Applied Structural and
Mechanical Vibrations

Applied Structural and Mechanical Vibrations

Theory, methods and measuring instrumentation

Paolo L. Gatti and Vittorio Ferrari

First published 1999
by E & FN Spon
11 New Fetter Lane, London EC4P 4EE

Simultaneously published in the USA and Canada
by Routledge
29 West 35th Street, New York, NY 10001

E & FN Spon is an imprint of the Taylor & Francis Group

© 1999 Paolo L. Gatti and Vittorio Ferrari

Typeset in 10/12pt Sabon by
Mathematical Compostion Setters Ltd, Salisbury
Printed and bound in Great Britain by
TJ International, Padstow, Cornwall

British Library Cataloguing in Publication Data
A catalogue record for this book is available from the British Library

Library of Congress Cataloging in Publication Data
Gatti, Paolo L., 1959–
    Applied structural and mechanical vibrations : theory, methods,
and measuring instrumentation / Paolo L. Gatti and Vittorio Ferrari.
        p.   cm.
    Includes bibliographical reference and index.
    1. Structural dynamics. 2. Vibration. 3. Vibration--Measurement.
I. Ferrari, Vittorio, 1962–   . II. Title.
TA654.G34   1999
620.3--dc21   98-53028
                                                           CIP

ISBN 0-419-22710-5

To my wife Doria, for her patience and understanding, my parents Paolina and Remo, and to my grandmother Maria Margherita

(Paolo L. Gatti)

To my wife and parents

(V. Ferrari)

Contents

PART II
Measuring instrumentation
V. FERRARI

Appendices

P.L. GATTI

Preface

This book deals primarily with fundamental aspects of engineering vibrations within the framework of the linear theory. Although it is true that in practical cases it is sometimes not easy to distinguish between linear and nonlinear phenomena, the basic assumption throughout this text is that the principle of superposition holds.

Without claim of completeness, the authors' intention has been to discuss a number of important topics of the subject matter by bringing together, in book form, a central set of ideas, concepts and methods which form the common background of real-world applications in disciplines such as structural dynamics, mechanical, aerospace, automotive and civil engineering, to name a few.

In all, the authors claim no originality for the material presented. However, we feel that a book such as this one can be published at the end of the 1990s because, while it is true that the general theory of linear vibrations is well established (Lord Rayleigh's book *Theory of Sound* is about a century old), this by no means implies that the subject is 'closed' and outside the mainstream of ongoing research. In fact, on the one hand, the general approach to the subject has significantly changed in the last 30 years or so. On the other hand, the increasing complexity of practical problems puts ever higher demands on the professional vibration engineer who, in turn, should acquire a good knowledge in a number of disciplines which are often perceived as distinct and separate fields.

Also, in this regard, it should be considered that the computer revolution of recent years, together with the development of sophisticated algorithms and fully automated testing systems, provide the analyst with computation capabilities that were unimaginable only a few decades ago. This state of affairs, however – despite the obvious advantages – may simply lead to confusion and/or erroneous results if the phenomena under study and the basic assumptions of the analysis procedures are not clearly understood.

The book is divided into two parts. Part I (Chapters 1 to 12) has been written by Paolo L. Gatti and is concerned with the theory and methods of linear engineering vibrations, presenting the topics in order of increasing difficulty – from single-degree-of-freedom systems to random vibrations and stochastic processes – and also including a number of worked examples in every chapter. Within this part, the first three chapters consider, respectively, some basic definitions and concepts to be used throughout the book (Chapter 1), a number of important aspects of mathematical nature (Chapter 2) and a concise treatment of analytical mechanics (Chapter 3). In a first reading, if the reader is already at ease with Fourier series, Fourier and Laplace transforms, Chapter 2 can be skipped without loss of continuity. However, it is assumed that the reader is familiar with fundamental university calculus, matrix analysis (although Appendix A is dedicated to this topic) and with some basic notions of probability and statistics.

Part II (Chapters 13 to 15) has been written by Vittorio Ferrari and deals with the measurement of vibrations by means of modern electronic instrumentation. The reason why this practical aspect of the subject has been included as a complement to Part I lies in the importance – which is sometimes overlooked – of performing valid measurements as a fundamental requirement for any further analysis. Ultimately, any method of analysis, no matter how sophisticated, is limited by the quality of the raw measurement data at its input, and there is no way to fix a set of poor measurements. The quality of measurement data, in turn, depends to a large extent on how properly the available instrumentation is used to set up a measuring chain in which each significant source of error is recognized and minimized. This is especially important in the professional world where, due to a number of reasons such as limited budgets, strict deadlines in the presentation of results and/or real operating difficulties, the experimenter is seldom given a second chance.

The choice of the topics covered in Part II and the approach used in the exposition reflect the author's intention of focusing the attention on basic concepts and principles, rather than presenting a set of notions or getting too much involved in inessential technological details. The aim and hope is, first, to help the reader – who is only assumed to have a knowledge of basic electronics – in developing an understanding of the essential aspects related to the measurement of vibrations, from the proper choice of transducers and instruments to their correct use, and, second, to provide the experimenter with guidelines and advice on how to accomplish the measurement task.

Finally, it is possible that this book, despite the attention paid to reviewing all the material, will contain errors, omissions, oversights and/or misprints. We will be grateful to readers who spot any of the above or who have any comment for improving the book. Any suggestion will be received and considered.

Milan 1998

Paolo Luciano Gatti,
Vittorio Ferrari

Email addresses:
pljgatti@tin.it
ferrari@bsing.ing.unibs.it

Acknowledgements

I wish to thank Dr G. Brunetti at Tecniter s.r.l. (Cassina de' Pecchi, Milan) for allowing me to take some time off work and complete the manuscript (almost) on time, Eng. R. Giacchetti at the University of Ancona for introducing me (a nuclear physicist) to the fascinating subject of engineering vibrations, and my long time friend and electronics expert Dr V. Ferrari for his important contribution to this project. Last but not least, I wish to thank Professor Valz-Gris (Department of Physics, State University of Milan) for valuable mathematical advice.
Paolo L. Gatti

I would like to thank Professor A. Taroni and Professor D. Marioli at the University of Brescia for their encouragement whilst I was writing this book.
Vittorio Ferrari

Both authors wish to thank everybody at Routledge (London) for their cooperation, competence and efficiency.

Part I
Theory and methods
Paolo L. Gatti

1 Review of some fundamentals

1.1 Introduction

It is now known from basic physics that force and motion are strictly connected and are, by nature, inseparable. This is not an obvious fact; it has taken almost two millennia of civilized human history and the effort of many great minds to understand. At present, it is the starting point of almost every branch of known physics and engineering. One of these branches is **dynamics**: the study that relates the motion of physical bodies to the forces acting on them. Within certain limitations, this is the realm of Newton's laws, in the framework of the theory that is generally referred to as **classical physics**.

Mathematically, the fact that force causes a change in the motion of a body is written

$$F = \frac{d}{dt}(mv) \tag{1.1}$$

This is Newton's second law which defines the unit of force once the fundamental units of mass and distance are given.

An important part of dynamics is the analysis and prediction of vibratory motion of physical systems, in which the system under study oscillates about a stable equilibrium position as a consequence of a perturbing disturbance which, in turn, starts the motion by displacing the system from such a position.

This type of behaviour and many of its aspects – wanted or unwanted – is common everyday experience for all of us and is the subject of this book. However, it must be clear from the outset that we will only restrict our attention to 'linear vibrations' or, more precisely, to situations in which vibrating systems can be modelled as 'linear' so that the principle of superposition applies. Future sections of this chapter and future chapters will clarify this point in stricter detail.

1.2   The role of modelling (linear and nonlinear, discrete and continuous systems, deterministic and random data)

In order to achieve useful results, one must resort to models. This is true in general and applies also to all the cases of our concern. Whether these models be mathematical or nonmathematical in nature, they always represent an idealization of the actual physical system, since they are based on a set of assumptions and have limits of validity that must be specified at some point of the investigation. So, for the same system it is possible to construct a number of models, the 'best' being the simplest one that retains all the essential features of the actual system under study.

Generally speaking, the modelling process can be viewed as the first step involved in the analysis of problems in science and engineering: the so-called 'posing of the problem'. Many times this first step presents considerable difficulties and plays a key role to the success or failure of all subsequent procedures of symbolic calculations and statement of the answer. With this in mind, we can classify oscillatory systems according to a few basic criteria. They are not absolute but turn out to be useful in different situations and for different types of vibrations.

First, according to their behaviour, systems can be **linear** or **nonlinear.** Formally, linear systems obey differential equations where the dependent variables appear to the first power only, and without their cross products; the system is nonlinear if there are powers greater than one, or fractional powers. When the equation contains terms in which the independent variable appears to powers higher than one or to fractional powers, the equation (and thus the physical system that the equation describes) is with variable coefficients and not necessarily nonlinear. The fundamental fact is that for linear system the principle of superposition applies: the response to different excitations can be added linearly and homogeneously. In equation form, if $f(x)$ is the output to an input x, then the system is linear if for any two inputs x_1 and x_2, and any constant a,

$$f(x_1 + x_2) = f(x_1) + f(x_2) \qquad \text{(additive property)} \qquad (1.2)$$

$$f(ax) = af(x) \qquad \text{(homogeneous property)} \qquad (1.3)$$

The distinction is **not** an intrinsic property of the system but depends on the range of operation: for large amplitudes of vibration **geometrical nonlinearity** ensues and – in structural dynamics – when the stress–strain relationship is not linear **material nonlinearity** must be taken into account.

Our attention will be focused on linear systems. For nonlinear ones it is the author's belief that there is no comprehensive theory (it may be argued that this could be their attraction), and the interested reader should refer to specific literature.

Second, according to the physical characteristics – called **parameters** – systems can be **continuous** or **discrete**. Real systems are generally continuous since their mass and elasticity are distributed. In many cases, however, it is useful and advisable to replace the distributed characteristics with discrete ones; this simplifies the analysis because ordinary differential equations for discrete systems are easier to solve than the partial differential equations that describe continuous ones.

Discrete-parameter systems have a finite number of **degrees of freedom**, i.e. only a finite number of independent coordinates is necessary to define their motion. The well-known finite-element method, for example, is in essence a discretization procedure that retains aspects of either continuous and discrete systems and exploits the calculation capabilities of high-speed digital computers. Whatever discretization method is used, one advantage is the possibility to improve the accuracy of the analysis by increasing the number of degrees of freedom.

Also in this case, the distinction is more apparent than real; continuous systems can be seen as limiting cases of discrete ones and the connection of one formulation to the other is very close. However, a detailed treatment of continuous systems probably gives more physical insight in understanding the 'standing-wave–travelling-wave duality', intrinsic in every vibration phenomenon.

Third, in studying the response of a system to a given excitation, sometimes the type of excitation dictates the analysis procedure rather than the system itself. From this point of view, a classification between **deterministic** and **random** (or stochastic or nondeterministic) data can be made. Broadly speaking, deterministic data are those that can be described by an explicit mathematical relationship, while there is no way to predict an exact value at a future instant of time for random data. In practice, the ability to reproduce the data by a controlled experiment is the general criterion to distinguish between the two. With random data each observation will be unique and their description is made only in terms of statistical statements.

1.3 Some definitions and methods

As stated in the introduction, the particular behaviour of a particle, a body or a complex system that moves about an equilibrium position is called oscillatory motion. It is natural to try to describe such a particle, body or system using an appropriate function of time $x(t)$. The physical meaning of $x(t)$ depends on the scope of the investigation and, as often happens in practice, on the available measuring instrumentation: it might be displacement, velocity, acceleration, stress or strain in structural dynamics, pressure or density in acoustics, current or voltage in electronics or any other quantity that varies with time.

A function that repeats itself exactly after certain intervals of time is called **periodic**. The simplest case of periodic motion is the harmonic (or sinusoidal) that can be defined mathematically by a sine or cosine function:

$$x(t) = X \cos(\omega t - \theta) \tag{1.4}$$

where:

> X is the **maximum**, or **peak amplitude** (in the appropriate units);
> $(\omega t - \theta)$ is the **phase angle** (in radians);
> ω is the **angular frequency** (in rad/s);
> θ is the **initial phase angle** (in radians), which depends on the choice of the time origin and can be taken equal to zero if there is no relative reference to other sinusoidal functions.

The time between two identical conditions of motion is the **period** T. It is measured in seconds and is the inverse of the **frequency** ν whose unit is the hertz (Hz, with dimensions of s^{-1}) and, in turn, represents the number of cycles per unit time. The following relations hold:

$$\omega = 2\pi\nu \tag{1.5}$$

$$T = \frac{1}{\nu} = \frac{2\pi}{\omega} \tag{1.6}$$

A plot of eq (1.4), amplitude versus time, is shown in Fig. 1.1 where the peak amplitude is equal to unity.

A vector **x** of modulus X that rotates with angular velocity ω in the xy plane is a useful representation of sinusoidal motion: $x(t)$ is now the instantaneous projection of **x** on the x-axis (on the y-axis for a sine function).

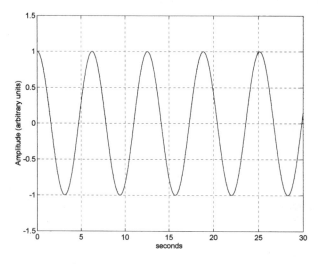

Fig. 1.1 Cosine harmonic oscillation of unit amplitude.

Other representations are possible, each one with its own particular advantages; however, the use of complex numbers for oscillatory quantities gives probably the most elegant and compact way of dealing with the problem.

Recalling from basic calculus the Euler equations

$$e^{-iz} = \cos z - i \sin z$$

$$\cos z = \frac{1}{2}(e^{iz} + e^{-iz})$$

$$\sin z = \frac{1}{2i}(e^{iz} - e^{-iz}) \qquad (1.7)$$

where i is the imaginary unit ($i \equiv \sqrt{-1}$) and $e = 2.71828 \ldots$ is the well-known basis of Naperian logarithms, an oscillatory quantity can be conveniently written as the complex number

$$x(t) = Ce^{-i\omega t} \qquad (1.8)$$

where C is the **complex amplitude**, i.e. a complex number that contains both magnitude and phase information and can be written as $(a + ib)$ or $Xe^{i\theta}$ with magnitude $|C| = X = \sqrt{a^2 + b^2}$ and phase angle θ, where $\tan\theta = b/a$, $\cos\theta = a/X$ and $\sin\theta = b/X$.

The number $C^* = a - ib$ is the complex conjugate of C and the square of the magnitude is given by $|C|^2 = CC^* = X^2 = a^2 + b^2$.

The idea of eq (1.8) – called the **phasor representation** – is the temporary replacement of a **real** physical quantity by a **complex** number for purposes of calculation; the usual convention is to assign physical significance only to the real part of eq (1.8), so that the oscillatory quantity $x(t)$ can be expressed in any of the four ways

$$x(t) = a\cos(\omega t) + b\sin(\omega t) = X\cos(\omega t - \theta) \qquad (1.9)$$

or

$$x(t) = Ce^{-i\omega t} = Xe^{-i(\omega t - \theta)} \qquad (1.10)$$

where only the real part is taken of the expressions in eq (1.10).

A little attention must be paid when we deal with the energy associated with these oscillatory motions. The various forms of energy (energy, energy density, power or intensity) depend quadratically on the vibration amplitudes, and since $\text{Re}(x^2) \neq [\text{Re}(x)]^2$ we need to **take the real part first and then square** to find the energy.

Furthermore, it is often useful to know the time-averaged energy or power and there is a convenient way to extract this value in the general case.

Suppose we have the two physical quantities expressed in the real form of eq (1.4)

$$x_1(t) = X_1 \cos(\omega_1 t - \theta_1)$$

$$x_2(t) = X_2 \cos(\omega_2 t - \theta_2)$$

It is easy to show that $\langle x_1 x_2 \rangle$, i.e. the average value of the product $x_1 x_2$, is different from zero only if $\omega_1 = \omega_2$ and in this case we get

$$\langle x_1 x_2 \rangle \equiv \frac{1}{T} \int_0^T x_1(t) x_2(t) dt = \frac{1}{2} X_1 X_2 \cos(\theta_1 - \theta_2) \tag{1.11}$$

where $T = 2\pi/\omega$ and the factor 1/2 comes from the result $\langle \cos^2 \rangle = 1/2$.

If we want to use phasors and represent the physical quantities as

$$x_1(t) = X_1 e^{-i(\omega_1 t - \theta_1)}$$

$$x_2(t) = X_2 e^{-i(\omega_2 t - \theta_2)}$$

we see that in order to get the correct result of eq (1.11) we must calculate the quantity

$$\tfrac{1}{2} \operatorname{Re}(x_1 x_2^*) \tag{1.12}$$

In the particular case of $x_1 = x_2 \equiv x$ (where these terms are expressed in the form of eq (1.8)), our convention says that the average value of x squared is given by

$$\tfrac{1}{2} \operatorname{Re}(xx^*) = \tfrac{1}{2} CC^* = \tfrac{1}{2} |C|^2 = \tfrac{1}{2} X^2 = \tfrac{1}{2} (a^2 + b^2) \tag{1.13}$$

Phasors are very useful for representing oscillating quantities obeying linear equations; other authors (especially in electrical engineering books) use the letter j instead of i and $e^{i\omega t}$ instead of $e^{-i\omega t}$ and some other authors use the positive exponential notation $e^{i\omega t}$. Since we mean to take the real part of the result, the choice is but a convention; any expression is fine as long as we are consistent. The negative exponential is perhaps more satisfactory when dealing with wave motion, but in any case it is possible to change the formulas to the electrical engineering notation by replacing every i with $-j$.

Periodic functions in general are defined by the relation

$$x(t) = x(t + nT) \qquad n = 1, 2, 3, \ldots \tag{1.14}$$

and will be considered in subsequent chapters where the powerful tool of Fourier analysis will be introduced.

1.3.1 The phenomenon of beats

Let us consider what happens when we add two sinusoidal functions of slightly different frequencies ω_1 and ω_2, with $\omega_2 = \omega_1 + \varepsilon$ and ε being a small quantity compared to ω_1 and ω_2. In phasor notation, assuming for simplicity equal magnitude and zero initial phase for both oscillations x_1 and x_2, we get

$$x_1(t) + x_2(t) = Xe^{-i\omega_1 t} + Xe^{-i\omega_2 t} \tag{1.15}$$

that can be written as

$$Xe^{-i(\omega_1 + \omega_2)t/2}[e^{i(\omega_2 - \omega_1)t/2} + e^{-i(\omega_2 - \omega_1)t/2}]$$

with a real part given by

$$2X \cos(\bar{\omega}t)\cos(\omega_{av}t) \tag{1.16}$$

where $\omega_{av} = (\omega_1 + \omega_2)/2$ and $\bar{\omega} = (\omega_2 - \omega_1)/2$. We can see eq (1.16) as an oscillation of frequency ω_{av} and a time-dependent amplitude $2X \cos(\bar{\omega}t)$. A graph of this quantity is shown in Figs 1.2 and 1.3 where $\omega_1 = 8.0$ rad/s and $\omega_2 - \omega_1 = 0.6$ (Fig. 1.2) and 1.0 (Fig. 1.3), respectively.

Physically, the two original waves remain nearly in phase for a certain time and reinforce each other; after a while, however, the crests of the first wave correspond to the troughs of the other and they practically cancel out. This pattern repeats on and on and the result is the **phenomenon of beats** illustrated in Figs 1.2 and 1.3. Maximum amplitude occurs when $\bar{\omega}t = n\pi$ ($n = 0, 1, 2, ...$), that is, every $\pi/\bar{\omega}$ seconds. Therefore, the frequency of the

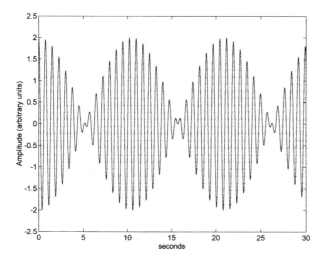

Fig. 1.2 Beat phenomenon ($\omega_2 - \omega_1 = 0.6$).

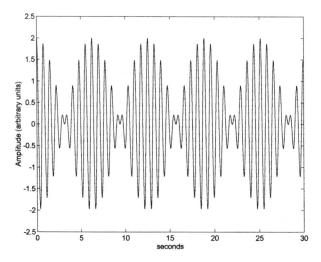

Fig. 1.3 Beat phenomenon $(\omega_2 - \omega_1 = 1.0)$.

beats is $(\bar{\omega}/\pi) = \nu_2 - \nu_1$, equal to the difference in frequency between the two original signals.

When the signals have unequal amplitudes (say A and B) the total amplitude does not become zero, it oscillates between $A + B$ and $A - B$, but the general pattern can still be easily identified.

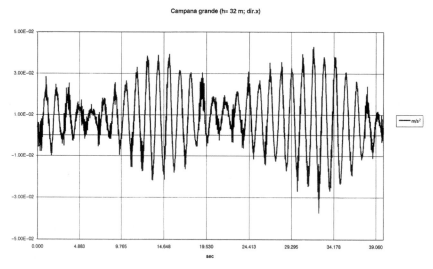

Fig. 1.4 Beat phenomenon measured on an Italian belltower (oscillation of largest bell).

In acoustics, for example, beats are heard as a slow rise and fall in sound intensity (at the beat frequency $\nu_2 - \nu_2$) when two notes are slightly out of tune. Many musicians exploit this phenomenon for tuning purposes, they play the two notes simultaneously and tune one until the beats disappear.

Figure 1.4 illustrates a totally different situation. It is an actual vibration measurement performed on an ancient belltower in Northern Italy during the oscillation of the biggest bell. The measurement was made at about two-thirds of the total height of about 50 m on the body of the tower in the transverse direction. There is a clear beat phenomenon between the force imposed on the structure by the oscillating bell (at about 0.8 Hz, with little variations of a few percent) and the first flexural mode of the tower (0.83 Hz). Several measurements were made and the beat frequency was shown to vary between 0.03 and 0.07 Hz, indicating a situation close to **resonance**. This latter concept, together with the concepts of forced oscillations and modes of a vibrating system will be considered in future chapters.

1.3.2 Displacement, velocity and acceleration

If the oscillating quantity $x(t)$ in eq (1.8) is a displacement and we recall the usual definitions of velocity and acceleration

$$v(t) \equiv \dot{x}(t) = \frac{dx(t)}{dt}$$

$$a(t) \equiv \ddot{x}(t) = \frac{d^2x(t)}{dt^2}$$

we get from the phasor representation

$$v(t) = -i\omega C e^{-i\omega t} = \omega C e^{-i(\omega t + \pi/2)}$$
$$a(t) = \omega^2 C e^{-i(\omega t + \pi)} \tag{1.17}$$

since the phase angle of $-i$ is $\pi/2$ and the phase angle of -1 is π. The velocity leads the displacement of 90°, the acceleration leads the velocity of 90° and all three rotate clockwise in the Arland–Gauss plane (abscissa = real part, ordinate = imaginary part) as time passes. Moreover, from eqs (1.17) we note that the maximum velocity amplitude is $V_{max} = \omega C$, while the maximum acceleration amplitude is $A_{max} = \omega^2 C$.

In theory it should not really matter which one of these three quantities is measured; the necessary information and frequency content of a signal is the same whether displacement, velocity or acceleration is considered and any one quantity can be obtained from any other one by integration or differentiation. However, physical considerations on the nature of vibrations themselves and on the electronic sensors and transducers used to measure them somehow make one parameter preferred over the others.

Physically, it will be seen that displacement measurements give most weight to low-frequency components and, conversely, acceleration give most weight to high-frequency components. So, the frequency range of the expected signals is a first aspect to consider; when a wide-band signal is expected, velocity is the appropriate parameter to select because it weights equally low- and high-frequency components. Furthermore, velocity (rms values, see next section), being directly related to the kinetic energy, is preferred to quantify the **severity**, i.e. the destructive effect, of vibration.

On the other hand, acceleration sensitive transducers (accelerometers) are commonly used in practice because of their versatility: small physical dimensions, wide frequency and dynamic ranges, easy commercial availability and the fact that analogue electronic integration is more reliable than electronic differentiation are important characteristics that very often make acceleration the measured parameter. All these aspects play an important role but, primarily, it must be the final scope and aim of the investigation that dictates the choice to make for the particular problem at hand.

Let us make some heuristic considerations from the practical point of view of the measurement engineer. Suppose we have to measure a vibration phenomenon which occurs at about $\nu = 1$ Hz with an expected displacement amplitude (eq (1.17)) of $C = \pm 1$ mm. It is not difficult to find on the market a cheap displacement sensor with, say, a total range of 10 mm and a sensitivity of 0.5 V/mm. In our situation, such a sensor would produce an output signal of 1 V, meaning a good signal-to-noise ratio in most practical situations. On the other hand, the acceleration amplitude in the above conditions is about $(2\pi\nu)^2 C = (39.48)(2 \times 10^{-3}) = 7.9 \times 10^{-2}$ m/s$^2 \cong 8.1 \times 10^{-3} g$ so that a standard general-purpose accelerometer with a sensitivity of, say, 100 mV/g would produce an output signal of 0.81 mV $= 8.1 \times 10^{-4}$ V, which is much less favourable from a signal-to-noise ratio viewpoint.

By contrast – for example in heavy machinery – forces applied to massive elements generally result in small displacements which occur at relatively high frequencies. So, for purpose of illustration, suppose that a machinery element vibrates at about 100 Hz with a displacement amplitude of ± 0.05 mm. The easiest solution in this case would be an acceleration measurement because the acceleration amplitude is now 39.5 m/s$^2 \cong 4g$, so that a general purpose (and relatively unexpensive) 100 mV/g accelerometer would produce an excellent peak-to-peak signal of about 400 mV. In order to measure such small displacements at those values of frequency, we would probably have to resort to more expensive optical sensors.

1.3.3 *Quantification of vibration level and the decibel scale*

The most useful descriptive quantity – which is related to the power content of the of the vibration and takes the time history into account – is the **root**

mean square value (x_{rms}), defined by

$$x_{rms} \equiv \sqrt{\lim_{T \to \infty} \frac{1}{T} \int_0^T x^2(t)dt} \qquad (1.18)$$

For sinusoidal motion it is easily seen that $x_{rms} = X/\sqrt{2} = 0.707X$. In the general case X/x_{rms} is called the the **form factor** and gives some indication of the waveshape under study when impulsive components are present or the waveform is extremely jagged. It is left as an easy exercise to show that for a triangular wave (see Fig. 1.5) $x_{rms} = 0.577X$.

In common sound and vibration analysis, amplitudes may vary over wide ranges (the so-called dynamic range) that span more than two or three decades. Since the dynamic range of electronic instrumentation is limited and the graphical presentation of such signals can be impractical on a linear scale, the **logarithmic decibel (dB) scale** is widely used.

By definition, two quantities differ by one bel if one is 10 times (10^1) greater than the other, two bels if one is 100 times (10^2) greater than the other and so on. One tenth of a bel is the decibel and the dB level (L) of a quantity x, with respect to a reference value x_0, is given by

$$L(\text{dB}) \equiv 10 \log_{10} \left(\frac{x}{x_0} \right) \qquad (1.19)$$

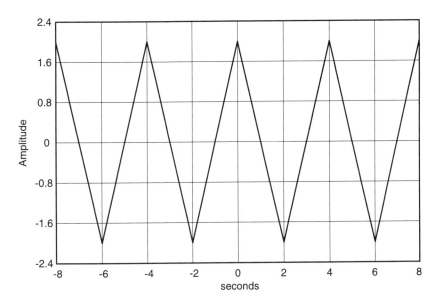

Fig. 1.5 Triangular wave.

So, for example, 3 dB represent a doubling of the relative quantity, i.e. $x \cong 2.0x_0$. Decibels, like radians, are dimensionless; they are not 'units' in the usual sense and consistency dictates that the reference value must be universally accepted. When this is not the case, the reference value must be clearly specified. Sound intensity levels (SIL), in acoustics, are given by eq (1.19) and the reference intensity value is $I_0 = 10^{-12}$ W/m^2.

A more commonly used version of this logarithmic scale requires the proportionality factor 20 instead of 10, and is related to the fact – already mentioned in the preceding paragraph – that energies, powers, rms values, etc. depend quadratically on the amplitudes.

We have

$$L(\text{dB}) = 20 \, \log_{10} \left(\frac{x}{x_0} \right) \tag{1.20}$$

In acoustics, eq (1.20) defines the sound pressure level (SPL) where the reference value (in air) is the pressure $p_0 = 2 \times 10^{-5}$ Pa. In vibration measurements one can have a displacement level L_d, a velocity level L_v, an acceleration level L_a or a force level L_f. These are obtained, respectively, by the relations

$$L_d = 20 \log_{10}(d/d_0)$$
$$L_v = 20 \log_{10}(v/v_0)$$
$$L_a = 20 \log_{10}(a/a_0)$$
$$L_f = 20 \log_{10}(f/f_0)$$

The reference values are not universally accepted, but usually we have

$$d_0 = 10^{-11} \text{ m}$$
$$v_0 = 10^{-9} \text{ m/s}$$
$$a_0 = 10^{-6} \text{ m/s}^2$$
$$f_0 = 10^{-6} \text{ N}$$

Another choice that is commonly adopted in instrumentation, e.g. spectrum analysers, is to take 1 V or the input voltage range (IVR) as the reference value. Vibration sensors, in fact, feed to the analyser an output voltage which depends – through the sensitivity of the sensor – on the vibration level to be measured. The analyser, in turn, allows the operator to preset an input voltage range (IVR = ± 0.5 V, ± 1 V, ± 5 V, ± 10 V for instance) which is sometimes used as a reference when displaying power spectra on a dB scale. So, the dB value of a voltage signal fed to the analyser is given by

$$L(\text{dBV}) = 20 \log_{10} \left(\frac{V_{in}}{\text{IVR}} \right)$$

and in this case one always gets negative dB levels, because always $V_{in} \leqslant \text{IVR}$.

From the definition, it is clear that decibels are not added and subtracted linearly. If we have two decibel levels L_1 and L_2 (obviously referred to the **same** reference value) that must be added, the total dB level, say L_T, is not $L_1 + L_2$ but

$$L_T = 10 \log_{10}(10^{L_1/10} + 10^{L_2/10}) \tag{1.21}$$

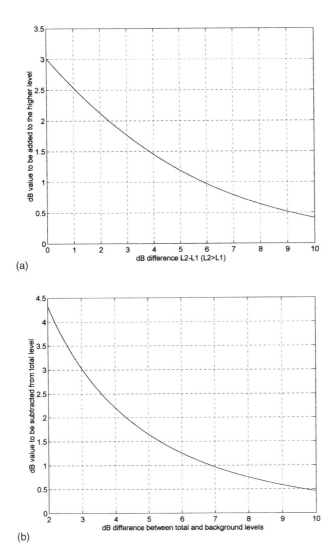

(a)

(b)

Fig. 1.6 (a) Addition and (b) subtraction of dB levels (dB levels calculated as in eq (1.19)).

since

$$10^{L_1/10} = \frac{x_1}{x_0} \quad \text{and} \quad 10^{L_2/10} = \frac{x_2}{x_0}$$

and the total level is by definition

$$L_T = 10 \log_{10} \left(\frac{x_1 + x_2}{x_0} \right)$$

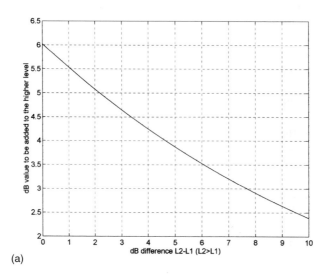

(a)

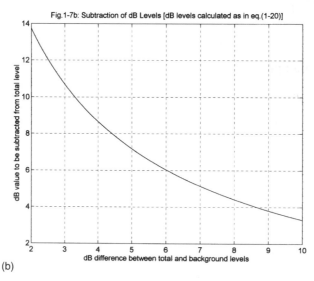

(b)

Fig. 1.7 (a) Addition and (b) subtraction of dB levels (dB levels calculated as in eq (1.20)).

Similar considerations apply in the case of subtractions of dB levels, the most common case being the subtraction of a background level L_b from a total measured level L_T, in order to obtain the actual level L_a. Of course, there is now a minus sign in the parentheses of (1.21) and in this case $L_1 = L_T$ and $L_2 = L_b$.

The graphs in Fig. 1.6(a), (b) (dB levels calculated according to eq (1.19)) and Fig. 1.7(a), (b) (dB levels calculated according to eq (1.20)) are useful when adding or subtracting dB levels and there is no pocket calculator at hand. For example, we want to add the two levels $L_1 = 60$ dB and $L_2 = 63$ dB. Referring to Fig. 1.6(a), the difference $\Delta L \equiv L_2 - L_1$ is 3 dB and this value gives on the ordinate axis a dB increment of $1.76 \cong 1.8$ to be added to the higher level (63 dB). The result of the addition is then 64.8 dB.

For an example of subtraction, suppose that we measure a total level of 70 dB on a background level of 63 dB. Referring to Fig. 1.6(b), for the abscissa entry $\Delta L = 7$ dB one obtains for the ordinate value a dB decrement $= 0.97 \cong 1.0$ to be subtracted from the total level. The actual value is then 69.0 dB.

A final word on the calculation of the average dB level of a series of N measurements: this is given by

$$L_{avg} = 10 \log_{10} \left(\frac{1}{N} \sum_{i=1}^{N} 10^{L_i/10} \right) \tag{1.22}$$

and **not** by the familiar

$$L_{avg} = \frac{1}{N} \sum_{i=1}^{N} L_i$$

1.4 Springs, dampers and masses

Almost any physical system possessing elasticity and mass can vibrate. The simplest realistic model considers three basic discrete elements: a spring, a damper and a mass. Mathematically, they relate applied forces to displacement, velocity and acceleration, respectively. Let us consider them more closely.

The restoring force that acts when a system is slightly displaced from equilibrium is supplied by internal elastic forces that tend to bring the system back to the original position. In solids, these forces are the macroscopic manifestation of short-range microscopic forces that arise when atoms or molecules are displaced from the equilibrium position they occupy in some organized molecular structure.

These phenomena are covered by the theory of elasticity. In addition, in complex structures, macroscopic restoring forces occur when different structural elements meet. In general, it is not easy to predict the performance of a such a joint, even under laboratory conditions, because the stiffness is dependent on very local distortions. Here probably lies the main reason of the fallibility of predictions of stiffnesses in many cases.

This may come as a surprise to many engineering graduates and it is due to the mistaken belief that 'everything' can be calculated with a high-speed computer, using finite-element techniques with very fine resolution. As it turns out, it is frequently cheaper to test than to predict.

The easiest way to model such characteristics relating forces to displacements is by means of a simple stiff element: the **linear massless spring**, shown in Fig. 1.8. The assumption of zero mass assures that a force F acting on one end must be balanced by a force $-F$ on the other end so that, in such conditions, the spring undergoes an elongation equal to the difference between the displacements x_2 and x_1 of the endpoints. For small elongations it is convenient and also correct to assume a linear relation, i.e.

$$F = k(x_2 - x_1) \tag{1.23}$$

where k is a constant (the spring **stiffness**, with units N/m) that represents the force required to produce a unit displacement in the specified direction.

Often, one end of the spring is considered to be fixed, the displacement of the other end is simply labelled x and – since F is an elastic restoring force – eq (1.23) is written $F = -kx$, with the minus sign indicating that the force opposes the displacement. The reciprocal of stiffness, i.e. $1/k$, is frequently used and is called **flexibility** or **compliance**.

In real-world systems, energy is always dissipated by some means. In the cases concerning us the energies of interest are the kinetic energy of motion and the potential strain energy due to elasticity. Friction mechanisms of different kinds are often considered to represent this phenomenon of energy 'damping' that ultimately results in production of heat and is probably the main uncertainty in most engineering problems. In fact, a word of caution is necessary: any claim that damping in structures can be predicted with accuracy should be treated with scepticism.

On a first approach, we will see that damping can be neglected (undamped systems) without losing the physical insight to the problem at hand; nevertheless, it must be kept in mind that this is a representation of an unreal situation or an approximation that can be accepted only in certain circumstances.

Fig. 1.8 Ideal massless spring.

The simplest model of damping mechanism is the **massless viscous damper** that relates forces to velocities. An example can be a piston fitting loosely in a cylinder filled with oil so that the oil can flow around the piston as it moves inside the cylinder. The graphical symbol usually adopted is the dashpot shown in Fig. 1.9. Once again, a force F on one end must be balanced by a force $-F$ on the other end and linearity is assumed, i.e.

$$F = c(\dot{x}_2 - \dot{x}_1) \tag{1.24}$$

where the dot indicates the time derivative and c is the coefficient of viscous damping, with units N s/m.

If one end is fixed and the velocity of the other end is labelled $\dot{x}$, eq (1.24) is written $F = -c\dot{x}$, with the minus sign indicating that the damping force resists an increase in velocity.

The quantity that relates forces to accelerations is the mass and the fundamental relation is Newton's second law, which can be written, with respect to an inertial frame of reference,

$$F = m\ddot{x} \tag{1.25}$$

In the SI system of units the mass is measured in kilograms and eq (1.25) defines the unit of force (see also Section 1.1). The mass represents the inertia properties of physical bodies that, under the action of a given applied force F, are set in motion with an acceleration that is inversely proportional to their mass.

Finally, it is interesting to consider a few examples of 'equivalent springs' k_{eq}, meaning by this term the replacement of one or more combination of stiff elements with a single spring that takes into account and represents – for the problem at hand – the stiffness of such combination.

Two springs can be connected in series or in parallel, as shown in Fig. 1.10(a) and (b). (However, for a different approach see Section 13.7.3) In the first case we have

$$F = k_2(x_2 - x_1)$$
$$F = k_1 x_1 \tag{1.26}$$

For an equivalent spring of stiffness k_{eq} connected between the fixed point and x_2 we would have $F = k_{eq}x_2$. It is then easy to show that it must be

$$\frac{1}{k_{eq}} = \frac{1}{k_1} + \frac{1}{k_2} = \frac{k_1 + k_2}{k_1 k_2} \tag{1.27}$$

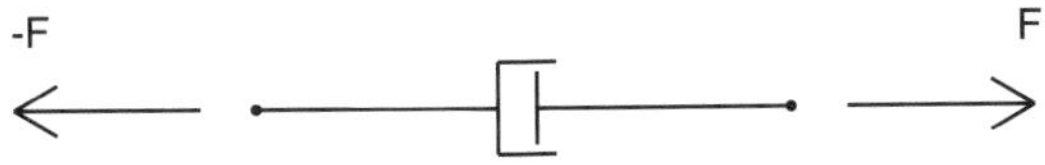

Fig. 1.9 Ideal massless dashpot.

and extending to the case of N springs connected in series we obtain

$$k_{eq} = \left(\sum_{i=1}^{N} \frac{1}{k_i} \right)^{-1} \tag{1.28}$$

For the connection in parallel of two springs (Fig. 1.10(b)) the force F divides into F_1 and F_2 (with the obvious condition $F = F_1 + F_2$) and the relations $F_1 = k_1 x$, $F_2 = k_2 x$. The equivalent spring obeys $F = k_{eq} x$, from which it follows that $k_{eq} = k_1 + k_2$. The generalization to N springs in parallel is straightforward and we get

$$k_{eq} = \sum_{i=1}^{N} k_i \tag{1.29}$$

Considering the real spring of Fig. 1.11(a), under the action of an axial tension or compression its stiffness constant is

$$k = \frac{Gd^4}{64nR^3} \tag{1.30}$$

where G (in N/m^2) is the modulus of elasticity in shear of the material of which the spring is made, n is the number of turns and the other symbols are shown in Fig. 1.11a. Different values are obtained for torsion or bending actions and Young's modulus, in these latter cases, also plays a part.

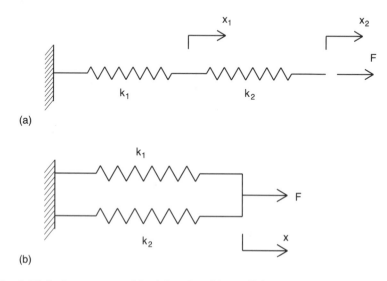

Fig. 1.10 Springs connected in: (a) series; (b) parallel.

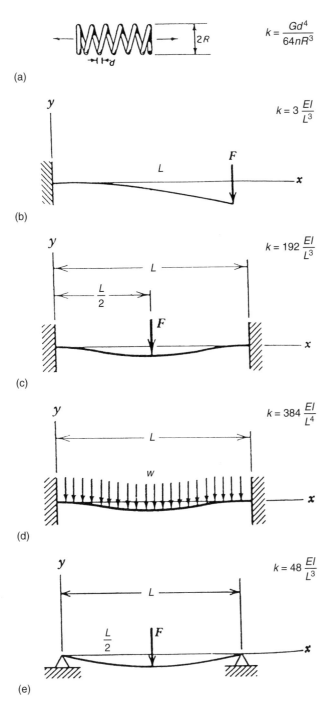

$$k = \frac{Gd^4}{64nR^3}$$

(a)

$$k = 3\frac{EI}{L^3}$$

(b)

$$k = 192\frac{EI}{L^3}$$

(c)

$$k = 384\frac{EI}{L^4}$$

(d)

$$k = 48\frac{EI}{L^3}$$

(e)

Fig. 1.11 (a) Typical helical spring. (b)–(e) Bending of bars.

Other examples are shown in:

- Fig. 1.11(b): cantilever with fixed-free boundary conditions and force F applied at the free end, k is the local stiffness at the point of application of the force,
- Fig. 1.11(c): fixed–fixed bar of length L with transverse localized force F at $L/2$; k is the local stiffness at the point of application of the force;
- Fig. 1.11(d): fixed–fixed bar of length L with uniform transverse load w (N/m), k is the local stiffness at the point of maximum displacement;
- Fig. 1.11(e): bar simply supported at both ends with force F at $L/2$, k is the local stiffness at the point of application of the force.

The result of Fig. 1.11(b) comes from the fact that we know from basic theory of elasticity that the vertical displacement x of the cantilever free end under the action of F is given by

$$x = \frac{FL^3}{3EI}$$

so that

$$k \equiv \frac{F}{x} = \frac{3EI}{L^3}$$

Similar considerations apply to the other examples, E being Young's modulus in N/m^2 and I being the cross-sectional moment of inertia in m^4.

1.5   Summary and comments

Chapter 1 introduces and reviews some basic notions in engineering vibrations with which the reader should already have some familiarity. At the very beginning, it is important to point out that the modelling process is always the first step to be taken in every approach to problems in science and engineering. The model must be able to reproduce, within an acceptable degree of accuracy, the essential features of the physical system under investigation. The definitions of the terms 'acceptable degree of accuracy' and 'essential features' depend on the particular problem at hand and, in general, should be based on decisions that specify: the results that are needed, the error one is willing to accept, the money cost involved, how and what to measure if measurements must be taken. When the budget is fixed and time deadlines are short, a compromise between cost and attainable level of accuracy must be made and agreed upon.

More specifically, a few classifications are given which may help in setting the guidelines of the type of modelling schemes that can be adopted for a wide class of problems.

First and above all is the distinction between **linear** and **nonlinear analysis**. Linearity or nonlinearity are **not** intrinsic properties of the system under study, but different behaviours of mechanical and structural systems under different conditions. Small amplitudes of motion, in general, are the range where linearity holds and the cornerstone of linearity is the principle of superposition.

Second, **discrete** and **continuous systems** can be distinguished, or, equivalently, **finite** or **infinite-degree-of-freedom** systems. Continuous distributed parameters are often substituted by discrete localized parameters to deal with ordinary differential equations rather than with partial differential equations and perform the calculations via computer.

Third, signals encountered in the field of linear vibrations can be classified as **deterministic** or **random**: an analytical form can be written for the former, while statistical methods must be adopted for the latter. The type of signals encountered in a particular problem often dictates the method of analysis.

The other sections of the chapter introduce some basic definitions and methods that will be used throughout the text. A few examples are simple sinusoidal motion, its complex (**phasor**) representation and the **decibel scale**. The phenomenon of beats is then considered, for its own intrinsic interest and as an application of phasors, and, finally, an examination of the parameters that make systems vibrate or prevent them from vibrating – namely mass, stiffness and damping – is made, together with their simplest schematic representations.

2   Mathematical preliminaries

2.1   Introduction

The purpose of this chapter is twofold: (1) to introduce and discuss a number of concepts and fundamental results of a mathematical nature which will be used whenever necessary in the course of our analysis of linear vibrations, and (2) to provide the reader with some notions which are important for a more advanced and more mathematically oriented approach to the subject matter of this text. In this light, some sections of this chapter can be skipped in a first reading and considered only after having read the chapters that follow, in particular Chapters 6–9.

It is important to point out that not all the needed results will be considered in this chapter. More specifically, matrix analysis is considered separately in Appendix A, while the whole of Chapter 11 is dedicated to the discussion of some basic concepts of probability and statistics which, in turn, serve the purpose of introducing the subjects of stochastic processes and random vibrations (Chapter 12). Also, when short mathematical remarks do not significantly interfere with the main line of reasoning of the subject being considered, brief digressions are made whenever needed in the course of the text.

So, without claim of completeness and/or extreme mathematical rigour, this chapter is intended mainly for reference purposes. We simply hope that it can be profitably used by readers of this and/or other books on the specific field of engineering vibrations and related technical subjects.

2.2   Fourier series and Fourier transforms

In general terms, Fourier analysis is a mathematical technique that deals with two problems:

1. the addition of several sinusoidal oscillations to form a resultant (harmonic synthesis);
2. the reverse problem, i.e. given a resultant, the problem of finding the sinusoidal oscillations from which it was formed (harmonic analysis).

As a trivial opening example, it is evident that adding two harmonic oscillations of the same frequency

$$x_1 = A_1 \sin \omega_0 t + B_1 \cos \omega_0 t$$
$$x_2 = A_2 \sin \omega_0 t + B_2 \cos \omega_0 t \qquad (2.1)$$

leads to a harmonic oscillation $x_1 + x_2$ of the same frequency with a sine amplitude $A_1 + A_2$ and a cosine amplitude $B_1 + B_2$. If the two oscillations have different frequencies, the resultant will not be harmonic. However, if the frequencies being summed are all multiples of some fundamental frequency ω_1, then the resultant oscillation will repeat itself after a time $T = 2\pi/\omega_1$ and we say that it is periodic with a period of T seconds. For example, the function

$$x(t) = \cos(3t) + 1.4 \sin(5t) + 2.1 \cos(6t) \qquad (2.2)$$

shown in Fig. 2.1, repeats itself with period $T = 2\pi/1 = 2\pi$ seconds, so that $x(t) = x(t + T)$ (in Fig. 2.1, note that the time axis extends from $t = 0$ to $t = 6\pi \cong 18.85$).

If, on the other hand, the frequencies being summed have no common factor, the resultant waveform is not periodic and never repeats itself. As an example, the function

$$x(t) = \cos(t) + 1.4 \sin(\sqrt{2}t) + 1.3 \cos(\sqrt{17}t) \qquad (2.3)$$

is shown in Fig. 2.2 from $t = 0$ to $t = 50$ seconds.

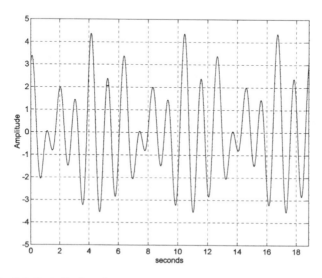

Fig. 2.1 Periodic function.

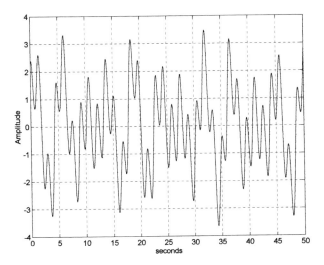

Fig. 2.2 Nonperiodic function.

In short, the process of harmonic synthesis can be expressed mathematically as

$$x(t) = \sum_{n=1}^{N} (A_n \cos \omega_n t + B_n \sin \omega_n t) \qquad (2.4)$$

where the quantities A_n, and B_n are, respectively, the cosine and sine amplitudes at the frequency ω_n. When a fundamental frequency ω_1 exists, then $\omega_n = n\omega_1$ and we call ω_n the nth harmonic frequency.

2.2.1 *Periodic functions: Fourier series*

As stated before, harmonic analysis is, in essence, the reverse process of harmonic synthesis. J.B. Fourier's real achievement, however, consists of the fact that he showed how to solve the problem by dealing with an infinite number of frequencies, that is, for example, with an expression of the type

$$x(t) = \sum_{n=1}^{\infty} [A_n \cos(n\omega_1 t) + B_n \sin(n\omega_1 t)] \qquad (2.5)$$

where it is understood that the meaning of eq (2.5) is that the value of $x(t)$ at any instant can be obtained by finding the partial sum S_N of the first N harmonics and defining $x(t)$ as the limiting value of S_N when N tends to infinity. As an example, it is left to the reader to show that eq (2.5) with the coefficients $A_n = 0$, and $B_n = 1/n$ is the well-known periodic 'sawtooth' oscillation (see Fig. 2.3 representing the partial sum S_6 with $\omega_1 = 1$ rad/s).

Incidentally, two short comments can be made at this point.

First, the example of Fig. 2.3 gives us the opportunity to note that a function which executes very rapid changes must require that high-frequency components have appreciable amplitudes: more generally it can be shown that a function with discontinuous jumps (the 'sawtooth', for example) will have A and B coefficients whose general trend is proportional to n^{-1}. By contrast, any continuous function that has jumps in its first derivative (for example, the triangular wave of Fig. 1.5) will have coefficients that behave asymptotically as n^{-2}.

Second, although our notation reflects this particular situation, the term 'oscillation' does not necessarily imply that we have to deal with time-varying physical quantities: for example, time t could be replaced by a space variable, say z, so that the frequency ω would then be replaced by a 'spatial frequency' (the so-called wavenumber, with units of rad/m and usually denoted by k or κ), meaning that $x(z)$ has a value dependent on position. Moreover, in this case, the quantity $2\pi/k$ (with the units of a length) is the wavelength λ of the oscillation.

So, returning to the main discussion, we can say that a periodic oscillation $x(t)$ whose fluctuations are not too pathological (in a sense that will be clarified in the following discussion) can be written as

$$x(t) = \frac{1}{2} A_0 + \sum_{n=1}^{\infty} [A_n \cos(n\omega_1 t) + B_n \sin(n\omega_n t)]$$

$$= \frac{1}{2} A_0 + \sum_{n=1}^{\infty} \left[A_n \cos\left(\frac{2\pi n t}{T}\right) + B_n \sin\left(\frac{2\pi n t}{T}\right) \right] \qquad (2.6a)$$

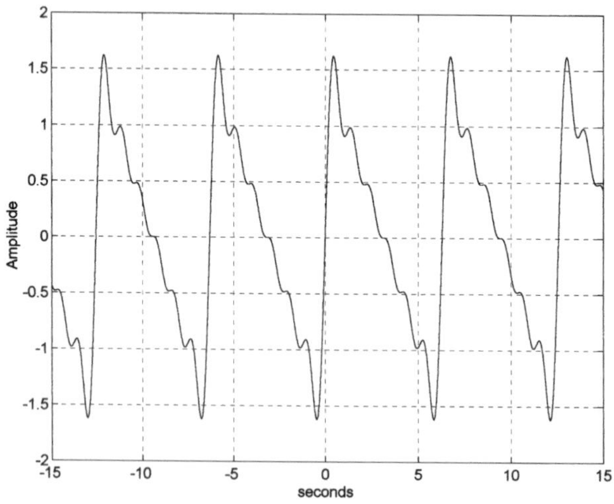

Fig. 2.3 Harmonic synthesis of sawtooth oscillation (first six terms).

and the Fourier coefficients A_n, and B_n are given by

$$A_n = \frac{2}{T} \int_{t_1}^{t_2} x(t)\cos(n\omega_1 t)dt$$

$$B_n = \frac{2}{T} \int_{t_1}^{t_2} x(t)\sin(n\omega_1 t)dt \qquad (2.6b)$$

where $2\pi/\omega_1 = T$ is the period of the oscillation and the limits of integration t_1, t_2 can be chosen at will provided that $t_2 - t_1 = T$ (the most frequent choices being obviously $t_1 = 0, t_2 = T$ or $t_1 = -T/2$, $t_2 = T/2$). Note that the 'static' term $(1/2)A_0$ has been included to allow the possibility that $x(t)$ oscillates about some value different from zero; furthermore, we write this term as $(1/2)A_0$ only for a matter of convenience. By so doing, in fact, the first of eqs (2.6b) applies correctly even for this term and reads

$$\frac{A_0}{2} = \frac{1}{T} \int_0^T x(t)dt \qquad (2.6c)$$

where we recognize the expression on the r.h.s. as the average value of our periodic oscillation.

As noted in Chapter 1, it often happens that complex notation can provide a very useful tool for dealing with harmonically varying quantities. In this light, it is not difficult to see that, by virtue of Euler's equations (1.7), the term $A\cos\omega t + B\sin\omega t$ can also be written as

$$C^+ e^{i\omega t} + C^- e^{-i\omega t} \qquad (2.7a)$$

where the complex amplitudes C^+ and C^- are given by

$$C^+ = \frac{1}{2}(A - iB)$$

$$C^- = \frac{1}{2}(A + iB) \qquad (2.7b)$$

and no convention of taking the real or imaginary part of eq (2.7a) is implied because C^+ and C^- combine to give a real resultant. Then, eq (2.6a) can also be written as

$$x(t) = \frac{1}{2}A_0 + \sum_{n=1}^{\infty} (C_n^+ e^{in\omega_1 t} + C_n^- e^{-in\omega_1 t})$$

$$= \sum_{n=-\infty}^{\infty} C_n e^{in\omega_1 t} \qquad (2.8)$$

where in the last expression the complex coefficients C_n are such that $C_n = C^*_{-n}$,

$$C_n = \frac{1}{2}(A_n - iB_n)$$

$$C_0 = \frac{1}{2}A_0 \qquad (2.9)$$

$$C_{-n} = \frac{1}{2}(A_n + iB_n)$$

and eqs (2.6b) translate into the single formula

$$C_n = \frac{1}{T}\int_0^T x(t)e^{-in\omega_1 t}dt \qquad (2.10)$$

Now, if we become a bit more involved with mathematical details we can turn our attention to some issues of interest. The first issue is how we obtain eqs (2.6b) (or eq. (2.10)). Assuming that our periodic function $x(t)$ can be written in the form

$$x(t) = \sum_{n=-\infty}^{\infty} C_n e^{in\omega_1 t}$$

let us multiply both sides of this equation by $e^{-im\omega_1 t}$ and integrate over one period, to get

$$\int_0^T x(t)e^{-im\omega_1 t}dt = \sum_{n=-\infty}^{\infty}\int_0^T C_n e^{in\omega_1 t}e^{-im\omega_1 t}dt \qquad (2.11)$$

Noting that

$$\int_0^T e^{im\omega_1 t}e^{-im\omega_1 t}dt = T\delta_{mn}$$

where δ_{mn} is the Kronecker delta defined by

$$\delta_{mn} = \begin{cases} 0 & m \neq n \\ 1 & m = n \end{cases} \qquad (2.12)$$

then

$$\int_0^T x(t)e^{-im\omega_1 t}dt = TC_n\delta_{mn} = TC_m$$

which is precisely eq (2.10).

The second issue we consider has to do with the conditions under which the series (2.6a) (or (2.8)) converges to the function $x(t)$. As a matter of fact, three basic assumptions have been tacitly made in the foregoing discussion: (1) the expansion (2.6a) is possible, (2) the function $x(t)e^{-im\omega_1 t}$ under the integral in eq (2.11) is integrable over one period and (3) the order of integration and summation can be reversed (r.h.s. of eq (2.11)) when calculating the Fourier coefficients (eqs (2.6b) or (2.10)). Various sets of conditions have been discovered which ensure that these assumptions are justified, and the Dirichlet theorem that follows expresses one of these possibilities:

Dirichlet theorem. If the periodic function $x(t)$ is single valued, has a finite number of maxima and minima and a finite number of discontinuities, and if $\int_0^T |x(t)|\,dt$ is finite, then the Fourier series (2.6a) converges to $x(t)$ at all the points where $x(t)$ is continuous. At jumps (discontinuities) the Fourier series converges to the midpoint of the jump. Moreover, if $x(t)$ is complex (a case of little interest for our purposes), the conditions apply to its real and imaginary parts separately.

Two things are worthy of note at this point. First, the usefulness of this theorem lies mainly in the fact that we do not need to test the convergence of the Fourier series. We just need to check the function to be expanded, and if the Dirichlet conditions are satisfied, then the series (when we get it) will converge to the function $x(t)$ as stated. Second, the Dirichlet conditions are sufficient for a periodic function to have a Fourier series representation, but not necessary. In other words, a given function may fail to satisfy the Dirichlet conditions but it may still be expandable in a Fourier series.

The third and last issue we consider is the relation between the mean squared value of $x(t)$, i.e.

$$\langle x^2(t) \rangle \equiv \frac{1}{T} \int_0^T x^2(t)\,dt \tag{2.13}$$

and the coefficients of its Fourier series. The result we will obtain is called Parseval's theorem and is very important in many practical applications where energy and/or power are involved. If, for example, we use the expansion in exponential terms (eq (2.8)) it is not difficult to see that

$$\langle x^2(t) \rangle = \frac{1}{T} \int_0^T |x(t)|^2\,dt$$

$$= \frac{1}{T} \int_0^T \left(\sum_n C_n e^{in\omega_1 t} \right) \left(\sum_m C_m^* e^{-im\omega_1 t} \right) dt$$

$$= \sum_n \sum_m C_n C_m^* \delta_{nm} = \sum_{n=-\infty}^{\infty} |C_n|^2 \tag{2.14a}$$

which shows that each Fourier component of $x(t)$, independently of the other Fourier components, makes its own separate contribution to the mean squared value. In other words, no cross (or 'interference') terms of the form $C_n C_m^*$ ($n \neq m$) appear in the expression of the mean squared value of $x(t)$. If, on the other hand, we use the series expansion in sine and cosine terms, it is immediate to determine that Parseval's theorem reads

$$\langle x^2(t) \rangle = \frac{A_0^2}{4} + \frac{1}{2} \sum_{n=1}^{\infty} (A_n^2 + B_n^2) \qquad (2.14b)$$

2.2.2   *Nonperiodic functions: Fourier transforms*

Nonperiodic functions cannot be represented in terms of a fundamental component plus a sequence of harmonics, and the series of harmonic terms must be generalized to an integral over all values of frequency. This can be done by means of the Fourier transform, which, together with the Laplace transform, is probably the most widely adopted integral transform in many branches of physics and engineering.

Now, given a sufficiently well-behaved nonperiodic function $f(t)$ (i.e. the counterpart of the function $x(t)$ of Section 2.2.1) the generalization of the Fourier series to a continuous range of frequencies can be written as

$$f(t) = \int_{-\infty}^{\infty} F(\omega) e^{i\omega t} d\omega \qquad (2.15)$$

provided that the integral exists. Then, the counterpart of eq (2.10) becomes

$$F(\omega) = \frac{1}{2\pi} \int_{-\infty}^{\infty} f(t) e^{-i\omega t} dt \qquad (2.16)$$

where the function $F(\omega)$ is called the Fourier transform of $f(t)$ and, conversely, $f(t)$ is called the inverse Fourier transform of $F(\omega)$. Also, the two functions $f(t)$ and $F(\omega)$ are called a Fourier transform pair and it can be said that eq (2.15) addresses the problem of Fourier synthesis, while eq (2.16) addresses the problem of Fourier analysis.

A set of conditions for the validity of eqs.(2.15) and (2.16) are given by the Fourier Integral theorem, which can be stated as follows:

Fourier integral theorem.   If a function $f(t)$ satisfies the Dirichlet conditions on every finite interval and if $\int_{-\infty}^{\infty} |f(t)| \, dt < \infty$, then eqs (2.15) and (2.16) are correct. In other words, if we calculate $F(\omega)$ as shown in eq (2.16) and substitute the result in eq (2.15) this integral gives the value of $f(t)$ where $f(t)$ is continuous, while at jumps, say at $t = t_0$, it gives the value of the midpoint of the jump, i.e. the value $[(f(t_0^-) + f(t_0^+)]/2$.

Without giving a rigorous mathematical proof (which can be found in many excellent mathematics texts), let us try to justify the formulas above. Starting from the Fourier series of the preceding section, it may seem reasonable to represent a nonperiodic function $f(t)$ by letting the interval of periodicity $T \to \infty$. So, let us rewrite eq (2.8) as

$$f(t) = \sum_{-\infty}^{\infty} C_n e^{in\omega_1 t} = \sum_{-\infty}^{\infty} C_n e^{i\omega_n t} \tag{2.17}$$

where $\omega_n \equiv n\omega_1$ and we know that the fundamental frequency ω_1 is related to the period T by $\omega_1 = 2\pi/T$. In this light, if we define $\Delta\omega \equiv \omega_{n+1} - \omega_n$, it is seen immediately that $\Delta\omega = 2\pi/T$ so that eq (2.10) becomes

$$C_n = \frac{\Delta\omega}{2\pi} \int_{-T/2}^{T/2} f(t) e^{-i\omega_n t} dt \tag{2.18}$$

and can be substituted in eq (2.17) to give

$$f(t) = \frac{1}{2\pi} \sum_{-\infty}^{\infty} \left(\int_{-T/2}^{T/2} f(\alpha) e^{-i\omega_n \alpha} d\alpha \right) e^{i\omega_n t} \Delta\omega$$

$$= \frac{1}{2\pi} \sum_{-\infty}^{\infty} \left(\int_{-T/2}^{T/2} f(\alpha) e^{i\omega_n(t-\alpha)} d\alpha \right) \Delta\omega = \frac{1}{2\pi} \sum_{-\infty}^{\infty} g(\omega_n) \Delta\omega \tag{2.19a}$$

where we define (note that α is a dummy variable of integration)

$$g(\omega_n) \equiv \int_{-T/2}^{T/2} f(\alpha) e^{i\omega_n(t-\alpha)} d\alpha \tag{2.19b}$$

Now, if we let $T \to \infty$, then $\Delta\omega \to 0$ and, in the limit

$$\frac{1}{2\pi} \sum_{-\infty}^{\infty} g(\omega_n) \Delta\omega \to \frac{1}{2\pi} \int_{-\infty}^{\infty} g(\omega) d\omega$$

$$g(\omega_n) \to \int_{-\infty}^{\infty} f(\alpha) e^{i\omega(t-\alpha)} d\alpha$$

so that eq (2.19a) becomes

$$f(t) = \frac{1}{2\pi} \int_{-\infty}^{\infty} g(\omega)d\omega = \frac{1}{2\pi} \int_{-\infty}^{\infty} \int_{-\infty}^{\infty} f(\alpha)e^{i\omega(t-\alpha)} d\alpha d\omega$$

$$= \int_{-\infty}^{\infty} \left(\frac{1}{2\pi} \int_{-\infty}^{\infty} f(\alpha)e^{-i\omega\alpha} d\alpha \right) e^{i\omega t} d\omega = \int_{-\infty}^{\infty} F(\omega)e^{i\omega t} d\omega \qquad (2.20)$$

where we have defined

$$F(\omega) \equiv \frac{1}{2\pi} \int_{-\infty}^{\infty} f(\alpha)e^{-i\omega\alpha} d\alpha \qquad (2.21)$$

Equations (2.20) and (2.21) are the same as (2.15) and (2.16). Equation (2.20) also justifies the different notations frequently encountered in various texts where, for example, the multiplying factor $1/(2\pi)$ is attached to the inverse transform. The difference is irrelevant from a mathematical point of view, but care must be exercised in practical situations when using tables of Fourier transform pairs. Some commonly encountered forms (other than our definition given in eqs (2.15) and (2.16)) are as follows:

$$f(t) = \frac{1}{2\pi} \int_{-\infty}^{\infty} F(\omega)e^{i\omega t} d\omega$$

$$(2.22a)$$

$$F(\omega) = \int_{-\infty}^{\infty} f(t)e^{-i\omega t} dt$$

$$f(t) = \frac{1}{\sqrt{2\pi}} \int_{-\infty}^{\infty} F(\omega)e^{i\omega t} d\omega$$

$$(2.22b)$$

$$F(\omega) = \frac{1}{\sqrt{2\pi}} \int_{-\infty}^{\infty} f(t)e^{-i\omega t} dt$$

$$f(t) = \int_{-\infty}^{\infty} F(\nu)e^{i2\pi\nu t} d\nu$$

$$(2.22c)$$

$$F(\nu) = \int_{-\infty}^{\infty} f(t)e^{-i2\pi\nu t} dt$$

where in eqs (2.22c) the ordinary frequency ν (in Hz) is used rather than the angular frequency $\omega = 2\pi\nu$ (in rad/s).

Example 2.1 Given a rectangular pulse in the form

$$f(t) = \begin{cases} 1/2 & -1 < t < 1 \\ 0 & \text{otherwise} \end{cases} \tag{2.23}$$

(a so-called 'boxcar' function) we want to investigate its frequency content. From eq (2.16) we have

$$F(\omega) = \frac{1}{2\pi} \int_{-1}^{1} \frac{1}{2} e^{-i\omega t} dt = -\frac{1}{4\pi i \omega} (e^{-i\omega t}\big|_{-1}^{1})$$

$$= \frac{1}{4\pi i \omega} (e^{i\omega} - e^{-i\omega}) = \frac{\sin \omega}{2\pi \omega} \tag{2.24a}$$

and the graph of $F(\omega)$ is sketched in Fig. 2.4.

Furthermore, if we want the Fourier transform in terms of ordinary frequency, from the condition $F(\omega)d\omega = F(\nu)d\nu$ and the fact that $d\omega = 2\pi d\nu$ we get

$$F(\nu) = 2\pi F(\omega) = \frac{\sin(2\pi\nu)}{2\pi\nu} \tag{2.24b}$$

Note that in this particular example $F(\omega)$ is a real function (and this is because the real function $f(t)$ is even). In general, this is not so, as the reader can verify, for instance, by transforming the simple exponential function

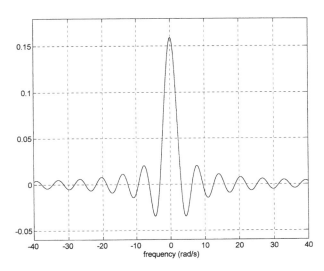

Fig. 2.4 Fourier transform of even 'boxcar' function.

$f(t) = \exp(-t)$, which we assume to be zero for $t < 0$. The result is

$$F(\omega) = \frac{1 - i\omega}{2\pi(1 + \omega^2)}$$

Alternatively, the reader can consider the Fourier transform of the pulse function $f(t) = 1/2$ for $0 < t < 2$ and zero otherwise (which is just a time-shifted version of (2.23)) and note that the result can be written as

$$F(\omega) = \frac{e^{-i\omega}}{2\pi\omega} \sin \omega$$

with exactly the same magnitude as the expression (2.24a) but different phase (because the phase depends on the arbitrary definition of time zero).

Before considering some properties of Fourier transforms, a word on notation is necessary: in what follows – and whenever convenient in the course of future chapters – the symbol $F(\omega) = \mathcal{F}\{f(t)\}$ will indicate that $F(\omega)$ is the Fourier transform of $f(t)$. Conversely, the symbol $f(t) = \mathcal{F}^{-1}\{F(\omega)\}$ will indicate that $f(t)$ is the inverse Fourier transform of $F(\omega)$. With this notation, the first easily verified property is linearity, that is

$$\mathcal{F}\{af(t) + bg(t)\} = a\mathcal{F}\{f(t)\} + b\mathcal{F}\{g(t)\} \tag{2.25}$$

where a and b are two constants and $f(t)$ and $g(t)$ are two Fourier-transformable functions.

A second important property has to do with the Fourier transform of the function $df(t)/dt$. Integrating by parts and taking into account the fact that – for the transform to exist – necessarily we must have $f(t) \to 0$ as $t \to \pm\infty$, we get

$$\mathcal{F}\left\{\frac{df}{dt}\right\} = \frac{1}{2\pi} \int_{-\infty}^{\infty} \frac{df}{dt} e^{-i\omega t} \, dt$$

$$= \frac{1}{2\pi} [fe^{-i\omega t}]_{-\infty}^{\infty} + \frac{i\omega}{2\pi} \int_{-\infty}^{\infty} fe^{-i\omega t} \, dt = i\omega\mathcal{F}\{f\} \tag{2.26a}$$

which is just a special case of the relation

$$\mathcal{F}\left\{\frac{d^n f(t)}{dt^n}\right\} = (i\omega)^n \mathcal{F}\{f(t)\} \tag{2.26b}$$

Clearly, eq (2.26b) is true only if $d^n f/dt^n$ is a Fourier-transformable function in its own right. By similar arguments, if $f(t)$ is absolutely integrable then the function $I(t) \equiv \int_{t_0}^{t} f(\gamma)d\gamma$ is continuous. If it is also absolutely integrable, then

$$\mathcal{F}\{I(t)\} = \frac{1}{2\pi} \int_{-\infty}^{\infty} I e^{-i\omega t} dt$$

$$= -\frac{1}{2\pi i \omega} [I e^{-i\omega t}]_{-\infty}^{\infty} + \frac{1}{2\pi i \omega} \int_{-\infty}^{\infty} f e^{-i\omega t} dt = \frac{1}{i\omega} \mathcal{F}\{f(t)\} \qquad (2.27)$$

Consider now two Fourier-transformable functions $f(t)$ and $g(t)$ with Fourier transforms $F(\omega)$ and $G(\omega)$, respectively. The function $w(t)$ defined by

$$w(t) \equiv f(t) * g(t) \equiv \int_{-\infty}^{\infty} f(t - \tau)g(\tau)d\tau \qquad (2.28)$$

is called the convolution of $f(t)$ and $g(t)$ and it has considerable importance when calculating the time response of linear systems. Provided that $\int_{-\infty}^{\infty} |f(t)g(t)| \, dt < \infty$ we can take the transform of $w(t)$ and use Fubini's theorem (see any book on calculus) to get

$$W(\omega) \equiv \mathcal{F}\{w(t)\} = \frac{1}{2\pi} \int_{-\infty}^{\infty} \int_{-\infty}^{\infty} f(t - \tau)g(\tau)e^{-i\omega t} d\tau dt$$

$$= \frac{1}{2\pi} \int_{\infty}^{\infty} g(\tau) \left[\int_{-\infty}^{\infty} f(t - \tau)e^{-i\omega t} dt \right] d\tau$$

Then, by making the change of variable $\varphi = t - \tau$, $d\varphi = dt$ in the t integral, it follows that

$$W(\omega) = \int_{-\infty}^{\infty} g(\tau)e^{-i\omega \tau} \left[\frac{1}{2\pi} \int_{-\infty}^{\infty} f(\varphi)e^{-i\omega \varphi} d\varphi \right] d\tau$$

$$= F(\omega) \int_{\infty}^{\infty} g(\tau)e^{-i\omega \tau} d\tau = 2\pi F(\omega)G(\omega) \qquad (2.29a)$$

which expresses the **convolution theorem**: in words, the transform of the convolution $f(t) * g(t)$ is 2π times the product of the transforms of $f(t)$ and $g(t)$. Again, the position of the 2π factor depends on the definition of Fourier transform that we use (eqs (2.22a–c)). By the same token, it is left to the reader to show that

$$\mathcal{F}^{-1}\{F(\omega) * G(\omega)\} = f(t)g(t) \qquad (2.29b)$$

The Parseval formula for Fourier transforms reads

$$\int_{-\infty}^{\infty} |f(t)|^2 dt = 2\pi \int_{-\infty}^{\infty} |F(\omega)|^2 d\omega \tag{2.30}$$

and can be derived as follows. Consider the integral

$$I \equiv \frac{1}{2\pi} \int_{-\infty}^{\infty} \int_{-\infty}^{\infty} f(t) F^*(\omega) e^{-i\omega t} dt d\omega$$

Assuming that the order of integration is immaterial (again Fubini's theorem), we can write it as

$$I = \int_{-\infty}^{\infty} F^*(\omega) \left[\frac{1}{2\pi} \int_{-\infty}^{\infty} f(t) e^{-i\omega t} dt \right] d\omega$$

$$= \int_{-\infty}^{\infty} F^*(\omega) F d\omega = \int_{-\infty}^{\infty} |F(\omega)|^2 d\omega \tag{2.31}$$

or, noting that $f(t) = \mathscr{F}^{-1}\{F(\omega)\}$ and hence $f^*(t) = \int_{-\infty}^{\infty} F^*(\omega) e^{-i\omega t} d\omega$, as

$$I = \frac{1}{2\pi} \int_{-\infty}^{\infty} f(t) \left[\int_{-\infty}^{\infty} F^*(\omega) e^{-i\omega t} d\omega \right] dt$$

$$= \frac{1}{2\pi} \int_{-\infty}^{\infty} f(t) f^*(t) dt = \frac{1}{2\pi} \int_{-\infty}^{\infty} |f(t)|^2 dt \tag{2.32}$$

so that eqs (2.31) and (2.32) together lead exactly to the Parseval theorem expressed by eq (2.30). Like its counterpart for periodic functions (eqs (2.14a) or (2.14b)), eq (2.30) is very important in practical applications when dealing with the squares of physical time-varying signals, i.e. quantities that have to do with the energy and/or the power associated with the signal itself.

2.2.3   *The bandwidth theorem (uncertainty principle)*

The idea of the uncertainty principle comes from the development of quantum mechanics during the 1920s. The principle, in its most famous form given by the physicist W. Heisenberg, reads $\Delta x \Delta p \geqslant \hbar$ and, broadly speaking, means that the product of the uncertainties in position (Δx) and momentum (Δp) for a particle is always greater than Planck's constant $\hbar = 1.05 \times 10^{-34}$ J s. More generally, the principle states that any phenomenon described by a pair of conjugate (or complementary) variables must obey some form of uncertainty relation.

For our purposes the conjugate variables are time t and frequency ν and the uncertainty principle in the field of signal analysis relates the duration of a transient function (a pulse) to the range of frequencies that are present in its transform. So, if we call Δt the duration of the transient $f(t)$ and $\Delta\nu$ the range of frequencies spanned by $F(\nu)$, then

$$\Delta t \Delta \nu \cong 1 \tag{2.33}$$

which is the so-called **bandwidth theorem**. The approximation sign in eq (2.33) means that the product lies usually in the range $0.5-3$ for most simple transients (try, for instance, the boxcar function of Example 2.1, where $\Delta\nu$ can be approximately taken as the width at the basis of the peak centred at $\nu = 0$), but its precise value is not really important for the essence of the argument. Stated simply, eq (2.33) shows that the two members of a Fourier transform pair – each one in its appropriate domain– cannot both be of short duration. The implications of this fact pervade the whole subject of signal analysis and have important consequences in both theory and practice.

In the course of this text, we will encounter many situations that agree with eq (2.33): for example, a lightly damped structure in free vibration will oscillate for a long time (large Δt) at its natural frequency ω_n, so that the spectrum of the recorded signal will show a very narrow peak at $\omega = \omega_n$ (small $\Delta\omega$); by contrast, if we want to excite many modes of a structure in a broad band of frequencies (large $\Delta\omega$), we can do so by a sudden blow of very short duration (small Δt) or by means of a random, highly erratic time signal, and so on. So, the essence of this short section is that the uncertainty principle represents an inescapable law of nature (or, at least, of the way in which we perceive the phenomena of nature) with which we must come to terms either by looking for trade-offs or by using different methods to extract all the desired information from a given signal.

Finally, if we observe that for slowly decaying signals it may not be so easy to define the quantities Δt and $\Delta\nu$, we can give the following definitions in order to make these concepts more precise:

$$(\Delta t)^2 = \frac{\displaystyle\int_{-\infty}^{\infty} t^2\,|f(t)|^2 dt}{\displaystyle\int_{-\infty}^{\infty} |f(t)|^2 dt}$$

$$\tag{2.34}$$

$$(\Delta \nu)^2 = \frac{\displaystyle\int_{-\infty}^{\infty} \nu^2\,|F(\nu)|^2 d\nu}{\displaystyle\int_{-\infty}^{\infty} |F(\nu)|^2 d\nu}$$

These integrals may not be easy to calculate, but they give a prescription applicable to a wide variety of signals: with these definitions it can be shown that the bandwidth theorem becomes $\Delta t \Delta \nu \geqslant 1/(4\pi)$.

2.3   Laplace transforms

From a general mathematical viewpoint, the Fourier transform is just a special case of a class of linear integral transforms which can be formally written as

$$T\{f(t)\} = \int_a^b K(t, u) f(t)dt \tag{2.35}$$

where the function $K(t, u)$ is called the kernel of the transform and $T\{f(t)\}$ is the integral transform of $f(t)$ with respect to the kernel $K(t, u)$. The various transforms are characterized by their kernels and their limits a and b so that, for example, for the Fourier transform $K(t, u) = (1/(2\pi))e^{-iut}$ and $a = -\infty$, $b = \infty$. Other types of integral transforms are, to name just a few, the Laplace transform, the Hankel transform and the Mellin transform.

Mathematically, their utility lies in the fact that, with certain kernels, ordinary differential equations yield algebraic expressions in the transformed variable (partial differential equations also lead to more tractable problems) and the solution in the transformed space is found more easily. Clearly, the problem of finding the solution in the original variable remains and the calculation of the inverse transformation is often the most difficult part of the whole procedure.

In its own right – together with the Fourier transform – the Laplace transform is the most popular integral transform. It is defined as

$$F(s) \equiv \mathscr{L}\{f(t)\} \equiv \int_0^\infty f(t)e^{-st}dt \tag{2.36}$$

where s is a complex quantity. Equation (2.36) is particularly useful when we are interested in functions whose Fourier transform does not exist, i.e. for example the function $f(t) = e^{bt}$, where $b > 0$. In many cases the trouble at $t \to \infty$ can be fixed by multiplication with a factor e^{-ct} if c is real and larger than some minimum value a (which, in turn, depends on the function to be transformed: for $f(t) = e^{bt}$, for example, we must have $c > b$), while the 'bad' behaviour at $t \to -\infty$ can be taken care of by noting that our interest often lies in functions $f(t)$ that are zero for $t < 0$. In these circumstances, the function $g(t) \equiv f(t)e^{-ct}$ does have a Fourier transform and we get

$$\int_0^\infty f(t)e^{-ct}e^{-i\omega t}dt = \int_0^\infty f(t)e^{-(c+i\omega)t}dt = F(c + i\omega)$$

Introducing the complex variable $s = c + i\omega$, we obtain exactly the integral of eq (2.36) which, in turn, exists only in the right half of the s-plane where $c = \text{Re}(s) > a$ (a is the minimum value for c mentioned above).

Therefore, broadly speaking, we can say that the Laplace kernel is a damped version of the Fourier kernel or, more properly, that the Fourier kernel is an undamped version of the Laplace kernel. The inverse transform, in turn, can be written as

$$f(t)e^{-ct} = \frac{1}{2\pi} \int_{-\infty}^{\infty} F(c + i\omega)e^{i\omega t} d\omega \tag{2.37}$$

Now, since $s = c + i\omega$ and hence $d\omega = ds/i$, eq (2.37) becomes

$$f(t)e^{-ct} = \frac{1}{2\pi i} \int_{c-i\infty}^{c+i\infty} F(s)e^{(s-c)t} ds = \frac{1}{2\pi i} e^{-ct} \int_{c-i\infty}^{c+i\infty} F(s)e^{st} ds$$

from which it follows that

$$\mathcal{L}^{-1}\{F(s)\} \equiv f(t) = \frac{1}{2\pi i} \int_{c-i\infty}^{c+i\infty} F(s)e^{st} ds \tag{2.38}$$

where the integral on the r.h.s. is known as Laplace inversion (or Bromwich) integral and is understood in the principal-value sense. It converges to $f(t)$ where $f(t)$ is continuous, while at jumps it converges to the midpoint of the jump; in particular, for $t = 0$ the integral converges to $(1/2)f(0^+)$.

Note that we have made the factor $1/(2\pi)$ appear in the inverse transform and not in the forward transform; this is merely a matter of convenience due to the fact that the Laplace transform is almost universally defined as in eq (2.36).

Although in practical situations one generally refers to tables of Laplace transform pairs (e.g. Erdélyi [1]), the integral of eq (2.38) can be evaluated as a complex (contour) integral by resorting to the theorem of residues and the following understanding:

1. For $t > 0$ the integration is calculated along the straight vertical line $\text{Re}(s) = c$, where c is large enough so that all the poles of the function $F(s)e^{st}$ lie to the left of this line. The contour is closed, forming a large semicircle around the complex plane to the left of the line.
2. For $t < 0$, since $f(t)$ must be zero, the contour is closed, forming a semicircle that extends to the right of the line (no poles in the contour). More details on complex integration can be found, for example, in Mathews and Walker [2] or Sidorov *et al.* [3].

Now, if we turn to some properties of Laplace transforms, it is not difficult to show that, first of all, the Laplace transform is a linear transformation and that the first derivative of $f(t)$ transforms into

$$\mathscr{L}\left\{\frac{df(t)}{dt}\right\} = sF(s) - f(0) \tag{2.39}$$

Also, a double integration by parts leads to

$$\mathscr{L}\left\{\frac{d^2f(t)}{dt^2}\right\} = s^2 F(s) - sf(0) - \left.\frac{df(t)}{dt}\right|_{t=0} \tag{2.40}$$

or, if we write

$$f^{(0)}(t) \equiv f(t), f^{(1)}(t) \equiv df(t)/dt, ..., f^{(n)}(t) \equiv d^n f(t)/dt^n$$

then, more generally

$$\mathscr{L}\left\{f^{(n)}(t)\right\} = s^n F(s) - \left[\sum_{j=0}^{n-1} s^j f^{(n-j-1)}(0)\right] \tag{2.41}$$

where – in eqs (2.39), (2.40) and (2.41) – it must be noted that 0 really means 0^+, the limit as zero is approached from the positive side.

If $f(t)$ is piecewise continuous, then the function $I(t) \equiv \int_0^t f(\alpha)d\alpha$ is continuous, $I(0) = 0$ and we get

$$\mathscr{L}\{I(t)\} \equiv \mathscr{L}\left\{\int_0^t f(\alpha)d\alpha\right\} = \frac{1}{s}\mathscr{L}\{f(t)\} \tag{2.42}$$

Furthermore, given two functions $f(t)$ and $g(t)$, the convolution theorem for Laplace transforms reads

$$\mathscr{L}\{f(t) * g(t)\} = \mathscr{L}\{f(t)\}\mathscr{L}\{g(t)\} \equiv F(s)G(s) \tag{2.43}$$

where in the rightmost expression we have defined $F(s)$ and $G(s)$, the Laplace transforms of $f(t)$ and $g(t)$, respectively.

Example 2.2.   With reference to the problem of finding the inverse Laplace transform of a function it is of interest to note that one often has to deal with functions which can be written as the ratio of two polynomials in s, i.e.

$$F(s) = \frac{P(s)}{Q(s)} \tag{2.44a}$$

where $Q(s)$ is of higher degree than $P(s)$. If we suppose that $Q(s)$ is a polynomial of order n with the n distinct roots $q_1, q_2, ..., q_n$, then the polynomial in the denominator can be written as $Q(s) = (s - q_1)(s - q_2) ... (s - q_n)$ and the function $F(s)$ as

$$F(s) = \sum_{j=1}^{n} \frac{A_j}{s - q_j} \tag{2.44b}$$

where the coefficients A_j are obtained from

$$A_j = \lim_{s \to q_j} [(s - q_j)F(s)] \tag{2.44c}$$

Since

$$\mathcal{L}^{-1}\left\{\frac{1}{s - q_j}\right\} = \exp(q_j t) \tag{2.45}$$

the inverse transform of $F(s)$ is

$$f(t) \equiv \mathcal{L}^{-1}\{F(s)\} = \sum_{j=1}^{n} A_j \exp(q_j t) \tag{2.46}$$

As a simple example, consider the function

$$F(s) = \frac{s + 1}{s(s + 2)}$$

where the roots of the denominator are $q_1 = 0, q_2 = -2$. From eq (2.44c) we get $A_1 = A_2 = 1/2$; therefore

$$F(s) = \frac{1}{2s} + \frac{1}{2(s + 2)}$$

and

$$f(t) = \frac{1}{2} + \frac{1}{2} e^{-2t}$$

As for the case of repeated roots and/or more difficult cases in general, the reader can find more details on partial-fraction theory in any basic text on calculus or algebra.

In the light of the foregoing discussion, we may observe that the main advantages of the Laplace transform over the Fourier transform are that:

1. The Laplace integral converges for a large class of functions for which the Fourier integral is divergent.
2. By virtue of eq (2.41), the Laplace transform provides automatically for the initial conditions at $t = 0$. This latter characteristic will be considered (and exploited) in the following examples and in future chapters.

Example 2.3. For this example we consider the ordinary (homogeneous) differential equation

$$\frac{d^2 f(t)}{dt^2} + a^2 f(t) = 0 \tag{2.47}$$

where a is a constant and we are given the initial conditions $f(0) = f_0$ and $df/dt \,|_{t=0} = f_0'$. Taking the Laplace transform on both sides gives

$$[s^2 F(s) - sf_0 - f_0'] + a^2 F(s) = 0$$

so that the solution in the s-domain is easily obtained as

$$F(s) = \frac{sf_0}{s^2 + a^2} + \frac{f_0'}{s^2 + a^2}$$

Then, from a table of Laplace transforms, we get $\mathcal{L}^{-1}\{s/(s^2 + a^2)\} = \cos at$ and $\mathcal{L}^{-1}\{(s^2 + a^2)^{-1}\} = a^{-1} \sin at$ and hence

$$f(t) = f_0 \cos at + \frac{f_0'}{a} \sin at \tag{2.48}$$

which is precisely the result that we will obtain in eqs (4.8) and (4.9) by standard methods.

On the other hand, if we consider the nonhomogeneous differential equation

$$\frac{d^2 f(t)}{dt^2} + a^2 f(t) = g(t) \tag{2.49}$$

where, for example, the forcing function is $g(t) = \hat{g} \cos \omega t$ and the initial conditions are still given by $f(0) = f_0$, $df/dt \,|_{t=0} = f_0'$, Laplace transformation of both sides leads, after a little manipulation, to

$$F(s) = \frac{\hat{g}s}{(s^2+a^2)(s^2+\omega^2)} + \frac{sf_0}{(s^2+a^2)} + \frac{f_0'}{(s^2+a^2)}$$

because $\mathcal{L}\{\cos\omega t\} = s/(s^2+\omega^2)$. In order to return to the time domain, we already know the inverse transform of the last two terms, while for the first term we can use the convolution theorem and write

$$\mathcal{L}^{-1}\left\{\left(\frac{s}{s^2+\omega^2}\right)\left(\frac{1}{s^2+a^2}\right)\right\} = \mathcal{L}^{-1}\left\{\frac{s}{s^2+\omega^2}\right\} * \mathcal{L}^{-1}\left\{\frac{1}{s^2+a^2}\right\}$$

$$= [\cos\omega t] * [a^{-1}\sin at] = \frac{1}{a}\int_0^t \cos(\omega\tau)\sin a(t-\tau)d\tau \quad (2.50)$$

so that the time domain solution of eq (2.49) is

$$f(t) = f_0\,\cos at + \frac{f_0'}{a}\sin at + \frac{1}{a}\int_0^t \cos\omega\tau\,\sin a(t-\tau)d\tau \quad (2.51)$$

(see also eq (5.19) with the appropriate modifications).

Partial differential equations can also be solved with the aid of Laplace transforms, the effect of Laplace transformation being the reduction of independent variables by one. Also, Laplace and Fourier transforms can be used together, as in the following example.

Example 2.4. (Initial-value problem of an infinitely long flexible string). This problem will be considered in detail in Section 8.2; see this section for the meaning of the symbols.
The equation of motion for the small oscillations of a vibrating string is

$$\frac{\partial^2 y}{\partial x^2} = \frac{1}{c^2}\frac{\partial^2 y}{\partial t^2} \quad (2.52)$$

Let the initial conditions of the problem be specified by the two functions

$$y(x,0) = u(x)$$

$$\left.\frac{\partial y(x,t)}{\partial t}\right|_{t=0} = w(x) \quad (2.53)$$

If we call $Y(x, s)$ the Laplace transform of $y(x, t)$ with respect to the time variable and transform eq (2.52), we get

$$\frac{\partial^2 Y}{\partial t^2} = \frac{1}{c^2} [s^2 Y - su(x) - w(x)] \tag{2.54}$$

Now, let us take the Fourier transform of eq (2.54) with respect to the space variable x and define

$$\Psi(k, s) \equiv \mathcal{F}\{Y(x, s)\}$$
$$U(k) \equiv \mathcal{F}\{u(x)\}$$
$$W(k) \equiv \mathcal{F}\{w(x)\}$$

where the variable k (the wavenumber) is the conjugate variable of x. We have

$$-k^2 \Psi(k, s) = \frac{s^2}{c^2} \Psi(k, s) - \frac{s}{c^2} U(k) - \frac{1}{c^2} W(k)$$

from which it follows that

$$\Psi(k, s) = \frac{sU(k) + W(k)}{s^2 + k^2 c^2} \tag{2.55}$$

Taking the inverse Laplace transform of (2.55) gives

$$\chi(k, t) \equiv \mathcal{L}^{-1}\{\Psi(k, s)\} = U(k)\cos(kct) + \frac{W}{kc} \sin(kct) \tag{2.56}$$

and the final solution can be obtained by inverse Fourier transformation of eq (2.56). The inverse Fourier transform of the first term on the r.h.s. is

$$\mathcal{F}^{-1}\{U(k)\cos(kct)\} = \frac{1}{2} \int_{-\infty}^{\infty} U(k)[e^{ikct} + e^{-ikct}]e^{ikx} dk$$

$$= \frac{1}{2} \int_{-\infty}^{\infty} U(k)e^{ik(x + ct)} dk + \frac{1}{2} \int_{-\infty}^{\infty} U(k)e^{ik(x - ct)} dk$$

$$= \frac{1}{2} [u(x + ct) + u(x - ct)] \tag{2.57a}$$

while for the second term we may note that

$$\mathscr{F}^{-1}\left\{W(k)\,\frac{\sin(kct)}{kc}\right\} = \mathscr{F}^{-1}\left\{W(k)\int_0^t \cos(kc\alpha)d\alpha\right\}$$

where α is a dummy variable of integration. Hence

$$\mathscr{F}^{-1}\left\{W(k)\,\frac{\sin(kct)}{kc}\right\} = \frac{1}{2c}\int_{x-ct}^{x+ct} w(\xi)d\xi \tag{2.57b}$$

and putting eqs (2.57a) and (2.57b) back together yields

$$y(x,t) = \mathscr{F}^{-1}\{\chi(k,t)\} = \frac{1}{2}\,[u(x+ct)+u(x-ct)] + \frac{1}{2c}\int_{x-ct}^{x+ct} w(\xi)d\xi \tag{2.58}$$

which is the same result we will obtain in eq (8.7a) by a different method.

Incidentally, note that for a string of finite length L, we can proceed from eq (2.54) by expanding the functions $Y(x,s)$, $u(x)$ and $w(x)$ in terms of Fourier series rather than taking their Fourier transforms.

2.4   The Dirac delta function and related topics

In physics and engineering it is often convenient to envisage some finite effect achieved in an arbitrarily small interval of time or space. For example, the cases with which we will have to deal in future chapters are impulsive forces which act for a very short time or localized actions applied at a given point on a structure.

In this regard, the general attitude consists of using mathematics as a means for handling relations reasonably and efficiently, rather than as a free-standing discipline. Under the guide of mathematical reasonableness and, hopefully, physical insight, we often interchange the order of summations of infinite series (implemented as appropriate limits when necessary) with integration or differentiation, without asking whether the interchange is rigorously warranted from a mathematical viewpoint. The final results are then checked *a posteriori* for physical sense and consistency, and only at that stage, in doubtful cases, can we reassess our mathematics. In their own right, the ideas of the Dirac delta function, the Heaviside function, etc. – which have been widely used since the beginning of this century without a rigorous mathematical justification – reflect very well this 'practical' attitude.

Before getting a bit more involved with the mathematical details, let us consider briefly the standard (practical) approach.

Intuitively, the Dirac delta function $\delta(t)$ (we write it as a function of time only for future convenience) is defined to be zero when $t \neq 0$ and infinite at $t = 0$ in such a way that the area under it is unity. So, if a and b are two positive numbers, we can write

$$\delta(t) = 0 \qquad t \neq 0 \tag{2.59a}$$

and

$$\int_{-a}^{b} \delta(t)dt = \int_{-\infty}^{\infty} \delta(t)dt = 1 \tag{2.59b}$$

which evidently implies $\int_{-a}^{-b} \delta(t)dt = \int_{a}^{b} \delta(t)dt = 0$. The weakness of the definition given by eqs (2.59a and b) is evident: strictly speaking, one can object that, first, $\delta(t)$ is not a function and, second, that eqs (2.59a) and (2.59b) are incompatible (if a function is zero everywhere except at one point, its integral, no matter what definition of integral is used, is necessarily zero).

Nonetheless, one generally ignores these objections and proceeds by stating that the definition of eqs (2.59a and b) can be replaced by the following and equivalent definition: let $g(t)$ be any well-behaved function which we may call a 'test function', then

$$\int_{-a}^{b} g(t)\delta(t)dt = g(0) \tag{2.60}$$

the rationale behind eq (2.60) being that, by virtue of (2.59a), there is no contribution to the integral from anywhere where $g(t) \neq g(0)$ so that $g(t)$ can be replaced by the constant $g(0)$ which, in turn, can be moved outside the integral sign. The equality in (2.60) then follows from eq. (2.59b).

Equation (2.60) is the usual definition commonly found in textbooks, which is sometimes called the 'weak definition' of $\delta(t)$ because it leaves undefined the value of the integral when one of the limits of integration is precisely zero. As a matter of fact, eq (2.60) and its consequences are compatible with many different assignments of the integrals

$$\int_{left} \delta(t)dt \equiv \int_{-a}^{0} \delta(t)dt$$

$$\int_{right} \delta(t)dt \equiv \int_{0}^{b} \delta(t)dt \tag{2.61}$$

which are only subject to the condition $\int_{left} \delta(t)dt + \int_{right} \delta(t)dt = 1$. Therefore, any definition that assigns values to the integrals of eqs (2.61) can be called a 'strong' definition of $\delta(t)$. In practice, however, provided that no integration stops at zero (as is often the case), no strong definition is necessary and we can simply ignore the difference. On the other hand, when one of the integration limits is zero, we definitely need a strong definition, also in the light of the fact that there exist equations that are 'weakly true but strongly false'.

On physical grounds, the delta function may arise, for example, in a situation in which we consider a sudden impulsive blow applied to a mass m. If this impulsive force $f(t)$ lasts from $t = t_0$ to $t = t_1$, from Newton's second law we get

$$\int_{t_0}^{t_1} f(t)dt = \int_{t_0}^{t_1} m\frac{dv}{dt}dt = \int_{v_0}^{v_1} mdv = m(v_1 - v_0) \tag{2.62}$$

stating that the impulse of $f(t)$ equals the change in momentum of the mass. Also, note that the precise shape of $f(t)$ is irrelevant, the relevant quantity being its area between the two instants of time. So, let this area be unity and let, for simplicity, the initial velocity of the mass be $v_0 = 0$. If $t_1 - t_0$ is very small we can simply ignore the motion of the mass during this time and assume that its momentum changed abruptly from zero to mv_1. As $t_1 - t_0$ gets smaller and smaller, the graph of momentum versus time approaches a step Heaviside function and – since (see eq (2.62)) the function $f(t)$ is the slope of this graph – we are practically requiring $f(t)$ to be the derivative of such a step function, i.e. infinite at $t = t_0$ and zero otherwise. No ordinary function possesses these properties; however, since we are not so much interested in $f(t)$ itself but in the effect it produces, it can be convenient to introduce an entity – identified by the symbol $\delta(t - t_0)$ – that represents this infinitely high and infinitely narrow 'unnatural' force. In this light, $\delta(t)$ can physically represent a sudden blow applied at $t_0 = 0$.

So, returning to the general discussion, it is not difficult to show that the following (weakly true) properties of $\delta(t)$ hold:

1. $$\int_a^b g(t)\delta(t - c)dt = \begin{cases} g(c) & a < c < b \\ 0 & \text{otherwise} \end{cases} \tag{2.63}$$

Also

2. $$\delta(\alpha t) = \frac{1}{|\alpha|}\delta(t) \tag{2.64}$$

where $\alpha \neq 0$. From the evenness of $\delta(t)$ (i.e. $\delta(-t) = \delta(t)$ or $\delta(t - \tau) = \delta(\tau - t)$) it follows that

3. $$\int_{-\infty}^{\infty} g(\tau)\delta(t - \tau)d\tau = \int_{-\infty}^{\infty} g(\tau)\delta(\tau - t)d\tau = g(t) \tag{2.65}$$

with the consequence that $g(t) * \delta(t) = g(t)$, meaning that $\delta(t)$ is the identity element in the operation of convolution. Next, if we consider the step Heaviside function

$$\theta(t) \equiv \begin{cases} 0 & t < 0 \\ 1 & t > 0 \end{cases} \tag{2.66}$$

we can formally write

4. $\delta(t) = \dfrac{d}{dt}\,\theta(t)$ (2.67a)

and also

5. $\theta(t) = \displaystyle\int_{-\infty}^{t} \delta(\tau)d\tau$ (2.67b)

Then, if we turn our attention to the derivative $\delta'(t)$ of $\delta(t)$, we may note that this function is not directly determined by the definition of $\delta(t)$ so that, consequently, it is meaningless to ask what this derivative is. Instead, in order for the ordinary rules of integration by parts remain valid, the approach is to define $\delta'(t)$ in such a way that $\delta'(t) = 0$ when $t \neq 0$. In symbols, if $g(t)$ is a differentiable test function, we get

$$\int_a^b g(t)\,\frac{d\delta(t)}{dt}\,dt = g(t)\delta(t)\big|_a^b - \int_a^b \frac{dg(t)}{dt}\,g(t) = -\frac{dg(t)}{dt}\bigg|_{t=0}$$

and hence

6. $\displaystyle\int_{-\infty}^{\infty} g(t)\delta'(t)dt = -g'(0)$ (2.68)

so that, by extension, the nth derivative $\delta^{(n)}(t)$ is required to vanish at $t \neq 0$, and for a well-behaved function $g(t)$ we get

7. $\displaystyle\int_{-\infty}^{\infty} g(t)\delta^{(n)}(t)dt = (-1)^n g^{(n)}(0)$ (2.69)

The formal 'proofs' of all the above properties are not difficult and are left to the reader. Other properties will be considered if and whenever needed in the course of future chapters.

Nonetheless, it is worth noting that the above properties do not follow from the definition of $\delta(t)$, but they are mere assumptions consistent with the formal properties of the integral.

Similarly, the fact that the Dirac function can be considered as the limit of a sequence of functions $\chi_\varepsilon(t)$ (where ε is a 'smallness' parameter) by writing

$$\delta(t) = \lim_{\varepsilon \to 0} \chi_\varepsilon(t) \tag{2.70}$$

is not justified in the ordinary sense. Let us ignore this for the moment and proceed in our discussion by giving some examples of how we can see $\delta(t)$ as a limit of this kind. If we take a sequence of functions $\chi_\varepsilon(t)$ whose integral $\int_{-\infty}^{\infty} \chi_\varepsilon(t)dt$ equals unity for any value of ε, some common examples of eq (2.70) are as follows:

1. Gaussian

$$\chi_\varepsilon(t) = \frac{1}{\varepsilon\sqrt{\pi}} e^{-t^2/\varepsilon^2} \tag{2.71a}$$

2. Lorentzian (resonance shape)

$$\chi_\varepsilon(t) = \frac{1}{\pi} \frac{\varepsilon}{(t^2 + \varepsilon^2)} \tag{2.71b}$$

3. Dirichlet (diffraction peak)

$$\chi_\varepsilon(t) = \frac{\sin(t/\varepsilon)}{\pi t} \tag{2.71c}$$

4. Square step (boxcar function of width 2ε and height $1/(2\varepsilon)$)

$$\chi_\varepsilon(t) = \frac{1}{2\varepsilon} \theta(\varepsilon - |t|) \tag{2.71d}$$

where $\theta(t)$ is the Heaviside function of eq (2.66).

In the limit of eq (2.70), these functions formally satisfy the defining property of $\delta(t)$. So, for example, for the Gaussian form of $\chi_\varepsilon(t)$ we can verify that eq (2.60) holds, i.e.

$$\lim_{\varepsilon \to 0} \int_{-\infty}^{\infty} g(t) \frac{1}{\varepsilon\sqrt{\pi}} e^{-t^2/\varepsilon^2} dt = \lim_{\varepsilon \to 0} \frac{1}{\varepsilon\sqrt{\pi}} \int_{-\infty}^{\infty} \varepsilon g(\varepsilon\xi) e^{-\xi^2} d\xi$$

$$= \frac{1}{\sqrt{\pi}} g(0) \int_{-\infty}^{\infty} e^{-\xi^2} d\xi = \frac{1}{\sqrt{\pi}} g(0) \sqrt{\pi} = g(0)$$

where in the second integral we introduced the variable $\xi = t/\varepsilon$ and the well-behaved test function $g(t)$ is assumed to vanish fast enough at infinity to ensure the convergence of any integral in which it occurs.

For purpose of illustration, Figs 2.5(a) and 2.5(b) show a sequence of such functions for three decreasing values of ε: Gaussian functions with unit area in Fig. 2.5(a) and Lorentzian functions with unit area in Fig. 2.5(b). From these graphs it is evident that, as ε gets smaller, the functions $\chi_\varepsilon(t)$ become taller and narrower and approach the delta function in the limit of $\varepsilon \rightarrow 0$.

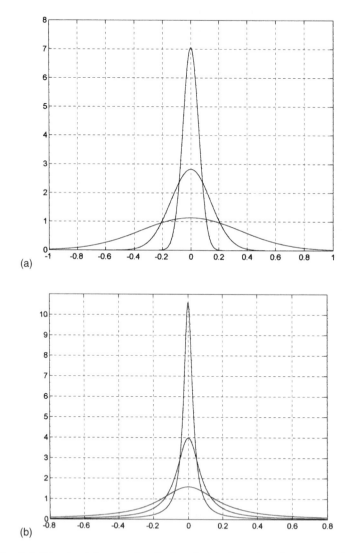

(a)

(b)

Fig. 2.5 Dirac's delta as the limit of: (a) Gaussian functions; (b) Lorentzian functions.

(Incidentally, note that all the functions $\chi_\varepsilon(t)$ are symmetrical (even) and imply the strong definition $\int_{left} \delta(t)dt = \int_{right} \delta(t)dt = 1/2$. This is not strictly necessary and, for example, the lopsided functions

$$\chi_\varepsilon(t) = \frac{2\theta(t)}{\varepsilon\sqrt{\pi}} e^{-t^2/\varepsilon^2}$$

have a limit (for $\varepsilon \to 0$) that satisfies the weak definition of $\delta(t)$. However, in this case the strong definition $\int_{left}\delta(t)dt = 0$ and $\int_{right}\delta(t)dt = 1$ is implied.)

In the light of the discussion above, we can also obtain a Fourier integral representation of $\delta(t)$. In fact, if $A \equiv 1/\varepsilon$, eq (2.70) and (2.71c) give

$$\delta(t) = \lim_{A \to \infty} \frac{\sin At}{\pi t} \tag{2.72}$$

Then, noting that

$$\frac{\sin At}{\pi t} = \frac{1}{2\pi} \int_{-A}^{A} e^{i\omega t} d\omega$$

we can write eq (2.72) as

$$\delta(t) = \lim_{A \to \infty} \frac{\sin At}{\pi t} = \lim_{A \to \infty} \frac{1}{2\pi} \int_{-A}^{A} e^{i\omega t} d\omega = \frac{1}{2\pi} \int_{-\infty}^{\infty} e^{i\omega t} d\omega \tag{2.73}$$

which also tells us that, formally, $\delta(t) = \mathscr{F}^{-1}\{1/(2\pi)\}$. This, in turn, implies $\mathscr{F}\{\delta(t)\} = 1/(2\pi)$, and in fact, from eq (2.60) we get

$$\mathscr{F}\{\delta(t)\} = \frac{1}{2\pi} \int_{-\infty}^{\infty} \delta(t)e^{-i\omega t} dt = \frac{1}{2\pi} \tag{2.74}$$

More generally

$$\mathscr{F}\{\delta(t-\tau)\} = \frac{1}{2\pi} \int_{-\infty}^{\infty} \delta(t-\tau)e^{-i\omega t} dt = \frac{e^{-i\omega\tau}}{2\pi} \tag{2.75}$$

which leads to the following Fourier representation of $\delta(t-\tau)$:

$$\delta(t-\tau) = \frac{1}{2\pi} \int_{-\infty}^{\infty} e^{i\omega(t-\tau)} d\omega \tag{2.76}$$

The representation (2.76), in turn, is often used to obtain many important properties of Fourier transforms. As an example, we can obtain Parseval's theorem. If we suppose that $f(t) = \mathscr{F}^{-1}\{F(\omega)\}$, by introducing the dummy variables of integration α and β, we can formally write

$$\int_{-\infty}^{\infty} |f(t)|^2 dt = \int_{-\infty}^{\infty} \left(\int_{-\infty}^{\infty} F(\alpha)e^{i\alpha t} d\alpha \right) \left(\int_{-\infty}^{\infty} F^*(\beta)e^{-i\beta t} d\beta \right) dt$$

$$= \int_{-\infty}^{\infty} \int_{-\infty}^{\infty} \int_{-\infty}^{\infty} F(\alpha)F^*(\beta)e^{it(\alpha - \beta)} d\alpha d\beta dt$$

$$= \int_{-\infty}^{\infty} \int_{-\infty}^{\infty} F(\alpha)F^*(\beta) \left(\int_{-\infty}^{\infty} e^{it(\alpha - \beta)} dt \right) d\alpha d\beta$$

$$= 2\pi \int_{-\infty}^{\infty} F^*(\beta) \left(\int_{-\infty}^{\infty} F(\alpha)\delta(\alpha - \beta)d\alpha \right) d\beta$$

$$= 2\pi \int_{-\infty}^{\infty} F^*(\beta)F(\beta)d\beta$$

$$= 2\pi \int_{-\infty}^{\infty} |F(\beta)|^2 d\beta$$

which is precisely the Parseval formula of eq (2.30).

2.4.1 Brief introduction to the theory of distributions

According to the developments of the preceding section, we note that, ultimately, the delta function and its derivatives make sense only when multiplied by a sufficiently well-behaved function and integrated over some finite or infinite domain. As a matter of fact, it is only in this context that they are ever required in practice, and equations where they appear without an integral sign should be considered just as convenient half-way stages that allow a larger degree of flexibility in calculations. Since, on a number of occasions in future chapters, we will need and use mathematical entities such as the Dirac function, the Heaviside function, etc. in connection with differential equations, the scope of this short section of a mathematical nature is – without claim of completeness – to give the reader a general idea of how the Dirac delta function itself and the various relations involving this function can be justified on the basis of the rigorous theory of distributions. We assume that the interested reader has some familiarity with mathematical notations, mathematical terminology and with the Lebesgue theory of measure and integration (for most practical purposes there is no difference between Lebesgue and Riemann integrals; however, the Lebesgue theory is more useful for theoretical purposes because certain important theorems hold for Lebesgue integrals but not for Riemann integrals).

The reason for including this section has to do with the fact that an improper use of entities such as the delta function can lead to serious errors which can go unnoticed if the 'usual' definitions are taken too literally. For example, at the beginning of Section 2.4 – following the customary non-rigorous approach – we 'defined' $\delta(t)$ by means of the two relations

$$\delta(t) = \begin{cases} 0 & t \neq 0 \\ \infty & t = 0 \end{cases}$$

and

$$\int_{-\infty}^{\infty} \delta(t)dt = 1$$

Now, consider the function $2\delta(t)$. In accordance with the second relation we must have $\int_{-\infty}^{\infty} 2\delta(t)dt = 2$ but, owing to the first relation, the function $2\delta(t)$ must be equal to infinity at $t = 0$. We are then led to the absurd result $2 = 1$. This simple but meaningful example shows that care must be exercised in these cases and that it is of great help to have a general idea of the way in which mathematics deals (rigorously) with this subject so that, when in doubt, there is the possibility of double-checking the results and the mathematical manipulations which are often obtained and performed by following the non-rigorous approach.

Returning to the main discussion of this section, we can say that, in essence, ordinary differential calculus sometimes runs into difficulties because of the existence of nondifferentiable functions. The theory of distributions – initially developed by Schwartz during the late 1940s-early 1950s – frees differential calculus from these difficulties by extending it to a class of objects called distributions or generalized functions, the main ideas behind this extension being as follows:

1. Every continuous function is a distribution.
2. Every distribution has partial derivatives which are themselves distributions; moreover, for differentiable functions, the new notion of derivative coincides with the old one.
3. The usual formal rules of calculus hold.
4. There is a supply of convergence theorems that is adequate for handling the usual limit process.

In this light, it is clear that the class of distributions is larger than the class of differentiable functions to which calculus applies in its original form.

For example, if we consider functions of one variable and integrals are taken with respect to the Lebesgue measure, we say that a function $f(x)$ is locally integrable if it is measurable and $\int_K |f| \, dx < \infty$ for every compact $K \subset \mathfrak{R}$ ($\mathfrak{R}$ being the real line). Now, if we consider f as something that assigns the number $\int f(x)\varphi(x)dx$ to every suitably chosen 'test function' φ rather than as something that assigns the number $f(x)$ to each $x \in \mathfrak{R}$, we adopt a new point of view that turns out to be particularly useful in physics and engineering (in fact, physicists have been improperly using distributions long before the mathematical theory was constructed). Obviously, the first thing to do is to specify the class of test functions.

So, let Ω be an open, nonempty set of $\mathfrak{R}^n$ (for our purposes we generally have $\Omega = \mathfrak{R}$) and let $\mathscr{D}(\Omega)$ be the vector space of all $\varphi \in C^\infty(\Omega)$ whose support is compact. Recall that supp(φ), i.e. the support of the function φ, is the closure of the set $\{x \in W : \varphi(x) \neq 0\}$. In simpler terms, these latter statements mean that (1) φ is continuous in Ω with all its derivatives and (2) φ is identically zero outside its support which, in turn, is a compact (a finite interval in $\mathfrak{R}$) subset of Ω.

At this point mathematicians introduce an appropriate topology τ on $\mathscr{D}(\Omega)$ and indicate with $\mathscr{D}(\Omega)$ the topological vector space $(\mathscr{D}(\Omega), \tau)$. We simply take this for granted and only point out two important facts:

1. In $\mathscr{D}(\Omega)$, every Cauchy sequence (see Section 2.5 on Hilbert spaces for the definition of a Cauchy sequence) converges.
2. The convergence of the sequence $\{\varphi_j\}$ to φ (as $j \to \infty$) implies that (a) there is a compact set $K \subset \Omega$ such that supp(φ_j) $\subset K$ for every $j = 1, 2, \ldots$, supp(φ) $\subset K$ and (b) the derivatives $D^\alpha \varphi_j$ converge uniformly to $D^\alpha \varphi$ for every multi-index α.

(A word on notation to explain the shorthand symbol $D^\alpha \varphi$: we define a multi-index as an ordered n-tuple of non-negative numbers $\alpha = (\alpha_1, \alpha_2, \ldots, \alpha_n)$ whose 'order' is $|\alpha| \equiv \alpha_1 + \ldots + \alpha_n$ and we write

$$D^a \varphi \equiv \frac{\partial^{|\alpha|} \varphi}{\partial x_1^{\alpha_1} \ldots \partial x_n^{\alpha_n}}$$

with the assumption that $D^0 \varphi = \varphi$. In this light, condition (b) means that the derivatives of φ_j of every order converge uniformly to the derivatives of φ of the same order.)

We are now in a position to give the definition of distribution: a continuous linear functional on $\mathscr{D}(\Omega)$ is called a distribution in Ω where, by the term 'functional' it is customary to denote any operator that maps a function belonging to an appropriate linear space ($\mathscr{D}(\Omega)$ in this case) to a real or complex number.

The space of all distributions in Ω is denoted by $\mathscr{D}'(\Omega)$. In other words, a distribution $T \in \mathscr{D}'(\Omega)$ is a mapping $T: \mathscr{D}(\Omega) \to \mathscr{C}$ or $\mathfrak{R}$ ($\mathscr{C}$ is the complex field and $\mathfrak{R}$ is the real field) which is continuous with respect to the topology τ, this latter statement meaning that $\{\varphi_j\} \to \varphi$ in $\mathscr{D}(\Omega)$ implies $T(\varphi_j) \to T(\varphi)$.

Note that, since it is often customary to indicate the action of a linear functional $T \in \mathscr{D}'(\Omega)$ as $T(\varphi) \equiv \langle T, \varphi \rangle$ ($T(\varphi)$ is a complex or real number) and since linearity means that

$$
\begin{aligned}
\langle T_1, \varphi_1 + \varphi_2 \rangle &= \langle T, \varphi_1 \rangle + \langle T, \varphi_2 \rangle \\
\langle T, a\varphi \rangle &= a\langle T, \varphi \rangle
\end{aligned}
\tag{2.77}
$$

we can immediately consider $\mathscr{D}'(\Omega)$ as a linear space by means of the 'natural' definitions

$$
\begin{aligned}
\langle T_1 + T_2, \varphi \rangle &\equiv \langle T_1, \varphi \rangle + \langle T_2, \varphi \rangle \\
\langle aT, \varphi \rangle &\equiv a\langle T, \varphi \rangle
\end{aligned}
\tag{2.78}
$$

which, in mathematical terminology, means that $\mathscr{D}'(\Omega)$ is the 'dual space' of $\mathscr{D}(\Omega)$.

At this point, we can make some considerations which are of direct importance for our purposes:

- If $f(x)$ is a locally integrable function and dx denotes the Lebesgue measure, the relation

$$
\langle T_f, \varphi \rangle \equiv \int_{-\infty}^{\infty} f(x)\varphi(x)dx
\tag{2.79}
$$

  defines a linear functional T_f so that it is customary to identify the distribution T_f with f, write (improperly) f in place of T_f and say that such distributions 'are' functions.

- Every $x \in \Omega$ determines a linear functional δ_x on $\mathscr{D}(\Omega)$ defined by

$$
\langle \delta_x, \varphi \rangle \equiv \varphi(x)
\tag{2.80}
$$

  and called the Dirac distribution. If $x = 0$ one simply writes $\delta_0 = \delta$ and we have $\langle \delta, \varphi \rangle \equiv \varphi(0)$.

Distributions that can be represented in the form of eq (2.79) are generally called 'regular', while distributions which cannot be reduced to such a form – δ for example, and its derivatives – are called 'singular' (or, sometimes 'symbolic functions'). In this regard, when we write an equation such as eq (2.60), we are treating the Dirac distribution as if it were a regular distribution, i.e. we introduce a (nonexistent) locally integrable function $\delta(x)$ and write

$$
\int_{-\infty}^{\infty} \delta(x)\varphi(x)dx = \varphi(0)
\tag{2.81}
$$

which is a formal justification of eqs (2.59a and b).

Without getting into details that are beyond our scope, we can also take for granted that the topology defined on $\mathcal{D}(\Omega)$ induces a topology on $\mathcal{D}'(\Omega)$ (called the weak* ('weak-star') topology by mathematicians) and the statement $T_j \rightarrow T$ in $\mathcal{D}'(\Omega)$ means that

$$\lim_{j \to \infty} \langle T_j, \varphi \rangle = \langle T, \varphi \rangle \qquad \varphi \in \mathcal{D}(\Omega)$$

Then, it can be shown that every distribution can be obtained as a limit of a sequence of regular distributions, i.e. for any $T \in \mathcal{D}'(\Omega)$ we can find a sequence of functions $f_j(x)$ such that, as $j \rightarrow \infty$

$$\int_{-\infty}^{\infty} f_j(x)\varphi(x) \rightarrow \langle T, \varphi \rangle \tag{2.82}$$

and it is noteworthy that such a sequence can always be a sequence of functions that belong to $C^\infty(\Omega)$ ($C^\infty(\mathcal{R})$ in the case of eq (2.82)). A typical example is given by sequences of functions that approximate the Dirac distribution (Section 2.4); it is only in the sense of distributions that the passage to the limit as $\varepsilon \rightarrow 0$ of eqs (2.71a, b and c) is justified and we can see the Dirac delta function as a limit of ordinary functions.

We are now in a position to consider the subject of calculus with distributions. Given a distribution $T \in \mathcal{D}'(\Omega)$ we define the derivatives of T by means of the relation

$$\langle D^\alpha T, \varphi \rangle \equiv (-1)^{|\alpha|} \langle T, D^\alpha \varphi \rangle \tag{2.83a}$$

noting that $D^\alpha T \in \mathcal{D}'(\Omega)$ because the r.h.s. of eq (2.83a) is always a well-defined quantity. In its simpler form (first derivative) eq (2.83a) reads

$$\left\langle \frac{\partial T}{\partial x_j}, \varphi \right\rangle = -\left\langle T, \frac{\partial \varphi}{\partial x_j} \right\rangle \tag{2.83b}$$

From this definition it can be shown that if f is a continuous function which admits a continuous derivative $\partial f/\partial x_j \equiv f'$ in the ordinary sense, and if T_f is the (regular) distribution associated with f, then, in the sense of distributions $\partial T_f/\partial x_j = T_{f'}$. In fact, integration by parts gives

$$\left\langle \frac{\partial T_f}{\partial x_j}, \varphi \right\rangle = \int_\Omega \frac{\partial f(x)}{\partial x_j} \varphi(x)dx$$

$$= -\int_\Omega f(x) \frac{\partial \varphi(x)}{\partial x_j} dx = -\left\langle T_f, \frac{\partial \varphi}{\partial x_j} \right\rangle$$

because $\varphi = 0$ on the boundary of Ω. (In essence, the definition of eq (2.83a) is intentionally tailored on the property of integration by parts: since eq (2.83a) holds for continuous functions with continuous derivatives, we extend it to distributions by taking it as the definition of generalized derivative.) It should be noted, however, that it is not always true that the distribution derivative $\partial T_f / \partial x_j$ equals $T_{f'}$ (or more generally, it is not always true that $D^\alpha T_f = T_{D^\alpha f}$) even when both quantities have a meaning. In particular, if Ω is a segment of $\mathcal{R}$, it can be shown that $DT_f = T_{Df}$ (or, in usual notation $T_f' = T_{f'}$) if and only if the function $f(x)$ is absolutely continuous.

At this point it is important to note the considerable implications of eq (2.83a) which can be summarized as follows:

- Distributions always possess derivatives of any order with respect to any variable.
- If a sequence of distributions $\{T_j\}$ converges to T, then the sequence of derivatives $\{T_j'\}$ converges to T' (or, equivalently, every convergent series of distributions can be differentiated term by term any number of times).

Furthermore, when ordinary derivatives exist and are well-behaved, the derivatives in the sense of distributions (the so-called generalized derivatives) coincide with the ordinary derivatives; if, on the other hand, ordinary derivatives do not exist, then the generalized derivatives with all their desirable properties (say, for example, $D^\alpha D^\beta T = D^\beta D^\alpha T$ for all multi-indices α and β, a property which may not be valid for ordinary derivatives) 'take over' and allow a much larger degree of flexibility, while at the same time maintaining the formal rules of ordinary calculus.

Let us now consider a few examples. The first typical example is given by the locally integrable Heaviside function $\theta(x)$ (eq (2.66)) for which the equation (see also eq (2.67a))

$$\frac{d\theta(x)}{dx} = \delta(x) \tag{2.84}$$

has a meaning only in the sense of distributions. In fact,

$$\int_{-\infty}^{\infty} \frac{d\theta(x)}{dx} \varphi(x)dx = -\int_{-\infty}^{\infty} \theta(x)\frac{d\varphi(x)}{dx}dx$$

$$= -\int_0^{\infty} \frac{d\varphi(x)}{dx}dx = \varphi(0) = \int_{-\infty}^{\infty} \delta(x)\varphi(x)dx$$

where in the last equality we adopted the notation of eq (2.81). Note that, strictly speaking, the distribution θ is the functional defined by the relation

$$\langle \theta, \varphi \rangle = \int_0^\infty \varphi(x)\,dx$$

The second example is given by the derivatives of $\delta(x)$ which, in the distribution sense, are well-defined functionals. In fact, in the case of the first derivative, for example, we have

$$\int_{-\infty}^\infty \delta'(x)\varphi(x)\,dx = -\int_{-\infty}^\infty \delta(x)\varphi'(x)\,dx = -\varphi'(0)$$

which justifies eq (2.68) and, by extension, eq (2.69).

Now let $f(x)$ be a function which is continuous everywhere except at the origin where it has a jump discontinuity of a_1, that is

$$a_1 \equiv f(0^+) - f(0^-)$$

and suppose further that $f(x)$ is differentiable everywhere except at $x = 0$. Its ordinary derivative is defined for $x < 0$ and $x > 0$, but it is undefined at $x = 0$. Now, since the r.h.s. of eq (2.83b) reads

$$-\langle f, \varphi' \rangle = -\int_{-\infty}^\infty f(x)\varphi'(x)\,dx$$

$$= -\int_{-\infty}^0 f(x)\varphi'(x)\,dx - \int_0^\infty f(x)\varphi'(x)\,dx$$

$$= \int_{-\infty}^\infty f'(x)\varphi(x)\,dx + a_1\varphi(0) = \int_{-\infty}^\infty [f'(x) + a_1\delta(x)]\varphi(x)\,dx$$

it follows from eq (2.83b) that the generalized derivative $\{f'(x)\}$ is given by

$$\{f'(x)\} = f'(x) + a_1\delta(x) \qquad (2.85a)$$

or

$$\{f'(x)\} = f'(x) + a_1\delta(x - x_1) \qquad (2.85b)$$

if the jump occurs at $x = x_1$. In other words, the generalized derivative is the ordinary derivative plus a_1 times a delta function. Similarly, for the second derivative we get

$$\{f''(x)\} = f'(x) + a_1\delta'(x) + b_1\delta(x) \qquad (2.86a)$$

where b_1 is the jump on the first derivative $f'(x)$. More generally, if $f(x)$ has jumps of magnitude $a_1, a_2, ..., a_n$ at the points $\alpha_1, \alpha_2, ..., \alpha_n$ and $f'(x)$ has jumps of magnitude $b_1, b_2, ..., b_m$ at the points $\beta_1, \beta_2, ..., \beta_m$, we have

$$\{f''(x)\} = f''(x) + a_1\delta'(x - \alpha_1) + ... + a_n\delta'(x - \alpha_n)$$
$$+ b_1\delta(x - \beta_1) + ... + b_m\delta(x - \beta_m) \tag{2.86b}$$

so that, with these considerations in mind, one generally drops the notation $\{f'(x)\}, \{f''(x)\}$ etc. and simply writes $f'(x), f''(x)$ etc.

Example 2.5.   As a preliminary result, it is left to the reader to show that, in the sense of distributions, the following equality holds:

$$f(x)\delta(x) = f(0)\delta(x) \tag{2.87}$$

where $f(x)$ is any continuous function.

Then, using this result and considering an absolutely continuous function $g(x)$, let us determine that the distribution derivative of $g(x)\theta(x)$ is

$$g'(x)\theta(x) + g(0)\delta(x) = g'(x)\theta(x) + g(x)\delta(x)$$

i.e. the result that we obtain by applying the ordinary differentiation rule of a product. In fact, from the definition of generalized derivative we get

$$-\langle g\theta, \varphi' \rangle = -\int_{-\infty}^{\infty} g(x)\theta(x)\varphi'(x)dx$$

$$= -\int_{0}^{\infty} g(x)\varphi'(x)dx = -g(x)\varphi(x)|_{0}^{\infty} + \int_{0}^{\infty} g'(x)\varphi(x)dx$$

$$= g(0)\varphi(0) + \int_{-\infty}^{\infty} g'(x)\theta(x)\varphi(x)dx$$

$$= \int_{-\infty}^{\infty} [g'(x)\theta(x) + g(0)\delta(x)]\varphi(x)dx$$

so that the term within brackets in the last integral defines the distribution derivative $\{(g\theta)'\}$ and we get, as expected

$$\{(g\theta)'\} = g'(x)\theta(x) + g(0)\delta(x) = g'(x)\theta(x) + g(x)\delta(x) \tag{2.88}$$

where eq (2.87) has been taken into account. This result is just a specific example which shows that the usual product rule for differentiation holds (if the products involved have a meaning).

Finally, in just the same way in which we can introduce a generalized derivative, we can introduce the notion of generalized antiderivative – i.e. indefinite integrals – by saying that the distribution Z is the indefinite integral of T if $\langle Z', \varphi \rangle = \langle T, \varphi \rangle$ or $\langle T, \varphi \rangle = -\langle Z, \varphi' \rangle$ and it is worth noting the fact that the distribution derivative coincides with the usual derivative in all cases in which we can recover the original function by integration. As soon as the ordinary derivative becomes of little practical utility and does not allow us to go back to the original function, the two concepts differ. This is one of the major strengths of the theory of distributions.

We will not pursue the subject any further and we close this section with a few final comments:

1. It is possible to extend to distributions a large number of operations which are commonly used with ordinary functions. Among these operations we find, for example, the translation $f(x - a)$, the multiplication by a function $g(x) \in C^\infty$, the change of variable, the convolution, etc. So, for example, if $\Omega \subset \mathcal{R}^n$, $g \in C^\infty(\Omega)$ and T is a distribution on Ω, then by the product gT we understand the distribution satisfying

$$\langle (gT), \varphi \rangle = \langle T, g\varphi \rangle \qquad \text{for all } \varphi \in \mathcal{D}(\Omega) \tag{2.89}$$

2. There exists an important theorem which states that every distribution can be expressed as the derivative $D^\alpha f$ (for some multi-index α) of some continuous function f.

3. In the light of comments (1) and (2), the reader can refer to a number of excellent books to see how the theory of distribution has important consequences in the study of ordinary differential equations as well as partial differential equations (e.g. Friedman [4]) because, as a matter of fact, the development of the theory of distributions was to a large extent motivated by problems involving differential equations. For example, suppose that we want to find the distribution g that satisfies

$$g' = f \tag{2.90a}$$

for a given distribution f, on some interval $[a, b]$ on the real line. If f and g were ordinary functions (say, for example, $f \in C[a, b]$ and $g \in C^1[a, b]$, i.e. f is continuous on $[a, b]$ and g continuous with continuous first derivative on $[a, b]$) then eq (2.90) would be a simple first order differential equation. However, since f and g are actually distributions, we must go back to the definition of generalized derivative and eq (2.90a) reads $\langle g', \varphi \rangle = \langle f, \varphi \rangle$ or

$$-\langle g, \varphi' \rangle = \langle f, \varphi \rangle \qquad \text{for all } \varphi \in \mathcal{D}(a, b) \tag{2.90b}$$

If by eq (2.90a) we understand eq (2.90b), then eq (2.90a) is said to be a distributional differential equation. Clearly, this same procedure applies to more general differential equations, for example to

$$Ag = f \tag{2.91a}$$

where A is a (generalized) differential operator given by

$$A = a_0(x) \frac{d^k}{dx^k} + a_1(x) \frac{d^{k-1}}{dx^{k-1}} + \ldots + a_k(x)$$

($a_0(x), \ldots, a_k(x) \in C^\infty$) and we interpret eq (2.91) as a differential equation involving generalized derivatives of g so that we seek g such that

$$\langle Ag, \varphi \rangle = \langle f, \varphi \rangle \qquad \text{for all } \varphi \in \mathcal{D}(a, b) \tag{2.91b}$$

which is equivalent to

$$\langle g, \hat{A}\varphi \rangle = \langle f, \varphi \rangle \qquad \text{for all } \varphi \in \mathcal{D}(a, b) \tag{2.91c}$$

where the operator $\hat{A}$ results from repeated application of (2.89) and of the definition of generalized derivative. Explicitly

$$\hat{A}\varphi = (-1)^k \frac{d^k}{dx^k} (a_0\varphi) + \ldots + a_k\varphi$$

Naturally, one would expect that if f is continuous (i.e. a distribution generated by a continuous function) then the solution g should be a function that is k times continuously differentiable. This is indeed so. In other words, when the distributions involved are generated by sufficiently differentiable functions, we recover the classical concept of differential equation and, in this case, g is called a classical solution. On the other hand, if f is a regular distribution generated by a function which is locally integrable but not continuous – or it is a singular distribution – then eq. (2.91a) cannot be expected to have any meaning in the classical sense and the solution is called a weak or generalized solution.

4. There exist many possible choices of the test space other than the space of infinitely differentiable finite functions (i.e. the space $\mathcal{D}(\Omega)$), for example, in $\mathcal{R}$ we can choose the test space to be the space of all infinitely differentiable functions which, together with all their derivatives, approach zero faster than any power of $1/|x|$. This test space is usually denoted by $\mathcal{S}(\mathcal{R})$ and it is used in connection with

Fourier transforms (e.g. Vladimirov [5]). The space $\mathscr{S}'$ of all continuous linear functionals on $\mathscr{S}$ defines the space of the so-called 'tempered' distributions. This shows that, as a matter of fact, there is no need to commit oneself to any definite choice of test space; rather, it is better to choose a test space which appears to be the most suitable for the problem, (or the class of problems) at hand. In this regard, however, it should be noted that the smaller the test space, the greater the freedom in performing various analytical operations (differentiation, passage to the limit, etc.) and the larger the number of functionals. Nonetheless, the test space should not be too small, otherwise there will not be enough test functions to allow us to 'tell ordinary functions (i.e. regular distributions) apart', this latter statement meaning that given any two continuous functions f_1 and f_2 where $f_1 \neq f_2$ there must be a test function φ such that $\langle f_1, \varphi \rangle \neq \langle f_2, \varphi \rangle$.

2.5   The notion of Hilbert space

One of the most important concepts in mathematics and in its applications to physics and engineering is that of linear or vector space, a concept with which we assume that the reader has some familiarity (if this is not the case, see the appendix on matrix analysis for a refresher on the fundamental properties of linear spaces). Broadly speaking, linear spaces can be seen as extensions of the three-dimensional space $\mathscr{R}^3$ of usual vectors and, in this light, we can consider finite-dimensional linear spaces with an arbitrary number n of dimensions. For our purposes, it should be mentioned that these spaces are the 'natural' setting for the study of vibrating systems with a finite number of degrees of freedom. In these spaces, the notion of inner or scalar product – which is a symmetric, linear, positive definite operation and provides a means of measuring both the length (or norm) of a vector and the distance between two points – plays a predominant role. These concepts can be generalized to spaces with an infinite number of dimensions (in a sense that will be clear from the following discussion).

More specifically, if we let X be a complex vector space, an inner product $\langle u \,|\, v \rangle$ of $u, v \in X$ is any operation that satisfies the following axioms for all $u, v, w \in X$ and $a, b \in \mathscr{C}$ (i.e. any two complex numbers)

1. $\langle u \,|\, v \rangle \in \mathscr{C}$
2. $\langle u \,|\, v \rangle = \langle v \,|\, u \rangle^*$
3. $\langle au + bv \,|\, w \rangle = a \langle u \,|\, w \rangle + b \langle v \,|\, w \rangle$
4. $\langle u \,|\, u \rangle \geqslant 0$ and $\langle u \,|\, u \rangle = 0$ if and only if $u = 0$

where the asterisk denotes complex conjugation.

A vector space endowed with the inner product $\langle \cdot \,|\, \cdot \rangle$ is called an **inner product space**. Two observations can be made immediately: first, although the inner product is in general a complex number (axiom 1), axiom 2 shows that

$\langle u \mid u \rangle$ is always a real number, so that the axiom of positive definiteness (axiom 4) makes complete sense and, second, it should be noted that in our definition linearity (i.e. the properties of additivity and homogeneity) applies to the first slot of the inner product and for complex inner product spaces it makes a difference whether linearity is defined with respect to the first or the second slot. In fact, from the axioms above it follows that additivity applies also to the second slot, i.e. $\langle u \mid v + w \rangle = \langle u \mid v \rangle + \langle u \mid w \rangle$, but homogeneity does not apply, i.e. we have $\langle u \mid av \rangle = a^* \langle u \mid v \rangle$.

It is possible to define an inner product on real vector spaces as well, and this is an important special case which will often serve our purposes. In this case the axioms read (note that now $a, b \in \mathcal{R}$)

1'. $\langle u \mid v \rangle \in \mathcal{R}$
2'. $\langle u \mid v \rangle = \langle v \mid u \rangle$
3'. $\langle au + bv \mid w \rangle = a \langle u \mid w \rangle + b \langle v \mid w \rangle$
4'. $\langle u \mid u \rangle \geqslant 0$ and $\langle u \mid u \rangle = 0$ if and only if $u = 0$

and linearity in the second slot can be verified immediately.

A simple example of inner product is given by the product of two column vectors $\mathbf{x}, \mathbf{y}$ of an n-dimensional vector space, i.e. the product, $\langle \mathbf{x} \mid \mathbf{y} \rangle \equiv \mathbf{x}^T \mathbf{y} = x_1 y_1 + x_2 y_1 + x_2 y_2 + \ldots + x_n y_n$, where $\mathbf{x} = (x_1, x_2, \ldots, x_n)^T$ and $\mathbf{y} = (y_1, y_2, \ldots, y_n)^T$.

A less intuitive example of inner product is defined in the space of square-integrable (in the Lebesque sense) functions defined on an interval $\alpha < x < \beta$: a space which is commonly denoted by mathematicians and physicists as $L^2(\alpha, \beta)$. It is left to the reader to show that this is a linear space and to verify that for any two functions f and $g \in L^2(\alpha, \beta)$ the expression

$$\langle f \mid g \rangle \equiv \int_\alpha^\beta f(x) g^*(x) dx \tag{2.92}$$

defines an inner product in $L^2(\alpha, \beta)$.

With the concept of inner product at our disposal, it is straightforward to introduce the notion of orthogonality between two members of an inner product space $X : u, v \in X$ are said to be orthogonal if

$$\langle u \mid v \rangle = 0 \tag{2.93}$$

and one often writes $u \perp v$. For example, in the real inner product space $L^2(-\pi, \pi)$ the two functions $u(x) = \sin x$ and $v(x) = \cos x$ are orthogonal because

$$\langle u \mid v \rangle = \int_{-\pi}^{\pi} \sin x \, \cos x \, dx = 0$$

An important property which holds in any inner product space is the so-called Cauchy–Schwartz inequality which reads

$$| \langle u \mid v \rangle | \leqslant \langle u \mid u \rangle^{1/2} \langle v \mid v \rangle^{1/2} \tag{2.94}$$

Furthermore, for the interested reader it is worth noting that, from a mathematical point of view, an inner product space is structured at three levels because the inner product generates a norm (i.e. a notion of length) $\| \cdot \|$ according to the definition

$$\| u \| \equiv \langle u \mid u \rangle^{1/2} \tag{2.95a}$$

and the norm, in turn, generates a metric (i.e. a notion of distance) $d(\cdot , \cdot)$ by means of the definition

$$d(u, v) \equiv \| u - v \| \tag{2.95b}$$

In terms of the norm generated by the inner product, the Cauchy–Schwartz inequality can be written as $| \langle u \mid v \rangle | \leqslant \| u \| \| v \|$.

It should be noted that both the norm and the metric are primitive concepts whose definitions are given by a number of axioms and do not require the existence of an inner product (in fact, not all norms are generated by an inner product and not all metrics are generated by a norm), however, our main interest lies in inner product spaces and this justifies the definitions and use of eqs (2.95a and b).

In normed and inner product spaces the notion of completeness of the space plays an important role in many circumstances. In finite- (say n-) dimensional linear spaces completeness has to do with the fact that any vector of the space can be expressed as a linear combination of a set of n linearly independent vectors (i.e. a basis of the space) and that the dimensionality of the space is defined by means of the maximum number of linearly independent vectors which may be chosen from among the vectors of the space. Furthermore, when an inner product is defined between vectors, we can always choose a orthonormal set $\{x_j\}_{j=1}^{n}$ such that $\langle x_k \mid x_m \rangle = \delta_{km}$. These concepts can be extended to infinite-dimensional linear inner product spaces with the only difference that in the infinite-dimensional case the completeness of the space is not always guaranteed.

In order to properly define the notion completeness of a space in a general case, we need the notion of Cauchy sequence. A sequence $\{u_n\}$ in a subset (either a proper subset or the entire space) Y of a normed space X is called a Cauchy sequence if

$$\lim_{n, m \to \infty} \| u_m - u_n \| = 0$$

or, more formally, for any given $\varepsilon > 0$ there exists a number N such that $\| u_m - u_n \| < \varepsilon$ whenever $m, n > N$. Every convergent series is clearly a

Cauchy sequence but the point is that not every Cauchy sequence is convergent because the limit may not belong to the space. When this is the case, we say that the space is incomplete. By contrast we have the following definition of completeness: a subset (proper or not) Y of a normed space X is complete if every Cauchy sequence in Y converges to an element of Y.

As mentioned before, completeness is not a problem in finite-dimensional spaces because there is a theorem which states that every finite-dimensional normed space is complete.

When a (infinite-dimensional) space is incomplete, however, the situation can be remedied by adding to the original space all those elements that are the limits of Cauchy sequences; this process is called completion of the space and a few examples will help clarify these points.

Example 2.6. The easiest way to construct an infinite-dimensional linear space is to generalize the space of n-tuples of real or complex numbers to the space l^2 of all infinite sequences $u = (u_1, u_2, ..., u_k, ...)$ which satisfy the convergence condition $\sum_{k=1}^{\infty} |u_k|^2 < \infty$, where the sum and multiplication by a scalar are defined as $(u_1, ..., u_k, ...) + (v_1, ..., v_k, ...) = (u_1 + v_1, ..., u_k + v_k, ...)$ and $a(u_1, ..., u_k, ...) = (au_1, ..., au_k, ...)$. This space becomes an infinite-dimensional inner product space when equipped with the inner product

$$\langle u \mid v \rangle = \sum_{k=1}^{\infty} u_k v_k$$

and, generalizing from the finite-dimensional case, the simplest orthonormal 'basis' (we will return on this aspect later in this section) in this space consists of the vectors

$$e_1 = (1, 0, 0, ...)$$
$$e_2 = (0, 1, 0, 0, ...)$$
$$e_3 = (0, 0, 1, 0, ...)$$

The proof that this space is complete can be found in any textbook on advanced calculus.

Example 2.7. Let us now consider the space of all real-valued continuous functions defined on the interval $[0, 1]$, i.e. the space $C[0, 1]$ equipped with the L^2-norm $\| u \|^2 \equiv \int_0^1 u^2 dx$, which is the norm induced by the inner product $\langle u \mid v \rangle = \int_0^1 u(x)v(x)dx$. It is readily shown that this space is not complete; in fact, consider the Cauchy (as the reader can verify) sequence $\{u_n\}$ defined by

$$u_n = \begin{cases} 0 & 0 \leqslant x < 1/2 \\ (x - 1/2)^{1/n} & 1/2 \leqslant x \leqslant 1 \end{cases}$$

Its 'limit' is the discontinuous function

$$u(x) = \begin{cases} 0 & 0 \leqslant x < 1/2 \\ 1 & 1/2 \leqslant x \leqslant 1 \end{cases}$$

which obviously does not belong to $C[0,1]$. Hence $C[0,1]$, and more generally the space $C[a,b]$, with the L^2-norm is not complete. Furthermore, it can be shown that if we complete this space by adding all the limits of Cauchy sequences, we obtain the space $L^2[a,b]$. By contrast, the space $C[a,b]$ equipped with the norm $\|u\|_\infty \equiv \sup\{|u(x)|; x \in [a,b]\}$ is indeed complete.

Example 2.8.   The final example – the most important for our purposes – is the space $L^2(\Omega)$, where $\Omega \subset \Re^n$. A famous theorem of functional analysis states and proves that this space is complete; following, for example, Vladimirov ([5], Chapter 1) the Riesz–Fisher theorem states that:

> If a sequence of functions u_n, $n = 1, 2, \ldots$ belonging to $L^2(\Omega)$ is a Cauchy sequence in $L^2(\Omega)$ there is a function $f \in L^2(\Omega)$ such that $\|u_n - f\| \to 0$ as $n \to \infty$; the function f is unique to within values on a set of measure zero.

This latter statement has to do with the fact that Lebesque integration is essential for the completeness of $L^2(\Omega)$, in fact the space of functions that are Riemann square integrable is not complete.

We are now in a position to give a precise definition of a Hilbert space: a Hilbert space is a complete inner-product linear space. Clearly, all finite-dimensional inner-product linear spaces are Hilbert spaces, but the most interesting cases of Hilbert spaces in physics and engineering are spaces of functions such as the aforementioned $L^2(\Omega)$.

At this point, the problem arises of extending the familiar notion of orthonormal basis to a general (infinite-dimensional) Hilbert space, that is, we are faced with the problem of representing a given function as a linear combination of some given set of functions or, in other words, with the problem of series expansions of functions in terms of a given set. Needless to say, the prototype of such series expansions is the Fourier series. The additional question of how one generates bases in infinite-dimensional spaces is partially answered by considering Sturm–Liouville problems (Section 2.5.1) which, in turn, are special cases of eigenvalue problems with a number of interesting properties. The most relevant of these properties is that their eigenfunctions form orthonormal bases in $L^2(\Omega)$, where Ω is the domain in which the Sturm–Liouville problem is formulated, i.e. for our purposes (see Chapter 8), a limited interval $[a,b]$ of the real line $\Re$.

Some explanatory remarks are in order. Now, if we take as a representative example the Hilbert space $L^2(\Omega)$, it is obvious that this space is not finite dimensional, i.e. it is not possible to find a finite set of functions in $L^2(\Omega)$ that spans $L^2(\Omega)$. The best that can be done is to construct an infinite set of functions with the property that any member of the space can be approximated arbitrarily closely by a finite linear combination of these functions, provided that a sufficiently large number of functions is used. This, in essence, is the idea that leads to the notion of a basis consisting of a countably infinite set and, in the end, to the counterpart of orthonormal bases of a finite-dimensional linear space, i.e. sets of functions $\{\phi_1, \phi_2, ..., \phi_k, ...\}$ for which

$$\langle \phi_i \mid \phi_j \rangle = \delta_{ij} \equiv \begin{cases} 1 & i = j \\ 0 & \text{otherwise} \end{cases} \tag{2.96}$$

In this light, the following definition applies: let X be any inner-product space and let $\Phi \equiv \{\phi_k\}_{k=1}^{\infty}$ be an orthonormal set in X. Then we say that Φ is a **maximal orthonormal set** in X if there is no other non-zero member ϕ in X that is orthogonal to all the ϕ_k. In other words, Φ is maximal if $\langle \phi \mid \phi_k \rangle = 0$ for all k implies $\phi = 0$, meaning that it is not possible to add to Φ a further nonzero element that is orthogonal to all existing members of Φ. Furthermore, a maximal orthonormal set in a Hilbert space H is called an **orthonormal basis** (or complete orthonormal set; note that completeness in this case is not the same as completeness of the space, although the two concepts are intimately connected) for H. It is obvious that we are generalizing from finite-dimensional inner-product spaces and, as a matter of fact, this definition coincides with the familiar idea of orthonormal basis in the finite-dimensional case.

Example 2.9. It can be shown that the set

$$\Phi = \{1/\sqrt{2}\} \cup \{\sin \pi kx, \cos \pi kx\}_{k=1}^{\infty}$$

is a maximal orthonormal set in $L^2(-1, 1)$. Since $L^2(-1, 1)$ is complete (i.e. a Hilbert space), Φ is an orthonormal basis.

Before completing our brief discussion on Hilbert spaces, we need a further definition and two important theorems which underlie the developments of the discussion of Chapter 8. Let $\{\phi_k\}$ be an orthonormal set in an inner-product space X. Then, for any $u \in X$ the numbers $\langle u \mid \phi_k \rangle$ are called the Fourier coefficients of u with respect to $\{\phi_k\}$. Clearly, these are the infinite-dimensional counterparts of the components u_k of an element u of a finite-dimensional space. In this case, we know that if $\{\phi_1, ..., \phi_n\}$ is an orthonormal basis for an inner-product space X with dimension n, then for any $u \in X$

$$u = \sum_{k=1}^{n} u_k \phi_k$$

where $u_k = \langle u \mid \phi_k \rangle$. The two theorems, which we state without proof, are as follows.

The best approximation theorem. Let X be an inner product space and $\{\phi_k\}_{k=1}^{\infty}$ an orthonormal set in X. Let u be a member of X and let s_n and t_n denote the partial sums $s_n = \sum_{k=1}^{n} u_k \phi_k$, $t_n = \sum_{k=1}^{n} c_k \phi_k$ where $u_k = \langle u \mid \phi_k \rangle$ is the kth Fourier coefficient of of u and the c_k are arbitrary real or complex numbers. Then:

1. Best approximation:

$$|| u - s_n || \leq || u - t_n || \tag{2.97}$$

2. Bessel's inequality: the series $\sum_{k=1}^{\infty} | u_j |^2$ converges and

$$\sum_{k=1}^{\infty} | u_k |^2 \leq || u ||^2 \tag{2.98}$$

3. Parseval's formula:

$$\sum_{k=1}^{\infty} | u_k |^2 = || u ||^2 \qquad \text{if and only if } \lim_{n \to \infty} || u - s_n || = 0 \tag{2.99}$$

The Fourier series theorem. Let H be a Hilbert space and let $\Phi = \{\phi_k\}_{k=1}^{\infty}$ be an orthonormal set in H. Then any $u \in H$ can be expressed in the form

$$u = \sum_{k=1}^{\infty} \langle u \mid \phi_k \rangle \phi_k \tag{2.100}$$

if and only if Φ is an orthonormal basis, i.e. a maximal orthonormal set.

Moreover, another important theorem of functional analysis guarantees the existence of a countable orthonormal basis in any (separable) Hilbert space (we will never consider nonseparable Hilbert spaces, so the definition of 'separable Hilbert space' is not really relevant for our purposes).

2.5.1   *Sturm–Liouville problems*

This section lays the mathematical framework in which we will be able to deal with the 'modal approach' to the vibrations of strings and rods, a topic that will be considered in Chapter 8. As a preliminary remark, it must be said that the method of separation of variables is a common technique used for solving a number of linear initial-boundary-value problems. This method leads to a differential eigenvalue problem whose solution is essential to the solution of the problem as a whole, and it is such eigenvalue problems that are examples of Sturm–Liouville problems, the eigenfunctions of which are candidates for ortho-normal bases of the Hilbert space of (Lebesque) square integrable functions.

As a matter of fact, it can be shown that the eigenfunctions of Sturm–Liouville problems constitute orthonormal bases for $L^2(\Omega)$, where Ω is an appropriate subset of $\mathcal{R}$.

A Sturm–Liouville operator L is a linear differential operator of the form

$$Lu \equiv \frac{1}{\rho}\left[-\frac{d}{dx}\left(p\,\frac{du}{dx}\right) + qu\right] \tag{2.101a}$$

defined on an interval $[a, b]$ of the real line. The basic assumptions are that $p(x), dp/dx, q(x)$ and $\rho(x)$ are continuous real-valued functions on $[a, b]$ that satisfy

$$p(x) > 0$$
$$q(x) \geqslant 0$$
$$\rho(x) > 0 \tag{2.101b}$$

on $[a, b]$. Furthermore, we have the linear operators B_1 and B_2 that specify the boundary values of a continuous function and are defined by

$$B_1 u \equiv \alpha_1 u(a) + \beta_1 u'(a)$$
$$B_2 u \equiv \alpha_2 u(b) + \beta_2 u'(b) \tag{2.101c}$$

where by the symbol u' we mean du/dx and the constants in eq (2.101c) satisfy $(i = 1, 2)$

$$\alpha_i \geqslant 0$$
$$\beta_i \geqslant 0 \tag{2.101d}$$
$$\alpha_i + \beta_i > 0$$

Then, we define a 'regular' Sturm–Liouville problem on a finite interval $[a, b]$ as an eigenvalue problem of the form

$$Lu = \lambda u$$
$$B_1 u = 0 \tag{2.102}$$
$$B_2 u = 0$$

where λ is a complex number. The first of eqs (2.102) is a differential equation which is often written explicitly in the form

$$-(pu')' + qu = \lambda \rho u \tag{2.103}$$

and the second and third of eqs (2.102) represent the boundary conditions. If any of the requirements in the definitions differ from those given here (limited interval, eqs (2.101b) or (2.101d)), the problem is known as a 'singular' Sturm–Liouville problem.

More generally, if for a moment we abstract the discussion from the present special case, the defining features of an eigenvalue problem (we will have to deal with such problems in Chapters 6–8), which is expressed synthetically as $Au = \lambda u$ where A is a linear operator and u is whatever kind of object A can operate on (a column vector if A is a matrix, a function if A is a differential or an integral operator), are that: (1) $u = 0$ is a solution known as the trivial solution and (2) there are special nonzero values of λ called **eigenvalues,** for which the equation $Au = \lambda u$ has nontrivial solutions. In the context of matrix problems these solutions are called **eigenvectors,** whereas for differential or integral equations they are known as **eigenfunctions.** In either case they are determined only up to a multiplicative constant (i.e. if u is an eigenvector or eigenfunction, then so is αu for any constant α) and, because of this indeterminacy, it is customary to 'normalize' the eigenvectors or eigenfunctions in some convenient manner. When A is a differential operator defined on a domain Ω with boundary S, it is necessary to specify the boundary conditions which are, in their own right, an essential part of the eigenvalue problem.

Returning to our main discussion, the problem (2.102) is considered in the space $L^2(a, b)$ endowed with the inner product

$$\langle u \mid v \rangle = \int_a^b u(x)v^*(x)\rho(x)dx \tag{2.104}$$

and the function $\rho(x)$, because of its role, is called a weighting function. The first issue to consider is the domain $D(L)$ of the operator L because, clearly, not all members of $L^2(a, b)$ have derivatives in the classical sense. For our purposes it suffices to take

$$D(L) = \{u \in L^2(a, b) \cap C^2[a, b]; B_1u = B_2u = 0\} \tag{2.105}$$

which is a proper subspace of $L^2(a, b)$ and, broadly speaking, is a 'sufficiently large' subspace of $L^2(a, b)$ (the correct mathematical terminology is that $D(L)$ is 'dense' in $L^2(a, b)$).

Now, it turns out that Sturm–Liouville operators are examples of what are known as **symmetrical operators,** and symmetrical operators have many of the nice properties that symmetrical matrices possess in linear algebra. In general, if L is a linear operator defined on a Hilbert space H with domain $D(L)$, L is said to be symmetrical if

$$\langle Lu \mid v \rangle = \langle u \mid Lv \rangle \qquad \text{for all } u, v \in D(L) \tag{2.106}$$

where it must always be kept in mind that the definition applies to members of the domain of L, and since boundary conditions play a role in the choice of the domain (as in (2.105)), these will be crucial in determining whether a given operator is symmetrical. In this light, the results that follow are direct generalizations of the situation that pertains to symmetrical matrices.

Lemma 1. The eigenvalues of a symmetrical linear operator are real.

The proof of this lemma is straightforward if we consider that given an arbitrary complex number λ, we have $(\lambda - \lambda^*) = 2\text{Im}(\lambda)$. Then, since $Lu = \lambda u$, we get

$$(\lambda - \lambda^*)\langle u \mid u \rangle = \lambda\langle u \mid u \rangle - \lambda^*\langle u \mid u \rangle$$
$$= \langle \lambda u \mid u \rangle - \langle u \mid \lambda u \rangle = \langle Lu \mid u \rangle - \langle u \mid Lu \rangle = 0$$

where the last equality is obtained by virtue of eq (2.106).

Lemma 2. Let L be a symmetrical linear operator defined on a Hilbert space H. Then, the eigenfunctions corresponding to two distinct eigenvalues are orthogonal (the proof is easy and is left to the reader).

Finally, the following two theorems summarize all the important results for Sturm-Liouville operators.

Theorem 1.
1. The Sturm–Liouville operator is symmetrical.
2. The Sturm–Liouville operator L is positive, that is $\langle Lu \mid u \rangle \geq 0$ for all $u \in D(L)$.

An important corollary to Theorem 1 is that the eigenvalues of L are all non-negative and form a countable set. This means that they can be arranged in the sequence $0 \leq \lambda_1 \leq \lambda_2 \leq ...$; in addition, it can be shown that $\lambda_n \to \infty$ as $n \to \infty$.

Theorem 2. The eigenfunctions of a regular Sturm–Liouville problem form an orthonormal basis for $L^2(a, b)$, that is, recalling the Fourier series theorem of the preceding section, given a Sturm–Liouville problem defined on $D(L)$ and any function $f \in L^2(a, b)$, the eigenfunctions $\{\phi_k\}_{k=1}^{\infty}$ are such that

$$f = \sum_{k=1}^{\infty} \langle f \mid \phi_k \rangle \phi_k$$

in the sense that $\lim_{n \to \infty} \| f - s_n \| = 0$, where s_n is the finite sum

$$s_n = \sum_{k=1}^{n} \langle f \mid \phi_k \rangle \phi_k$$

Finally, a further significant result can be considered if we introduce the **Rayleigh quotient** R, i.e. a functional defined on $D(L)$ by

$$R(u) \equiv \frac{\langle Lu \mid u \rangle}{\langle u \mid u \rangle} \qquad \text{for all } u \in D(L) \tag{2.107}$$

where it should be noted that $R(u) \geqslant 0$ for the positivity of L.

Theorem 3. The minimum of $R(u)$ over all functions $u \in D(L)$ that are orthogonal to the first n eigenfunctions is λ_{n+1}. That is

$$\lambda_{n+1} = \min\{R(u) : u \in D(L); \ \langle u \mid \phi_1 \rangle = \ldots = \langle u \mid \phi_n \rangle = 0\}$$

The proofs of the above theorems are not given here and can be found in any book on advanced analysis. However, because of its importance in future discussions, we will just show here the proof of the fact that the Sturm–Liouville operator is symmetrical. For any $u, v \in D(L)$, indicating for simplicity the complex conjugate with an overbar rather than the usual asterisk, we have

$$\langle Lu \mid v \rangle - \langle u \mid Lv \rangle = \int_a^b [-(pu')'\bar{v} - qu\bar{v} + u(p\bar{v}')' + qu\bar{v}]dx$$

$$= \int_a^b [-(pu')'\bar{v} + (p\bar{v}')'u]dx = \int_a^b [(p\bar{v}'u)' - (p\bar{u}v')']dx$$

$$= [p\bar{v}'u - pu'\bar{v}]_a^b$$

$$= p(b)[u(b)\bar{v}'(b) - u'(b)\bar{v}(b)] - p(a)[u(a)\bar{v}'(a) - u'(a)\bar{v}(a)]$$

$$\tag{2.108}$$

Now, since $v \in D(L)$, so does $\bar{v}$. It follows that $B_1 u = B_1 \bar{v} = 0$, or in matrix form, using eq (2.101c)

$$\begin{bmatrix} u(a) & u'(a) \\ \bar{v}(a) & \bar{v}'(a) \end{bmatrix} \begin{bmatrix} \alpha_1 \\ \beta_1 \end{bmatrix} = \begin{bmatrix} 0 \\ 0 \end{bmatrix}$$

From the conditions (2.101d) at least one of α_1 and β_1 must be nonzero, and this is only possible if the matrix is singular, i.e. if

$$u(a)\bar{v}'(a) - u'(a)\bar{v}(a) = 0$$

The same line of reasoning applied to the boundary condition $B_2 u = B_2 \bar{v} = 0$ leads to

$$u(b)\bar{v}'(b) - u'(b)\bar{v}(b) = 0$$

and from these two equations it follows that the right-hand side of eq (2.108) is zero and L is symmetric, as stated in Theorem 1.

2.5.2 A few additional remarks

As is apparent from the foregoing sections, the theoretical background needed for the treatment of the eigenvalue problems of our concern constitutes a vast and delicate subject in its own right. In particular, we will be interested in the vibration of one, two- or three-dimensional continuous systems extending in a limited domain of space Ω. Clearly, a detailed discussion of the mathematical theory of these aspects is far beyond the scope of this book. However, we believe that a brief section with some additional remarks to what has been said up to this point is not out of place, because it may be of help to the interested reader who wants to refer to more advanced texts on the subject.

Consider the differential operator of order m with continuous complex-valued coefficients $p_0, p_1, ..., p_m$ $(p_m \neq 0)$

$$Lu \equiv \sum_{j=0}^{m} p_j(x) u^{(j)}(x) \qquad (2.109a)$$

where $a \leqslant x \leqslant b$ and $u \in C^m[a, b] \cap L^2(a, b)$. In addition, consider the m boundary operators of order $\leqslant m - 1$

$$B_l u \equiv \sum_{i=0}^{m-1} [\alpha_{li} u^{(i)}(a) + \beta_{li} u^{(i)}(b)] \qquad (l = 0, 1, ..., m-1) \qquad (2.109b)$$

with complex coefficients α_{li}, β_{li} and suppose further that, for all $u, v \in C^m[a, b] \cap L^2(a, b)$ satisfying the boundary conditions $B_l u = B_l v = 0$ $(l = 0, 1, ..., m - 1)$ we have (we return to the asterisk to indicate complex conjugation and the following condition, in words, states that the operator is symmetric)

$$\int_b^a (Lu)v^* dx = \int_a^b u(Lv)^* dx \qquad (2.110)$$

and, lastly, that $\rho(x) > 0$ is a continuous, real-valued function on $[a, b]$. Now if we consider the eigenvalue problem on $[a, b]$

$$Lu = \lambda \rho u$$
$$B_l u = 0 \qquad\qquad (2.111)$$

the following theorem can be proven.

Theorem 4.   There exists a sequence λ_k, u_k $(k = 1, 2, ...)$ of eigenpairs of problem   (2.111)   such   that $u_k \in C^m[a, b] \cap L^2(a, b)$, $B_l u_k = 0$ $(l = 1, ..., m - 1;\ k = 1, 2, ...)$ is a orthonormal basis in $L^2(a, b)$ endowed with the inner product

$$\langle u \mid v \rangle = \int_a^b \rho(x) u(x) v^*(x) dx$$

that is, for all $f \in L^2(a, b)$

$$f = \sum_{k=1}^{\infty} \langle f \mid \phi_k \rangle \phi_k \qquad\qquad (2.112)$$

with the series converging in the L^2-sense. Moreover, all eigenvalues are real and have finite multiplicity (meaning that a multiple eigenvalue can only be repeated a finite number of times). Note that the Sturm–Liouvulle problem of the preceding section is a special case of the symmetrical problem (2.111).

A more general approach which deals with the free vibrations of three-dimensional structures with sufficiently smooth boundaries considers a variational formulation of the problem in the bounded domain $\Omega \subset \mathcal{R}^3$ (the space occupied by the structure) with a smooth boundary S. The general point of Ω is denoted by $\mathbf{x} = (x_1, x_2, x_3)$ and $\mathbf{u}(\mathbf{x})e^{i\omega t}$ denotes the displacement field at time t at point $\mathbf{x}$. Further, it is supposed that a part of the boundary Γ_0 is fixed and the remaining part Γ is free. Clearly $S = \Gamma_0 \cup \Gamma$.

The formulation of the problem is rather involved and we do not consider it here. However, in its final steps this formulation considers an appropriate space $\mathscr{C}$ of 'sufficiently differentiable' functions defined on Ω with values in $\mathcal{R}^3$ and leads to the introduction of two linear operators K and M which are called the stiffness and mass operator, respectively. In terms of these operators the problem is formulated as

$$K\mathbf{u} = \lambda M\mathbf{u} \qquad \mathbf{u} \in \mathscr{C}_0 \qquad\qquad (2.113)$$

where $\lambda = \omega^2$ and $\mathscr{C}_0$ (the so-called 'space of admissible functions') is the space of all functions that belong to $\mathscr{C}$ for which $\mathbf{u} = 0$ on Γ_0. Since Ω is a bounded domain and the two operators can be shown to be symmetrical and positive definite, the fundamental results are that:

1. There exists an increasing sequence of positive eigenvalues

$$0 < \lambda_1 \leqslant \lambda_2 \leqslant \ldots \leqslant \lambda_k \leqslant \ldots \tag{2.114}$$

   and every eigenvalue has a finite multiplicity;
2. The eigenfunctions $\{u_k\}$ satisfy the orthogonality conditions

$$\langle Mu_j \mid u_k \rangle = a_j \delta_{jk}$$
$$\langle Ku_j \mid u_k \rangle = a_j \lambda_j \delta_{jk} = a_j \omega_j^2 \delta_{jk} \tag{2.115}$$

   form a complete set in $\mathscr{C}_0$ and are called, in engineering terminology, **elastic modes** (the meaning of this term will be clearer in future chapters).

When the structure is free – i.e. $\Gamma_0 = \varnothing$, where $\varnothing$ is the empty set – the eigenvalue problem (2.113) is posed in $\mathscr{C}$, the stiffness operator is only positive semidefinite and there exist solutions of the type $\{\lambda = 0, \mathbf{u} \neq 0\}$. The eigenfunctions corresponding to these solutions are called **rigid-body modes** and can be indicated by the symbol $\mathbf{u}_{\text{rig}}$. In this case it can be shown that:

1. any displacement field $\mathbf{u} \in \mathscr{C}$ can be expanded in terms of rigid-body modes and elastic modes;
2. the elastic modes and the rigid-body modes are mutually orthogonal.

References

1. Erdélyi, A., Magnus, W., Oberhettinger, F. and Tricomi, F.G., *Tables of Integral Transforms*, 2 vols, McGraw-Hill, New York, 1953.
2. Mathews, J. and Walker, R.L., *Mathematical Methods of Physics*, 2nd edn, Addison-Wesley, Reading, Mass., 1970.
3. Sidorov, Yu. V., Fedoryuk, M.V. and Shabunin, M.I., *Lectures on the Theory of Functions of a Complex Variable*, Mir Publishers, Moscow, 1985.
4. Friedman, A., *Generalized Functions and Partial Differential Equations*, Prentice Hall, Englewood Cliffs, NJ, 1963.
5. Vladimirov, V.S., *Equations of Mathematical Physics*, Mir Publishers, Moscow, 1984.

3 Analytical dynamics – an overview

3.1 Introduction

In order to describe the motion of a physical system, it is necessary to specify its position in space and time. Strictly speaking, only relative motion is meaningful, because it is always implied that the description is made with respect to some **observer** or **frame of reference.**

In accordance with the knowledge of his time, Newton regarded the concepts of length and time interval as absolute, which is to say that these quantities are the same in all frames of reference. Modern physics showed that Newton's assumption is only an approximation but, nevertheless, an excellent one for most practical purposes. In fact, Newtonian mechanics, vastly supported by experimental evidence, is the key to the explanation of the great majority of everyday facts involving force and motion.

If one introduces as a fundamental entity of mechanics the convenient concept of **material particle** – that is, a body whose position is completely defined by three Cartesian coordinates x, y, z and whose dimension can be neglected in the description of its motion – **Newton's second law** reads

$$m\mathbf{a} = \mathbf{F} \tag{3.1}$$

where $\mathbf{F}$ is the resultant (i.e. the vector sum) of all the forces applied to the particle, $\mathbf{a} = d^2\mathbf{x}/dt^2$ is the particle acceleration and the quantity m characterizes the material particle and is called its mass. Obviously, $\mathbf{x}$ is here the vector of components x, y, z.

Equation (3.1) must not be regarded as a simple identity, because it establishes a form of interaction between bodies and thereby describes a law of nature; this interaction is expressed in the form of a differential equation that includes only the second derivatives of the coordinates with respect to time. However, eq (3.1) makes no sense if the frame of reference to which it is referred is not specified. A difficulty then arises in stating the cause of acceleration: it may be either the interaction with other bodies or it may be due to some distinctive properties of the reference frame itself. Taking a step further, we can consider a set of material particles and suppose that a frame of reference exists such that all accelerations of the particles are a result of their mutual interaction. This can be verified if the forces satisfy **Newton's third law**, that is they are equal in magnitude and opposite in sign for any given pair of particles.

Such a frame of reference is called **inertial**. With respect to an inertial frame of reference a free particle moves uniformly in a straight line and every observer in uniform rectilinear motion with respect to an inertial frame of reference is an inertial observer himself.

3.2   Systems of material particles

Let us consider a system of N material particles and an inertial frame of reference. Each particle is subjected to forces that can be classified either as: **internal forces**, due to the other particles of the system or **external forces**, due to causes that are external to the system itself. We can write eq (3.1) for the kth particle as

$$m_k \ddot{\mathbf{x}}_k = \mathbf{F}_k^{(ext)} + \mathbf{F}_k^{(int)} \tag{3.2}$$

where $k = 1, 2, ..., N$ is the index of particle, $\mathbf{F}_k^{(ext)}$ and $\mathbf{F}_k^{(int)}$ are the resultants of external and internal forces, respectively. In addition, we can write the resultant of internal forces as

$$\mathbf{F}_k^{(int)} = \sum_{j=1}^{N} \mathbf{F}_{kj} \tag{3.3}$$

which is the vector sum of the forces due to all the other particles, $\mathbf{F}_{kj}$ being the force on the kth particle due to the jth particle. Newton's third law states that

$$\mathbf{F}_{kj} = -\mathbf{F}_{jk} \tag{3.4}$$

hence

$$\sum_{k=1}^{N} \mathbf{F}_k^{(int)} = \sum_{k=1}^{N} \sum_{j=1}^{N} \mathbf{F}_{kj} = 0 \tag{3.5}$$

and eq (3.2), summing on the particle index k, leads to

$$\sum_{k=1}^{N} m_k \ddot{\mathbf{x}}_k = \mathbf{F}_k^{(ext)} \tag{3.6}$$

These results are surely well known to the reader but, nevertheless, they are worth mentioning because they show the possibility of writing equations where internal forces do not appear.

3.2.1   *Constrained systems*

Proceeding further in our discussion, we must account for the fact that, in many circumstances, the particles of a given system are not free to occupy any arbitrary position in space, the only limitation being their mutual influences. In other words, we must consider **constrained systems**, where the positions and/or the velocities of the particles are connected by a certain number of relationships that limit their motion and express mathematically the equations of **constraints**.

A perfectly rigid body is the simplest example: the distance between any two points remains unchanged during the motion and $3N - 6$ equations (if $N \geqslant 3$ and the points are not aligned) must be written to satisfy this condition.

In every case, a constraint implies the presence of a force which may be, *a priori*, undetermined both in magnitude and direction; these forces are called **reaction forces** and must be considered together with all other forces. For the former, however, a precise law for their dependence on time, coordinates or velocities (of the point on which they act or of other points) is not given; when we want to determine the motion or the equilibrium of a given system, the information about them is supplied by the constraints equations.

Constraints, in turn, may be classified in many ways according to their characteristics and to the mathematical form of the equations expressing them, we give the following definitions: if the derivatives of the coordinates do not appear in a constraint equation we speak of **holonomic constraint** (with the further subdivision in **rheonomic** and **scleronomic**), their general mathematical expression being of the type

$$f(\mathbf{x}_1, \mathbf{x}_2, ..., t) = 0 \qquad (3.7)$$

where time t appears explicitly for a rheonomic constraint and does not appear for a scleronomic one. In all other cases, the term **nonholonomic** is used.

For example, two points rigidly connected at a distance L must satisfy

$$(x_1 - x_2)^2 + (y_1 - y_2)^2 + (z_1 - z_2)^2 - L^2 = 0 \qquad (3.8)$$

A point moving in circle in the x–y plane must satisfy

$$x^2 + y^2 - R^2 = 0 \qquad (3.9a)$$

or, in parametric form,

$$\left. \begin{array}{l} x = R\cos\theta \\ y = R\sin\theta \\ z = 0 \end{array} \right\} \qquad (3.9b)$$

where the angle θ (the usual angle of polar coordinates) is the parameter and shows how this system has only one **degree of freedom** (the angle θ), a concept that will be defined soon.

Equations (3.8) and (3.9) are two typical examples of holonomic (scleronomic) constraints. A point moving on a sphere whose radius increases linearly with time $(R = at)$ is an example of rheonomic constraint, the constraint equation being now

$$x^2 + y^2 + z^2 - a^2 t^2 = 0 \tag{3.10a}$$

or, in parametric form

$$\left. \begin{array}{l} x = at(\sin\theta\cos\varphi) \\ y = at(\sin\theta\sin\varphi) \\ z = at\cos\varphi \end{array} \right\} \tag{3.10b}$$

where the usual angles for spherical coordinates have been used.

From the examples above, it can be seen that holonomic constraints reduce the number of independent coordinates necessary to describe the motion of a system to a number that defines the **degrees of freedom** of the system. Thus, a perfectly rigid body of N material points can be described by only six independent coordinates; in fact, a total of $3N$ coordinates identify the N points in space but the constraints of rigidity are expressed by $3N - 6$ equations, leaving only six degrees of freedom.

Consider now a sphere of radius R that rolls without sliding on the $x–y$ plane. Let X and Y be the coordinates of the centre with respect to the fixed axes x and y and let ω_x and ω_y be the components of the sphere angular velocity along the same axes. The constraint of rolling without sliding is expressed by the equations (the dot indicates the time derivative)

$$\begin{array}{l} \dot{X} - \omega_y R = 0 \\ \dot{Y} - \omega_x R = 0 \end{array} \tag{3.11a}$$

which are the projections on the x and y axes of the vector equation

$$\mathbf{v} + \omega \times (\mathbf{C} - \mathbf{O}) = 0$$

where $\mathbf{v}$ and ω are the sphere linear and angular velocities, the symbol $\times$ indicates the vector product, $\mathbf{C}$ is the position vector of the point of contact of the sphere with the $x–y$ plane and $\mathbf{O}$ is the position vector of the centre of the sphere. Making use of the Euler angles ϕ, θ and ψ (e.g. Goldstein [1]), eqs (3.11a) become

$$\begin{array}{l} \dot{X} + R(\dot{\psi}\cos\phi\sin\theta - \dot{\theta}\sin\phi) = 0 \\ \dot{Y} + R(\dot{\psi}\sin\phi\sin\theta - \dot{\theta}\cos\phi) = 0 \end{array} \tag{3.11b}$$

which express the nonholonomic constraint of zero velocity of the point C.

A disc rolling without sliding on a plane is another classical example – which can be found on most books of mechanics – of nonholonomic constraint. These constraints do not reduce the number of degrees of freedom but only limit the way in which a system can move in order to go from one given position to another.

In essence, they are constraints of 'pure mobility': they do not restrict the possible configurations of the system, but how the system can reach them.

Obviously, a holonomic constraint of the kind (dropping the vector notation and using scalar quantities for simplicity)

$$f(x_1, x_2) = 0$$

implies the relation on the time derivative

$$\frac{\partial f}{\partial x_1} \dot{x}_1 + \frac{\partial f}{\partial x_2} \dot{x}_2 = 0$$

but when we have a general relation of the kind

$$A(x_1, x_2)\dot{x}_1 + B(x_1, x_2)\dot{x}_2 = 0$$

which cannot be obtained by differentiation (i.e. it is not an exact differential) the constraint is nonholonomic. In the equation above, A and B are two functions of the variables x_1 and x_2. Incidentally, it may be interesting to note that the assumption of relativistic mechanics stating that the velocity of light in vacuum ($c \cong 3 \times 10^8$ m/s) is an upper limit for the velocities of physical bodies is, as a matter of fact, a good example of nonholonomic constraint.

Thus, in the presence of constraints:

1. The coordinates x_k are no longer independent (being connected by the constraint equations).
2. The reaction forces appear as unknowns of the problem; they can only be determined *a posteriori*, that is, they are part of the solution itself. This 'indetermination' is somehow the predictable result of the fact that we omit a microscopical description of the molecular interactions involved in the problem and we make up for this lack of knowledge with information on the behaviour of constraints – the reaction forces – on a macroscopic scale. So, unless we are specifically interested in the determination of reaction forces, it is evident the interest in writing, if possible, a set of equations where the reaction forces do not appear.

For every particle of our system we must now write

$$m\mathbf{a} = \mathbf{F} + \Phi \tag{3.12}$$

where $\mathbf{F}$ is the resultant of active (internal and external) forces, Φ is the resultant of reactive (internal and external) forces and the incomplete knowledge on Φ is supplied by the equation(s) of constraint.

The nature of holonomic constraints itself allows us to tackle point (1) by introducing a set of generalized independent coordinates; in addition, we are led to equations where reaction forces disappear if we restrict our interest to reactive forces that, under certain circumstances of motion, do no work. These are the subjects of the next section.

3.3 Generalized coordinates, virtual work and d'Alembert principles: Lagrange's equations

If m holonomic constraints exist between the $3N$ coordinates of a system of N material particles, the number of degrees of freedom is reduced to $n = 3N - m$. It is then possible to describe the system by means of n configuration parameters $q_1, q_2, ..., q_n$, usually called **generalized coordinates**, which are related to the Cartesian coordinates by a transformation of the form

$$\left. \begin{array}{l} \mathbf{x}_1 = \mathbf{x}_1(q_1, q_2, ... q_n, t) \\ \quad \vdots \\ \mathbf{x}_N = \mathbf{x}_N(q_1, q_2, ... q_n, t) \end{array} \right\} \tag{3.13}$$

and time t does not appear explicitly if the constraints are not time dependent. The advantage lies obviously in the possibility of choosing a convenient set of generalized coordinates for the particular problem at hand.

From eq (3.13) we note that the velocity of the kth particle is given by

$$\mathbf{v}_k \equiv \frac{d\mathbf{x}_k}{dt} = \sum_{j=1}^{n} \frac{\partial \mathbf{x}_k}{\partial q_j} \dot{q}_j + \frac{\partial \mathbf{x}_k}{\partial t} \tag{3.14}$$

Let us now define the kth ($k = 1, 2, ..., N$) **virtual displacement** $\delta\mathbf{x}_k$ as an infinitesimal displacement of the kth particle compatible with the constraints. In performing this displacement we assume both active and reactive forces to be 'frozen' in time at the instant t, that is to say that they do not change as the system passes through this infinitesimal change of its configuration. This justifies the term 'virtual', as opposed to a 'real' displacement $d\mathbf{x}_k$, which occurs in a time dt. Similarly, we can define the kth virtual work done by active and reactive forces as

$$(\mathbf{F}_k + \Phi_k) \cdot \delta\mathbf{x}_k$$

and the total work on all of the N particles is

$$\sum_{k=1}^{N} (\mathbf{F}_k + \Phi_k) \cdot \delta \mathbf{x}_k \qquad (3.15)$$

If the system is in equilibrium, eq (3.15) is zero, because each one of the N terms in parentheses is zero; if in addition we restrict our considerations to reactive forces whose virtual work is zero, eq (3.15) becomes

$$\sum_{k=1}^{N} \mathbf{F}_k \cdot \delta \mathbf{x}_k = 0 \qquad (3.16)$$

which expresses the **principle of virtual work**. We point out that only active forces appear in eq (3.16) and, in general, $F_k \neq 0$ because the virtual displacements are not all independent, being connected by the constraints equations. The method leads to a number of equations $< N$, but equal – for holonomic constraints – to the number of degrees of freedom.

The assumption on the constraints leading to eq (3.16) is not very restrictive and, in practice, is valid for all holonomic constraints without friction. When the constraints are not frictionless, the equation is still valid if we count the tangential components of friction forces as active forces themselves. It could be added that if the principle of virtual work allows to obtain the equilibrium condition with frictionless constraints, the same conditions must apply when friction is present.

Since holonomic constraints are our major concern, we consider the transformation (3.13) and write

$$\delta \mathbf{x}_k = \sum_{j=1}^{n} \frac{\partial \mathbf{x}_k}{\partial q_j} \delta q_j \qquad (3.17)$$

where time does not appear because of the definition of virtual displacement. Substitution in eq (3.16) gives the principle of virtual work

$$\sum_{k,j} \mathbf{F}_k \cdot \frac{\partial \mathbf{x}_k}{\partial q_j} \delta q_j = \sum_{j=1}^{n} Q_j \delta q_j = 0 \qquad (3.18)$$

in terms of the **generalized forces**, defined as

$$Q_j = \sum_{k=1}^{N} \mathbf{F}_k \cdot \frac{\partial \mathbf{x}_k}{\partial q_j} \qquad (3.19)$$

The generalized forces do not necessarily have the dimensions of a force themselves, but the product $Q_j \delta q_j$ has the dimension of work. Now all the δq_js are independent and eq (3.18) implies

$$Q_j = 0 \qquad (3.20)$$

for every $j = 1, 2, ..., n$. Equation (3.20) expresses the equilibrium condition of our system and, as such, applies to the static case.

The extension to the dynamic case – with the difference that now we want to determine the equations of motion – is made by means of **d'Alembert's principle**, which sees eq (3.1) as an equilibrium equation by writing

$$\mathbf{F} - m\mathbf{a} = 0 \qquad (3.21)$$

which means that a material particle is in equilibrium if we consider the inertia force $(-m\mathbf{a})$ together with all other forces whose resultant is $\mathbf{F}$. Under the same assumptions that lead to eq (3.16), we can rewrite it in the form

$$\sum_{k=1}^{N} (\mathbf{F}_k - m_k \ddot{\mathbf{x}}_k) \cdot \delta \mathbf{x}_k = 0 \qquad (3.22)$$

and transform it into an equation where the virtual displacements of the generalized coordinates appear, so that we are allowed to say that every multiplicative coefficient of these virtual displacements is individually equal to zero.

We have already considered the term $\sum \mathbf{F}_k \cdot \delta \mathbf{x}_k$; for the other term we can substitute eq (3.17) into eq (3.22) and obtain

$$\sum_{j,k} m_k \ddot{\mathbf{x}}_k \cdot \frac{\partial \mathbf{x}_k}{\partial q_j} \qquad (3.23)$$

For the summation on the index k we can write

$$\sum_{k=1}^{N} m_k \ddot{\mathbf{x}}_k \cdot \frac{\partial \mathbf{x}_k}{\partial q_j} = \sum_{k=1}^{N} \left[\frac{d}{dt} \left(m_k \dot{\mathbf{x}}_k \cdot \frac{\partial \mathbf{x}_k}{\partial q_j} \right) - m_k \dot{\mathbf{x}}_k \cdot \frac{d}{dt} \left(\frac{\partial \mathbf{x}_k}{\partial q_j} \right) \right] \qquad (3.24)$$

and since it is not difficult to verify that (eq (3.14))

$$\frac{d}{dt} \left(\frac{\partial \mathbf{x}_k}{\partial q_j} \right) = \frac{\partial \mathbf{v}_k}{\partial q_j}$$

$$\frac{\partial \mathbf{x}_k}{\partial q_j} = \frac{\partial \mathbf{v}_k}{\partial \dot{q}_j}$$

eq (3.24) becomes

$$\sum_{k=1}^{N} m_k \ddot{\mathbf{x}}_k \cdot \frac{\partial \mathbf{x}_k}{\partial q_j} = \sum_{k=1}^{N} \left[\frac{d}{dt} \left(m_k \mathbf{v}_k \cdot \frac{\partial \mathbf{v}_k}{\partial \dot{q}_j} \right) - m_k \mathbf{v}_k \cdot \left(\frac{\partial \mathbf{v}_k}{\partial q_j} \right) \right]$$

Substituting back in eq (3.23), we get the expression

$$\sum_{j=1}^{n} \left\{ \frac{d}{dt} \left[\frac{\partial}{\partial \dot{q}_j} \left(\frac{1}{2} \sum_{k=1}^{N} m_k v_k^2 \right) \right] - \frac{\partial}{\partial q_j} \left(\frac{1}{2} \sum_{k=1}^{N} m_k v_k^2 \right) \right\} \delta q_j$$

where we recognize the total kinetic energy of the system

$$T = \frac{1}{2} \sum_{k=1}^{N} m_k v_k^2 \qquad (3.25)$$

which, in general, is a function of the kind $T = T(q_1, ..., q_n, \dot{q}_1, ..., \dot{q}_n, t)$, or, for short, $T = T(q, \dot{q}, t)$.

Putting all the pieces back together we finally get the expression we were looking for, i.e.

$$\sum_{j=1}^{n} \left[\frac{d}{dt} \left(\frac{\partial T}{\partial \dot{q}_j} \right) - \frac{\partial T}{\partial q_j} - Q_j \right] \delta q_j = 0$$

from which it follows that

$$\frac{d}{dt} \left(\frac{\partial T}{\partial \dot{q}_j} \right) - \frac{\partial T}{\partial q_j} = Q_j \qquad (j = 1, 2, ..., n) \qquad (3.26)$$

owing to the independence of the δq_js. Equations (3.26) are called **Lagrange's equations of the second kind** and are valid for holonomic systems with frictionless constraints. It can be shown that they form a system of n second-order differential equations which can be solved for the second derivatives and written as

$$\ddot{q}_j = \ddot{q}_j(q, \dot{q}, t) \qquad (j = 1, 2, ..., n) \qquad (3.27)$$

The solution is determined completely by introducing $2n$ constants of integration, obtained by imposing the initial conditions at $t = 0$.

When the system is **conservative** (see next section for more details), the forces can be obtained from a scalar function $V = V(\mathbf{x}_1, ..., \mathbf{x}_N)$ as

$$\mathbf{F}_k = -\text{grad}_k V \equiv -\nabla_k V \qquad (3.28)$$

The elementary work, i.e. the work done as the system passes through an infinitesimal displacement, is an exact differential and eq (3.18) can be written as

$$\sum_{j=1}^{n} Q_j \delta q_j = - \sum_{j=1}^{n} \frac{\partial V}{\delta q_j} \delta q_j$$

which implies

$$Q_j = - \frac{\partial V}{\partial q_j} \tag{3.29}$$

Lagrange's equations (3.26) thus become

$$\frac{d}{dt}\left(\frac{\partial T}{\partial \dot{q}_j} \right) - \frac{\partial T}{\partial q_j} = - \frac{\partial V}{\partial q_j}$$

and defining the **Lagrangian function** (or simply Lagrangian) as the difference between kinetic and potential energies, i.e.

$$L \equiv T - V \tag{3.30}$$

we obtain **Lagrange's equations of the first kind** for conservative systems as

$$\frac{d}{dt}\left(\frac{\partial L}{\partial \dot{q}_j} \right) - \frac{\partial L}{\partial q_j} = 0 \qquad (j = 1, 2, ..., n) \tag{3.31}$$

where $L = L(q, \dot{q}, t)$ and we exploited the fact that $\partial V/\partial \dot{q}_j = 0$.

When some of the forces acting on the system are conservative and some others are not, it is worth noting that Lagrange's equations can always be written in the form of eq (3.31), where now a term Q_j appears on the right-hand side. In this case, the potential V accounts for conservative forces and the Q_js represent all the forces that cannot be derived from a potential function.

A fundamental property of Lagrange's equations (3.26) and (3.31) is that they are invariant under an arbitrary transformation of generalized coordinates; in fact, it can be proven that an invertible transformation of the type

$$\left. \begin{aligned} q_1 &= q_1(\bar{q}_1, \bar{q}_2, ..., \bar{q}_n, t) \\ &\vdots \\ q_n &= q_n(\bar{q}_1, \bar{q}_2, ..., \bar{q}_n, t) \end{aligned} \right\} \tag{3.32}$$

converts, for example, eqs (3.31) into

$$\frac{d}{dt}\left(\frac{\partial L_1}{\partial \dot{q}_j}\right) - \frac{\partial L_1}{\partial q_j} = 0 \qquad (j = 1, 2, ..., n) \tag{3.33}$$

where L_1 is the appropriate Lagrangian in the new coordinates.

3.3.1 *Conservative forces*

It is well known that the work done by a force **F** on a material particle which undergoes an infinitesimal displacement $d\mathbf{x}$ is

$$\delta W \equiv \mathbf{F} \cdot d\mathbf{x} \tag{3.34}$$

Suppose now that the particle moves along some curve in space (say, from a point A to a point B), with the force varying as the particle moves. On a curve, x, y and z are related by the equations of the curve and in three dimensions two equations are needed. Thus, along a curve there is only one independent variable and the total work from A to B, i.e. the integral of eq (3.34) is an ordinary integral of a function of one variable, more precisely, it is a line integral. To evaluate a line integral, we must write it as a single integral using one independent variable.

For example, in two dimensions (the x–y plane), given the force field

$$\mathbf{F} = xy\mathbf{i} - y^2\mathbf{j}$$

where **i** and **j** are, respectively, the usual unit vectors in the positive x and y directions, let us find the work from $A = (0, 0)$ to $B = (2, 1)$ along the two paths:

1. straight line $y = x/2$ $(dy = dx/2)$;
2. parabola $y = x^2/4$ $(dy = x\,dx/2)$.

It is not difficult to see that

$$W = \int (xy\mathbf{i} - y^2\mathbf{j}) \cdot (\mathbf{i}\,dx + \mathbf{j}\,dy)$$

becomes an integral in dx and leads to the following results: $W = 1$ in case (1), and $W = 2/3$ in case (2). So, the work done may depend on the path the particle follows; in fact, it usually will when there is friction.

A force field for which the quantity

$$W = \int \mathbf{F} \cdot d\mathbf{x} \tag{3.35}$$

depends upon the path as well as the endpoints is called **nonconservative**. Physically this means that energy has been dissipated, for example by friction.

However, there are **conservative fields** for which the integral above is the same between two given points, regardless of what path we calculate it along.

It can be shown from calculus that, ordinarily,

$$\nabla \times \mathbf{F} \equiv \text{curl } \mathbf{F} = 0 \tag{3.36}$$

is a necessary and sufficient condition for the integral (3.35) to be independent of path, that is $\nabla \times \mathbf{F} = 0$ for conservative fields and $\nabla \times \mathbf{F} \neq 0$ for nonconservative fields. Explicitly, the components of the vector $\nabla \times \mathbf{F}$ can be obtained from the determinant

$$\nabla \times \mathbf{F} = \begin{vmatrix} \mathbf{i} & \mathbf{j} & \mathbf{k} \\ \dfrac{\partial}{\partial x} & \dfrac{\partial}{\partial y} & \dfrac{\partial}{\partial z} \\ F_x & F_y & F_z \end{vmatrix}$$

$$= \mathbf{i}\left(\frac{\partial F_z}{\partial y} - \frac{\partial F_y}{\partial z}\right) + \mathbf{j}\left(\frac{\partial F_x}{\partial z} - \frac{\partial F_z}{\partial x}\right) + \mathbf{k}\left(\frac{\partial F_y}{\partial x} - \frac{\partial F_x}{\partial y}\right)$$

It is not difficult to justify the considerations above. Suppose that for a given $\mathbf{F}$ there is a function $W(x, y, z)$ such that

$$\mathbf{F} = \nabla W \equiv \mathbf{i}\,\frac{\partial W}{\partial x} + \mathbf{j}\,\frac{\partial W}{\partial y} + \mathbf{k}\,\frac{\partial W}{\partial z}$$

then, from the fact that $\partial^2 W/\partial x \partial y = \partial^2 W/\partial y \partial x$ etc., we see that the components of $\nabla \times \mathbf{F}$ are all zero. Then, if $\mathbf{F} = \nabla W$ it follows that $\nabla \times \mathbf{F} = 0$. Conversely, if $\nabla \times \mathbf{F} = 0$, then we can find a function $W(x, y, z)$ for which $\mathbf{F} = \nabla W$. In this case we can write

$$\mathbf{F} \cdot d\mathbf{x} = \nabla W \cdot d\mathbf{x} = \frac{\partial W}{\partial x}\,dx + \frac{\partial W}{\partial y}\,dy + \frac{\partial W}{\partial z}\,dz = dW \tag{3.37}$$

and

$$\int_A^B \mathbf{F} \cdot d\mathbf{x} = \int_A^B dW = W(B) - W(A) \tag{3.38}$$

where $W(B)$ and $W(A)$ are the values of the function W at the endpoints of the path of integration. Since the integral does not depend on the path but only on the endpoints *A* and *B*, then **F** is conservative, *W* is the **scalar potential** of the force **F** and it is customary to define the **potential energy** as $V = -W$.

From its definition, it is clear that *V* can be changed by adding any constant; this corresponds to the choice of the zero level of the potential energy and has no effect on **F**.

The differential *dW* of eq (3.37) is an exact differential. In the light of the discussion above, we can say that $\nabla \times \mathbf{F} = 0$ is a necessary and sufficient condition for $\mathbf{F} \cdot d\mathbf{x}$ to be an exact differential. More generally, the following theorem of vector calculus can be proven:

If the components of **F** have continuous first partial derivatives in a simply connected region, then any one of the following five conditions implies all the others:

1. $\nabla \times \mathbf{F} = 0$ at every point of the region.
2. $\oint \mathbf{F} \cdot d\mathbf{x} = 0$ around every simple closed curve in the region.
3. **F** is conservative, that is, $\int_A^B \mathbf{F} \cdot d\mathbf{x}$ does not depend on the path of integration from point *A* to point *B* (the path, obviously, must lie entirely in the region).
4. $\mathbf{F} \cdot d\mathbf{x}$ is an exact differential.
5. There exists a function *V* such that $\mathbf{F} = -\nabla V$, where *V* is single-valued.

Generally speaking, a region is 'simply connected' if any simple closed curve in the region can be shrunk to a point without encountering any points not in the region.

Two examples of conservative forces that will be of interest to us are gravitational forces and elastic forces. It is very well known to the reader that the potential energy of a body of mass *m* lifted above the surface of the earth to a height *h* is *mgh* (where $h = 0$ is the choice for zero potential energy) and the potential energy of a stretched (or compressed) spring within its linear range is $k(\Delta l)^2/2 = F(\Delta l)/2$. In fact, $F = -k(\Delta l)$, where Δl is the displacement from the unstretched position, which is commonly assumed as the position of zero potential energy.

The elastic potential energy is also called **strain energy**; in general, for an elastically deformed body it can be written as

$$E_{strain} = \int_V w_{strain} \, dV$$

where w_{strain} is the strain energy per unit volume of the body, *dV* is the element of volume and we have used here the letter *E* for the energy to avoid ambiguities between the usual notation *V* for the potential energy and *V* for volume. Again, zero potential energy is assigned to the undeformed body.

Example 3.1.   Illustratively, we can calculate the potential energy of a rod of length L and cross-section A under the action of an axial tensile force $F(x, t)$. The infinitesimal element dx of the rod undergoes an elongation

$$\frac{\partial u(x, t)}{\partial x} dx = \varepsilon(x, t) dx$$

where $u(x, t)$ is the displacement and $\varepsilon(x, t)$ is the strain at point x and time t. The strain energy of the volume element $A dx$ is equal in magnitude to the work done by the force F, i.e.

$$dE_{strain} = \frac{1}{2} F(x, t)\varepsilon(x, t) dx$$

from the definition of axial stress $\sigma(x, t) = F/A$ and the assumption to remain within the elastic range – so that $\sigma = E\varepsilon$ ($E =$ Young's modulus) – we get

$$dE_{strain} = \frac{1}{2} EA\varepsilon^2 dx$$

from which the strain energy of the rod follows as

$$E_{strain} = \frac{1}{2} \int_0^L EA\varepsilon^2 dx = \frac{1}{2} \int_0^L EA \left(\frac{\partial u}{\partial x}\right)^2 dx \tag{3.39}$$

Obviously, the action of an axial compressive force $F(x, t)$ leads to the same result.

3.3.2   A few generalizations of Lagrange's equations

One of the most important nonconservative forces is **viscous damping**. The viscous force is proportional to the particle velocity, so that for the kth particle

$$\mathbf{F}_k^{(visc)} = -c\mathbf{v}_k \tag{3.40a}$$

or, explicitly, for the single components

$$F_{k\alpha}^{(visc)} = -cv_{k\alpha} \tag{3.40b}$$

where k is the particle index ($k = 1, 2, ..., N$) and α is the component index ($\alpha = 1, 2, 3$; $\alpha = 1$ indicates the x component, $\alpha = 2$ the y component and $\alpha = 3$ the z component).

These forces can be derived from a scalar function D – a kind of generalized potential – of the form

$$D = -\frac{1}{2}\sum_{k=1}^{N} c_k (\mathbf{v}_k \cdot \mathbf{v}_k) = -\frac{1}{2}\sum_{k} c_k \sum_{\alpha=1}^{3} v_{k\alpha}^2 \tag{3.40c}$$

from which it follows that $F_{k\alpha}^{(visc)} = -\partial D/\partial v_{k\alpha}$, or

$$\mathbf{F}_k^{(visc)} = -\nabla_{v_k} D \tag{3.41}$$

where the subscript v_k indicates that the gradient is taken with respect to velocities. For example, in the case of a single particle moving in one direction with velocity v the equations above simply state that $F = -cv$ and $D = cv^2/2$, so that $F = -dD/dv$.

From eq (3.41) we get

$$\sum_{k} \mathbf{F}_k^{(visc)} \cdot \delta \mathbf{x}_k = -\sum_{j,k} \nabla_{v_k} D \cdot \frac{\partial \mathbf{x}_k}{\partial q_j} \delta q_j$$

and since $\partial \mathbf{x}_k/\partial q_j = \partial \mathbf{v}_k/\partial \dot{q}_j$ it follows that

$$Q_j^{(visc)} = -\frac{\partial D}{\partial \dot{q}_j} \tag{3.42}$$

Lagrange's equation may then be written as

$$\frac{d}{dt}\left(\frac{\partial L}{\partial \dot{q}_j}\right) - \frac{\partial L}{\partial q_j} + \frac{\partial D}{\partial \dot{q}_j} = 0 \tag{3.43}$$

in the case of conservative and viscous forces, or as

$$\frac{d}{dt}\left(\frac{\partial L}{\partial \dot{q}_j}\right) - \frac{\partial L}{\partial q_j} + \frac{\partial D}{\partial \dot{q}_j} = Q_j \tag{3.44}$$

when conservative, viscous and nonconservative (other than viscous) forces are acting on our system. The first two terms on the left-hand side account for conservative forces (through the function V which is part of the Lagrangian L), the third term accounts for viscous forces and the term on the right-hand side accounts for all other nonconservative forces.

Sometimes it may be convenient to work with constrained coordinates. Let us suppose that we have an n-degrees of freedom system and $n+1$ coordinates (which we will call $q_1, q_2, ..., q_{n+1}$), i.e. we have one coordinate in excess with respect to the minimum number required to describe our system. Therefore, they will be connected by one constraint equation of the general form

$$f(q_1, q_2, ..., q_n, q_{n+1}) = 0 \qquad (3.45)$$

from which it follows that

$$\sum_{i=1}^{n+1} \frac{\partial f}{\partial q_i} \delta q_i = 0 \qquad (3.46)$$

and implies that only n out of the $n+1$ δq_is are independent, the $(n+1)$th being determined by eq (3.46). The application of d'Alembert's principle to the principle of virtual work leads again to the expression

$$\sum_{j=1}^{n+1} \left[\frac{d}{dt} \left(\frac{\partial T}{\partial \dot{q}_j} \right) - \frac{\partial T}{\partial q_j} - Q_j \right] \delta q_j = 0 \qquad (3.47)$$

but now this does not imply that the coefficients in brackets are individually zero. Let us then add λ times eq (3.46) to eq (3.47), where λ is an unknown arbitrary parameter called the **Lagrangian multiplier**,

$$\sum_{j=1}^{n+1} \left[\frac{d}{dt} \left(\frac{\partial T}{\partial \dot{q}_j} \right) - \frac{\partial T}{\partial q_j} - Q_j + \lambda \frac{\partial f}{\partial q_j} \right] \delta q_j = 0 \qquad (3.48)$$

and let us choose λ so that

$$\frac{d}{dt} \left(\frac{\partial T}{\partial \dot{q}_{n+1}} \right) - \frac{\partial T}{\partial q_{n+1}} - Q_{n+1} + \lambda \frac{\partial f}{\partial q_{n+1}} = 0 \qquad (3.49)$$

The sum (3.48) is then a sum of n terms implying the n Lagrange equations

$$\frac{d}{dt} \left(\frac{\partial T}{\partial \dot{q}_j} \right) - \frac{\partial T}{\partial q_j} = Q_j$$

which, together with eqs (3.49) and (3.45), are $n+2$ equations in the $n+2$ unknowns $q_1, q_2, ..., q_{n+1}, \lambda$. The extension to the case of more than one, say N $(N > n)$, coordinates is straightforward; we will then have $m = N - n$ constraint equations and m Lagrangian multipliers $\lambda_1, \lambda_2, ..., \lambda_m$, thus obtaining $N + m$ equations in $N + m$ unknowns, i.e. the N coordinates plus the m multipliers.

The reader has probably noticed that in the foregoing discussion on Lagrangian multipliers we assumed one (or more) holonomic constraint (eq (3.45)). The method can be used also in the case of nonholonomic constraints which, very often, have the general form

$$\sum_{j=1}^{n} b_{lj}\dot{q}_j + b_l = 0 \qquad\qquad (3.50)$$

where $l = 1, 2, ..., m$ in the case of m nonholonomic constraints. The generalized coordinates are now n in number, $b_{lj} = b_{lj}(q_1, ..., q_n)$ and $b_l = b_l(t)$. In this case it is not possible to find m functions such that eqs (3.50) are equivalent to $f_l = const$, but referring to virtual displacements we can write eq (3.50) in differential form

$$\sum_{j=1}^{n} b_{lj}\delta q_j = 0$$

multiply it by λ_l, sum it to the usual expression (3.47), where now $j = 1, 2, ..., n$, choose the m multipliers so that, say, the last m terms are zero, i.e.

$$\frac{d}{dt}\left(\frac{\partial T}{\partial \dot{q}_j}\right) - \frac{\partial T}{\partial q_j} - Q_j + \sum_{l=1}^{m} \lambda_l b_{lj} = 0 \qquad (j = n - m + 1, ..., n) \qquad (3.51)$$

and obtain explicitly the Lagrange multipliers; it follows that

$$\frac{d}{dt}\left(\frac{\partial T}{\partial \dot{q}_j}\right) - \frac{\partial T}{\partial q_j} - Q_j + \sum_{l=1}^{m} \lambda_l b_{lj} = 0 \qquad (j = 1, 2, ..., n - m) \qquad (3.52)$$

Equations (3.50) and (3.52), are now the n equations that can be solved for the n coordinates q_j. Equivalently, we can say that eqs (3.50), (3.51) and (3.52) are $n + m$ equations in $n + m$ unknowns.

The method is particularly useful in problems of statics because the parameters $\lambda_1, \lambda_2, ..., \lambda_m$ represent the generalized reactive forces which automatically appear as a part of the solution itself.

3.3.3 Kinetic energy and Rayleigh's dissipation function in generalized coordinates. Energy conservation

From the foregoing discussion it is clear that Lagrange's equations are based on the calculation of the derivatives of the kinetic energy with respect to the generalized coordinates. Since we know that

$$T = \frac{1}{2}\sum_{k=1}^{N} m_k(\mathbf{v}_k \cdot \mathbf{v}_k)$$

from eq (3.14) we get the kinetic energy as the sum of three terms

$$T = \frac{1}{2}A + \sum_j A_j \dot{q}_j + \frac{1}{2}\sum_{ij} A_{ij}\dot{q}_i\dot{q}_j \qquad (3.53)$$

where

$$A = \sum_k m_k \left(\frac{\partial \mathbf{x}_k}{\partial t}\right)^2 \qquad (3.53\text{a})$$

$$A_j = \sum_k m_k \frac{\partial \mathbf{x}_k}{\partial q_j}\frac{\partial \mathbf{x}_k}{\partial t} \qquad (3.53\text{b})$$

$$A_{ij} = \sum_k m_k \frac{\partial \mathbf{x}_k}{\partial q_i}\frac{\partial \mathbf{x}_k}{\partial q_j} \qquad (3.53\text{c})$$

and a noteworthy simplification occurs in the case of scleronomic (time independent) constraints, i.e. when time does not appear explicitly in the coordinate transformation (3.13). In fact, in this case, only the third term in eq (3.53) is different from zero and the kinetic energy is a quadratic form in the generalized velocities.

First we note that

$$A_{ij} = A_{ji} \qquad (3.54)$$

and the term in question can also be written in matrix notation as

$$T = \frac{1}{2}\sum_{ij} A_{ij}\dot{q}_i\dot{q}_j = \frac{1}{2}\dot{\mathbf{q}}^T \mathbf{A}\dot{\mathbf{q}} \qquad (3.55)$$

where

$$\dot{\mathbf{q}} = \begin{bmatrix} \dot{q}_1 \\ \dot{q}_2 \\ \cdots \\ \dot{q}_n \end{bmatrix} \qquad \mathbf{A} = \begin{bmatrix} A_{11} & A_{12} & \cdots & A_{1n} \\ A_{21} & A_{22} & \cdots & A_{2n} \\ \cdots & \cdots & \cdots & \cdots \\ A_{n1} & A_{n2} & \cdots & A_{nn} \end{bmatrix}$$

$\mathbf{A}$ is symmetrical and $\dot{\mathbf{q}}^T$ is simply the transpose of matrix $\dot{\mathbf{q}}$.

Furthermore, by virtue of Euler's theorem on homogeneous functions – which we state briefly without proof – T is a homogeneous function of degree 2 in the $\dot{q}$s and can be written as

$$T = \frac{1}{2} \sum_{j=1}^{n} \dot{q}_j \frac{\partial T}{\partial \dot{q}_j} \tag{3.56}$$

a form that may turn out to be useful in many circumstances.

Euler's theorem on homogeneous functions. A general function $f(x, y, z)$ is homogeneous of degree m if

$$f(\alpha x, \alpha y, \alpha z) = \alpha^m f(x, y, z)$$

Euler's theorem says that if f is homogeneous of degree m then

$$x \frac{\partial f}{\partial x} + y \frac{\partial f}{\partial y} + z \frac{\partial f}{\partial z} = mf$$

By looking at the general expression of Rayleigh's dissipation function (eq (3.40c)) it is not difficult to see that this function too is a quadratic form in the generalized velocities which can be written as

$$D = \frac{1}{2} \sum_{ij} C_{ij} \dot{q}_i \dot{q}_j \tag{3.57}$$

where

$$C_{ij} = \sum_{k} c_k \frac{\partial \mathbf{x}_k}{\partial \dot{q}_i} \frac{\partial \mathbf{x}_k}{\partial \dot{q}_j} \tag{3.57a}$$

and

$$C_{ij} = C_{ji} \tag{3.57b}$$

From eq (3.56) it follows that

$$2 \frac{dT}{dt} = \sum_{j} \ddot{q}_j \frac{\partial T}{\partial \dot{q}_j} + \sum_{j} \dot{q}_j \frac{d}{dt} \left(\frac{\partial T}{\partial \dot{q}_j} \right) \tag{3.58}$$

On the other hand, since $T = T(q, \dot{q})$, we can also write

$$\frac{dT}{dt} = \sum_{j} \ddot{q}_j \frac{\partial T}{\partial \dot{q}_j} + \sum_{j} \dot{q}_j \frac{\partial T}{\partial q_j} \tag{3.59}$$

subtracting eq (3.59) from (3.58) and using Lagrange's equations we get

$$\frac{dT}{dt} = \sum_j \dot{q}_j \left[\frac{d}{dt} \left(\frac{\partial T}{\partial \dot{q}_j} \right) - \left(\frac{\partial T}{\partial q_j} \right) \right] = \sum_j \dot{q}_j Q_j \qquad (3.60)$$

If the forces are conservative, i.e. $Q_j = -\partial V/\partial q_j$, the term on the right-hand side is nothing but $-dV/dt$ so that

$$\frac{d}{dt}(T + V) = 0 \qquad (3.61)$$

or

$$T + V = const \equiv E \qquad (3.62)$$

which states the conservation of energy for a conservative system with scleronomic constraints.

If the nonconservative forces are viscous in nature, the term on the right-hand side of eq (3.60) is the power dissipated by such forces; following the same line of reasoning that leads to eq (3.56) we have

$$P \equiv \sum_j \dot{q}_j Q_j = - \sum_j \dot{q}_j \frac{\partial D}{\partial \dot{q}_j} = -2D$$

hence

$$\frac{d}{dt}(T + V) = -2D \qquad (3.63)$$

and, as expected, energy is not conserved in this case, $2D$ being the rate of energy dissipation due to the frictional forces.

3.3.4 Hamilton's equations

Lagrange's equations for a n-degree-of-freedom system are n second-order equations. In many cases, such a set of equations is equivalent to a set of $2n$ first order equations; in analytical mechanics these are Hamilton's equations, which we introduce briefly. Whenever a Lagrangian function exists, we can define the **conjugate momenta** as

$$p_j \equiv \frac{\partial L}{\partial \dot{q}_j} \qquad (3.64)$$

and a Hamiltonian function

$$H \equiv \sum_j \frac{\partial L}{\partial \dot{q}_j} \dot{q}_j - L = \sum_j p_j \dot{q}_j - L \qquad (3.65)$$

where the $\dot{q}_j$s are expressed as functions of the generalized momenta and then H is, in general, a function $H = H(p, q, t)$. By differentiating eq (3.65) and taking into account Lagrange's equations, we get

$$\sum_j \left(\frac{\partial H}{\partial p_j} dp_j + \frac{\partial H}{\partial q_j} dq_j \right) = \sum_j (\dot{q}_j dp_j - \dot{p}_j dq_j)$$

which is valid for any choice of the differentials dp_j and dq_j; it then follows that

$$\dot{q}_j = \frac{\partial H}{\partial p_j}$$

$$\dot{p}_j = - \frac{\partial H}{\partial q_j} \qquad (3.66)$$

These are Hamilton's canonical equations, showing a particular formal elegance and symmetry. It is not difficult to show that in the case of scleronomic constraints the function H is the total energy $T + V$ of the system, which is a constant of motion when the Lagrangian does not depend explicitly on time t. Since we will make little use of Hamilton's equations, we do not pursue the subject any further and the interested reader can find references for additional reading at the end of Part I.

3.4 Hamilton's principle of least action

Many problems of physics and engineering lend themselves to mathematical formulations that belong to the specific subject called 'calculus of variations'. In principle, the basic problem is the same as finding the maximum and minimum values of a function $f(x)$ in ordinary calculus. We calculate df/dx and set it equal to zero; the values of x that we find are called **stationary points** and they correspond to maximum points, minimum points or points of inflection with horizontal tangent. In the calculus of variation the quantity to make stationary is an integral of the form

$$I = \int_{x_1}^{x_2} f(x, y, \dot{y}) dx \qquad (3.67)$$

where $\dot{y} = dy/dx$. The statement of the problem is as follows: given the two points (x_1, y_1) and (x_2, y_2) and the form of the function f, find the curve $y = y(x)$, passing through the given points, which makes the integral I have the smallest value (or stationary value). Such a function $y(x)$ is called an **extremal**. Then let $\eta(x)$ be an arbitrary function whose only properties are to be zero at x_1 and x_2 and have continuous second derivatives. We define $Y(x)$ as

$$Y(x) = y(x) + \varepsilon\eta(x) \tag{3.68}$$

where $y(x)$ is the desired (unknown) extremal and ε is a parameter. Due to the arbitrariness of $\eta(x)$, $Y(x)$ represents any curve that can be drawn through the points (x_1, y_1) and (x_2, y_2). Then

$$I(\varepsilon) = \int_{x_1}^{x_2} f(x, Y, \dot{Y})dx$$

and we want $dI/d\varepsilon = 0$ when $\varepsilon = 0$. By differentiating under the integral sign and substituting eq (3.68) we are led to

$$\left(\frac{dI}{d\varepsilon}\right)_{\varepsilon=0} = \int_{x_1}^{x_2}\left[\frac{\partial f}{\partial y}\eta(x) + \frac{\partial f}{\partial \dot{y}}\dot{\eta}(x)\right]dx = 0 \tag{3.69}$$

because $Y = y$ when $\varepsilon = 0$. The second term can be integrated by parts as

$$\int_{x_1}^{x_2}\frac{\partial f}{\partial \dot{y}}\dot{\eta}(x)dx = \frac{\partial f}{\partial \dot{y}}\eta(x)\Bigg|_{x_1}^{x_2} - \int_{x_1}^{x_2}\frac{d}{dx}\left(\frac{\partial f}{\partial \dot{y}}\right)\eta(x)dx$$

and the first term on the right-hand side is zero because of the requirements on $\eta(x)$. We obtain

$$\left(\frac{dI}{d\varepsilon}\right)_{\varepsilon=0} = \int_{x_1}^{x_2}\left[\frac{\partial f}{\partial y} - \frac{d}{dx}\left(\frac{\partial f}{\partial \dot{y}}\right)\right]\eta(x)dx = 0$$

and for the arbitrariness of $\eta(x)$

$$\frac{d}{dx}\left(\frac{\partial f}{\partial \dot{y}}\right) - \frac{\partial f}{\partial y} = 0 \tag{3.70}$$

which is called the **Euler–Lagrange equation**. The extension to more than one dependent variable (but always one independent variable) is straightforward: one gets as many Euler–Lagrange equations as the number of dependent variables.

A word of warning about the mathematical symbols: the reader will often find δI for the differential $(dI/d\varepsilon)_{\varepsilon=0}d\varepsilon$, where the δ symbol reads 'the variation of' and indicates that ε and not x is the differentiation variable. Similarly, $\delta y = (\partial Y/\partial\varepsilon)_{\varepsilon=0}d\varepsilon = \eta(x)d\varepsilon$ and $\delta\dot{y} = (\partial\dot{Y}/\partial\varepsilon)_{\varepsilon=0}d\varepsilon = \dot{\eta}(x)d\varepsilon$.

The formal structure of eq (3.70) already suggests the connection with the problems of interest to us. In fact, if we replace the independent variable x with time and if the function f is the Lagrangian of some mechanical system, eq (3.70) is a Lagrange equation for a single-degree-of-freedom conservative system where $y(x)$ – the dependent variable – is nothing but the generalized coordinate $q(t)$.

In the case of n degrees of freedom we have n generalized coordinates and the procedure outlined above leads exactly to the n Lagrange equations (3.31). The integral to be made stationary is called **action**; it is usually denoted by the symbol S and is defined as

$$S \equiv \int_{t_1}^{t_2} L(q, \dot{q}, t)dt \tag{3.71}$$

where L is the Lagrangian of the system under consideration. **Hamilton's principle of least action** can then be written

$$\delta S = \delta \int_{t_1}^{t_2} L(q, \dot{q}, t)dt = 0 \tag{3.72}$$

Thus, for a conservative system with frictionless constraints the natural motion has the property of making the Hamiltonian action stationary with respect to all varied motions for which $\delta q(t_1) = \delta q(t_2) = 0$.

More specifically, the class of varied motions considered in this case are called 'synchronous' to make the increments δq correspond to the virtual displacements introduced in preceding sections; other types of increments can be considered but this is beyond the scope of the present discussion.

In essence, Hamilton's principle is an integral version of the virtual work principle because it considers the entire motion of the system between two instants t_1 and t_2. Furthermore, Hamilton's principle is a necessary and sufficient condition for the validity of Lagrange's equations; as such, it can be obtained by starting from d'Alembert's principle, i.e. the equation

$$\sum_k (\mathbf{F}_k - m_k\ddot{\mathbf{x}}_k) \cdot \delta\mathbf{x}_k = 0$$

(e.g. Meirovitch [2]). The advantages – as with Lagrange's and Hamilton's equations – are the possibility of working with a scalar function, the invariance with respect to the coordinate system used and their validity for linear and nonlinear systems.

An extended, or **generalized Hamilton principle** can be stated as

$$\delta \int_{t_1}^{t_2} (T - V)dt + \int_{t_1}^{t_2} \delta W_{nc} dt = 0 \qquad (3.73)$$

where

$$\delta W_{nc} = \sum_j Q_j \delta q_j$$

is the virtual work done by nonconservative forces (including damping forces and forces not accounted for in V).

Example 3.2.   As an example, we can use the generalized Hamilton principle to derive the governing equation of motion for the transverse vibration of an Euler–Bernoulli beam (shear deformation and rotary inertia ignored, see Chapter 8 for more details) in the general case of variable mass density and bending stiffness $EI(x)$, as shown in Fig. 3.1.

The kinetic energy is

$$T = \frac{1}{2} \int_0^L \hat{m}(x) \left(\frac{\partial y}{\partial t} \right)^2 dx \qquad (3.74)$$

where $\hat{m}(x)$ is the mass per unit length and $y = y(x, t)$ is the transverse displacement from the neutral axis; the potential strain energy is in this case

$$V = \frac{1}{2} \int_0^L EI(x) \left(\frac{\partial^2 y}{\partial x^2} \right)^2 dx \qquad (3.75)$$

where E is Young's modulus and $I(x)$ is the cross-sectional area moment of inertia of the beam about its neutral axis. Finally, the virtual work done by the

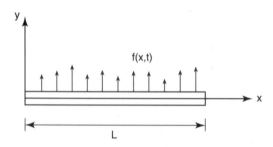

Fig. 3.1 Transverse vibration of a slender (Euler–Bernoulli) beam.

nonconservative distributed external force $f(x, t)$ is

$$\delta W_{nc} = \int_0^L f(x, t)\delta y(x, t)dx \tag{3.76}$$

where $\delta y = \delta y(x, t)$ is an arbitrary virtual displacement.

Inserting the expressions above in the generalized Hamilton principle, under the assumptions that the operator δ commutes with the time and spatial derivatives and that the time and spatial integrations are interchangeable, we obtain, for example, for the term $\int \delta T dt$,

$$\frac{1}{2} \int_{t_1}^{t_2} \int_0^L \hat{m} \left(\frac{\partial y}{\partial t}\right)^2 dxdt = \int_0^L \int_{t_1}^{t_2} \hat{m} \frac{\partial y}{\partial t} \frac{\partial}{\partial t}(\delta y)dtdx$$

Integration by parts with respect to time gives

$$\int_0^L \left[\left(\hat{m} \frac{\partial y}{\partial t} \delta y \right) \Big|_{t_1}^{t_2} - \int_{t_1}^{t_2} \hat{m} \left(\frac{\partial^2 y}{\partial t^2}\right) \delta y dt \right] dx$$

$$= -\int_{t_1}^{t_2} \int_0^L \hat{m} \left(\frac{\partial^2 y}{\partial t^2}\right) \delta y dxdt \tag{3.77}$$

because δy vanishes at $t = t_1$ and $t = t_2$. Similar calculations apply for the strain energy term, where now two integrations by part with respect to x must be performed and we get

$$-\int_{t_1}^{t_2} \delta V dt = \int_{t_1}^{t_2} \left[-\left(EI \frac{\partial^2 y}{\partial x^2} \right) \delta \frac{\partial y}{\partial x} \Big|_0^L + \frac{\partial}{\partial x}\left(EI \frac{\partial^2 y}{\partial x^2} \right) \delta y \Big|_0^L \right.$$

$$\left. - \int_0^L \frac{\partial^2}{\partial x^2}\left(EI \frac{\partial^2 y}{\partial x^2} \right) dx \right] dt$$

Putting all the pieces back together in Hamilton's principle we obtain

$$\int_{t_1}^{t_2} \left\{ \int_0^L \left[-\hat{m} \frac{\partial^2 y}{\partial t^2} - \frac{\partial^2}{\partial x^2}\left(EI \frac{\partial^2 y}{\partial x^2} \right) + f(x, t) \right] \delta y dx \right.$$

$$\left. + \frac{\partial}{\partial x}\left(EI \frac{\partial^2 y}{\partial x^2} \right) \delta y \Big|_0^L - EI \left(\frac{\partial^2 y}{\partial x^2}\right) \delta \frac{\partial y}{\partial x} \Big|_0^L \right\} dt = 0$$

Because of the arbitrariness of δy, the equation above is satisfied only if the integrand vanishes for all t and all x in the domain $0 < x < L$, i.e.

$$\frac{\partial^2}{\partial x^2}\left(EI\frac{\partial^2 y}{\partial x^2}\right) + \hat{m}\frac{\partial^2 y}{\partial t^2} = f(x, t) \tag{3.78}$$

The other terms should also vanish, so that

$$\frac{\partial}{\partial x}\left(EI\frac{\partial^2 y}{\partial x^2}\right)\delta y \bigg|_0^L = 0 \tag{3.79}$$

$$EI\frac{\partial^2 y}{\partial x^2}\delta\frac{\partial y}{\partial x}\bigg|_0^L = 0 \tag{3.80}$$

Equation (3.79) can be satisfied if either the displacement is zero – which implies $\delta y = 0$ – or the shearing force

$$\frac{\partial}{\partial x}\left(EI\frac{\partial^2 y}{\partial x^2}\right)$$

is zero at either end; eq (3.80), in turn, can be satisfied if either the slope is zero, i.e. $\delta(\partial y/\partial x) = 0$, or the bending moment

$$EI\frac{\partial^2 y}{\partial x^2}$$

is zero at either end.

It is not difficult to identify eq (3.78) as the differential equation of motion for the transverse (or flexural or bending) vibrations of a slender beam and eqs (3.79) and (3.80) as the essential and natural boundary conditions, which automatically appear from Hamilton's principle.

For the moment, it suffices to say that in any case of beam vibration, four boundary conditions must exist: two at each end. The specific problem under investigation dictates which particular set of conditions apply in that particular case.

3.5 The general problem of small oscillations

In application of mechanics, our interest lies specifically in a special form of motion known as 'small oscillations'.

It is well explained in many textbooks how the problem of an oscillating simple pendulum leads, in the general case, to an equation relating the deflection angle θ to time which is a nonelementary (elliptic) integral. However, the problem is greatly simplified if $\theta \ll 1$, i.e. the deflection angle is small compared to a radian. The fundamental point in this case is that the frequency of small oscillations does not depend on the amplitude of oscillation.

A slightly more complicated example is given by the double pendulum, which is composed of two masses m_1 and m_2 as shown in Fig. 3.2. The masses are connected to each other by a light string of length l_2 and the first mass is suspended from a fixed point by means of a light string of length l_1.

It is obvious that this is a two-degree-of-freedom system and a proper choice of generalized coordinates are the two angles θ and ϕ. The Lagrangian function is in this case

$$L = \frac{1}{2}(m_1 + m_2)l_1^2\dot{\theta}^2 + \frac{1}{2}m_2l_2^2\dot{\phi}^2 + m_2l_1l_2\dot{\theta}\dot{\phi}\cos(\theta - \phi)$$
$$+ (m_1 + m_2)gl_1\cos\theta + m_2gl_2\cos\phi \tag{3.81}$$

where the origin of the coordinates is taken at the suspension point and the y axis is directed downwards.

Using Lagrange's equations, the reader is invited to derive the equations of motion. It will soon be clear that the equations are nonlinear and, in order to proceed further, some simplification is necessary (unless we want to resort to numerical integration).

The procedure of 'linearization' (or **small amplitude approximation**) is based on the assumption that θ and ϕ remain small and consists of neglecting all higher-order terms in a Taylor series expansion of the cosine terms in eq (3.81). Then, the two equations of motion become linear and can be

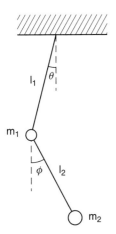

Fig. 3.2 Double pendulum.

tackled with much less effort. For our example, the Lagrangian becomes

$$L = \frac{1}{2}(m_1 + m_2)l_1^2\dot{\theta}^2 + \frac{1}{2}m_2l_2^2\dot{\phi}^2 + m_2l_1l_2\dot{\theta}\dot{\phi}$$

$$- \frac{1}{2}(m_1 + m_2)gl_1\theta^2 - \frac{1}{2}m_2gl_2\phi^2 \qquad (3.82)$$

and the equations of motion are easily obtained as

$$(m_1 + m_2)l_1\ddot{\theta} + m_2l_2\ddot{\phi} + (m_1 + m_2)g\theta = 0 \qquad (3.83a)$$

$$l_2\ddot{\phi} + l_1\ddot{\theta} + g\theta = 0 \qquad (3.83b)$$

These introductory remarks are much more general than the two simple cases above may suggest and, as a matter of fact, they are fundamental to all the problems considered in this book.

We note immediately that it makes more sense to examine and perform the appropriate modifications to the terms T and V **before** Lagrange's equations are formulated rather than simplify the equations of motion once they have been obtained. This is, in essence, the central idea to the discussion that follows.

All oscillations occur about a position of equilibrium and since force is equal to the derivative (with respect to the coordinate) of potential energy, the equilibrium condition for a single-degree-of-freedom system can be written as

$$\frac{\partial V}{\partial q} = 0 \qquad (3.84)$$

Let q_0 be the solution of eq (3.84), representing the equilibrium configuration of our system. From basic physics we know that the equilibrium is stable if V has a minimum at $q = q_0$. When we consider small deviations from the equilibrium we can expand V in a Taylor series as

$$V(q) - V(q_0) = \left(\frac{\partial V}{\partial q}\right)_{q=q_0}(q - q_0) + \frac{1}{2}\left(\frac{\partial^2 V}{\partial q^2}\right)_{q=q_0}(q - q_0)^2 + \dots \quad (3.85)$$

The first term on the right-hand side is zero in accordance with eq (3.84) and, without loss of generality, we can assume $V(q_0) = 0$ so that

$$V(q) \cong \frac{1}{2}ku^2 \qquad (3.86)$$

where we have defined

$$k \equiv \left(\frac{\partial^2 V}{\partial q^2} \right)_{q = q_0}$$

(3.87)

and $u = q - q_0$. The condition of stable equilibrium (force acting in the opposite direction of displacement) clearly implies $k > 0$. The kinetic energy, in its turn, can be written in the general form

$$T = \frac{1}{2} a(q) \dot{q}^2 = \frac{1}{2} a(q) \dot{u}^2$$

(3.88)

(note that $a(q) = m$ if u is a Cartesian coordinate) and expanding the coefficient $a(q)$ in a Taylor series we get

$$T \cong \frac{1}{2} a(q_0) \dot{q}^2 + \frac{1}{2} \left(\frac{\partial a}{\partial q} \right)_{q = q_0} (q - q_0) \dot{q}^2 + \ldots$$

$$= \frac{1}{2} a_0 \dot{u}^2 + \frac{1}{2} \left(\frac{\partial a}{\partial u} \right)_{u = 0} u \dot{u}^2 + \ldots$$

where we define $a(q_0) \equiv a_0$. The zero-order term of T is already of the same order as the second-order term in the expansion of V; we retain this term only and obtain the Lagrangian function

$$L = \frac{1}{2} a_0 \dot{u}^2 - \frac{1}{2} k u^2$$

(3.89)

which leads to the equation of motion

$$a_0 \ddot{u} + k u = 0$$

or

$$\ddot{u} + \omega^2 u = 0$$

(3.90)

where we defined $\omega^2 \equiv k/a_0$.

The solution of eq (3.90) is discussed in detail in Chapter 4, but its essential feature is that ω is the characteristic frequency of oscillation, which is **completely defined by the physical properties of the system itself and does not**

depend on the initial conditions of motion. This is a consequence of the small-oscillation assumption and is no longer valid when higher-order approximations are considered; mathematically, this important property of ω is strictly connected to the fact that V is a quadratic function of the generalized coordinate.

The considerations above can be generalized to n-degree of freedom conservative systems with scleronomic constraints. As before, the system is in equilibrium when the generalized applied forces are zero, i.e.

$$\frac{\partial V}{\partial q_j} = 0 \qquad (j = 1, 2, ..., n) \tag{3.91}$$

Let the equilibrium configuration be identified by the coordinates $q_{01}, q_{02}, ..., q_{0n}$; stability requires that the stationary point of V defined by eqs (3.91) is a minimum and if we confine our attention to small oscillations about the equilibrium position, all the relevant functions may be expanded in a Taylor series. Letting $u_i \equiv q_i - q_{0i}$ be the variation of the ith generalized coordinate from equilibrium, we have

$$V(q_1, q_2, ..., q_n) - V(q_{01}, q_{02}, ..., q_{0n})$$

$$= \sum_i \left(\frac{\partial V}{\partial q_i}\right)_{q=q_0} u_i + \frac{1}{2} \sum_{i,j} \left(\frac{\partial^2 V}{\partial q_i \partial q_j}\right)_{q=q_0} u_i u_j + ... \tag{3.92}$$

where the subscript $q = q_0$ refers now to the entire set of coordinates. As before, we retain only the lowest order nonzero term, i.e.

$$V = \frac{1}{2} \sum_{i,j} \left(\frac{\partial^2 V}{\partial q_i \partial q_j}\right)_{q=q_0} u_i u_j = \frac{1}{2} \sum_{i,j} k_{ij} u_i u_j \tag{3.93}$$

where we defined the constants

$$k_{ij} \equiv \left(\frac{\partial^2 V}{\partial q_i \partial q_j}\right)_{q=q_0}$$

which are clearly symmetrical in the indices i and j.

An analogous line of reasoning applies to the kinetic energy which, in the case of scleronomic constraints, can be written as (eq (3.55))

$$T = \frac{1}{2} \sum_{i,j} A_{ij} \dot{q}_i \dot{q}_j = \frac{1}{2} \sum_{i,j} A_{ij} \dot{u}_i \dot{u}_j \tag{3.94}$$

In general, the coefficients A_{ij} are functions of the generalized coordinates and can be expanded in a Taylor series about equilibrium:

$$A_{ij}(q_1, q_2, ..., q_n) = m_{ij} + \sum_k \left(\frac{\partial A_{ij}}{\partial q_k}\right)_{q=q_0} u_k + ...$$

Only the first term of the expansion is retained, so that

$$T = \frac{1}{2} \sum_{i,j} m_{ij} \dot{u}_i \dot{u}_j \tag{3.95}$$

where m_{ij} represents the constant coefficients $A_{ij}(q_{01}, q_{02}, ..., q_{0n})$. From eqs (3.93) and (3.95) we obtain the Lagrangian

$$L = \frac{1}{2} \sum_{i,j} (m_{ij} \dot{u}_i \dot{u}_j - k_{ij} u_i u_j) \tag{3.96}$$

Performing the appropriate differentiations, Lagrange's equations lead (since $m_{ij} = m_{ji}$ and i and j are dummy indices of summation) to the n equations of motion

$$\sum_{j=1}^{n} (m_{ij} \ddot{u}_j + k_{ij} u_j) = 0 \qquad (i = 1, 2, ..., n) \tag{3.97}$$

which are equivalent to the matrix expression

$$\mathbf{M\ddot{u} + Ku = 0} \tag{3.98}$$

where the mass, stiffness, displacement and acceleration matrices have been introduced:

$$\mathbf{M} = \begin{bmatrix} m_{11} & m_{12} & \cdots & m_{1n} \\ m_{21} & m_{22} & \cdots & m_{2n} \\ \cdots & \cdots & \cdots & \cdots \\ m_{n1} & m_{n2} & \cdots & m_{nn} \end{bmatrix}$$

$$\mathbf{K} = \begin{bmatrix} k_{11} & k_{12} & \cdots & k_{1n} \\ k_{21} & k_{22} & \cdots & k_{2n} \\ \cdots & \cdots & \cdots & \cdots \\ k_{n1} & k_{n2} & \cdots & k_{nn} \end{bmatrix} \tag{3.98a}$$

$$\mathbf{u} = \begin{bmatrix} u_1 \\ u_2 \\ \cdots \\ u_n \end{bmatrix} \qquad \ddot{\mathbf{u}} = \begin{bmatrix} \ddot{u}_1 \\ \ddot{u}_2 \\ \cdots \\ \ddot{u}_n \end{bmatrix} \tag{3.98b}$$

In the language of matrices, the symmetry of the coefficients m_{ij} and k_{ij} translates into

$$\mathbf{M} = \mathbf{M}^T$$

$$\mathbf{K} = \mathbf{K}^T \tag{3.99}$$

If dissipative forces of viscous nature are also acting on our n-degree of freedom system, the generalization is straightforward. The formal analogy of the Rayleigh dissipation function D (see eq (3.57)) with the kinetic energy term – a symmetrical quadratic form in the generalized velocities – suggests that the coefficients C_{ij} in the function D can also be expanded in a Taylor series as

$$C_{ij}(q_1, q_2, ..., q_n) = c_{ij} + \sum_k \left(\frac{\partial C_{ij}}{\partial q_k}\right)_{q=q_0} u_k + ...$$

where $c_{ij} = C_{ij}(q_{01}, q_{02}, ..., q_{0n})$ and only the first term on the right-hand side is retained. Using Lagrange's equations in the form of eq (3.43) leads now to the n equations of motion

$$\sum_{j=1}^{n} (m_{ij}\ddot{u}_j + c_{ij}\dot{u}_j + k_{ij}u_j) = 0 \qquad (i = 1, 2, ..., n) \tag{3.100}$$

or, in matrix form

$$\mathbf{M}\ddot{\mathbf{u}} + \mathbf{C}\dot{\mathbf{u}} + \mathbf{K}\mathbf{u} = 0 \tag{3.101}$$

where

$$\mathbf{C} = \begin{bmatrix} c_{11} & c_{12} & \cdots & c_{1n} \\ c_{21} & c_{22} & \cdots & c_{2n} \\ \cdots & \cdots & \cdots & \cdots \\ c_{n1} & c_{n2} & \cdots & c_{nn} \end{bmatrix}$$

is called the **damping matrix** and the symmetry of the coefficients implies $\mathbf{C} = \mathbf{C}^T$. The addition of viscous forces in eq (3.90) – i.e. the single degree of freedom system – follows easily as a particular case of eqs (3.100).

Finally, generalized external forces can be taken into account by a term Q_i on the right-hand side of eqs (3.97) or (3.100), which in matrix notation means the addition of the column vector **f** on the right-hand side of eqs (3.98) or (3.101) to give

$$\mathbf{M\ddot{u} + Ku = f} \tag{3.102}$$

in the case of an 'undamped' (no dissipative forces) system, or

$$\mathbf{M\ddot{u} + C\dot{u} + Ku = f} \tag{3.103}$$

in the case of a viscously damped system. Explicitly $\mathbf{f} = [Q_1 \; Q_2 \; \cdots \; Q_n]^T$ and we have used the symbol **f** to avoid ambiguities with the column vector **q** of the generalized coordinates that we used in eq (3.55).

The solutions of eqs (3.98) and (3.101) – free vibration problem – and of eqs (3.102) and (3.103) – forced vibration problem – will be obtained and analysed in future chapters.

In the double pendulum example considered at the beginning of this section, the explicit quadratic expressions of the kinetic and potential energy are, in matrix form

$$T = \frac{1}{2} [\dot{\theta} \;\; \dot{\phi}] \begin{bmatrix} (m_1 + m_2)l_1^2 & m_2 l_1 l_2 \\ m_2 l_1 l_2 & m_2 l_2^2 \end{bmatrix} \begin{bmatrix} \dot{\theta} \\ \dot{\phi} \end{bmatrix}$$

$$V = \frac{1}{2} [\theta \;\; \phi] \begin{bmatrix} (m_1 + m_2)gl_1 & 0 \\ 0 & m_2 gl_2 \end{bmatrix} \begin{bmatrix} \theta \\ \phi \end{bmatrix}$$

where the elements of the mass and stiffness matrices are, respectively

$$m_{11} = (m_1 + m_2)l_1^2 \qquad k_{11} = (m_1 + m_2)gl_1$$

$$m_{12} = m_{21} = m_2 l_1 l_2 \qquad k_{12} = k_{21} = 0$$

$$m_{22} = m_2 l_2^2 \qquad k_{22} = m_2 gl_2$$

3.5.1 A qualitative discussion of Lagrange's method in small-oscillation problems

In the light of the discussion above, some qualitative observations can be made. They will not add anything substantial to what have been already said but they may be helpful in providing further insight in the formulation of the problem. The simplification procedure of small oscillation consists of performing appropriate modifications in the terms T and V so that only linear terms appear in Lagrange's equations. Let us then rewrite the kinetic energy (for a *n*-degree-of-freedom system with scleronomic constraints) and the *i*th Lagrange equation as

$$T = \frac{1}{2} \sum_{l,k} A_{lk}(q)\dot{q}_l \dot{q}_k \tag{3.104}$$

$$\frac{d}{dt}\left(\frac{\partial T}{\partial \dot{q}_i}\right) = \frac{\partial T}{\partial q_i} - \frac{\partial V}{\partial q_i} \tag{3.105}$$

The first term on the right-hand side of eq (3.105) will always be at least of second order, since the two $\dot{q}$s in the expression of T are unaffected by the $\partial/\partial q_i$ operation; this term becomes

$$\frac{1}{2} \sum_{l,k} \frac{\partial A_{lk}}{\partial q_i} \dot{q}_l \dot{q}_k$$

and will therefore be neglected.

On the left-hand side, the term $\partial T/\partial \dot{q}_i$ will be linear in the generalized velocities, i.e. we get

$$\frac{\partial T}{\partial \dot{q}_i} = \frac{1}{2} \sum_{l,k} A_{lk}(\delta_{li}\dot{q}_k + \dot{q}_l \delta_{ki}) = \frac{1}{2} \sum_k A_{ik}\dot{q}_k + \frac{1}{2} \sum_l A_{li}\dot{q}_l = \sum_k A_{ik}\dot{q}_k$$

where δ is the usual Kronecker symbol and the symmetry of the coefficients has been used. The operator d/dt applies both to the q and $\dot{q}$ variables of this term. The time differentiation of any q results in the appearance of a second $\dot{q}$ and therefore in a term that will be neglected; the time differentiation of the $\dot{q}$s will produce an expression linear in the $\ddot{q}$s with coefficients (the A_{ij}) that in general depend on some of the qs. In formula

$$\frac{d}{dt}\left(\sum_k A_{ik}\dot{q}_k\right) = \sum_{k,j} \frac{\partial A_{ik}}{\partial q_j} \dot{q}_k \dot{q}_j + \sum_{k,j} A_{ik}\delta_{kj}\ddot{q}_j$$

$$= \sum_{k,j} \frac{\partial A_{ik}}{\partial q_j} \dot{q}_k \dot{q}_j + \sum_j A_{ij}\ddot{q}_j$$

The required linear term is obtained by assigning to these qs their equilibrium values (the coefficients m_{ij} in the Taylor expansion of the A_{ij}).

In other words, it is clear that the qs in the kinetic energy term do not play a significant role when small oscillations are considered: every time they are differentiated the term is neglected and when they are not differentiated they are given their equilibrium values. It follows that it is reasonable to give the qs in T their equilibrium values before substitution in Lagrange's equations.

For the potential energy, we can observe that linear terms in $\partial V / \partial q_j$ will come from the second-order terms in a Taylor expansion of V. Thus, V is required correct to the second-order in the qs and no first-order terms appear since $\partial V / \partial q_i$ is zero at equilibrium.

Summarizing:

1. In T, assign to the qs their equilibrium values.
2. Obtain V to the second-order in the qs.
3. Apply Lagrange's equations to the modified T and V.

Returning again to the double pendulum example, the kinetic energy is

$$T = \frac{1}{2} m_1 l_1^2 \dot{\theta}^2 + \frac{1}{2} m_2 [l_1^2 \dot{\theta}^2 + l_2^2 \dot{\phi}^2 + 2 l_1 l_2 \dot{\theta} \dot{\phi} \cos(\theta - \phi)]$$

and assigning to the generalized coordinates their values at equilibrium ($\theta = 0$ and $\phi = 0$) we get

$$T = \frac{1}{2} (m_1 + m_2) l_1^2 \dot{\theta}^2 + \frac{1}{2} m_2 l_2^2 \dot{\phi}^2 + m_2 l_1 l_2 \dot{\theta} \dot{\phi}$$

which, together with the Taylor expansion of the cosine terms that appear in the potential energy, adds up to form the 'small oscillations' Lagrangian of eq (3.82).

3.6 Lagrangian formulation for continuous systems

At the end of Section 3.4 (Example 3.2) we used Hamilton's principle to obtain the differential equation of motion for the transverse vibrations of an Euler–Bernoulli beam. We assumed without much explanation that the function $y(x, t)$ describes the beam transverse displacement from its neutral axis and we were led by Hamilton's principle to the equation of motion (and boundary conditions!) in a straightforward manner.

A few words of explanation are needed at this point. The assumption above is common in continuum mechanics; the equations of motion are expressed in terms of scalar or vector (or tensor) fields – i.e. physical quantities that depend, in general, on the spatial coordinates x, y, z and on time t – and owing to the fact that the space variables vary in a continuous way, the system is said to possess **infinite degrees of freedom**. In addition, Example 3.2 suggested the fundamental importance of space boundary conditions in the continuous case: in any given problem, in fact, their formulation is an essential part of the problem itself.

Now, since it is reasonable, and in general correct, to consider continuous systems as limiting cases of discrete systems and since we have seen that Lagrange's equations can be derived from Hamilton's principle, it is instructive to obtain a Lagrangian formulation for continuous systems (in the foregoing discussion, Lagrange's equations have only been given for finite-degree of freedom systems) and consider briefly the nature of the modifications that occur in the transition from discrete to continuous systems.

Restricting ourselves for the moment to one dimensional continuous systems – for example a string or a beam in longitudinal or transverse vibrations – the key aspect of the transition is in the role played by the position coordinate x. This is **not** a generalized coordinate of the problem, but a continuous index replacing the index i (or j, or whichever discrete index we decide to use) of the discrete case. For purposes of illustration, think for example of a slender beam of length L undergoing longitudinal vibrations; its discrete counterpart could be a chain of n equal mass points spaced a distance L/n apart and connected by uniform massless springs. Just as each value of i ($i = 1, 2, ..., n$) corresponds to a different generalized coordinate $v_i(t)$ of the system, for each value of x there is a generalized coordinate $v(x, t)$, so that x, like t, can be considered as a parameter entering into the Lagrangian. If the system is two- or three-dimensional, the generalized coordinate is labelled by two or three continuous 'space indices' and written as $v(x, y, t)$ or $v(x, y, z, t)$. The indices x, y, z and t are completely independent of each other and this is the reason why many authors write the derivative of v with respect to any of them as a total derivative rather than a partial derivative; we prefer to adopt the partial derivative notation.

We start from Hamilton's principle

$$\delta \int_{t_1}^{t_2} L \, dt = 0$$

where L appears now as a space integral (over an appropriate domain Ω) of a Lagrangian density $\mathscr{L}$. In the case of the beam of Example 3.2 we have

$$\mathscr{L} = \frac{1}{2}\left[\hat{m}\left(\frac{\partial y}{\partial t}\right)^2 - EI\left(\frac{\partial^2 y}{\partial x^2}\right)^2\right] \tag{3.106}$$

and the space integration is in dx on the domain $W = [0, l]$. We used here the lowercase letter l for the beam length to avoid ambiguities with the uppercase L which stands for the Lagrangian function.

For a general one-dimensional continuum, the Lagrangian density will be a function of the kind

$$\mathscr{L} = \mathscr{L}(v, v', v'', \dot{v}, \dot{v}') \tag{3.107}$$

where the overdots designate time derivatives and the primes designate space derivatives. Expressing the variation δL in Hamilton's principle we get

$$\int_{t_1}^{t_2} \int_0^l \left(\frac{\partial \mathscr{L}}{\partial v} \delta v + \frac{\partial \mathscr{L}}{\partial v'} \delta v' + \frac{\partial \mathscr{L}}{\partial v''} \delta v'' + \frac{\partial \mathscr{L}}{\partial \dot{v}} \delta \dot{v} + \frac{\partial \mathscr{L}}{\partial \dot{v}'} \delta \dot{v}' \right) dx\,dt = 0 \quad (3.108)$$

and a rather tedious symbolic integration by parts (remember that Hamilton's principle implies $\delta v(t_1) = \delta v(t_2) = 0$) can be performed in order to factor out the term δv. For example, the second and the fifth terms of eq (3.108) give

$$\int_0^l \frac{\partial \mathscr{L}}{\partial v'} \delta v'\,dx = \left(\frac{\partial \mathscr{L}}{\partial v'} \delta v \right) \Bigg|_0^l - \int_0^l \frac{\partial}{\partial x} \left(\frac{\partial \mathscr{L}}{\partial v'} \right) \delta v\,dx$$

$$\int_{t_1}^{t_2} \int_0^l \frac{\partial \mathscr{L}}{\partial \dot{v}'} \delta \dot{v}'\,dx\,dt = \int_0^l \left[\left(\frac{\partial \mathscr{L}}{\partial \dot{v}'} \delta v' \right) \Bigg|_{t_1}^{t_2} - \int_{t_1}^{t_2} \frac{\partial}{\partial t} \left(\frac{\partial \mathscr{L}}{\partial \dot{v}'} \right) \delta v'\,dt \right] dx$$

$$= -\int_{t_1}^{t_2} \left[\left(\frac{\partial}{\partial t} \left(\frac{\partial \mathscr{L}}{\partial \dot{v}'} \right) \delta v \right) \Bigg|_0^l \right] dt + \int_{t_1}^{t_2} \int_0^l \frac{\partial}{\partial x} \left(\frac{\partial}{\partial t} \left(\frac{\partial \mathscr{L}}{\partial \dot{v}'} \right) \right) \delta v\,dx\,dt$$

In conclusion, we arrive at the somewhat lengthy expression

$$\int_{t_1}^{t_2} \int_0^l \frac{\partial \mathscr{L}}{\partial v} \delta v\,dx\,dt + \int_{t_1}^{t_2} \left(\frac{\partial \mathscr{L}}{\partial v'} \delta v' \right) \Bigg|_0^l dt - \int_{t_1}^{t_2} \int_0^l \frac{\partial}{\partial x} \left(\frac{\partial \mathscr{L}}{\partial v'} \right) \delta v\,dx\,dt$$

$$+ \int_{t_1}^{t_2} \left[\frac{\partial \mathscr{L}}{\partial v''} \delta v' \Bigg|_0^l - \frac{\partial}{\partial x} \left(\frac{\partial \mathscr{L}}{\partial v''} \right) \delta v \Bigg|_0^l \right] dt + \int_{t_1}^{t_2} \int_0^l \frac{\partial^2}{\partial x^2} \left(\frac{\partial \mathscr{L}}{\partial v''} \right) \delta v\,dx\,dt$$

$$- \int_{t_1}^{t_2} \int_0^l \frac{\partial}{\partial t} \left(\frac{\partial \mathscr{L}}{\partial \dot{v}} \right) \delta v\,dx\,dt - \int_{t_1}^{t_2} \frac{\partial}{\partial t} \left(\frac{\partial \mathscr{L}}{\partial \dot{v}'} \right) \delta v \Bigg|_0^l dt$$

$$+ \int_{t_1}^{t_2} \int_0^l \frac{\partial}{\partial x} \left(\frac{\partial}{\partial t} \left(\frac{\partial \mathscr{L}}{\partial \dot{v}'} \right) \right) \delta v\,dx\,dt = 0$$

and using the usual argument on the arbitrariness and independence of the variations δv, the above equation gives the governing differential equation and the boundary conditions as follows:

$$
\left[\frac{\partial \mathcal{L}}{\partial v} - \frac{\partial}{\partial x} \left(\frac{\partial \mathcal{L}}{\partial v'} \right) + \frac{\partial^2}{\partial x^2} \left(\frac{\partial \mathcal{L}}{\partial v''} \right) \right.
$$

$$
\left. - \frac{\partial}{\partial t} \left(\frac{\partial \mathcal{L}}{\partial \dot{v}} \right) + \frac{\partial}{\partial x} \left(\frac{\partial}{\partial t} \left(\frac{\partial \mathcal{L}}{\partial \dot{v}'} \right) \right) \right] = 0 \quad (3.109)
$$

$$
\left[\frac{\partial \mathcal{L}}{\partial v'} - \frac{\partial}{\partial x} \left(\frac{\partial \mathcal{L}}{\partial v''} \right) - \frac{\partial}{\partial t} \left(\frac{\partial \mathcal{L}}{\partial \dot{v}'} \right) \right] \delta v = 0 \quad \text{at} \ \begin{cases} x = 0 \\ x = l \end{cases} \quad (3.110)
$$

$$
\frac{\partial \mathcal{L}}{\partial v''} \, \delta v' = 0 \quad\quad\quad\quad\quad\quad\quad \text{at} \ \begin{cases} x = 0 \\ x = l \end{cases} \quad (3.111)
$$

The procedure above lends itself to further generalizations; an elegant and detailed discussion, together with some interesting applications, can be found, for example in the text of Junkins and Kim [3]. Their formulation applies to a wide family of multibody hybrid discrete/distributed systems (collections of interconnected rigid and elastic bodies), resulting in a hybrid system of ordinary and partial integrodifferential equations.

The authors explicitly say that the method 'is especially attractive for more mature analysts because, on the average, integration by parts ceases to be fun sometime around age forty-five'. Their treatment, however, is beyond the scope of this book and the interested reader can find additional information also in Lee and Junkins [4].

Two examples will now illustrate the Lagrangian formulation discussed above.

Example 3.3.   Let us go back to the transverse vibrations of the beam considered in Example 3.2. In this case, the Lagrangian density is given by eq (3.106), i.e.

$$
\mathcal{L}(\dot{v}, v'') = \frac{1}{2} [\hat{m}\dot{v}^2 - EIv''^2] \quad (3.112)
$$

which is obviously a particular case of the symbolic expression (3.107). The equation of motion can be obtained from eq (3.109) which now becomes

$$
\left[\frac{\partial^2}{\partial x^2} \left(\frac{\partial \mathcal{L}}{\partial v''} \right) - \frac{\partial}{\partial t} \left(\frac{\partial \mathcal{L}}{\partial \dot{v}} \right) \right] = 0 \quad (3.113)
$$

with the boundary conditions

$$\frac{\partial \mathscr{L}}{\partial v''} \, \delta v' \, \Big|_0^l = 0 \tag{3.114}$$

$$\frac{\partial}{\partial x} \left(\frac{\partial \mathscr{L}}{\partial v''} \right) \delta v \, \Big|_0^l = 0 \tag{3.115}$$

By performing on the Lagrangian density of eq (3.112) the appropriate derivatives prescribed by Lagrange's equation, it is easy to obtain the explicit expression

$$\frac{\partial^2}{\partial x^2} (EIv'') + \hat{m}\ddot{v} = 0 \tag{3.116}$$

for the differential equation of motion and

$$\frac{\partial}{\partial x} (EIv'')\delta v \, \Big|_0^l = 0 \tag{3.117}$$

$$EIv''\delta v' \, \Big|_0^l = 0 \tag{3.118}$$

for the boundary conditions. As expected, since no external forces have been considered in this example, eq (3.116) is the homogeneous counterpart of eq (3.78) and eqs (3.117) and (3.118) are exactly the same as eqs (3.79) and (3.80).

Example 3.4.   If, on the other hand, we consider the longitudinal vibrations of the same beam, the Lagrangian density has the form

$$\mathscr{L}(\dot{v}, v') = \frac{1}{2} [\hat{m}\dot{v}^2 - EAv'^2] \tag{3.119}$$

where A is the cross-sectional area of the beam. The reader is invited to verify that the same line of reasoning as in the preceding example leads to the equation of motion

$$-\frac{\partial}{\partial x} (EAv') + \hat{m}\ddot{v} = 0 \tag{3.120}$$

and the boundary conditions

$$EAv'\delta v \, \Big|_0^l = 0 \tag{3.121}$$

References

1. Goldstein, H., *Classical Mechanics*, Addison Wesley, Reading, Mass., 1965.
2. Meirovitch, L., *Analytical Methods in Vibrations*, Macmillan, New York, 1967.
3. Junkins, J.L. and Kim, Y., *Introduction to Dynamics and Control of Flexible Structures*, AIAA Education Series, 1993.
4. Lee, S. and Junkins, J.L., Explicit generalizations of Lagrange's equations for hybrid coordinate dynamical systems, Department of Aerospace Engineering, Technical Report AERO 91-0301 Texas A&M University, College Station TX, March 1991.

4 Single-degree-of-freedom systems

4.1 Introduction

This chapter deals with the simplest system capable of vibratory motion: the **single-degree-of-freedom (SDOF) system** which, in its discrete-parameters form, is often called **harmonic oscillator**.

Despite its apparent simplicity, this system contains and exhibits most of the essential features of vibrating systems and its analysis is a necessary prerequisite to any further investigation in vibration theory and practice. In addition, many complex systems behave and can be considered, under certain circumstances, as SDOF systems, thus considerably simplifying the procedures of measurement and analysis. It is often a matter of the modelling scheme that we choose to adopt for the system under examination and of the degree of approximation that we are willing to accept.

The second-order ordinary differential equation which these SDOF obey is commonly found in many branches of physics and engineering – acoustics, mechanical, structural engineering and electronics, to name a few – and all its physical and mathematical aspects are worth considering for their own sake since they are the basis of useful analogies between these different fields.

This basic equation, in the general form that is of interest to us, is the equation of motion of the linear harmonic oscillator shown in Fig. 4.1 under an external applied force $f(t)$. It can be written as (Chapter 3)

$$m \frac{d^2x(t)}{dt^2} + c \frac{dx(t)}{dt} + kx(t) = f(t) \qquad (4.1)$$

where m, c and k are the mass, the damping and elastic constant of our SDOF system (see also Section 1.4). Unless otherwise specified, we will assume that these quantities are time-independent.

When the motion is pure rotational, there is a whole formal analogy with the translational case for which eq (4.1) applies. Obviously, the quantities to consider now are angular displacements, angular velocities and angular accelerations. The analogies are listed in Table 4.1 and, as a consequence of this analogy, only translational systems will be discussed in the following

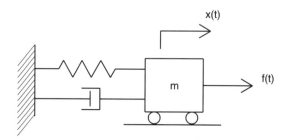

Fig. 4.1 Harmonic oscillator.

Table 4.1 Analogies between translational and rotational systems

Translation	Rotation
Linear displacement x	Angular displacement α
Force f	Torque M
Spring constant k	Spring constant k_r
Damping constant c	Damping constant c_r
Mass m	Moment of inertia J
Spring law $F = k(x_1 - x_2)$	Spring law $M = k_r(\alpha_1 - \alpha_2)$
Damping law $F = c(\dot{x}_1 - \dot{x}_2)$	Damping law $M = c_r(\dot{\alpha}_1 - \dot{\alpha}_2)$
Inertia law $F = m\ddot{x}$	Inertia law $M = J\ddot{\alpha}$

sections. The treatment of rotational systems is obtained by substitution of the appropriate quantities with the consistent units.

As a word of caution, it must be said that rotational quantities cannot be measured as easily as translational quantities and therefore the experimental part may turn out to be more critical.

4.2 The harmonic oscillator I: free vibrations

When friction forces are absent (or negligible), any elastic system that, in some way, is slightly displaced from its equilibrium position and is subsequently let free by removing the cause of the initial disturbance, executes an oscillatory motion and continues to vibrate forever unless we decide to interfere with it again. This particular condition is called **undamped free vibrations**. The frequency characteristics of the oscillatory motion depend on the parameters of the system itself, that is, on its mass and elasticity; the amplitude characteristics, on the contrary, depend on the initial conditions and the vibration does not die out because no energy is lost during the motion.

When some kind of damping is present, energy is lost during the motion and the amplitude of vibration decreases as time passes, until it stops completely; this is the case of **damped free vibrations**.

Once again, however, the frequency characteristics of the oscillatory motion depend on the parameters of the system, and not on the initial conditions that started the motion. No musical instrument could be played in tune if this general rule did not apply.

In the case of 'strong damping', the system does not vibrate at all but loses quickly its initial energy and simply returns to its equilibrium position without oscillating. We will determine quantitatively the meaning of the term 'strong damping' in the following sections.

4.2.1 Undamped free vibrations

Let us now consider the simple ideal system of Fig. 4.2. It consists of a mass m and a massless spring k; the mass can only move in the x direction and no friction of any kind acts during the motion. The equation of motion is eq (4.1) with $c = 0$, $f = 0$ and can be written as

$$\ddot{x}(t) + w_n^2 x(t) = 0 \qquad (4.2)$$

where we have defined

$$w_n^2 = \frac{k}{m} \qquad (4.3)$$

and $x = 0$ determines the equilibrium position of the mass.

It is interesting to note that if the mass was suspended in a vertical position, eq (4.2) would still hold because the weight mg (gravity acceleration $g = 9.81$ m/s^2) cancels out with the term $k\delta_{st}$, which represents the restoring force acting on the mass in the vertical equilibrium position, in which the spring is stretched by the amount δ_{st}. The subscript 'st' is for static, and at equilibrium $mg = k\delta_{st}$.

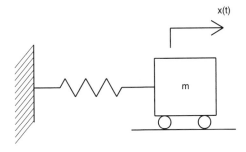

Fig. 4.2 Undamped harmonic oscillator.

This is a general principle that applies to all linear elastic systems and for this reason the static equilibrium position is chosen as a reference to eliminate gravity forces from the equations of motion. The displacements thus determined are the dynamic response; total deflections, stresses etc. are obtained by adding the relevant static quantities to the result of the dynamic analysis.

The general solution of eq (4.2) can be expressed in exponential form as $x(t) = Ae^{\alpha t}$; the characteristic equation is $\alpha^2 + w_n^2 = 0$ with solutions $\alpha = \pm iw_n$. Using complex numbers we get

$$x(t) = Ae^{+iw_n t} + Be^{-iw_n t} \tag{4.4}$$

where A and B are two complex constants determined by the initial conditions. Since $x(t)$ is a real quantity, we must have $A = B^*$. Moreover, if the initial conditions for $t = 0$ are given by

$$x(0) = x_0$$
$$\dot{x}(0) = v_0 \tag{4.5}$$

we obtain

$$x(t) = \frac{1}{2}\left(x_0 - i\frac{v_0}{w_n}\right)e^{iw_n t} + \frac{1}{2}\left(x_0 + i\frac{v_0}{w_n}\right)e^{-iw_n t} \tag{4.6}$$

Equation (4.4) represents a pure oscillation at the frequency w_n, which is the **natural frequency of the undamped system** (or undamped natural frequency). It is easy to show that eq (4.4) can also be alternatively written in the forms

$$x(t) = C\cos(w_n t - \theta) \tag{4.7}$$
$$x(t) = D\cos w_n t + E\sin w_n t \tag{4.8}$$

where the constants of integration are now C and θ in the first case, D and E in the second case and all of them can be expressed in terms of x_0 and v_0.
Explicitly:

$$A + B = D = C\cos\theta = x_0$$

$$i(A - B) = E = C\sin\theta = \frac{v_0}{w_n}$$

$$C = \sqrt{D^2 + E^2} = \sqrt{x_0^2 + \left(\frac{v_0}{w_n}\right)^2} \tag{4.9}$$

$$\tan\theta = \frac{E}{D} = \frac{v_0}{x_0 w_n}$$

Another possibility is to use the convention on complex numbers explained in Chapter 1 and write the solution in the form $x(t) = C e^{(-i\omega_n t)}$, where C is complex with magnitude $|C|$ and phase angle θ, and we agree to take the real part (or the imaginary part) to represent the quantity $x(t)$.

Two points need to be emphasized:

- Only the initial displacement x_0 and the initial velocity v_0 need be given to determine completely the subsequent motion of the oscillator. The mathematical counterpart of this statement is that the solution of any second-order differential equation has two arbitrary constants in it
- The frequency of the oscillation depends only on k and m and not at all on x_0 and v_0. With a given spring, an increase of mass results in a decrease of the natural frequency and vice versa; with a given mass, an increase of the spring stiffness results in an increase of the natural frequency and vice versa.

Returning for a moment to the case in which the mass is suspended vertically and the equilibrium position corresponds to a certain amount of stretching of the spring, we can obtain the natural frequency of the system by the static deflection (δ_{st}) only. In fact, at equilibrium $mg = k\delta_{st}$ and so, from (4.3), we get

$$\omega_n = \sqrt{g/\delta_{st}} \tag{4.10}$$

The system we are considering is conservative: once the motion has been started no energy is lost through friction and no energy is supplied by an external source. The total energy E_T does not depend on time and is the sum of the kinetic and potential energies which, in turn, are given respectively by

$$E_k = \frac{1}{2} m v^2(t)$$

$$E_p = \int_0^x ky(t)dy = \frac{1}{2} kx^2(t)$$

where $v(t) \equiv \dot{x}(t)$ and y is a dummy variable of integration. Using for example the form of eq (4.7), we obtain for the total energy

$$E_T = \tfrac{1}{2} kC^2 = \tfrac{1}{2} m\omega_n^2 C^2 \tag{4.11}$$

where $\omega_n C \equiv V$ is the **velocity amplitude** of the motion. So, the total energy is equal to the potential energy at maximum displacement or to the kinetic energy at maximum velocity and $E_{k,max} = E_{p,max}$. In terms of x_0 and v_0, we see that E_T depends on the square of these two quantities.

If we consider average values over a time period T, it is left as an easy exercise to show that

$$\langle E_k \rangle = \langle E_p \rangle = \tfrac{1}{4} k C^2 = \tfrac{1}{4} m V^2 \tag{4.12}$$

4.2.2   Damped free vibrations

The equation to solve is now

$$m \frac{d^2 x(t)}{dt^2} + c \frac{dx(t)}{dt} + kx(t) = 0 \tag{4.13}$$

which corresponds to the system of Fig. 4.1 where $f = 0$. Different forms of damping other than viscous are possible and we will consider this aspect later. To solve eq (4.13) we make use of the exponential function again; the characteristic equation is now $m\alpha^2 + c\alpha + k = 0$, with solutions α_1 and α_2 given by

$$\alpha_{1,2} = -\frac{c}{2m} \pm \sqrt{\left(\frac{c}{2m}\right)^2 - \frac{k}{m}} \tag{4.14}$$

so that the general solution of eq (4.13) is written

$$x(t) = A e^{\alpha_1 t} + B e^{\alpha_2 t} \tag{4.15}$$

where A and B are two constants to be evaluated from the initial conditions. Three cases are possible, depending on the sign of the term under the square root in eq (4.14). Let us introduce the definitions

$$c_{cr} \equiv 2\sqrt{km} = 2m\omega_n \tag{4.16}$$

$$\zeta \equiv \frac{c}{c_{cr}} \tag{4.17}$$

where:

- c_{cr} is the value of damping that makes the radical in eq (4.14) equal to zero and is called **critical damping**;
- ζ is a nondimensional quantity known as the **damping ratio**.

With these definitions the equation of motion (4.13) becomes

$$\ddot{x} + 2\omega_n \zeta \dot{x} + \omega_n^2 x = 0 \tag{4.18}$$

and the three cases correspond to $\zeta = 1, \zeta > 1$ and $\zeta < 1$ since the roots $\alpha_{1,2}$ are written

$$\alpha_{1,2} = (-\zeta \pm \sqrt{\zeta^2 - 1})\omega_n \qquad (4.19)$$

Case 1. Critically damped motion: $\zeta = 1$ $(c = c_{cr})$

In this case $\alpha_1 = \alpha_2 = -\omega_n$ and in order to have two linearly independent solutions, $x(t)$ has the form

$$x(t) = (A + tB)e^{-\omega_n t} \qquad (4.20)$$

With the usual initial conditions of eqs (4.5), we get for the constants A and B

$$A = x_0$$
$$B = v_0 + x_0\omega_n \qquad (4.21)$$

Equation (4.20) represents an exponentially decaying response; the system does not oscillate and returns to rest in the shortest time possible. For a given system, Fig. 4.3 shows three types of responses $x(t)$ which differ by the value of initial velocity (the initial displacement being $x_0 = 0.1$ m in all of the three cases).

In many applications that employ meters and measuring instrumentation in general, the physical systems or some of their moving parts are critically damped on purpose in order to avoid unwanted overshoot and oscillations.

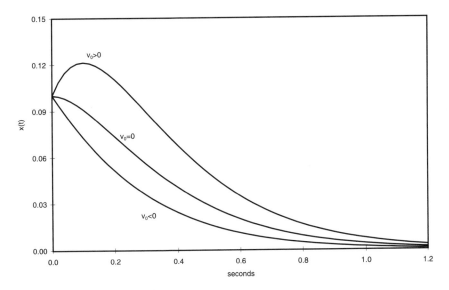

Fig. 4.3 Critically damped motion.

Case 2. Overdamped motion: $\zeta > 1$ $(c > c_{cr})$

The two roots are separate and real; the general solution is

$$x(t) = Ae^{(-\zeta + \sqrt{\zeta^2 - 1})\omega_n t} + Be^{(-\zeta - \sqrt{\zeta^2 - 1})\omega_n t} \tag{4.22}$$

which is an exponentially decaying function similar to the ones shown in Fig. 4.3. The constants A and B determined by the initial conditions and substitution of the latter in eq (4.20) gives

$$A = \frac{v_0 + \omega_n x_0(\zeta + \sqrt{\zeta^2 - 1})}{2\omega_n\sqrt{\zeta^2 - 1}}$$

$$B = \frac{-v_0 - \omega_n x_0(\zeta - \sqrt{\zeta^2 - 1})}{2\omega_n\sqrt{\zeta^2 - 1}} \tag{4.23}$$

Once again, the system returns to rest without oscillating but it takes longer than in case 1; how much longer, for fixed initial conditions, depends on how much ζ is greater than unity.

Case 3. Underdamped motion: $0 < \zeta < 1$ $(c < c_{cr})$

This is the case of greatest interest in vibration problems. Substituting eq (4.19) into eq (4.15), the general solution becomes

$$x(t) = e^{-\zeta\omega_n t}(Ae^{i\sqrt{1 - \zeta^2}\omega_n t} + Be^{-i\sqrt{1 - \zeta^2}\omega_n t}) \tag{4.24}$$

which represents an oscillation at frequency $\omega_n\sqrt{1 - \zeta^2}$ with an exponentially decaying amplitude. The curves $\pm Xe^{-\zeta\omega_n t}$ envelop the displacement–time relationship and touch it at the points where $\cos(\omega_d t - \theta) = \pm 1$. These are not the points of maximum displacement: the actual maxima lie a bit to their left and can be determined by equating the derivative of $x(t)$ to zero; obviously, they depend on the value of ζ.

Looking at the general solution, it is common to define

$$\omega_d \equiv \omega_n\sqrt{1 - \zeta^2} \tag{4.25}$$

as the **frequency of damped free oscillation**. The oscillation is **not** periodic because the motion never repeats itself; however, if ζ is small compared to unity the motion is very nearly periodic and ω_d is very nearly equal to ω_n. In this case, the expression for ω_d can be expanded and all but the first two terms can be neglected so that

$$\omega_d = \omega_n - \tfrac{1}{2}\omega_n\zeta^2 + \ldots$$

To evaluate the effect of damping on the frequency it may be useful to note that a plot of ω_d/ω_n versus ζ is a circle of unit radius in the first quadrant (since $0 \leqslant \omega_d/\omega_n \leqslant 1$ and $0 \leqslant \zeta \leqslant 1$).

We point out that, once again, the frequency is independent of the amplitude of motion and the decay properties, which in turn, are independent of the way the SDOF system is started into motion. Strictly speaking, the word 'frequency' used in this case of nonperiodic motion is improper, but when the damping is small the word still has some meaning. This can be seen in Fig. 4.4 where we have plotted the displacement time history of a underdamped system with $\omega_n = 5$ rad/s and $\zeta = 0.1$ started into motion by the initial conditions $x_0 = 0.1$m and $v_0 = 0$ m/s.

The constants A and B can be obtained from the initial conditions. We get

$$A = \frac{x_0\omega_d - i(v_0 + \zeta x_0\omega_n)}{2\omega_d}$$

$$B = \frac{x_0\omega_d + i(v_0 + \zeta x_0\omega_n)}{2\omega_d}$$

(4.26)

with $A = B^*$ since $x(t)$ must be a real quantity.

Other equivalent forms can be used to express the general solution (4.24); we can write

$$x(t) = e^{-\zeta\omega_n t}(D \cos \omega_d t + E \sin \omega_d t)$$

$$= Xe^{-\zeta\omega_n t} \cos(\omega_d t - \theta)$$

(4.27)

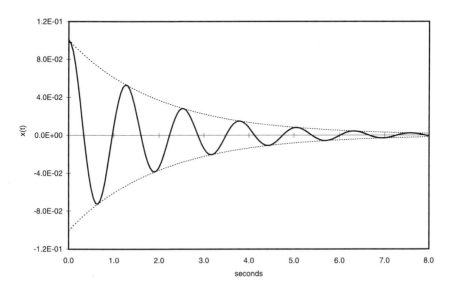

Fig. 4.4 Underdamped motion ($\zeta = 0.1$).

The second expression is the real part of $e^{-\zeta\omega_n t}Ce^{-i\omega_d t}$ where C is complex with magnitude X and phase angle θ; in this case the initial conditions determine X and θ that are given by

$$X = \sqrt{D^2 + E^2} = \sqrt{x_0^2 + \left(\frac{v_0 + \zeta x_0 \omega_n}{\omega_d}\right)^2}$$

$$\tan\theta = \frac{E}{D} = \frac{v_0 + \zeta x_0 \omega_n}{x_0 \omega_d} \tag{4.28}$$

With reference to the decaying amplitude, one often find references to the **decay time** (τ) which is defined as the time it takes to have an amplitude that is $1/e$ of its initial value. This is $\tau = 1/\zeta\omega_n$. Furthermore, if one writes the general solution in complex form as

$$x(t) = Ce^{-i(\omega_d - i\zeta\omega_n)t}$$

the complex quantity

$$\hat\omega_d = \omega_d - i\zeta\omega_n \tag{4.29}$$

is defined **complex damped frequency** of free oscillation and contains information on both the damped frequency of the system and its decay time.

Energy is not conserved in this case, since friction forces are at work during the motion. We can calculate the rate of loss of energy and consider the subject from this point of view. Using, for example, the second form of eq (4.27) for the general solution to the equation of motion, we can first calculate the energy of the system at any instant. We have

$$E(t) = \frac{1}{2}mv^2 + \frac{1}{2}kx^2$$

$$= \frac{1}{2}m\omega_d^2 A^2 - m\omega_d A\left(\frac{dA}{dt}\right)\sin(\omega_d t - \theta)\cos(\omega_d t - \theta)$$

$$+ \frac{1}{2}m\left(\frac{dA}{dt}\right)^2 \cos^2(\omega_d t - \theta)$$

where the time dependent amplitude $Xe^{-\zeta\omega_n t}$ has been written for convenience of notation in the general form $A(t)$. If we average this quantity over one period, the second term (with the product $\sin(\cdots)\cos(\cdots)$) goes to zero. If, in addition, $A(t)$ is slowly varying so that dA/dt is small compared to $\omega_d A$, the third term can be neglected and we obtain the approximate expression for the energy

$$E(t) \cong \tfrac{1}{2} k[A(t)]^2 = \tfrac{1}{2} m[V(t)]^2$$
$$= \tfrac{1}{2} kX^2 e^{-2\zeta\omega_n t} = \tfrac{1}{2} m\omega_n^2 X^2 e^{-2\zeta\omega_n t} \qquad (4.30)$$

where $V(t)$ is the time-dependent velocity amplitude and in the last expression ω_n can be substituted for ω_d because of the requirement of slowly varying amplitude.

The rate of energy loss is given by the product of the friction force $c\dot{x}$ times the velocity, i.e. $c\dot{x}^2$ (force times velocity is power, the rate of change of energy), and with the same approximations as above we get the average energy loss per second which, by definition, is the average dissipated power

$$P \equiv -\frac{dE}{dt} \cong \frac{1}{2} \omega_n^2 X^2 e^{-2\zeta\omega_n t} \qquad (4.31)$$

4.2.3 Logarithmic decrement

The free oscillation response of a given SDOF system can be used to determine the amount of damping when this is not known. We are referring to the underdamped case with viscous damping.

It must be noted that the true damping characteristics of physical systems are in general difficult to define, but it is often common practice to consider equivalent viscous damping ratios ζ for systems in free vibration that show similar decay rates. So, it is important to see how this damping ratio can be obtained by an experimental measurement of the free oscillation time history.

Considering two successive positive (or negative) peaks x_1 and x_2 that occur at times t_1 and t_2 respectively, we can calculate the ratio

$$\frac{x_1}{x_2} = \frac{e^{-\zeta\omega_n t_1} C e^{-i\omega_d t_1}}{e^{-\zeta\omega_n t_2} C e^{-i\omega_d t_2}}$$

and, since $t_2 = t_1 + T_d = t_1 + 2\pi/\omega_d$ (where T_d is the period of damped oscillation), we get

$$\frac{x_1}{x_2} = e^{+\zeta\omega_n (2\pi/\omega_d)}$$

The logarithmic decrement δ is defined as the natural logarithm of this amplitude ratio, i.e.

$$\delta \equiv \ln\frac{x_1}{x_2} = 2\pi\zeta \frac{\omega_n}{\omega_d} = \frac{2\pi\zeta}{\sqrt{1-\zeta^2}} \qquad (4.32)$$

If, as it is often the case, the damping is low, eq (4.32) can be approximated by

$$\delta \cong 2\pi\zeta \qquad (4.33)$$

and the exponential in the ratio x_1/x_2 can be expanded in series retaining only the first two terms; since $\omega_d \cong \omega_n$ this leads to

$$\zeta \cong \frac{x_1 - x_2}{2\pi x_2} \qquad (4.34)$$

Figure 4.5 illustrates a plot of the exact and the approximate values of δ, as given by eqs (4.32) and (4.33) respectively, as a function of ζ. Figure 4.6, in turn, is a graph of practical use where, as a function of the approximate ζ given by eq (4.34), ζ_{exact} is given. From the recorded decaying time history the approximate value ζ_{approx} is obtained by using eq (4.34) and then, from Fig. 4.6, the exact value of ζ can be determined.

In case of very low damping – i.e. when the difference between the amplitudes of two successive peaks is small – it is more convenient (and more accurate) to consider two peaks that are a few cycles apart, say x_i and x_{i+m} (m cycles apart), so that

$$\ln \frac{x_i}{x_{i+m}} = 2\pi m \zeta \frac{\omega_n}{\omega_d} \qquad (4.35)$$

and the same approximations as before give

$$\zeta \cong \frac{x_i - x_{i+m}}{2\pi m x_{i+m}} \qquad (4.36)$$

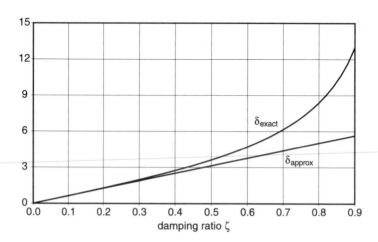

Fig. 4.5 Exact and approximate logarithmic decrement vs damping ratio.

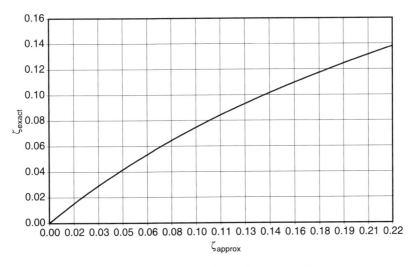

Fig. 4.6 Exact damping ratio versus approximate damping ratio.

It is left as an exercise to the reader to determine the number of cycles, as a function of ζ, required to reduce the oscillation amplitude by 50%. This can be done by using eq (4.35) and, as a quick rule of thumb, it can be convenient to remember that for 10% critical damping (i.e. $\zeta = 0.10$) the amplitude is reduced by 50% in one cycle, for 5% critical damping (i.e. $\zeta = 0.05$) the amplitude is reduced by 50% in about two cycles, more precisely in 2.2 cycles.

The value of the logarithmic decrement δ can be expressed in terms of energy considerations as well. From the definition of δ we get

$$\frac{x_2}{x_1} = e^{-\delta}$$

and, since the energy of the system at the relative maxima x_1 and x_2 is given by

$$E_{p_1} = \tfrac{1}{2} k x_1^2 \qquad E_{p_2} = \tfrac{1}{2} k x_2^2$$

it follows that

$$\frac{\Delta E_p}{E_{p_1}} \equiv \frac{E_{p_1} - E_{p_2}}{E_{p_1}} = 1 - \left(\frac{x_2}{x_1}\right)^2 = 1 - e^{-2\delta}$$

Table 4.2 Analogies between mechanical and electrical quantities*

Mechanical quantity	Electrical quantity
Mass m (kg)	Inductance L (H)
Compliance $1/k$ (m/N)	Capacitance C (F)
Damping coefficient c (Ns/m)	Resistance R (Ω)
Force f (N)	Voltage V (V)
Displacement x (m)	Charge Q (C)
Velocity dx/dt (m/s)	Current $i = dQ/dt$ (A)
Acceleration d^2x/dt^2 (m/s^2)	Rate of change of current di/dt (A/s)

* See Section 13.7.3 for more details.

For low damping the exponential term can be expanded giving

$$\delta = \frac{\Delta E_p}{2E_{p_1}} \tag{4.37}$$

4.2.4 Further analogies

Now that we have gained some initial insight in the behaviour of mechanical systems, it is interesting to go back to the introduction to this chapter and show some analogies between these systems and other systems that, in their own right, belong to different branches of physics and engineering. The importance lies in the fact that a good understanding of the essential aspects and phenomena in one particular field can be taken over to another field of interest, once the appropriate substitutions are made. It is then possible to compare an unfamiliar system with one that is better known and extend the line of reasoning to the former system. The analogy between translational and rotational systems – although both in the realm of mechanical vibrations – is a good example.

The electrical circuit is a widely exploited vibrating system in which the kinetic, potential energy and dissipation may be expressed by equations similar to the ones we have considered so far. Table 4.2 shows the dynamic analogies between mechanical and electrical elements.

The analogy above is sometimes called the **voltage–force analogy** and is constructed on the basis of Kirchhoff's voltage law, which states that the algebraic sum of all the voltages around any closed circuit is equal to zero in any network. Kirchhoff's current law can be used to establish the **current–force analogy** and other analogies with acoustical systems can be given. The interested reader can refer, for example, to Section 13.7.3 or to Olson [1] and Seto [2].

4.3 The harmonic oscillator II: forced vibrations

It often happens that a system is set into vibration because it is subjected in some way to an external excitation that supplies the energy to keep it oscillating. In these cases one speaks of **forced oscillations** because the external excitation – the input to our system – is the 'forcing' cause of the vibration. The characteristics of the system's response (i.e. the output) depend on its physical properties and on the type of excitation as well. Figure 4.7 illustrates schematically this situation in the case of a single-input single-output linear system.

Common examples can be the diaphragm of a microphone that vibrates because it is linked, by means of sound waves, to the vibrations of a violin string; a tall and slender structure that vibrates under the forcing action of the wind or a building under the action of an earthquake or the passage of heavy traffic in its vicinity.

The input excitation and the output response can be in the form of displacement, velocity, acceleration or force and in a number greater than one; in addition, there can be some form of feedback between some of the outputs and inputs. However, in many cases of interest, the system under investigation does not feed back any appreciable amount of energy to the forcing (or 'driving') system that supplies the excitation, either because the linkage between the two is weak or because the driving system has so great a reserve of energy that the amount fed back is negligible in comparison.

So, for the present, we need not be concerned with these complications and consider as the only relevant property of the driving system the fact that it supplies an excitation that persists for a certain interval of time; furthermore, linearity assures that the principle of superposition (Section 1.2) holds.

A simple case of a single-input single-output linear system is the harmonic oscillator under the action of an external time-varying force $f(t)$ as shown in Fig. 4.1. The equation of motion is eq (4.1), which we can rewrite here in the form

$$m\ddot{x}(t) + c\dot{x}(t) + kx(t) = f(t) \tag{4.38}$$

If damping is absent or negligible, we put $c = 0$ in eq (4.38) and we speak of **undamped forced vibrations**; when damping is taken into account the term **damped forced vibrations** is commonly used.

In both cases the excitation force can be:

1. applied to the mass of our SDOF system;
2. a motion of the 'foundation' that supports the system.

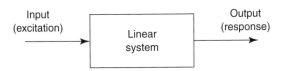

Fig. 4.7 Single-input single-output linear system (schematic 'black-box' representation).

In case 1 the response may be expressed either as the amplitude of motion of the mass (**motion response**) or the fraction of the applied force transmitted to the support (**force transmissibility**).

In case 2 the response is usually given in terms of the amplitude of mass motion divided by amplitude of support motion, and one speaks of **motion transmissibility**.

All these quantities depend on the frequency of the forcing excitation and vary for different types and degrees of damping; the excitation, in turn, can be harmonic, periodic or nonperiodic in time or, more generally, deterministic or random. This latter aspect – bearing in mind the use to be made of the result – somehow dictates the procedure of analysis.

Because of its fundamental nature and for the multitude of practical applications, we will start with the simplest case of excitation: a harmonically varying function of time.

4.3.1   Forced vibrations

The system we consider is illustrated in Fig. 4.1 and the relevant equation of motion is eq (4.1) or, which is the same, eq (4.38). This is, mathematically speaking, an inhomogeneous linear differential equation with constant coefficients and its solution can be expressed as the sum of two parts: a **complementary function** and a **particular integral**.

The complementary function is obtained by solving the homogeneous equation, i.e. eq (4.13), where $f(t) = 0$. This part was discussed in the previous sections and we saw that the solution involves two arbitrary constants to satisfy the initial conditions. Furthermore, complementary solutions are transient in nature because, as a result of damping, they decay with time.

On the other hand, a particular solution (or integral) depends on the exciting force and does not involve any arbitrary constant; by itself, it will not satisfy the initial conditions of displacement and velocity at time $t = 0$. The particular solution represents the steady-state condition of motion because it persists as long as the exciting force does. It can be obtained by a process of trial and error when $f(t)$ is a simple function of time and by more complex techniques when this is not the case. In brief, steady-state motion is the motion of a system that has forgotten how it started.

We begin with a mass-applied harmonic forcing function that we write for convenience in complex exponential form as

$$f(t) = f_0 e^{-i\omega t} \tag{4.39}$$

with magnitude f_0 and zero phase angle (Note that, as stated in Chapter 1, the positive exponential convention – i.e. $f(t) = f_0 e^{i\omega t}$ – could be adopted without affecting the essence of the results; the choice is subjective and either is fine as long as consistency is maintained.)

The general solution can be written, as stated above, as

$$x(t) = x_1(t) + x_2(t) \tag{4.40}$$

where:

- $x_1(t)$ is the transient complementary function that dies out with time because of damping and can be expressed in any one of the forms discussed in the preceding section. (An important observation, however, must be made at this point. Although we can write $x_1(t)$ as, say, in the first of eqs (4.27), it should be noted that now the two constants D and E are no longer given by eq (4.28) because in the forced vibration case they must be evaluated by taking into account also the particular solution. In this regard, see, e.g. eqs. (4.60) and (4.61).)
- $x_2(t)$ is the particular solution. When $t \gg 1/(\zeta \omega_n)$ only this part remains, $x(t) = x_2(t)$ and the motion takes place at the forcing frequency with amplitude and phase characteristics that depend on the particular value of the forcing frequency.

We can assume the particular solution to be of the form

$$x_2(t) = Xe^{-i\omega t} = |X| e^{-i(\omega t - \phi)} \tag{4.41}$$

where X is the complex amplitude with magnitude $|X|$ and phase angle ϕ (with respect to zero, or more generally, to the phase angle of $f(t)$). Calculating the first and second time derivatives of eq (4.41) and substituting in eq (4.38) we obtain after a few stages

$$X = \frac{f_0}{k - m\omega^2 - ic\omega} \tag{4.42}$$

which can be written more conveniently as

$$X = \frac{f_0}{k} \left(\frac{1}{1 - \beta^2 - 2i\zeta\beta} \right) \tag{4.43}$$

since $c = \zeta c_{cr} = 2\zeta m\omega_n$, $\omega_n^2 = k/m$ and we have defined $\beta \equiv \omega/\omega_n$.

The magnitude of X, its real and imaginary parts and the phase angle ϕ (in radians) can be easily obtained from complex algebra and are given respectively by

$$|X| = \left(\frac{f_0}{k} \right) \frac{1}{\sqrt{(1 - \beta^2)^2 + (2\zeta\beta)^2}} \tag{4.44}$$

$$\operatorname{Re} X = \left(\frac{f_0}{k}\right) \frac{1 - \beta^2}{(1 - \beta^2)^2 + (2\zeta\beta)^2} \tag{4.45}$$

$$\operatorname{Im} X = \left(\frac{f_0}{k}\right) \frac{2\zeta\beta}{(1 - \beta^2)^2 + (2\zeta\beta)^2} \tag{4.46}$$

$$\tan\phi \equiv \frac{\operatorname{Im} X}{\operatorname{Re} X} = \frac{2\zeta\beta}{1 - \beta^2} \tag{4.47}$$

The explicit form of the steady-state solution is then

$$x_2(t) = \frac{f_0/k}{\sqrt{(1 - \beta^2)^2 + (2\zeta\beta)^2}} e^{-i(\omega t - \phi)} \tag{4.48}$$

For the moment, we concentrate our attention on eqs (4.44) and (4.47). The ratio of the amplitude response to the static displacement which would be produced by the force f_0 – i.e. the ratio f_0/k – is a nondimensional quantity often called **dynamic magnification factor** $(D \equiv k\,|X|/f_0)$ and, as the phase angle ϕ, is a function only of the frequency ratio β and of the damping factor ζ. Before making some general comments on the behaviour of our system, it is interesting to study the graphical representations of D and ϕ as functions of β, with ζ as a parameter. They are shown in Figs 4.8 and 4.9.

It is evident that the damping factor ζ has a large influence on the amplitude and phase angle in the region where $\beta \cong 1$ (i.e. when $\omega \cong \omega_n$). The condition

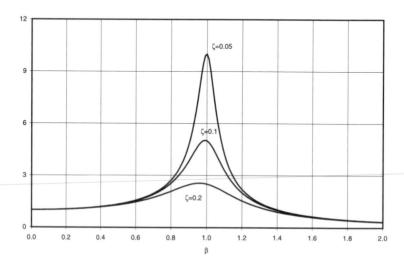

Fig. 4.8 Dynamic magnification factor versus β.

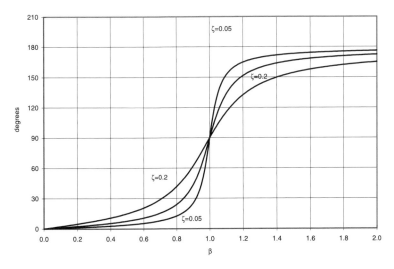

Fig. 4.9 Phase angle versus β.

$\omega = \omega_n$ is called **resonance**; for small damping it is characterized by violent vibration and may lead to disastrous effects. It can be shown that the maximum value of amplitude (amplitude resonance) does not occur at exactly $\beta = 1$, but at

$$\beta = \sqrt{1 - 2\zeta^2} \tag{4.49a}$$

and the maximum value of the dynamic magnification factor is

$$D_{max} = \frac{1}{2\zeta\sqrt{1 - \zeta^2}} \tag{4.50a}$$

For light damping, such as when $\zeta \leqslant 0.05$, eqs (4.49a) and (4.50a) can be approximated by

$$\beta \cong 1 \tag{4.49b}$$

and

$$D_{max} \cong \frac{1}{2\zeta} \tag{4.50b}$$

without significant loss of accuracy (for $\zeta = 0.05$ the relative error, in percentage with respect to the true value, is 0.25% on β and 0.13% on D_{max}). A brief point to note is that eq (4.50b) – from electrical engineering terminology – defines the **quality factor** Q ($Q \equiv 1/(2\zeta)$) which describes the sharpness of the response at resonance and can be used as a measure of the system's damping characteristics. Low damping corresponds to high values of Q and to a strongly peaked response, high damping corresponds to low values of Q and to smaller amplitudes at resonance.

As damping increases, the amplitude of motion decreases and the value of D_{max} is shifted to the left of the vertical line $\beta = 1$. If $\zeta > 1/\sqrt{2} \cong 0.707$, the amplitude response shows no peak, D is a monotonous decreasing function of β and the maximum value of D occurs at $\omega = 0$.

With these considerations in mind, we can go back to Fig. 1.4 and understand what was meant by saying that the oscillating bell imposes on the structure (the belltower) a transverse force with frequency close to resonance for the first flexural mode of the tower. In the time domain, the phenomenon of beats is of great use in detecting such phenomena, especially when the frequency resolution of the instrumentation at hand is not adequate and cannot resolve – in the frequency domain – the two peaks of the driving force and of the natural frequency of the vibrating system.

In general applications, harmonically driven systems can be used in two different ways. One type requires the system to respond strongly only to particular values of frequency (for example an acoustic resonator or the tuning circuit of a radio), and in this case friction must be small in order to have a large response only at the natural frequency of the driven system. The other type requires the system to respond more or less equally well to all frequencies in a certain range (for example the diaphragm of a microphone or a vibration measuring device such as an accelerometer or a seismometer) and this requirement can be met by an appropriate selection of the values of friction, stiffness and mass.

The phase angle ϕ describes the time shift ($t = \phi/\omega$) between the output displacement $x_2(t)$ and the force excitation $f(t)$; from Fig. 4.9 it can be seen to vary, as a function of β, between $0°$ and $180°$, passing through the point $\phi = 90°$ at $\beta = 1$ for all values of ζ. The transition between the two extremes becomes more and more gradual as the damping factor increases. This shows that the motion is not in phase with the force, the angle of lag of the displacement behind the force being given by ϕ, which is zero when $\beta \ll 1$, $90°$ when $\beta = 1$ and $180°$ when $\beta \gg 1$ and approaches infinity (indicating that the displacement is opposite in direction to the force).

Let us examine more closely these three regions that correspond, respectively, to the cases $\omega \ll \omega_n$, $\omega \cong \omega_n$ and $\omega \gg \omega_n$.

- When the driving frequency is much smaller than the natural frequency of our system, the displacement is in phase with the force and the amplitude and phase angle are small because both the inertia and damping forces are small.
- At resonance, the inertia force is balanced by the spring force, the amplitude is large (if damping is small), the displacement lags behind the force of $90°$ and the velocity is in phase with the force.

- Finally, when the driving frequency is much greater than the natural frequency, the amplitude becomes very small (approaching zero as ω tends to infinity), the displacement is opposed to the force which, in turn, is now in phase with acceleration and is expended almost entirely in overcoming the large inertia force.

The displacement response of eq (4.44) can be approximated, at these frequency conditions, as

$$|X| \cong \left(\frac{f_0}{k}\right) e^{-i\omega t} \qquad \omega \ll \omega_n \qquad (4.51a)$$

$$|X| \cong \left(\frac{f_0}{c\omega_n}\right) e^{-i(\omega t + \pi/2)} \qquad \omega \cong \omega_n \qquad (4.51b)$$

$$|X| \cong \left(\frac{f_0}{m\omega^2}\right) e^{-i(\omega t + \pi)} \qquad \omega \gg \omega_n \qquad (4.51c)$$

From eqs (4.51) – depending on which element is primarily responsible for the system's behaviour – the oscillator is often described as **stiffness controlled, resistance (or damping) controlled** and **mass controlled** in the three cases above, respectively.

It is left as a very useful exercise to the reader to actually draw the rotating vectors and see how they add up to give the displacement response and phase relations according to the equation

$$\mathbf{x}(t) = \frac{\mathbf{f}(t)}{k} - \frac{m}{k}\ddot{\mathbf{x}}(t) - \frac{c}{k}\dot{\mathbf{x}}(t)$$

where the bold type indicates a vector.

So far in this section we have considered the displacement amplitude response; however, velocity and acceleration amplitude responses can be considered as well, especially if we remember that these are the two quantities more frequently measured in experimental practice. They show different characteristics in the limits of $\beta \to 0$ and $\beta \to \infty$ and also the peak values occur at slightly different forcing frequencies. If the resonant frequency is defined as the frequency for which the response is maximum, our SDOF system has three resonant frequencies, all of them different from the damped natural frequency.

Equation (4.49) defines the **displacement resonant frequency**, which is $\omega = \omega_n \sqrt{1 - 2\zeta^2}$.

The velocity resonant frequency can be obtained by the standard techniques of calculus if we consider that (Section 1.3.2)

$$|V| \equiv |\dot{X}| = \omega |X|$$

Equating to zero the derivative of the velocity amplitude with respect to β leads to the condition

$$\frac{1}{\beta}|X| = -\frac{d|X|}{d\beta}$$

and using eq (4.44), the **velocity resonant frequency** can be determined; this is

$$\omega = \omega_n$$

An analogous procedure gives the **acceleration resonant frequency** which is

$$\omega = \frac{\omega_n}{\sqrt{1 - 2\zeta^2}}$$

In practice, for the degree of damping of common physical systems, the difference among the three resonant frequencies is often negligible; nevertheless, the difference is worth noting for sake of completeness.

4.3.2   *Force transmissibility and harmonic motion of the support*

At the beginning of this chapter we defined the concepts of force and motion transmissibility: the first quantity considers the fraction of driving force (applied to the mass) transmitted to the support, the second quantity expresses the amplitude of mass motion with respect to an excitation due to the motion of the 'foundation' that supports our SDOF system. From the definitions themselves, it is not difficult to understand that both these quantities are of fundamental importance in the broad field of vibration isolation where two general problems are of interest:

- the transmission of as little vibration as possible to the base that supports a vibrating system;
- the isolation of sensitive instruments from possible vibrations of the base that supports them.

An extensive treatment of this subject is beyond the scope of this book but the basic aspect can be outlined as follows.

When the mass is subjected to a time-varying force $f(t)$, the force transmitted to the support through the spring and the damper is

$$f_T(t) = c\dot{x}(t) + kx(t) \tag{4.52}$$

The two quantities on the right-hand side add vectorially to give a magnitude

$$|f_T| = \sqrt{(c\dot{x})^2 + (kx)^2}$$

which, assuming $x(t)$ in the form $|X| \exp(\omega t - \phi)$, can be written more explicitly as

$$|f_T| = |X| \sqrt{c^2\omega^2 + k^2}$$

The force transmissibility itself, defined by the ratio $T \equiv f_T/f$, can be expressed as a harmonically varying quantity with magnitude $|T|$ and phase angle (with respect to f) ϕ_T. From the considerations above and eq (4.44) we get

$$|T| = \sqrt{\frac{1 + (2\zeta\beta)^2}{(1 - \beta^2)^2 + (2\zeta\beta)^2}} \qquad (4.53)$$

The phase angle can be obtained by observing that velocity and displacement are 90° out of phase and for this reason f_T leads the displacement of an angle θ given by

$$\tan\theta = 2\zeta\beta$$

The displacement, in turn, lags behind the driving force of the angle ϕ given by eq (4.47); it follows that the phase angle of T is $\phi_T = \phi - \theta$, and remembering from trigonometry that

$$\tan(\gamma - \delta) = \frac{\tan\gamma - \tan\delta}{1 + \tan\gamma \tan\delta}$$

we obtain

$$\tan\phi_T = \frac{2\zeta\beta^3}{1 - \beta^2 + 4\zeta^2\beta^2} \qquad (4.54)$$

When the support is vibrating, it is easy to see that the equation of motion has the form

$$m\ddot{x} + c(\dot{x} - \dot{y}) + k(x - y) = 0 \qquad (4.55)$$

where the function $y(t)$ describes the support motion and $x(t)$ is the displacement of the mass relative to a fixed frame of reference. Equation (4.55) can be rearranged to

$$m\ddot{x} + c\dot{x} + kx = c\dot{y} + ky \tag{4.56}$$

Using complex algebra, we let the support motion be of the form $y(t) = Ye^{-i\omega t}$ and the complex motion transmissibility is found to be

$$\frac{X}{Y} = \frac{1 - 2i\zeta\beta}{1 - \beta^2 - 2i\zeta\beta} \tag{4.57}$$

with magnitude and phase angle ϕ_{xy} given by

$$\left|\frac{X}{Y}\right| = \sqrt{\frac{1 + (2\zeta\beta)^2}{(1 - \beta^2)^2 + (2\zeta\beta)^2}}$$

$$\tan\phi_{xy} = \frac{2\zeta\beta^3}{1 - \beta^2 + 4\zeta^2\beta^2}$$

which are exactly the same as the magnitude and phase angle of the force transmissibility T of equations (4.53) and (4.54).

Graphs of eqs (4.53) and (4.54) are shown in Figs 4.10 and 4.11 for the two values of damping $\zeta = 0.05$ and $\zeta = 0.2$.

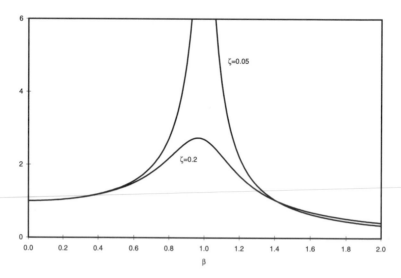

Fig. 4.10 Transmissibility (magnitude) versus β.

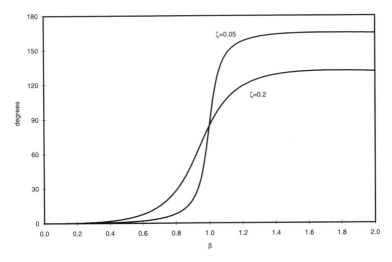

Fig. 4.11 Transmissibility (phase angle) versus β.

So, the problem of isolating a mass from the motion of the support is identical to that of limiting the force transmission to the base of a vibrating system. Figure 4.10 shows that the transmissibility amplitude is smaller than one only in the region $\beta > \sqrt{2}$ and that all curves, at $\beta = \sqrt{2}$, have the same value equal to unity. Surprisingly, damping does not make the situation any better in the effective range of vibration isolation and an undamped system is superior to a damped one in reducing the transmissibility, zero damping giving the smallest motion or force transmission. However, zero damping represents an unreal condition and a small amount of damping is desirable to keep the response within reasonable limits if the system has to go through the resonant region.

In the design of isolation systems, when the frequency ratio β is greater than $\sqrt{2}$, the damping is usually kept small, the transmissibility can be approximated by

$$|T| \cong \frac{1}{\beta^2 - 1} \tag{4.58}$$

and it is often convenient to speak in terms of isolation effectiveness, rather than transmissibility. The isolation effectiveness is defined mathematically by the quantity $1 - T$.

A general consideration on base motion can be made by going back to eq (4.55). The relative motion of the mass with respect to the base is

$$w \equiv x - y$$

By subtracting the quantity $m\ddot{y}$ from both sides of eq (4.55) we obtain

$$m\ddot{w} + c\dot{w} + kw = -m\ddot{y} \equiv f_{eff}(t) \tag{4.59}$$

where we can see that the base motion has the effect of adding a reversed inertia force $f_{eff}(t) = -m\ddot{y}$ to the equation of relative motion. Equation (4.59) is useful because in many applications relative motion is more important than absolute motion and also because the base acceleration is relatively easy to measure. The quantity $f_{eff}(t)$ is the effective support excitation loading, i.e. the system responds to the base acceleration as it would to an external load equal to the product (mass) × (base acceleration). The minus sign indicates that the effective force opposes the direction of the base acceleration; this has little importance in practice, since the base motion is generally assumed to act in an arbitrary direction.

4.3.3 *Resonant response of damped and undamped SDOF systems*

Immediately after the driving force is turned on it is not reasonable to expect that the oscillator response is given by eq (4.43). In fact, the force has not been acting long enough even to establish what its frequency is and it takes a while for the motion to settle into the steady state. The mathematical counterpart to this statement is, as we have seen before, that the general solution is the sum of two parts: the transient complementary function and a particular integral which represents the steady-state term. Explicitly, we can write

$$x(t) = e^{-\zeta\omega_n t}(A \cos \omega_d t + B \sin \omega_d t) + |X| \cos(\omega t - \phi) \tag{4.60}$$

The amplitude and phase angle of the steady-state term are still given by eqs (4.44) and (4.47), but the initial conditions $x(t=0) = x_0$ and $\dot{x}(t=0) = v_0$ now lead to

$$A = x_0 - |X| \cos \phi$$

$$B = \frac{1}{\omega_d}(v_0 - \omega |X| \sin \phi + \zeta\omega_n A) \tag{4.61}$$

With the intention to investigate what happens in resonance conditions (i.e. when $\beta = 1$), we assume that the system starts from rest ($x_0 = v_0 = 0$) and we get for A and B the values

$$A = 0$$

$$B = -\frac{\omega f_0}{2\zeta k \omega_d}$$

since $\phi = \pi/2$ and $|X| = f_0/2\zeta k$ when $\beta = 1$. With the further assumption of small damping, the damped frequency is nearly equal to the undamped frequency, we can write $\omega/\omega_d \cong 1$ and eq (4.60) becomes

$$x(t) \cong \frac{f_0}{2\zeta k} (1 - e^{-\zeta\omega_n t}) \sin \omega t \qquad (4.62)$$

from which it is evident that the response rapidly builds up asymptotically to its maximum value $f_0/2\zeta k$. It is left to the reader to draw a graph of eq (4.62), and also to determine, for different values of ζ, how many cycles are needed to practically reach the maximum response amplitude.

The case of an undamped SDOF system can be easily worked out from the considerations of the preceding sections by letting $\zeta \to 0$. In this case the magnitude of the response is given by (eq (4.44))

$$|X| = \left(\frac{f_0}{k}\right) \frac{1}{1 - \beta^2}$$

the phase angle is given by $\phi = 0$ and eq (4.60) becomes

$$x(t) = A \cos \omega_n t + B \sin \omega_n t + \frac{f_0}{k(1 - \beta^2)} \cos \beta\omega_n t \qquad (4.63)$$

Again, we assume for simplicity that our system starts from rest and from the initial conditions we get

$$A = -\frac{f_0}{k(1 - \beta^2)}$$

$$B = 0$$

Substituting in eq (4.63), the displacement response is

$$x(t) = \frac{f_0}{k(1 - \beta^2)} (\cos \beta\omega_n t - \cos \omega_n t)$$

which becomes indeterminate at resonance, i.e. when we let $\beta \to 1$. Using L'Hospital's rule we finally obtain

$$x(t)_{res} = \lim_{\beta \to 1} \frac{f_0 \omega_n t}{2k\beta} \sin \beta\omega_n t = \frac{f_0 \omega_n t}{2k} \sin \omega_n t \qquad (4.64)$$

The response builds up linearly with time and it is evident that after a few cycles the linear equations considered so far are no longer valid: the amplitude of oscillation increases indefinitely until disruptive effects ensue. Figure 4.12 illustrates the undamped resonant response given by eq (4.64).

4.3.4   Energy considerations

In the case of forced vibrations of a viscously damped system energy is dissipated because of damping and energy is supplied to the system by the driving force. The energy input per cycle can be obtained considering the infinitesimal work dW_f done by $f(t)$ as the system moves through a small distance dx and integrating over one cycle, i.e.

$$W_f \equiv \int_0^{2\pi/\omega} f dx = -\omega X f_0 \int_0^{2\pi/\omega} \cos \omega t \, \sin(\omega t - \phi) dt$$

where we have taken the harmonically varying force in the form $f(t) = f_0 \cos \omega t$ and the displacement in the form $x(t) = X \cos(\omega t - \phi)$, from which it follows $dx = -\omega X \sin(\omega t - \phi) dt$.

The integration gives after a few passages

$$W_f = \pi X f_0 \sin \phi \tag{4.65}$$

It is instructive to see how the same result can be obtained by using phasors if we remember the convention explained in Section 1.3 (eq (1.12)). We have now to consider the product force times velocity (which is the input supplied

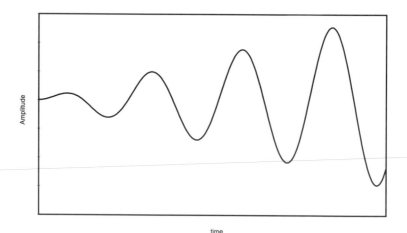

Fig. 4.12 Undamped resonant response.

power) where force and velocity are in the complex form

$$f(t) = f_0 e^{-i\omega t}$$

$$\dot{x}(t) = -i\omega X e^{-i(\omega t - \phi)}$$

respectively. Note that we have temporarily dropped the notation $|X|$ for the magnitude of the complex displacement and we are using X throughout this section to be consistent with the 'sinusoidal' notation of eq (4.65).

The integrated value over one cycle is exactly W_f and is obtained by calculating the quantity

$$W_f = \left(\frac{2\pi}{\omega}\right) \frac{1}{2} \operatorname{Re}[f(t)\dot{x}^*(t)]$$

$$= \frac{\pi}{\omega} \operatorname{Re}[i\omega X f_0 e^{-i\phi}] = \pi X f_0 \sin\phi$$

where we had to multiply by the period $T = 2\pi/\omega$ because eq (1.12) gives the average over one cycle and incorporates the division by T.

The same procedure can be used to calculate the work done by the damping force ($f_D = c\dot{x}$) per cycle of motion. We have now in sinusoidal notation

$$W_D = \int_0^{2\pi/\omega} f_D dx = c\omega^2 X^2 \int_0^{2\pi/\omega} \sin^2(\omega t - \phi)dt = \pi c\omega X^2 \qquad (4.66)$$

or, using phasors,

$$W_D = \left(\frac{2\pi}{\omega}\right) \frac{1}{2} \operatorname{Re}[c\dot{x}(t)\dot{x}^*(t)] = \pi c\omega X^2$$

It is not difficult at this point to show that $W_f = W_D$; since $\sin\phi = \operatorname{Im} X/|X|$ we get from eqs (4.44), (4.46) and (4.47)

$$\sin\phi = \frac{2\zeta\beta}{\sqrt{(1 - \beta^2)^2 + (2\zeta\beta)^2}} = \frac{2\zeta\beta k}{f_0} X^2$$

that must be substituted in eq (4.65) to give

$$W_f = 2\pi\zeta\beta k X^2 \qquad (4.67)$$

The known relations $k = m w_n^2$, $w = \beta w_n$ and $c = 2\zeta m w_n$ can be used to rearrange the result of eq (4.66) to

$$W_D = 2\pi \zeta \beta k X^2$$

which is the same as eq (4.67) and proves that the energy delivered by the driving force just equals the energy lost by friction. This fact implies that the work done per cycle by the spring and inertial forces is zero. In fact, the inertia and spring forces are related to the displacement by

$$f_I \equiv m\ddot{x} = -m w^2 x$$

$$f_S = kx$$

and a plot of f_I (or f_S) versus x over one cycle is a straight line enclosing a zero area. If we remember that the area enclosed in a graph of this kind represents the work done by the force over one cycle, we have justified the statement above. Obviously, this same statement can be proven by performing the calculations in sinusoidal or phasor notation. With regard to the damping force we have already determined that the work done by f_D over one cycle is different from zero (eq (4.66)), but a graphical representation may also be useful. We can write

$$f_D \equiv c\dot{x} = -c w X \sin(w t - \phi)$$

squaring and rearranging leads to

$$\left(\frac{f_D}{c w X}\right)^2 = 1 - \cos^2(w t - \phi) = 1 - \left(\frac{x}{X}\right)^2$$

i.e.

$$\left(\frac{f_D}{c w X}\right)^2 + \left(\frac{x}{X}\right)^2 = 1 \tag{4.68}$$

which relates force and displacement and is the equation of an ellipse with area equal to $\pi c w X^2$. We note that at resonance the phase angle ϕ is $\pi/2$ radians and eq (4.65) reduces to

$$W_f = \pi X f_0 \tag{4.69}$$

4.4   Damping in real systems, equivalent viscous damping

Damping is an inherent property of every real system; its effect is to remove energy from the system by dissipating it as heat or by radiating it away. There are many mechanisms which can cause damping in materials and structures:

internal friction, fluid resistance, sliding friction at joints and interfaces within a structure and at its connections and supports. Therefore, the basic physical characteristics of damping are seldom fully understood and many different types – besides viscous damping which we have considered so far – can be encountered in practice. One often finds reference to structural (hysteretic), Coulomb (dry-friction) or velocity-squared (aerodynamic drag) damping. They are all damping mechanisms based on some modelling assumptions that try to explain and fit the experimental data from vibration analysis. Unfortunately, in real systems damping is rarely of a viscous nature even if, on the other hand, most systems are lightly damped and the difference is insignificant in regions away from resonance. It is then possible to obtain approximate models of nonviscous damping in terms of equivalent viscous dampers and exploit this simple vibration model in different situations.

The concept of **equivalent viscous damping** is based on the equivalence of energy dissipated per cycle by a viscous damping mechanism and by the given nonviscous real situation. We have seen in the preceding section that the energy loss per cycle (eq (4.66)) is directly proportional to the frequency of motion, the damping coefficient c and the square of the amplitude. However, experimental tests show that the actual energy loss per cycle of stress is directly proportional to the square of the amplitude, but independent of frequency over wide ranges of frequency and temperature; this suggests a relation of the type

$$W_D = \alpha X^2 \tag{4.70}$$

where α is a constant for a given frequency and temperature range.

This type of damping is called **structural (or hysteretic) damping** and is attributed to the hysteresis phenomenon observed in cyclic stress of elastic materials, where the energy loss per cycle is equal to the area inside the hysteresis loop. Equating eqs (4.66) and (4.70) we get

$$\pi c_{eq} \omega X^2 = \alpha X^2$$

The equivalent viscous damping coefficient can be defined in this case as

$$c_{eq} = \frac{\alpha}{\pi \omega} \tag{4.71}$$

Our structurally damped system subjected to harmonic excitation can thus be treated as if it were viscously damped with a coefficient given by eq (4.71). By introducing this result in the equation of motion we obtain the complex amplitude response

$$X = \frac{f_0}{k} \left(\frac{1}{1 - \beta^2 - i\gamma} \right) \tag{4.72}$$

with magnitude, real and imaginary parts and phase angle given by

$$|X| = \left(\frac{f_0}{k}\right) \frac{1 - \beta^2}{\sqrt{(1 - \beta^2)^2 + \gamma^2}} \tag{4.73}$$

$$\text{Re } X = \left(\frac{f_0}{k}\right) \frac{1 - \beta^2}{(1 - \beta^2)^2 + \gamma^2} \tag{4.74}$$

$$\text{Im } X = \left(\frac{f_0}{k}\right) \frac{\gamma}{(1 - \beta^2)^2 + \gamma^2} \tag{4.75}$$

$$\tan \phi = \frac{\gamma}{1 - \beta^2} \tag{4.76}$$

where we have defined the **structural damping factor** (or **loss factor**) $\gamma \equiv \alpha/(\pi k)$. A plot of eqs (4.73) and (4.76) is similar to Figs 4.8 and 4.9 but there are differences worthy of note:

- The amplitude response is always maximum at $\beta = 1$ (irrespective of the value of γ) and for very low values of β the response depends on γ.
- The phase angle tends to the value $\arctan \gamma$ for $\beta \to 0$, while $\phi(\beta \to 0) = 0$ for viscous damping.

By comparing the denominators of eqs (4.43) and (4.72) we can see that γ corresponds to the quantity $2\zeta\beta$ of the viscous case and since damping factors are usually small and are effective only in the vicinity of resonance, we have

$$\gamma \cong 2\zeta \tag{4.77}$$

Another equivalent way to introduce structural damping is to incorporate in the complex equation of motion a term which is proportional to displacement but in phase with velocity, i.e.

$$m\ddot{x} + k(1 - i\gamma)x = f_0 e^{-i\omega t} \tag{4.78}$$

where γ is as before; $k(1 - i\gamma)$ (or $k(1 + i\gamma)$ if we adopt the positive exponential form $e^{+i\omega t}$) is called **complex stiffness** and was introduced for the calculation of the flutter speeds of airplane wings and tail surfaces [3]. One word of caution is necessary: the analogy between structural and viscous damping is valid only for harmonic excitation, because a driving force at frequency ω is implied in the foregoing discussion.

Other damping models that are frequently used and encountered in practice are Coulomb and velocity-squared damping. We limit the discussion to some fundamental results.

Coulomb damping arises from sliding of two dry surfaces; to start the motion the force must overcome the resistance due to friction, i.e. it must be greater than $\mu_s mg$, where $0 < \mu_s < 1$ is the static friction coefficient and mg is the weight of the sliding mass. When this happens, the resistance force suddenly drops to $\mu_k mg$, where μ_k is the kinetic friction coefficient and is generally smaller than μ_s. The friction force opposes velocity (i.e. $f_D = \pm \mu_k mg$, with the appropriate sign for every half cycle of motion) and remains constant as long as the forces acting on m are sufficient to overcome the dry friction. The motion stops when this is no longer the case.

It is easy to see that a graph of force versus displacement is a rectangle in this case (with sides $2X$ and $2\mu_k mg$) and the energy lost per cycle is given by $W_D = 4\mu_k mgX$, so that the following equivalent damping coefficient is obtained:

$$c_{eq} = \frac{4\mu_k mg}{\pi \omega X} \qquad (4.79)$$

Other characteristics of Coulomb damping are:

1. The free vibration decay still occurs at the 'frequency' ω_n (remember, however, that this is an improper term because the decaying motion is not strictly periodic) but is linear (and not exponential) in time with an amplitude reduction of $4\mu_k mg/k$ per cycle of motion (this can be easily verified by solving the homogeneous equation of motion).
2. In forced vibration conditions the damping does not limit the amplitude at resonance and the quantity $\tan \phi$ is independent of β, but its sign changes abruptly as β passes through 1.

Bodies moving with moderate speed in a fluid (for example air, water or oil) experience a resisting force that is proportional to the square of the speed (aerodynamic damping), i.e.

$$f_D = \pm a\dot{x}^2$$

where the minus sign is used when $\dot{x}$ is positive and vice versa. It is left to the reader to determine that the energy lost per cycle is given by

$$W_D = \tfrac{8}{3} a\omega^2 X^3 \qquad (4.80)$$

and the equivalent viscous damping is given in this case by

$$c_{eq} = \frac{8}{3\pi} a\omega X \qquad (4.81)$$

The constant a is in general related to the drag coefficient, to the exposed surface area of the body and to the density of the fluid in which the body is immersed.

It should be noted that both coefficients (4.79) and (4.81) are nonlinear, since they are functions of the amplitude of the vibration.

4.4.1   Measurement of damping

There are many ways to quantify and measure the damping of a system. The ideal situation would be that all these quantities were consistent with each other and that a linear relation would hold between any two of them. This is not always the case, and care must be taken to ensure that the chosen quantity is clearly specified. Further complications arise because it appears that there is not a unified set of symbols to describe damping and because the generalization to systems with a high degree of damping or to systems with more than one degree of freedom is not always straightforward.

In the following, we will describe a few of the most common ways to measure this parameter, based on the insight we have gained on the behaviour of the harmonically excited SDOF system considered so far.

Free vibration decay

This method has already been explained in Section 4.2.3 (eqs (4.34) and (4.36)). In practice, the system is excited by any convenient means and then allowed to vibrate freely. The time history of the free vibration is recorded and the displacement amplitudes of successive peaks can be used for the calculation of damping. For instance, with this technique mean values for the logarithmic decrement of concrete have been quoted in the order of 0.03 (uncracked) and 0.1 (cracked). Obviously, the constituents of the concrete and the amount and type of reinforcement influence the amount of damping present.

The equipment and instrumentation required in this case are minimal and the 'convenient means' to impart the excitation may sometimes be very simple. For example, the damping characteristics of tall buildings have been obtained by observing the free vibration caused by wind, nondamaging impact or the release of a taut cable connecting the building to the ground. This latter technique, with minor modifications, was used to determine the damping of offshore platforms [4] and bridges [5]. The author has measured the damping of a 300 kg subway two-way switch by standing on it and suddenly jumping down, recording the signal with only one accelerometer connected to a digital oscilloscope. Also, the damping of a few ancient Italian belltowers has been measured by suddenly stopping the oscillation of the bell on top (by means of a braking system that electrically controls the bell oscillations).

Resonant response

We have seen that at resonance the phase angle is $\phi = \pi/2$. If the phase angle can be measured, one can detect resonance by adjusting the exciting frequency until the condition above is attained. The measured displacement is then given by

$$|X|_{res} = \frac{f_0/k}{2\zeta}$$

which is eq (4.44) with $\beta = 1$; it follows that

$$\zeta = \frac{f_0/k}{2|X|_{res}} \tag{4.82}$$

However, it may not be easy to apply the exact resonance frequency and the measurement of the phase may also be somewhat difficult. An alternative is to obtain the amplitude response curve in the vicinity of resonance and measure the peak value; for viscous damping this is given by (eq (4.50))

$$|X|_{max} = \frac{f_0/k}{2\zeta\sqrt{1 - \zeta^2}}$$

from which ζ can be easily obtained. In ordinary structures the term ζ^2 can be neglected with respect to unity and we get

$$\zeta = \frac{f_0/k}{2|X|_{max}} \tag{4.83}$$

If the damping is hysteretic in nature we have

$$|X|_{res} = |X|_{max} = \frac{f_0/k}{\gamma} \tag{4.84}$$

In any case the static displacement must be known or measured by some means and this may present a problem with this technique.

Half-power (bandwidth)

This technique assumes that the frequency response curve is available from experimental measurements and avoids the need for the static response. A sinusoidal excitation at a closely spaced sequence of frequencies in the resonance region is applied to the structure and the resulting displacement curve is plotted as a function of frequency. The points where the amplitude response (or the dynamic magnification factor) is reduced to $(1/\sqrt{2})$ of its peak value are used to calculate damping. There are two such points; they are often defined as **half-power points** (because power and energy are proportional to the square of the amplitude) or -3 dB points [because $20\log_{10}(1/\sqrt{2}) = -3$] and they can be obtained by the condition

$$\left(\frac{f_0}{k}\right)\frac{1}{2\zeta\sqrt{2}} = \left(\frac{f_0}{k}\right)\frac{1}{\sqrt{(1-\beta^2)^2 + (2\zeta\beta)^2}}$$

Upon squaring and rearranging we get the equation

$$\beta^4 - 2\beta^2(1 - 2\zeta^2) + 1 - 8\zeta^2 = 0$$

whose roots are given by

$$\beta_{1,2}^2 = (1 - 2\zeta^2) \pm 2\zeta\sqrt{1 + \zeta^2}$$

and can be simplified as follows:

$$\beta_{1,2}^2 \cong 1 \pm 2\zeta - 2\zeta^2$$

The two values of β can be finally obtained using the binomial expansion

$$\beta_{1,2} \cong 1 \pm \zeta - \zeta^2 \tag{4.85}$$

thus giving

$$\zeta = \frac{1}{2}(\beta_1 - \beta_2) = \frac{\omega_1 - \omega_2}{2\omega_n} \tag{4.86}$$

This same result can be obtained by measuring the frequencies at which $\phi = \pm\pi/4$ (or $\tan\phi = \pm1$). From eq (4.47) we get the two values of β from the conditions

$$\frac{2\zeta\beta_1}{1 - \beta_1^2} = +1 \qquad \frac{2\zeta\beta_2}{1 - \beta_2^2} = -1$$

which give

$$1 - \beta_1^2 - 2\zeta\beta_1 = 0 \tag{4.87a}$$

$$1 - \beta_2^2 + 2\zeta\beta_2 = 0 \tag{4.87b}$$

and the result of eq (4.86) can be obtained by subtracting eq (4.87a) from eq (4.87b).

This latter procedure relies again on the possibility to measure phase angles between input force and output displacement which, as we said before, may not be an easy task.

Energy loss per cycle

When force–displacement measurements can be made by running a harmonic excitation test at a specified frequency over a whole cycle, the damping can be determined from a plot of their relationship. We have seen (eq (4.68)) that a graph of the viscous force versus displacement is an ellipse that intercepts the ordinate axis at $c\omega X$. The same is true when the total force

$$f = f_I + f_D + f_S$$

versus displacement is plotted, the only difference being the fact that now the ellipse is inclined at an angle with respect to the coordinate axes. Since ω and X are both known, the damping c can be obtained by the intercept on the ordinate axis.

If the damping is not viscous, the graph will not in general be an ellipse, but an equivalent viscous damping can be determined by measuring the area W_D enclosed by the force–displacement relationship and equating it to the value obtained in the viscous case, i.e. $\pi c_{eq}\omega X^2$. We get

$$c_{eq} = \frac{W_D}{\pi\omega X^2} \tag{4.88}$$

and a damping ratio

$$\zeta_{eq} = \frac{W_D}{\pi\omega X^2}\left(\frac{1}{2\sqrt{km}}\right) = \frac{W_D}{2\pi X^2}\left(\frac{1}{k\beta}\right) \tag{4.89}$$

which requires us to estimate or measure m and k. This latter parameter can be obtained by running the test at a very low frequency, which in practice corresponds to almost static conditions. The force–displacement diagram f_S versus x is a straight line with slope k. Alternatively, the area W_S under the diagram can be used to determine k as

$$k = \frac{2W_S}{X^2}$$

thus giving ζ as a function of the ratio of the damping energy loss per cycle divided by the strain energy at maximum displacement, i.e.

$$\zeta_{eq} = \left(\frac{1}{4\pi\beta}\right)\frac{W_D}{W_S} \tag{4.90}$$

Running the test at resonance makes β:

1. equal to unity in the denominator of eq (4.90);
2. equal to zero the inclination angle – with respect to the coordinate axes – of the principal axes of the ellipse obtained from the diagram f versus x.

Condition 2 applies because at resonance the inertia force exactly balances the spring force and a diagram of the applied force versus displacement is identical to the damping force versus displacement diagram.

Frequency response function

The **frequency response function** (FRF) is a widely used quantity in many fields of engineering and is of fundamental importance in many applications of vibration analysis. Here we briefly introduce the basic definitions in order to proceed with the topic we are presently discussing, i.e. the determination of damping. However, due to its importance, the subject will be considered in detail in subsequent chapters.

For linear systems, the FRF – usually indicated by $H(\omega)$ – establishes a relationship between the Fourier transform of the input signal $X(\omega)$ and the Fourier transform output signal $Y(\omega)$. The general relationship is

$$Y(\omega) = H(\omega)X(\omega) \tag{4.91}$$

i.e.

$$H(\omega) = \frac{Y(\omega)}{X(\omega)} \tag{4.92}$$

provided that $X(\omega) \neq 0$. Equation (4.92) is often given as the definition of $H(\omega)$. Since, in general, the FRF is a complex-valued function of a real-valued independent variable ω (thus involving three quantities: the frequency ω, the real and imaginary part of $H(\omega)$), two $x - y$ graphs are needed for complete information. The graphical display is a matter of choice, depending on the information required. Obviously, for the concept of FRF to make sense, the input and output signals must be Fourier transformable – a condition that is met by all physically realizable systems – and the input signal must be non-zero at all frequencies of interest.

If now we refer to our harmonically excited SDOF system, the input signal is the sinusoidal force and the output signal is the displacement response; both signals can be taken in the complex-valued form and the complex FRF of this system can be found by solving the equation of motion for an arbitrary Fourier component, i.e.

$$(-m\omega^2 - ic\omega + k)Xe^{-i\omega t} = fe^{-i\omega t}$$

thus giving

$$H(\omega) \equiv \frac{X}{f} = \frac{1}{k - m\omega^2 - i\omega c} \tag{4.93}$$

where we recognize eq (4.42) and we understand that the FRF representation gives information on both the magnitude of response and the phase angle of lag between response and excitation. This particular form of FRF (displacement response–driving force) is called **receptance**; other forms can be obtained by considering velocity or acceleration as the measured response. Table 4.3 gives a brief list of these various forms with the corresponding names that are commonly used.

The real part of the function in eq (4.93) is obtained from eq (4.45) simply dividing by the force and is plotted in Fig 4.13 for two different values of ζ. It is not difficult to determine that the two extrema occur at the values

$$\beta_{1,2} = \sqrt{1 \pm 2\zeta}$$

so that the damping ratio is usually calculated by

$$\zeta = \frac{1}{2} \left(\frac{(\beta_1/\beta_2)^2 - 1}{(\beta_1/\beta_2)^2 + 1} \right) = \frac{1}{2} \left(\frac{(\omega_1/\omega_2)^2 - 1}{(\omega_1/\omega_2)^2 + 1} \right) \tag{4.94}$$

The same expressions on the right-hand side give the value of γ in the case of hysteretic damping.

The last method that we consider is based on a form of display of the FRF called the **Nyquist plot**. This is a plot of the imaginary part of the FRF versus its real part; as such it does not contain the frequency information explicitly but it is a useful representation in view of the generalization of FRFs to multiple-degree-of-reedom systems.

The particular form of FRF we consider now for our viscously damped SDOF system is **mobility**, which is the ratio of the velocity response function divided by the driving force (Table 4.3) and which we indicate for the present

Table 4.3 Different types of frequency response functions

Relationship	Function
Displacement–force	Receptance, admittance, dynamic compliance
Velocity–force	Mobility
Acceleration–force	Accelerance, inertance
Force–displacement	Dynamic stiffness
Force–velocity	Mechanical impedance
Force–acceleration	Apparent mass

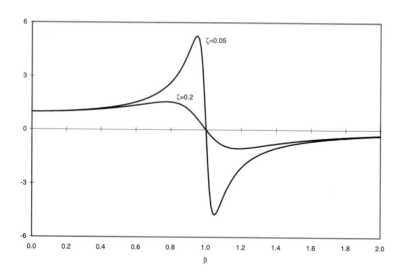

Fig. 4.13 Re[$H(\omega)$] versus β.

by $M(\omega)$. By taking the time derivative of receptance, the real and imaginary part of $M(\omega)$ can be obtained, i.e.

$$\text{Re}[M(\omega)] = \frac{c\omega^2}{(k + m\omega^2)^2 + (c\omega)^2} \tag{4.95}$$

$$\text{Im}[M(\omega)] = \frac{-\omega(k - m\omega^2)}{(k + m\omega^2)^2 + (c\omega)^2} \tag{4.96}$$

so that defining the quantities U and V as

$$U \equiv \text{Re}[M(\omega)] - \frac{1}{2c} \qquad V \equiv \text{Im}[M(\omega)]$$

we get

$$U^2 + V^2 = \frac{1}{4c^2} \tag{4.97}$$

which is the equation of a circle of radius $1/(2c)$ and centre at the origin of the $U - V$ plane. Referring back to the plane with axes $\text{Re}[M(\omega)] - \text{Im}[M(\omega)]$ (which is horizontally translated to the left of the $U - V$ plane of the quantity $1/(2c)$) the centre is now at the point $(1/(2c), 0)$. So, a Nyquist plot of

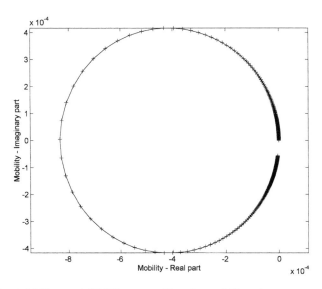

Fig. 4.14 Damped SDOF system: Nyquist mobility plot.

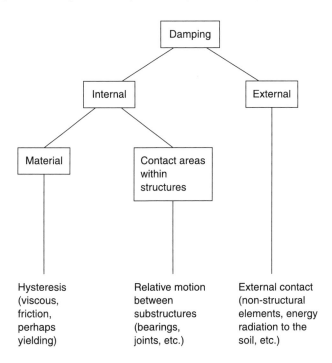

Fig. 4.15 Different types of damping. (Reproduced with permission from H. Bachmann, W.J. Ammann *et al.*, *Vibration Problems in Structures – Practical Guidelines*, Birkhäuser-Verlag, Basel, Boston, Berlin, 1995.)

mobility traces out an exact circle as the frequency ω sweeps from zero to infinity and therefore the measurement of damping reduces to a measure of the radius of such a circle.

In the same way it can be determined that for a hysteretically damped system a Nyquist plot of the receptance will form a circle of radius $1/(2\gamma)$ and centre at the point $(0, -1/(2\gamma))$.

The Nyquist plot of mobility, for an arbitrary value of ζ of a viscously damped SDOF, is shown in Fig. 4.14 where the small crosses are equal increments in frequency.

Finally, Fig. 4.15 [6] is a useful schematic representation of the different types and sources of damping encountered in structural dynamics.

4.5   Summary and comments

In Chapter 4 the general model of **single-degree-of-freedom** (SDOF) system is considered. A single mass with a spring and a viscous damper – the so-called harmonic oscillator – is the simplest model which, despite its simplicity, shows many of the fundamental characteristics of vibrating systems in general. The equation of motion for such a system has been obtained in Chapter 2 and here it is solved in the **undamped** and **damped** cases, considering **free vibrations** first (homogeneous equation: no forcing external excitation) and then **forced vibrations** (nonhomogeneous equation) under the action of an external sinusoidal excitation. Energy considerations are made in both cases.

The section on free vibrations introduces the concepts of **natural frequency, overdamped, critically damped** and **underdamped** systems, showing how and when a system can vibrate depending on the values of its parameters. The underdamped case is the most important in vibration analysis and leads to the definitions of **frequency of damped free oscillations** and **logarithmic decrement**.

The section on forced vibrations sheds light on the important phenomenon of **resonance**, which plays a major role in so many applications in physics and engineering. Three frequency ranges are considered according to a comparison between forcing frequency and natural frequency of the system and the phase relationship between input (excitation) and output (motion: in the form of displacement, velocity or acceleration) is shown.

In some circumstances, when the transmission of force or motion from support to mass (or vice versa) is of interest, one speaks of **force** or **motion transmissibility**, obtaining the surprising result that, away from resonance, damping does not seem to be of any help in limiting the amplitude of motion. Zero damping, however, is an unreal condition; a small amount of damping is always desirable because the transient part of the vibration must be considered when a mechanical or structural system is set into motion and disruptive effects may ensue if damping is too low. Such effects do appear if an undamped SDOF system is excited at resonance: the motion increases without bounds and the range of linearity is soon exceeded.

Furthermore, damping represents the mechanism of energy dissipation of the system, and in real situations this mechanism is often very hard to define analytically; the viscous model is adopted in general for convenience but sometimes leads to results that do not agree with experimental measurements. To overcome this difficulty, the concept of **equivalent viscous damping** is introduced by comparing the energy loss per cycle in different situations (hysteretic, Coulomb and velocity-squared damping). Nevertheless, an experimental measurement of this parameter is often necessary because, unlike mass and stiffness, it cannot be predicted with an acceptable degree of accuracy from theoretical considerations alone.

With this in mind, five different methods to measure damping are then considered at the end of the chapter. The instrumentation requirements to accomplish this task vary considerably: in some cases an appropriate vibration sensor and an oscilloscope may do, but only highly (and sometimes costly) sophisticated electronic instruments may do the job in others. In general, the choice is dictated by the desired accuracy and by the operating conditions of measurement; sometimes – for example in hostile industrial environments – the speed of the measurement may be paramount and the results obtained can be good enough for all practical purposes.

References

1. Olson, H.F., *Dynamical Analogies*, D. Van Nostrand, Princeton, NJ, 1958.
2. Seto, W.W., *Mechanical Vibrations*, Shaum's Outline Series in Engineering, McGraw-Hill, New York, 1964.
3. Theodorsen, Th. and Garrick, I.E., *Mechanisms of Flutter. A Theoretical and Experimental Investigation of the Flutter Problem*, NACA Report 685, 1940.
4. Black J.L., Method for determining the damping coefficients of an offshore platform, *Eurodyn 93, Proceedings of the 2nd European Conference on Structural Dynamics*, Trondheim, Norway, 21–23 June 1993.
5. Larssen, R.M. *et al.*, Reliability updating of a cable-stayed bridge during construction based on measured dynamic response, *Eurodyn 93, Proceedings of the 2nd European Conference on Structural Dynamics*, Trondheim, Norway, 21–23 June 1993.
6. Bachmann, H. *et al.*, *Vibration Problems in Structures – Practical Guidelines*, Birkhäuser Verlag, 1995.

5 More SDOF – transient response and approximate methods

5.1 Introduction

The harmonic excitation considered so far is a special kind of deterministic dynamic loading that only in a few cases can approximate a real situation. Nevertheless, its consideration is a necessary prerequisite for any further analysis, and not only for didactical purposes. If we remember that for linear systems the principle of superposition holds, from the fact that any reasonably well-behaved excitation function can be written as the sum or integral of a series of simple functions, it follows that the total response is then the sum (or integral) of the individual responses. So, in principle, the complications seem to be more of a mathematical nature rather than a physical nature, and such a statement of the problem does not seem to add anything substantial to the understanding of the behaviour of a linear SDOF system under the action of a complex exciting load. However, things are not so simple; a number of different approaches and techniques are available to deal with this problem and the choice is partly a subjective matter and partly dictated by the complexity of the situation, the final results that one wants to achieve and the mathematical tractability of the calculations by means of analytical or computer-based methods.

The first and main distinction can be made between:

- time-domain techniques
- frequency-domain techniques.

As the name itself implies, the first approach relies on the manipulation of the functions involved (generally the loading and the response functions) in the domain of time as the independent variable of interest. The important concepts are ultimately the **impulse response function** and the **convolution, or Duhamel's, integral**.

On the other hand, the second technique is based on the powerful tool of mathematical transforms: the manipulations are made in the domain of an appropriate independent variable (frequency, for example, hence the name) and then, if necessary, the result is transformed back to the domain of the original variable.

he two approaches, as one might expect, are strictly connected and the result does not depend on the particular technique adopted for the problem at hand. Unfortunately, except for simple cases, both techniques involve evaluations of integrals that are not always easy to solve and their practical application must often rely heavily on numerical methods which, in their turn, require the relevant functions to be 'sampled' at regular intervals of the independent variable. This 'sampling' procedure introduces further complications that belong to the specific field of **digital signal analysis,** but they cannot be ignored when measurements are taken and computations via electronic instrumentation are performed. Their basic aspects will be dealt in future chapters.

Until a few decades ago the computations involved in frequency-domain techniques were no less than those in a direct evaluation of the discrete convolution in the time domain. The development of a special algorithm called the fast Fourier transform [1] has completely changed this situation, cutting down computational time of orders of magnitude and making frequency techniques more effective.

Both the convolution integral and the transform methods apply when linearity holds; for nonlinear systems recourse must be made to a direct numerical integration of the equations of motion, a technique which, obviously, applies to linear systems as well.

When the predominant frequency of vibration is the most important parameter and the system is relatively complex, the Rayleigh 'energy method' and other techniques with a similar approach turn out to be useful to obtain such a parameter. The simplest application represents a multiple- (or infinite-) degree-of-freedom system as a 'generalized' SDOF system after an educated guess of the vibration pattern has been made. The method is approximate (but so are numerical methods, and in general are much more time consuming) and its accuracy depends on how well the estimated vibration pattern matches the true one. Its utility lies in the fact that even a crude but reasonable guess often results in a frequency estimate which is good enough for most practical purposes.

5.2 Time domain – impulse response, step response and convolution integral

Let us refer back to Fig. 4.7. The SDOF system considered so far is a particular case of the situation that this figure illustrates, i.e. a single-input single-output linear system where the output $x(t)$ and the input $f(t)$ (written as a force for simplicity, but it need not necessarily be so) are related through a linear differential equation of the general form

$$a_n \frac{d^n x}{dt^n} + a_{n-1} \frac{d^{n-1} x}{dt^{n-1}} + \ldots + a_1 \frac{dx}{dt} + a_0 x$$

$$= b_r \frac{d^r f}{dt^r} + b_{r-1} \frac{d^{r-1} f}{dt^{r-1}} + \ldots + b_1 \frac{df}{dt} + b_0 f \tag{5.1}$$

The coefficients a_i and b_j $(i = 1, 2, 3, \cdots, n; \; j = 1, 2, 3, \cdots, r)$, that is the parameters of the problem, may also be functions of time, but in general we shall consider only cases when they are constants. On physical grounds, this assumption means that the system's parameters (mass, stiffness and damping) do not change, or change very slowly, during the time of occurrence of the vibration phenomenon. This is the case, for example, for our spring- -mass–damper SDOF system whose equation of motion is eq (4.13), which is just a particular case of eq (5.1).

Very common sources of excitation are **transient phenomena** and **mechanical shocks**, both of which are obviously nonperiodic and are characterized by an energy release of short duration and sudden occurrence. Broadly speaking, we can define a mechanical shock as a transmission of energy to a system which takes place in a short time compared with the natural period of oscillation of the system, while transient phenomena may last for several periods of vibration of the system.

An impulse disturbance, or shock loading, may be for example a 'hammer blow': a force of large magnitude which acts for a very short time. Mathematically, the Dirac delta function (Chapter 2) can be used to represent such a disturbance as

$$f(t) = \hat{f}\delta(t) \tag{5.2}$$

where $\hat{f}$ has the dimensions newton-seconds and describes an impulse (time integral of the force) of magnitude

$$\int_{-\infty}^{+\infty} f(t)dt = \hat{f} \int_{-\infty}^{+\infty} \delta(t)dt = \hat{f} \tag{5.3}$$

One generally speaks of **unit impulse** when $\hat{f} = 1$.

From Newton's second law $fdt = mdv$, assuming the system at rest before the application of the impulse, the result on our system will be a sudden change in velocity equal to $\hat{f}/m$, without an appreciable change in its displacement. Physically, it is the same as applying to the free system the initial conditions $x(0) = 0$ and $\dot{x}(0) = \hat{f}/m$. The response can thus be written (eq (4.8))

$$x(t) = \frac{\hat{f}}{m\omega_n} \sin \omega_n t \tag{5.4}$$

for an undamped system, and (eq (4.27))

$$x(t) = \frac{\hat{f}}{m\omega_d} e^{-\zeta\omega_n t} \sin \omega_d t \tag{5.5}$$

for a damped system. In both cases it is convenient to write the response as

$$x(t) = \hat{f} h(t) \tag{5.6}$$

where $h(t)$ is called the **unit impulse response** (some authors also call it the weighting function) and is given by

$$h(t) \equiv \frac{1}{m\omega_n} \sin \omega_n t$$

$$\tag{5.7a}$$

$$h(t) \equiv \frac{e^{-\zeta\omega_n t}}{m\omega_d} \sin \omega_d t$$

for the undamped and damped case, respectively. Equations (5.7a) represent the response to an impulse applied at time $t = 0$; if the impulse is applied at time $t = \tau$ ($\tau \neq 0$), they become

$$h(t - \tau) \equiv \frac{1}{m\omega_n} \sin \omega_n(t - \tau)$$

$$\tag{5.7b}$$

$$h(t - \tau) \equiv \frac{e^{-\zeta\omega_n(t - \tau)}}{m\omega_d} \sin \omega_d(t - \tau)$$

for $t > \tau$ and zero for $0 < t < \tau$, since the change of variable from t to $t - \tau$ is geometrically a simple translation of the coordinate axes to the right by an amount τ seconds. In practice, an impact of duration Δt of the order of 10^{-3} s (and Δt is short compared to the system's period $T = 2\pi/\omega_n$) is a common occurrence in vibration testing of structures. In these cases, the considerations above apply.

Figure 5.1 illustrates a graph of $h(t - \tau)$ for a damped system with unit mass, damping ratio $\zeta = 0.2$ and damped natural frequency $\omega_d = 2.0$ rad/s.

Example 5.1. Let us consider the response of an undamped system to an impulse of a constant force f_0 that acts for the short (compared to the system's period) interval of time $0 < t < t_1$. We assume the system to be initially at rest and we have

$$m\ddot{x} + kx = \begin{cases} f_0 & 0 < t \leqslant t_1 \\ 0 & t > t_1 \end{cases} \tag{5.8}$$

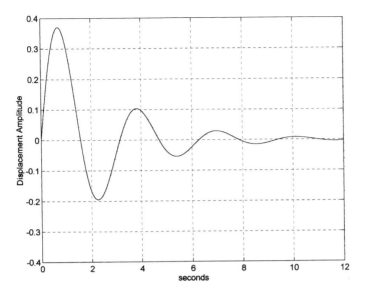

Fig. 5.1 Impulse response function $h(t)$ – damped system.

During the 'forced vibration era' $(0 < t < t_1)$ the response of the system is given by eq (4.63) where $\beta = 0$ (because $\omega = 0$) and the particular integral is given by f_0/k, i.e.

$$x(t) = A \cos \omega_n t + B \sin \omega_n t + \frac{f_0}{k} \qquad (5.9)$$

the initial conditions $x(0) = 0$ and $\dot{x}(0) = 0$ determine the constants A and B and we get

$$x(t) = \frac{f_0}{k} (1 - \cos \omega_n t) \qquad (5.10)$$

In the 'free vibration era' $(t > t_1)$ the excitation is no longer active and the response is given by eq (4.8) with initial conditions determined by the state of the system at the instant $t = t_1$. Explicitly this is written

$$x(t) = x(t_1)\cos \omega_n(t - t_1) + \frac{v(t_1)}{\omega_n} \sin \omega_n(t - t_1) \qquad (5.11)$$

where

$$x(t_1) = \frac{f_0}{k}(1 - \cos \omega_n t_1) \cong \frac{f_0}{2k}(\omega_n t_1)^2$$

$$v(t_1) \equiv \dot{x}(t_1) = \frac{f_0 \omega_n}{k} \sin \omega_n t_1 \cong \frac{f_0 \omega_n^2 t_1}{k}$$

(5.12)

and the approximations above hold for $t_1 \to 0$ (or, better, $t_1/T \to 0$). By noting that $\omega_n^2/k = 1/m$ and that the impulse has a value $\hat{f} = f_0 t_1$, from eq (5.11) we obtain in the limit

$$x(t) = \frac{\hat{f}}{m\omega_n} \sin \omega_n t$$

(5.13)

which is, as expected, the impulse response of the undamped system (eq (5.4)). It is left to the reader to determine how considerations similar to the ones above lead to eq (5.5) for a damped system.

A general transient loading such as the one shown in Fig. 5.2 can be regarded as a superposition of impulses; each impulse is applied at time $t = \tau$ and has a magnitude given by $f(\tau)\Delta\tau$, with τ varying along the time axis (shaded area in Fig. 5.2).

Mathematically we can write

$$\Delta f(t, \tau) = f(\tau)\Delta\tau\delta(t - \tau)$$

(5.14)

and the function $f(t)$ can be approximated by a superposition of these impulses as

$$f(t) \cong \sum \Delta f(t, \tau) = \sum f(\tau)\Delta\tau\delta(t - \tau)$$

(5.15)

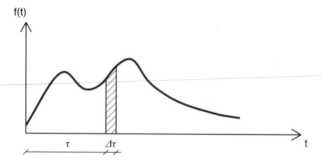

Fig. 5.2 General loading as a series of impulses.

The response to the impulse of eq (5.14) is

$$\Delta x(t, \tau) = f(\tau)\Delta\tau h(t - \tau)$$

and the total response from time $\tau = 0$ to time $\tau = t$ is obtained by summing the effects of all the impulses up to the instant t, i.e.

$$x(t) \cong \sum \Delta x(t, \tau) = \sum f(\tau)\Delta\tau\delta(t - \tau) \tag{5.16}$$

Passing to the limit of $\Delta\tau \to 0$, the summation becomes an integral and the response from $\tau = 0$ to $\tau = t$ is finally given by

$$x(t) = \int_0^t f(\tau)h(t - \tau)d\tau \tag{5.17}$$

The integral of eq (5.17) is known as **Duhamel's integral** or **convolution integral** and may be used to determine the response to an arbitrary input as long as it satisfies certain mathematical conditions. For our damped SDOF system the explicit expression of eq (5.17) is given by

$$x(t) = \frac{1}{m\omega_d} \int_0^t f(\tau)e^{-\zeta\omega_n(t - \tau)} \sin\omega_d(t - \tau) \tag{5.18}$$

if the system is initially at rest. If this is not the case, the complementary function must be added, thus obtaining

$$x(t) = e^{-\zeta\omega_n t}(x_0 \cos\omega_d t + \frac{\dot{x}_0 + \zeta\omega_n x_0}{\omega_d} \sin\omega_d t)$$

$$+ \frac{1}{m\omega_d} \int_0^t f(\tau)e^{-\zeta\omega_n(t - \tau)} \sin\omega_d(t - \tau)d\tau \tag{5.19}$$

It should be noted that all linear systems can be completely characterized by their impulse response function (functions if there is more than one input and more than one output) and furthermore, the response to any input is given by the input function's convolution with the system's impulse response function. The importance of these concepts deserves a few comments and observations.

The nature of the convolution integral can be 'visualized' by considering that its evaluation is performed through multiplication of $f(\tau)$ by each incremental shift in $h(t - \tau)$. As the present time t varies the impulse response $h(t - \tau)$ scans the excitation function, producing a weighted sum of past inputs up to the present instant; so, in terms of the superposition principle the system's response $x(t)$ may be interpreted as being the weighted superposition of past input $f(\tau)$ values 'weighted' or multiplied by $h(t - \tau)$. In other words, to find the response $x(t_0)$ for some $t = t_0$, we form the function $h(-\tau)$ and we shift it to $h(t_0 - \tau)$; the area of the product $f(\tau)h(t_0 - \tau)$ yields $x(t_0)$.

In the foregoing discussion we have assumed that the force $f(t)$ starts at $t = 0$; since this may not be case and $f(t)$ can extend to negative values of t, we can write eq (5.17) in the more general form

$$x(t) = \int_{-\infty}^{t} f(\tau)h(t - \tau)d\tau \tag{5.20}$$

In order for a constant-parameter linear system to be **physically realizable** (causal), it is necessary that it responds only to past inputs; this requires

$$h(t - \tau) = 0 \qquad \tau > t \tag{5.21}$$

In fact, if we recall that $h(t - \tau)$ is the response to an impulse applied at time $t = \tau$ (i.e. $t - \tau = 0$), for $t - \tau < 0$ (i.e. $\tau > t$) there is no response because no impulse has been applied. This justifies eq (5.21) and allows us to extend the upper limit of integration in eq (5.20) from $\tau = t$ to $\tau = +\infty$ without changing the result; we can then write

$$x(t) = \int_{-\infty}^{+\infty} f(\tau)h(t - \tau)d\tau \tag{5.22}$$

We note in passing that a physical system is always causal if the independent variable is the real time t; however, not all physical systems are causal (for example, if the independent variable is a space variable).

Consider now, in eq (5.20), the change of variable obtained by defining $\lambda = t - \tau$, where λ can be interpreted as the time delay between the occurrence of the input and the instant when its result is calculated; we get

$$x(t) = \int_{-\infty}^{t} f(\tau)h(t - \tau)d\tau = -\int_{+\infty}^{0} f(t - \lambda)h(\lambda)d\lambda$$

where the minus sign in the second integral comes from the fact that $d\tau = -d\lambda$. Changing over the limits of integration to dispense with the minus sign we get

$$x(t) = \int_{-\infty}^{t} f(\tau)h(t - \tau)d\tau = \int_{0}^{+\infty} f(t - \lambda)h(\lambda)d\lambda \tag{5.23}$$

which is an alternative form of the convolution integral.

Another alternative form can be obtained either by putting $\lambda = t - \tau$ in eq (5.20) or by noting – in the second integral of eq (5.23) – that $h(\lambda) = 0$ for $\lambda < 0$, since there is no response before the impulse occurs: this yields

$$x(t) = \int_{-\infty}^{+\infty} f(t - \lambda)h(\lambda)d\lambda = \int_{-\infty}^{+\infty} f(\tau)h(t - \tau)d\tau \equiv f(t) * h(t) \qquad (5.24)$$

The equality between the two integrals is obvious and in the last expression we have used the common symbol * to indicate the convolution between two functions. Equation (5.24) is the most general form of the convolution integral for our purposes.

An analogous line of reasoning, starting from eq (5.17) leads to

$$x(t) = \int_0^t f(t - \lambda)h(\lambda)d\lambda = \int_0^t f(\tau)h(t - \tau)d\tau \qquad (5.25)$$

which, with eq (5.24), shows the symmetry of the convolution integral in the excitation $f(t)$ and the impulse response $h(t)$. Simplicity suggests that the simpler of the two functions be shifted in performing the calculations.

A constant-parameter linear system is **stable** if any possible bounded input produces a bounded output. Since

$$|x(t)| = \left| \int_{-\infty}^{+\infty} f(\tau)h(t - \tau)d\tau \right| \leqslant \int_{-\infty}^{+\infty} |f(\tau)|\,|h(t - \tau)|\,d\tau \qquad (5.26)$$

if the input is bounded, i.e. there exists some finite constant K such that

$$|f(t)| \leqslant K \qquad \text{for all } t \qquad (5.27)$$

then

$$|x(t)| \leqslant K \int_{-\infty}^{+\infty} |h(\tau)|\,d\tau \qquad (5.28)$$

It follows that if the impulse response function is absolutely integrable, i.e.

$$\int_{-\infty}^{+\infty} |h(t)|\,dt < \infty \qquad (5.29)$$

then the output will be bounded and the system is stable. Note that $x(t)$ does not need to satisfy

$$\int_{-\infty}^{+\infty} |x(t)| \, dt < \infty \tag{5.30}$$

which is necessary in the classical Fourier transform (frequency response) approach.

A final comment to show the important property of **frequency preservation** in linear systems. Let $h(t)$ be the impulse response of a constant-parameter linear system; for an arbitrary, well behaved input $f(t)$ the nth derivative of the output $x(t)$ is given by

$$\frac{d^n x(t)}{dt^n} = \frac{d^n}{dt^n} \int_{-\infty}^{+\infty} h(\lambda) f(t - \lambda) d\lambda = \int_{-\infty}^{+\infty} h(\lambda) \frac{d^n f(t - \lambda)}{dt^n} \, d\lambda \tag{5.31}$$

If we assume a sinusoidal input of the general form $f(t) = f_0 \sin(\omega t - \theta)$, we obtain, for example, for its second derivative $d^2 f(t)/dt^2 = -\omega^2 f(t)$. Inserting this result in eq (5.31) we get

$$\frac{d^2 x(t)}{dt^2} = -\omega^2 \int_{-\infty}^{+\infty} h(\lambda) f(t - \lambda) d\lambda = -\omega^2 x(t) \tag{5.32}$$

so that $x(t)$ must be sinusoidal with the same frequency as $f(t)$. It follows that a constant-parameter linear system cannot cause any frequency translation, but can only modify the amplitude and phase of an applied input.

Example 5.2.   Let us now determine the response of a damped system to the **step function** given by eq (5.33), which is often called the Heaviside function when $f_0 = 1$. In real situations this could model some kind of machine operation or a car running over a surface that changes level abruptly (i.e. a curb).

$$f(t) = \begin{cases} f_0 & t \geqslant 0 \\ 0 & t < 0 \end{cases} \tag{5.33}$$

For $t \geqslant 0$, assuming that the system starts from rest, the response can be written as the convolution integral

$$x(t) = \frac{f_0}{m\omega_d} \int_0^t e^{-\zeta \omega_n(t - \tau)} \sin \omega_d(t - \tau) d\tau \tag{5.34}$$

which can be easily calculated if we remember that

$$\int e^{at} \sin bt \, dt = \frac{e^{at}}{a^2 + b^2} (a \sin bt - b \cos bt)$$

Since $a = -\zeta\omega_n$, $b = \omega_d$ and $a^2 + b^2 = \omega_n^2$; after some manipulations we get

$$x(t) = \frac{f_0}{k}\left[1 - e^{-\zeta\omega_n t}\left(\frac{\zeta\omega_n}{\omega_d}\sin\omega_d t + \cos\omega_d t\right)\right]$$

$$= \frac{f_0}{k} - \frac{f_0 e^{-\zeta\omega_n t}}{k\sqrt{1 - \zeta^2}}\cos(\omega_d t - \theta) \qquad (5.35)$$

where in the second expression

$$\theta = \arctan\left(\frac{\zeta}{\sqrt{1 - \zeta^2}}\right) \qquad (5.36)$$

It is worth noting how the same result can be obtained by writing (for $t \geq 0$) the general solution of the equation of motion. This is

$$x(t) = \frac{f_0}{k} + e^{-\zeta\omega_n t}(D\cos\omega_d t + E\sin\omega_d t) \qquad (5.37)$$

The initial conditions $x(0) = 0$ and $\dot{x}(0) = 0$ give for the constants

$$D = -\frac{f_0}{k}$$

$$E = -\frac{\zeta\omega_n f_0}{\omega_d k}$$

which, after substitution in eq (5.37) and a little algebra, give exactly the solution of eq (5.35). Figure 5.3 is a graph of eq (5.35) with: $f_0 = 50$ N, $k = 1000$ N/m, $\zeta = 0.1$ and $\omega_n = 3.14$ rad/s.

For an undamped system eq (5.37) becomes

$$x(t) = \frac{f_0}{k}(1 - \cos\omega_n t) \qquad (5.38)$$

The response to a unit step input (sometimes called **indicial response**) and the unit step function (Heaviside) itself are also very important in linear vibration theory and are often indicated with a symbol of their own: as we did in Chapter 2, we will write $\theta(t)$ for the Heaviside function applied at $t = 0$ while we will use the symbol $s(t)$ for the response to a Heaviside input. We can write symbolically eq (4.13) for our SDOF linear system as

$$D[x(t)] = f(t) \qquad (5.39)$$

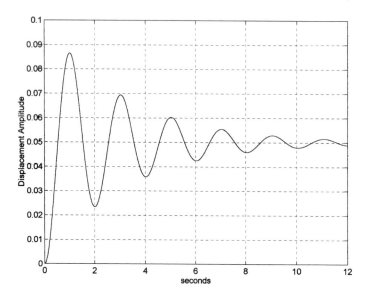

Fig. 5.3 Response to step function.

where D is the linear differential operator

$$D \equiv m \frac{d^2}{dt^2} + c \frac{d}{dt} + k \tag{5.40}$$

Following this notation, we can then write the symbolic relations

$$D[h(t)] = \delta(t)$$
$$D[s(t)] = \theta(t) \tag{5.41}$$

and since – in the sense of distributions (Chapter 2) – it can be shown that

$$\theta(t) = \int_0^t \delta(\lambda) d\lambda \qquad \text{or} \qquad \frac{d\theta(t)}{dt} = \delta(t)$$

it is left to the reader to determine the following important relations between the impulse response and the Heaviside response

$$s(t) = \int_0^t h(\lambda) d\lambda \qquad \text{or} \qquad \frac{ds(t)}{dt} = h(t) \tag{5.42}$$

which can be verified in eqs (5.7a), (5.35) and (5.38) (see also eq (5.34)).

Two more examples will now show further applications of what has been said up to this point. Both examples refer to an undamped SDOF system, since this system is often considered – for comparison purposes – in the analysis of **shock loading**. In these cases, the excitation is short compared to the natural period of the structure and as a result, the response is not significantly influenced by the presence of damping. We point out that the response of structures to shock is vital in design, particularly preliminary design, to select the system's parameters in a manner that limits within a specified range a certain response quantity – e.g. maximum response amplitude, maximum stress etc. – in order to prevent undesired vibration or damage of the structure. The concept of **shock** (or **response**) **spectrum** has been devised to deal with these problems and, due to its importance, it will be considered separately in the following section.

Example 5.3. The first example is the response to a rectangular pulse of duration t_1 and amplitude f_0. This problem has already been considered in Example 5.1 (eqs (5.10) and (5.11)), but now we proceed from eq (5.11) by substituting the appropriate initial conditions (5.12) to obtain an explicit expression for the 'free vibration era' $t > t_1$. The substitution gives

$$x(t) = \frac{f_0}{k} (1 - \cos \omega_n t_1) \cos \omega_n(t - t_1) + \frac{f_0}{k} \sin \omega_n t_1 \sin \omega_n(t - t_1)$$

which, after some manipulations, yields the response for $t > t_1$

$$x(t) = \frac{f_0}{k} [\cos \omega_n(t - t_1) - \cos \omega_n t] \tag{5.43}$$

Alternatively, the rectangular pulse $f(t)$ can be seen as the superposition of two step functions, i.e. $f(t) = f_1(t) + f_2(t)$, where $f_1(t)$ is given by eq (5.8), $f_2(t)$ is a step of magnitude $-f_0$ applied at $t = t_1$ and the response can be determined accordingly as a superposition of the two responses. Needless to say, the results are again eq (5.10) for the 'forced vibration era' and eq (5.43) for the 'free vibration era'. The reader is invited to draw a graph of the response for different values of t_1.

Example 5.4. In this second example we consider the response of an undamped system to a half-sine shock excitation, that is, to an excitation function of the form

$$f(t) = \begin{cases} f_0 \sin \omega t & 0 < t \leqslant t_1 \\ 0 & t > t_1 \end{cases} \tag{5.44}$$

During the 'forced vibration era' $(0 \leqslant t \leqslant t_1)$ the exciting function is a sinusoid with circular frequency $\omega = \pi/t_1$ rad/s. In order to calculate the response, we notice that $f(t)$ can be written in phasor form as $f_0 e^{i\omega t}$ if we agree to take the imaginary part as described in Chapter 1 (recall that the form of the exponential and the choice of the real or imaginary part is a matter of convenience, it is only important to be consistent). The response is then given by

$$x(t) = \text{Im} \left[\frac{f_0}{m\omega_n} \int_0^t e^{i\omega(t - \tau)} \sin \omega_n \tau d\tau \right] = \text{Im} \left[\frac{f_0 e^{i\omega t}}{m\omega_n} \int_0^t e^{-i\omega\tau} \sin \omega_n \tau d\tau \right]$$

and the integral can be easily solved using again

$$\int e^{at} \sin bt dt = \frac{e^{at}}{a^2 + b^2} (a \sin bt - b \cos bt)$$

where now $a = -i\omega$, $b = \omega_n$ and $a^2 + b^2 = \omega_n^2 - \omega^2 = \omega_n^2(1 - \beta^2)$. The calculation of the definite integral in brackets yields

$$\frac{f_0 e^{i\omega t}}{k\omega_n(1 - \beta^2)} [e^{-i\omega t}(-i\omega \sin \omega_n t - \omega_n \cos \omega_n t) + \omega_n]$$

$$= \frac{f_0}{k\omega_n(1 - \beta^2)} (-i\omega \sin \omega_n t - \omega_n \cos \omega_n t) + \frac{f_0 e^{i\omega t}}{k(1 - \beta^2)}$$

Taking the imaginary part of the expression above we get the system's response for $0 \leqslant t \leqslant t_1$, which is

$$x(t; \ t \leqslant t_1) = \frac{f_0}{k(1 - \beta^2)} (\sin \omega t - \beta \sin \omega_n t) \tag{5.45}$$

The response in the 'free vibration era' $(t > t_1)$ is obtained from eq (4.8) with initial conditions that are determined by the state of the system at $t = t_1$: this is

$$x(t; \ t > t_1) = x(t_1) \cos \omega_n(t - t_1) + \frac{\dot{x}(t_1)}{\omega_n} \sin \omega_n(t - t_1) \tag{5.46}$$

where now

$$x(t_1) = \frac{f_0}{k(1 - \beta^2)} (\sin \omega t_1 - \beta \sin \omega_n t_1)$$

$$\dot{x}(t_1) = \frac{f_0}{k(1 - \beta^2)} (\omega \cos \omega t_1 - \beta \omega_n \cos \omega_n t_1)$$

$$\tag{5.47}$$

Substitution of eqs (5.47) into eq (5.46) – considering that $\omega = \pi/t_1$ – yields

$$x(t;\ t > t_1) = \frac{f_0\beta}{k(\beta^2 - 1)}\ [\sin \omega_n t + \sin \omega_n(t - t_1)] \tag{5.48}$$

Figures 5.4(a) and 5.4(b) illustrate how the same undamped SDOF system ($k = 1.5$ N/m, $\omega_n = 3.14$ rad/s) responds to two different half-sine excitations.

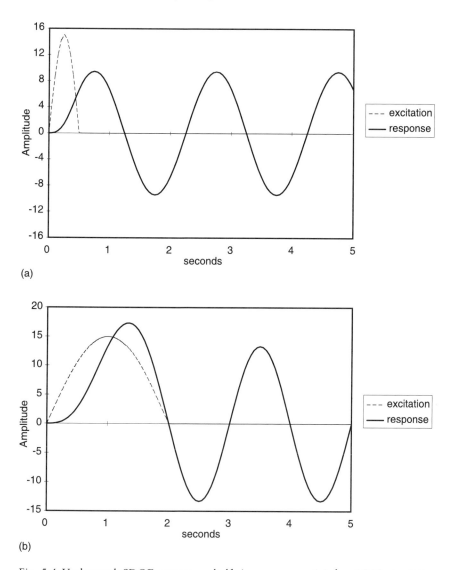

(a)

(b)

Fig. 5.4 Undamped SDOF system – half-sine response. (a) $f_0 = 15$ N, $t = 0.5$ s, $\omega = 6.28$ rad/s, $\beta = 2$. (b) $f_0 = 15$ N, $t = 2.0$ s, $\omega = 1.57$ rad/s, $\beta = 0.5$.

Both the excitation and the response are shown in the two graphs even if the measurement units are obviously different (newtons for the excitation function and metres for the response displacement function).

5.2.1   Shock response

We defined a shock as a sudden application of a force (or other form of excitation) that results in a transient response of a system. Roughly speaking, a shock takes place in a short time (t_1) compared with the natural period of the system, i.e. $t_1/T < 1$, and the vibration severity that it causes can be categorized by analysing, as a standard reference, the maximum value of the response of an undamped SDOF system. The whole time history of the response is not considered for this purpose, and significance is given to its maximum value plotted as a function of t_1/T or, alternatively, as a function of ω_n/ω when the quantity ω (frequency of the exciting pulse) can be defined.

Such curves, extensively used in engineering practice, are called **response** or **shock spectra**. Several kinds of maxima are important, but in general the so-called **maximax response** is considered, which is the maximum of the response attained at any time following the onset of the shock pulse.

Mathematically speaking, given an exciting force $f(t)$ of duration t_1, we want to determine the quantity

$$x_{max}(t_1/T) = \frac{1}{m\omega_n} \left| \int_0^t f(\tau) h(t - \tau) d\tau \right|_{max} \tag{5.49}$$

Let us consider a rectangular pulse of amplitude f_0 and duration t_1. The response to such excitation is given by eqs (5.10) and (5.43) in the 'forced' and 'free' vibration era, respectively. The maximum value of response will be attained in either the first or second era, depending on the value of t_1: when $t < t_1$, equating the derivative of eq (5.10) to zero gives the condition $\sin \omega_n t = 0$, which is verified at a time t_m given by $\omega_n t_m = n\pi$, provided that

$$t_m = \frac{n\pi}{\omega_n} < t_1$$

$$\tag{5.50}$$

$$\frac{t_1}{T} > \frac{1}{2}$$

where n is an integer and we have taken $n = 1$ because we are interested in the first maximum and we know that $\omega_n = 2\pi/T$. Substitution of t_m in eq (5.10) gives the maximum value

$$x_{max} = \frac{2f_0}{k} \qquad (t < t_1) \tag{5.51}$$

and this result can be stated by saying that the maximum value of the ratio of dynamic response to static deformation (**maximum dynamic magnification factor**) is equal to 2.

In the free vibration era the response is given by eq (5.43). Equating its derivative to zero gives the condition

$$\sin \omega_n(t - t_1) = \sin \omega_n t$$

which – considering again the first maximum – is verified at a time t_m given by

$$\omega_n(t_m - t_1) = \pi - \omega_n t_m$$

i.e.

$$t_m = \frac{T}{4} + \frac{t_1}{2} \tag{5.52}$$

provided that $t_m > t_1$, i.e. $t_1/T < 1/2$. Substitution of eq (5.52) in eq (5.43) yields after some manipulation

$$x_{max} = \frac{2f_0}{k} \sin\left(\frac{\pi t_1}{T}\right) \qquad (t > t_1) \tag{5.53}$$

The response spectrum for a rectangular pulse excitation is then obtained by combining eqs (5.51) and (5.53) and is shown in Fig. 5.5.

We turn now to the half-sine excitation considered in Example 5.4. The response to this kind of pulse is given by eqs (5.45) and (5.48) for the forced and free vibration era, respectively and, once again, the maximum value can be attained for $0 \leqslant t < t_1$ or $t > t_1$, depending on the value of t_1. The two cases are illustrated in Figs 5.4(a) and 5.4(b).

Let us suppose that the maximum response occurs in the forced vibration era; equating the derivative of eq (5.45) to zero gives

$$\cos \omega t = \cos \omega_n t$$

which is satisfied at a time t_m determined by

$$\omega t_m = 2\pi n \pm \omega_n t_m \qquad n = 0, 1, 2, 3, \cdots \tag{5.54}$$

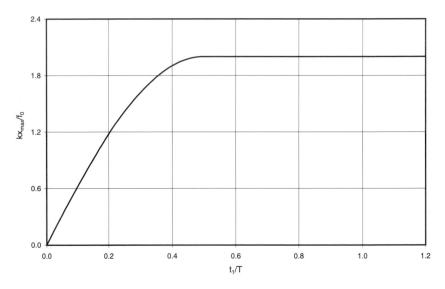

Fig. 5.5 Rectangular pulse – shock spectrum.

The first maximum is attained by considering the minus sign and $n = 1$ in eq (5.54) and we get

$$t_m = \frac{2\pi}{\omega + \omega_n} \qquad (5.55)$$

provided that $t_m < t_1$, which means $t_1/T > 1/2$ (or $\beta < 1$) if we remember that $t_1 = \pi/\omega$, $T = 2\pi/\omega_n$ and hence $t_1/T = 1/(2\beta)$. Substitution of eq (5.55) in eq (5.45) yields after some manipulation

$$x_m = \frac{f_0}{k(1 - \beta)} \sin\left(\frac{2\pi\beta}{1 + \beta}\right) \qquad (t < t_1) \qquad (5.56a)$$

or, defining $\eta \equiv t_1/T$

$$x_m = \frac{2f_0\eta}{k(2\eta - 1)} \sin\left(\frac{2\pi}{1 + 2\eta}\right) \qquad (t < t_1) \qquad (5.56b)$$

So, for example, if $\beta = 3/4$ it follows $t_1/T = 2/3$ and the maximum value of the dynamic magnification factor D is given by

$$D_{max} \equiv \left(\frac{x}{f_0/k}\right)_{max} = 1.7355$$

If the maximum occurs for $t > t_1$, equating the derivative of eq (5.48) to zero leads to

$$\cos \omega_n t = -\cos \omega_n(t - t_1)$$

which is satisfied at the time t_m given by

$$\omega_n t_m = n\pi \pm \omega_n(t_m - t_1) \tag{5.57a}$$

$$t_m = \frac{T}{4} + \frac{t_1}{2} \tag{5.57b}$$

provided that $t_m > t_1$, i.e. $t_1/T < 1/2$ (or $\beta > 1$) and we have considered the first maximum ($n = 1$ and the minus sign) passing from eq (5.57a) to eq (5.57b).

Substitution of the explicit expression for t_m in eq (5.48) yields for the maximum value of displacement

$$x_{max} = \frac{2f_0\beta}{k(\beta^2 - 1)} \cos\left(\frac{\pi}{2\beta}\right) \qquad (t > t_1) \tag{5.58a}$$

or, as a function of $t_1/T \equiv \eta$

$$x_{max} = \frac{4f_0\eta}{k(1 - 4\eta^2)} \cos(\pi\eta) \qquad (t > t_1) \tag{5.58b}$$

for example, if $\beta = 4/3$ ($t_1/T = 3/8$) the maximum value of D is found to be

$$D_{max} \equiv \left(\frac{x}{f_0/k}\right)_{max} = 1.312$$

The shock spectrum for a half-sine pulse excitation is shown in Fig. 5.6, where D_{max} is plotted as a function of t_1/T.

At resonance, i.e. when $\beta = 1$ (or $t_1/T = 1/2$), the expressions given by eqs (5.56a or 5.56b) and (5.58a or 5.58b) become indeterminate but the value of D_{max} can be obtained by calculating the limit for $\beta \to 1$ and using L'Hospital's rule in either one of the above equations. It is not difficult to show that

$$D_{max} = \lim_{\beta \to 1} \frac{1}{1 - \beta} \sin\left(\frac{2\pi\beta}{1 + \beta}\right)$$

$$= \lim_{\beta \to 1} \frac{2\beta}{\beta^2 - 1} \cos\left(\frac{\pi}{2\beta}\right) = \frac{\pi}{2} = 1.571$$

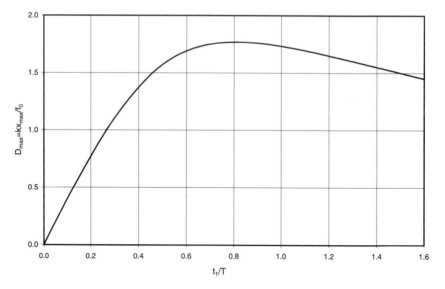

Fig. 5.6 Half-sine pulse – shock spectrum.

Following a similar line of reasoning, shock spectra for other impulsive loading conditions can be worked out (e.g. Harris [2], Jacobsen [3]).

The calculations become more and more laborious, especially if damping is taken into account, and other techniques are available (for example the Laplace transform, which will be considered later, and the plane-phase or other graphical methods), but often one has to make use of a computer for an extensive investigation of the problem.

However, the following general conclusions can be drawn for step-type and pulse-type excitations, where damping is of relatively less importance unless the system is highly damped.

The excitation – a function of time only – may be a force applied to the mass (as in the examples above) or a ground motion acting on the spring anchorage. The ground motion, in its turn, can be in the form of displacement, velocity or acceleration. All these cases are mathematically similar and the general equation

$$\frac{m}{k}\ddot{v} + v = \xi(t) \tag{5.59}$$

can be used, where v and ξ are the response and the excitation, respectively, in the appropriate form. For example, if $y(t)$ is the ground displacement with respect to a fixed frame of reference, the equation of motion is (4.55) with

$c = 0$, i.e.

$$m\ddot{x} + k(x - y) = 0$$

which, differentiating twice with respect to time, can be rearranged to

$$\frac{m}{k}\left(\frac{d^2\ddot{x}}{dt^2}\right) + \ddot{x} = \ddot{y}(t)$$

and can be treated as a second-order equation in $\ddot{x}$ when the ground acceleration $\ddot{y}(t)$ is given. Note the formal analogy with eq (5.59) where $\nu = \ddot{x}$ and $\xi = \ddot{y}$.

For **step-type excitations** x_{max} occurs after the step has risen to its full value (t_1 being the rise time) and the extreme values of D_{max} are 1 and 2. As the ratio t_1/T approaches zero, D_{max} approaches the upper limit of 2; when t_1/T approaches infinity (low rise time compared to the system's natural period), the step loses its character of dynamic excitation and D_{max} approaches the lower limit of 1, which means that x_{max} tends to the static value f_0/k.

The reader is invited to verify that for some shapes of the step rise time, D_{max} is equal to unity at certain finite values of t_1/T: for example, for a step with constant slope rise, $D_{max} = 1$ when $t_1/T = 1, 2, 3, \cdots$. Different values are obtained for different shapes of the slope, but the minimum value of t_1/T for which this occurs is 1. When $D_{max} = 1$, the amplitude of motion with respect to the **final** value of the excitation as a base (the so-called **residual response**) is zero and this fact is sometimes exploited in the practical design situation in order to achieve the smallest possible residual response.

For **pulse-type excitations**, when the ratio of pulse duration to system natural period t_1/T is less than $1/2$, the shape of certain types of pulses of equal area (equal impulse) is of secondary significance in determining the maxima of the system's response. If $t_1/T < 1/4$ the pulse shape has little significance in almost all cases; only when $t_1/T > 1/2$ must the pulse shape be considered carefully. Furthermore, the maximum response usually occurs for $1/2 < t_1/T < 1$ and D_{max} has values between 1.5 and 1.8. If the pulse has a vertical rise, D_{max} attains asymptotically the value of 2 and in the particular case of rectangular pulse $D_{max} = 2$ for $t_1/T \geqslant 1/2$.

Again, the reader is invited to verify that the residual response amplitude is zero for certain values of t_1/T and if the pulse has a vertical rise or a vertical decay (but not both) there are no zero values except at $t_1/T = 0$. In the case of a rectangular pulse, the residual response is zero for $t_1/T = 1, 2, 3, \cdots$. As an example, Fig. 5.7 shows the response of the undamped SDOF to a rectangular pulse of duration $t_1 = 4$ s in the case of $t_1/T = 3$.

Note that Fig. 5.7 illustrates a particular case; for different values of t_1/T the oscillation of the system continues after the end of the pulse with

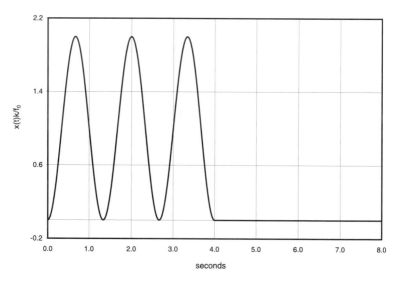

Fig. 5.7 Rectangular pulse response ($t_1/T = 3$; $t_1 = 4$ s).

amplitude characteristics that depend on the value of t_1/T considered in the specific case under study.

For several shapes of pulse the minimum value of t_1/T for zero residual response is 1 and occurs for an exciting rectangular pulse; for a sine pulse the lowest value for zero residual response is $t_1/T = 1.5$ and for a symmetrical triangular pulse is $t_1/T = 2$.

5.3   Frequency and *s*-domains. Fourier and Laplace transforms

Fourier analysis is a mathematical technique that deals with the addition of several harmonic oscillations to form a resultant and with the opposite problem. Any branch of physical sciences in which harmonically varying quantities play a part makes use of the theory of Fourier analysis at some stage of its theoretical development. The subject of engineering vibrations is no exception, and we have already given a review of the fundamental mathematical aspects in Chapter 2. We move on from there, recalling some of these fundamentals when needed in the course of our investigation. With respect to Chapter 2, small differences in notation will be noted in a few cases. They have been adopted to suit our present needs and, in any case, are irrelevant to the essence of the discussion as long as consistency is maintained from the beginning to the end of the specific argument being discussed.

5.3.1 Response to periodic excitation

A periodic signal is a particular type of deterministic (i.e. that can be expressed by an explicit mathematical relationship) signal that repeats itself in time every T seconds; T is called the **period** and for all values of t

$$f(t) = f(t \pm nT)$$

where $n = 1, 2, 3, \cdots$. Provided that the fluctuations in $f(t)$ are not too pathological, a periodic function can be represented by a convergent series of harmonic functions whose frequencies are integral multiples of a certain fundamental frequency $\omega_0 \equiv 2\pi/T$.

A periodic exciting function $f(t)$ (written in the form of a force for our present convenience) can be written as

$$f(t) = \sum_{n=0}^{\infty} (a_n \cos n\omega_0 t + b_n \sin n\omega_0 t) \tag{5.60a}$$

or, alternatively

$$f(t) = X_0 + \sum_{n=1}^{\infty} X_n \cos(n\omega_0 t - \theta_n)$$

$$= X_0 + \mathrm{Re} \left[\sum_{n=1}^{\infty} X_n e^{-i(n\omega_0 t - \theta_n)} \right] \tag{5.60b}$$

or, in complex notation

$$f(t) = \sum_{n=-\infty}^{+\infty} C_n e^{-in\omega_0 t} \tag{5.60c}$$

where

$$X_0 = a_0, \qquad X_n = \sqrt{a_n^2 + b_n^2} \qquad \tan \theta_n = \frac{b_n}{a_n} \qquad (n = 1, 2, 3, \cdots)$$

establish the relation between eqs (5.60a) and (5.60b), and

$$C_0 = a_0 \qquad C_n = (a_n + ib_n)/2 \qquad C_{-n} = (a_n - ib_n)/2$$

give the relation between eq (5.60a) and the complex form.

If all the Cs are real, then all the bs are zero, $f(t)$ is real and an even function of t, i.e. $f(-t) = f(t)$; if $C_{-n} = C_n^*$ (i.e. all as and bs are real, as in our case), then $f(t)$ is real but not necessarily symmetrical about $t = 0$. We point out that eq (5.60c) is not a phasor representation of the input signal (the second expression of eq (5.60b) is!), it is a different way of writing the periodic function $f(t)$ in its own right, where no convention of taking only the real or imaginary part is implied.

The conditions for the convergence of the series are extremely general and cover the majority of engineering situations; one important restriction to be kept in mind is that, when $f(t)$ is discontinuous, the series gives the average value of $f(t)$ at the discontinuity. The **Fourier coefficients** of the series are given by (eq (2.6b) or (2.10))

$$a_n = \frac{2}{T} \int_0^T f(t)\cos(n\omega_0 t)dt$$

$$b_n = \frac{2}{T} \int_0^T f(t)\sin(n\omega_0 t)dt$$

(5.61)

for the expression of eq (5.60a) and

$$C_n = \frac{1}{T} \int_0^T f(t)e^{in\omega_0 t}dt$$

(5.62)

for the complex representation of eq (5.60c).

The mean (over one period) and mean-square values of $f(t)$ and of its time derivative can be expressed in terms of the Fourier coefficients as (Parseval's theorem)

$$\langle f(t) \rangle = X_0 = a_0$$

$$\langle f^2(t) \rangle = X_0^2 + \frac{1}{2} \sum_{n=1}^{\infty} X_n^2 = \sum_{n=-\infty}^{\infty} |C_n|^2$$

(5.63)

and

$$\langle \dot{f}(t) \rangle = 0$$

$$\langle \dot{f}^2(t) \rangle = \frac{1}{2} \sum_{n=1}^{\infty} n^2 \omega_0^2 X_n^2 = \sum_{n=-\infty}^{\infty} n^2 \omega_0^2 |C_n|^2$$

(5.64)

Obviously, the periodic function could be a displacement and in this case eqs (5.64) would give the mean and mean square of velocity.

Under such an excitation, we have the equation of motion for our damped SDOF system

$$m\ddot{x} + c\dot{x} + kx = a_0 + \sum_{n=1}^{\infty} (a_n \cos n\omega_0 t + b_n \sin n\omega_0 t) \tag{5.65}$$

Owing to the principle of superposition, the steady-state response to each harmonic component can be calculated separately and the results added to obtain – following the notation of eq (4.40) – the particular solution $x_2(t)$. This is

$$x_2(t) = \frac{a_0}{k} + \sum_{n=1}^{\infty} \frac{(a_n/k)}{\sqrt{(1-n^2\beta^2)^2 + (2\zeta n\beta)^2}} \cos(n\omega_0 t - \phi_n)$$

$$+ \sum_{n=1}^{\infty} \frac{(b_n/k)}{\sqrt{(1-n^2\beta^2)^2 + (2\zeta n\beta)^2}} \sin(n\omega_0 t - \phi_n) \tag{5.66}$$

which can be recognized as a Fourier series itself where

$$\phi_n = \arctan\left(\frac{2\zeta n\beta}{1 - n^2\beta^2}\right) \tag{5.67}$$

and $\beta = \omega_0/\omega_n$.

Alternatively, if the excitation is written in the more compact form of eq (5.60c), x_2 too can be taken in the form of a series with summation from $-\infty$ to $+\infty$. Since we have determined in Chapter 4 (eqs (4.48) and (4.92)) that the steady-state response of our system to a sinusoidal load $fe^{-i\omega_0 t}$ is written

$$x_2(t) = fH(\omega_0)e^{-i\omega_0 t} \tag{5.68}$$

where $H(\omega_0)$ is the complex frequency response function (receptance in this case, see also subsequent sections) and is given by

$$H(\omega_0) = \frac{1}{k - m\omega_0^2 - ic\omega_0} = \frac{1/k}{1 - \beta^2 - 2i\zeta\beta} \tag{5.69}$$

the response to our periodic load can be written as

$$x(t) = \sum_{n=-\infty}^{+\infty} C_n H(n\omega_0)e^{-in\omega_0 t} = \sum_{n=-\infty}^{+\infty} \left(\frac{C_n/k}{1 - n^2\beta^2 - 2i\zeta n\beta}\right) e^{-in\omega_0 t} \tag{5.70}$$

from which it follows that

$$\langle x^2(t) \rangle = \sum_{n=-\infty}^{\infty} |H(n\omega_0)|^2 |C_n|^2 \qquad (5.71)$$

The complementary function $x_1(t)$ must be added to eq (5.66) or (5.70) in order to obtain the general solution $x(t) = x_1(t) + x_2(t)$. Again (Chapter 4), the complementary function dies out with time and contains two arbitrary constants that must be determined by the initial conditions. From the discussion above we see that if the frequency of one of the harmonics in the excitation is close to the system's natural frequency ω_n, its contribution to the response will be relatively large (especially for light damping) and a condition of resonance may exist when $n\omega_0 = \omega_n$ for some value of n (it is clear from the context, but we note that the subscript n in ω_n is for 'natural' frequency of the system and is **not** an integer $n = 1, 2, 3, \cdots$).

The undamped case can be easily obtained as a particular case of the equations above with $c = 0$.

5.3.2   *Transform methods I: preliminary remarks*

All the problems of transient response that we have considered up to the present can be solved by means of other methods by, so to speak, looking at them from a different angle. These techniques – which involve integral transforms – may seem only mathematical complications when applied to relatively simple cases but their usefulness and power become clear when more complicated problems of nonperiodic excitations are encountered. Nonperiodic signals cannot be represented in terms of a fundamental component plus a sequence of harmonics, but must be considered as a superposition of signals of all frequencies. In other words, the series must be generalized into an integral over all values of ω. In Chapter 2 we introduced the Fourier transform of a function $y(t)$ as (eq (2.16))

$$Y(\omega) = \mathcal{F}\{y(t)\} \equiv \frac{1}{2\pi} \int_{-\infty}^{+\infty} y(t) e^{-i\omega t} dt \qquad (5.72)$$

and the inverse Fourier transform (Fourier integral representation) as

$$y(t) = \mathcal{F}^{-1}[Y(\omega)] = \int_{-\infty}^{+\infty} Y(\omega) e^{+i\omega t} d\omega \qquad (5.73)$$

Equation (5.73) represents the synthesis of $y(t)$ from complex exponential harmonics while eq (5.72) represents analysis of $y(t)$ into its frequency components, each frequency having for its amplitude the magnitude $|Y(\omega)|$ of the complex quantity $Y(\omega)$ and for its phase the phase angle of $Y(\omega)$. If $y(t)$ is real then (since t is real) $Y(-\omega)$ must be the complex conjugate of $Y(\omega)$, i.e. $Y(-\omega) = Y^*(\omega)$. Obviously, eqs (5.72) and (5.73) have meaning only when the integrals converge.

As stated in Chapter 2, we are often interested in functions whose Fourier transforms do not exist and are zero for negative values of t (for example, the Heaviside function). There, we determined that the (one-sided) Laplace transform of a function $y(t)$ is obtained as (eq (2.36))

$$Y(s) = \mathscr{L}[y(t)] = \int_0^\infty y(t)e^{-st}dt \qquad (5.74)$$

Moreover, we also determined that the inverse transform is

$$y(t) = \mathscr{L}^{-1}[Y(s)] = \frac{1}{2\pi i} \int_{-i\infty+\varepsilon}^{i\infty+\varepsilon} Y(s)e^{st}ds \qquad (5.75)$$

and that this integral can be evaluated as a contour integral in the complex plane (the so-called Bromwich integral).

The utility of integral transforms lies in the fact that their application to linear differential equations yields algebraic expressions in the transformed variable. By treating the transformed problem, solutions may be found more easily. However, these solutions are in the transformed space and the problem of inverting these solutions remains. Obtaining the inverse transformation usually turns out to be the most formidable part of the entire problem. Luckily, extensive tables of Fourier and Laplace transforms are available in the literature (e.g. Erdélyi *et al.* [4], Thomson [5], Graff [6] and Inman [7]).

If now we turn our attention to the relationship between the input (excitation) and the output (response) of a linear SDOF system, we can summarize the situation as follows. By virtue of the superposition principle – and we refer here to the cases of our interest, i.e. linear, physically realizable and stable systems – frequency components in the frequency band from ω to $\omega + d\omega$ in the input (excitation) will correspond with components in the same frequency band in the output (response) signal. More specifically, we determined in Chapter 4 that a sinusoidal excitation at frequency ω produces a (steady-state) response at the same frequency. Further evidence of this has been obtained by the 'frequency preservation property' discussed in connection with the convolution integral (eq (5.32)) and in Section 5.3.1. In other words, the quantities that are affected – and change – in 'passing through the system' are the amplitude and the phase (with respect to the input), not the frequency. So, if our sinusoidal excitation is written in the form $f(t) = F(\omega)e^{i\omega t}$, where the function $F(\omega)$ provides the information on the input amplitude and phase at the frequency ω, then the system's response will be in the form $x(t) = X(\omega)e^{i\omega t}$, where $X(\omega)$ contains the necessary information on the amplitude and phase of the output at the same frequency. Hence, the effect of the system can be represented by means of a (complex) function $H(\omega)$ which (1) depends only on the system's characteristics and (2) provides the information on how the system itself affects (at the frequency ω) the amplitude and phase of the input signal. In mathematical terms we have

$$X(\omega)e^{i\omega t} = H(\omega)F(\omega)e^{i\omega t}$$

from which it follows that

$$X(\omega) = H(\omega)F(\omega) \tag{5.76a}$$

and $H(\omega)$ is called the system's 'frequency response function' (FRF). Equation (5.76a) is, as a matter of fact, the fundamental input–output relationship for linear systems in the frequency domain and is strictly related to the excitation-response relationship in the time domain obtained in section 5.2. The nature of this relation – which clearly has to do with Fourier transforms of the time signals – will be considered in a later section.

As far as the Laplace domain is concerned, it will be shown that the fundamental input–output relation in the s-domain is given by

$$X(s) = H(s)F(s) \tag{5.76b}$$

although less physical meaning can be attached to the Laplace variable s. The term **transfer function** (or 'generalized impedance', from electrical engineering terminology) is used for the function $H(s)$ and, incidentally, we note that the term 'transfer function' is also sometimes used for $H(\omega)$. Since $H(s)$ and $H(\omega)$ are different, we will not follow this loose terminology.

5.3.3   Transform methods [II]: applications

As useful applications of the preceding discussion, we consider in this section a number of problems which we have already solved by means of time-domain techniques.

Example 5.5.   The equation of motion of an undamped system, initially at rest, excited by a delta function at $t = 0$ is

$$m\ddot{x} + kx = \delta(t)$$

taking the Laplace transform of both sides yields

$$(ms^2 + k)X(s) = 1 \tag{5.77}$$

where $X(s) = \mathcal{L}[x(t)]$. Solving for $X(s)$ we have

$$X(s) = \frac{1/m}{s^2 + w_n^2} \tag{5.78}$$

where we recognize the symbolic notation of eq (5.76b) with

$$H(s) = \frac{1/m}{s^2 + w_n^2} \quad \text{and} \quad F(s) = 1$$

From any table of Laplace transforms one finds (a is a constant)

$$\mathcal{L}^{-1}\left[\frac{1}{s^2 + a^2}\right] = \frac{1}{a}\sin(at)$$

and we finally get

$$x(t) = \frac{1}{m w_n}\sin w_n t \tag{5.79}$$

in agreement with the first of eqs (5.7a) which gives the unit impulse response for an SDOF undamped system.

The introduction of viscous damping in our system would obviously lead to the second of eqs (5.7a) because the function

$$X(s) = \frac{1/m}{s^2 + w_n^2 + 2\zeta w_n s}$$

transforms back to

$$x(t) = \frac{1}{m w_d} e^{-\zeta w_n t}\sin w_d t$$

where $w_d = w_n\sqrt{1 - \zeta^2}$ and $\zeta < 1$.

Example 5.6. If the forcing function to our undamped SDOF system is a unit step Heaviside function $\theta(t)$, the transformed equation with zero initial conditions is

$$(ms^2 + k)X(s) = \frac{1}{s} \tag{5.80}$$

from which follows the indicial response

$$x(t) = \mathcal{L}^{-1}[X(s)] = \mathcal{L}^{-1}\left[\frac{1/m}{s(s^2 + \omega_n^2)}\right] = \frac{1}{k}(1 - \cos \omega_n t) \tag{5.81}$$

in agreement with eq (5.38).

Example 5.7. Let the exciting function be a unit rectangular impulse of duration t_1, i.e.

$$f(t) = \begin{cases} 1 & 0 < t \leqslant t_1 \\ 0 & t > t_1 \end{cases}$$

which can be written as

$$f(t) = \theta(t) - \theta(t - t_1) \tag{5.82}$$

The transformed equation of our undamped system is now

$$(ms^2 + k)X(s) = \frac{1 - e^{-t_1 s}}{s}$$

where the right-hand side is the Laplace transform of eq (5.82). Solving for $X(s)$ yields

$$X(s) = \frac{1/m}{s(s^2 + \omega_n^2)} - \left(\frac{1}{m}\right)\frac{e^{-t_1 s}}{s(s^2 + \omega_n^2)} \tag{5.83}$$

We already know the inverse transform of the first term on the right-hand side (eq (5.81)) and we concentrate our attention on the second term. This can be seen as the product

$$\left(\frac{e^{-t_1 s}}{s}\right)\left(\frac{1}{s^2 + \omega_n^2}\right)$$

and we know from Chapter 2 that the inverse transform of such a product is the convolution integral of the two inverse transformed functions, i.e.

$$\mathcal{L}^{-1}[G_1(s)G_2(s)] = \int_0^t g_1(t - \tau)g_2(\tau)d\tau \equiv g_1 * g_2 \tag{5.84}$$

where $G_1(s)$ and $G_2(s)$ are the transforms of $g_1(t)$ and $g_2(t)$, respectively. From a table of transforms we get

$$\mathcal{L}^{-1}\left[\frac{e^{-t_1 s}}{s}\right] = \begin{cases} 1 & t > t_1 > 0 \\ 0 & t < t_1 \end{cases}$$

so that the convolution integral is zero for $t < t_1$, leaving only the first term in eq (5.83) and

$$\frac{1}{\omega_n}\int_{t_1}^{t} \sin \omega_n(t - \tau)d\tau = \frac{1}{\omega_n^2}[1 - \cos \omega_n(t - t_1)]$$

for $t > t_1$. The inverse transformation of eq (5.83) finally yields

$$x(t) = \frac{1}{k}(1 - \cos \omega_n t) \qquad\qquad t < t_1$$

$$x(t) = \frac{1}{k}[\cos \omega_n(t - t_1) - \cos \omega_n t] \qquad t > t_1$$

which, aside from the constant f_0, are exactly eqs (5.10) and (5.43).

So far, we have not yet considered the possibility of obtaining directly the final solution satisfying given initial conditions. This is one advantage of solving linear differential equations with constant coefficients by the Laplace transform method. By standard methods one finds a general solution containing arbitrary constants, and further calculations for the values of the constants are needed to solve a particular problem. The Laplace transforms of derivatives given in Chapter 2 (eqs (2.39) and (2.40)) will now be used to clarify this point.

Let us consider the general equation of motion for a damped SDOF system

$$m\ddot{x} + c\dot{x} + kx = f(t) \tag{5.85}$$

with initial conditions $x(0) = x_0$ and $\dot{x}(0) = v_0$. The Laplace transformation of both sides gives

$$m(s^2 X - sx_0 - v_0) + c(sX - x_0) + kx = F(s) \tag{5.86}$$

where, as customary, we are using lower-case letters for functions in the time domain and capital letters for functions in the transformed domain.

Solving for $X(s)$ and rearranging leads to

$$X(s) = \frac{F(s)/m}{s^2 + \omega_n^2 + 2\zeta\omega_n s} + \frac{x_0(s + 2\zeta\omega_n)}{s^2 + \omega_n^2 + 2\zeta\omega_n s} + \frac{v_0}{s^2 + \omega_n^2 + 2\zeta\omega_n s} \tag{5.87}$$

The first term on the right-hand side (product of two functions of s) transforms back to the convolution integral

$$\frac{1}{m\omega_d} \int_0^t f(\tau)e^{-\zeta\omega_n(t-\tau)} \sin\omega_d(t-\tau)d\tau \qquad (5.88\text{a})$$

the second term transforms back to (see any list of Laplace transforms)

$$x_0 e^{-\zeta\omega_n t}\left(\cos\omega_d t + \frac{\zeta\omega_n}{\omega_d}\sin\omega_d t\right) \qquad (5.88\text{b})$$

and we have already considered the third term whose inverse transform is

$$\frac{v_0}{\omega_d}e^{-\zeta\omega_n t}\sin\omega_d t \qquad (5.88\text{c})$$

The sum of the three expressions (5.88a, b and c) finally gives the general response

$$x(t) = \frac{1}{m\omega_d}\int_0^t f(\tau)e^{-\zeta\omega_n(t-\tau)}\sin\omega_d(t-\tau)d\tau$$

$$+ e^{-\zeta\omega_n t}\left(x_0\cos\omega_d t + \frac{v_0 + \zeta\omega_n x_0}{\omega_d}\sin\omega_d t\right)$$

which is, as expected, eq (5.19) and shows explicitly how this method takes directly into account the initial conditions in the calculation of the response to external excitation.

5.4   Relationship between time-domain response and frequency-domain response

From the discussion and the examples of preceding sections it appears that both $h(t)$, the impulse response function (IRF), and the frequency response function (FRF) $H(\omega)$ (or the transfer function $H(s)$) completely define the dynamic characteristics of a linear system. This fact suggests that we should be able to derive one from the other and *vice versa*. The key connection between the two domains is established by the convolution theorem and by the Fourier (or Laplace) transform of the Dirac delta function. In Chapter 2 (eq (2.29a)) we determined that the Fourier transform of the convolution of two functions $g_1(t)$ and $g_2(t)$ – provided that $\int_{-\infty}^{+\infty} |g_1(t)g_2(t)|\, dt < \infty$ – is the product of the two transformed functions. With our definition of the Fourier transform, as a formula this statement reads

$$\mathcal{F}\{g_1(t) * g_2(t)\} = 2\pi\mathcal{F}\{g_1(t)\}\mathcal{F}\{g_2(t)\} \equiv 2\pi G_1(\omega)G_2(\omega)$$

and results, as we have seen, from an application of Fubini's theorem.

Now, since we know from Section 5.2 that the time-domain response $x(t)$ of a linear system is given by the convolution (Duhamel's integral) between the forcing function $f(t)$ and the system's IRF $h(t)$, i.e.

$$y(t) \atop x(t) = h(t) * f(t) \equiv \int_{-\infty}^{\infty} h(t - \tau)f(\tau)d\tau$$

we can Fourier transform both sides of this equation to get the input–output relationship in the frequency domain

$$\mathcal{F}\{x(t)\} = 2\pi\mathcal{F}\{h(t)\}\mathcal{F}\{f(t)\} \tag{5.89}$$

Equation (5.89) justifies eq (5.76a) from a more rigorous mathematical point of view. In fact the two equations (5.76a) and (5.89) are the same if we define

$$H(\omega) \equiv 2\pi\mathcal{F}\{h(t)\} = \int_{-\infty}^{\infty} h(t)e^{-i\omega t}dt \tag{5.90a}$$

and

$$F(\omega) \equiv \mathcal{F}[f(t)] = \frac{1}{2\pi} \int_{-\infty}^{\infty} f(t)e^{-i\omega t}dt$$

$$\tag{5.90b}$$

$$X(\omega) \equiv \mathcal{F}[x(t)] = \frac{1}{2\pi} \int_{-\infty}^{\infty} x(t)e^{-i\omega t}dt$$

In this light, note that the functions $F(\omega)$ and $X(\omega)$ are the Fourier transforms of the functions $f(t)$ and $x(t)$, respectively, but the FRF $H(\omega)$ (eq (5.90a)) differs slightly from the definition of the Fourier transform of the IRF $h(t)$ (there is no $1/(2\pi)$ multiplying factor). This is a consequence of our definition of the Fourier transform (eqs (2.15) and (2.16)) and of the fact that the fundamental input–output relationship for linear systems is almost always found in the form given by eq (5.76a). However, this is only a minor inconvenience since (Chapter 2) the position of the factor $1/(2\pi)$ is optional so long as it appears in either the Fourier transform equation or the inverse Fourier transform equation. Hence, the inverse transform equation corresponding to eq (5.90a) is

$$h(t) = \frac{1}{2\pi} \int_{-\infty}^{\infty} H(\omega)e^{i\omega t}d\omega \tag{5.91}$$

which, conforming to our definition of the inverse transform can be written $h(t) = (2\pi)^{-1}\mathscr{F}^{-1}\{H(\omega)\}$. No such inconvenience arises in the case of Laplace transforms because in this case, by virtue of the convolution theorem (eq (2.43)), the Laplace transform of the Duhamel integral leads to

$$X(s) = H(s)F(s) \equiv \mathscr{L}\{h(t)\}\mathscr{L}\{f(t)\} \tag{5.92}$$

where we defined $X(s)$ as the Laplace transform of the response $x(t)$.

With the above general developments in mind, we may now recall that, for a linear system, the function $h(t)$ represents the system's response to a delta input, i.e. $x(t) = h(t)$ when $f(t) = \delta(t)$. Consequently, since $\mathscr{F}\{\delta(t)\} = 1/(2\pi)$ (eq (2.74)), and $\mathscr{L}\{\delta(t)\} = 1$, eqs (5.89) and (5.92) give respectively

$$X(\omega) = H(\omega)$$
$$\tag{5.93}$$
$$X(s) = H(s)$$

which tell us that, in the case of δ-excitation, the Fourier (Laplace) transform of the system's response is precisely its frequency response (transfer) function, a circumstance which is also exploited in experimental practice.

With the definitions above, it is now not difficult to verify eq (5.76a) (and hence eq (5.93)) for the case of, say, a viscously damped SDOF system. In fact, in this case we know from the second of eqs (5.7a) that

$$h(t) = \frac{1}{m\omega_d}e^{-\zeta\omega_n t}\sin\omega_d t$$

so that, with the aid of a table of integrals from which we get

$$\int e^{ax}\sin bx\, dx = \frac{e^{ax}}{a^2 + b^2}(a\sin bx - b\cos bx)$$

we can calculate the term

$$H(\omega) = 2\pi\mathscr{F}\{h(t)\} = \frac{1}{m\omega_d}\int_0^\infty e^{-\zeta\omega_n t}e^{-i\omega t}\sin\omega_d t\, dt$$

by noting that, in our specific case $a = -(\zeta\omega_n + i\omega)$ and $b = \omega_d = \omega_n\sqrt{1 - \zeta^2}$. It is then left to the reader to show that the actual calculation leads to

$$H(\omega) = \frac{1}{m\omega_n^2 - m\omega^2 + 2im\zeta\omega\omega_n} = \frac{1}{k(1 - \beta^2 + 2i\zeta\beta)} \tag{5.94}$$

where, as usual, $\beta = \omega/\omega_n$. If we multiply this result by $F(\omega) = F\{\delta(t)\} = 1/(2\pi)$ we obtain explicitly the right-hand side of eq (5.76a) for our viscously damped SDOF system. Then, the left hand-side can be obtained by virtue of eq (5.5) and it can be determined immediately that, as expected,

$$X(\omega) = \mathscr{F}\{x(t)\} = \frac{1}{2\pi k}\left(\frac{1}{1 - \beta^2 + 2i\zeta\beta}\right)$$

In the light of these considerations we can write the response of a system to an input $f(t)$, whose Fourier and Laplace transforms are $F(\omega)$ and $F(s)$, as

$$x(t) = \mathscr{F}^{-1}\{H(\omega)F(\omega)\} = \int_{-\infty}^{+\infty} H(\omega)F(\omega)e^{+i\omega t}d\omega \qquad (5.95)$$

or, when more convenient, as the inverse Laplace transform of the product $H(s)F(s)$. Thus the following three equivalent definitions of the FRF $H(\omega)$ can be given:

1. $H(\omega)$ is 2π times the Fourier transform of $h(t)$.
2. For a sinusoidal input, i.e. $f(t) = e^{i\omega t}$, $H(\omega)$ is the coefficient of the resulting sinusoidal response $H(\omega)e^{i\omega t}$.
3. Provided that $F(\omega) \neq 0$ in the frequency range of interest, $H(\omega)$ equals the ratio $X(\omega)/F(\omega)$, where $X(\omega)$ is the Fourier transform of $x(t)$.

Figure 5.8. is a frequently found schematic representation of the fact that the dynamic characteristics of a 'single-input single-output' system are fully defined by either $h(t)$ or $H(\omega)$

A final note of interest concerns two systems connected in cascade as shown in Fig. 5.9. Denoting by $h(t)$ and $H(\omega)$ the IRFs and FRFs of the combined

f(t)	h(t)	x(t)
input		output
F(ω)	H(ω)	X(ω)

Fig. 5.8 Symbolic representation of a linear system.

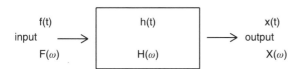

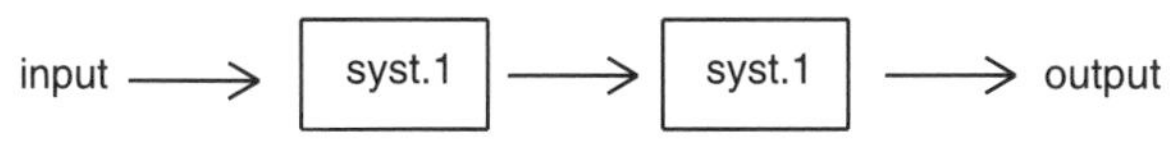

Fig. 5.9 Systems in cascade.

system, by $h_1(t)$, $h_2(t)$ and $H_1(\omega)$ and $H_2(\omega)$ the relevant functions of the two subsystems we have

$$h(t) = h_1(t) * h_2(t) \tag{5.96a}$$

and

$$H(\omega) = H_1(\omega)H_2(\omega) \tag{5.96b}$$

5.5   Distributed parameters: generalized SDOF systems

Up to the present we have always considered the simplest type of SDOF system, i.e. a system where all the parameters of interest – mass, damping and elasticity – are represented by discrete localized elements. This is, of course, an idealized view. However, even when more complicated modelling is required for the case under study, we have to remember that the key aspect of SDOF systems is that **only one** generalized coordinate is sufficient to describe their motion; if this characteristic is maintained it can be shown that the equation of motion of our SDOF system, no matter how complex, can always be written in the form

$$m^* \ddot{z}(t) + c^* \dot{z}(t) + k^* z(t) = f^*(t) \tag{5.97}$$

where $z(t)$ is the single generalized coordinate and the symbols with asterisks (not to be confused with complex conjugation) represent generalized physical properties – the generalized parameters – of our system with respect to the coordinate $z(t)$. This latter statement means that a different choice of this coordinate leads to different values of the generalized parameters. The possibility of writing an equation such as (5.97) allows us to extend all the considerations that we have made so far to a broader class of systems: assemblages of rigid bodies with localized spring elements, systems with distributed mass, damping and elasticity (bars, plates, etc.) or combinations thereof can be analysed in this way once the relevant generalized parameters have been determined. Their values can be obtained, in general, from energy principles such as Hamilton's or the principle of virtual displacements (Chapter 3) but general standardized forms of these expressions can be given for practical use.

It is important to note that the SDOF behaviour of the system under investigation may sometimes correspond very closely to the real situation but, more often, is merely an assumption based on the consideration that only a single vibration pattern (or deflected shape in case of continuous systems) is developed. For example, a beam that deforms in flexure is, as a matter of fact, a system with an infinite number of degrees of freedom, but in certain circumstances a SDOF analysis can be good and accurate enough for all practical purposes.

For continuous systems in particular, the success of this procedure – which is a particular case of the **assumed modes method** (Chapter 9) – depends on the validity of the assumption above and on an appropriate choice of a **shape** (or **trial) function** $\Psi(x)$ which, in turn, depends on the physical characteristics of the system and also on the type of loading. Ideally, the selected shape function should satisfy all the boundary conditions of the problem. At a minimum, it should satisfy the essential boundary conditions.

In this light, a few definitions are given here and then some examples will clarify the considerations above.

- An **essential** (or **geometric**) **boundary condition** is a specified condition placed on displacements or slopes on the boundary of a physical body (e.g. at the clamped end of a cantilever bar both displacement and slope must be zero).
- A **natural** (or **force**) **boundary condition** is a condition on bending moment and shear (e.g. at the free end of a cantilever bar the bending moment and the shear force must be zero).
- A **comparison function** is a function of the space coordinate(s) satisfying **all** the boundary conditions – essential and natural – of the problem at hand, plus appropriate conditions of continuity up to an appropriate order.
- An **admissible function** is a function that satisfies the essential boundary conditions and is continuous with its derivatives up to an appropriate order. For a specific problem, the class of comparison functions is a subset of the class of admissible functions.
- An **assumed mode** (or **shape function**) is a comparison or an admissible admissible function used to approximate the deformation of a continuous body.

Example 5.8. Let us consider the rigid bar of length L metres and mass m kilograms shown in Fig. 5.10. The angle θ of rotation about the hinge, where $\theta = 0$ at static equilibrium, can be chosen as the generalized coordinate. The vertical displacement $z(t)$ of the tip of the bar can be another choice; for small oscillations $z = L\theta$ as shown in the figure.

Since the bar is considered rigid, the system has distributed mass (along the length of the bar), localized stiffness, damping (the spring and the dashpot) and is subjected to a localized force $f(t)$.

For small oscillations about the equilibrium, we assume the shape function $u(x, t) = x\theta(t)$. The virtual displacement is then given by $\delta u(x, t) = x\delta\theta$ and from the principle of virtual displacements it is not difficult to obtain

$$\left(\frac{mL^2}{3}\ddot{\theta} + cb^2\dot{\theta} + kL^2\theta - f(t)a \right) \delta\theta = 0$$

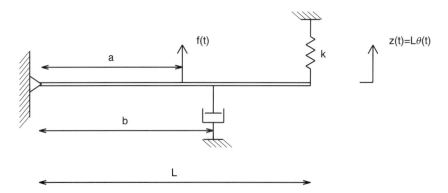

Fig. 5.10 Hinged rigid bar with localized stiffness (spring) and damping (dashpot).

since $\delta\theta \neq 0$ we get

$$\frac{mL^2}{3}\ddot{\theta} + cb^2\dot{\theta} + kL^2\theta = f(t)a \tag{5.98}$$

which is of the form (5.97) where the generalized parameters are

$$m^* = mL^2/3 \tag{5.99a}$$

$$c^* = cb^2 \tag{5.99b}$$

$$k^* = kL^2 \tag{5.99c}$$

$$f^*(t) = f(t)a \tag{5.99d}$$

The reader can easily verify that the choice of $z(t)$ as a generalized coordinate leads to different values for the generalized parameters.

Example 5.9.   As a second example we can consider the model of a beam of length L that deforms in flexure and is supported by an elastic foundation; we assume the following schematized beam and foundation characteristics:

- a mass $\hat{m}(x)$ per unit length and a flexural rigidity EI (beam);
- a spring constant of $\hat{k}(x)$ per unit length (foundation);
- a damping constant of $\hat{c}(x)$ per unit length (foundation).

In addition, the beam is subjected to a distributed transverse external force $\hat{f}(x,t)$ per unit length.

This is obviously a one-dimensional system where all the relevant parameters are distributed and can be represented for purpose of illustration by the drawing of Fig. 5.11. The direction of the x-axis is also shown.

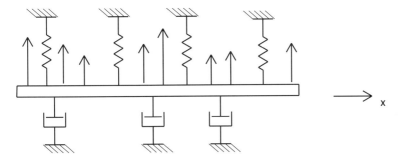

Fig. 5.11 Schematized beam on elastic foundation.

Our method assumes that only one mode is developed during the motion and represents the deflected shape $u(x, t)$ of the beam as the product

$$u(x, t) = \Psi(x)z(t) \tag{5.100}$$

where $\Psi(x)$ is the chosen admissible function and $z(t)$ is the unknown generalized coordinate. The principle of virtual displacements considers all the forces that do work and reads

$$-\delta V_{strain} + \delta W_{inertia} + \delta W_{damping} + \delta W_{spring} + \delta W_{ext} = 0 \tag{5.101}$$

where $-\delta V_{strain} = \delta W_{strain}$ from the definition of potential energy V. It will be shown in later chapters that the strain potential energy of a beam undergoing a transverse deflection $u(x, t)$ is given by

$$V = \frac{1}{2} \int_0^L EI(u'')^2 dx \tag{5.102}$$

where we indicated for simplicity of notation $u''(x, t) \equiv \partial^2 u/\partial x^2$. It follows from eq (5.102)

$$\delta V = \int_0^L EI(u'')^2 \delta u'' dx = z(t)\delta z \int_0^L EI[\Psi''(x)]^2 dx \tag{5.103}$$

since $u'' = z\Psi''$, $\delta u'' = \delta z\Psi''$ and $\Psi'' \equiv d^2\Psi/dx^2$.

For the inertia forces we get

$$\delta W_{inertia} = -\int_0^L \hat{m}\ddot{u}(x, t)\delta u dx = -\ddot{z}(t)\delta z \int_0^L \hat{m}[\Psi(x)]^2 dx \tag{5.104}$$

since $\ddot{u} = \ddot{z}\Psi$ and $\delta u = \delta z\Psi$.

Similarly, for damping forces we get

$$\delta W_{damping} = -\int_0^L \hat{c}\dot{u}(x, t)\delta u dx = -\dot{z}(t)\delta z \int_0^L \hat{c}[\Psi(x)]^2 dx \qquad (5.105)$$

and for the distributed spring and external forces

$$\delta W_{spring} = -\int_0^L \hat{k}u(x, t)\delta u dx = -z(t)\delta z \int_0^L \hat{k}[\Psi(x)]^2 dx \qquad (5.106)$$

$$\delta W_{ext} = \int_0^L \hat{f}(x, t)\delta u dx = \delta z \int_0^L \hat{f}(x, t)[\Psi(x)]dx \qquad (5.107)$$

Addition of the various terms leads to

$$\left[-z \int EI(\Psi'')^2 dx - \ddot{z} \int \hat{m}\Psi^2 dx - \dot{z} \int \hat{c}\Psi^2 dx - z \int \hat{k}\Psi^2 dx + \int \hat{f}\Psi dx \right]\delta z = 0$$

where all the integrals are taken between 0 and L. The virtual displacement δz is arbitrary and therefore can be cancelled out, leaving the equation of motion of our system in the form of eq (5.97) where, after rearranging, the generalized parameters are given by

$$m^* = \int_0^L \hat{m}\Psi^2 dx \qquad (5.108a)$$

$$c^* = \int_0^L \hat{c}\Psi^2 dx \qquad (5.108b)$$

$$k^* = \int_0^L EI(\Psi'')^2 dx + \int_0^L \hat{k}\Psi^2 dx \qquad (5.108c)$$

$$f^* = \int_0^L \hat{f}(x, t)\Psi dx \qquad (5.108d)$$

It is now evident how the particular choice of the shape function affects the generalized parameters of the system. Moreover, it is also clear that the application of Hamilton's principle leads the same expressions of eqs (5.108a–d).

The most general case of the type shown above consists of a system which is a combination of distributed and localized masses, springs, dampers and external forces. Again, the displacement is assumed of the form

$u(x, t) = \Psi(x)z(t)$ where $z(t)$ is the unknown generalized coordinate and the following standardized expressions for the generalized parameters can be given:

$$m^* = \int_0^L \hat{m}\Psi^2 dx + \sum_j M_j \Psi_j^2(x_j) + \sum_j I_{0j}[\Psi'(x_j)]^2 \qquad (5.109)$$

$$c^* = \int_0^L \hat{c}\Psi^2 dx + \sum_j c_j \Psi_j^2(x_j) \qquad (5.110)$$

$$k^* = \int_0^L EI[\Psi'']^2 dx + \int_0^L \hat{k}\Psi^2 dx + \sum_j k_j \Psi_j^2(x_j) \qquad (5.111)$$

$$f^* = \int_0^L \hat{f}\Psi dx + \sum_j F_j \Psi(x_j) \qquad (5.112)$$

where the integrals take into account the distributed parameters of the system under investigation (the symbols with a caret ($\wedge$) are intended per unit length) and the summations take into account the discrete elements. For example, the contribution to the generalized damping of a localized dashpot of constant c_1 at a distance x_1 from the origin of the axes is given by $c_1\Psi^2(x_1)$ in the summation of eq (5.110). With regard to the generalized mass, the second summation accounts for the rotation effects of localized rigid-body masses: I_{0j} is the mass moment of inertia of the jth mass and the first derivative of Ψ at the point x_j represents the rotation at that point. Referring back to the example of Fig. 5.10, it is not difficult to see that the generalized parameters of eqs (5.99a–d) are particular cases of eqs (5.109)–(5.112) where the assumed motion was $u(x, t) = x\theta(t)$ and the connection to the general case is given by

$$\Psi(x) = x$$

$$z(t) = \theta(t)$$

and the generalized mass accounts for translational motion of the centre of mass (at $x = L/2$) and the rotational motion around the centre of mass (mass moment of inertia $I_0 = mL^2/12$).

Care must be taken in the calculation of the generalized stiffness when 'destabilizing' forces are acting, since they add a further contribution to eq (5.111). Destabilizing forces may arise in different situations and may be of various nature. Gravity, for example, can be such a force in the case of an inverted pendulum, where a mass M is mounted on the tip of a light rigid bar as in Fig. 5.12.

The reader is invited to determine that the effective stiffness of the system can be written as $k - k_G$ where k_G depends on the weight Mg. When $k - k_G < 0$ the system becomes unstable.

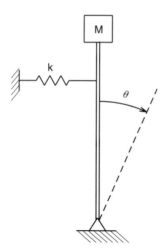

Fig. 5.12 Inverted pendulum.

For rotating systems, the centrifugal force can be destabilizing and, again, the generalized stiffness shows an additional term due to its effect.

However, for our purposes, compressive or tensile axial forces are the main concern. They tend to reduce (compressive forces) or increase (tensile forces) the stiffness of a structure and can play a significant role in determining the system's response under dynamic loading. Suppose that the beam of Fig 5.11 is subjected to an axial compressive loading $P(x)$. A deformed infinitesimal portion of the beam (exaggerated for purposes of illustration) is shown in Fig. 5.13, where it is assumed that the element remains dx in length as it rotates of α and that the axial force does not change direction as the structure moves.

As before, we choose $u(x,t) = \Psi(x)z(t)$, from which it follows that $\delta u = \Psi \delta z$. From the figure we get $dx = dx \cos \alpha + \varepsilon$, i.e. $\varepsilon = dx(1 - \cos \alpha)$. As the element dx passes through a virtual displacement, we have $\delta \varepsilon = dx \sin \alpha \delta \alpha$, and since for small angles $\sin \alpha \cong \alpha$, $\alpha \cong u'(x,t) = \Psi' z$ and

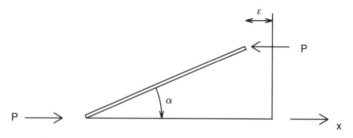

Fig. 5.13 Beam element under axial compressive loading.

$\delta\alpha = \Psi'\delta z$ we obtain

$$\delta\varepsilon = [\Psi']^2 z\delta z dx \qquad (5.113)$$

The virtual work done by $P(x)$ is finally obtained as

$$\delta W_{axial} = z\delta z \int_0^L P(x)[\Psi'(x)]^2 dx \qquad (5.114)$$

The term of eq (5.114) must now be added to eq (5.101) to give a generalized stiffness in the form $k^* - k_G^*$ where k^* is as before and

$$k_G^* = \int_0^L P(x)[\Psi'(x)]^2 dx \qquad (5.115)$$

In the case of a simple beam under the action of the axial compressive load only – if P does not depend on x and can be taken out of the integral – we can obtain the critical buckling load from the condition $k^* - k_G^* = 0$ as

$$P_{buckling} = \frac{\displaystyle\int EI[\Psi(x)]^2 dx}{\displaystyle\int [\Psi'(x)]dx} \qquad (5.116)$$

which is, obviously, relative to the assumed shape $\Psi(x)$.

As the simple examples above show, and as the word itself implies, destabilizing forces lead to stability problems. Stability is a broad subject in its own right and extends outside our scope. Some of its basic aspects will be considered when and if appropriate in the course of the book. For the moment, it suffices to say that stability is in general connected to situations in which the physical parameters (m, k or c or their generalized counterparts; one or more of them) become negative. The motion is not well-behaved in these cases and may diverge, i.e. increase without bounds, with or without oscillating. An example of diverging motion, even if no destabilizing forces are active, is the undamped oscillator excited at resonance.

One final word to point out that the assumed mode procedure can be extended to more complicated elements. For example, if the element undergoing flexure is two dimensional – i.e. a rectangular membrane or a plate – we can assume the displacement of the centre as the generalized coordinate $z(t)$ and write

$$u(x, y, t) = \Psi(x, y)z(t)$$

where, again, $\Psi(x, y)$ is a reasonable shape function consistent with the boundary conditions. However, a good choice of the shape function becomes more and more difficult as the number of dimensions increases and, as a consequence, the method may lead to unreliable results.

5.5.1 Rayleigh (energy) method and the improved Rayleigh method

Often, in practical situations, the quantity of main concern is the fundamental frequency of vibration of a given structural or mechanical system. When the system is complex, the exact determination of such a quantity may not be an easy task, long computation time and difficult calculations being involved in the process. The basis of a class of approximate methods to obtain the needed result is the so-called **Rayleigh's method**.

When an undamped elastic system vibrates at its fundamental frequency, each part of the system executes simple harmonic motion about its equilibrium position. The principle of conservation of energy applies for such a system and during the motion two extreme situations occur:

- All the energy is in the form of potential strain energy at maximum displacement.
- All the energy is in the form of kinetic energy when the system passes through its equilibrium position (maximum velocity).

Conservation of total energy requires that the potential energy at maximum displacement must equal the kinetic energy at maximum velocity, i.e.

$$E_{p, max} = E_{k, max} \tag{5.117}$$

The Rayleigh method calculates these maximum values, equates them and solves for frequency, since this quantity always appears in the kinetic energy term as a consequence of the simple harmonic motion of the system.

The undamped harmonic oscillator of Fig. 4.2 is the simplest example: the motion of such a system can be written as (Chapter 4)

$$x(t) = C \cos(\omega_n t - \theta)$$

the potential and kinetic energies are then

$$E_p \equiv \frac{1}{2} k x^2 = \frac{1}{2} k C^2 \cos^2(\omega_n t - \theta)$$

$$E_k \equiv \frac{1}{2} m \dot{x}^2 = \frac{1}{2} m \omega_n^2 C^2 \sin^2(\omega_n t - \theta)$$

Their maximum values are

$$E_{p,max} = \frac{1}{2} kC^2$$

$$E_{k,max} = \frac{1}{2} m\omega_n^2 C^2$$

Equation (5.117) follows because, individually, the two terms above must equal the total energy. Solving for ω_n we get the well-known result

$$\omega_n^2 = \frac{k}{m} \tag{5.118}$$

Another example can be the beam of Fig. 5.10 in free vibration; we make it conservative by considering $c = 0$ and $f(t) = 0$: no energy is removed from the system and no energy is fed into it. We assume as before $u(x,t) = x\theta$ and a harmonic motion of the generalized coordinate given by $\theta(t) = \theta_0 \sin \omega t$. The maximum values of the potential and kinetic energies are now

$$E_{p,max} = \frac{1}{2} kL^2\theta_0^2$$

$$E_{k,max} = \frac{1}{2} m\omega_n^2 \left(\frac{L}{2}\right)^2 \theta_0^2 + \frac{1}{2} I_0 \omega_n^2 \theta_0^2$$

$$= \frac{1}{2} m\omega_n^2\theta_0^2 \left(\frac{L}{2}\right)^2 + \frac{1}{2}\left(\frac{mL^2}{12}\right)\omega_n^2\theta_0^2 = \frac{1}{2}\omega_n^2\theta_0^2\frac{mL^2}{3}$$

equating the two energies and solving for frequency gives

$$\omega_n^2 = \frac{kL^2}{mL^2/3} = \frac{k^*}{m^*} \tag{5.119}$$

where the generalized stiffness and mass are the same as in eqs (5.99a) and (5.99c).

It is clear at this point that Rayleigh's method is strictly related to the assumed modes method (which is used to obtain the equation of motion), and the generality of the symbolic relation $\omega_n^2 = k^*/m^*$ – often referred to as the **Rayleigh quotient** – can be appreciated.

Further generalization will be given in later chapters when appropriate.

As a last example of SDOF system we consider a simple cantilever beam (i.e. a beam that is clamped at one end and free at the other) that undergoes flexural vibrations without energy loss during its motion. We assume the x-axis in the horizontal direction.

The assumption $u(x, t) = \Psi(x)z(t)$ characterizes the SDOF behaviour of this system and, again, $z(t) = Z_0 \sin \omega_n t$ characterizes the harmonic time dependence of this motion. The maximum potential and kinetic energies are

$$E_{p, max} = \frac{1}{2} \int_0^L EI(u'')^2 dx = \frac{Z_0^2}{2} \int_0^L EI[\Psi'']^2 dx \qquad (5.120)$$

$$E_{k, max} = \frac{1}{2} \int_0^L \hat{m}(\dot{u})^2 dx = \frac{Z_0^2 \omega_n^2}{2} \int_0^L \hat{m}[\Psi]^2 dx \qquad (5.121)$$

respectively, where EI is the flexural rigidity, $\hat{m}$ is the mass per unit length. Equating and solving for the frequency gives

$$\omega_n^2 = \frac{\displaystyle\int_0^L EI[\Psi'']^2 dx}{\displaystyle\int_0^L \hat{m}[\Psi]^2 dx} = \frac{k^*}{m^*} \qquad (5.122)$$

At this point it is interesting to test the effect of different choices for $\Psi(x)$ on the calculated frequency. Since the exact deformation shape can only be obtained by solving the equation of motion (but in this case the value of the fundamental frequency would be determined also) and therefore it is not known, we will try three trial functions

$$\Psi_1(x) = \left(\frac{x}{L}\right)^2 \qquad (5.123)$$

$$\Psi_2(x) = 1 - \cos\left(\frac{\pi x}{2L}\right) \qquad (5.124)$$

$$\Psi_3(x) = \frac{1}{3L^4}(x^4 - 4x^3 L + 6x^2 L^2) \qquad (5.125)$$

All of these functions give zero displacement at the clamped end ($x = 0$) and unit displacement at the free end ($x = L$) of the cantilever, the rationale being as follows: Ψ_1 is the simplest function one can think of in the given situation, Ψ_2 is a reasonable sinusoidal function and Ψ_3 is the static deflection curve of a cantilever beam under uniform load. In all of the cases above we can calculate the Rayleigh quotient and obtain an approximate value for the fundamental frequency of our system.

After some calculation that the reader is invited to try, we get the following results:

$$w_n^{(1)} = 4.472 \sqrt{\frac{EI}{\hat{m}L^4}} \tag{5.126}$$

$$w_n^{(2)} = 3.662 \sqrt{\frac{EI}{\hat{m}L^4}} \tag{5.127}$$

$$w_n^{(3)} = 3.530 \sqrt{\frac{EI}{\hat{m}L^4}} \tag{5.128}$$

respectively, where the superscript in parentheses refers to the chosen trial function. It will be determined in Chapter 8 that the exact fundamental frequency of the system we are considering is

$$w_n^{exact} = 3.516 \sqrt{\frac{EI}{\hat{m}L^4}} \tag{5.129}$$

The first consideration is that all of the trial functions produce a result that overestimates the exact value; this is a fundamental characteristic of the Rayleigh quotient and will be proven rigorously on a mathematical basis. On physical grounds, one can observe that additional constraints must be applied to the system if it is forced to vibrate in a shape that is different from its natural one; these constraints add stiffness to the system and hence an increase in frequency. Obviously, if the exact shape function (the lowest order eigenfunction) is used for $\Psi(x)$, the result is eq (5.129).

In addition, we note that the degree of approximation is rather crude (27% high) for the first function, but satisfactory for the other two. Qualitatively, we can say that the more the trial function resembles the true deflection shape, the more accurate the result will be. A closer examination requires the analysis of the boundary conditions. For the cantilever beam the following boundary conditions must be satisfied:

1. zero displacement and slope at the clamped end $(x = 0)$, i.e.

$$\begin{aligned} \Psi(0) &= 0 \\ \Psi'(0) &= 0 \end{aligned} \tag{5.130}$$

which we recognize as essential boundary conditions;

2.   zero bending moment and shear force at the free end $(x = L)$, i.e.

$$EI[\Psi''(L)] = 0$$
$$EI[\Psi'''(L)] = 0$$

(5.131)

which we recognize as natural boundary conditions.

All the trial functions satisfy the conditions of eqs (5.130) but only $\Psi_3(x)$ satisfies all four; $\Psi_1(x)$ does not satisfy the first of eqs (5.131) and $\Psi_2(x)$ does not satisfy the second of eqs (5.131).

In general, the deflection produced by a static load is a good candidate for $\Psi(x)$ because it automatically satisfies all the necessary boundary conditions and simplifies the calculation of the potential strain energy that can be obtained as the work done by the static load to produce the desired deflection. Only the function $\Psi(x)$ appears in this calculation and not its second derivative.

A common assumption is to choose the deflection shape that results from the application of the gravity load due to the mass of the structure. In this case, the direction of gravity must be chosen to match the probable deformation shape: in the analysis of the free vibrations of a vertical cantilever for example, the direction of gravity must be horizontal if we are interested in lateral motions of the structure. Obviously, this does not correspond to anything real, it is just a useful expedient.

There are two are the reasons that justify the assumption above:

1.   It is not necessary to spend much time in the choice of an assumed shape because any reasonable function compatible with the essential boundary conditions leads to acceptable results. It will be shown in Chapter 9 that the error on the calculated frequency is of the order of ε^2, if ε is the error of the assumed shape with respect to the exact one.
2.   The displacements in free vibration result, as a matter of fact, from the application of inertia forces and these forces depend, in their turn, on the mass distribution in our system.

This latter consideration, together with the serious disadvantage that the method does not allow us to estimate ε if the exact $\Psi(x)$ is not known, leads to the **improved Rayleigh method**, whose line of reasoning is as follows. Suppose that the true deflection $\Psi(x)$ is the same as the deflection produced by an external load $f(x)$. Then, deflection $u(x, t) = z\Psi(x)$ is produced by a load $zf(x)$ and the potential strain energy is the work done by this force to give the displacement $z\Psi$, i.e.

$$E_p = \frac{1}{2} \int_0^L (zf)(z\Psi)dx = \frac{1}{2} z^2 \int_0^L f(x)\Psi(x)dx$$

If, as before, $z(t) = Z_0 \sin \omega_n t$ we get

$$E_{p,max} = \frac{1}{2} Z_0^2 \int_0^L f(x)\Psi(x)dx$$

Equating to the maximum kinetic energy of eq (5.121) gives

$$f(x) = \omega_n^2 \hat{m}(x)\Psi(x) \tag{5.132}$$

which states that the load of eq (5.132), where we recognize inertia forces, produces the exact vibration shape. Equation (5.132) is true if $\Psi(x)$ is the true shape. Our assumed shape, which we call now $\Psi_0(x)$, is probably different from the true one and hence the load $\omega_n^2 \hat{m}(x)\Psi_0(x)$ will produce a shape different from $\Psi_0(x)$, let us call it $\bar{\Psi}_1(x)$. This function cannot be calculated because of the unknown ω_n^2 factor, but intuition suggests that it is likely to be a better approximation than Ψ_0 for the true deflected shape. Nevertheless, the function $\Psi_1(x) \equiv \bar{\Psi}_1/\omega_n^2$ (not to be confused with the function of eqs (5.123) which was a particular Ψ_0 for the cantilever problem) can be obtained from

$$\Psi_1(x) = \hat{m}(x)\Psi_0(x) \tag{5.133}$$

and we can write the maximum potential energy as

$$E_{p,max} = \frac{1}{2} Z_0^2 \int_0^L (\omega_n^2 \hat{m}\Psi_0)(\omega_n^2 \Psi_1)dx = \frac{1}{2} Z_0^2 \omega_n^4 \int_0^L \hat{m}\Psi_0\Psi_1 dx \tag{5.134}$$

Equating to the maximum kinetic energy gives

$$\omega_n^2 = \frac{\displaystyle\int_0^L \hat{m}[\Psi_0]^2 dx}{\displaystyle\int_0^L \hat{m}\Psi_0\Psi_1 dx} \tag{5.135}$$

At this point it seems reasonable to proceed further and use Ψ_1 also for the kinetic energy, so that

$$E_{k,max} = \frac{1}{2}\omega_n^2 Z_0^2 \int_0^L \hat{m}[\omega_n^2 \Psi_1]^2 dx = \frac{1}{2}\omega_n^6 Z_0^2 \int_0^L \hat{m}\Psi_1^2 dx \tag{5.136}$$

Equating to $E_{p,\,max}$ of eq (5.134) gives now

$$\omega_n^2 = \frac{\displaystyle\int_0^L \hat{m}\Psi_0\Psi_1 dx}{\displaystyle\int_0^L \hat{m}[\Psi_1]^2 dx} \tag{5.137}$$

Further iteration – that is, the use of Ψ_1 to obtain an even better approximate function Ψ_2 and use the latter to calculate the frequency – is generally not worth it.

We still do not have an estimate for the error ε, but indirectly we can have an idea by looking at the difference between the frequencies obtained from eq (5.122) and eq (5.135) or (5.137). If this difference is large, the function Ψ_0 is not a very good approximation for the true deflected shape and ε is large as well; if it is small, ε is small as well and Ψ_0 is a good choice.

Now, going back to the cantilever problem, we show an application of the improved Rayleigh method. We start from the function $\Psi_0 = (x/L)^2$ – which produced the result of eq (5.126) – and use the inertia forces $\hat{m}\Psi_0$ to calculate a better deflected shape. We know from beam theory that

$$EI \frac{d^4\Psi_1}{dx^4} = \hat{m}\frac{x^2}{L^2}$$

where Ψ_1 is the shape that results from the application of the forces on the right-hand side. By integrating four times and calculating the constants of integration from the boundary conditions of eqs (5.130) and (5.131) we get

$$\Psi_1(x) = \frac{\hat{m}}{EI}\left(\frac{1}{360L^2}x^6 - \frac{L}{18}x^3 + \frac{L^2}{8}x^2\right) \tag{5.138}$$

which can be used in eq. (5.135) to obtain

$$\omega_n = 3.530\sqrt{\frac{EI}{\hat{m}L^4}} \tag{5.139}$$

or in eq (5.137) to get

$$\omega_n = 3.5164\sqrt{\frac{EI}{\hat{m}L^4}} \tag{5.140}$$

These values are much better estimates of the exact frequency (given by eq (5.129)) and the large difference between the frequencies obtained from eq (5.126) and (5.139) – 26.7% with respect to the lower value – indirectly suggests that the assumed Ψ_0 was not a good approximation for the true deflected shape and hence the relative error on the frequency must have been large as well.

5.6   Summary and comments

Chapter 5 continues the discussion on SDOF systems. When the excitation is not a simple sinusoidal function, the response of the system can be obtained by means of various techniques, which obviously apply to harmonic excitation as well. The main distinction is between time-domain and frequency-domain techniques.

If the functions involved are analysed in the time domain, a fundamental concept is the **impulse response function** $h(t)$, whose convolution with the forcing exciting function provides the time response of our SDOF system. This particular form of convolution is known as **Duhamel integral**, which, in turn, can be visualized as a sum of the input excitation 'weighted' by an appropriately shifted form of the impulse response $h(t)$. As far as dynamic aspects are concerned, the function $h(t)$ is an inherent property of the system and characterizes it completely.

In this light, the response to the frequently encountered situation of loadings of short duration that may release a considerable amount of energy can be considered. One generally speaks of **transient** or **shock loading**, depending on a comparison between the time duration of the input load and the system's period of oscillation $2\pi/\omega_n$. The ratio between these two latter quantities is the natural abscissa axis for the representation of **shock spectra**, where the maximum response of the system is plotted on the ordinate axis without regard to the entire time history of the event. Shock spectra are obtained considering an undamped SDOF system as a standard reference and are widely used for design and comparison purposes in order to assess the potential disruptive effects of various forms of shock.

Another class of loadings is given by periodic (i.e. with a repetitive pattern in time) functions. Fourier's theorem states that a general periodic (and well-behaved) signal is the superposition of an infinite number of simple sinusoidal functions with frequencies that are all integral multiples of a value ω_0. It follows that its mathematical form is a convergent Fourier series of such functions and, owing to the principle of superposition, the response of a linear system is a similar Fourier series as well. Amplitudes and phases are modified between input and output (excitation and response), but it is not so for the frequency content and even if, rigorously, an infinite number of terms appear in the mathematical representation of the above series, a finite and limited number of terms often suffices for all practical purposes.

The generalization of Fourier series to nonperiodic signals leads to the Fourier and Laplace transformation integrals, which constitute the basis of the frequency-domain approach. Besides the fact that they often allow a simplification of the mathematics required to solve specific problems, their importance cannot be overstated and the fundamental concepts of **frequency response function** $H(\omega)$ and **transfer function** $H(s)$ of a linear system follow directly from their application.

These latter functions play a crucial role in almost every aspect of linear vibration analysis. They completely characterize a linear system in the frequency domain and – as $h(t)$ in the time domain – they are inherent properties of the system under study. Given these similarities, logic dictates that it must be possible to obtain $H(\omega)$ – or $H(s)$ – from $h(t)$ and *vice versa*. The connection is the Fourier (or Laplace) transform: $h(t)$ and $H(\omega)$ are a Fourier transform pair and, likewise, $h(t)$ and $H(s)$ are a Laplace transform pair.

Unfortunately, both the time-domain and the frequency-domain approaches often lead to integral expressions which cannot be evaluated analytically and, therefore, recourse must then be made to computer calculations on 'sampled' versions of the original signals. This sampling process is not harmless and its effects will be considered in Part II of the book which deals with electronic instrumentation. It is, however, important to point out right away that some care must be exercised in these cases if we do not want to run into undesirable consequences.

The last part of the chapter shows how, in some circumstances, an SDOF analysis can be extended to a wide class of more complex systems when some basic assumptions on the system's behaviour can be made or when the needed results can be accepted with a reasonable degree of approximation. The concept of **generalized parameters** is introduced in order to obtain a SDOF equation of motion or to calculate an approximate value of the fundamental frequency by means of the **Rayleigh energy method**. The assumption that only one vibration pattern (or 'shape') is developed during the motion is particularly useful in an approximate examination of continuous systems, where the static deflection under an appropriate load – often their own weight – is a good choice for the assumed vibration shape in most cases. The **assumed shape** then satisfies automatically the **essential** and **natural boundary conditions** of the problem. Nevertheless, simpler shapes satisfying only the essential boundary conditions can be chosen if great accuracy is not needed. Needless to say, the exact value of frequency is obtained if one has the luck (or the physical insight) to choose the correct deformed shape.

The **improved Rayleigh method** provides more accuracy in the calculation of the fundamental frequency – which is always overestimated when the assumed shape is not correct – and allows an indirect qualitative evaluation of the error with respect to the (unknown) exact value by looking at the improvement of the first one or two iterations involved in the process. Further iterations are, in general, not needed.

References

1. Cooley, J.W. and Tukey, J.W., An algorithm for the machine calculation of complex Fourier series, *Mathematics of Computation*, **19**, 297–301, 1965.
2. Harris, C.M. (ed.), *Shock and Vibration Handbook*, 3rd edn, McGraw-Hill, New York, 1988.
3. Jacobsen, L.S. and Ayre, R.S., *Engineering Vibrations*, McGraw-Hill, New York, 1958.
4. Erdélyi, A., Magnus, W., Oberhettingher, F. and Tricomi, F.G., *Tables of Integral Transforms*, 2 vols, McGraw-Hill, New York, 1953.
5. Thomson, W.T., *Laplace Transformation*, 2nd edn, Prentice Hall, Englewood Cliffs, NJ, 1960.
6. Graff, K.F., *Wave Motion in Elastic Solids*, Dover, New York, 1991.
7. Inman, D.J., *Engineering Vibrations*, Prentice Hall, Englewood Cliffs, NJ, 1994.

6 Multiple-degree-of-freedom systems

6.1 Introduction

In one way or another, almost any vibrating system could be modelled and represented as an SDOF system. In general, the quality of this assumption depends on the type of system and excitation, on the accuracy required and on the scope of the investigation. In some circumstances the assumption is correct; in some other cases it may lead to a description of the system's dynamic behaviour within an acceptable degree of accuracy but, in many other cases, the assumption is just an extreme oversimplification leading to inaccurate results which have almost nothing to do with the real situation. Since, *a priori*, the true behaviour of a real system is in general not known, the assessment of the validity of the results obtained from a SDOF analysis may not be an easy task. Therefore, in order to obtain a meaningful description for a wide class of systems, more complex representations are needed from the outset.

When one coordinate is not sufficient to characterize the motion of a given system, one speaks rightfully of two-, three-, ... *n*- or, in general, multiple-degree-of-freedom (MDOF) systems; where the number refers to the independent coordinates necessary to describe completely the vibration phenomenon.

The SDOF model enables us to explain – without particular mathematical difficulties – many fundamental concepts such as free and forced vibrations, natural frequency and resonance. Broadly speaking, all of these concepts can be extended to MDOF models. However, some important differences will appear; in anticipation we can say that the natural vibration of an MDOF system may occur at a number of different frequencies. Each one of them corresponds to a particular pattern (or 'shape' to give a pictorial view) of the system's motion and these different configurations, known as **natural** or **normal modes of vibration**, play a crucial role in almost every aspect of further analysis.

As discussed in Chapter 3, a set of *n* simultaneous ordinary differential equations of motion – one for each degree of freedom – must now be obtained in order to mathematically describe our system and a proper choice of the generalized coordinates is one of the key points of the whole process. In fact, the theory of finite-dimensional vector spaces and the closely related subject of linear matrix algebra are the most convenient mathematical tools required to deal with MDOF systems: it is well-known that the choice of a basis in a vector space determines the particular matrix representation of operators and vectors and may considerably simplify the problem at hand.

6.2   A simple undamped 2-DOF system: free vibration

As a starting point, let us consider the simple system of Fig. 6.1. It consists of two masses M and m connected by a spring k_2, with the mass M suspended from a fixed point by a spring k_1. We assume that this system is constrained to move only in the vertical direction; in this case the coordinates x_1 and x_2 that specify the position of the two masses at any instant are sufficient to describe the motion completely and we have a 2-DOF system.

If we take the position of static equilibrium as a reference, the terms representing the weights of the masses cancel out with the spring tensions at equilibrium, exactly as stated in Section 4.2.1; as the masses move at x_1 and x_2, respectively, the tension in the lower spring is $k_2(x_2 - x_1)$. The (coupled) equations of motion are

$$M\ddot{x}_1 = -k_1 x_1 + k_2(x_2 - x_1)$$
$$m\ddot{x}_2 = -k_2(x_2 - x_1) \tag{6.1}$$

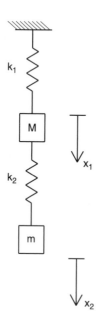

Fig. 6.1 Simple 2-DOF system.

as can also be determined by using Lagrange's equations. Let us consider now if such a system can vibrate so that all the masses move with the same frequency $\omega/(2\pi)$ and, if this is possible, at how many different frequencies this vibration can occur. Let us suppose further that, for example, $M = 3m$, $k_2 = k$ and $k_1 = 4k$. Equations (6.1) become

$$3m\ddot{x}_1 + 5kx_1 - kx_2 = 0$$
$$m\ddot{x}_2 - kx_1 + kx_2 = 0 \qquad (6.2)$$

By analogy with the behaviour of SDOF systems, we look for harmonic solutions. Using for example the sinusoidal notation we write

$$x_1 = A \cos(\omega t - \theta)$$
$$x_2 = B \cos(\omega t - \theta) \qquad (6.3)$$

where A, B and ω are constants.
Substituting eqs. (6.3) in eqs. (6.2) leads to

$$-3m\omega^2 A + 5kA - kB = 0$$
$$-m\omega^2 B - kA + kB = 0 \qquad (6.4)$$

We note in passing that the same phase angle is used because, were it not so, only the trivial case $A = 0$ and $B = 0$ (i.e. no motion at all) would be a solution; the case in which the phase angles differ by π is equivalent to a change of the sign of B, which is, as yet, still arbitrary.
A nontrivial solution of eqs (6.4) exists if the determinant

$$\begin{vmatrix} 5k - 3m\omega^2 & -k \\ -k & k - m\omega^2 \end{vmatrix}$$

vanishes. For reasons that will become clear in the following sections, the problem of determining the values of frequency for which eqs (6.4) admit nontrivial solutions is called the **eigenvalue** (or **characteristic value**) problem. Equating to zero the determinant above leads to the **frequency equation**

$$3m^2\omega^4 - 8km\omega^2 + 4k^2 = 0 \qquad (6.5)$$

with solutions

$$\omega_1^2 = 2k/(3m) \quad \text{i.e.} \quad \omega_1 = \sqrt{2k/(3m)}$$
$$\omega_2^2 = 2k/m \quad \text{i.e.} \quad \omega_2 = \sqrt{2k/m} \qquad (6.6)$$

since ω is a positive quantity. For each of these frequencies the ratio A/B can be determined by the following relations:

$$A/B = \frac{k}{5k - 3m\omega^2}$$

$$A/B = \frac{k - m\omega^2}{k}$$

(6.7)

which have been obtained from eqs (6.4) (two equations for three unknowns).

So, for each ω given by eqs (6.6), the phase θ and one of the constants A and B can be arbitrarily assigned: the choice of θ specifies the instant at which the motion is started and the choice of A (or B) determines the amplitude of vibration.

The two values of frequency at which harmonic motions of the type (6.3) exist are called the **natural frequencies** of the system. They are inherent characteristics of the system under study and correspond, as we will see shortly, to well-determined configurations in space of the system itself, the so-called **normal modes of vibration.**

For the value of frequency $\sqrt{2k/(3m)}$ rad/s, we obtain from eqs (6.7) $A/B = 1/3$. This means that at any instant of time both masses are above or below their equilibrium position (in phase), the displacement of mass m being three times the displacement of mass M. Likewise, for the frequency $\omega_2 = \sqrt{2k/m}$ rad/s, we get $A/B = -1$: so that at any instant the masses have the same displacement with respect to their equilibrium positions, but on opposite sides of them (opposition of phase).

Thus, each normal frequency 'belongs' to a well-defined pattern of motion – its relevant normal mode – and vice versa, each normal mode represents a pattern of motion that occurs at a well-defined value of frequency. The natural frequencies and the **mode shapes** (i.e. the pattern of motion of the normal modes) are unique for a given system, the amplitude of the mode shapes is not. However, for a given mode, the ratio of amplitudes is determined uniquely.

This degree of arbitrariness is a mathematical consequence of the fact that the eigenvalue problem (6.4) is homogeneous and multiplication of a mode shape by a constant does not change the mode shape itself.

Taking the analysis of our simple 2-DOF system a little further, we can write explicitly the solution corresponding to the frequency ω_1. This is

$$x_1 = A_1 \cos(\omega_1 t - \theta_1)$$
$$x_2 = 3A_1 \cos(\omega_1 t - \theta_1)$$

For this solution, we note that the quantity $(3x_1 - x_2)$ vanishes at all times and the same happens to the quantity $(x_1 + x_2)$ for the solution relative to ω_2.

Now, subtracting the second of eqs (6.2) from the first gives

$$m(3\ddot{x}_1 - \ddot{x}_2) + 2k(3x_1 - x_2) = 0 \tag{6.8}$$

and adding the first equation of (6.2) to three times the second gives

$$3m(\ddot{x}_1 + \ddot{x}_2) + 2k(x_1 + x_2) = 0 \tag{6.9}$$

If we define the new coordinates

$$\eta_1 = 3x_1 - x_2$$
$$\eta_2 = x_1 + x_2 \tag{6.10}$$

eq (6.8) and (6.9) become

$$m\ddot{\eta}_1 + 2k\eta_1 = 0$$
$$3m\ddot{\eta}_2 + 2k\eta_2 = 0 \tag{6.11}$$

which are two **uncoupled** equations in the new coordinates and, as such, can be solved independently. The simplification is noteworthy, because eqs (6.2) were coupled and had to be solved simultaneously. This new set of coordinates, or any multiples of them, are referred to as **normal coordinates**. In a normal mode only one normal coordinate is nonzero and the solutions to eqs (6.11) can be immediately written following the procedure described in Section 4.2.1.

The four arbitrary constants appearing in the expressions of $\eta_1(t)$ and $\eta_2(t)$ can be determined from the initial conditions and then, if needed, x_1 and x_2 may be recovered. It is desirable and easier at this point to fit the initial conditions before x_1 and x_2 are separately determined, otherwise we get four simultaneous equations for the four constants.

If now we turn our attention to the energy of our system, it is easy to write the potential and kinetic energies:

$$V = \tfrac{1}{2}k_1 x_1^2 + \tfrac{1}{2}k_2(x_2 - x_1)^2 \tag{6.12a}$$
$$T = \tfrac{1}{2}M\dot{x}_1^2 + \tfrac{1}{2}m\dot{x}_2^2 \tag{6.13a}$$

which become, in our example $(M = 3m; \ k_2 = k; \ k_1 = 4k)$

$$V = \tfrac{5}{2}kx_1^2 + \tfrac{1}{2}kx_2^2 - kx_1x_2 \tag{6.12b}$$
$$T = \tfrac{3}{2}m\dot{x}_1^2 + \tfrac{1}{2}m\dot{x}_2^2 \tag{6.13b}$$

These same quantities can be expressed in terms of the normal coordinates. Inverting eqs (6.10) gives

$$x_1 = \tfrac{1}{4}(\eta_1 + \eta_2)$$

$$x_2 = \tfrac{1}{4}(3\eta_2 - \eta_1)$$

(6.14)

and substitution in the energy expressions (6.12b) and (6.13b) leads to

$$V = \tfrac{1}{4} k(\eta_1^2 + \eta_2^2)$$

(6.15)

$$T = \tfrac{1}{8} m(\dot{\eta}_1^2 + 3\dot{\eta}_2^2)$$

(6.16)

Note that, when they are written in normal coordinates, no cross-product terms appear in the potential and kinetic energy expressions, the only nonzero terms being squares of the coordinates and their time derivatives. It is obviously not so in the original coordinates, where the cross-product term kx_1x_2 appears in the potential energy (eq (6.12b)). Furthermore, if we write Lagrange's equation for η_1

$$\tfrac{1}{4} m\ddot{\eta}_1 + \tfrac{1}{2} k\eta_1 = 0$$

(6.17)

an energy equation can be obtained; multiplying eq (6.17) by $\dot{\eta}_1$ and integrating with respect to time, we get

$$\tfrac{1}{8} m\dot{\eta}_1^2 + \tfrac{1}{4} k\eta_1^2 = const.$$

(6.18)

which shows that the η_1 contributions to V and T have a constant sum. The same applies to the coordinate η_2, as the reader can immediately verify. So, the energy associated with any particular normal coordinate remains constant and this result – a refinement of the energy equation for the whole system – implies that the total energy of the system is divided into parts which are separately constant. **There is no energy interchange between any two different normal coordinates.** This is clearly not true for the original coordinates, because the cross-product term in V provides the mechanism for an energy interchange between x_1 and x_2.

The results above are quite general and this particular example has been given in order to illustrate some basic aspects of a simple undamped MDOF system without getting too much involved in the specific techniques adopted in the solution of such problems. In the case of more complex systems, it is obvious that the procedure above leads to arduous and cumbersome calculations and one has to resort to other more tractable methods.

6.3   Undamped *n*-DOF systems: free vibration

In Chapter 3, making use of Lagrange's equations, we obtained the equations
of motion of a *n*-DOF system without damping as (eq (3.97))

$$\sum_{j=1}^{n} (m_{ij}\ddot{u}_j + k_{ij}u_j) = 0 \qquad i = 1, 2, \dots , n \tag{6.19}$$

Equations (6.19) are equivalent to the matrix expression

$$\mathbf{M\ddot{u} + Ku = 0} \tag{6.20}$$

where $\mathbf{M}$ and $\mathbf{K}$ are symmetrical (i.e. equal to their transpose) $n \times n$ matrices
and $\mathbf{u}$ is an $n \times 1$ column vector. The solution of the n coupled equations
(6.19) – or (6.20) – represents the free vibration of our system, where the
initial conditions are now given by a set of n initial displacements and n initial
velocities.

As an example, we can refer back to the simple 2-DOF system of Fig. 6.1
and write the equations of motion (6.1) in matrix form as

$$\begin{bmatrix} M & 0 \\ 0 & m \end{bmatrix} \begin{bmatrix} \ddot{x}_1 \\ \ddot{x}_2 \end{bmatrix} + \begin{bmatrix} k_1 + k_2 & -k_2 \\ -k_2 & k_2 \end{bmatrix} \begin{bmatrix} x_1 \\ x_2 \end{bmatrix} = \begin{bmatrix} 0 \\ 0 \end{bmatrix} \tag{6.21}$$

(It is worth noting that the emphasis on matrix notation is due to the fact that
this 'language' is convenient and particularly appropriate for computer
implementation; its brevity is a virtue and it has proven to be an extremely
useful organizational tool to keep track of complicated sequences of
operations.)

The coupling of the equations arises from the fact that, in general, the mass
(or inertia) matrix $\mathbf{M}$ or the stiffness matrix $\mathbf{K}$, or both, may not be diagonal:
when $\mathbf{M}$ is not diagonal the system is called **inertially**, or dynamically,
coupled, when $\mathbf{K}$ is not diagonal we speak of **elastic** (or stiffness, or static)
coupling, see for example, eqs. (6.21). Obviously, the system is inertially and
elastically coupled when both $\mathbf{M}$ and $\mathbf{K}$ are not diagonal.

However, it is very important to note that the type of coupling is not an
intrinsic property of the system, but depends on the choice of coordinates used
to express the equations of motion. In fact, it is not difficult – always referring
to the simple system of Fig. 6.1 as an example – to choose a set of coordinates
in which the stiffness matrix is diagonal and the mass matrix is not. In this
light, it could be inferred that the most desirable situation in which both $\mathbf{M}$
and $\mathbf{K}$ are diagonal – and the equations of motion are uncoupled, which
means much easier to solve – can eventually be achieved with an appropriate
choice of coordinate system. In the following we will see that this is always
the case for an undamped MDOF system.

Returning to eq (6.20), we investigate the possibility of having solutions of the form

$$\mathbf{u}(t) = \mathbf{z}e^{-i\omega t} \tag{6.22}$$

where $\mathbf{z}$ is a 'shape' vector and the time dependency has been separated out. Physically, this means that all the coordinates execute a synchronous motion and the system configuration changes its amplitude but not its shape during the motion. Substitution in eq (6.20) gives

$$(\mathbf{K} - \omega^2 \mathbf{M})\mathbf{z} = 0 \tag{6.23}$$

or, equivalently,

$$\mathbf{Kz} = \omega^2 \mathbf{Mz} \tag{6.24}$$

Equation (6.23), or (6.24), represents a set of n homogeneous linear equations and it is known that a nontrivial (different from zero, i.e. no motion at all) solution exists if and only if the determinant of the coefficients vanishes, i.e.

$$\left| \mathbf{K} - \omega^2 \mathbf{M} \right| = 0 \tag{6.25}$$

Mathematically, eq (6.24) represents a problem commonly known as **generalized** (or 'linearized' for other authors) **eigenvalue problem**. We may refer to it also as the 'real eigenvalue problem', to distinguish it from the 'complex eigenvalue problem', which is relative to systems with damping.

The standard, or canonical, form of the eigenvalue problem (see also the Appendix A on matrix analysis for more details) is

$$\mathbf{Az} = \lambda \mathbf{z} \tag{6.26a}$$

or

$$(\mathbf{A} - \lambda \mathbf{I})\mathbf{z} = 0 \tag{6.26b}$$

where $\mathbf{A}$ is a square $(n \times n)$ matrix, $\mathbf{z}$ is a right $n \times 1$ eigenvector and λ is the corresponding eigenvalue (or proper value). The solution of the 'eigenproblem', given the matrix $\mathbf{A}$, consists of finding the values of λ and the vectors for which eqs (6.26) hold. Eigenvalue problems occur in so many areas of physics and engineering that a large volume of literature is available on the subject, and this is also due to the fact that, for large systems, its solution may present considerable computational difficulties.

The expansion of the determinant of eq (6.25) leads to an algebraic equation of order n in ω^2. This is known as the characteristic or **frequency equation** (remember eq (6.5)) and its n roots are the eigenvalues of the undamped free vibration problem. These n roots are not necessarily all different and we must consider their algebraic multiplicities, i.e. the number of times a repeated root occurs as a solution of the frequency equation. However, we leave for later the case of repeated roots, the so-called eigenvalue degeneracy, which is, for the most part, only a mathematical complication.

Physically, the positive square roots of the eigenvalues represent the natural frequencies of our system. We denote them as $\omega_1, \omega_2, \ldots, \omega_n$, where the subscript increases as the value of frequency increases and the lowest frequency, ω_1, is called the **fundamental frequency**.

When the natural frequencies ω_j $(j = 1, 2, \ldots, n)$ are known, we can go back to eq (6.23) and solve for the eigenvectors $\mathbf{z}_j$ $(j = 1, 2, \ldots, n)$, which, in turn, can only be determined within a multiplicative arbitrary constant because of the homogeneous nature of the problem. In other words, the eigenvectors are arbitrary to the extent of an indeterminate multiplier: if $\mathbf{z}_j$ is a solution of the problem, $\alpha \mathbf{z}_j$ is also a solution, α being a constant. Fixing the value of α by some convention determines the eigenvector completely, or, alternatively, assigning a definite value to one element of the eigenvector determines its remaining $n - 1$ elements uniquely. This process is known as 'normalization' and more than one convention can be used, as we shall see later.

Whatever normalization is chosen to fix its amplitude, each eigenvector $\mathbf{z}_j$ represents a unique pattern or 'shape' of vibration of our MDOF system, a vibration pattern which takes place at the frequency ω_j; this occurrence suggests for the $\mathbf{z}_j$ the common names of **natural modes of vibration, modal shapes** or **modal vectors**.

The general solution of the problem, given its linearity, is a sum of the $2n$ solutions (note that, mathematically, the roots of the frequency equation are $\pm \omega_1, \pm \omega_2, \ldots, \pm \omega_n$)

$$\mathbf{z}_j \, e^{\pm i\omega_j t} \qquad j = 1, 2, \ldots, n$$

i.e.

$$\mathbf{u} = \sum_{j=1}^{n} (A_j e^{+i\omega_j t} + B_j e^{-i\omega_j t})\mathbf{z}_j \qquad (6.27\text{a})$$

or, alternatively

$$\mathbf{u} = \sum_{j=1}^{n} C_j \mathbf{z}_j \, \cos(\omega_j t - \theta_j) \qquad (6.27\text{b})$$

$$\mathbf{u} = \sum_{j=1}^{n} (D_j \cos \omega_j t + E_j \sin \omega_j t)\mathbf{z}_j \qquad (6.27\text{c})$$

where the set of $2n$ constants – A_j and B_j for the case of eq (6.27a), C_j and θ_j for the case of eq (6.27b) and D_j and E_j for eq (6.27c) – can be obtained from the initial conditions

$$\mathbf{u}(t=0) = \mathbf{u}_0$$

$$\dot{\mathbf{u}}(t=0) = \dot{\mathbf{u}}_0$$

(6.28)

We may note that, up to this point, we have re-expressed in matrix notation and extended to a number of n degrees of freedom many of the concepts introduced with the simple example of Section 6.2. The matrix approach, however, goes deeper into the fundamental aspects of the problem and some useful properties of the eigenvectors need to be pointed out before proceeding further.

6.3.1   *Orthogonality relationships of eigenvectors and normalization*

For our convenience, let us rewrite the eigenvalue problem of eq (6.24) as

$$\mathbf{Kz} = \lambda \mathbf{Mz}$$

(6.29)

where $\lambda = \omega^2$. Now, let $\mathbf{z}_i$ and $\mathbf{z}_j$ be any two eigenvectors whose corresponding eigenvalues are, respectively, λ_i and λ_j and we assume $\lambda_i \neq \lambda_j$. Then the two equations

$$\mathbf{Kz}_i = \lambda_i \mathbf{Mz}_i$$

$$\mathbf{Kz}_j = \lambda_j \mathbf{Mz}_j$$

(6.30)

are identically satisfied. Premultiplication of the first of eqs (6.30) by $\mathbf{z}_j^T$ gives

$$\mathbf{z}_j^T \mathbf{Kz}_i = \lambda_i \mathbf{z}_j^T \mathbf{Mz}_i$$

(6.31)

and transposing both sides, since $\mathbf{K} = \mathbf{K}^T$ and $\mathbf{M} = \mathbf{M}^T$, we get

$$\mathbf{z}_i^T \mathbf{Kz}_j = \lambda_i \mathbf{z}_i^T \mathbf{Mz}_j$$

(6.32)

On the other hand, premultiplication of the second of eqs (6.30) by $\mathbf{z}_i^T$ yields

$$\mathbf{z}_i^T \mathbf{Kz}_j = \lambda_j \mathbf{z}_i^T \mathbf{Mz}_j$$

(6.33)

and subtracting eq (6.33) from eq (6.32) leads to

$$(\lambda_i - \lambda_j)\mathbf{z}_i^T \mathbf{Mz}_j = 0$$

which implies

$$\mathbf{z}_i^T \mathbf{M} \mathbf{z}_j = 0 \tag{6.34}$$

because of the assumption of distinct eigenvalues. Equation (6.34) generalizes the usual concept of orthogonality of two vectors $\mathbf{x}$ and $\mathbf{y}$, which can be written as

$$\mathbf{x}^T \mathbf{y} = 0$$

where $\mathbf{x}$ and $\mathbf{y}$ are two elements of a Euclidean space (i.e. a vector space of finite dimension where an inner product has been defined). In this case one speaks of orthogonality of two vectors when their inner product is zero.

In the case of eq (6.34) we speak of orthogonality with respect to the mass matrix or, in brief, **mass-orthogonality**. It is now straightforward to show that the eigenvectors $\mathbf{z}_i$ and $\mathbf{z}_j$ are also orthogonal with respect to the stiffness matrix: by virtue of eq (6.34), eq (6.32) gives

$$\mathbf{z}_i^T \mathbf{K} \mathbf{z}_j = 0 \tag{6.35}$$

which is precisely the **stiffness-orthogonality** condition of the two eigenvectors.

When $i = j$, the left-hand side of eqs. (6.34) and (6.35) is different from zero and we can write

$$\mathbf{z}_i^T \mathbf{M} \mathbf{z}_i = M_{ii}$$
$$\mathbf{z}_i^T \mathbf{K} \mathbf{z}_i = K_{ii} \tag{6.36}$$

where the scalars M_{ii} and K_{ii} are called the modal or generalized mass and the modal or generalized stiffness of the ith mode, respectively. These terms are rather misleading because the eigenvectors are subjected to an arbitrary scaling factor which, as yet, has been left undetermined; nevertheless the ratio K_{ii}/M_{ii} is uniquely determined because, by virtue of eq (6.29), we have

$$\frac{K_{ii}}{M_{ii}} \equiv \frac{\mathbf{z}_i^T \mathbf{K} \mathbf{z}_i}{\mathbf{z}_i^T \mathbf{M} \mathbf{z}_i} = \lambda_i \frac{\mathbf{z}_i^T \mathbf{M} \mathbf{z}_i}{\mathbf{z}_i^T \mathbf{M} \mathbf{z}_i} = \lambda_i \equiv \omega_i^2 \tag{6.37}$$

The formal analogy with the usual $\omega_n^2 = k/m$ for an SDOF system is evident.

Going back in our discussion on eigenvectors, we tacitly assumed that the frequency equation has precisely n roots, counting multiplicities. Strictly speaking, this is true in the complex field (which is algebraically closed) and little can be said if we limit ourselves to real numbers. However, we note that both $\mathbf{M}$ and $\mathbf{K}$ are symmetrical and symmetrical matrices are just Hermitian – or self-adjoint – matrices with real entries. A direct consequence of their

Hermitian property is that all eigenvalues are real. In fact, consider the second of eqs (6.30), its adjoint equation, i.e. its transposed complex conjugate, is

$$\mathbf{z}_j^H \mathbf{K} = \lambda_j^* \mathbf{z}_j^H \mathbf{M} \tag{6.38}$$

because $\mathbf{K} = \mathbf{K}^H$ and $\mathbf{M} = \mathbf{M}^H$, where $\mathbf{K}^H$ and $\mathbf{M}^H$ indicate the Hermitian adjoint matrices of $\mathbf{K}$ and $\mathbf{M}$, respectively. Now, multiply from the right of eq (6.38) by $\mathbf{z}_i$, multiply from the left of the first of eqs (6.30) by $\mathbf{z}_j^H$ and subtract one of the resulting equations from the other. We get

$$(\lambda_i - \lambda_j^*)\mathbf{z}_j^H \mathbf{M} \mathbf{z}_i = 0 \tag{6.39a}$$

and the particular case of $i = j$ gives

$$(\lambda_i - \lambda_i^*)\mathbf{z}_i^H \mathbf{M} \mathbf{z}_i = 0 \tag{6.39b}$$

Once again, the Hermitian property of $\mathbf{M}$ can be used to show that the matrix product $\mathbf{z}_i^H \mathbf{M} \mathbf{z}_i$ is real. (Hint: write explicitly the real and imaginary parts of the eigenvector as $\mathbf{z}_i = \mathbf{a}_i + i\mathbf{b}_i$, perform the multiplication and note that the imaginary part of the result must vanish because $\mathbf{a}_i^T \mathbf{M} \mathbf{b}_i = \mathbf{b}_i^T \mathbf{M} \mathbf{a}_i$; it follows that $\text{Im}(\lambda_i) = 0$, i.e. λ_i is real.) On physical grounds, it could be argued that an imaginary part of ω different from zero would introduce an exponential decreasing or increasing factor in our solution (6.22) and this factor, in its turn, would lead to changes of the total energy as time passes. This is in contradiction with the fact that we are dealing with a conservative system.

Equations (6.34), (6.35) and (6.37), the othogonality conditions, can be stated more concisely if we form the modal matrix $\mathbf{Z}$ by assembling the eigenvectors side by side as columns of an $n \times n$ square matrix $[z_{ij}]$, where the first subscript is the component index and the second subscript is the mode or eigenvector index.

Explicitly, the modal matrix is written

$$\mathbf{Z} \equiv [\mathbf{z}_1 \quad \mathbf{z}_2 \quad \dots \quad \mathbf{z}_n] = \begin{bmatrix} z_{11} & z_{12} & \dots & z_{1n} \\ z_{21} & z_{22} & \dots & z_{2n} \\ \dots & \dots & \dots & \dots \\ n_{z1} & z_{n2} & \dots & z_{nn} \end{bmatrix} \tag{6.40}$$

and the orthogonality conditions become

$$\mathbf{Z}^T \mathbf{M} \mathbf{Z} = \text{diag}(M_{11}, M_{22}, \dots, M_{nn}) = M_{ii}\mathbf{I}$$
$$\mathbf{Z}^T \mathbf{K} \mathbf{Z} = \text{diag}(K_{11}, K_{22}, \dots, K_{nn}) = K_{ii}\mathbf{I} \tag{6.41}$$

where $\mathbf{I}$ is the unit $n \times n$ matrix.

The process of normalization removes the indeterminacy on the amplitude of the eigenvectors. In principle, any scaling factor will do as long as consistency is maintained throughout the process of analysis, but the most usual conventions are:

1. Set the largest component of each eigenvector equal to unity and determine the remaining $n-1$ components accordingly.
2. Mass normalization: scale each eigenvector so that

$$\mathbf{p}_i^T \mathbf{M} \mathbf{p}_i = 1 \qquad (i = 1, 2, \dots, n) \tag{6.42a}$$

or, in general

$$\mathbf{p}_i^T \mathbf{M} \mathbf{p}_j = \delta_{ij} \tag{6.42b}$$

where $\mathbf{p}_i$ is the mass-normalized ith eigenvector and δ_{ij} is the usual Kronecker symbol, equal to unity when $i = j$ and zero when $i \neq j$. The relationship between the mass-normalized eigenvector and its more general form is obviously

$$\mathbf{p}_i = \frac{1}{\sqrt{M_{ii}}} \mathbf{z}_i \tag{6.43}$$

and the immediate consequence of eq (6.42) is

$$\mathbf{p}_i^T \mathbf{K} \mathbf{p}_j = \lambda_i \delta_{ij} = \omega_i^2 \, \delta_{ij} \tag{6.44}$$

so that the concise form of eqs (6.42b) and (6.44) now reads

$$\mathbf{p}^T \mathbf{M} \mathbf{P} = \mathbf{I}$$
$$\tag{6.45}$$
$$\mathbf{P}^T \mathbf{K} \mathbf{P} = \mathrm{diag}(\lambda_1, \lambda_2, \dots, \lambda_n) \equiv \mathbf{L}$$

where $\mathbf{P}$ is the modal matrix formed with the vectors $\mathbf{p}_i$ (sometimes called the weighted modal matrix) and we have defined the $n \times n$ matrix of eigenvalues $\mathbf{L}$: its only entries different from zero lie on the main diagonal and are equal to the eigenvalues. This normalization process is the most relevant in the field of modal testing, which we shall consider in later chapters.

3. Another possibility is to normalize the eigenvectors so that all the modal masses are equal to M, where M is some convenient value, for example the total mass of the system.
4. Set each mode vector length (its norm in vector space terminology) equal to unity.

It is left to the reader to write down the counterpart of eqs (6.45) for cases 3 and 4.

6.3.2   *General solution of the undamped free vibration problem, degeneracy and normal coordinates*

The discussion of the preceding section has shown that the eigenvectors z_i ($i = 1, 2, \ldots, n$) are independent and mutually orthogonal with respect to the mass matrix. It is then straightforward to consider this set of vectors as a complete basis spanning an n-dimensional vector space and write, by virtue of the expansion theorem, any nth order vector $\mathbf{x}$ as

$$\mathbf{x} = \sum_{j=1}^{n} \alpha_j \mathbf{z}_j \tag{6.46}$$

where the α_j are scalar multipliers which can be easily obtained by making use of the othogonality conditions after premultiplication of both sides of eq (6.46) by $\mathbf{z}_i^T \mathbf{M}$, i.e.

$$\mathbf{z}_i^T \mathbf{M} \mathbf{x} = \sum_j \alpha_j \mathbf{z}_i^T \mathbf{M} \mathbf{z}_j = \alpha_i M_{ii}$$

from which it follows that

$$\alpha_i = \frac{\mathbf{z}_i^T \mathbf{M} \mathbf{x}}{M_{ii}} = \frac{\mathbf{z}_i^T \mathbf{M} \mathbf{x}}{\mathbf{z}_i^T \mathbf{M} \mathbf{z}_i} \tag{6.47}$$

The expression

$$\mathbf{x} = \sum_{j=1}^{n} \frac{\mathbf{z}_j^T \mathbf{M} \mathbf{x}}{M_{jj}} \mathbf{z}_j = \sum_{j=1}^{n} \frac{\mathbf{z}_j \mathbf{z}_j^T \mathbf{M}}{M_{jj}} \mathbf{x} \tag{6.48a}$$

is often called the modal expansion of $\mathbf{x}$. From the second form of the expansion given in eq (6.48) the spectral expansion of the unit matrix can be deduced

$$\mathbf{I} = \sum_j \frac{\mathbf{z}_j \mathbf{z}_j^T \mathbf{M}}{\mathbf{z}_j^T \mathbf{M} \mathbf{z}_j} \tag{6.49a}$$

which turns out to be a useful calculation check on the matrices of the right-hand side. Incidentally, we may note that eq (6–49a) expresses the completeness of the orthogonal set of vectors $\mathbf{Z}_j$.

The same considerations apply if we consider the mass-normalized eigenvectors $\mathbf{p}_i$; we now get

$$\mathbf{x} = \sum_{j=1}^{n} \mathbf{p}_j^T \mathbf{M} \mathbf{x} \mathbf{p}_j = \sum_{j=1}^{n} \mathbf{p}_j \mathbf{p}_j^T \mathbf{M} \mathbf{x} \tag{6.48b}$$

and

$$\mathbf{I} = \sum_{j} \mathbf{p}_j \mathbf{p}_j^T \mathbf{M} \tag{6.49b}$$

Therefore, the general solution of the undamped free vibration problem can be written in any one of the forms of eqs (6.27a–c) by expanding the initial conditions (6.28) on the basis of eigenvectors and using the orthogonality conditions as shown above. For example, these calculations performed on the expression (6.27c) yield

$$\mathbf{u} = \sum_{j=1}^{n} \left(\frac{\mathbf{z}_j^T \mathbf{M} \mathbf{u}_0}{M_{jj}} \cos \omega_j t + \frac{\mathbf{z}_j^T \mathbf{M} \dot{\mathbf{u}}_0}{\omega_j M_{jj}} \sin \omega_j t \right) \mathbf{z}_j \tag{6.50}$$

or, in terms of the mass-normalized eigenvectors

$$\mathbf{u} = \sum_{j=1}^{n} \left(\mathbf{p}_j^T \mathbf{M} \mathbf{u}_0 \cos \omega_j t + \frac{\mathbf{p}_j^T \mathbf{M} \dot{\mathbf{u}}_0}{\omega_j} \sin \omega_j t \right) \mathbf{p}_j \tag{6.51}$$

and it is not difficult to see that the relationships among the constants in the different forms (6.27a–c), by analogy with eqs (4.9), are

$$A_j + B_j = D_j = C_j \cos \theta_j$$
$$i(A_j - B_j) = E_j = C_j \sin \theta_j \qquad (j = 1, 2, \ldots, n) \tag{6.52}$$

Alternatively, we can write eq (6.50) as

$$\mathbf{u} = \left(\sum_{j} \frac{\mathbf{z}_j \mathbf{z}_j^T \mathbf{M}}{M_{jj}} \cos \omega_j t \right) \mathbf{u}_0 + \left(\sum_{j} \frac{\mathbf{z}_j \mathbf{z}_j^T \mathbf{M}}{\omega_j M_{jj}} \sin \omega_j t \right) \dot{\mathbf{u}}_0 \tag{6.53}$$

where the vectors $\mathbf{u}_0$ and $\dot{\mathbf{u}}_0$ have now been put into evidence. Equations (6.50) and (6.53) are equivalent, and it is a matter of convenience which one we decide to use.

In any case, for each coordinate u_i – i.e. the ith component of the vector $\mathbf{u}$ – the solution is in general a sum of simple harmonic oscillations in all of the frequencies satisfying the frequency equation and it is worth noting that, unless it happens that all of the frequencies are commensurable (rational fractions of each other), u_i never repeats its initial value and therefore it is not a periodic function of time. However, in the specific case in which the initial displacement vector resembles a particular modal vector (say, for example, $\mathbf{u}_0 = \beta\mathbf{p}_r$, where β is a constant) and the initial velocity vector is zero, we get from eq (6.51)

$$\mathbf{u} = \beta\mathbf{p}_r \cos\omega_r t \tag{6.54}$$

so that the system executes a harmonic oscillation at the natural frequency ω_r and its configuration resembles the rth mode at all times. This implies that any one natural mode can be excited independently of the others by an appropriate choice of the initial conditions. This occurrence can be useful in some practical applications.

Eigenvalue degeneracy

When one or more of the roots of the frequency equation are repeated, the argument leading to eq (6.34) ceases to be valid for $\lambda_i = \lambda_j$. Hence, this case needs to be discussed separately.

We recall that, in the general case, the characteristic polynomial $f(\lambda) \equiv |\mathbf{K} - \lambda\mathbf{M}|$ can be written as

$$f(\lambda) = (\lambda - \lambda_1)^{d_1}(\lambda - \lambda_2)^{d_2} \dots (\lambda - \lambda_m)^{d_m}$$

where $d_1 + d_2 + \dots + d_m = n$, λ_i $(i = 1, 2, \dots, m)$ are the distinct roots of the polinomial and d_i is called, in mathematical terms, the algebraic multiplicity of λ_i.

If $d_i > 1$ we speak, in the terminology of physics and engineering, of d_i-fold degeneracy of the eigenvalue λ_i; obviously $d_1 = d_2 = \dots = d_m = 1$ and $m = n$ is the nondegenerate case of n distinct eigenvalues considered so far. On the other hand, the dimension of the subspace corresponding to the eigenvalue λ_i is called its geometric multiplicity. Put simply, the geometric multiplicity is just the minimum number of linearly independent eigenvectors associated with an eigenvalue and it is, in general, different from its algebraic multiplicity. However, it can be proven that Hermitian matrices (and, more generally, normal matrices, i.e. matrices such that $AA^H = A^H A$) are 'nondefective', which is to say that the geometric multiplicity is the same as the algebraic multiplicity for each eigenvalue. The direct consequence of this statement is that it is always possible to generalize the orthogonality relationships shown above by a proper choice of d_i orthogonal eigenvectors in the d_i-dimensional subspace corresponding to the d_i-fold degenerate eigenvalue λ_i.

Without loss of generality, let us suppose, for example, that λ is a double root of the frequency equation; any two of the eigenvector components may be chosen arbitrarily, the rest being fixed by the eigenvalue equations. Any pair of randomly chosen eigenvectors – say $\tilde{z}_k$ and $\tilde{z}_l$, which we assume to have been mass normalized – will not, in general, be mutually orthogonal but the linear combination

$$z_l \equiv c_1 \tilde{z}_k + c_2 \tilde{z}_l$$

is also an eigenvector and we want to choose the constants c_1 and c_2 so that z_l is orthogonal to $\tilde{z}_k$. The orthogonality condition requires that

$$z_l^T \, \mathbf{M} \tilde{z}_k = c_1 + c_2 \tilde{z}_l^T \, \mathbf{M} \tilde{z}_k = 0 \qquad (6.55)$$

and eq (6.55) together with the mass-normalization condition for z_l provides the necessary equations to fix c_1 and c_2. The eigenvectors z_l and $\tilde{z}_k$ are precisely what we were looking for: they are automatically orthogonal to the eigenvectors of the other distinct eigenvalues and they have been constructed to be mutually orthogonal.

The process is completely analogous to the Gram–Schmidt algorithm used to replace a set of linearly independent vectors by an orthonormal set. It is obvious that this replacement may be carried out in infinitely many ways, but the point we are trying to make here is that it is always possible to obtain a set of n eigenvectors in order to form the modal matrix. This justifies the statement above that the presence of multiple roots is, for our purposes, only a mathematical complication.

Normal coordinates

Let us define the new set of coordinates y related to the original coordinates u by

$$\mathbf{u} = \mathbf{P}\mathbf{y} = \sum_{j=1}^{n} \mathbf{p}_j y_j(t) \qquad (6.56a)$$

or, in terms of components

$$u_i = \sum_{j=1}^{n} p_{ij} y_j \qquad (6.56b)$$

where $\mathbf{P} = [p_{ij}]$ is the modal matrix of eqs (6.45).

The equations of motion (6.20) now become

$$\mathbf{M}\mathbf{P}\ddot{\mathbf{y}} + \mathbf{K}\mathbf{P}\mathbf{y} = 0$$

$$\ddot{\mathbf{y}} + \mathbf{L}\mathbf{y} = 0 \qquad (6.57a)$$

or, in terms of components

$$\ddot{y}_j + \lambda_j y_j \equiv \ddot{y}_j + \omega_j^2 \, y_j = 0 \qquad (j = 1, 2, \dots, n) \tag{6.57b}$$

which represent n uncoupled SDOF equations of motion, one equation for each mode. Their solution is simply harmonic and can be easily obtained by the methods of Chapter 4. Each coordinate y_j corresponds to a vibration of the system with only one frequency and the complete motion is built up out of the sum of these component oscillations, each one of them weighted with appropriate magnitude and phase factors which can be obtained from the initial condition. In fact, premultiplication by $\mathbf{P}^T$ yields, by virtue of eqs (6.45)

$$\begin{aligned} \mathbf{y}_0 &= \mathbf{P}^T \mathbf{M} \mathbf{u}_0 \\[4pt] \dot{\mathbf{y}}_0 &= \mathbf{P}^T \mathbf{M} \dot{\mathbf{u}}_0 \end{aligned} \tag{6.58}$$

The new generalized coordinates which uncouple the equations of motion are called **normal coordinates** and the procedure of coordinate transformation leading to eqs (6.57a or b) is known as modal, or mode superposition, analysis. It is worth pointing out that modal analysis plays a central role in many aspects of vibration problems.

Pursuing further this line of reasoning, we can consider the potential and kinetic energies of our MDOF system which, in the case of small oscillations, can be written (eqs (3.93) and (3.95)) in matrix form as

$$V = \mathbf{u}^T \mathbf{K} \mathbf{u} \tag{6.59}$$

$$T = \dot{\mathbf{u}}^T \mathbf{M} \dot{\mathbf{u}} \tag{6.60}$$

Since $\mathbf{u} = \mathbf{P} \mathbf{y}$, it follows $\mathbf{u}^T = \mathbf{y}^T \mathbf{P}^T$ and substitution in eqs (6.59) and (6.60) gives

$$V = \tfrac{1}{2} \mathbf{y}^T \mathbf{L} \mathbf{y} = \tfrac{1}{2} \sum_{j=1}^{n} \omega_j^2 \, y_j^2 \tag{6.61}$$

$$T = \tfrac{1}{2} \dot{\mathbf{y}}^T \mathbf{I} \dot{\mathbf{y}} = \tfrac{1}{2} \sum_{j=1}^{n} \dot{y}_j^2 \tag{6.62}$$

so that the normal coordinates expressions of the potential and kinetic energies are sum of squares only, without any cross-product terms.

It is then straightforward to write the Lagrangian of the system as the sum of Lagrangians for n harmonic oscillators

$$L = \tfrac{1}{2} \sum_{j} (\dot{y}_j^2 - \omega_j^2 \, y_j^2) \tag{6.63}$$

obtain the Lagrange equations (i.e. eqs (6.57b)) and follow, for each normal coordinate, the same procedure that led to eq (6.18) in order to determine that the total energy associated with any particular y_j does not change with time and that there is no energy interchange between any two different normal coordinates. This is not true for the original set of coordinates where cross-product terms appear.

These important considerations extend and generalize to any undamped MDOF system the results obtained in the particular case of the simple 2-DOF system of Fig. 6.1.

Example 6.1. We are now in a position to discuss the system of Fig. 6.1 along the lines of the formal development above. Substitution of the appropriate values in eqs (6.21) yields

$$
\begin{bmatrix} 3m & 0 \\ 0 & m \end{bmatrix} \begin{bmatrix} \ddot{x}_1 \\ \ddot{x}_2 \end{bmatrix} + \begin{bmatrix} 5k & -k \\ -k & k \end{bmatrix} \begin{bmatrix} x_1 \\ x_2 \end{bmatrix} = \begin{bmatrix} 0 \\ 0 \end{bmatrix}
\tag{6.64}
$$

Assuming a harmonic time-dependent solution leads to the eigenvalue problem

$$
\begin{bmatrix} 5k & -k \\ -k & k \end{bmatrix} \begin{bmatrix} x_1 \\ x_2 \end{bmatrix} = \omega^2 \begin{bmatrix} 3m & 0 \\ 0 & m \end{bmatrix} \begin{bmatrix} x_1 \\ x_2 \end{bmatrix}
\tag{6.65}
$$

which is the explicit form of eq (6.24) for our particular problem. We have nontrivial solutions when $|K - \omega^2 M| = 0$, i.e. when the frequency equation (6.5) is satisfied; this means

$$
\omega_1 = \sqrt{2k/3m}
$$
$$
\tag{6.66}
$$
$$
\omega_2 = \sqrt{2k/m}
$$

Substitution of ω_1 in eq (6.65) gives

$$
\begin{bmatrix} 3k & -k \\ -k & k/3 \end{bmatrix} \begin{bmatrix} x_1 \\ x_2 \end{bmatrix} = \begin{bmatrix} 0 \\ 0 \end{bmatrix}
$$

from which it follows that $3x_1 = x_2$; the same procedure for ω_2 gives $x_1 = -x_2$.

We can normalize according to convention 1 of Section 6.3.1 and get the eigenvectors

$$
z_1 = \begin{bmatrix} 1/3 \\ 1 \end{bmatrix} \qquad z_2 = \begin{bmatrix} 1 \\ -1 \end{bmatrix}
\tag{6.67a}
$$

or normalize with respect to the mass matrix (convention 2 of Section 6.3.1, eq (6–42a)) and get

$$
\mathbf{p}_1 = \begin{bmatrix} \dfrac{1}{\sqrt{12m}} \\[2ex] \dfrac{3}{\sqrt{12m}} \end{bmatrix} \qquad \mathbf{p}_2 = \begin{bmatrix} \dfrac{1}{2\sqrt{m}} \\[2ex] -\dfrac{1}{2\sqrt{m}} \end{bmatrix} \tag{6.67b}
$$

It is then easy to verify the orthogonality relationships, i.e.

$$
\mathbf{z}_1^T \mathbf{M} \mathbf{z}_2 = \mathbf{z}_1^T \mathbf{K} \mathbf{z}_2 = 0
$$

or

$$
\mathbf{p}_1^T \mathbf{M} \mathbf{p}_2 = \mathbf{p}_1^T \mathbf{K} \mathbf{p}_2 = 0
$$

and – if we decide to use the eigenvectors of eqs (6.67a) – to obtain the modal masses and stiffnesses as

$$
M_{11} \equiv \mathbf{z}_1^T \mathbf{M} \mathbf{z}_1 = \tfrac{4}{3} m \qquad M_{22} \equiv \mathbf{z}_2^T \mathbf{M} \mathbf{z}_2 = 4m
$$

$$
K_{11} \equiv \mathbf{z}_1^T \mathbf{K} \mathbf{z}_1 = \tfrac{8}{9} k \qquad K_{22} \equiv \mathbf{z}_2^T \mathbf{K} \mathbf{z}_2 = 8k
$$

or, in case we want to use the mass-normalized eigenvectors, to determine that

$$
\mathbf{p}_i^T \mathbf{K} \mathbf{p}_i = \omega_i^2 \qquad (i = 1, 2)
$$

and

$$
\frac{K_{11}}{M_{11}} = \omega_1^2 \qquad \frac{K_{22}}{M_{22}} = \omega_2^2
$$

We can now form the mass-orthonormal modal matrix

$$
\mathbf{P} = \begin{bmatrix} \dfrac{1}{\sqrt{12m}} & \dfrac{1}{2\sqrt{m}} \\[2ex] \dfrac{3}{\sqrt{12m}} & -\dfrac{1}{2\sqrt{m}} \end{bmatrix}
$$

and obtain the normal coordinates. Since $\mathbf{x} = \mathbf{P}\mathbf{y}$, it follows that $\mathbf{y} = \mathbf{P}^T \mathbf{M} \mathbf{x}$.

In our case, explicitly,

$$
\begin{bmatrix} y_1 \\ y_2 \end{bmatrix} = \begin{bmatrix} \dfrac{1}{\sqrt{12m}} & \dfrac{3}{\sqrt{12m}} \\ \dfrac{1}{2\sqrt{m}} & -\dfrac{1}{2\sqrt{m}} \end{bmatrix} \begin{bmatrix} 3m & 0 \\ 0 & m \end{bmatrix} \begin{bmatrix} x_1 \\ x_2 \end{bmatrix}
$$

which is to say

$$
y_1 = \left(\frac{\sqrt{3m}}{2} \right) x_1 + \left(\frac{\sqrt{3m}}{2} \right) x_2
$$

$$
y_2 = \left(\frac{3}{2}\sqrt{m} \right) x_1 - \left(\frac{1}{2}\sqrt{m} \right) x_2
$$

(6.68)

Suppose now that the system is started into motion with the following initial conditions:

$$
\mathbf{x}_0 = [1 \quad 1]^T \qquad \dot{\mathbf{x}}_0 = [0 \quad 0]^T
$$

(6.69)

we can write, for example, the general solution in the form of eq (6.50) or (6.53), i.e.

$$
\mathbf{x} = \sum_j (\mathbf{p}_j^T \mathbf{M} \mathbf{x}_0 \cos \omega_j t) \mathbf{p}_j = \sum_j (\mathbf{p}_j \mathbf{p}_j^T \mathbf{M} \cos \omega_j t) \mathbf{x}_0
$$

which means

$$
\begin{bmatrix} x_1 \\ x_2 \end{bmatrix} = \begin{bmatrix} 1/2 \\ 3/2 \end{bmatrix} \cos \omega_1 t + \begin{bmatrix} 1/2 \\ -1/2 \end{bmatrix} \cos \omega_2 t
$$

(6.70)

Figures 6.2(a) and 6.2(b) show the first 30 s of motion of x_1 and x_2; for purposes of illustration, k and m have been assigned the values $k = 20$ N/m and $m = 5$ kg, so that $\omega_1 = 1.633$ rad/s and $\omega_2 = 2.828$ rad/s.

It is left to the reader to transform the initial conditions of eqs (6.69) into the initial conditions for the normal coordinates and to write explicitly the general solution for y_1 and y_2.

We point out that, in its schematic simplicity, the system considered in this example may be considered as representative of a wide class of undamped systems, which are essentially the same when we adopt some simplifying but

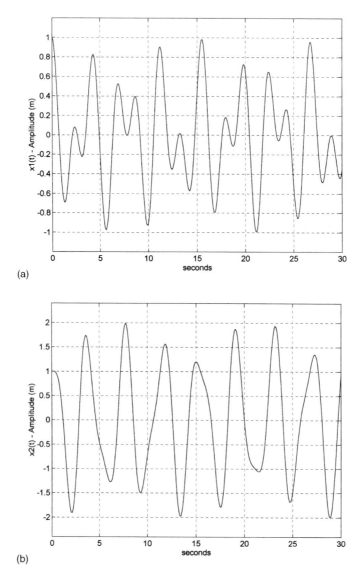

Fig. 6.2 First 30 s of time history for: (a) $x_1(t)$; (b) $x_2(t)$.

reasonable assumptions in the modelling process. The simplest and obvious example, for reasons explained in Chapter 4, is the same system which moves horizontally, i.e. the system of Fig. 6.3.

A second example is the two-storey shear building shown in Fig. 6.4. The basic assumptions are as follows: the frame is constrained to move in the

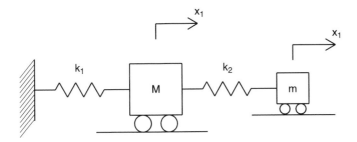

Fig. 6.3 Undamped 2-DOF translational system.

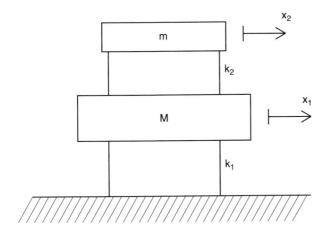

Fig. 6.4 Schematic (undamped) model of a two-storey building.

horizontal direction with no joint rotations, each floor is considered as a lumped mass at the floor level and the stiffness is associated with the lateral flexure of the weightless columns. This commonly adopted model can be extended to a *n*-storey frame building in a straightforward manner. Furthermore, on a qualitative basis, it can be (correctly) argued that the stiffness of the *i*th storey must depend on the flexural rigidity *EI* of all the columns in the storey and on the storey height *l*. In future chapters this statement will be made quantitatively clear.

The third example is shown in Fig. 6.5: two discs are mounted on a light (massless) shaft and have moments of inertia J_1 and J_2 about the shaft. The shaft, in turn, is clamped at one end. The generalized coordinates which come naturally in setting up the equations of motions are now θ_1 and θ_2, i.e. the angles through which the discs rotate about the shaft; these coordinates are the counterpart of x_1 and x_2 of the original problem. Physically, the torque exerted by the shaft on a disc is equal to the rate of change of the angular momentum of the disc about the shaft. Referring back to the analogy between

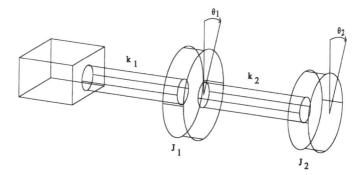

Fig. 6.5 Undamped 2-DOF rotational system: discs on a massless shaft.

translational and rotational systems of Table 4.1, it is not difficult to realize that the stiffness considered here is torsional stiffness, which is directly proportional to the angle and in a direction tending to return the discs to their equilibrium positions.

It is then left to the reader to show that, after the appropriate substitutions have been made, all the systems above lead to a formally identical set of equations of motion which can be solved with the matrix methods of the preceding sections.

Example 6.2.   A schematic vehicle model which accounts for bounce (up and down) and pitch (angular) motions is the one depicted in Fig. 6.6. This is a 2-DOF model; we assume the vertical coordinate x to be positive in the upward direction and the angle of pitch θ to be positive in the clockwise direction.

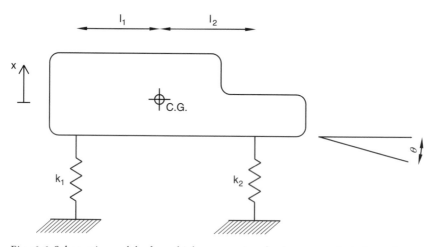

Fig. 6.6 Schematic model of a vehicle accounting for bounce and pitch motions.

The Lagrangian function is

$$L = \tfrac{1}{2}\left[m\dot{x}^2 + J\dot{\theta}^2 - k_1(x + l_1\theta)^2 - k_2(x - l_2\theta)^2 \right] \tag{6.71}$$

where J is the moment of inertia about the centre of gravity (C.G. in the figure). Lagrange's equations lead to the equations of motion

$$\begin{bmatrix} m & 0 \\ 0 & J \end{bmatrix}\begin{bmatrix} \ddot{x} \\ \ddot{\theta} \end{bmatrix} + \begin{bmatrix} k_1 + k_2 & k_1 l_1 - k_2 l_2 \\ k_1 l_1 - k_2 l_2 & k_1 l_1^2 + k_2 l_2^2 \end{bmatrix}\begin{bmatrix} x \\ \theta \end{bmatrix} = \begin{bmatrix} 0 \\ 0 \end{bmatrix} \tag{6.72}$$

which have already been written in matrix form; note that the static coupling disappears if $k_1 l_1 = k_2 l_2$. If we assume the following reasonable values for a car: $m = 1000$ kg; $J = 600$ kg m^2; $k_1 = k_2 = 10\,000$ N/m; $l_1 = 1.0$ m; $l_2 = 1.5$ m, it is not difficult to determine the two eigenvalues

$$\begin{array}{ll} \lambda_1 = 18.821 & \omega_1 = 4.338 \\[2mm] \lambda_2 = 55.346 & \omega_2 = 7.439 \end{array} \tag{6.73}$$

and the corresponding eigenvectors (normalized in accordance to the convention (1) of Section 6.3.1)

$$\mathbf{z}_1 = \begin{bmatrix} 1.000 \\ 0.236 \end{bmatrix} \qquad \mathbf{z}_2 = \begin{bmatrix} -0.142 \\ 1.000 \end{bmatrix} \tag{6.74}$$

It may be pointed out that the first mode is mainly translational, while the second mode is, on the contrary, mainly rotational. It can be argued *a posteriori* – on the basis of this consideration – that a reasonable estimate of the two frequencies could have been obtained by simply calculating the ratios

$$\omega_1^2 \approx \frac{k_1 + k_2}{m} = 20.0 \qquad \omega_2^2 \approx \frac{k_1 l_1^2 + k_2 l_2^2}{J} = 54.167$$

where the translational stiffness is divided by the mass to give the translational frequency and the rotational stiffness is divided by the moment of inertia to give the rotational frequency. The argument is, in principle, correct but it must be said that such circumstances are very seldom known in advance, especially when more complicated models are involved.

It is now left to the reader to verify the mass and stiffness orthogonality of the two eigenvectors, obtain the mass orthonormal eigenvectors, form the modal matrix and determine the normal coordinates. In addition, it may be useful to note how the accuracy of the orthogonality conditions and of eq (6.49a), i.e. the spectral expansion of the unit matrix, can be improved by retaining a different number of significant decimal figures in the entire process of calculation.

6.4   Eigenvalues and eigenvectors: sensitivity analysis

In the preceding sections we have seen that, for a given MDOF system, the eigenvalues and eigenvectors are physical quantities which strongly characterize the system under investigation. Consequently, before any further consideration can be made, it is interesting to have an idea of how these quantities change when the system is subjected to small modifications of its structural parameters, typically its mass and/or its stiffness. These modifications may reflect a number of different situations (for example, the addition or removal of springs or masses, a variation in the elasticity coefficients of materials or in their cross-sectional area, etc.) and being aware of their effects on the dynamical properties of the system may sometimes prevent – among other things – costly and ineffective system modifications. Furthermore, the effects of small errors in the measurement of the system properties can also be investigated.

This whole subject is known as 'sensitivity analysis' (or 'conditioning of the eigenvalue problem'); here we will consider here the case of a nondegenerate system subjected to a time-independent (stationary) small perturbation of its mass and/or stiffness matrix.

Let the unperturbed (zeroth-order) problem be

$$\mathbf{K}_0 \mathbf{p}_i^{(0)} = \lambda_i^{(0)} \mathbf{M}_0 \mathbf{p}_i^{(0)} \tag{6.75}$$

where we assume that the unperturbed eigenvalues $\lambda_i^{(0)}$ and mass orthonormal eigenvectors $\mathbf{p}_i^{(0)}$ are known. Let $\mathbf{K}_1$ and $\mathbf{M}_1$ be small perturbations of the stiffness and mass matrices, respectively. The perturbed problem can be written as

$$(\mathbf{K}_0 + \mathbf{K}_1)\mathbf{p}_i = \lambda_i(\mathbf{M}_0 + \mathbf{M}_1)\mathbf{p}_i \tag{6.76}$$

This assumption of small perturbations suggests that we expand the perturbed eigenvalues and eigenvectors in terms of a parameter γ such that the zeroth, first, etc. powers of γ correspond to the zeroth, first, etc. orders of the perturbation calculation. We then replace $\mathbf{K}_1$ and $\mathbf{M}_1$ by $\gamma\mathbf{K}_1$ and $\gamma\mathbf{M}_1$ and express λ_i and $\mathbf{p}_i$ as (well-behaved and rapidly converging) power series in γ. In the final result γ can be set equal to 1.

The perturbed eigenvalues and eigenvectors are written

$$\lambda_i = \lambda_i^{(0)} + \gamma\lambda_i^{(1)} + \gamma^2\lambda_i^{(2)} + \dots$$

$$\mathbf{p}_i = \mathbf{p}_i^{(0)} + \gamma\mathbf{p}_i^{(1)} + \gamma^2\mathbf{p}_i^{(2)} + \dots \tag{6.77}$$

and substitution into the eigenvalue problem gives

$$(\mathbf{K}_0 + \gamma \mathbf{K}_1)(\mathbf{p}_i^{(0)} + \gamma \mathbf{p}_i^{(1)} + \dots)$$
$$= (\lambda_i^{(0)} + \gamma \lambda_i^{(1)} + \dots)(\mathbf{M}_0 + \gamma \mathbf{M}_1)(\mathbf{p}_i^{(0)} + \gamma \mathbf{p}_i^{(1)} + \dots)$$

We can now equate the coefficients of equal powers of γ on both sides to obtain a series of equations that represent successively higher orders of the perturbation. We shall limit ourselves to the first order, since this is the most important perturbative term and it seldom makes sense to consider orders higher than the first or, at most, the second. We get

$$\mathbf{K}_0 \mathbf{p}_i^{(0)} = \lambda_i^{(0)} \mathbf{M}_0 \mathbf{p}_i^{(0)}$$

$$\mathbf{K}_0 \mathbf{p}_i^{(1)} + \mathbf{K}_1 \mathbf{p}_i^{(0)} = \lambda_i^{(0)} \mathbf{M}_0 \mathbf{p}_i^{(1)} + \lambda_i^{(0)} \mathbf{M}_1 \mathbf{p}_i^{(0)} + \lambda_i^{(1)} \mathbf{M}_0 \mathbf{p}_i^{(0)}$$

(6.78)

As expected, the zeroth-order problem (the first of eqs (6.78)) is the unperturbed problem. Let us now consider the first-order problem (the second of eqs (6.78)) and expand the vector $\mathbf{p}_i^{(1)}$ on the basis of the unperturbed eigenvectors, i.e.

$$\mathbf{p}_i^{(1)} = \sum_r c_{ir} \mathbf{p}_r^{(0)}$$

(6.79)

It is not difficult to see that substitution of eq (6.79) in the first-order problem leads to

$$\sum_r c_{ir}(\lambda_r^{(0)} - \lambda_i^{(0)})\mathbf{M}_0 \mathbf{p}_r^{(0)} + \mathbf{K}_1 \mathbf{p}_i^{(0)} = \lambda_i^{(0)} \mathbf{M}_1 \mathbf{p}_i^{(0)} + \lambda_i^{(1)} \mathbf{M}_0 \mathbf{p}_i^{(0)}$$

and premultiplication of this equation by $\mathbf{p}_k^{(0)T}$ gives

$$c_{ik}(\lambda_k^{(0)} - \lambda_i^{(0)}) + \mathbf{p}_k^{(0)T} \mathbf{K}_1 \mathbf{p}_i^{(0)} = \lambda_i^{(0)} \mathbf{p}_k^{(0)T} \mathbf{M}_1 \mathbf{p}_i^{(0)} + \lambda_k^{(1)}$$

(6.80)

where we used the orthogonality condition $\mathbf{p}_k^{(0)T} \mathbf{M}_0 \mathbf{p}_i^{(0)} = \delta_{ki}$. When $k = i$, eq (6.80) yields

$$\lambda_k^{(1)} = \mathbf{p}_k^{(0)T}(\mathbf{K}_1 - \lambda_k^{(0)} \mathbf{M}_1)\mathbf{p}_k^{(0)}$$

(6.81)

and when $k \neq i$, we get from eq (6.80)

$$c_{ik} = \frac{\mathbf{p}_k^{(0)T}(\mathbf{K}_1 - \lambda_i^{(0)} \mathbf{M}_1)\mathbf{p}_i^{(0)}}{\lambda_i^{(0)} - \lambda_k^{(0)}}$$

(6.82)

Now, in order to determine completely the first order perturbation of the eigenvector, only the coefficient c_{kk} is left; by imposing the normalization condition $\mathbf{p}_k^{(1)T}\mathbf{M}\mathbf{p}_k^{(1)} = 1$ and retaining only the first power of γ, we get

$$\mathbf{p}_k^{(0)T}\mathbf{M}_0\mathbf{p}_k^{(1)} + \mathbf{p}_k^{(0)T}\mathbf{M}_1\mathbf{p}_k^{(0)} + \mathbf{p}_k^{(1)T}\mathbf{M}_0\mathbf{p}_k^{(0)} = 0$$

and by virtue of the expansion (6.79) we are finally led to

$$c_{kk} = -\tfrac{1}{2}\mathbf{p}_k^{(0)T}\mathbf{M}_1\mathbf{p}_k^{(0)} \tag{6.83}$$

We can now explicitly write the result of the first-order perturbation calculation for the ith eigenvalue and the ith eigenvector as

$$\lambda_i = \lambda_i^{(0)} + \mathbf{p}_i^{(0)T}(\mathbf{K}_1 - \lambda_i^{(0)}\mathbf{M}_1)\mathbf{p}_i^{(0)} \tag{6.84}$$

$$\mathbf{p}_i = \mathbf{p}_i^{(0)} - \tfrac{1}{2}\mathbf{p}_i^{(0)T}\mathbf{M}_1\mathbf{p}_i^{(0)} + \sum_{\substack{r=1 \\ (r \neq i)}}^{n} \left(\frac{\mathbf{p}_r^{(0)T}(\mathbf{K}_1 - \lambda_i^{(0)}\mathbf{M}_1)\mathbf{p}_i^{(0)}}{\lambda_i^{(0)} - \lambda_r^{(0)}} \right) \tag{6.85}$$

From the expressions above, it may be noted that only the ith unperturbed parameters enter into the calculation of the perturbed eigenvalue, while the complete unperturbed eigensolution is required to obtain the perturbed eigenvector. Roughly speaking, we could say that the perturbation has the effect of 'mixing' the ith unperturbed eigenvector $\mathbf{p}_i^{(0)}$ with all the other eigenvectors $\mathbf{p}_r^{(0)}$ ($r \neq i$) for which the term in brackets of eq (6.85) is different from zero. Furthermore, a quick glance at the same equation suggests that the closer eigenvectors (to $\mathbf{p}_i^{(0)}$) give a greater contribution, because, for these vectors, the term $\lambda_i^{(0)} - \lambda_r^{(0)}$ is smaller.

It is evident that – with only minor modifications – the same results apply if we use the $\mathbf{z}$ eigenvectors; we only have to take into account the fact that in this case $\mathbf{z}_k^{(0)T}\mathbf{M}_0\mathbf{z}_i^{(0)} = M_{ki}\delta_{ki}$. Also, the perturbed $\mathbf{p}$ vectors are orthonormal with respect to the new mass matrix, not with respect to $\mathbf{M}_0$.

Example 6.3.   Let us go back to the system of Fig. 6.1, whose eigensolution has been considered in Example 6.1, and make the following modifications: we increase the first mass by $0.25m$ and decrease the second mass by $0.1m$. The total mass of the system changes from $4.0m$ into $4.15m$, which corresponds to an increase of 3.75% with respect to the original situation. Also, let us increase the stiffness of the first spring of $0.1k$ so that the term $k_1 + k_2$ changes from $5.0k$ into $5.1k$, i.e. an increase of 2.0% with respect to the original situation. These modifications can be considered small and we expect accurate results from our perturbative calculations.

The perturbation terms are

$$\mathbf{M}_1 = \begin{bmatrix} 0.25m & 0 \\ 0 & -0.1m \end{bmatrix} \qquad \mathbf{K}_1 = \begin{bmatrix} 0.1k & 0 \\ 0 & 0 \end{bmatrix}$$

We remember from Example 6.1 that: $\lambda_1^{(0)} = 2k/3m$, $\lambda_2^{(0)} = 2k/m$

$$\mathbf{M}_0 = \begin{bmatrix} 3m & 0 \\ 0 & m \end{bmatrix} \qquad \mathbf{K}_0 = \begin{bmatrix} 5k & -k \\ -k & k \end{bmatrix}$$

and

$$\mathbf{p}_1^{(0)} = \begin{bmatrix} \dfrac{1}{\sqrt{12\,m}} & \dfrac{3}{\sqrt{12\,m}} \end{bmatrix}^T \qquad \mathbf{p}_2^{(0)} = \begin{bmatrix} \dfrac{1}{2\sqrt{m}} & -\dfrac{1}{2\sqrt{m}} \end{bmatrix}^T$$

so that the first-order perturbative terms for the eigenvalues can be obtained as (eq (6.84))

$$\lambda_1^{(1)} = \mathbf{p}_1^{(0)T}(\mathbf{K}_1 - \lambda_1^{(0)}\mathbf{M}_1)\mathbf{p}_1^{(0)} = 0.0444 \ (k/m)$$

$$\lambda_2^{(1)} = \mathbf{p}_2^{(0)T}(\mathbf{K}_1 - \lambda_2^{(0)}\mathbf{M}_1)\mathbf{p}_2^{(0)} = -0.0500 \ (k/m)$$

and hence

$$\lambda_1 = \frac{2k}{3m} + 0.04444\,\frac{k}{m} = 0.711\,\frac{k}{m}$$

$$\lambda_2 = \frac{2k}{m} - 0.050\,\frac{k}{m} = 1.950\,\frac{k}{m}$$

(6.86)

For the first eigenvector, the expansion coefficients are given by

$$c_{11} = -\tfrac{1}{2}\mathbf{p}_1^{(0)T}\mathbf{M}_1\mathbf{p}_1^{(0)} = 0.0271$$

$$c_{12} = \frac{\mathbf{p}_2^{(0)T}(\mathbf{K}_1 - \lambda_1^{(0)}\mathbf{M}_1)\mathbf{p}_1^{(0)}}{\lambda_1^{(0)} - \lambda_2^{(0)}} = 0.0289$$

from which it follows that

$$\mathbf{p}_1 = \mathbf{p}_1^{(0)} + c_{11}\mathbf{p}_1^{(0)} + c_{12}\mathbf{p}_2^{(0)} = \begin{bmatrix} \dfrac{0.311}{\sqrt{m}} & \dfrac{0.875}{\sqrt{m}} \end{bmatrix}^T$$

(6.87a)

and the same procedure for the second eigenvector leads to

$$\mathbf{p}_2 = \begin{bmatrix} \dfrac{0.459}{\sqrt{m}} & -\dfrac{0.584}{\sqrt{m}} \end{bmatrix}^T$$

(6.87b)

Because of the simplicity of this example, these results can be compared to the exact calculation for the modified system, which can be performed with small effort. The exact eigenvalues are $\lambda_1^{(exact)} = 0.712(k/m)$ and $\lambda_2^{(exact)} = 1.968(k/m)$, which corresponds to a relative error of 0.07% on the first frequency and a relative error of 0.46% on the second. The exact eigenvectors are

$$
\mathbf{p}_1^{(exact)} = \begin{bmatrix} \dfrac{0.313}{\sqrt{m}} & \dfrac{0.871}{\sqrt{m}} \end{bmatrix}^T \qquad \mathbf{p}_2^{(exact)} = \begin{bmatrix} \dfrac{0.458}{\sqrt{m}} & -\dfrac{0.594}{\sqrt{m}} \end{bmatrix}^T
$$

and they must be compared, respectively, to eqs (6.87a and b).

Some complications appear in the case of degenerate eigenvalues. We will not deal with this subject in detail but a few qualitative considerations can be made. Suppose, for example, that two independent eigenvectors $\mathbf{p}_{i1}^{(0)}$ and $\mathbf{p}_{i2}^{(0)}$ correspond to the unperturbed eigenvalue $\lambda_i^{(0)}$ (twofold degeneracy). In general, the perturbation will split this eigenvalue into two different values, say λ_{i1} and λ_{i2}; as the perturbation tends to zero, the eigenvectors will tend to two unperturbed eigenvectors $\tilde{\mathbf{p}}_{i1}^{(0)}$ and $\tilde{\mathbf{p}}_{i2}^{(0)}$, which will be in general two linear combinations of $\mathbf{p}_{i1}^{(0)}$ and $\mathbf{p}_{i2}^{(0)}$. The additional problem is, as a matter of fact, the determination of $\tilde{\mathbf{p}}_{i1}^{(0)}$ and $\tilde{\mathbf{p}}_{i2}^{(0)}$: this particular pair – out of the infinite number of combinations of $\mathbf{p}_{i1}^{(0)}$ and $\mathbf{p}_{i2}^{(0)}$ – will depend on the perturbation itself.

For instance, let $\lambda_i^{(0)}$ be an *m*-fold degenerate eigenvalue and let $\mathbf{p}_{i1}^{(0)}, \mathbf{p}_{i2}^{(0)}, \dots, \mathbf{p}_{im}^{(0)}$ be a possible choice of mass-orthonormal eigenvectors (i.e. a basis in the subspace relative to the *i*th eigenvalue). We can then write the expansions

$$
\mathbf{p}_i^{(0)} = \sum_{j=1}^{m} c_j^{(0)} \mathbf{p}_{ij}^{(0)} \tag{6.88}
$$

and

$$
\mathbf{p}_i^{(1)} = \sum_{r=1}^{n} \sum_{j=1}^{m} c_{rj}^{(1)} \mathbf{p}_{rj}^{(0)} \tag{6.89}
$$

substitute them in the first-order problem and project the resulting equation successively on the eigenvectors $\mathbf{p}_{i1}^{(0)}, \mathbf{p}_{i2}^{(0)}, \dots, \mathbf{p}_{im}^{(0)}$ (this is done by premultiplying, respectively, by $\mathbf{p}_{i1}^{(0)T}, \mathbf{p}_{i2}^{(0)T}, \dots, \mathbf{p}_{im}^{(0)T}$). We obtain, after some manipulation, a system of *m* homogeneous equations, which admits nontrivial solutions if the determinant of the coefficients is equal to zero. This condition results in an algebraic equation of degree *m* in $\lambda_i^{(1)}$ and its *m* solutions $\lambda_{i1}^{(1)}, \lambda_{i2}^{(1)}, \dots, \lambda_{im}^{(1)}$ represent the first-order corrections to $\lambda_i^{(0)}$. Substitution of each one of these values into the homogeneous system allows the calculation of the zeroth-order coefficients $c_j^{(0)}$ ($j = 1, 2, \ldots, m$) for the relevant eigenvector. We have thus obtained the desired m linear combinations of the unperturbed eigenvectors (i.e. the $\tilde{\mathbf{p}}_{ij}^{(0)}$); once these are known, the coefficients $c_{rj}^{(1)}$ can be obtained by projecting the first-order equation on different eigenvectors.

It is interesting to note that, in many cases, the effect of the perturbation is to completely or partially 'remove the degeneracy' by splitting the degenerate eigenvalue into a number of different frequencies that were indistinguishable in the original system. This circumstance can be useful in some practical applications, and it is worth pointing out that similar procedures apply – with only minor modifications – in the case of distinct but closely spaced eigenvalues.

The subject of sensitivity analysis is much broader than shown in our discussion; in general, we can say that some linear systems are extremely sensitive to small changes in the system, and others are not. Sensitive systems are often said to be 'ill-conditioned', whereas insensitive systems are said to be 'well-conditioned'.

We will see that the generalized eigenvalue problem of eq (6.24) (or (6.29), which is the same) can be transformed into a standard eigenvalue problem (eq (6–26a)), where $\mathbf{A}$ is an appropriate matrix whose form and entries depend on how the transformation is carried out. The key point is that the eigenvalues are continuous functions of the entries of $\mathbf{A}$, so we have reason to believe that a small perturbation matrix will correspond to a small change of the eigenvalues. But one often needs precise bounds to know how small is 'small' in each case.

We will not pursue this subject further here for two reasons: first, a detailed discussion is beyond the scope of this book and, second, it would lead us too far away from the main topic of this chapter. For the moment, it suffices to say that if $\mathbf{A}$ is diagonalizable (see Appendix A on matrix analysis), it can be shown that it is possible to define a 'condition number' that represents a quantitative measure of ill-conditioning of the system and provides an upper bound on the perturbation of the eigenvalues due to a unit norm change in the system matrix; furthermore, it may be of interest to note that normal matrices are well-conditioned with respect to eigenvalue computations, that the condition number is generally conservative and that a better bound can be obtained if both $\mathbf{A}$ and the perturbing matrices are Hermitian. (The interested reader is referred to Horn and Johnson [1] and Junkins and Kim [2].)

6.4.1 Light damping

The free vibration of a damped system is governed by eq (3.101), i.e.

$$\mathbf{M}\ddot{\mathbf{u}} + \mathbf{C}\dot{\mathbf{u}} + \mathbf{K}\mathbf{u} = 0 \qquad (6.90)$$

As in the undamped case there are $2n$ independent solutions which can be superposed to meet $2n$ initial conditions. Assuming a trial solution of the form

$$\mathbf{u} = \mathbf{z}e^{\lambda t} \tag{6.91}$$

leads to

$$(\lambda^2 \mathbf{M} + \lambda \mathbf{C} + \mathbf{K})\mathbf{z} = 0 \tag{6.92}$$

which admits a nontrivial solution if the matrix in parentheses on the left-hand side is singular. Equation (6.92) represents what is commonly called a complex (or quadratic) eigenvalue problem because the eigenvalue λ and the elements of the eigenvector $\mathbf{z}$ are, in general, complex numbers; if λ and $\mathbf{z}$ satisfy eq (6.92), then so also do λ^* and $\mathbf{z}^*$, where the asterisk denotes complex conjugation. In general, the complex eigenvalue problem is much more difficult than its undamped counterpart and much less attention has been given to efficient numerical procedures for its solution, but we will return to these aspects later.

For the moment, let us make the following assumptions: the solution of the undamped problem is known and the system is lightly damped. The damping term can then be considered a small perturbation of the original undamped system and we are in a position to investigate its effect on the eigensolution of the conservative system.

Let $\pm i\omega_j$ and $\mathbf{p}_j$ ($j = 1, 2, \ldots, n$) be the eigenvalues and the mass-orthonormal eigenvectors of the conservative system (i.e. when $\mathbf{C} = 0$ in eq (6.92)); under the assumption of light damping we can write

$$\begin{aligned} \lambda_j &= i\omega_j + \Delta\lambda_j \\ \mathbf{z}_j &= \mathbf{p}_j + \Delta\mathbf{p}_j \end{aligned} \tag{6.93}$$

Substitute these expressions in eq (6.92) and retain only the first-order terms (note that the terms in $\Delta\lambda\mathbf{C}$ and $\mathbf{C}\Delta\mathbf{z}$ are neglected because they are second-order for light damping). After some manipulation we arrive at

$$(\mathbf{K} - \omega_j^2\mathbf{M})\Delta\mathbf{p}_j + i\omega_j(\mathbf{C} + 2\Delta\lambda_j\mathbf{M})\mathbf{p}_j = 0 \tag{6.94}$$

Now, as we did in eq (6.79), we expand $\Delta\mathbf{p}$ on the basis of the unperturbed eigenvectors, i.e.

$$\Delta\mathbf{p}_j = \sum_r a_{jr}\mathbf{p}_r \tag{6.95}$$

Substitute eq (6.95) in (6.94) and premultiply the resulting expression by $\mathbf{p}_k^T$ to get

$$\sum_r a_{jr}(\omega_r^2 - \omega_j^2)\mathbf{p}_k^T\mathbf{M}\mathbf{p}_r + i\omega_j\mathbf{p}_k^T(\mathbf{C} + 2\Delta\lambda_j\mathbf{M})\mathbf{p}_j = 0 \tag{6.96}$$

Since $\mathbf{p}_k^T\mathbf{M}\mathbf{p}_r = \delta_{kr}$, for $k = j$ we get the first-order perturbation of the jth eigenvalue

$$\Delta\lambda_j = -\tfrac{1}{2}\mathbf{p}_j^T\,\mathbf{C}\mathbf{p}_j \tag{6.97}$$

Note that a term M_{jj} appears in the denominator of the right-hand side of eq (6.97) if we do not use mass-orthonormal vectors in the calculation.

From eq (6.97) two observations can be made:

- Each correction to the unperturbed eigenvalues takes the form of a real negative part (matrix $\mathbf{C}$ is generally positive definite) which transforms the solution into a damped oscillatory motion and accounts for the fact that the free vibration of real systems dies out with time because there is always some loss of energy.
- The first-order correction involves only the diagonal terms of the matrix $\mathbf{P}^T\mathbf{C}\mathbf{P}$ which is, in general, nondiagonal unless some assumptions are made on the damping matrix (remember that both $\mathbf{M}$ and $\mathbf{K}$ become diagonal under the similarity transformation $\mathbf{P}^T\mathbf{M}\mathbf{P}$ and $\mathbf{P}^T\mathbf{K}\mathbf{P}$). Off-diagonal terms have only a second-order effect on the unperturbed eigenvalues.

When $k \neq j$ eq (6.96) gives

$$a_{jk} = i\omega_j\,\frac{\mathbf{p}_k^T\mathbf{C}\mathbf{p}_j}{(\omega_j^2 - \omega_k^2)} \tag{6.98a}$$

Again, a term M_{kk} appears at the denominator on the right-hand side if the calculation is made with eigenvectors that are not mass-orthonormal; note also that a minus sign appears on the right-hand side if we start with $\lambda_j = -i\omega_j + \Delta\lambda_j$. The perturbed eigenvector is then

$$\mathbf{z}_j = \mathbf{p}_j + \sum_{\substack{k \\ k \neq j}} i\omega_j\,\frac{\mathbf{p}_k\mathbf{C}\mathbf{p}_j}{(\omega_j^2 - \omega_k^2)}\,\mathbf{p}_k \tag{6.98b}$$

showing that the perturbation splits the original real eigenvector into a pair of complex vectors having the same real part as the undamped mode (remember that, in vibration terminology, the term mode is analogous to eigenvector:

more precisely, a mode is a particular pattern of motion which is mathematically represented by an eigenvector) but having small conjugate imaginary parts.

On physical grounds – unless damping has some desirable characteristics which will be considered in a later section – this occurrence translates into the fact that, in a particular mode, each coordinate has a relative amplitude and a relative phase with respect to any other coordinate. In other words, the free vibration of a generally damped system oscillating in a particular mode is no longer a synchronous motion of the whole system: the individual degrees of freedom no longer move in phase or antiphase and they no longer reach their extremes of motion together.

For obvious reasons, this pattern of motion is usually called a 'complex mode', as opposed to the 'real mode' of the undamped system where each coordinate does have an amplitude, but a phase angle which is either $0°$ or $180°$ and real numbers suffice for a complete description.

6.5 Structure and properties of matrices M, K and C: a few considerations

A fundamental part of the analysis of MDOF systems – and of any physical phenomenon in general – is the solution of the appropriate equations of motion. However, as we stated in Chapter 1, the first step in any investigation is the formulation of the problem; this step involves the selection of a mathematical model which has to be both effective and reliable, meaning that we expect our model to reproduce the behaviour of the real physical system within an acceptable degree of accuracy and, possibly, at the least cost. We always must keep in mind that, once the mathematical model has been chosen, we solve that particular model and the solution can never give more information than that implicitly contained in the model itself. These observations become more important when we consider that:

- Numerical procedures implemented on digital computers play a central role (think, for example, to the finite-element method) in the analysis of systems with more than three or four degrees of freedom.
- Matrix algebra is the 'natural language' of these procedures
- The effectiveness and reliability of numerical techniques depend on the structure and properties of the input matrices.
- Continuous systems (i.e. systems with an infinite number of degrees of freedom) are very often modelled as MDOF systems.

As in the case of an SDOF system, the principal forces acting on an MDOF system are (1) the inertia forces, (2) the elastic forces, (3) the damping forces and (4) the externally applied forces. We will not consider, for the moment, the forces of type (4). Under the assumption of small amplitude vibrations, we have seen in Chapter 3 that matrices **M**, **K** and **C** are symmetrical. Symmetry is a desirable property and results in significant computational advantages. In essence, the symmetry property of **M** and **K** depends on the form of the kinetic and potential energy functions and the symmetry **C** of depends on the existence of the Rayleigh dissipation function.

Unfortunately, for most systems the damping properties are very difficult, if not impossible, to define. For this reason the most common choices for the treatment of its effects are (1) neglect damping altogether (this is often a better assumption than it sounds), (2) assume 'proportional damping' (Section 6.7) or (3) use available experimental information on the damping characteristics of a typical similar structure or on the structure itself.

We know from Chapter 3 that both kinetic and potential energies can be written as quadratic forms and we know from basic physics that they are essentially positive quantities. If none of the degrees of freedom has zero mass, $\dot{\mathbf{u}}^T \mathbf{M} \dot{\mathbf{u}}$ (eq (3.95)) is never zero unless $\dot{\mathbf{u}}$ is a zero vector and hence **M**, besides being symmetrical, is also positive definite; if some of the degrees of freedom have zero mass then **M** is a positive semidefinite matrix.

Similar considerations apply for the stiffness matrix; unless the system is unrestrained and capable of rigid-body modes, **K** is a positive definite matrix. When this is not the case, i.e. when rigid-body modes are possible (Section 6.6), the stiffness matrix is positive semidefinite. It is worth pointing out that if a matrix **A** is symmetrical and positive definite, then $\mathbf{A}^{-1}$ always exists (i.e. **A** is nonsingular) and is a symmetrical positive definite matrix itself. The fact that either **M**, or **K**, or both, are nonsingular is useful when we want to transform the generalized eigenvalue problem (eq (6.29)) into a standard eigenvalue problem (eq (6.26a)), which is the form required by some numerical eigensolvers (section 6.8).

6.5.1 Mass properties

The simplest procedure for defining the mass properties of a structure is by concentrating, or lumping, its mass at the points where the displacements are defined. This is certainly not a problem for a simple system such as the one in Fig. 6.1 where mass is, as a matter of fact, localized, but a certain degree of arbitrariness is inevitable for more complex systems. In any case, whatever the method we use to concentrate the masses of a given structure, if we choose the coordinates as the absolute displacement of the masses we obtain a diagonal mass matrix. In fact, the off-diagonal terms are zero because an acceleration at one point produces an inertia force at that point only; this is not strange if we consider that m_{ij} is the force that must be applied at point i to equilibrate the inertia forces produced by a unit acceleration at point j, so that $m_{ii} = m_i$ and $m_{ij} = 0$ for $i \neq j$.

A diagonal matrix is certainly desirable for computational purposes, but a serious disadvantage of this approach is the fact that the mass associated with rotational degrees of freedom is zero because a point has no rotational inertia. This means that – when rotational degrees of freedom must be considered in a specific problem – the mass matrix is singular. In principle, the problem could be overcome by assigning some rotational inertia to the masses associated with rotational degrees of freedom (in which case the diagonal mass coefficient would be the rotational inertia of the mass), but this is easier said than done. The general conclusion is that the lumped mass matrix is a diagonal matrix with nonzero elements for each translational degree of freedom and zero diagonal elements for each rotational degree of freedom.

A different approach is based on the **assumed-modes method**, a far-reaching technique developed along the line of reasoning of Section 5.5 (see also Chapter 9). In that section, a distributed parameter system was modelled as an SDOF system by an appropriate choice of a shape, or trial, function under the assumption that only one vibration pattern is developed during the motion. This basic idea can be improved by superposing n shape functions $\Psi_1(x), \Psi_2(x), \dots, \Psi_n(x)$ so that

$$u(x, t) = \sum_{i=1}^{n} \Psi_i(x) z_i(t) \tag{6.99}$$

where the $z_i(t)$ constitute a set of n generalized time-dependent coordinates. (Note that we considered the trial functions in eq (6.99) to depend on one spatial coordinate only, thus implying a one-dimensional problem (for example, an Euler–Bernoulli beam); this is only for our present convenience, and the extension to two or three spatial coordinates is straightforward.)

In essence, eq (6.99) represents an n-DOF model of a continuous system, and since the kinetic energy of a continuous system is an integral expression depending on the partial derivative $\partial u / \partial t$, we can substitute eq (6.99) into this expression to arrive at the familiar form

$$T = \frac{1}{2} \sum_{i,j=1}^{n} m_{ij} \dot{z}_i \dot{z}_j = \frac{1}{2} \dot{z}^T \mathbf{M} \dot{z} \tag{6.100}$$

where the coefficients m_{ij} will now depend on the mass distribution of the system and on the trial functions Ψ_i. Consider, for example, the axial vibration of an elastic bar of length L and mass $\hat{m}$ per unit length; the kinetic energy is given by

$$T = \frac{1}{2} \int_0^L \hat{m} \left(\frac{\partial u}{\partial t} \right)^2 dx \tag{6.101}$$

Inserting eq (6.99) into eq (6.101) leads to

$$T = \frac{1}{2} \sum_{i,j=1}^{n} \left(\int_0^L \hat{m} \, \Psi_i \Psi_j \, dx \right) \dot{z}_i \dot{z}_j \tag{6.102}$$

which is formally equivalent to eq (6.100) when we define the coefficients m_{ij} as

$$m_{ij} = \int_0^L \hat{m} \, \Psi_i \Psi_j dx \qquad (6.103)$$

and it is evident that $m_{ij} = m_{ji}$, i.e. the mass matrix is symmetrical. Note that if we define the row vector of shapes $\mathbf{y} \equiv [\Psi_1 \ldots \Psi_n]$ and the column vector of generalized coordinates $\mathbf{z} \equiv [z_1 \ldots z_n]^T$ we can write eqs (6.99) and (6.102) and the mass matrix, respectively, as

$$u(x, t) = \mathbf{yz}$$

$$T = \tfrac{1}{2} \int_0^L \hat{m} \mathbf{z}^T \mathbf{y}^T \mathbf{yz} dx \qquad (6.104)$$

$$\mathbf{M} = \int_0^L \hat{m} \mathbf{y}^T \mathbf{y} dx$$

For purposes of illustration, let us consider a clamped–free bar in axial vibration: we can model this continuous system as a 2-DOF system and express the displacement by means of the two shape functions $\Psi_1 = x/L$ and $\Psi_2 = (x/L)^2$. From eq (6.103) we get

$$m_{11} = \int_0^L \hat{m} \left(\frac{x}{L}\right)^2 dx = \frac{\hat{m}L}{3}$$

$$m_{12} = m_{21} = \int_0^L \hat{m} \left(\frac{x}{L}\right)^3 dx = \frac{\hat{m}L}{4} \qquad (6.105a)$$

$$m_{22} = \int_0^L \hat{m} \left(\frac{x}{L}\right)^4 dx = \frac{\hat{m}L}{5}$$

and hence the mass matrix

$$\mathbf{M} = \hat{m}L \begin{bmatrix} 1/3 & 1/4 \\ 1/4 & 1/5 \end{bmatrix} \qquad (6.105b)$$

6.5.2 Elastic properties

The determination of the elastic properties of a given system is a standard problem of structural analysis. This task is accomplished by means of the influence coefficients which express the relation between the displacement at a point and the forces acting at that and other points of the system. The flexibility influence coefficient a_{ij} is defined as the displacement at point $x = x_i$ due to a unit load applied at point $x = x_j$ with all other loads equal to zero. The principle of superposition for linearly elastic systems allows one to write the displacement u_i at point i as

$$u_i = \sum_{j=1}^{n} a_{ij} f_j \qquad (6.106)$$

where f_j is the force applied at $x = x_j$. The units of the flexibility coefficients are metres per newton when the relation (6.106) is between linear displacements and forces, if angular displacements and torques are considered, the units change accordingly. In the case of a single spring it is evident that the flexibility coefficient is simply $a = 1/k$, where k is the spring constant.

Equation (6.106) can be written in matrix form as $\mathbf{u} = \mathbf{Af}$, where $\mathbf{A}$ is called the flexibility matrix and the other symbols are obvious.

The stiffness influence coefficient k_{ij} is defined as the load required at $x = x_i$ to produce a unit displacement at $x = x_j$ when all other displacements are held to zero. This definition is more involved than the definition of the flexibility coefficient; nevertheless the stiffness coefficients are sometimes easier to determine than the flexibility coefficients. Consider, for example a one-dimensional 3-DOF system, if $x = x_1$ is given a unit displacement (i.e. $u_1 = 1$) and $u_j = 0$ for $j \neq 1$, the forces at points 1, 2 and 3 required to maintain this displacement configuration are exactly k_{11}, k_{21} and k_{31}, i.e. the first column of the stiffness matrix; moreover, these coefficients must be considered with the appropriate sign: positive in the sense of positive force and negative otherwise. By a line of reasoning parallel to that used for a_{ij}, we can write

$$f_i = \sum_{j=1}^{n} k_{ij} u_j \qquad (6.107)$$

or, equivalently, $\mathbf{f} = \mathbf{Ku}$, where $\mathbf{K}$ is the stiffness matrix. The conservative nature of the linearly elastic systems we consider (meaning that the sequence of load applications is unimportant) leads to Clapeyron's law for the total strain energy which reads

$$E_{strain} = \tfrac{1}{2} \sum_{j=1}^{n} f_j u_j \qquad (6.108)$$

and is valid for a structure which is initially stress free and not subjected to temperature changes. Furthermore, Maxwell's reciprocity theorem holds and it can be invoked to prove that $a_{ij} = a_{ji}$ and $k_{ij} = k_{ji}$, or, in matrix form

$$\mathbf{A} = \mathbf{A}^T$$
$$\mathbf{K} = \mathbf{K}^T$$

(6.109)

i.e. the flexibility matrix and the stiffness matrix are symmetrical.

The structure of eqs (6.106) and (6.107) suggests that the two matrices should be related; it is so, and it is not difficult to prove that

$$\mathbf{K} = \mathbf{A}^{-1}$$
$$\mathbf{A} = \mathbf{K}^{-1}$$

(6.110)

The interested reader can find excellent discussions of this topic in many textbooks (e.g. Bisplinghoff *et al.* [3]).

Depending on the specific problem, in some cases it may be easier to obtain directly the stiffness matrix whereas in some other cases it may be more convenient to obtain the flexibility matrix and then invert it. We must remember, however, that for an unrestrained system – where rigid body motions are possible – the stiffness matrix $\mathbf{K}$ is singular, hence it has no inverse. This is consistent with the fact that for an unrestrained system we cannot define flexibility influence coefficients.

Example 6.4.   Using the definitions given above, the reader is invited to determine that the flexibility and stiffness matrices for the one-dimensional 3-DOF system of Fig. 6.7 are, respectively

$$\mathbf{A} = \begin{bmatrix} \dfrac{1}{k_1} & \dfrac{1}{k_1} & \dfrac{1}{k_1} \\ \dfrac{1}{k_1} & \dfrac{1}{k_1} + \dfrac{1}{k_2} & \dfrac{1}{k_1} + \dfrac{1}{k_2} \\ \dfrac{1}{k_1} & \dfrac{1}{k_1} + \dfrac{1}{k_2} & \dfrac{1}{k_1} + \dfrac{1}{k_2} + \dfrac{1}{k_3} \end{bmatrix}$$

$$\mathbf{K} = \begin{bmatrix} k_1 + k_2 & -k_2 & 0 \\ -k_2 & k_2 + k_3 & -k_3 \\ 0 & -k_3 & k_3 \end{bmatrix}$$

It is then straightforward to show that $\mathbf{KA} = \mathbf{I}$.

Example 6.5.   Consider now the system of Fig. 6.8. The mass m is connected to a base of mass M which can undergo translational and rotational motion.

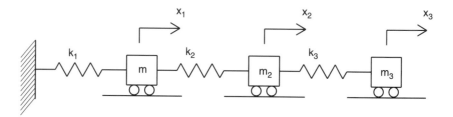

Fig. 6.7 One-dimensional 3-DOF system.

The mass moment of inertia of the base is J_G, k_r is a rotational spring and all the other symbols are shown in the figure. The system has three degrees of freedom and the equations of free motion can be written in the usual matrix form as $\mathbf{M\ddot{u}} + \mathbf{Ku} = 0$, where $\mathbf{u} = [x_1 \quad x_2 \quad \theta]^T$. Using the definition of stiffness and flexibility influence coefficients, the reader is invited to determine that

$$\mathbf{K} = \begin{bmatrix} k_1 + k_2 & -k_2 & 0 \\ -k_2 & k_2 & 0 \\ 0 & 0 & k_r \end{bmatrix} \qquad \mathbf{A} = \begin{bmatrix} \dfrac{1}{k_1} & \dfrac{1}{k_1} & 0 \\ \dfrac{1}{k_1} & \dfrac{1}{k_1} + \dfrac{1}{k_2} & 0 \\ 0 & 0 & \dfrac{1}{k_r} \end{bmatrix}$$

and that the stiffness matrix can be also (and maybe more easily) determined by making use of Lagrange's equations, the Lagrangian of this system being

$$L = \tfrac{1}{2} M\dot{x}_1^2 + \tfrac{1}{2} m(\dot{x}_2 + l\dot{\theta})^2 + \tfrac{1}{2} J_G \dot{\theta}^2 - \tfrac{1}{2} k_1 x_1^2 - \tfrac{1}{2} k_2 (x_2 - x_1)^2 - \tfrac{1}{2} k_r \theta^2$$

As expected, Lagrange's equations lead to the equations of motion and hence also to the mass matrix. Note that this choice of coordinates results in both dynamic and static coupling of the equations of motion.

In principle, the above definition of influence coefficients may be used whenever convenient. However, the assumed-modes method can be useful in many circumstances. Suppose, for example, that we are dealing with a continuous system; as for the kinetic energy, its potential energy is an integral expression depending on the partial spatial derivatives of the function $u(x,t)$. The basic assumption of eq (6.99) generates a n-DOF model of our system and can be substituted in the potential energy expression to arrive at the familiar form

$$V = \tfrac{1}{2} \sum_{i,j=1}^{n} k_{ij} z_i z_j = \tfrac{1}{2} \mathbf{z}^T \mathbf{K} \mathbf{z} \qquad (6.111)$$

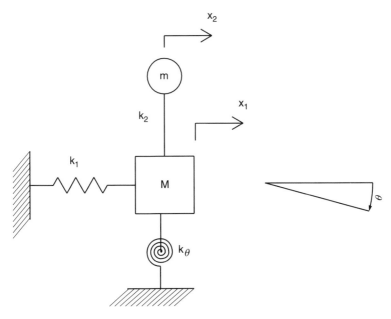

Fig. 6.8 3-DOF system: mass *m* elastically supported on a base *M*.

where the coefficients k_{ij} will depend on the stiffness distribution and on the trial functions Ψ_i. If we consider again the axial vibration of an elastic bar of length L and axial rigidity EA (where A is the cross-sectional area), its potential strain energy is given by

$$V = \frac{1}{2} \int_0^L EA \left(\frac{\partial u}{\partial x} \right)^2 dx \qquad (6.112)$$

Substitution of eq. (6.99) in (6.112) yields

$$V = \frac{1}{2} \sum_{i,j=1}^{n} \left(\int_0^L EA \frac{d\Psi_i}{dx} \frac{d\Psi_j}{dx} dx \right) z_i z_j \qquad (6.113)$$

which is formally equal to eq (6.111) if we define

$$k_{ij} = \int_0^L EA(\Psi_i' \, \Psi_j') \, dx \qquad (6.114)$$

where, as is customary, the prime indicates the derivative with respect to the spatial variable, i.e. $\Psi_i' \equiv d\Psi_i/dx$. To be more specific, let the bar be clamped at one end and free at the other end; we choose a 2-DOF model by choosing the 'reasonable' functions $\Psi_1 = x/L$ and $\Psi_2 = (x/L)^2$, then

$$k_{11} = \int_0^L EA \frac{1}{L^2} \, dx = \frac{EA}{L}$$

$$k_{12} = k_{21} = \int_0^L EA \frac{2x}{L^3} \, dx = \frac{EA}{L} \tag{6.115a}$$

$$k_{22} = \int_0^L EA \frac{4x^2}{L^4} \, dx = \frac{4}{3} \frac{EA}{L}$$

where we have assumed a uniform axial rigidity. In matrix form, eqs (6.115a) are written

$$K = \frac{EA}{L} \begin{bmatrix} 1 & 1 \\ 1 & 4/3 \end{bmatrix} \tag{6.115b}$$

Note that the mass matrix of eq (6.105b) refers to the same system as the stiffness matrix (6.115b) and both have been obtained by using the same shape functions: the common terminology in this case is that the mass matrix is 'consistent' with the stiffness matrix of eq (6.115b). Strictly speaking, the term applies to the finite-element approach, but the latter can be considered as an application of the assumed-modes method where the shape functions represent deflection patterns of limited portions (the so-called finite elements) of a given structural system. In the end, in order to construct a mathematical model of the whole structure, these elements are assembled together in a common (or global) frame of reference.

The assumed-modes method leaves open the question of what constitutes a judicious choice of the shape functions. We have already faced this problem in Chapter 5 where we stated the importance of boundary conditions for continuous systems and we defined, for a given system, the classes of admissible and comparison functions (incidentally, note that $\Psi_1 = x/L$ and $\Psi_2 = (x/L)^2$ are admissible functions for the clamped–free bar in axial vibration). We will not pursue this subject here; for the moment we adhere to the considerations of Section 5.5 with only one additional observation due to the fact that now we must choose more than one function: the trial functions should be linearly independent and form a complete set. We will be more specific about this and about finite-element modelling whenever needed in the course of the discussion.

6.5.3 More othogonality conditions

The conditions of eqs (6.42b) and (6.44) are only a particular case of a broader class of othogonality properties involving the eigenvectors $\mathbf{p}_i$ and the matrices $\mathbf{M}$ and $\mathbf{K}$. If $\mathbf{p}_i$ is an eigenvector, the eigenvalue problem $\mathbf{K}\mathbf{p}_i = \lambda_i \mathbf{M}\mathbf{p}_i$ is identically satisfied. Let us premultiply both sides of the eigenvalue problem by $\mathbf{p}_j^T \mathbf{K}\mathbf{M}^{-1}$, we get, by virtue of (6.42b) and (6.44)

$$\mathbf{p}_j^T \mathbf{K}\mathbf{M}^{-1}\mathbf{K}\mathbf{p}_i = \lambda_i \mathbf{p}_j^T \mathbf{K}\mathbf{M}^{-1}\mathbf{M}\mathbf{p}_i = \lambda_i \mathbf{p}_j^T \mathbf{K}\mathbf{p}_i = \lambda_i^2 \delta_{ji} \qquad (6.116)$$

Premultiplication of the eigenvalue problem by $\mathbf{p}_j^T \mathbf{K}\mathbf{M}^{-1}\mathbf{K}\mathbf{M}^{-1}$ yields

$$\mathbf{p}_j^T \mathbf{K}\mathbf{M}^{-1}\mathbf{K}\mathbf{M}^{-1}\mathbf{K}\mathbf{p}_i = \lambda_i \mathbf{p}_j^T \mathbf{K}\mathbf{M}^{-1}\mathbf{K}\mathbf{p}_i = \lambda_i^3 \delta_{ji} \qquad (6.117)$$

where the result of eq (6.116) has been taken into account. The process can be repeated to give

$$\mathbf{p}_j^T (\mathbf{K}\mathbf{M}^{-1})^a \mathbf{K}\mathbf{p}_i = \lambda_i^{a+1} \delta_{ji} \qquad (a = 0, 1, 2, \dots) \qquad (6.118)$$

which can be rewritten in the equivalent form

$$\mathbf{p}_j^T \mathbf{M}(\mathbf{M}^{-1}\mathbf{K})^b \mathbf{p}_i = \lambda_i^b \delta_{ji} \qquad (b = 0, 1, 2, \dots, +\infty) \qquad (6.119)$$

just by premultiplying the term in parentheses on the left-hand side of eq (6.118) by $\mathbf{M}\mathbf{M}^{-1}$. The cases $b = 0$ and $b = 1$ in eq (6.119) correspond, respectively, to eqs (6.42b) and (6.44).

An analogous procedure can be started by premultiplying both sides of the eigenvalue problem by $\mathbf{p}_j^T \mathbf{M}\mathbf{K}^{-1}$ to give

$$\delta_{ji} = \mathbf{p}_j^T \mathbf{M}\mathbf{K}^{-1}\mathbf{K}\mathbf{p}_i = \lambda_i \mathbf{p}_j^T \mathbf{M}\mathbf{K}^{-1}\mathbf{M}\mathbf{p}_i$$

which means, provided that $\lambda_i \neq 0$

$$\mathbf{p}_j^T \mathbf{M}\mathbf{K}^{-1}\mathbf{M}\mathbf{p}_i = \mathbf{p}_j^T \mathbf{M}(\mathbf{M}^{-1}\mathbf{K})^{-1}\mathbf{p}_i = \lambda_i^{-1}\delta_{ji} \qquad (6.120)$$

and the term in the centre is due to the fact that $\mathbf{K}^{-1}\mathbf{M} = (\mathbf{M}^{-1}\mathbf{K})^{-1}$.

Now, premultiplying both sides of the eigenvalue problem by $\mathbf{p}_j^T \mathbf{M}\mathbf{K}^{-1}\mathbf{M}\mathbf{K}^{-1}$ and taking eq (6.120) into account we get

$$\mathbf{p}_j^T \mathbf{M}\mathbf{K}^{-1}\mathbf{M}\mathbf{K}^{-1}\mathbf{M}\mathbf{p}_i = \mathbf{p}_j^T \mathbf{M}(\mathbf{M}^{-1}\mathbf{K})^{-2}\mathbf{p}_i = \lambda_i^{-2}\delta_{ji}$$

Repeated application of this procedure leads to

$$\mathbf{p}_j^T \mathbf{M}(\mathbf{M}^{-1}\mathbf{K})^b \mathbf{p}_i = \lambda_i^b \delta_{ji} \qquad (b = -1, -2, \dots, -\infty) \qquad (6.121)$$

Equations (6.119) and (6.121) can be put together to express the family of orthogonality conditions

$$\mathbf{p}_j^T \mathbf{M} (\mathbf{M}^{-1}\mathbf{K})^b \mathbf{p}_i = \lambda_i^b \delta_{ji} \qquad (-\infty < b < +\infty) \tag{6.122}$$

6.6 Unrestrained systems: rigid-body modes

Suppose that, for a given system, there is a particular vector $\mathbf{r}$ ($\mathbf{r} \neq 0$) for which the relationship

$$\mathbf{Kr} = 0 \tag{6.123}$$

holds. First of all, this occurrence implies that the matrix $\mathbf{K}$ is singular and that the potential energy is now a positive semidefinite quadratic form. Furthermore, this vector can be substituted in the eigenvalue problem of eq (6.29) to give $\lambda \mathbf{Mr} = \mathbf{Kr} = 0$, i.e. $\lambda \mathbf{Mr} = 0$ and hence, since $\mathbf{M}$ is generally positive definite, $\lambda \equiv \omega^2 = 0$. It follows that $\mathbf{r}$ can be considered an eigenvector corresponding to the eigenvalue $\lambda = 0$. At first sight, an oscillation at zero frequency may appear surprising, but the key point is that this solution does not correspond to an oscillatory motion at all: we speak of oscillatory motion because eq (6.29) has been obtained by assuming a time dependence of the form $e^{-i\omega t}$. However, if we assume a more general solution of the form

$$\mathbf{u} = \mathbf{r}f(t) \tag{6.124}$$

and we substitute it in eq (6.20) we get $\ddot{f}(t) = 0$, which corresponds to a uniform motion $f(t) = at + b$ where a and b are two constants. In practice, the system moves as a whole and there is no change in the potential energy because such rigid displacement does not produce any elastic restoring force. In other words, we are dealing with a system with a neutrally stable equilibrium position. Some examples could be: an aeroplane in rectilinear flight, a shaft supported at both ends by frictionless sleeves or, in general, any structure supported on springs that are very soft compared to its stiffness. This latter situation is of considerable importance in the field of dynamic structural testing and it is usually referred to as the 'free' condition (see also Chapter 10).

The maximum number of eigenvectors which correspond to the eigenvalue $\lambda = 0$ (the so-called **rigid-body modes**; this is why we used the letter $\mathbf{r}$) is six because a three-dimensional body has a maximum of six rigid-body degrees of freedom (three translational and three rotational): moreover, for a given problem, it is generally not difficult to identify the rigid-body modes by inspection.

For example, suppose that we model an aeroplane body and wings as a flexible beam with three lumped masses, M being the mass of the fuselage and m being the mass of each wing: if we only consider motions in the plane of the page, the first two modes occur at zero frequency and they correspond to a rigid-body translation and a rigid-body rotation as shown in Fig. 6.9. The higher modes are elastic modes representing flexural deformations of our simple structure.

The arguments used to arrive at the orthogonality relationships retain their validity; hence, the rigid-body modes are **M**- and **K**-orthogonal to the elastic eigenvectors (because they are associated with distinct eigenvalues) and the rigid-body modes can always be assumed mutually **M**-orthogonal (because they just represent a particular case of eigenvalue degeneracy). The only difference is the stiffness orthogonality condition for rigid-body modes that now reads, because of eq (6.123)

$$\mathbf{r}_i \mathbf{K} \mathbf{r}_j = 0 \tag{6.125}$$

for every index i and j.

As in the case of the 'usual' orthogonality conditions, rigid-body modes do not represent a difficulty in all the aspects of the foregoing discussions. In order to be more specific, we can consider an n-DOF system with m rigid-body modes and write eq (6.56a) as

$$\mathbf{u} = \sum_{j=1}^{m} \mathbf{r}_j w_j(t) + \sum_{k=1}^{n-m} \mathbf{p}_k y_k(t) = \mathbf{R} \mathbf{w} + \mathbf{P} \mathbf{y} \tag{6.126}$$

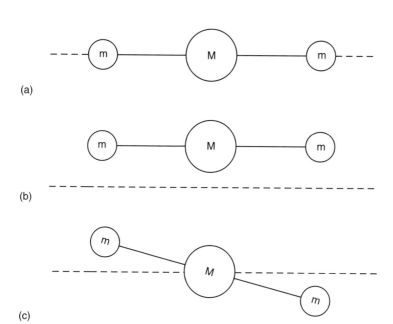

(a)

(b)

(c)

Fig. 6.9 (a) Simple aeroplane model. (b) Rigid-body translation. (c) Rigid-body rotation.

where we introduced the rigid-body $n \times m$ matrix $\mathbf{R} = [\mathbf{r}_1 \ldots \mathbf{r}_m]$ and the $m \times 1$ matrix $\mathbf{w} = [w_1 \ldots w_m]^T$. of normal coordinates associated with rigid-body modes. The matrices $\mathbf{P}$ and $\mathbf{y}$ are associated with elastic modes and retain their original meaning; however, their dimensions are now $n \times (n-m)$ and $(n-m) \times 1$, respectively. Substitution of eq (6.126) into $\mathbf{M\ddot{u}} + \mathbf{Ku} = 0$ leads to

$$\mathbf{MR\ddot{w}} + \mathbf{MP\ddot{y}} + \mathbf{KPy} = 0 \tag{6.127}$$

because $\mathbf{KRw} = 0$. Premultiplying eq (6.127) by $\mathbf{R}^T$ and $\mathbf{P}^T$, by virtue of the othogonality conditions, gives

$$\ddot{\mathbf{w}} = 0$$
$$\ddot{\mathbf{y}} + \mathbf{Ly} = 0 \tag{6.128}$$

where $\mathbf{L}$ is the $(n-m) \times (n-m)$ diagonal matrix of eigenvalues different from zero. The expressions (6.128) show that the equation for the elastic modes remains unchanged, while the rigid-body normal equations have solutions of the form $w_j = at + b$ ($j = 1, \ldots, m$). The general solution, for example in the form of eq (6.27c), can then be written as the eigenvector expansion

$$\mathbf{u} = \sum_{j=1}^{m}(a_j t + b_j)\mathbf{r}_j + \sum_{k=1}^{n-m}(D_k \cos\omega_k t + E_k \sin\omega_k t)\mathbf{p}_k \tag{6.129}$$

where the $2n$ constants are determined by the initial conditions

$$\mathbf{u}_0 = \sum_{j=1}^{m} b_j\mathbf{r}_j + \sum_{k=1}^{n-m} D_k\mathbf{p}_k$$

$$\dot{\mathbf{u}}_0 = \sum_{j=1}^{m} a_j\mathbf{r}_j + \sum_{k=1}^{n-m} \omega_k E_k\mathbf{p}_k$$

By using once again the orthogonality conditions, we arrive at the explicit expression

$$\mathbf{u} = \sum_{j=1}^{m}\left[(\mathbf{r}_j^T\mathbf{M\dot{u}}_0)t + \mathbf{r}_j^T\mathbf{Mu}_0\right]\mathbf{r}_j$$

$$+ \sum_{k=1}^{n-m}\left[(\mathbf{p}_k^T\mathbf{Mu}_0)\cos\omega_k t + \frac{\mathbf{p}_k^T\mathbf{M\dot{u}}_0}{\omega_k}\sin\omega_k t\right]\mathbf{p}_k \tag{6.130a}$$

or, equivalently

$$\mathbf{u} = \left(\sum_{j=1}^{m} \mathbf{r}_j \mathbf{r}_j^T + \sum_{k=1}^{n-m} \mathbf{p}_k \mathbf{p}_k^T \cos \omega_k t \right) \mathbf{M} \mathbf{u}_0$$

$$+ \left(\sum_{j=1}^{m} \mathbf{r}_j \mathbf{r}_j^T t + \sum_{k=1}^{n-m} \frac{\mathbf{p}_k \mathbf{p}_k^T}{\omega_k} \sin \omega_k t \right) \mathbf{M} \dot{\mathbf{u}}_0 \qquad (6.130b)$$

Equations (6.130a) and (6.130b) are the counterpart of eqs (6.51) and (6.53) when the system admits m rigid-body modes. Furthermore, by virtue of eq (6.126), it is not difficult to see that the potential and kinetic energy are now given by

$$V = \frac{1}{2} \mathbf{u}^T \mathbf{K} \mathbf{u} = \frac{1}{2} \mathbf{y}^T \mathbf{L} \mathbf{y} = \frac{1}{2} \sum_{k=1}^{n-m} \omega_k^2 y^2$$

$$\qquad (6.131)$$

$$T = \frac{1}{2} \dot{\mathbf{u}}^T \mathbf{M} \dot{\mathbf{u}} = \frac{1}{2} \sum_{j=1}^{m} \dot{w}_j^2 + \frac{1}{2} \sum_{k=1}^{n-m} \dot{y}_k^2$$

where $\mathbf{L}$ is the $(n - m) \times (n - m)$ matrix of eq (6.128).

The expressions above show that the elastic motion and the rigid-body motion are completely uncoupled and that the rigid-body modes, as expected, give no contribution to the potential energy. Besides the fact that the general solution must take rigid-body modes into account, another aspect that deserves attention is the fact that $\mathbf{K}$ is now singular. This is mostly a problem of a computational nature because some important numerical techniques require the inversion of the stiffness matrix. The highly specialized subject of the numerical solution of the eigenproblem is well beyond the scope of the book but it is worth knowing that there are simple ways to circumvent this problem.

One solution is the addition of a small fictitious stiffness along an adequate number of degrees of freedom of the unrestrained system. This is generally done by adding restraining springs to prevent rigid-body motions and make the stiffness matrix nonsingular. If the additional springs are very 'soft' (that is, they have a very low stiffness) the modified system will have frequencies and mode shapes that are very close to those of the unrestrained system. It is apparent that this procedure involves a certain approximation because, as a matter of fact, the original system has been modified. In practice, however, a satisfactory degree of accuracy can be achieved in most cases.

A second possibility, extensively used by eigensolvers, is called **shifting**. We calculate the shifted matrix

$$\hat{\mathbf{K}} = \mathbf{K} - \rho \mathbf{M} \qquad (6.132)$$

where ρ is the shift, and solve the eigenproblem

$$\hat{\mathbf{K}}\mathbf{z} = \mu\mathbf{M}\mathbf{z} \tag{6.133}$$

Since the solution of the eigenproblem is unique, it is not difficult to see that the original eigenvalues are given by

$$\lambda_i = \rho + \mu_i \tag{6.134}$$

and that the original eigenvectors are left unchanged by the shifting process.

Suppose, for example, that a given system leads to the eigenproblem

$$\begin{bmatrix} 5 & -5 \\ -5 & 5 \end{bmatrix}\mathbf{z} = \lambda\begin{bmatrix} 2 & 0 \\ 0 & 1 \end{bmatrix}\mathbf{z}$$

with eigenvalues $\lambda_1 = 0$, $\lambda_2 = 15/2$ and mass-normalized eigenvectors

$$\mathbf{p}_1 = \begin{bmatrix} \dfrac{1}{\sqrt{3}} & \dfrac{1}{\sqrt{3}} \end{bmatrix}^T \qquad \mathbf{p}_2 = \begin{bmatrix} \dfrac{1}{\sqrt{6}} & -\dfrac{2}{\sqrt{6}} \end{bmatrix}^T$$

Imposing a shift of, say, $\rho = -3$ leads to the shifted eigenproblem of eq (6.133)

$$\begin{bmatrix} 11 & -5 \\ -5 & 8 \end{bmatrix}\mathbf{z} = \mu\begin{bmatrix} 2 & 0 \\ 0 & 1 \end{bmatrix}\mathbf{z}$$

and to the characteristic equation $2\mu^2 - 27\mu - 63 = 0$ which admits the solutions $\mu_1 = 3$ and $\mu_2 = 21/2$. Equation (6.134) is verified and it is easy to determine that the eigenvectors are the same as before.

This procedure does not involve any approximation, so that, in principle, it may be sufficient to have solution algorithms for eigenvalues different from zero. In fact, it is always possible to operate on the shifted matrix $\hat{\mathbf{K}}$ which, in turn, can always be made nonsingular by an appropriate choice of the shifting value ρ.

A third possibility lies in the fact that, as stated above, rigid-body modes can often be identified by inspection. Thus, by imposing the condition of orthogonality between rigid-body modes and the elastic modes we can obtain as many constraint equations as there are rigid-body modes. The constraint equations are then used to perform a coordinate transformation between $(n - m)$ independent coordinates and the original n coordinates. This leads to a reduced eigenvalue problem where the rigid-body modes have been eliminated; the reduced system is positive definite and can be solved by means of any standard eigensolver. This procedure, for the reasons explained above, is often called **sweeping of rigid-body modes**.

For example, we can consider the system of Fig. 6.10 where the three masses can only move along the x-axis. Choosing the coordinates shown in the figure, we obtain the following mass and stiffness matrices:

$$\mathbf{M} = \begin{bmatrix} m_1 & 0 & 0 \\ 0 & m_2 & 0 \\ 0 & 0 & m_3 \end{bmatrix} \qquad \mathbf{K} = \begin{bmatrix} k_1 & -k_1 & 0 \\ -k_1 & k_1 + k_2 & -k_2 \\ 0 & -k_2 & k_2 \end{bmatrix}$$

The rigid-body mode represents a translation of the whole system in the x direction and, besides normalization, it can be written as $\mathbf{r} = \begin{bmatrix} 1 & 1 & 1 \end{bmatrix}^T$. Any elastic mode $\mathbf{z} = \begin{bmatrix} x_1 & x_2 & x_3 \end{bmatrix}^T$ must be orthogonal to $\mathbf{r}$, i.e.

$$\mathbf{r}^T \mathbf{M} \mathbf{z} = m_1 x_1 + m_2 x_2 + m_3 x_3 = 0 \tag{6.135}$$

Equation (6.156), as a matter of fact, is a holonomic constraint ensuring that the centre of mass remains fixed at the origin; we can use this equation to reduce by one the number of degrees of freedom of the system by expressing one of the coordinates as a function of the other two. Which coordinate to eliminate is only a matter of choice; for example we can eliminate x_1 and write the coordinate transformation

$$\begin{bmatrix} x_1 \\ x_2 \\ x_3 \end{bmatrix} = \begin{bmatrix} -m_2/m_1 & -m_3/m_1 \\ 1 & 0 \\ 0 & 1 \end{bmatrix} \begin{bmatrix} x_2 \\ x_3 \end{bmatrix}$$

where the first equation is obtained from the constraint (6.135) and the other two are simple identities. In matrix form the transformation above can be written as

$$\mathbf{z} = \mathbf{T}\hat{\mathbf{z}} \tag{6.136}$$

where we call $\mathbf{z}$ the 'constrained' 3×1 vector and $\hat{\mathbf{z}}$ the 2×1 vector of independent coordinates. At this stage we can form the reduced eigenproblem

$$\hat{\mathbf{K}}\hat{\mathbf{z}} = \lambda \hat{\mathbf{M}}\hat{\mathbf{z}} \tag{6.137a}$$

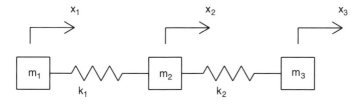

Fig. 6.10 Unrestrained 3-DOF system.

where $\hat{\mathbf{K}}$ and $\hat{\mathbf{M}}$ are the 2×2 matrices

$$\hat{\mathbf{K}} = \mathbf{T}^T \mathbf{K} \mathbf{T}$$

$$\hat{\mathbf{M}} = \mathbf{T}^T \mathbf{M} \mathbf{T}$$

(6.137b)

As expected, the reduced eigenproblem yields the two eigenvalues different from zero of the original problem and two 2×1 eigenvectors $\hat{\mathbf{z}}_1$ and $\hat{\mathbf{z}}_2$ which, in turn, can be used to obtain the full eigenvectors by means of the coordinate transformation (6.136). Obviously, substitution of $\hat{\mathbf{z}}_1$ into eq (6.136) will yield the first elastic mode and substitution of $\hat{\mathbf{z}}_2$ will yield the second elastic mode and, at the end of the process, we must not forget the rigid-body mode that has been removed by the reduction procedure. In general, for large systems this method is not numerically very efficient for two reasons: first because the reduction is not significant (the rigid-body modes are at most six) and second because the transformation of eq (6.137b) destroys some computationally useful properties of the original matrices (for instance, in our simple example above the mass matrix $\mathbf{M}$ is diagonal while $\hat{\mathbf{M}}$ is not).

6.7   Damped systems: proportional and nonproportional damping

6.7.1   *Proportional damping*

In Section 6.4.1 we considered the inclusion of a small amount of viscous damping in the equations of motion as a perturbative term of an originally undamped n-DOF system. Our intention was twofold: to have a general idea of what to expect when some dissipative effects are taken into account and to investigate the behaviour of lightly damped structures, a frequently occurring situation in practical vibration analysis. The starting point was eq (6.90), which we rewrite here for our present convenience:

$$\mathbf{M}\ddot{\mathbf{u}} + \mathbf{C}\dot{\mathbf{u}} + \mathbf{K}\mathbf{u} = 0$$

(6.138)

When the undamped solution is known it is always possible to perform the coordinate transformation (6.56a), premultiply by the modal matrix $\mathbf{P}^T$ and arrive at

$$\mathbf{I}\ddot{\mathbf{y}} + \mathbf{P}^T \mathbf{C} \mathbf{P}\dot{\mathbf{y}} + \mathbf{L}\mathbf{y} = 0$$

(6.139)

which is not a set of n uncoupled equations unless the matrix $\mathbf{P}^T\mathbf{C}\mathbf{P}$ is diagonal. When this is the case, the n uncoupled equations of motion can be written as

$$\ddot{y}_j + 2\omega_j \zeta_j \dot{y} + \omega_j^2 y = 0 \qquad (j = 1, 2, \dots, n)$$

(6.140)

where ζ_j is the *j*th **modal damping ratio,** in analogy with the SDOF equation (4.18). The solution of eq (6.140) is given in Chapter 4 and for $0 < \zeta_j < 1$ we are already familiar with its oscillatory character at the frequency

$$\omega_j \sqrt{1 - \zeta_j^2}$$

and its exponentially decaying amplitude. At this point the usefulness of investigating the condition under which the damping matrix can be diagonalized by the coordinate transformation (6.56a) is evident. Some common assumptions are as follows:

$$C = a\mathbf{M}$$

$$C = b\mathbf{K} \tag{6.141}$$

$$C = a\mathbf{M} + b\mathbf{K}$$

i.e. a damping matrix which is proportional either to the mass matrix or to the stiffness matrix, or both. The damping distributions of eqs (6.141) are classified as **proportional damping** and the last of eqs (6.141), which covers the two cases before, is often referred to as **Rayleigh damping.** Some justifications can be given for the assumptions above; for example:

- The first case may represent a situation in which each mass is connected to a viscous damper whose other end is connected to 'the ground' and every coefficient c_{ij} is in the same proportion a to the mass coefficient m_{ij}.
- A damping element in parallel with each spring element with a constant ratio $b = c_{ij}/k_{ij}$ can be invoked for the second case, but the use of eqs (6.141) is mostly a matter of convenience which turns out to be adequate in many practical situations. In these circumstances the damped eigenvectors are the same as the undamped ones and it is evident that eqs (6.141) take advantage of their orthogonality properties to arrive at (eqs (6.139) and (6.140))

$$\mathbf{p}_i^T C \mathbf{p}_j = 2\omega_i \zeta_i \delta_{ij} \tag{6.142}$$

which allows us to determine the proportionality coefficient(s) when one (or two in the third case of (6.141)) damping ratio(s) has been specified or measured for the system under investigation.

For example, suppose we assumed mass proportional damping and we measured ζ_k from a free-vibration amplitude decay test performed by imposing appropriate initial conditions. Then, provided that we know the value of ω_k, we get

$$a\mathbf{p}_k^T \mathbf{M}\mathbf{p}_k = a = 2\omega_k \zeta_k$$

and the other damping ratios can be obtained from (6.142) as

$$\zeta_j = \left(\frac{1}{\omega_j}\right) \omega_k \zeta_k \qquad (j \neq k) \tag{6.143}$$

On the other hand, the assumption of stiffness proportional damping leads to

$$b\mathbf{p}_k^T \mathbf{K} \mathbf{p}_k = b = \frac{2\zeta_k}{\omega_k}$$

and hence

$$\zeta_j = \left(\frac{\zeta_k}{\omega_k}\right) \omega_j \qquad (j \neq k) \tag{6.144}$$

Equations (6.143) and (6.144), respectively, show that mass proportional damping assigns a higher damping to low-frequency modes, while stiffness proportional damping assigns higher damping to high-frequency modes.

A greater degree of control on the damping ratios can be achieved if we assume Rayleigh damping, where we can specify the damping ratios for any two modes, say the kth and the mth, to get

$$\mathbf{p}_k^T(a\mathbf{M} + b\mathbf{K})\mathbf{p}_k = a + b\omega_k^2 = 2\omega_k\zeta_k$$

$$\mathbf{p}_m^T(a\mathbf{M} + b\mathbf{K})\mathbf{p}_m = a + b\omega_m^2 = 2\omega_m\zeta_m \tag{6.145a}$$

so that a and b can be obtained and substituted in eq (6.142) to determine the jth damping ratio when j is different from k and m. Equation (6.145a) in matrix form reads

$$\frac{1}{2}\begin{bmatrix} 1/\omega_k & \omega_k \\ 1/\omega_m & \omega_m \end{bmatrix}\begin{bmatrix} a \\ b \end{bmatrix} = \begin{bmatrix} \zeta_k \\ \zeta_m \end{bmatrix} \tag{6.145b}$$

Note that Rayleigh damping results approximately in a constant damping ratio for the middle-frequency modes and an increasing damping ratio for the low- and high-frequency modes. However, the situation may vary depending on the specific values of a and b; the reader is invited to consider a few reasonable cases and draw a graph of the function $\zeta = \zeta(\omega)$ for each case.

The foregoing procedure can be extended if we take into account the additional orthogonality conditions of Section 6.5.1. In fact, assume for example that r damping ratios, say $\zeta_1, \zeta_2, \dots, \zeta_r$, are given; then a damping matrix satisfying eq (6.142) can be obtained by using the Caughey summation

$$\mathbf{C} = \sum_{j=0}^{r-1} a_j \mathbf{M}(\mathbf{M}^{-1}\mathbf{K})^j \tag{6.146}$$

where the r coefficients can be obtained from the r simultaneous equations

$$\frac{1}{2}\begin{bmatrix} 1/\omega_1 & \omega_1 & \omega_1^3 & \dots & \omega_1^{2r-3} \\ 1/\omega_2 & \omega_2 & \omega_2^3 & \dots & \omega_2^{2r-3} \\ 1/\omega_3 & \omega_3 & \omega_3^3 & \dots & \omega_3^{2r-3} \\ \dots & \dots & \dots & \dots & \dots \\ 1/\omega_r & \omega_r & \omega_r^3 & \dots & \omega_r^{2r-3} \end{bmatrix} \begin{bmatrix} a_0 \\ a_1 \\ a_2 \\ \dots \\ a_{r-1} \end{bmatrix} = \begin{bmatrix} \zeta_1 \\ \zeta_2 \\ \zeta_3 \\ \dots \\ \zeta_r \end{bmatrix} \tag{6.147}$$

Rayleigh damping is obviously a particular case of eq (6.146) where we have $a = a_0$ and $b = a_1$.

From the foregoing considerations, we can see the assumption of proportional damping leads, in the end, to a diagonal damping matrix

$$\hat{\mathbf{C}} \equiv \mathbf{P}^T\mathbf{C}\mathbf{P} = \text{diag}\left[2\zeta_j\omega_j\right] \tag{6.148}$$

If some technique other than the mode superposition (for example, a direct numerical integration of the equations of motion) is needed for a specific problem, it is evident that the explicit damping matrix can be obtained as $(\mathbf{P}^T)^{-1}\hat{\mathbf{C}}\mathbf{P}^{-1}$. However, this expression is not computationally very convenient, especially for large systems. The inversion of the modal matrix can be avoided if we consider that $\mathbf{I} = \mathbf{P}^T\mathbf{M}\mathbf{P}$, from which it follows that $\mathbf{P}^{-1} = \mathbf{P}^T\mathbf{M}$, $(\mathbf{P}^T)^{-1} = \mathbf{M}\mathbf{P}$ and hence

$$\mathbf{C} = \mathbf{M}\mathbf{P}\hat{\mathbf{C}}\mathbf{P}^T\mathbf{M} \tag{6.149a}$$

which, because of eq (6.148), can be written as

$$\mathbf{C} = \mathbf{M}\left(\sum_{j=1}^{n} 2\zeta_j\omega_j\mathbf{p}_j\mathbf{p}_j^T \right)\mathbf{M} \tag{6.149b}$$

where the contribution of each mode has been put into evidence. Those modes which are not included in eq (6.149b) are considered as undamped (i.e. their damping ratio is zero).

Finally, it is interesting to note that the equations of motion of an MDOF system with hysteretic damping can also be uncoupled by using the normal modes (i.e. the eigenvectors) of the undamped system because it is customary to assume a hysteretic damping matrix which is proportional to the stiffness matrix. Nevertheless, the attention to this case is mainly focused on the analysis of the forced response because this type of damping presents some difficulties to a rigorous free-vibration analysis. In fact, it must be remembered (Section 4.4) that the equivalent damping coefficients are defined under the assumption of harmonic excitation forces at a given frequency w so that the equations of motion can be written as

$$\mathbf{M\ddot{u}} + (1 + i\gamma)\mathbf{Ku} = \mathbf{f}_0 e^{i\omega t} \qquad (6.150a)$$

However, for a proportional damping matrix as above it is left to the reader to show that, under the assumption $\mathbf{u} = \mathbf{z}e^{i\tilde{\omega}t}$, the free-vibration solution of eq (6.150a) consists of the same eigenvectors as for the undamped problem with eigenvalues that are given by

$$\tilde{\omega}_j^2 = \omega_j^2 \, (1 + i\gamma) \qquad (6.150b)$$

where ω_j is the undamped jth natural frequency.

6.7.2   *Nonproportional damping*

The assumption of proportional damping, despite its usefulness in many theoretical and practical analysis, is not always justified. Therefore, in order to gain some insight in the behaviour of real structures, one must also consider the more general case in which the equations of motion (6.138) do not uncouple under the transformation to normal coordinates. If we assume a solution of the form

$$\mathbf{u} = \mathbf{z}e^{\lambda t} \qquad (6.151)$$

and substitute it in eq (6.138) we arrive at the complex eigenvalue problem of eq (6.92) which admits a nontrivial solution if

$$\det(\lambda^2 \mathbf{M} + \lambda \mathbf{C} + \mathbf{K}) = 0 \qquad (6.152)$$

which results in a characteristic equation of order $2n$ with real coefficients. The fact that the coefficients are all real implies that the $2n$ solutions are either real or, if they are complex, they occur in complex conjugate pairs. Moreover, if the initial disturbance of the system results in an exponential decay (steady or oscillating, depending on the amount of damping) and in a progressive loss of energy, it is not difficult to see that the real eigenvalues must be negative and the complex eigenvalues must have a negative real part.

The general solution to the free vibration problem is again obtained by superposing the $2n$ solutions as

$$\mathbf{u} = \sum_{j=1}^{2n} A_j \mathbf{z}_j\, e^{\lambda_j t} \qquad (6.153)$$

where the constants A_j are determined from the initial conditions.

Each complex eigenvalue will be accompanied by a complex eigenvector and together they must satisfy the complex eigenvalue problem, meaning that the real and imaginary parts on the left-hand side of eq (6.92) must separately be zero. This implies that the complex eigenvectors also occur in complex conjugate pairs. A general eigenvalue can then be written explicitly as $\lambda_j = -\mu_j + i\kappa_j$ and it will be associated with the eigenvector $\mathbf{z}_j = \mathbf{a}_j + i\mathbf{b}_j$, while the eigenvalue $\lambda_{j+1} \equiv \lambda_j^* = -\mu_j - i\kappa_j$ will be associated with the eigenvector $\mathbf{z}_{j+1} = \mathbf{z}_j^* = \mathbf{a}_j - i\mathbf{b}_j$. Note that there is no loss of generality in assigning the $(j+1)$th index to the complex conjugate solution of the jth eigenpair. When this is the case, then $A_{j+1} = A_j^*$, because we are dealing with a physical problem and the general solution must be real.

Alternatively, we can assign the jth index to both eigenpairs, sum from 1 to n in eq (6.153) and consider the expression

$$\mathbf{u}_j = A_j(\mathbf{a} + i\mathbf{b}_j)e^{(-\mu_j + i\kappa_j)t} + A_j^*(\mathbf{a}_j - i\mathbf{b}_j)e^{(-\mu_j - i\kappa_j)t} \qquad (6.154)$$

as the jth contribution to the general solution. In order to take a closer look at the expression (6.154) let us drop for a moment the index j and assume for simplicity that $A_j = A_j^* = 1$; then (6.154) becomes

$$\mathbf{u} = e^{-\mu t}(\mathbf{z}e^{i\kappa t} + \mathbf{z}^* e^{-i\kappa t}) \qquad (6.155a)$$

or, for each one of the n elements of the column vector $\mathbf{u}$

$$u_m = e^{-\mu t}(z_m e^{i\kappa t} + z_m^* e^{-i\kappa t}) \qquad (6.155b)$$

where the subscript m must now be interpreted as a matrix element index and not a mode index (which has been temporarily dropped). In eq (6.155b) we recognize two phasors rotating in opposite directions of the Argand–Gauss plane. Their sum results in a real oscillating quantity with a decaying amplitude, which can also be written as

$$u_m = 2\,|z_m|\,e^{-\mu t}\cos(\kappa t + \theta_m) \qquad (6.155c)$$

where the phase angle is given by $\theta_m = \operatorname{Im}(z_m)/\operatorname{Re}(z_m)$ and is, in general, different from θ_k when $k \neq m$.

So, in a complex mode different coordinates no longer move in phase or in antiphase with each other (0° or 180° phase angles), and consequently they no longer pass through their equilibrium position or reach their extremes of motion simultaneously. In other words, there are phase differences between the various parts of the structure. An animated display of a complex mode shape shows a 'travelling wave' situation, with no stationary nodal points or nodal lines (points or lines of no motion). This is different from the animation of a real mode shape, in which well defined nodal points or lines can be identified and a 'standing wave' situation can be recognized.

Another important difference has to do with the orthogonality conditions of complex eigenvectors. Suppose that z_j is an eigenvector of the generally damped problem relative to the eigenvalue λ_j. Then the complex eigenproblem $(\lambda_j^2 \mathbf{M} + \lambda_j \mathbf{C} + \mathbf{K})z_j = 0$ is identically satisfied. Now, premultiply by z_k^T to get

$$z_k^T(\lambda_j^2 \mathbf{M} + \lambda_j \mathbf{C} + \mathbf{K})z_j = 0 \qquad (6.156)$$

write the complex eigenproblem for the kth eigenpair, transpose it and postmultiply by z_j. We get

$$z_k^T(\lambda_k^2 \mathbf{M} + \lambda_k \mathbf{C} + \mathbf{K})z_j = 0 \qquad (6.157)$$

Subtracting eq (6.156) from eq (6.157), provided that $\lambda_k \neq \lambda_j$, leads to the orthogonality condition

$$(\lambda_k + \lambda_j)z_k^T \mathbf{M}z_j + z_k^T \mathbf{C}z_j = 0 \qquad (6.158)$$

A second orthogonality condition can be obtained if we multiply eq (6.156) by λ_k and eq (6.157) by λ_j and subtract one of the resulting equations from the other; if $\lambda_k \neq \lambda_j$ we get

$$\lambda_k \lambda_j z_k^T \mathbf{M}z_j - z_k^T \mathbf{K}z_j = 0 \qquad (6.159)$$

Equations (6.158) and (6.159) are not as simple as their real mode counterparts but, as in that case, they still hold true for repeated eigenvalues provided that an appropriate choice is made for the eigenvectors associated with a repeated root. Furthermore, since it is often convenient to represent a complex eigenvalue as

$$\lambda_j = \omega_j(-\zeta_j + i\sqrt{1 - \zeta_j^2}) \qquad (6.160)$$

(the formal analogy with the SDOF case is evident: Chapter 4), when $\lambda_j = \lambda_k^*$

it is not difficult to obtain the relationships

$$2\zeta_k\omega_k = \frac{\mathbf{z}_k^T \mathbf{C} \mathbf{z}_k^*}{\mathbf{z}_k^T \mathbf{M} \mathbf{z}_k^*}$$

$$\omega_k^2 = \frac{\mathbf{z}_k^T \mathbf{K} \mathbf{z}_k^*}{\mathbf{z}_k^T \mathbf{M} \mathbf{z}_k^*}$$

(6.161)

where the kth damping ratio ζ_k and the kth frequency ω_k can be determined from the kth eigenvecctor and the damping, mass and stiffness matrices.

6.8 Generalized and complex eigenvalue problems: reduction to standard form

For conservative and nonconservative systems we need to find the solution of a particular eigenproblem: the so-called generalized eigenvalue problem in the first case and the complex eigenvalue problem in the second case. A widely adopted strategy of solution is to organize the equations of motions so that the eigenvalue problem can be represented in the standard form

$$\mathbf{A}\mathbf{v} = \lambda\mathbf{v} \qquad (6.162)$$

where the matrix $\mathbf{A}$ and the vector $\mathbf{v}$ depend on the particular transformation procedure. In general, the transformation is carried out for two reasons: first because a large number of effective solution algorithms are available for the standard eigenproblem, and second because all the properties of the eigenvalues and eigenvector of the original eigenproblem can be deduced from the properties of the standard eigensolution.

Our purpose here is simply to make the reader aware of the fact that several alternative methods are available for the solution of the various forms of eigenproblems and that many of these possibilities transform the original problem into the form (6.162). We merely outline some of the possibilities; for a detailed discussion the reader can refer to more advanced texts or to the wide body of specific literature. Broadly speaking, there is no such thing as 'the best method' and the selection among the alternative eigensolvers depends on the dimension of the problem, the number of eigenvalues and eigenvectors required and the structure (typically the symmetry and bandwidth) of the matrices involved.

6.8.1 Undamped systems

Let us consider the conservative case first. Both sides of the generalized eigenproblem $\mathbf{K}\mathbf{z} = \lambda\mathbf{M}\mathbf{z}$ can be premultiplied by $\mathbf{M}^{-1}$ (provided that $\mathbf{M}$ is nonsingular) to give

$$\mathbf{M}^{-1}\mathbf{K}\mathbf{z} \equiv \mathbf{A}\mathbf{z} = \lambda\mathbf{z} \qquad (6.163)$$

where we have defined the dynamic matrix $\mathbf{A} = \mathbf{M}^{-1}\mathbf{K}$. Alternatively we can premultiply the generalized problem by $\mathbf{K}^{-1}$ (provided that $\mathbf{K}$ is nonsingular) and arrive at

$$\mathbf{A}\mathbf{z} = \gamma\mathbf{z} \tag{6.164}$$

where the dynamic matrix is now defined as $\mathbf{A} \equiv \mathbf{K}^{-1}\mathbf{M}$ and $\gamma = 1/\lambda$. Numerical procedures known as iteration methods can be used both for (6.163) and (6.164); however, the form (6.164) is preferred because it can be shown that the iteration converges to the largest value of γ, i.e. to the fundamental frequency of the system under investigation (one generally speaks of inverse iteration in this case). Conversely, the iteration converges to the highest value of λ for eq (6.163), and higher frequencies are generally of less interest in vibration analysis. Once the first eigenpair has been obtained, the process can continue provided that the influence of the first eigenvector is removed from the trial vector that is chosen to start the iteration. The procedure is then repeated to obtain the second eigenpair and so on. Equation (6.163) leads to the determination of eigenpairs in decreasing order (i.e. $\omega_n, \omega_{n-1}, \dots, \omega_1$), while the reverse is true for eq. (6.164).

The main drawback of the methods above is that, in general, the dynamic matrix – $\mathbf{M}^{-1}\mathbf{K}$ or $\mathbf{K}^{-1}\mathbf{M}$, whichever is the case – is not symmetrical. However, the symmetry of $\mathbf{M}$ and $\mathbf{K}$ can be exploited to transform $\mathbf{A}$ into a symmetrical matrix before extracting its eigenvalues: for example the mass matrix can be factored into

$$\mathbf{M} = \mathbf{L}\mathbf{L}^T \tag{6.165}$$

where $\mathbf{L}$ – not to be confused with the diagonal matrix of eigenvalues used in preceding sections – is a lower triangular matrix (with positive diagonal terms because $\mathbf{M}$ is positive definite). Equation (6.165), known as the Cholesky factorization, expresses an important theorem of matrix algebra. Substitution of eq (6.165) into eq (6.164) and successive premultiplication by $\mathbf{L}^T$ leads to $\mathbf{L}^T\mathbf{K}^{-1}\mathbf{L}\mathbf{L}^T\mathbf{z} = \lambda\mathbf{z}$ which, defining the new set of coordinates $\mathbf{x} = \mathbf{L}^T\mathbf{z}$, turns into the standard form

$$\mathbf{S}\mathbf{x} = \gamma\mathbf{x} \tag{6.166}$$

where $\mathbf{S} \equiv \mathbf{L}^T\mathbf{K}^{-1}\mathbf{L}$ is a symmetrical matrix (because $\mathbf{K}^{-1}$ is symmetrical and hence $(\mathbf{L}^T\mathbf{K}^{-1}\mathbf{L})^T = \mathbf{L}^T\mathbf{K}^{-1}\mathbf{L}$). Solving the eigenproblem (6.166) leads to the same eigenvalues as the original starting problem and to eigenvectors $\mathbf{x}_j$ which satisfy the orthogonality conditions

$$\mathbf{x}_k^T\mathbf{x}_j = 0$$

$$\mathbf{x}_k^T\mathbf{S}\mathbf{x}_j = 0 \qquad (k \neq j) \tag{6.167}$$

Then, thanks to the upper triangular form of L^T, the original eigenvectors can be obtained by backward substitution.

Alternatively, instead of the Cholesky factorization, we can solve the standard eigenvalue problem for the matrix M (which is symmetrical and positive definite) and write its spectral decomposition as

$$M = RD^2R^T \tag{6.168}$$

where R is an orthogonal matrix ($RR^T = I$) and D^2 is the diagonal matrix of the (positive, this is why we write D^2) eigenvalues of M. Substitution of eq (6.168) in the generalized eigenproblem gives $Kz = \lambda RD^2R^Tz$, and since $DR^T = (RD)^T$ we arrive at

$$Kz = \lambda NN^Tz \tag{6.169}$$

where we have defined the matrix $N \equiv RD$. Now, premultiply eq (6.169) by N^{-1}, insert the identity matrix $N^{-T}N^T$ between K and z on the left-hand side and define the vector $x = N^Tz$. We have thus obtained a standard eigenvalue problem of the form (6.166), where the matrix S is now $S = N^{-1}KN^{-T}$.

The problem is symmetrical because $N^{-1}KN^{-T} \equiv (RD)^{-1}K(RD)^{-T} = D^{-1}R^TKRD^{-1}$ and it is easy to verify that $(D^{-1}R^TKRD^{-1})^T = D^{-1}R^TKRD^{-1}$.

Another possibility is to start from the generalized eigenvalue problem $Kz = \lambda Mz$, substitute the coordinate transformation $z = M^{-1/2}x$ in it and premultiply the resulting equation by $M^{-1/2}$ to arrive at the symmetrical standard eigenproblem

$$(M^{-1/2}KM^{-1/2})x = \lambda x \tag{6.170a}$$

which results in the same eigenvalues as the original problem and in n eigenvectors x_j ($j = 1, 2, \dots, n$) connected to the original mode shapes by

$$z_j = M^{-1/2}x_j \tag{6.170b}$$

Again, the symmetry of the problem is a computational advantage, but the numerical disadvantage is the calculation of the matrix $M^{-1/2}$. However, if M is a lumped mass matrix of the form

$$M = \mathrm{diag}(m_1, m_2, \dots, m_n)$$

then

$$M^{-1/2} = \mathrm{diag}(m_1^{-1/2}, m_2^{-1/2}, \dots, m_n^{-1/2})$$

Consider now the matrix equation (6.20), which constitutes a set of n second-order ordinary differential equations. Premultiplication by M^{-1} leads to

$$\ddot{u} + M^{-1}Ku = 0 \tag{6.171}$$

If now we define the $n \times 1$ vectors $\mathbf{x}_1$ and $\mathbf{x}_2$ as

$$\mathbf{x}_1 = \mathbf{u}$$
$$\mathbf{x}_2 = \dot{\mathbf{u}}$$

$$(6.172)$$

we get

$$\dot{\mathbf{x}}_1 = \dot{\mathbf{u}} = \mathbf{x}_2$$
$$\dot{\mathbf{x}}_2 = \ddot{\mathbf{u}} = -\mathbf{M}^{-1}\mathbf{K}\mathbf{x}_1$$

which can be written in matrix form as

$$\dot{\mathbf{x}} = \mathbf{A}\mathbf{x} \qquad (6.173)$$

where we have defined the $2n \times 1$ vector $\mathbf{x} = [\mathbf{x}_1, \mathbf{x}_2]^T$ and the matrix $\mathbf{A}$ is given by

$$\mathbf{A} = \begin{bmatrix} \mathbf{0} & \mathbf{I} \\ -\mathbf{M}^{-1}\mathbf{K} & \mathbf{0} \end{bmatrix}$$

Assuming a solution of eq (6.173) of the form $\mathbf{x} = \mathbf{s}e^{\lambda t}$ we arrive at the standard eigenproblem of order $2n$

$$\mathbf{A}\mathbf{s} = \lambda\mathbf{s} \qquad (6.174a)$$

Equation (6.174a) results in $2n$ eigenvalues which correspond to the natural frequencies of the system according to the relations

$$\lambda_1 = +i\omega_1 \qquad \lambda_2 = -i\omega_1$$
$$\lambda_3 = +i\omega_2 \qquad \lambda_4 = -i\omega_2$$
$$\vdots \qquad\qquad \vdots$$
$$\lambda_{2n-1} = +i\omega_n \qquad \lambda_{2n} = -i\omega_n$$

$$(6.174b)$$

where only the positive values of frequency have physical meaning and are used to obtain the $2n \times 1$ eigenvectors. These, in turn, have the form

$$\mathbf{s}_j = \begin{bmatrix} \mathbf{z}_j \\ \lambda_j \mathbf{z}_j \end{bmatrix} \qquad (6.174c)$$

where the $\mathbf{z}_j$ are the eigenvectors of the original problem. Note that this procedure, in essence, converts a set of n second-order ordinary differential equations into an equivalent set of $2n$ first-order ordinary differential equations by introducing velocities as an auxiliary set of variables. Equation (6.173) represents what is commonly called a state-space formulation of the equations of motions because, from a mathematical point of view, the set of $2n$ variables $u_1, u_2, \ldots, u_n, \dot{u}_1, \dot{u}_2, \ldots, \dot{u}_n$ defines the state (or phase) space of the system under investigation. We have already considered an analogous approach in Section 3.3.4, where – due to their importance in analytical mechanics – we briefly discussed Hamilton's canonical equations. In that case the conjugate momenta played the role of auxiliary variables.

6.8.2 Viscously damped systems

A state-space formulation of the eigenproblem can also be adopted for the case of viscously damped n-DOF systems when the transformation to normal coordinates fails to uncouple the equations of motion. This is done by combining the two equations

$$\mathbf{M\ddot{u}} + \mathbf{C\dot{u}} + \mathbf{Ku} = 0$$
$$\mathbf{M\dot{u}} - \mathbf{M\dot{u}} = 0 \tag{6.175a}$$

(i.e. the equations of motion of the damped system plus a simple identity) into the single matrix equation

$$\begin{bmatrix} \mathbf{C} & \mathbf{M} \\ \mathbf{M} & 0 \end{bmatrix} \begin{bmatrix} \dot{\mathbf{u}} \\ \ddot{\mathbf{u}} \end{bmatrix} + \begin{bmatrix} \mathbf{K} & 0 \\ 0 & -\mathbf{M} \end{bmatrix} \begin{bmatrix} \mathbf{u} \\ \dot{\mathbf{u}} \end{bmatrix} = \begin{bmatrix} 0 \\ 0 \end{bmatrix} \tag{6.175b}$$

which can be written in the form

$$\hat{\mathbf{M}}\dot{\mathbf{x}} + \hat{\mathbf{K}}\mathbf{x} = 0 \tag{6.175c}$$

where we have defined the $2n \times 1$ state vector $\mathbf{x} \equiv [\mathbf{u} \quad \dot{\mathbf{u}}]^T$ and the $2n \times 2n$ real symmetrical matrices

$$\hat{\mathbf{M}} = \begin{bmatrix} \mathbf{C} & \mathbf{M} \\ \mathbf{M} & 0 \end{bmatrix} \qquad \hat{\mathbf{K}} = \begin{bmatrix} \mathbf{K} & 0 \\ 0 & -\mathbf{M} \end{bmatrix}$$

which are not, in general, positive definite. Once again, assuming a solution of the form $\mathbf{x} = \mathbf{s}e^{\lambda t}$ leads to the generalized $2n$ symmetrical eigenvalue problem

$$\hat{\mathbf{K}}\mathbf{s} + \lambda\hat{\mathbf{M}}\mathbf{s} = 0 \tag{6.176}$$

whose characteristic equation and eigenvalues are the same as the ones of the complex eigenproblem (eq (6.92)); the coefficients of the characteristic equation are all real and the eigenvalues are real or in complex conjugate pairs. The $2n \times 1$ eigenvectors of eq (6.176) have the form

$$
\mathbf{s}_j = \begin{bmatrix} \mathbf{z}_j \\ \lambda_j \mathbf{z}_j \end{bmatrix}
\tag{6.177}
$$

where $\mathbf{z}_j$ the are the $n \times 1$ eigenvectors of the complex eigenproblem. Apart from the increased computational difficulty, the solution of eq. (6.176) develops along the same line of reasoning followed in the case of the generalized eigenvalue problem of undamped systems.

The advantage of the present formulation with respect to the complex eigenproblem is that the eigenvectors s are now orthogonal with respect to both $\hat{\mathbf{M}}$ and $\hat{\mathbf{K}}$, i.e.

$$
\mathbf{s}_i \hat{\mathbf{M}} \mathbf{s}_j = \hat{M}_{ii} \delta_{ij}
$$
$$
\mathbf{s}_i \hat{\mathbf{K}} \mathbf{s}_j = \hat{K}_{ii} \delta_{ij}
\tag{6.178}
$$

where the constants $\hat{M}_{ii}$ and $\hat{K}_{ii}$ are determined from the normalization. Note that the right-hand side of eq (6.178) is zero also when the two eigenvectors are complex conjugate of each other, because, as a matter of fact, they correspond to different eigenvalues. Moreover, eqs (6.178) can be taken a step further and the reader, by writing explicitly a complex eigenvector as $\mathbf{s}_j = \mathbf{a}_j + i\mathbf{b}_j$, is invited to obtain the orthogonality relationships between different eigenvectors in terms of their real and imaginary parts. These latter relationships are useful in the numerical solution of the complex eigenproblem.

Example 6.6. The reader is also invited to consider Example 6.1 (Fig. 6.1); assume the numerical values $m = 1$ kg, $k = 2$ N/m and add two viscous dampers of constants $c_1 = 0.02$ and $c_2 = 0.01$ N s/m in parallel with the springs k_1 and k_2s. The equations of motion (6.64) have now also a viscous term where the damping matrix is

$$
\mathbf{C} = \begin{bmatrix} c_1 + c_2 & -c_2 \\ -c_2 & c_2 \end{bmatrix} = \begin{bmatrix} 0.03 & -0.01 \\ -0.01 & 0.01 \end{bmatrix}
$$

Outline the formulation of the complex eigenproblem and of the generalized eigenproblem (6.176) and show that they both lead to the characteristic equation

$$
3.0000\lambda^4 + 0.0600\lambda^3 + 16.0002\lambda^2 + 0.1200\lambda + 16.0000 = 0
$$

whose solutions are

$$
\lambda_{1,2} = -0.0025 \pm 1.1547i
$$
$$
\lambda_{3,4} = -0.0075 \pm 2.0000i
$$

Determine the eigenvectors.

If needed, the generalized problem of eq (6.176) can be converted into the standard form by premultiplying both sides by $\hat{\mathbf{M}}^{-1}$ or $\hat{\mathbf{K}}^{-1}$ (when they exist), in close analogy to the procedure that led to eq (6.163) or (6.164). In this case it is not difficult to show that, for example, in terms of the original mass and damping matrices, $\hat{\mathbf{M}}^{-1}$ is given by

$$\hat{\mathbf{M}}^{-1} = \begin{bmatrix} 0 & \mathbf{M}^{-1} \\ \mathbf{M}^{-1} & -\mathbf{M}^{-1}\mathbf{C}\mathbf{M}^{-1} \end{bmatrix}$$

The fact that the symmetrical generalized eigenproblem (6.176) leads to complex eigenvalues may seem in contradiction with the statement of Section 6.3.1 that a symmetrical eigenvalue problem produces only real eigenvalues (a general conclusion that was obtained from eq (6.39b)). However, there is no contradiction because eq (6.176) is not a 'true' $2n$ generalized eigenproblem but a state-space formulation of a complex eigenproblem of order n. Substitution of the explicit expressions of s_j (eq (6.177)) and of $\hat{\mathbf{M}}$ into eq (6.39b) will show, after a few calculations, that the conclusion $\text{Im}(\lambda_j) = 0$ is no longer valid. In addition, note that the state formulation of eq (6.176) applies also for undamped systems; in this case, however, the matrix $\hat{\mathbf{M}}$ is given by

$$\hat{\mathbf{M}} = \begin{bmatrix} 0 & \mathbf{M} \\ \mathbf{M} & 0 \end{bmatrix}$$

because $\mathbf{C} = 0$. This formulation, as one might expect, leads to the same eigenvalues and eigenvectors of eqs (6.174b) and (6.174c). The fact that now we obtain purely imaginary eigenvalues can be once again proved with the help of eq (6.39b), which now reads

$$(\lambda_j - \lambda_j^*)\begin{bmatrix} \mathbf{z}_j^H & \lambda_i^* \mathbf{z}_j^H \end{bmatrix}\begin{bmatrix} 0 & \mathbf{M} \\ \mathbf{M} & 0 \end{bmatrix}\begin{bmatrix} \mathbf{z}_j \\ \lambda_i \mathbf{z}_j \end{bmatrix} = 0$$

The reader is invited to show that the calculation of the matrix product above and substitution of the explicit expressions $\mathbf{z}_j = \mathbf{a}_j + i\mathbf{b}_j$ and $\mathbf{z}_j^H = \mathbf{a}_j^T - i\mathbf{b}_j^T$ into the resulting equation leads to $\text{Re}(\lambda_j) = 0$.

The last possibility that we consider follows closely the procedure that led to eq (6.174a) for an undamped system. We start from $\mathbf{M}\ddot{\mathbf{u}} + \mathbf{C}\dot{\mathbf{u}} + \mathbf{K}\mathbf{u} = 0$, premultiply by $\mathbf{M}^{-1}$, define the vectors $\mathbf{x}_1$ and $\mathbf{x}_2$ as in (6.172) and obtain

$$\dot{\mathbf{x}}_1 = \dot{\mathbf{u}} = \mathbf{x}_2$$

$$\dot{\mathbf{x}}_2 = \ddot{\mathbf{u}} = -\mathbf{M}^{-1}\mathbf{K}\mathbf{x}_1 - \mathbf{C}\mathbf{x}_2$$

(6.179)

If now we define the $2n \times 1$ vector $\mathbf{x} = [\mathbf{x}_1 \quad \mathbf{x}_2]^T$ and the matrix

$$\mathbf{A} = \begin{bmatrix} 0 & \mathbf{I} \\ -\mathbf{M}^{-1}\mathbf{K} & -\mathbf{M}^{-1}\mathbf{C} \end{bmatrix} \tag{6.180a}$$

Equation (6.179) can be written in matrix form as $\dot{\mathbf{x}} = \mathbf{A}\mathbf{x}$, in formal analogy with eq (6.173). Once again, assuming a solution of the form $\mathbf{x} = \mathbf{s}e^{\lambda t}$ leads to a standard eigenvalue problem of order $2n$ for the matrix $\mathbf{A}$, i.e.

$$\mathbf{A}\mathbf{s} = \lambda\mathbf{s} \tag{6.180b}$$

The main drawback is that, in general, the problem is not symmetrical.

More details about the solution of nonsymmetrical standard eigenproblems are given in Appendix A. For the moment, it suffices to say that nonsymmetrical eigenproblems lead to two set of eigenvectors, the so-called right and left eigenvectors: the right eigenvectors are associated with the original eigenproblem (i.e. the matrix $\mathbf{A}$ in the case of the last formulation that led to eq (6.180)) while the left eigenvectors are associated with the transposed eigenproblem (i.e. the matrix $\mathbf{A}^T$ for the case of eq (6.180)). Then, an appropriate set of 'biorthogonality' conditions holds between left and right eigenvectors (in this regard, it may be noted that it is not necessary to solve the eigenvalue problem for $\mathbf{A}^T$ because the left eigenvectors can be obtained by inverting the matrix of right eigenvectors).

In general, the lack of symmetry is not a problem when the eigenvalues are distinct. However, some complications may arise with degenerate eigenvalues when the matrix is defective, that is when a degenerate eigenvalue has a strictly smaller geometric than algebraic multiplicity. In this case, the concept of a principal vector (or, less properly, generalized eigenvector) is introduced by performing linear operations on previously computed eigenvectors. We will deal with this very particular case as appropriate in the course of future chapters; the interested reader can refer, for example, to Newland [4] and Gantmacher [5].

6.9   Summary and comments

Chapter 6 has been entirely dedicated to the free vibration of undamped and damped n-DOF systems, where n is any finite number. Hardly any mention has been made of approximate methods, which will be deferred to later chapters.

As a preliminary, we started our analysis with a simple 2-DOF undamped system without making use of matrix algebra; some important conclusions were reached:

1. The equations of motion are, in general, coupled.
2. The system is capable of harmonic motion at two well-defined values of frequency (the natural frequencies).

3. Each frequency is associated with a specific, but not uniquely defined, pattern of motion (the normal modes of vibration), the indeterminacy on the amplitude of motion being a consequence of the homogeneous nature of the problem.
4. There exists a choice of coordinates (the normal coordinates) which uncouples the equations of motions and leads to potential- and kinetic-energy expressions with no cross-product terms.

A direct consequence of point 4 is that the coupling of the equations of motion is not an intrinsic property of the system but depends on the choice of coordinates.

All the conclusions above are then extended and generalized to the case of undamped systems with n degrees of freedom with the aid of matrix algebra, which is the natural mathematical 'tool' for computer implementation. The matrix formulation of the free-vibration problem has the form of a symmetrical generalized eigenvalue problem of order n whose solution determines the eigenvalues (the natural frequencies) and eigenvectors (the mode shapes) of the undamped system under investigation. The eigenvectors, in turn, satisfy the important conditions of mass and stiffness orthogonality and form a basis of an n-dimensional vector space. Hence, it follows from the linearity of the problem that its general solution can be obtained by superposition of the n 'normalized' eigenvectors. The process of normalization of eigenvectors is needed to remove their amplitude indeterminacy by an appropriate choice, made by the analyst, of a scaling factor. Moreover, the eigenvectors can be arranged in matrix form (the modal matrix) in a way that leads naturally to the normal coordinates, and hence to the uncoupling of the equations of motion, which is always possible for undamped systems. The simplification is noteworthy, because the problem is reduced to n independent SDOF equations of motion.

Once the intrinsic importance of the eigenvalues and eigenvectors has been established, we turn our attention to their sensitivity, i.e. the way in which they change as a consequence of small modifications of the mass and/or stiffness parameters of the system. This is done via a perturbation method, widely employed in many branches of physics and engineering. The calculations are limited to the first order because this is the most significant perturbation term.

Successively, the same perturbation technique is used to investigate the consequences of the addition of a small viscous damping term in the equations of motion; in brief, these consequences are:

- the appearance of complex eigenvalues and eigenvectors;
- the validity of the modal approach for lightly damped structures, where the influence of the off-diagonal terms of the matrix $\mathbf{P}^T\mathbf{CP}$ is, to the first order, negligible.

A whole section is then dedicated to a general discussion on the structure and properties of the mass and stiffness matrices because, in general, the effectiveness and reliability of a numerical solution (the only one possible for complex systems) depend on the structure and properties of the input matrices. As a matter of fact, there is no unique way to define the mass and stiffness properties of a given system, and the choice of one approach rather than a different one may reflect on the quality of the solution. For its conceptual importance and its far-reaching consequences in numerical analysis (especially finite-element methods), we introduce in this section the important approach of the assumed-modes method and show some examples for simple systems.

One particular property of the input matrices is their nonsingularity. However, situations occur when the stiffness matrix is singular because the system is capable of rigid-body motions. The name 'rigid-body modes', occurring at zero frequency, is a bit misleading because these are not oscillatory solutions of the equations of motion but correspond to a rigid translation of the system as a whole. Rigid-body modes do not involve any change in the potential elastic energy and are due to the fact that the system is unrestrained; they can often be identified by inspection and their effect is often removed by means of appropriate techniques in the solution of the eigenproblem. However, they must not be forgotten when we write the general free-vibration solution.

Damped systems are considered next. They lead to the formulation of the complex eigenvalue problem. In general, the transformation to normal coordinates does not uncouple the equations of motion because of the damping matrix. However, the assumption of 'proportional' damping leads to a set of n uncoupled equations of motion and to a set of n damping ratios, in analogy with the SDOF damped equation of motion. Then, the damped eigenvectors are the same as the undamped ones. A more general form of damping matrix which results in a set of n uncoupled equations of motion is given by the Caughey summation form, which exploits a general family of eigenvector orthogonality conditions.

Nevertheless, the assumption of these special forms of damping is not always justified and the case must be considered when the equations of motion cannot be uncoupled. We are led to a set of complex eigenvalues and eigenvectors. The complex eigenvectors represent motions of the system (the so-called complex modes) in which different parts of the system are no longer in phase or antiphase with each other and do not satisfy simple orthogonality relations as in the undamped case. When animation of the mode shapes is available, it is noted that complex modes depict waves travelling along the structure rather than stationary waves with well-defined nodal points or lines.

Finally, the last section deals with the possibility of transforming one particular type of eigenproblem into a different form, more amenable to numerical solution. Hence, we can transform a generalized eigenproblem into a standard form, or transform a complex eigenproblem into a generalized or a standard eigenproblem. A widely adopted strategy consists of formulating the equations in the state (or phase) space. The cost of the resulting simpler form of the problem is the doubling of its size (that is, we are led to an eigenproblem of order $2n$) and sometimes the loss of its symmetry. Nevertheless, the advantages generally outweigh the disadvantages and more effective numerical procedures can be employed to solve the transformed eigenproblem.

References

1. Horn, R.A. and Johnson, C.R., *Matrix Analysis*, Cambridge University Press, 1985.
2. Junkins, J.L. and Kim, Y., *Introduction to Dynamics and Control of Flexible Structures*, AIAA Education Series, 1993.
3. Bisplinghoff, R.L., Mar, J.W. and Pian, T.H.H., *Statics of Deformable Solids*, Dover, New York, 1965.
4. Newland, D.E., *Mechanincal Vibration Analysis and Computation*, Longman Scientific and Technical, 1989.
5. Gantmacher, F.R., *The Theory of Matrices*, Vol. 1, Chelsea Publishing Company, New York, 1977.

7 More MDOF systems – forced vibrations and response analysis

7.1 Introduction

The preceding chapter was devoted to a detailed discussion of the free-vibration characteristics of undamped and damped MDOF linear systems. In the course of the discussion, it has become more and more evident – both from a theoretical and from a practical point of view – that natural frequencies (eigenvalues) and mode shapes (eigenvectors) play a fundamental role. As we proceed further in our investigation, this idea will be confirmed.

Following Ewins [1], we can say that for any given structure we can distinguish between the **spatial model** and the **modal model**: the first being defined by means of the structure's physical characteristics – usually its mass, stiffness and damping properties – and the second being defined by means of its modal characteristics, i.e. a set of natural frequencies, mode shapes and damping factors. In this light we may observe that Chapter 6 led from the spatial model to the modal model; in Ewins' words, we proceeded along the 'theoretical route' to vibration analysis, whose third stage is the **response model**. This is the subject of the present chapter and concerns in the analysis of how the structure will vibrate under given excitation conditions.

The importance is twofold: first, for a given system, it is often vital for the engineer to understand what amplitudes of vibration are expected in prescribed operating conditions and, second, the modal characteristics of a vibrating system can be obtained by performing experimental tests in appropriate 'forced-vibration conditions', that is by exciting the structure and measuring its response. These measurements, in turn, often constitute the first step of the 'experimental route' to vibration analysis (again Ewins' definition), which proceeds in the reverse direction with respect to the theoretical route and leads from the measured response properties to the vibration modes and, finally, to a structural model.

Obviously, in common practice the theoretical and experimental approaches are strictly interdependent because, hopefully, the final goal is to arrive at a satisfactory and effective description of the behaviour of a given system; what to do and how to do it depends on the scope of the investigation, on the deadline and, last but not least, on the available budget.

In this chapter we pursue the theoretical route to its third stage, while the experimental route will be considered in later chapters.

7.2   Mode superposition

In the analysis of the dynamic response of a MDOF system, the relevant equations of motions are written in matrix form as

$$\mathbf{M\ddot{u} + C\dot{u} + Ku = f} \tag{7.1}$$

where $\mathbf{f} \equiv [f_1(t), f_2(t), ..., f_n(t)]^T$ is a time-dependent $n \times 1$ vector of forcing functions. In the most general case eqs (7.1) are a set of n simultaneous equations whose solution can only be obtained by appropriate numerical techniques, more so if the forcing functions are not simple mathematical functions of time.

However, if the system is undamped ($\mathbf{C} = 0$) we know that there always exists a set of normal coordinates y which uncouples the equations of motion. We pass to this set of coordinates by means of the transformation (6.56a), i.e.

$$\mathbf{u = Py} = \sum_{i=1}^{n} \mathbf{p}_i y_i \tag{7.2}$$

where $\mathbf{P}$ is the weighted modal matrix, that is the matrix of mass orthonormal eigenvectors. As for the free-vibration case, premultiplication of the transformed equations of motion by $\mathbf{P}^T$ gives

$$\mathbf{I\ddot{y} + Ly = P}^T\mathbf{f} \tag{7.3a}$$

where $\mathbf{L} = \text{diag}(\lambda_1, \lambda_2, ..., \lambda_n)$ is the diagonal matrix of eigenvalues and the term on the right-hand side is called the modal force vector. Equations (7.3a) represent a set of n uncoupled equations of motion; explicitly they read

$$\ddot{y}_j + \lambda_j y = \mathbf{p}_j^T\mathbf{f} \equiv \phi_j \qquad (j = 1, 2, ..., n) \tag{7.3b}$$

where we define the jth **modal participation factor** $\phi_j \equiv \mathbf{p}_j^T\mathbf{f}$, i.e. the jth element of the $n \times 1$ modal force vector, which clearly depends on the type of loading. In this regard, it is worth noting that the jth modal participation factor can be interpreted as the amplitude associated with the jth mode in the expansion of the force vector with respect to the inertia forces. In other words, if the vector $\mathbf{f}$ is expanded in terms of the inertia forces $\mathbf{Mp}_i$ generated by the eigenmodes, we have

$$\mathbf{f} = \sum_{i=1}^{n} \alpha_i \mathbf{Mp}_i \tag{7.4}$$

where the α_is are the expansion coefficients. Premultiplication of both sides of eq (7.4) by $\mathbf{p}_j^T$ leads to

$$\alpha_j = \mathbf{p}_j^T \mathbf{f}$$

and hence to the conclusion $\alpha_j = \phi_j$, which proves the statement above.

The equations of motion in the form (7.3b) can be solved independently with the methods discussed in Chapters 4 and 5: each equation is an SDOF equation and its general solution can be obtained by adding the complementary and particular solutions. The initial conditions in physical coordinates

$$\mathbf{u}(t = 0) = \mathbf{u}_0$$

$$\dot{\mathbf{u}}(t = 0) = \dot{\mathbf{u}}_0$$

are taken into account by means of the transformation to normal coordinates. The transformation (7.2) suggests that the initial conditions in normal coordinates could be obtained as

$$\mathbf{y}_0 = \mathbf{P}^{-1} \mathbf{u}_0$$

$$\dot{\mathbf{y}}_0 = \mathbf{P}^{-1} \dot{\mathbf{u}}_0$$

(7.5a)

However, as in eqs (6.58), it is preferable to use the orthogonality of eigenmodes and calculate

$$\mathbf{y}_0 = \mathbf{P}^T \mathbf{M} \mathbf{u}_0$$

$$\dot{\mathbf{y}}_0 = \mathbf{P}^T \mathbf{M} \dot{\mathbf{u}}_0$$

(7.5b)

The solution strategy considered above is often called the **mode superposition method** (or the normal mode method) and is based on the possibility to uncouple the equations of motion by means of an appropriate coordinate transformation. It is evident that the first step of the whole process is the solution of the free-vibration problem, because it is assumed that the eigenvalues and eigenvectors of the system under study are known.

The same method applies equally well to damped systems with proportional damping or, more generally, to damped systems for which the matrix $\mathbf{P}^T \mathbf{C} \mathbf{P}$ has either zero or negligible off diagonal elements. In this case the uncoupled equations of motion read

$$\mathbf{I}\ddot{\mathbf{y}} + \mathbf{P}^T \mathbf{C} \mathbf{P} \dot{\mathbf{y}} + \mathbf{L}\mathbf{y} = \mathbf{P}^T \mathbf{f}$$

(7.6a)

or, explicitly

$$\ddot{y}_j + 2\omega_j \zeta_j \dot{y}_j + \omega_j^2 y_j = \phi_j \qquad (j = 1, 2, ..., n)$$

(7.6b)

where damping can be more easily specified at the modal level by means of the damping ratios ζ_j rather than obtaining the damping matrix **C**. The initial conditions are obtained exactly as in eqs (7.5b) and the complete solution for the jth normal coordinate can be written in analogy with eq (5.19) as

$$
y_j = e^{-\zeta_j \omega_j t} \left(y_{j0} \cos \omega_{dj} t + \frac{\dot{y}_{j0} + \zeta_j \omega_j y_{j0}}{\omega_{dj}} \sin \omega_{dj} t \right)
$$

$$
+ \frac{1}{\omega_{dj}} \int_0^t \phi_j(\tau) e^{-\zeta_j \omega_j (t-\tau)} \sin [\omega_{dj}(t-\tau)] d\tau \qquad (7.7a)
$$

where we write y_{j0} and $\dot{y}_{j0}$ to mean the initial displacement and velocity of the jth normal coordinate and, in the terms ω_{dj}, the subscript d indicates 'damped'. As in the SDOF case, the damped frequency is given by

$$
\omega_{dj} = \omega_j \sqrt{1 - \zeta_j^2}
$$

and the exact evaluation of the Duhamel integral is only possible when the $\phi_j(t)$ are simple mathematical functions of time, otherwise some numerical technique must be used. It is evident that if we let $\zeta_j \to 0$, eq (7.7) leads immediately to the undamped solution. Also, we note in passing that for a system initially at rest (i.e. $y_{j0} = \dot{y}_{j0} = 0$) we can write the vector of normal coordinates in compact form as

$$
\mathbf{y} = \int_0^t \text{diag}[h_1(t-\tau), ..., h_n(t-\tau)] \mathbf{P}^T \mathbf{f}(\tau) d\tau \qquad (7.7b)
$$

where $\text{diag}[h_1(t), ..., h_n(t)]$ is a diagonal matrix of **modal** impulse response functions (eq (5.7a), where in this case $m = M_{jj} = 1$ because the eigenvectors are mass orthonormal)

$$
h_j(t) = \left(\frac{e^{-\zeta_j \omega_j t}}{\omega_{dj}} \right) \sin \omega_{dj} t \qquad (j = 1, 2, ..., n)
$$

Two important observations can be made at this point:

- If the external loading **f** is orthogonal to a particular mode $\mathbf{p}_k$, that is if $\mathbf{p}_k^T \mathbf{f} = 0$, that mode will not contribute to the response.
- The second observation has to do with the **reciprocity theorem** for dynamic loads, which plays a fundamental role in many aspects of linear vibration analysis. The theorem, a counterpart of Maxwell's reciprocal theorem for static loads, states that the response of the jth degree of

freedom due to an excitation applied at the kth degree of freedom is equal to the response of the kth degree of freedom when the same excitation is applied at the jth degree of freedom.

To be more specific, let us assume that the vibrating system is initially at rest, i.e. $\mathbf{u}_0 = \dot{\mathbf{u}}_0 = 0$ or, equivalently $y_{j0} = \dot{y}_{j0} = 0$ in eq (7.7) (this assumption is only for our present convenience and does not imply a loss of generality). From eq (7.2), the total response of the jth physical coordinate u_j is given by

$$u_j = \sum_{i=1}^{n} p_{ji} y_i \tag{7.8a}$$

Now, suppose that the structure is excited by a single force at the kth point, i.e. $\mathbf{f} = [0, 0, ..., 0, f_k, 0, ..., 0]^T$; the ith participation factor will be given by

$$\phi_i \equiv \mathbf{p}_i^T \mathbf{f} = p_{ki} f_k \qquad (i = 1, 2, ..., n) \tag{7.9a}$$

so that, by substituting eqs (7.7) (with zero initial conditions) and (7.9a) in eq (7.8a), we have

$$u_j = \sum_{i=1}^{n} p_{ji} p_{ki} \left(\frac{1}{\omega_{di}} \int_0^t f_k(\tau) e^{-\zeta_i \omega_i (t-\tau)} \sin[\omega_{di}(t-\tau)] d\tau \right) \tag{7.10a}$$

The same line of reasoning shows that the response of the kth physical coordinate is written as

$$u_k = \sum_{i=1}^{n} p_{ki} y_i \tag{7.8b}$$

and, under the assumption that we apply the same force as before at the jth degree of freedom (i.e. only the jth term of the vector $\mathbf{f}$ is different from zero), we have the following participation factors:

$$\phi_i \equiv \mathbf{p}_i^T \mathbf{f} = p_{ji} f_j \qquad (i = 1, 2, ..., n) \tag{7.9b}$$

Once again, substitution of the explicit expression of y_i and of eq (7.9b) into eq (7.8b) yields

$$u_k = \sum_{i=1}^{n} p_{ki} p_{ji} \left(\frac{1}{\omega_{di}} \int_0^t f_j(\tau) e^{-\zeta_i \omega_i (t-\tau)} \sin[\omega_{di}(t-\tau)] d\tau \right) \tag{7.10b}$$

which is equal to eq (7.10a) when the hypothesis of the reciprocity theorem is satisfied, that is, that the external applied load is the same in the two cases, the only difference being the point of application.

So, returning to the main discussion of this section, we saw that in order to obtain a complete solution we must evaluate n equations of the form (7.7) and substitute the results back in eq (7.2), where the response in physical coordinates is expressed as a superposition of the modal responses. For large systems, this procedure may involve a large computational effort. However, one major advantage of the mode superposition method for the calculation of dynamic response is that, frequently, only a small fraction of the total number of uncoupled equations need to be considered in order to arrive at a satisfactory approximate solution of eq (7.1). Broadly speaking, this is due to the fact that, in common situations, a large portion of the response is contained in only a few of the mode shapes, usually those corresponding to the lowest frequencies. Therefore, only the first s ($s \ll n$) equations need to be used in order to obtain a good approximate 'truncated' solution. This is written as

$$\mathbf{u}_{(s)} = \sum_{i=1}^{s} \mathbf{p}_i y_i(t) \tag{7.11}$$

How many modes must be included in the analysis (i.e. the value of s) depends, in general, on the system under investigation and on the type of loading, namely its spatial distribution and frequency content. Nevertheless, the significant saving of computation time can be appreciated if we consider, for example, that in wind and earthquake loading of structural systems we may have $n/s \geqslant 50$.

If not enough modes are included in the analysis, the truncated solution will not be accurate. On a qualitative basis, we can say that the lack of accuracy is due to the fact that – owing to the truncation process – part of the loading has not been included in the superposition. Since we can expand the external loading in terms of the inertia forces (eq (7.4)), we can calculate

$$\Delta \mathbf{f} = \mathbf{f} - \sum_{i=1}^{s} \phi_i \mathbf{M} \mathbf{p}_i \tag{7.12}$$

and note that a satisfactory accuracy is obtained when $\Delta \mathbf{f}$ corresponds, at most, to a static response. It follows that a good correction $\Delta \mathbf{u}$ – the so-called static correction – to the truncated solution $\mathbf{u}_{(s)}$ can be obtained from

$$\mathbf{K} \Delta \mathbf{u} = \Delta \mathbf{f} \tag{7.13}$$

Also, on physical grounds, lack of accuracy must be expected when the external loading has a frequency component which is close to one of the system's modes (say, the kth mode, where $k > s$) that has been neglected. In this case, in fact, the contribution of the *k*th mode to the response becomes important and an inappropriate truncation will fail to take this part of the response into account. This is a typical example of what we meant by saying that the frequency content of the input – together with its spatial distribution – determines the number of modes to be included in the sum (7.11).

From a more general point of view, it must also be considered that little or hardly any accuracy can be expected in both the theoretical (for example, by finite-element methods) calculation and the experimental determination (for example, by means of experimental modal analysis) of higher frequencies and mode shapes. Hence, for systems with a high number of degrees of freedom, modal truncation is almost a necessity.

A final note of practical use: frequently we may be interested in the maximum peak value of a physical coordinate u_j. An approximated value for this quantity, as a matter of fact, is based on the truncated mode summation and it reads

$$| u_j |_{max} = | p_{j1} y_{1,\,max} | + \sqrt{\sum_{k=2}^{s} | p_{jk} y_{k,\,max} |^2} \qquad (7.14)$$

where p_{jk} is the (jk)th element of the modal matrix or, in other words, the *j*th element of the *k*th eigenvector. Equation (7.14) is widely accepted and has been found satisfactory in most cases; the contribution of modes other than the first is taken into account by means of the term under the square root which, in turn, is a better expression than $\sum_k | p_{jk} y_{k,\,max} |$ because, statistically speaking, it is very unlikely that all maxima occur simultaneously.

7.2.1 *Mode displacement and mode acceleration methods*

The process of expressing the system response through mode superposition and restricting the modal expansion to a subset of *s* modes is often called the mode displacement method. Experience has shown that this method must be applied with care because, owing to convergence problems, many modes are needed to obtain an accurate solution. Suppose, for example that the applied load can be written in the form

$$\mathbf{f} = \mathbf{f}_0 g(t)$$

If we consider, for simplicity, the response of an undamped system initially at rest, we have the **mode displacement** solution

$$\mathbf{u}_{(s)} = \sum_{i=1}^{s} \mathbf{p}_i \mathbf{p}_i^T \mathbf{f}_0 \frac{1}{\omega_i} \int_0^t g(\tau) \sin[\omega_i(t - \tau)] d\tau \qquad (7.15)$$

which does not take into account the contribution of the modes that have been left out. Moreover – besides depending on the frequency content of the excitation and on the eigenvalues of the vibrating system, which are both taken into account in the convolution integral – the convergence of the solution depends also on how well the spatial part of the applied load $\mathbf{f}_0$ is represented on the basis of the s modes retained in the process. The mode acceleration method approximates the response of the missing modes by means of an additional pseudostatic response term. The line of reasoning has been briefly outlined in the preceding section (eqs (7.12) and (7.13)) and will be pursued a little further in this section.

We can rewrite the equations of motion of our undamped (and initially at rest) system in the form

$$\mathbf{Ku} = \mathbf{f} - \mathbf{M\ddot{u}}$$

premultiplicate both sides by $\mathbf{K}^{-1}$ (under the assumption of no rigid-body modes) and substitute the truncated expansion of the inertia forces to get the mode acceleration solution $\hat{\mathbf{u}}_{(s)}$ as

$$\hat{\mathbf{u}}_{(s)} = \mathbf{K}^{-1}\mathbf{f} - \mathbf{K}^{-1} \sum_{i=1}^{s} \ddot{y}_i \mathbf{Mp}_i \qquad (7.16)$$

and since $\mathbf{Mp}_i = \omega_i^{-2}\mathbf{Kp}_i$, we obtain

$$\hat{\mathbf{u}}_{(s)} = \mathbf{K}^{-1}\mathbf{f} - \sum_{i=1}^{s} \left(\frac{\ddot{y}_i}{\omega_i^2}\right)\mathbf{p}_i \qquad (7.17)$$

where the first term on the right-hand side of eq (7.17) is called the pseudo-static response and the name of the method is due to the $\ddot{y}_i$ in the second term. Moreover, note that if the loading is of the form $\mathbf{f} = \mathbf{f}_0 g(t)$, the term $\mathbf{K}^{-1}\mathbf{f}_0$ can be calculated only once. Then, it can be multiplied by $g(t)$ for each specific value of t for which the response is required.

Now, the expression

$$y_i = \frac{1}{\omega_i} \int_0^t \mathbf{p}_i^T \mathbf{f}(\tau)\sin[\omega_i(t-\tau)]d\tau$$

can be inserted in $\ddot{y}_i = \mathbf{p}_i^T \mathbf{f} - \omega_i^2 y_i$ which, in turn, is obtained from eq (7.3b); the result is then substituted in eq (7.17) to give

$$\hat{\mathbf{u}}_{(s)} = \sum_{i=1}^{s} \frac{\mathbf{p}_i \mathbf{p}_i^T}{\omega_i} \int_0^t \mathbf{f}(\tau)\sin[\omega_i(t-\tau)]d\tau + \left(\mathbf{K}^{-1} - \sum_{i=1}^{s} \frac{\mathbf{p}_i \mathbf{p}_i^T}{\omega_i^2}\right)\mathbf{f} \qquad (7.18)$$

Equation (7.18) can be put in its final form if we consider the spectral expansion of the matrix $\mathbf{K}^{-1}$. This is not difficult to obtain: we start from the spectral expansion of the identity matrix (eq (6.49b)), transpose both sides to obtain

$$\mathbf{I} = \sum_{i=1}^{n} \mathbf{M}\mathbf{p}_i\mathbf{p}_i^T$$

premultiply both sides by $\mathbf{K}^{-1}$ and consider that $\mathbf{M}\mathbf{p}_i = \omega_i^{-2}\mathbf{K}\mathbf{p}_i$. It follows

$$\mathbf{K}^{-1} = \sum_{i=1}^{n} \frac{\mathbf{p}_i\mathbf{p}_i^T}{\omega_i^2} \tag{7.19}$$

which is the expansion we were looking for. Inserting eq (7.19) into (7.18) leads to

$$\hat{\mathbf{u}}_{(s)} = \sum_{i=1}^{s} \frac{\mathbf{p}_i\mathbf{p}_i^T}{\omega_i} \int_0^t \mathbf{f}(\tau)\sin[\omega_i(t-\tau)]d\tau + \left(\sum_{i=s+1}^{n} \frac{\mathbf{p}_i\mathbf{p}_i^T}{\omega_i^2}\right)\mathbf{f} \tag{7.20}$$

where it is now evident the contribution of the $n-s$ modes that had been completely neglected in the mode displacement solution.

As opposed to the mode displacement method, the mode acceleration method shows better convergence properties and, in general, fewer eigenvalues and eigenvectors are needed to obtain a satisfactory solution. Nevertheless, some attention must always be paid to the number of modes employed in the superposition. In fact, if the highest (sth) eigenvalue is much larger than the highest frequency component ω_{max} of the applied load, say for example $\omega_s \geqslant 4\omega_{max}$, the response of modes $s+1, s+2, ..., n$ is essentially static because (Fig. 4.8)

$$\frac{\omega_{max}}{\omega_j} \ll 1 \qquad (j = s+1, s+2, ..., n)$$

and the pseudostatic term, as a matter of fact, is a proper representation of their contribution. On the other hand, if some frequency component of the loading is close to the frequency of a 'truncated' mode, the mode acceleration solution will be just as inaccurate as the mode displacement solution and no effective improvement should be expected in this case.

For viscously damped system with proportional damping, the mode acceleration solution can be obtained from

$$\mathbf{u} = \mathbf{K}^{-1}(\mathbf{f} - \mathbf{C}\dot{\mathbf{u}} - \mathbf{M}\ddot{\mathbf{u}})$$

and written as

$$\mathbf{u}_{(s)} = \mathbf{K}^{-1}\mathbf{f} - \sum_{i=1}^{s} \left(\frac{2\zeta_i \dot{y}_i}{\omega_i} \right) \mathbf{p}_i - \sum_{i=1}^{s} \left(\frac{\ddot{y}_i}{\omega_i^2} \right) \mathbf{p}_i \qquad (7.21)$$

where the last term on the right-hand side is exactly as in eq (7.17) and the second term has been obtained using the spectral expansion (7.19) and the expression $\mathbf{p}_i^T \mathbf{C} \mathbf{p}_i = 2\zeta_i \omega_i$ (i.e. eq (6.142)).

7.3   Harmonic excitation: proportional viscous damping

Suppose now that a viscously damped n-DOF system is excited by means of a set of sinusoidal forces with the same frequency ω but with various amplitudes and phases. We have

$$\mathbf{M}\ddot{\mathbf{u}} + \mathbf{C}\dot{\mathbf{u}} + \mathbf{K}\mathbf{u} = \mathbf{f}_0 e^{-i\omega t} \qquad (7.22)$$

and we assume that a solution exists in the form

$$\mathbf{u} = \mathbf{z} e^{-i\omega t} \qquad (7.23)$$

where $\mathbf{f}_0$ and $\mathbf{z}$ are $n \times 1$ vectors of time-independent complex amplitudes. Substitution of eq (7.23) into (7.22) gives

$$(-\omega^2 \mathbf{M} - i\omega \mathbf{C} + \mathbf{K})\mathbf{z} = \mathbf{f}_0$$

whose formal solution is

$$\mathbf{z} = (\mathbf{K} - i\omega \mathbf{C} - \omega^2 \mathbf{M})^{-1} \mathbf{f}_0 \equiv \mathbf{R}\mathbf{f}_0 \qquad (7.24)$$

where we define the receptance matrix (which is a function of ω) $\mathbf{R} = \mathbf{R}(\omega) \equiv (\mathbf{K} - i\omega \mathbf{C} - \omega^2 \mathbf{M})^{-1}$. The (jk)th element of this matrix is the displacement response of the jth degree of freedom when the excitation is applied at the kth degree of freedom only. Mathematically we can write

$$R_{jk} = \left(\frac{z_j}{f_{0k}} \right)_{f_{0m} = 0; \, m \neq k} \qquad (7.25)$$

The calculation of the response by means of eq (7.24) is highly inefficient because we need to invert a large (for large n) matrix for each value of frequency.

However, if the system is proportionally damped and the damping matrix becomes diagonal under the transformation $\mathbf{P}^T \mathbf{C} \mathbf{P}$ we can write

$$(\mathbf{K} - i\omega \mathbf{C} - \omega^2 \mathbf{M}) = \mathbf{R}^{-1}$$

premultiply both sides by $\mathbf{P}^T$ and postmultiply by $\mathbf{P}$ to get

$$[\mathbf{L} - i\omega\,\text{diag}(2\zeta_j\omega_j) - \omega^2\mathbf{I}] = \text{diag}(\omega_j^2 - 2i\omega\zeta_j\omega_j - \omega^2) = \mathbf{P}^T\mathbf{R}^{-1}\mathbf{P}$$

which we can write as $\mathbf{D} = \mathbf{P}^T\mathbf{R}^{-1}\mathbf{P}$, where we define for brevity of notation

$$\mathbf{D} \equiv \text{diag}(\omega_j^2 - 2i\omega\zeta_j\omega_j - \omega^2)$$

From the above it follows that $\mathbf{D}^{-1} = (\mathbf{P}^T\mathbf{R}^{-1}\mathbf{P})^{-1} = \mathbf{P}^{-1}\mathbf{R}\mathbf{P}^{-T}$ which, after pre- and postmultiplication of both sides by $\mathbf{P}$ and $\mathbf{P}^T$, respectively, leads to

$$\mathbf{R} = \mathbf{P}\mathbf{D}^{-1}\mathbf{P}^T = \mathbf{P}\,diag\left(\frac{1}{\omega_j^2 - \omega^2 - 2i\omega\zeta_j\omega_j}\right)\mathbf{P}^T \tag{7.26}$$

so that the solution (7.24) can be written as

$$\mathbf{z} = \mathbf{R}\mathbf{f}_0 = \mathbf{P}\,diag\left(\frac{1}{\omega_j^2 - \omega^2 - 2i\omega\zeta_j\omega_j}\right)\mathbf{P}^T\mathbf{f}_0 \tag{7.27}$$

and the (jk)th element of the receptance matrix can be explicitly written as

$$R_{jk} = \sum_{m=1}^{n}\left(\frac{1}{\omega_m^2 - \omega^2 - 2i\omega\zeta_m\omega_m}\right)p_{jm}p_{km} \tag{7.28}$$

Now, the term in brackets in eq (7.28) looks, indeed, familiar and a slightly different approach to the problem will clarify this point. For a proportionally damped system, the equations of motion (7.22) can be uncoupled with the aid of the modal matrix and written in normal coordinates as

$$\ddot{y}_j + 2\zeta_j\omega_j\dot{y}_j + \omega_j^2 y = \mathbf{p}_j^T\mathbf{f}_0 e^{-i\omega t} \tag{7.29}$$

Each equation of (7.29) is a forced SDOF equation with sinusoidal excitation. We assume a solution in the form

$$y_j(t) = \hat{y}_j e^{-i\omega t}$$

where $\hat{y}_j$ is the complex amplitude response. Following Chapter 4, we arrive at the steady-state solution (the counterpart of eq (4.42)),

$$\hat{y}_j = \frac{\mathbf{p}_j^T\mathbf{f}_0}{(\omega_j^2 - \omega^2 - 2i\omega\zeta_j\omega_j)} = \frac{\mathbf{p}_j^T\mathbf{f}_0}{\omega_j^2}\left(\frac{1}{1 - \beta_j^2 - 2i\zeta_j\beta_j}\right) \tag{7.30}$$

where $\beta_j \equiv (\omega/\omega_j)$.

By definition, the frequency response function (FRF) is the coefficient $H(\omega)$ of the response of a linear, physically realizable system to the input $e^{-i\omega t}$; with this in mind we recognize that

$$H_j(\omega) \equiv \left(\frac{1}{\omega_j^2 - \omega^2 - 2i\omega\zeta_j\omega_j} \right) \tag{7.31}$$

is the jth modal (because it refers to normal, or modal, coordinates) FRF. If we define the $n \times 1$ vector $\hat{\mathbf{y}} = [\hat{y}_1, \hat{y}_2, ..., \hat{y}_n]^T$ of response amplitudes we can put together the n equations (7.29) in the matrix expression

$$\hat{\mathbf{y}} = \text{diag}[H_j(\omega)]\mathbf{P}^T\mathbf{f}_0 \tag{7.32}$$

and the passage to physical coordinates is accomplished by the transformation (7.2), which, for sinusoidal solutions, translates into the relationship between amplitudes $\mathbf{z} = \mathbf{P}\hat{\mathbf{y}}$. Hence

$$\mathbf{z} = \mathbf{P}\,\text{diag}[H_j(\omega)]\mathbf{P}^T\mathbf{f}_0 \tag{7.33}$$

which must be compared to eq (7.27) to conclude that

$$\mathbf{R}(\omega) = \mathbf{P}\,\text{diag}[H_j(\omega)]\mathbf{P}^T \tag{7.34a}$$

Equation (7.34a) establishes the relationship between the FRF matrix ($\mathbf{R}$) of receptances in physical coordinates and the FRF matrix of receptances in modal coordinates. This latter matrix is diagonal because in normal (or modal) coordinates the equations of motion are uncoupled. This is not true for the equations in physical coordinates, and consequently $\mathbf{R}$ is not diagonal. Moreover, appropriate partitioning of the matrices on the right-hand side of eq (7.34a) leads to the alternative expression for the receptance matrix

$$\mathbf{R}(\omega) = \sum_{m=1}^{n} H_m(\omega)\mathbf{p}_m\mathbf{p}_m^T \tag{7.34b}$$

where the term $\mathbf{p}_m\mathbf{p}_m^T$ is an $(n \times 1)$ by $(1 \times n)$ matrix product and hence results in an $n \times n$ matrix. From eq (7.34a) or (7.34b) it is not difficult to determine that

$$\mathbf{R} = \mathbf{R}^T \tag{7.35}$$

i.e. $\mathbf{R}$ is symmetrical; this conclusion can also be reached by inspection of eq (7.28) where it is evident that $R_{jk} = R_{kj}$. This result is hardly surprising. In fact, owing to the meaning of the term R_{jk} (i.e. eq (7.25)), it is just a different statement of the reciprocity theorem considered in Section 7.2.

7.4   Time-domain and frequency-domain response

In Section 7.2, eq (7.7b) represents, in the time domain, the normal coordinate response of a proportionally damped system to a general set of applied forces. Since we pass to physical coordinates by means of the transformation $\mathbf{u} = \mathbf{P}\mathbf{y}$, we have

$$\mathbf{u} = \int_0^t \mathbf{P} \, \text{diag}[h_j(t - \tau)]\mathbf{P}^T\mathbf{f}(\tau)d\tau \qquad (7.36)$$

so that the $n \times n$ matrix of impulse response functions in physical coordinates is given by

$$\mathbf{h}(t) = \mathbf{P} \, \text{diag}[h_j(t)]\mathbf{P}^T = \sum_{m=1}^{n} h_m(t)\mathbf{p}_m\mathbf{p}_m^T \qquad (7.37a)$$

Explicitly, the (jk)th element of matrix (7.37a) is written

$$h_{jk}(t) = \sum_{m=1}^{n} h_m(t)p_{jm}p_{km} \qquad (7.37b)$$

and it is evident that $h_{jk} = h_{kj}$, or equivalently, $\mathbf{h}(t) = \mathbf{h}^T(t)$.

On the other hand, if we take the Fourier transform of both sides of eqs (7.6), we get

$$[\omega_j^2 - \omega^2 - 2i\omega\zeta_j\omega_j]Y_j(\omega) = \Phi_j(\omega) \qquad (7.38)$$

where we have called $Y_j(\omega)$ and $\Phi_j(\omega)$ the Fourier transforms of the functions $y_j(t)$ and $\mathbf{p}_j\mathbf{f}(t) = \phi_j(t)$, respectively. If we form the column vectors

$$\mathbf{Y}(\omega) = [Y_1(\omega), \, ..., \, Y_n(\omega)]^T$$

and

$$\mathbf{P}^T\mathbf{F}(\omega) = [\Phi_1(\omega), \, ..., \, \Phi_n(\omega)]^T$$

where $\mathbf{F}(\omega) \equiv [\mathcal{F}\{f_1(t)\}, \, ..., \, \mathcal{F}\{f_n(t)\}]^T$ is the (element by element) Fourier transform of $\mathbf{f}$, we obtain from eq (7.38)

$$\mathbf{Y}(\omega) = \text{diag}\left(\frac{1}{\omega_j^2 - \omega^2 - 2i\omega\zeta_j\omega_j}\right)\mathbf{P}^T\mathbf{F}(\omega)$$

$$= \text{diag}[H_j(\omega)]\mathbf{P}^T\mathbf{F}(\omega) \qquad (7.39)$$

Now, since $\mathbf{u}(t) = \mathbf{P}\mathbf{y}(t)$ it follows that $\mathscr{F}\{\mathbf{u}(t)\} \equiv \mathbf{U}(\omega) = \mathbf{P}\mathbf{Y}(\omega)$ and eq (7.39) leads to

$$\mathbf{U}(\omega) = \mathbf{P}\,\mathrm{diag}[H_j(\omega)]\mathbf{P}^T\mathbf{F}(\omega) \qquad (7.40)$$

which is the frequency-domain counterpart of the time-domain equation (7.36). Summarizing the results above and referring to the discussion of Chapter 5 about impulse-response functions and frequency-response functions, we can say that – as for the SDOF case – the modal coordinates functions $h_j(t)$ and $H_j(\omega)$ ($j = 1, 2, ..., n$) are a Fourier transform pair and fully define the dynamic characteristics of our n-DOF proportionally damped system.

In physical coordinates, the dynamic response of the same system is characterized by the matrices $\mathbf{h}(t)$ and $\mathbf{R}(\omega)$ whose elements are given, respectively, by eqs (7.37b) and (7.28). These matrices are also a Fourier transform pair (Section 5.4), i.e.

$$\mathbf{R}(\omega) = 2\pi\mathscr{F}\{\mathbf{h}(t)\}$$
$$\mathbf{h}(t) = (2\pi)^{-1}\mathscr{F}^{-1}\{\mathbf{R}(\omega)\} \qquad (7.41)$$

which is not unexpected if we consider that the Fourier transform is a linear transformation. Also, from the discussion of Chapter 5, it is evident that the considerations of this section apply equally well if ω is replaced by the Laplace operator s and the FRFs are replaced by transfer functions in the Laplace domain. Which transform to use is largely dictated by a matter of convenience.

A note about the mathematical notation

In general an FRF function is indicated by the symbol $H(\omega)$ and, consequently, a matrix of FRF functions can be written as $\mathbf{H}(\omega)$. However, as shown in Table 4.3, $H(\omega)$ can be a receptance, a mobility or an accelerance (or inertance) function; in the preceding sections we wrote $\mathbf{R}(\omega)$ because, specifically, we have considered only receptance functions, so that $\mathbf{R}(\omega)$ is just a particular form of $\mathbf{H}(\omega)$. Whenever needed we will consider also the other particular forms of $\mathbf{H}(\omega)$, i.e. the mobility and accelerance matrices and we will indicate them, respectively, with the symbols $\mathbf{V}(\omega)$ and $\mathbf{A}(\omega)$ which explicitly show that the relevant output is velocity in the first case and acceleration in the second case. Obviously, the general FRF symbol $\mathbf{H}(\omega)$ can be used interchangeably for any one of the matrices $\mathbf{R}(\omega)$, $\mathbf{V}(\omega)$ or $\mathbf{A}(\omega)$. By the same token, $\mathbf{H}(s)$ is a general transfer function and $\mathbf{R}(s)$, $\mathbf{V}(s)$ or $\mathbf{A}(s)$ are the receptance, mobility and accelerance transfer functions.

Finally, it is worth noting that some authors write FRFs as $H(i\omega)$ in order to remind the reader that, in general, FRFs are complex functions with a real and imaginary part or, equivalently, that they contain both amplitude and phase information. We do not follow this symbolism and write simply $H(\omega)$.

7.4.1 A few comments on FRFs

In many circumstances, one may want to consider an FRF matrix other than $\mathbf{R}(\omega)$. The different forms and definitions are listed in Table 4.3 and it is not difficult to show that, for a given system, the receptance, mobility and accelerance matrices satisfy the following relationships:

$$\mathbf{V}(\omega) = -i\omega\mathbf{R}(\omega)$$

$$\mathbf{A}(\omega) = -i\omega\mathbf{V}(\omega) = -\omega^2\mathbf{R}(\omega) \tag{7.42}$$

which can be obtained by assuming a solution of the form (7.23) and noting that

$$\dot{\mathbf{u}} \equiv \mathbf{v}e^{-i\omega t} = -i\omega\mathbf{z}e^{-i\omega t}$$

$$\ddot{\mathbf{u}} \equiv \mathbf{a}e^{-i\omega t} = -i\omega\mathbf{v}e^{-i\omega t} = -\omega^2\mathbf{z}e^{-i\omega t} \tag{7.43}$$

where we have defined the (complex) velocity and acceleration amplitudes $\mathbf{v}$ and $\mathbf{a}$. However, the definitions of Table 4.3 include also other FRFs, namely the dynamic stiffness, the mechanical impedance and the apparent mass which, for the SDOF case are obtained, respectively, as the inverse of receptance, mobility and accelerance. This is not so for an MDOF system.

Even if in this text we will generally use only $\mathbf{R}(\omega)$, $\mathbf{V}(\omega)$ or $\mathbf{A}(\omega)$, the reader is warned against, say, trying to obtain impedance information by calculating the reciprocals of mobility functions. In fact, the definition of a mobility function V_{jk}, in analogy with eq (7.25), implies that the velocity at point j is measured when a prescribed force input is applied at point k, with all other possible inputs being zero. The case of mechanical impedance is different because the definition implies that a prescribed velocity input is applied at point j and the force is measured at point k, with all other input points having zero velocity. In other words, all points must be fixed (grounded) except for the point to which the input velocity is applied.

Despite the fact that this latter condition is also very difficult (if not impossible) to obtain in practical situations, the general conclusion is that

$$V_{jk}(\omega) \neq \frac{1}{Z_{jk}(\omega)} \tag{7.44}$$

where we used for mechanical impedance the frequently adopted symbol Z. Similar relations hold between receptance and dynamic stiffness and between accelerance and apparent mass. So, in general [1], the FRF formats of dynamic stiffness, mechanical impedance and apparent mass are discouraged because they may lead to errors and misinterpretations in the case of MDOF systems.

Two other observations can be made regarding the FRF which are of interest to us:

- The first observation has to do with the reciprocity theorem. Following the line of reasoning of the preceding section where we determined (eq (7.35)) that the receptance matrix is symmetrical, it is almost straightforward to show that the same applies to the mobility and accelerance matrices.
- The second observation is to point out that only n out of the n^2 elements of the receptance matrix $\mathbf{R}(\omega)$ are needed to determine the natural frequencies, the damping factors and the mode shapes.

We will return to this aspect in later chapters but, in order to have an idea, suppose for the moment that we are dealing with a 3-DOF system with distinct eigenvalues and widely spaced modes. In the vicinity of a natural frequency, the summation (7.28) will be dominated by the term corresponding to that frequency so that the magnitude $|\ R_{jk}(\omega_i)\ |$ can be approximated by (eqs (7.28) and (7.34b))

$$|\ R_{jk}(\omega_i)\ | \cong \frac{|\ \mathbf{p}_i\mathbf{p}_i^T\ |_{jk}}{|\ \omega_i^2 - \omega_i^2 - 2\zeta_i\omega_i\ |} = \frac{|\ \mathbf{p}_i\mathbf{p}_i^T\ |_{jk}}{2\zeta_i\omega_i} \qquad (7.45)$$

where $j, k = 1, 2, 3$. Let us suppose further that we obtained an entire column of the receptance matrix, say the first column, i.e. the functions R_{11}, R_{21} and R_{31}; a plot of the magnitude of these functions will, in general, show three peaks at the natural frequencies ω_1, ω_2 and ω_3 and any one function can be used to extract these frequencies plus the damping factors ζ_1, ζ_2 and ζ_3.

Now, consider the first frequency ω_1: from eq (7.45) we get the expressions

$$|\ \mathbf{p}_1\mathbf{p}_1^T\ |_{11} = 2\zeta_1\omega_1^2\ |\ R_{11}(\omega_1)\ |$$

$$|\ \mathbf{p}_1\mathbf{p}_1^T\ |_{21} = 2\zeta_1\omega_1^2\ |\ R_{21}(\omega_1)\ | \qquad (7.46)$$

$$|\ \mathbf{p}_1\mathbf{p}_1^T\ |_{31} = 2\zeta_1\omega_1^2\ |\ R_{31}(\omega_1)\ |$$

where the terms on the right-hand side are known. If we write explicitly eqs (7.46), we obtain three equations in three unknowns which can be solved to obtain p_{11}, p_{21} and p_{31}, i.e. the components of the eigenvector $\mathbf{p}_1$. Then, the phase information on the three receptance functions can be used to assign a plus or minus sign to each component (phase at ω_1 is either $+90°$ or $-90°$) and determine completely the first eigenvector. The same procedure for ω_2

and w_3 leads, respectively, to $\mathbf{p}_2$ and $\mathbf{p}_3$ and, since the choice of the first column of the receptance matrix has been completely arbitrary, it is evident that any one column or row of an FRF matrix (receptance, mobility or acceleration) is sufficient to extract all the modal parameters. This is fundamental in the field of experimental modal analysis (Chapter 10) in which the engineer performs an appropriate series of measurements in order to arrive at a modal model of the structure under investigation.

Kramers–Kronig relations

Let us now consider a general FRF function. If we become a little more involved in the mathematical aspects of the discussion, we may note that FRFs, regardless of their origin and format, have some properties in common. Consider for example, an SDOF equation in the form (4.1) (this simplifying assumption implies no loss of generality and it is only for our present convenience). It is not difficult to see that a necessary and sufficient condition for a function $f(t)$ to be real is that its Fourier transform $F(\omega)$ have the symmetry property $F(-\omega) = F^*(\omega)$, which, in turn, implies that $\text{Re}[F(\omega)]$ is an even function of ω, while $\text{Im}[F(\omega)]$ is an odd function of ω. Since $H(\omega)$ is the Fourier transform of the real function $h(t)$, the same symmetry property applies to $H(\omega)$ and hence

$$H_{\text{Re}}(\omega) = H_{\text{Re}}(-\omega)$$

$$H_{\text{Im}}(\omega) = -H_{\text{Im}}(-\omega)$$

(7.47)

where, for brevity, we write H_{Re} and H_{Im} for the real and imaginary part of H, respectively. In addition, we can express $h(t)$ as

$$h(t) = \frac{1}{2\pi} \int_{-\infty}^{+\infty} H(\omega) e^{i\omega t} d\omega$$

(7.48)

divide the real and imaginary parts of $H(\omega)$ and, since $h(t)$ must be real, arrive at the expression

$$h(t) = \frac{1}{2\pi} \int_{-\infty}^{+\infty} H_{\text{Re}}(\omega) \cos \omega t d\omega - \frac{1}{2\pi} \int_{-\infty}^{+\infty} H_{\text{Im}}(\omega) \sin \omega t d\omega$$

$$= \frac{1}{\pi} \int_{0}^{+\infty} H_{\text{Re}}(\omega) \cos \omega t d\omega - \frac{1}{\pi} \int_{0}^{+\infty} H_{\text{Im}}(\omega) \sin \omega t d\omega$$

(7.49)

where the change of the limits of integration is permitted by the fact that, owing to eqs (7.47), the integrands in both terms on the r.h.s. are even functions of ω.

If we now introduce the **principle of causality** – which requires that the effect must be zero prior to the onset of the cause – and consider the cause to be an impulse at $t = 0$, it follows that $h(t)$ must be identically zero for negative values of time. The two terms of eq (7.49) are even and odd functions of time and so, if $h(t)$ is to vanish for all $t < 0$, we have

$$\int_0^\infty H_{\text{Re}}(\omega)\cos \omega t d\omega = - \int_0^\infty H_{\text{Im}}(\omega)\sin \omega t d\omega \qquad (7.50)$$

for all positive values of t. In other words, the two terms of eq (7.49) are each equal to $h(t)/2$ when t is positive but cancel out when t is negative.

Equation (7.50) constitutes another restriction on the mathematical properties of the real and imaginary parts of an FRF and means that if we know $H_{\text{Re}}(\omega)$, we can compute $H_{\text{Im}}(\omega)$ and *vice versa*.

The explicit relations between H_{Re} and H_{Im} can be found by writing the relation

$$H(\omega) = \int_{-\infty}^{+\infty} h(t)e^{-i\omega t}dt = \int_0^\infty h(t)e^{-i\omega t}dt$$

where the lower limit of integration can be set to zero because we assumed $h(t) = 0$ for $t < 0$. Next, by separating the real and imaginary parts of $H(\omega)$ we obtain

$$H_{\text{Re}}(\omega) = \int_0^\infty h(t)\cos \omega t dt$$

$$\qquad (7.51)$$

$$H_{\text{Im}}(\omega) = - \int_0^\infty h(t)\sin \omega t dt$$

In addition, from eq (7.49) we have

$$h(t) = \frac{2}{\pi} \int_0^\infty H_{\text{Re}}(\omega)\cos \omega t d\omega$$

which (introducing the dummy variable of integration $\tilde{\omega}$) can be substituted in the second of eqs (7.51) to give

$$H_{\text{Im}}(\omega) = -\frac{2}{\pi} \int_0^\infty \int_0^\infty H_{\text{Re}}(\tilde{\omega})\cos \tilde{\omega} t \sin \omega t dt d\tilde{\omega}$$

and hence, since it can be shown that

$$\int_0^\infty \sin at \, \cos btdt = \frac{a}{a^2 - b^2}$$

we can perform the time integration to obtain the result

$$H_{\mathrm{Im}}(\omega) = \frac{2\omega}{\pi} P \int_0^\infty \frac{H_{\mathrm{Re}}(\tilde{\omega})}{\tilde{\omega}^2 - \omega^2} \, d\tilde{\omega} \qquad (7.52)$$

where the symbol P indicates that it is necessary to take the Cauchy principal value of the integral because the integrand possesses a singularity.

By following a similar procedure and noting that from eq (7.50) we can also write $h(t) = -2\pi^{-1} \int_0^\infty H_{\mathrm{Im}}(\omega)\sin \omega t d\omega$, we can introduce this expression into the first of eqs (7.51) to obtain

$$H_{\mathrm{Re}}(\omega) = -\frac{2}{\pi} P \int_0^\infty \frac{\tilde{\omega}H_{\mathrm{Im}}(\tilde{\omega})}{\tilde{\omega}^2 - \omega^2} \, d\tilde{\omega} \qquad (7.53)$$

Equations (7.52) and (7.53) are known as Kramers–Kronig relations. Note that they are not independent but they are two alternative forms of the same restriction on $H(\omega)$ imposed by the principle of causality.

The conclusion is that for any given 'reasonable' choice of H_{Re} on the real axis there exists one and only one 'well-behaved' form of H_{Im}. The terms 'reasonable' and 'well-behaved' are deliberately vague because a detailed discussion involves considerations in the complex plane and would be out of place here: however, the reader can intuitively imagine that, for example, by 'reasonable' we mean continuous and differentiable and such as to allow the Kramers–Kronig integrals to converge.

We will not pursue this subject further because, in the field of our interest, the Kramers–Kronig relations are unfortunately of little practical utility. In fact, even with numerical integration, the integrals are very slowly convergent and experimental errors on, say, H_{Re} may produce anomalies in H_{Im} which can be easily misinterpreted and *vice versa*. Nevertheless, the significance of the Kramers–Kronig relations is mainly due to the fact that they exist and that their very existence reflects the fundamental relation between cause and effect, a concept of paramount importance in our quest for an increasingly refined and complete description of the physical world.

7.5 Systems with rigid-body modes

Consider now an undamped system with m rigid-body modes. From the equations of motion

$$\mathbf{M\ddot{u}} + \mathbf{Ku} = \mathbf{f}_0 e^{-i\omega t}$$

and the usual assumption of a harmonic solution in the form $\mathbf{u} = \mathbf{z}e^{-i\omega t}$ we get

$$(\mathbf{K} - \omega^2 \mathbf{M})\mathbf{z} = \mathbf{f}_0 \qquad (7.54)$$

whose formal solution is given by

$$\mathbf{z} = (\mathbf{K} - \omega^2 \mathbf{M})^{-1}\mathbf{f}_0 = \mathbf{R}(\omega)\mathbf{f}_0 \qquad (7.55)$$

where $\mathbf{R}(\omega) \equiv (\mathbf{K} - \omega^2 \mathbf{M})^{-1}$ is the receptance matrix of our undamped system. As in Section 7.3, our scope is to arrive at an explicit expression for this FRF matrix.

Referring back to Section 6.6, we can expand the vector $\mathbf{z}$ on the basis of the system's eigenvectors, which now include the m rigid-body modes: the expansion (whose coefficients must be determined) reads

$$\mathbf{z} = \sum_{i=1}^{m} \alpha_i \mathbf{r}_i + \sum_{l=1}^{n-m} \beta_l \mathbf{p}_l \qquad (7.56)$$

where we assume all modes to be mass orthonormal. Equation (7.56) can be substituted in eq (7.54) to obtain a somewhat lengthy expression which, in turn, can be premultiplied by $\mathbf{r}_j^T$ to give

$$\alpha_j = -\frac{\mathbf{r}_j^T \mathbf{f}_0}{\omega^2} \qquad (7.57a)$$

and premultiplied by $\mathbf{p}_j^T$ to give

$$\beta_j = \frac{\mathbf{p}_j^T \mathbf{f}_0}{\omega_j^2 - \omega^2} \qquad (7.57b)$$

so that eq (7.56) becomes

$$\mathbf{z} = \sum_{i=1}^{m} \left(-\frac{\mathbf{r}_i^T \mathbf{f}_0}{\omega^2} \right) \mathbf{r}_i + \sum_{l=1}^{n-m} \left(\frac{\mathbf{p}_l^T \mathbf{f}_0}{\omega_l^2 - \omega^2} \right) \mathbf{p}_l$$

$$= \left(-\frac{1}{\omega^2} \sum_{i=1}^{m} \mathbf{r}_i \mathbf{r}_i^T + \sum_{l=1}^{n-m} \frac{\mathbf{p}_l \mathbf{p}_l^T}{\omega_l^2 - \omega^2} \right) \mathbf{f}_0 \qquad (7.58)$$

which can be compared to eq (7.55) to conclude that the receptance matrix is written as

$$\mathbf{R}(\omega) = -\frac{1}{\omega^2} \sum_{i=1}^{m} \mathbf{r}_i \mathbf{r}_i^T + \sum_{l=1}^{n-m} \frac{\mathbf{p}_l \mathbf{p}_l^T}{\omega_l^2 - \omega^2} \qquad (7.59a)$$

and its (jk)th element is

$$R_{jk} = -\frac{1}{\omega^2} \sum_{i=1}^{m} r_{ji} r_{ki} + \sum_{l=1}^{n-m} \left(\frac{1}{\omega_l^2 - \omega^2}\right) p_{jl} p_{kl} \qquad (7.59b)$$

Note that the expansion (7.56) on the basis of modes which are not mass orthonormal results in a term M_{ii} in the denominator of the first sum on the right-hand side of eqs (7.59a) and (7.59b) and in a term M_{ll} in the denominator of the second sum.

Equations (7.59a) and (7.59b) are, respectively, the counterpart of eqs (7.34b) and (7.28) for an undamped system with rigid-body modes: the rigid-body modes contribution is evident and it is also evident that the function

$$\frac{1}{\omega_l^2 - \omega^2}$$

is the lth modal FRF $H_l(\omega)$ of an undamped system. In this light, the discussion of this section can be extended with only little effort to a proportionally damped system with m rigid-body modes. The reader is invited to do so.

As far as unrestrained systems are concerned, it is interesting to note that the mode displacement and the mode acceleration methods can also be used to determine their response. The mode displacement method does not present additional difficulties due to the presence of rigid-body modes, but the extension of the mode acceleration method is not straightforward. In essence, the reason lies in the fact that the stiffness matrix of an unrestrained system is singular and the method (Section 7.2.1) requires the calculation of $\mathbf{K}^{-1}$.

However, this difficulty can be circumvented; we do not pursue this subject here and for a detailed discussion the interested reader is referred, for example, to Craig [2].

7.6 The case of nonproportional viscous damping

The preceding sections have all dealt either with undamped systems or with systems whose damping matrix becomes diagonal under the transformation $\mathbf{P}^T \mathbf{CP}$. In these cases, the modal approach for the calculation of their response properties relies on the possibility to directly uncouple the equations of motion, solve each equation independently and superpose the individual responses.

As stated in Section 6.7.1, the assumption of proportional damping is not always justified and a general damping matrix leads, in the homogeneous case, to the complex eigenvalue problem (6.92). This, in turn, can either be solved directly as it is or can be tackled by adopting a state-space formulation, as shown in Section 6.8 (eqs (6.75a and b) or eqs (6.179)).

The nature of the problem itself leads to a complex eigensolution, but the eigenvectors that we obtain in the first case satisfy the 'undesirable' orthogonality conditions of eqs (6.158) and (6.159) which, in general, are of limited practical utility. By contrast, the state-space formulation results either in a generalized or in a standard eigenvalue problem – both of which forms are preferred for numerical solution – and in a set of much simpler orthogonality conditions. This approach is also more effective in the nonhomogeneous case.

Let us first consider the equations

$$\mathbf{M\ddot{u} + C\dot{u} + Ku = f}$$

$$\mathbf{M\dot{u} - M\dot{u} = 0}$$

(7.60a)

and write them in matrix form as

$$\begin{bmatrix} \mathbf{C} & \mathbf{M} \\ \mathbf{M} & \mathbf{0} \end{bmatrix} \begin{bmatrix} \dot{\mathbf{u}} \\ \ddot{\mathbf{u}} \end{bmatrix} + \begin{bmatrix} \mathbf{K} & \mathbf{0} \\ \mathbf{0} & -\mathbf{M} \end{bmatrix} \begin{bmatrix} \mathbf{u} \\ \dot{\mathbf{u}} \end{bmatrix} = \begin{bmatrix} \mathbf{f} \\ \mathbf{0} \end{bmatrix}$$

or

$$\mathbf{\hat{K}x + \hat{M}\dot{x} = q}$$

(7.60b)

where we define the matrix $\mathbf{q} = [\mathbf{f} \quad \mathbf{0}]^T$ and the matrices $\mathbf{\hat{M}}$, $\mathbf{\hat{K}}$ and $\mathbf{x}$ as in eq (6.175c). We are already familiar with the solution of the homogeneous counterpart of eq (7.60b); hence we can express the solution of (7.60b) as the superposition of eigenmodes

$$\mathbf{x} = \sum_{i=1}^{2n} y_i(t)\mathbf{s}_i$$

(7.61)

which can be substituted in eq (7.60b) and, taking eqs (6.178) into account, premultiplied by $\mathbf{s}_j^T$ to get the $2n$ independent first-order equations

$$\hat{K}_{jj} y_j + \hat{M}_{jj} \dot{y}_j = \mathbf{s}_j^T \mathbf{q}$$

or, equivalently

$$-\lambda_j y_j + \dot{y}_j = \phi_j(t)$$

(7.62)

where we defined

$$\phi_j(t) = \frac{\mathbf{s}_j^T \mathbf{q}}{\hat{M}_{jj}}$$

(7.63)

and took into account the relation $\hat{K}_{jj}/\hat{M}_{jj} = -\lambda_j$. Equations (7.62) can be easily solved by multiplying both sides by $e^{-\lambda_j t}$ and writing the result as

$$\frac{d}{dt}\left(y_j e^{-\lambda_j t}\right) = \phi_j(t) e^{-\lambda_j t}$$

Hence

$$y_j(t) = e^{\lambda_j t}\left(\int_0^t \phi_j(\tau) e^{-\lambda_j \tau} d\tau + y_j(0)\right) \tag{7.64}$$

It may now be useful to show how, with a more compact notation, we can arrive at the result (7.64) in matrix form. Let us write eq (7.61) as

$$\mathbf{x} = \mathbf{S}\mathbf{y} \tag{7.65}$$

where $\mathbf{S}$ is the $2n \times 2n$ matrix of eigenvectors and $\mathbf{y} = [y_1(t) \ldots y_{2n}(t)]^T$; now, substitute (7.65) in eq (7.60b) and premultiply by $\mathbf{S}^T$ to obtain

$$\mathbf{S}^T\hat{\mathbf{K}}\mathbf{S}\mathbf{y} + \mathbf{S}^T\hat{\mathbf{M}}\mathbf{S}\dot{\mathbf{y}} = \mathbf{S}^T\mathbf{q} \tag{7.66}$$

Without loss of generality, we can assume $\mathbf{S}^T\hat{\mathbf{M}}\mathbf{S} = \mathbf{I}$ and arrive at the matrix equation

$$\frac{d}{dt}\mathbf{y} = \text{diag}(\lambda_j) + \mathbf{S}^T\mathbf{q} \tag{7.67}$$

which is the matrix form of the $2n$ equations (7.62). If now we define the $2n \times 1$ matrix of the constants of integration $\mathbf{y}_0 \equiv [y_1(0) \ldots y_{2n}(0)]^T$, the $2n$ solutions of eq (7.64) can be combined into the single equation

$$\mathbf{y}(t) = \text{diag}(e^{\lambda_j t})\left(\int_0^t \text{diag}(e^{-\lambda_j t})\mathbf{S}^T\mathbf{q}(\tau)d\tau + \mathbf{y}_0\right) \tag{7.68}$$

and the solution in the original coordinates can be obtained by means of the transformation (7.65), so that

$$\mathbf{x} = \mathbf{S}\,\text{diag}(e^{\lambda_j t})\left(\int_0^t \text{diag}(e^{-\lambda_j \tau})\mathbf{S}^T\mathbf{q}(\tau)d\tau + \mathbf{y}_0\right) \tag{7.69}$$

Finally, if we remember that $\mathbf{x} = [\mathbf{u} \ \dot{\mathbf{u}}]^T$, it follows that the last n elements of $\mathbf{x}$ are the derivatives of the first n elements; this implies, as we know from the preceding chapter, that each eigenvector is in the form

$$\mathbf{s}_j = \begin{bmatrix} \mathbf{z}_j \\ \lambda_j \mathbf{z}_j \end{bmatrix} \tag{7.70}$$

By virtue of eq (7.70), the $2n \times 2n$ matrices $\mathbf{S}$ and $\mathbf{S}^T$ can be partitioned into

$$\mathbf{S} = \begin{bmatrix} \mathbf{Z} \\ \mathbf{Z} \ \mathrm{diag}(\lambda_j) \end{bmatrix} \tag{7.71a}$$

and

$$\mathbf{S}^T = \begin{bmatrix} \mathbf{Z}^T & \mathrm{diag}(\lambda_j)\mathbf{Z}^T \end{bmatrix} \tag{7.71b}$$

where the orders of $\mathbf{Z}$, $\mathbf{Z}^T$ and $\mathrm{diag}(\lambda_j)$ are $n \times 2n$, $2n \times n$ and $2n \times 2n$, respectively. With this in mind, noting that

$$\mathbf{S}^T\mathbf{q} = \begin{bmatrix} \mathbf{Z}^T & \mathrm{diag}(\lambda_j)\mathbf{Z}^T \end{bmatrix} \begin{bmatrix} \mathbf{f} \\ \mathbf{0} \end{bmatrix} = \mathbf{Z}^T\mathbf{f}$$

we can recover the displacement solution from eq (7.69) as

$$\mathbf{u} = \mathbf{Z} \ \mathrm{diag}(e^{\lambda_j t}) \left(\int_0^t \mathrm{diag}(e^{-\lambda_j \tau})\mathbf{Z}^T\mathbf{f}(\tau)d\tau + \mathbf{y}_0 \right) \tag{7.72}$$

which represents the response of our system to an arbitrary excitation.

7.6.1 Harmonic excitation and receptance FRF matrix

The solution for a harmonic excitation $\mathbf{f}_0 e^{i\omega t}$ can be worked out as a particular case of eq (7.64). The jth participation factor is now

$$\phi_j(t) = \frac{\mathbf{s}_j^T \mathbf{q}_0}{\hat{M}_{jj}} e^{i\omega t} \tag{7.73}$$

where $\mathbf{q}_0 = [\mathbf{f}_0 \ \ 0]^T$. Without loss of generality we can assume zero initial conditions and the normalization condition $\hat{M}_{jj} = 1$; then, eq (7.64) becomes

$$y_j(t) = e^{\lambda_j t} \int_0^t \mathbf{s}_j^T \mathbf{q}_0 e^{(i\omega - \lambda_j)\tau} d\tau = \frac{\mathbf{s}_j^T \mathbf{q}_0}{i\omega - \lambda_j} e^{i\omega t} - \frac{\mathbf{s}_j^T \mathbf{q}_0}{i\omega - \lambda_j} e^{\lambda_j t} \tag{7.74}$$

Since we are mainly interested in the steady-state solution, we can drop the second term on the right-hand side which (if the system is stable and all eigenvalues λ_j ($j = 1, 2, ..., 2n$) have negative real parts) dies away as $t \to \infty$ and arrive at the solution

$$\mathbf{x} = \mathbf{S} \operatorname{diag}\left(\frac{1}{i\omega - \lambda_j}\right) \mathbf{S}^T \mathbf{q}_0 e^{i\omega t} \tag{7.75a}$$

or, alternatively

$$\mathbf{x} = \sum_{m=1}^{2n} \left(\frac{\mathbf{s}_m \mathbf{s}_m^T}{i\omega - \lambda_m}\right) \mathbf{q}_0 e^{i\omega t} \tag{7.75b}$$

Next, once again by virtue of eq (7.70), we can partition the matrices $\mathbf{S}$ and $\mathbf{S}^T$ as in eqs (7.71a and b) and obtain

$$\mathbf{x} = \begin{bmatrix} \mathbf{Z} \\ \mathbf{Z} \operatorname{diag}(\lambda_j) \end{bmatrix} \operatorname{diag}\left(\frac{1}{i\omega - \lambda_j}\right) \mathbf{Z}^T \mathbf{f}_0 e^{i\omega t}$$

from which it follows that

$$\mathbf{u} = \mathbf{Z} \operatorname{diag}\left(\frac{1}{i\omega - \lambda_j}\right) \mathbf{Z}^T \mathbf{f}_0 e^{i\omega t} \tag{7.76a}$$

or, equivalently

$$\mathbf{u} = \sum_{m=1}^{2n} \left(\frac{\mathbf{z}_m \mathbf{z}_m^T}{i\omega - \lambda_m}\right) \mathbf{f}_0 e^{i\omega t} \tag{7.76b}$$

From the definition of receptance matrix and from eqs (7.76a) and (7.76b) we get the $n \times n$ matrix

$$\mathbf{R}(\omega) = \mathbf{Z} \operatorname{diag}\left(\frac{1}{i\omega - \lambda_j}\right) \mathbf{Z}^T = \sum_{m=1}^{2n} \left(\frac{1}{i\omega - \lambda_m}\right) \mathbf{z}_m \mathbf{z}_m^T \tag{7.77}$$

whose (jk)th element is obtained as

$$R_{jk}(\omega) = \sum_{m=1}^{2n} \left(\frac{1}{i\omega - \lambda_m}\right) z_{jm} z_{km} \tag{7.78a}$$

Furthermore, for the case in which we are mainly interested – i.e. underdamped systems – we know that both eigenvalues and eigenvectors appear in complex conjugate pairs; this implies that eq (7.78a) can be written as the sum of n terms

$$R_{jk}(\omega) = \sum_{m=1}^{n} \left(\frac{z_{jm}z_{km}}{i\omega - \lambda_m} + \frac{z^*_{jm}z^*_{km}}{i\omega - \lambda^*_m} \right)$$

$$= \sum_{m=1}^{n} \left(\frac{z_{jm}z_{km}}{\zeta_m\omega_m + i(\omega - \omega_m\sqrt{1-\zeta_m^2})} + \frac{z^*_{jm}z^*_{km}}{\zeta_m\omega_m + i(\omega + \omega_m\sqrt{1-\zeta_m^2})} \right)$$

(7.78b)

where the last expression was written by taking eq (6.160) into account.

So, as in the other cases, we have obtained an explicit expression for the receptance FRF matrix. This is precisely the response model for the system under study and, once again, we can see that the general element of the receptance matrix is the sum of the contributions of the different modes of vibration.

At this point, it is worth pointing out that in modal analysis terminology the eigenvalues λ_m are often called the poles and the term $z_{jm}z_{km}$ – referred to as the residue for mode m – is given a symbol in its own right: for example, the reader may find in current literature the symbols $_mA_{jk}$ or $r_{jk,m}$, both of which stand for $z_{jm}z_{km}$.

At this point it may be instructive to follow a similar line of reasoning as above to work out a response model (and an explicit expression for the receptance FRF matrix) by starting from the state-space formulation of eqs (6.179). As a useful – and not trivial – exercise, the reader is urged to do so by taking advantage of the guidelines that follow.

1. the homogeneous case leads to a standard $2n$ eigenvalue problem $\mathbf{As} = \lambda\mathbf{s}$ where, in general, the matrix $\mathbf{A}$ is not symmetrical.
2. The eigenvalues and eigenvectors occur in complex conjugate pairs (underdamped case) and the eigenvectors have the form $s_j = [\mathbf{z}_j \; \lambda_j\mathbf{z}_j]^T$.
3. If the matrix $\mathbf{A}$ is nondefective (which we assume to be the case), we can form the $2n \times 2n$ matrix $\mathbf{S}$ of column eigenvectors so that

$$\mathbf{S}^{-1}\mathbf{AS} = \text{diag}(\lambda_j)$$

(7.79)

4. The forced vibration equations can be cast in the form

$$\frac{d}{dt}\mathbf{x} = \mathbf{Ax} + \mathbf{q}$$

(7.80a)

where now we define

$$q = \begin{bmatrix} 0 \\ M^{-1}f \end{bmatrix} \tag{7.80b}$$

5. The transformation to normal coordinates $x = Sy$ can be substituted in eq (7.80a) in order to arrive at

$$\dot{y} = \mathrm{diag}(\lambda_j)y + S^{-1}q \tag{7.81}$$

which is formally similar to eq (7.67) and leads, in the end, to

$$x = S\,\mathrm{diag}(e^{\lambda_j t})\left(\int_0^t \mathrm{diag}(e^{-\lambda_j \tau})S^{-1}q\,d\tau + y_0 \right) \tag{7.82}$$

6. The matrix of right eigenvectors S can be partitioned into two $n \times 2n$ matrices as

$$S = \begin{bmatrix} S_{upper} \\ S_{lower} \end{bmatrix} = \begin{bmatrix} S_{upper} \\ S_{upper}\,\mathrm{diag}(\lambda_j) \end{bmatrix} \tag{7.83a}$$

and, by the same token, it is also convenient to partition S^{-1} into two $2n \times n$ matrices as

$$S^{-1} = \begin{bmatrix} S_{left}^{-1} & S_{right}^{-1} \end{bmatrix} \tag{7.83b}$$

Furthermore

$$S^{-1}q = \begin{bmatrix} S_{left}^{-1} & S_{right}^{-1} \end{bmatrix}\begin{bmatrix} 0 \\ M^{-1}f \end{bmatrix} = S_{right}^{-1}M^{-1}f \tag{7.84}$$

and hence

$$u = S_{upper}\,\mathrm{diag}(e^{\lambda_j t})\left(\int_0^t \mathrm{diag}(e^{-\lambda_j \tau})S_{right}^{-1}M^{-1}f(\tau)d\tau + y_0 \right) \tag{7.85}$$

which is the displacement response of our system to an arbitrary excitation.
7. Again, the case of harmonic excitation can be obtained as a particular case of eq (7.85). The jth participation factor is now

$$\phi_j(t) = v_j^T q_0 e^{i\omega t} \tag{7.86}$$

where we called $\mathbf{v}_j^T$ the left jth eigenvector of $\mathbf{A}$ (Appendix A). Note that $\mathbf{v}_j^T$ is a row $1 \times 2n$ vector (this is why we write the superscript T for transpose: because in our notation, as it is customary, vectors are arranged as columns) and its components form the jth row of matrix $\mathbf{S}^{-1}$ just like the components of $\mathbf{s}_j$ (the jth right eigenvector of $\mathbf{A}$) form the jth column of matrix $\mathbf{S}$.

8. The displacement response is given by

$$\mathbf{u} = \mathbf{S}_{upper} \ \mathrm{diag}\left(\frac{1}{i\omega - \lambda_j}\right) \mathbf{S}_{right}^{-1}\mathbf{M}^{-1}\mathbf{f}_0 e^{i\omega t} \tag{7.87}$$

and hence the $n \times n$ receptance FRF matrix is

$$\mathbf{R}(\omega) = \mathbf{S}_{upper} \ \mathrm{diag}\left(\frac{1}{i\omega - \lambda_j}\right) \mathbf{S}_{right}^{-1}\mathbf{M}^{-1} \tag{7.88}$$

9. The expression of a single receptance FRF function $R_{jk}(\omega)$ in terms of individual components is a bit involved; however, if we call b_{rs} the general (r, s)th element of the $2n \times n$ matrix $\mathbf{S}_{right}^{-1}\mathbf{M}^{-1}$, it is not difficult to determine that

$$R_{jk}(\omega) = \sum_{m=1}^{2n}\left(\frac{1}{i\omega - \lambda_j}\right)s_{jm}b_{mk} \tag{7.89}$$

7.7   MDOF systems with hysteretic damping

We stated in Section 6.7 that this type of damping does not lend itself easily to a rigorous free-vibration analysis because, strictly speaking, the concept of hysteretic (or structural) damping is based on an analogy with the viscous damping case when the system is excited by means of a harmonic forcing function. Nevertheless, experimental tests are often performed in a forced vibration condition and it is undoubtedly useful to obtain a response model for these systems in terms of eigenvalues and mode shapes, however questionable this free-vibration solution may be. Therefore, provided that the results are used judiciously, we justify the considerations that follow on the basis of physical sense.

In general, hysteretic damping is taken into account by expressing the equations of motion in the form (6.150), where the damping matrix is written as $i\gamma\mathbf{K}$. In the homogeneous case, assuming a solution in the form

$$\mathbf{u} = \mathbf{z}e^{i\lambda t} \tag{7.90}$$

leads to

$$[(1 + i\gamma)K - \lambda^2 M]z = 0$$

which admits a nontrivial solution if $\det[(1 + i\gamma)K - \lambda^2 M] = 0$. It is not difficult to see that we obtain now a set of n complex eigenvalues λ_j^2 (because the coefficients of the characteristic polynomial are complex) and that the set of n real eigenvectors z_j are the same as for the undamped case.

The eigenvalues contain information on both frequency and damping characteristics and they can be written as

$$\lambda_j^2 = \omega_j^2(1 + i\gamma) \tag{7.91}$$

where

$$\omega_j^2 = \frac{z_j^T K z_j}{z_j^T M z_j} \equiv \frac{K_{jj}}{M_{jj}} \tag{7.92}$$

Note that, as in the previous cases, the ω_js have well-defined values but the values of the K_{jj}s and M_{jj}s depend on the normalization that we choose. Once again, it is common practice to fix the indeterminacy on the eigenvectors by choosing, out of the many possibilities, the vectors p_j ($j = 1, 2, ..., n$) which satisfy the relations $p_i^T M p_j = \delta_{ij}$.

Incidentally, it may be worth noting that a more general case of proportional hysteretic damping can be considered by writing the equations of motion as

$$M\ddot{u} + iHu + Ku = f(t) \tag{7.93a}$$

and assuming that the hysteretic damping matrix H (not to be confused with a FRF matrix) is given by

$$H = aM + bK \tag{7.93b}$$

where a and b are two constants. We will not deal specifically with this case because, as the reader can verify, the nature of the eigensolution is the same as before and nothing is added to the essence of the problem.

With the above considerations in mind, it is now only a small effort to arrive at a response model in the case of the harmonic excitation $f(t) = f_0 e^{i\omega t}$.

We start from the equations of motion (6.150) and perform the change of coordinates

$$u = \sum_{i=1}^{n} p_i y_i(t) = Py$$

where P is the matrix of mass orthonormal eigenvectors. Next, we premultiply the resulting equation by $\mathbf{P}^T$ and arrive at

$$\mathbf{I}\ddot{\mathbf{y}} + \text{diag}(\lambda_j^2)\mathbf{y} = \mathbf{P}^T\mathbf{f} \tag{7.94a}$$

which represents a set of n uncoupled equations. The jth equation reads explicitly

$$\ddot{y}_j + \lambda_j^2 y_j = \mathbf{p}_j^T\mathbf{f} \tag{7.94b}$$

Assuming a harmonically varying response $y_j(t) = \hat{y}_j e^{i\omega t}$, it is not difficult to retrieve the solution in physical coordinates as

$$\mathbf{u} = \mathbf{P}\,\text{diag}\left(\frac{1}{\lambda_j^2 - \omega^2}\right)\mathbf{P}^T\mathbf{f}_0 e^{i\omega t} \tag{7.95}$$

and recognize the receptance matrix

$$\mathbf{R}(\omega) = \mathbf{P}\,\text{diag}\left(\frac{1}{\lambda_j^2 - \omega^2}\right)\mathbf{P}^T = \sum_{m=1}^{n}\left(\frac{1}{\lambda_m^2 - \omega^2}\right)\mathbf{p}_m\mathbf{p}_m^T \tag{7.96a}$$

whose (jk)th element is given by

$$R_{jk}(\omega) = \sum_{m=1}^{n}\left(\frac{p_{jm}p_{km}}{\lambda_m^2 - \omega^2}\right) = \sum_{m=1}^{n}\left(\frac{p_{jm}p_{km}}{\omega_m^2 - \omega^2 + i\gamma\omega_m^2}\right) \tag{7.96b}$$

and the last expression has been written by taking eq (7.91) into account.

The reader is invited to consider the more general case of proportional hysteretic damping expressed by eqs (7.93a) and (7.93b) and show that we arrive at

$$R_{jk}(\omega) = \sum_{m=1}^{n}\left(\frac{p_{jm}p_{km}}{\omega_m^2 - \omega^2 + i\gamma_m\omega_m^2}\right) \tag{7.97}$$

where now we have

$$\gamma_m = b + \frac{a}{\omega_m^2} \tag{7.98}$$

Despite the discussion at the beginning of this section, there seems to be no difficulty in the derivation of the response model of eqs (7.96a, b) and (7.97). There is, however, a subtle conceptual problem. Our FRFs must be the

Fourier transform of real impulse response functions, and this implies (Section 7.4) that the conditions (7.47) apply. This is not the case for the FRFs of eqs (7.96b) and (7.97); furthermore, if these latter are modified to agree with eqs (7.47) it follows that our FRFs do not satisfy the requirement of causality. We will not go proceed further in this discussion which is beyond the scope of this book, but it seems that these conceptual problems – although they can be ignored in many practical situations – are the price that we must pay for the inadequacy of a free vibration solution in the hysteretic case.

The interested reader can refer, for example, to Nashif *et al.* [3] and Newland [4].

7.8   A few remarks on other solution strategies: Laplace transform and direct integration

This chapter has dealt in some detail with the response properties of various types of MDOF system. However, special attention has been intentionally given to the so-called modal approach (or modal superposition, modal expansion techniques), where the dynamic response is expressed as a series expansion of eigenmodes. The reason is twofold: first of all, this text is mainly concerned with linear vibrations of structural and mechanical systems and, second, the modal approach has considerable importance in many aspects of experimental vibration measurements. Nevertheless, the reader would be right in assuming that other approaches are available in order to solve the forced vibration problem of MDOF systems.

7.8.1   Laplace transform method

At least in principle, the Laplace transform method can be directly applied to eq (7.1) to obtain

$$\mathbf{M}(s^2\mathbf{U} - s\mathbf{u}_0 - \dot{\mathbf{u}}_0) + \mathbf{C}(s\mathbf{U} - \mathbf{u}_0) + \mathbf{K}\mathbf{U} = \mathbf{F}(s) \tag{7.99}$$

where $\mathbf{U} = \mathbf{U}(s)$ and $\mathbf{F}(s)$ are, respectively, the Laplace transforms of $\mathbf{u}(t)$ and $\mathbf{f}(t)$, s is the Laplace operator and $\mathbf{u}_0$, $\dot{\mathbf{u}}_0$ are the vectors of initial displacements and velocities. For zero initial conditions eq (7.99) can be rewritten as

$$[s^2\mathbf{M} + s\mathbf{C} + \mathbf{K}]\mathbf{U}(s) \equiv \mathbf{G}(s)\mathbf{U}(s) = \mathbf{F}(s) \tag{7.100}$$

and consequently

$$\mathbf{U}(s) = \mathbf{G}^{-1}(s)\mathbf{F}(s) = \frac{\mathrm{adj}(\mathbf{G})}{\det(\mathbf{G})}\,\mathbf{F}(s) \tag{7.101}$$

where the last expression on the right-hand side for $\mathbf{G}^{-1}$ can be found in any book on matrix algebra: adj(**G**) is called classical adjoint (or adjugate, to avoid confusion with the Hermitian adjoint) of **G** and is the transposed matrix of cofactors of **G**. From eq (7.101) we recognize $\mathbf{G}^{-1}$ as the matrix of receptance transfer functions $R_{ij}(s)$ and we note that, in the inverse transformation of eq (7.101), the poles from the transfer functions are the eigenvalues of our system because they are obtained from the characteristic equation

$$\det(\mathbf{G}) \equiv \det[s^2\mathbf{M} + s\mathbf{C} + \mathbf{K}] = 0$$

This method also applies only for linear systems and may be useful for systems with nonproportional damping, where eqs (7.1) cannot be uncoupled by means of the classical modal matrix **P**. Note also that the terms 'poles' and 'residues' come directly from the Laplace transform approach.

In addition to what has been said above and in Section 5.3.3, we can briefly review the case of a SDOF system – thus keeping the mathematics extremely simple – and get an idea of how this technique works as far as transfer and frequency response functions are concerned.

If we take the Laplace transform of both sides of the equation $m\ddot{u} + c\dot{u} + ku = f(t)$ and assume zero initial conditions (which amounts to neglecting the solution of the homogeneous equation), we arrive at the SDOF counterpart of eq (7.100) which can be written as

$$H(s) = \frac{U(s)}{F(s)} \tag{7.102}$$

where the meaning of the symbols is obvious and

$$H(s) = \frac{1}{ms^2 + cs + k} \tag{7.103}$$

is the (complex-valued) receptance transfer function. The denominator of eq (7.103) is the characteristic equation, whose roots (the poles) can be written for an underdamped system as

$$s_{1,2} = \sigma \pm iw_d = -\zeta w_n \pm iw_n\sqrt{1-\zeta^2} \tag{7.104}$$

Now we note that $H(s)$ can be rewritten as

$$H(s) = \frac{1}{m(s - s_1)(s - s_2)} \tag{7.105a}$$

and expanded in partial fractions as

$$H(s) = \frac{A_1}{s - s_1} + \frac{A_2}{s - s_2} \tag{7.105b}$$

where the coefficients (residues) A_j can be obtained from

$$A_j = \lim_{s \to s_j} [(s - s_j)H(s)] \tag{7.106}$$

It is straightforward, in this case, to determine that $A_2 = A_1^*$ and obtain

$$A_1 = \frac{1}{2im\omega_d} \tag{7.107}$$

If now we consider that the frequency response function is simply the transfer function evaluated along the $i\omega$ axis, we obtain from eq (7.105b)

$$H(\omega) = \frac{A_1}{i\omega - s_1} + \frac{A_2}{i\omega - s_2} \tag{7.108a}$$

where we can substitute the explicit expressions for s_1, s_2, A_1 and A_2 to arrive at the familiar form

$$H(\omega) = \frac{1}{k - m\omega^2 + ic\omega} = \frac{1}{k(1 - \beta^2 + 2i\zeta\beta)} \tag{7.108b}$$

and, as usual, $\beta = \omega/\omega_n$.

Note that graphically, the transfer function is completely represented by two surfaces in the complex plane, i.e. its real and imaginary parts or its magnitude and phase. The corresponding FRF is obtained as the curve seen by an observer who looks down on a plane which cuts through the surface and whose normal is parallel to the σ-axis (Fig. 7.1).

7.8.2 Direct integration methods

A different approach to the solution of the forced-vibration problem for both SDOF and MDOF systems consists of a direct numerical integration of the equation(s) of motion in the time domain. The details of this approach belong rightfully to the subject of numerical techniques and are beyond the scope of this book; however, some general comments on the advantages and limitations of these methods are not out of place.

In particular, the reader is warned against the temptation to use direct integration as a 'black box', where you input the right equations and obtain the correct response time history.

The major advantage of direct integration is that it applies both to linear and nonlinear problems and, as a matter of fact, it is the only generally applicable method for the analysis of nonlinear systems. Nonetheless, as far as linear vibrations are concerned, direct integration methods may also be an effective alternative to the modal approach. For example, in the case of a

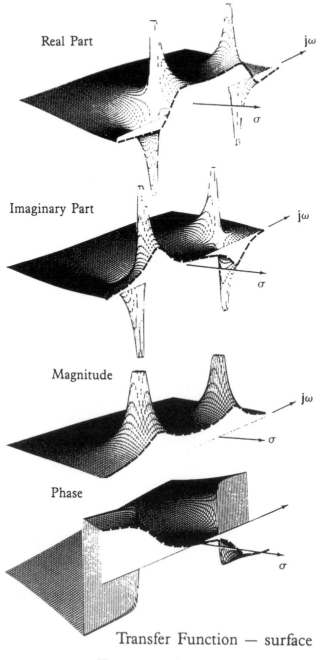

Transfer Function — surface

Frequency Response — dashed

Fig. 7.1 Representations of the transfer and frequency response functions. (Reprinted by permission of Hewlett-Packard.)

complex structure which undergoes the action of short-duration impulsive loads:

1. It is possible that many modes are excited and contribute significantly to the response.
2. Model idealizations very often result in inaccurate higher modes.

These factors make the modal approach less effective with respect to direct integration in which (1) the analyst does not need to transform the equations into a different form and (2) under the circumstances, only a short response time history is needed.

One drawback is that it is generally difficult, for an MDOF system, to define explicitly the damping matrix, but this potential difficulty is counterbalanced by an increase of flexibility in the choice of the damping characteristics, since there is no need to uncouple the equations of motion.

In essence, direct integration is based upon finite time differences; instead of trying to satisfy the equations of motion for any time t, these numerical methods consider the equilibrium between elastic, damping, inertia forces and applied loads only at discrete time intervals $t_n = n(\Delta t)$, $n = 1, 2, ..., N$ where Δt is the (usually constant) time step of integration and the response is calculated for the time interval $t_0, t_0 + N(\Delta t)$. This results in a 'sampled' time history response, as opposed to the continuous time solution which would be obtained from the exact integration of the equations of motion.

All the integration schemes assume appropriate variations of displacements, velocities and accelerations within each time step Δt and consist of expressions that relate these response parameters at a given time to their values at one or more previous time points. In other words, the procedure marches along the time dimension by assuming a general expression of the type

$$\mathbf{u}_{n+1} = \sum_{j=n-m}^{n} a_j \mathbf{u}_j + \sum_{j=n-m}^{n+1} b_j \dot{\mathbf{u}}_j + \sum_{j=n-m}^{n+1} c_j \ddot{\mathbf{u}}_j \qquad (7.109)$$

(where by $\mathbf{u}_{n+1}$ we mean the displacement at time t_{n+1} etc.) and substitute the relevant expressions in the equations of motion written either for time t_n or for time t_{n+1}. In this regard, an important subdivision exists between **explicit** and **implicit methods**, the former being when the equilibrium equations are expressed at time t_n, i.e.

$$\mathbf{M}\ddot{\mathbf{u}}_n + \mathbf{C}\dot{\mathbf{u}}_n + \mathbf{K}\mathbf{u}_n = \mathbf{f}_n$$

and the latter when the equilibrium equations are considered at time t_{n+1}. For a given time step Δt, implicit methods involve more computational effort than explicit methods but these latter cannot always be used effectively because of stability problems of the solution, a concept that will be made clearer in the following discussion.

Strictly speaking, eq (7.109) defines a direct 'multistep' integration method because the response at time t_{n+1} is calculated from the values of the response parameters at times $t_{n-m}, t_{n-m+1}, ..., t_{n+1}$; however, it must be noted that in most methods m is generally small. When $m = 1$ one speaks of single step methods.

Some of the most effective integration schemes are, to name a few:

- the central difference method (explicit),
- the linear acceleration method (implicit),
- the Houbolt method (implicit),
- the Wilson θ method (implicit)
- the family of Newmark methods (explicit or implicit depending on the choice of two parameters usually indicated with the symbols γ and β).

In general, whatever integration method we decide to adopt, there are two issues of fundamental importance: stability and accuracy of the solution. Stability has to do with the boundedness of the solution – which we do not want to grow indefinitely and become meaningless by being artificially amplified by the integration scheme – and accuracy has to do with the fact that, ideally, we want a solution with no (or small) amplitude and periodicity errors. Engineering common sense suggests that it may not be wise to integrate the response contribution of higher modes with a time step Δt that is larger than half their natural period T or, in other words, when the ratio $\Delta t/T$ is large. On the other hand, for large systems with many degrees of freedom, the highest period is so small that selection of an appropriate time step would make the whole procedure costly and impractical. So we must try to understand what kind of response is obtained when the ratio $\Delta t/T$ is large. In this regard we can distinguish between:

- **unconditionally stable** integration methods, where the solution remains bounded for any time step Δt and, in particular, when $\Delta t/T$ is large;
- **conditionally stable** methods, where the solution remains bounded only if Δt is smaller or equal to a certain critical value Δt_{cr}.

In particular, explicit methods are only conditionally stable, while most of the implicit methods are unconditionally stable. Therefore explicit methods can be used only provided that the restriction on the time step is observed and results in a reasonable value of Δt, otherwise one must resort to an implicit unconditionally stable method. In this case the solution does remain bounded but the selection of an appropriate time step (which can be generally much larger than in a conditionally stable case) reflects on the accuracy of the calculated response which, in turn, can be characterized in terms of amplitude accuracy and period accuracy. The first attribute refers to amplitude errors – specifically, artificial damping introduced by the numerical procedure – and the second refers to period elongations. If we want to represent appropriately the oscillating behaviour of the response, it is clear that both types of errors need to be avoided as much as possible.

We limit ourselves to these general considerations and we urge the interested reader to refer to specialized literature.

7.9 Frequency response functions of a 2-DOF system

The case of a simple 2-DOF system with proportional viscous damping will now be of help to illustrate from a more practical point of view some aspects of the preceding discussions. Let us consider the 2-DOF system of Fig. 7.2.

We assume the following characteristics:

$$\mathbf{M} = \begin{bmatrix} m_1 & 0 \\ 0 & m_2 \end{bmatrix} = \begin{bmatrix} 1000 & 0 \\ 0 & 500 \end{bmatrix} \text{kg}$$

$$\mathbf{K} = \begin{bmatrix} k_1 + k_2 & -k_2 \\ -k_2 & k_2 \end{bmatrix} = \begin{bmatrix} 1 \times 10^6 & -5 \times 10^5 \\ -5 \times 10^5 & 5 \times 10^5 \end{bmatrix} \text{N/m}$$

$$\mathbf{C} = \begin{bmatrix} c_1 + c_2 & -c_2 \\ -c_2 & c_2 \end{bmatrix} = \begin{bmatrix} 1000 & -500 \\ -500 & 500 \end{bmatrix} \text{N s/m}$$

and we want to arrive at the explicit expressions of the receptance FRF matrix.

Since the damping matrix is proportional to the stiffness matrix we know that the undamped modes uncouple the equations of motion, so we solve the undamped free-vibration problem and we obtain the following eigenvalues

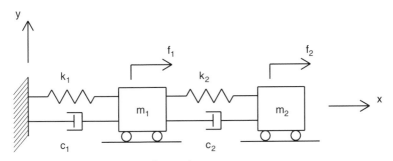

Fig. 7.2 Schematic 2-DOF translational system.

and mass-orthonormal eigenvectors:

$$\mathbf{L} = \begin{bmatrix} \lambda_1 & 0 \\ 0 & \lambda_2 \end{bmatrix} = \begin{bmatrix} 292.9 & 0 \\ 0 & 1707.1 \end{bmatrix} \qquad (7.110a)$$

$$\mathbf{P} = \begin{bmatrix} p_{11} & p_{12} \\ p_{21} & p_{22} \end{bmatrix} = \begin{bmatrix} 0.0224 & -0.0224 \\ 0.0316 & 0.0316 \end{bmatrix} \qquad (7.110b)$$

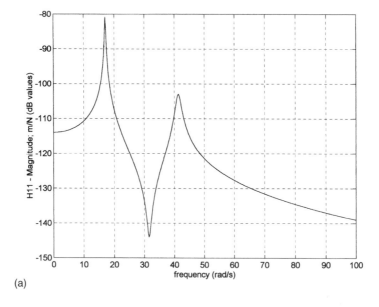

(a)

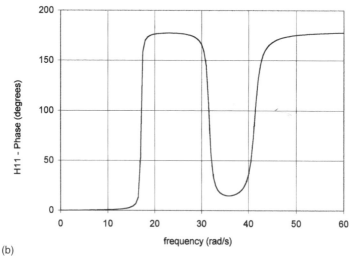

(b)

Fig. 7.3 Receptance $R_{11}(\omega)$: (a) magnitude, (b) phase, versus frequency.

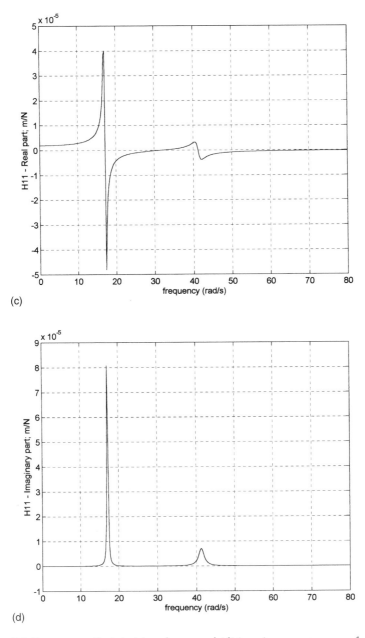

(c)

(d)

Fig. 7.3 Receptance $R_{11}(\omega)$: (c) real part and (d) imaginary part, versus frequency.

which have already been arranged in matrix form. The eigenmodes of our system occur at the frequencies $w_1 = 17.11$ and $w_2 = 41.32$ rad/s and referring to eq (6.142) it is not difficult to determine the modal damping ratios as $\zeta_1 = 0.0086$ and $\zeta_2 = 0.0207$. With this in mind we can now write the two modal receptance FRFs as

$$H_1(\omega) = \frac{1}{292.9 - w^2 - 0.293i\omega} \qquad (7.111\text{a})$$

$$H_2(\omega) = \frac{1}{1707.1 - w^2 - 1.707i\omega} \qquad (7.111\text{b})$$

and arrive at the receptance matrix in physical coordinates by virtue of eq (7.34b); we get

$$\mathbf{R}(\omega) = H_1(\omega)\mathbf{p}_1\mathbf{p}_1^T + H_2(\omega)\mathbf{p}_2\mathbf{p}_2^T$$

$$= H_1(\omega) \begin{bmatrix} 5.00 \times 10^{-4} & 7.07 \times 10^{-4} \\ 7.07 \times 10^{-4} & 1.00 \times 10^{-3} \end{bmatrix}$$

$$+ H_2(\omega) \begin{bmatrix} 5.00 \times 10^{-4} & -7.07 \times 10^{-4} \\ -7.07 \times 10^{-4} & 1.00 \times 10^{-3} \end{bmatrix} \qquad (7.112)$$

If now, for our convenience, we want to obtain each FRF by separating its real and imaginary parts, it only takes a little patience to arrive at

$$R_{11}(\omega) = \left(\frac{(w_1^2 - w^2)p_{11}p_{11}}{(w_1^2 - w^2)^2 + 4\zeta_1^2 w_1^2 w^2} + \frac{(w_2^2 - w^2)p_{12}p_{12}}{(w_2^2 - w^2)^2 + 4\zeta_2^2 w_2^2 w^2} \right)$$

$$+ i\left(\frac{(2\zeta_1 w_1 p_{11}p_{11})w}{(w_1^2 - w^2)^2 + 4\zeta_1^2 w_1^2 w^2} + \frac{(2\zeta_2 w_2 p_{12}p_{12})w}{(w_2^2 - w^2)^2 + 4\zeta_2^2 w_2^2 w^2} \right) \qquad (7.113\text{a})$$

$$R_{12}(\omega) = \left(\frac{(w_1^2 - w^2)p_{11}p_{21}}{(w_1^2 - w^2)^2 + 4\zeta_1^2 w_1^2 w^2} + \frac{(w_2^2 - w^2)p_{12}p_{22}}{(w_2^2 - w^2)^2 + 4\zeta_2^2 w_2^2 w^2} \right)$$

$$+ i\left(\frac{(2\zeta_1 w_1 p_{11}p_{21})w}{(w_1^2 - w^2)^2 + 4\zeta_1^2 w_1^2 w^2} + \frac{(2\zeta_2 w_2 p_{12}p_{22})w}{(w_2^2 - w^2)^2 + 4\zeta_2^2 w_2^2 w^2} \right) = R_{21}(\omega)$$

$$(7.113\text{b})$$

$$R_{22}(\omega) = \left(\frac{(\omega_1^2 - \omega^2)p_{21}p_{21}}{(\omega_1^2 - \omega^2)^2 + 4\zeta_1^2\omega_1^2\omega^2} + \frac{(\omega_2^2 - \omega^2)p_{22}p_{22}}{(\omega_2^2 - \omega^2)^2 + 4\zeta_2^2\omega_2^2\omega^2} \right)$$

$$+ i \left(\frac{(2\zeta_1\omega_1 p_{21}p_{21})\omega}{(\omega_1^2 - \omega^2)^2 + 4\zeta_1^2\omega_1^2\omega^2} + \frac{(2\zeta_2\omega_2 p_{22}p_{22})\omega}{(\omega_2^2 - \omega^2)^2 + 4\zeta_2^2\omega_2^2\omega^2} \right) \quad (7.113c)$$

in which we can substitute the appropriate values. From these expressions the magnitude and phase angle can be obtained as

$$|R_{jk}(\omega)| = \sqrt{[\text{Re}(H_{jk})]^2 + [\text{Im}(H_{jk})]^2} \qquad (7.114)$$

$$\phi_{jk}(\omega) = \arctan \frac{\text{Im}(R_{jk})}{\text{Re}(R_{jk})} \qquad (7.115)$$

It is now instructive to see how these functions look in graphic form; we must not be deceived by the simplicity of this example because many of the important characteristics of MDOF FRF (receptances in this case) are already present and, as a matter of fact, can be better appreciated in an example like this one rather than in a more complex case.

Since it is more convenient for the eye to visualize two-dimensional graphs and we are dealing with complex functions, our FRFs can only be completely represented if we draw two such graphs for each FRF. As for the SDOF case, the most common choices are two:

1. magnitude and phase as functions of frequency;
2. real and imaginary parts as functions of frequency.

Figures 7.3–7.5 show representations 1 and 2 for each receptance FRF of eqs (7.113a, b and c).

Note the dB scale on the graphs of magnitude and the fact that the phase angle is considered to vary from $0°$ to $360°$, with increasing angles in the counterclockwise direction. A quick look at these graphs shows two things right away:

- The first mode is much less damped than the second.
- Between the two resonances there is a considerable difference in the behaviour of the magnitude curves: on one hand the FRFs $R_{11}(\omega)$ and $R_{22}(\omega)$ show an evident 'antiresonance' slightly above 30 rad/s while, on the other hand, no such thing appears in the graphs of $R_{12}(\omega)$ and $R_{21}(\omega)$.

We will have more to say about the distinctive features of these graphs in later chapters; for the time being the reader is invited to draw the graphs of mobility and acceleration (representations 1 and 2) for this example.

In addition – although unnecessary for proportional damping – it may also be useful to adopt, for example, the second state-space formulation outlined in Section 7.6.1 to treat the above problem. In this case we form the dynamic matrix

$$\mathbf{A} = \begin{bmatrix} \mathbf{0} & \mathbf{I} \\ -\mathbf{M}^{-1}\mathbf{K} & -\mathbf{M}^{-1}\mathbf{C} \end{bmatrix} = \begin{bmatrix} 0 & 0 & 1 & 0 \\ 0 & 0 & 0 & 1 \\ -1000 & 500 & -1 & 0.5 \\ 1000 & -1000 & -1 & -1 \end{bmatrix}$$

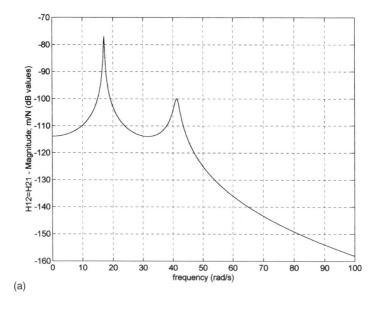

(a)

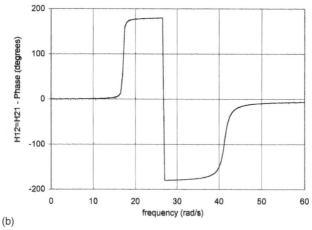

(b)

Fig. 7.4 Receptance $R_{12}(\omega)$: (a) magnitude, (b) phase, versus frequency.

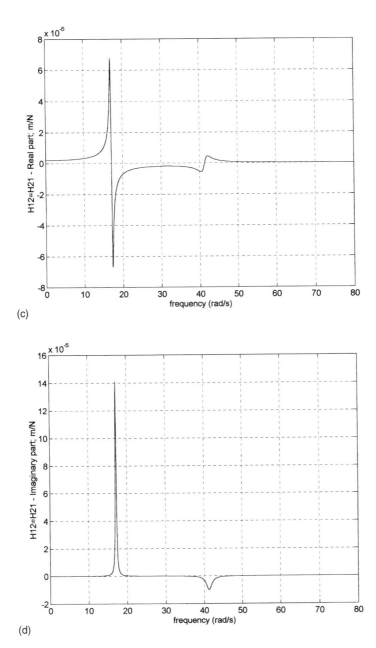

(c)

(d)

Fig. 7.4 Receptance $R_{12}(\omega)$: (c) real part and (d) imaginary part, versus frequency.

and obtain the matrix of eigenvalues and eigenvectors as

$$
\text{diag}(\lambda_j) = \begin{bmatrix} -0.854 + 41.308i & 0 & 0 & 0 \\ 0 & -0.854 - 41.308i & 0 & 0 \\ 0 & 0 & -0.146 + 17.114i & 0 \\ 0 & 0 & 0 & -0.146 - 17.114i \end{bmatrix}
$$

$$
S = \begin{bmatrix} 0.0138 + 0.0020i & 0.0138 - 0.0020i & 0.0232 - 0.0244i & 0.0232 + 0.0244i \\ -0.0196 - 0.0028i & -0.0196 + 0.0028i & 0.0328 - 0.0345i & 0.0328 + 0.0345i \\ -0.0948 + 0.5693i & -0.0948 - 0.5693i & 0.4140 + 0.4010i & 0.4140 - 0.4010i \\ 0.1340 - 0.8052i & 0.1340 + 0.8052i & 0.5855 + 0.5671i & 0.5855 - 0.5671i \end{bmatrix}
$$

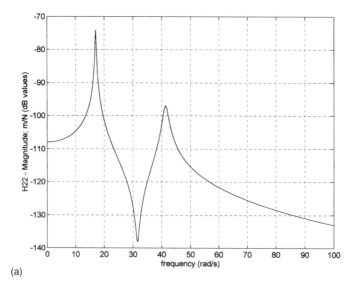

(a)

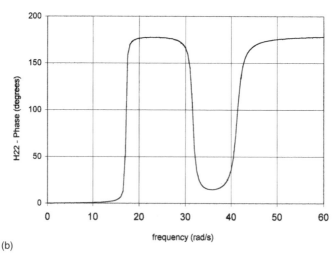

(b)

Fig. 7.5 Receptance $R_{22}(\omega)$: (a) magnitude, (b) phase, versus frequency.

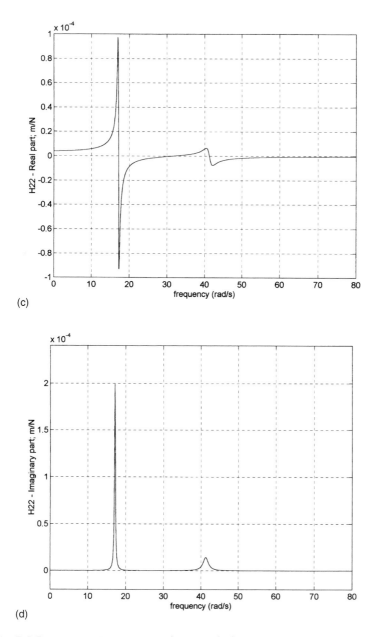

(c)

(d)

Fig. 7.5 Receptance $R_{22}(\omega)$: (c) real part and (d) imaginary part, versus frequency.

Next we form the matrix S_{upper} with the first two rows of S, obtain S^{-1}, form the matrix S_{right}^{-1} with its last two columns and calculate the product

$$
S_{right}^{-1}M^{-1} = \begin{bmatrix}
-0.0623 - 0.4287i & 0.0881 + 0.6163i \\
-0.0623 - 0.4287i & 0.0881 - 0.6063i \\
0.3141 - 0.2991i & 0.4443 - 0.4230i \\
0.3141 + 0.2991i & 0.4443 + 0.4230i
\end{bmatrix}
$$

Now, if we consider for example the function $R_{11}(\omega)$ we get

$$
R_{11}(\omega) = -\frac{6.052 \times 10^{-6}i}{i\omega - \lambda_1} + \frac{6.052 \times 10^{-6}i}{i\omega - \lambda_2}
$$

$$
-\frac{1.4608 \times 10^{-5}i}{i\omega - \lambda_3} + \frac{1.4608 \times 10^{-5}i}{i\omega - \lambda_4} \tag{7.116a}
$$

where the eigenvectors are ordered as in the matrix $\mathrm{diag}(\lambda_j)$. Since $\lambda_2 = \lambda_1^*$ and $\lambda_4 = \lambda_3^*$, we can substitute their values into eq (7.116a) to get

$$
R_{11}(\omega) = \frac{5.00 \times 10^{-4}}{1707.1 - \omega^2 + 1.707i\omega} + \frac{5.00 \times 10^{-4}}{292.9 - \omega^2 + 0.2929i\omega} \tag{7.116b}
$$

which is, as expected, the same as the $(1,1)$ element of matrix (7.112). The same applies for the other FRF functions $R_{12}(\omega) = R_{21}(\omega)$ and $R_{22}(\omega)$.

Note that the only difference between eqs (7.112) and (7.116b) is in the sign of the third (damping) term in both denominators: this is due to the fact that we can choose the harmonic excitation either in the form $f(t) = f_0 e^{-i\omega t}$ (as in Section 7.3) or in the form $f(t) = f_0 e^{i\omega t}$ (as in Section 7.6.1). It is obvious that the choice leads to no consequences as long as consistency is maintained.

Another useful exercise would be to consider the same 2-DOF system with a nonproportional damping matrix, for example

$$
C = \begin{bmatrix} 2500 & -2000 \\ -2000 & 2000 \end{bmatrix} \text{N s/m}
$$

The reader is invited to arrive at the matrix of receptances by following the state space formulation of eq (7.60a): the $2n \times 2n$ matrix S of M-orthonormal eigenvectors is in this case

$$
S = \begin{bmatrix}
0.0017 - 0.0018i & 0.0017 + 0.0018i & 0.0027 - 0.0027i & 0.0027 + 0.0027i \\
-0.0025 + 0.0024i & -0.0025 - 0.0024i & 0.0038 - 0.0039i & 0.0038 + 0.0039i \\
0.0680 + 0.0755i & 0.0680 - 0.0755i & 0.0453 + 0.0473i & 0.0453 - 0.473i \\
-0.0918 - 0.1112i & -0.0918 + 0.1112i & 0.0652 + 0.0656i & 0.0652 - 0.0656i
\end{bmatrix}
$$

and the matrix of eigenvalues is

$$
\text{diag}(\lambda_j) = \begin{bmatrix} -3.039 + 41.197i & 0 & 0 & 0 \\ 0 & -3.039 - 41.197i & 0 & 0 \\ 0 & 0 & -0.211 + 17.116i & 0 \\ 0 & 0 & 0 & -0.211 - 17.116i \end{bmatrix}
$$

Note that, as expected, $\mathbf{S}^T \hat{\mathbf{K}} \mathbf{S} = -\text{diag}(\lambda_j)$. Moreover, for the sake of completeness, it may be worth pointing out that the above eigenvectors and eigenvalues – provided that $\hat{\mathbf{M}}$ is nonsingular – can also be obtained from the standard eigenvalue problem

$$
\hat{\mathbf{M}}^{-1} \hat{\mathbf{K}} \mathbf{s} = -\lambda \mathbf{s}
$$

or from eq (6.180b). The reader is by now well aware of this fact which, nevertheless can be of help whenever a program that solves generalized eigenvalues problems is not available.

7.10 Summary and comments

This chapter has considered the response characteristics of MDOF systems. Intentionally, our approach has given more emphasis to the mode super-position solution strategy in view of future chapters which will consider the experimental part of the subject. For the moment, the discussion has followed what we may call a 'theoretical approach', in which the physical characteristics of the system under investigation (mass, stiffness and damping, i.e. the spatial model) are supposed to be known. This knowledge allows the analyst to obtain:

1. the system's eigenvalues and eigenvectors;
2. the system's response to an external excitation.

Point 1 has to do with the solution of the free-vibration problem and it has been considered in detail in Chapter 6.

As far as point 2 is concerned, we can distinguish between various types of systems and between various types of excitations: for example, the system under study may be undamped, viscously damped or hysteretically damped, and with or without rigid-body modes (here the analyst has sometimes a certain degree of control because he/she can choose to test the system in a restrained or an unrestrained condition). In turn, damping may be proportional or nonproportional, and the external excitations may or may not be simple functions of time.

Whatever solution strategy we decide to use, a general and important result is expressed by the reciprocity theorem (Section 7.2) which states that the response at point j of our system due to an excitation applied at point k is equal to the response at point k when the same excitation is applied at point j.

This occurrence has important consequences from an experimental point of view and, mathematically, translates into the fact that the matrices of IRF and of FRF are symmetrical.

Specifically, the mode superposition strategy for linear systems is to – where possible – uncouple the equations of motion by means of an appropriate coordinate transformation, solve the equations independently and, by virtue of the superposition principle, superpose the individual results to obtain the desired response. One of its major advantages has to do with the fact that the system's response can often be represented within a reasonable degree of accuracy by considering only a small fraction (say s, where $s < n$) of the n uncoupled equations of motion. This circumstance is exploited in the mode displacement and mode acceleration methods (Section 7.2.1) where, provided that some precautions on the number of modes to consider are observed, the response is obtained by means of a 'truncated' solution. In general, the mode acceleration method is preferred.

As for SDOF systems, the case of harmonic excitation is particularly important because it leads directly to the response model of the system under study, i.e. a set of FRFs, usually in the preferred forms of receptance, mobility or accelerance. For a given (linear) system, these FRFs are intrinsic characteristics which deserve special attention for the implications on any further stage of theoretical and experimental analysis. In Section 7.4 it is shown that there is a strict connection between the FRF matrix and the impulse response matrix and in each one of the following sections an explicit expression of the response model is obtained for various systems. In Section 7.4 we included also a note of theoretical importance, namely the Kramers–Kronig relations, which relate the real and imaginary part of a general FRF and express a fundamental consequence of the principle of causality.

For nonproportionally damped systems the price we have to pay in order to uncouple the equations of motion is the doubling of the order of our system. Section 7.6 describes two strategies that follow directly from the two state-space formulations of the homogeneous problem given in Section 6.8.

Despite the fact that the mathematical part is a bit more involved, the methods show no additional conceptual difficulty with respect to the 'standard' modal approach.

Some conceptual problems, however, do arise in the case of hysteretic damping, which is considered in Section 7.7. In fact, the solution of the homogeneous case is not fully justified in this case and only physical sense – and the need for a response model for future developments – can be invoked. The difficulty lies in the definition of hysteretic damping itself, which implies a harmonically forced vibration condition. Nevertheless, one could argue that experiments are generally performed by exciting the system and by measuring its response so that, as a matter of fact, we are in a forced-vibration condition despite the fact that the final result may involve only quantities that, mathematically, come directly from the homogeneous case. This latter may seem a proper justification, but the price we pay is more subtle: the FRFs that we obtain do not satisfy the requirement of causality.

Finally, Section 7.8 outlines other solution strategies that can be adopted to solve the forced vibration problem: the Laplace transform method and a direct integration of the equations of motion in the time domain. One word of caution about this latter technique: despite the computational capabilities of computers it must not be used as a 'black box', some care must always be exercised because of convergence and accuracy problems of the solution.

Specifically, many different integration schemes are available and a major distinction can be made between implicit and explicit integration procedures. In addition, it must be pointed out that direct integration is the only applicable technique for the analysis of nonlinear systems.

References

1. Ewins, D.J., *Modal Testing: Theory and Practice*, Research Studies Press, 1984.
2. Craig, R.R., Jr., *Structural Dynamics, An Introduction to Computer Methods*, John Wiley, New York, 1981.
3. Nashif, A.D., Jones, D.I.G. and Henderson, J.P., *Vibration Damping*, John Wiley, New York, 1985.
4. Newland, D.E., *Mechanical Vibration Analysis and Computation*, Longman Scientific and Technical, 1989.

8 Continuous or distributed parameter systems

8.1 Introduction

The representation of physical systems by means of discrete models in which properties like inertia, stiffness and damping are localized and identified with different elements like masses, springs and dampers is often very convenient and leads to satisfactory results in many circumstances. In reality, however, one has to deal with, say, aircraft structures, pipelines, car bodies, various types of buildings, etc.; in other words, with structures which generally comprise cables, rods, beams, plates and shells, all of which are neither rigid nor massless. Every material portion of the system may possess mass, stiffness and damping properties at the same time, and these properties may vary from point to point. In these cases, whenever possible, one can resort to continuous models (we already encountered some examples of such models in Chapters 3 and 5), where the displacement is a continuous function of both space and time and we are in presence of an infinite number of degrees of freedom.

Distributed parameter models are based on another idealization: the continuous elastic medium which, for its part, leads to a fundamental insight into the nature of mechanical vibrations: the so-called wave–mode duality. In everyday engineering problems we often tend to think of vibrations in terms of modes and of, say, acoustical phenomena in terms of waves. As a matter of fact, this distinction is somehow fictitious because we are just considering the same physical phenomena: that is, the propagation of a localized disturbance (mechanical in our case) which 'spreads' from one part of a medium into other parts of the same medium or into a different medium.

We are referring here to the propagation of mechanical waves which, as a matter of fact, represents a deeper level of explanation for mechanical vibrations. Normal modes of vibration are in fact particular motions of a system (for which the system is, let's say, particularly well suited) which ensue because of its finite physical dimensions in space and hence because of the presence of boundaries. In other words, the superposition of travelling waves reflected back and forth from the physical boundaries of the medium ultimately result in the appearance of standing waves which, in turn, represent the normal modes of vibration of our system.

This book is primarily concerned with aspects of mechanical vibrations that do not need a detailed discussion of the 'wave approach' and for this reason the subject of wave propagation and motion in elastic solids will only be touched on briefly whenever needed in the course of the discussion. For our purposes, the importance of the considerations above is mostly a matter of principle, but it must be pointed out that the wave-mode duality has also significant implications in all fields of engineering where the interest lies in the study of interactions between sound waves and solid structures.

In essence, distributed parameter systems present many conceptual analogies with MDOF systems. However, some important differences of mathematical nature will be clear from the outset. First of all, the motion of these systems is governed by partial differential equations and, second, these equations must be supplemented by an appropriate number of boundary conditions. Moreover, boundary conditions are just as important as the differential equations themselves and constitute a fundamental part of the problem; for short, one often says that the motion of a continuous system is governed by boundary value problems.

These problems are in general much more difficult to solve than their discrete counterpart (where boundary conditions enter only indirectly because they are implicitly included into the system's matrices) and hence, for continuous systems, exact solutions are available only for a limited number of cases. We will consider some of these cases and provide the exact solutions. Nonetheless, for more complex systems we have to resort to approximate solutions which, in turn, are often obtained through spatial discretization and in the end – despite the fact that the techniques of analysis may be highly sophisticated (e.g. the finite-element method) – bring us back to finite-degree-of-freedom systems.

On a more theoretical basis, we pass from the finite-dimensional vector spaces of the discrete case to infinite-dimensional vector spaces. More specifically, we have to deal with Hilbert spaces, i.e. infinite-dimensional vector spaces where an inner product has been defined and are complete with respect to the norm defined by means of the above inner product. The reader can find in Chapter 2 some introductory considerations on these theoretical aspects because it is important to be aware of the fact that the conceptual analogies with the discrete case rest on the fact that, broadly speaking, Hilbert spaces are the 'natural generalization' of the usual finite-dimensional vector spaces.

At the beginning of this chapter we will consider a simple system – the flexible string – in some detail in order to gain some insight on the fundamental aspects of wave propagation, natural frequencies and modes of a continuous system which, with the appropriate modifications, can be taken as representative for other types of distributed parameter systems. The beam in bending vibration will be considered next, before turning our attention to more general aspects of the differential eigenvalue problem and to the analysis of some two-dimensional systems.

8.2   The flexible string in transverse motion

The flexible string under tension – with some basic assumptions that will be considered soon – is the simplest model of continuous system where mass and elasticity are distributed over its whole extent. From the discussion developed in preceding chapters we can argue that, in principle, we could (and indeed we do) arrive at a satisfactory description of its motion by considering it as a linear array of oscillators where masses are lumped at discrete points and elasticity is introduced by means of massless springs connecting the masses. The greater the number of degrees of freedom, the better the approximation. Furthermore, we could also work out an asymptotic solution by increasing the number of masses indefinitely and by letting their mutual distance tend to zero. However, we adopt a different method of attack which contains more physical insight than the mathematical expedient of the limiting procedure; that is, we do not consider the motion of each one of the individual infinite number of points of the string but only concern ourselves with the shape of the string as a whole.

So, let us consider a string of indefinite length (we want to avoid for the moment a discussion of the boundary conditions) which is stretched by a tension of T_0 newtons and whose undisturbed position coincides with the x-axis.

Let us further assume that the displacements of each point of the string are wholly transverse in a direction parallel to the y-axis. It follows that the string motion is specified by a 'shape' function $y(x, t)$ where x and t play the role of independent variables: for a fixed time t_1 the graph of the function $y(x, t_1)$ depicts the shape of the string at that instant while the graph of the function

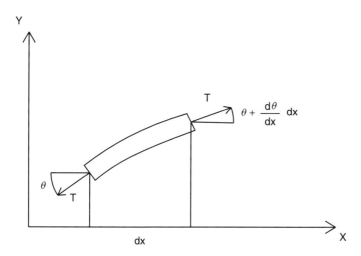

Fig. 8.1 Displaced differential element of taut string.

$y(x_1, t) - x_1$ fixed – represents the motion of the point located at x_1 as time passes, i.e. the time history of the particle at x_1.

Qualitatively, if we assume that any variation of the tension due to the transverse displacement of the string is negligible, we can apply Newton's second law to the differential element of the string shown in Fig. 8.1 to get the equation of motion in the vertical direction as

$$-T_0 \sin\theta + T_0 \sin\left(\theta + \frac{\partial\theta}{\partial x} dx\right) = \mu ds \frac{\partial^2 y}{\partial t^2} \tag{8.1}$$

where μ (kg/m) is the mass per unit length of the string, which for the present we assume uniform throughout the length of the string.

If we further assume that the slope of the string is everywhere small – i.e. $\partial y/\partial x \ll 1$ or, in other words, the inclination angle θ is always small compared with one radian, we can write

$$ds = dx \sqrt{1 + \left(\frac{\partial y}{\partial x}\right)^2} \cong dx$$

$$\sin\theta \cong \theta \cong \frac{\partial y}{\partial x}$$

so that eq (8.1) becomes

$$T_0 \frac{\partial^2 y}{\partial x^2} = \mu \frac{\partial^2 y}{\partial t^2} \tag{8.2a}$$

which we can choose to write in the form

$$c^2 \frac{\partial^2 y}{\partial x^2} = \frac{\partial^2 y}{\partial t^2} \tag{8.2b}$$

where $c = \sqrt{T_0/\mu}$ has the dimensions of a velocity and, in the approximations above, is independent of both x and t. Note that the small slopes approximation (or small-amplitude approximation) expressed by $\partial y/\partial x \ll 1$ allows us to neglect all quantities of second and higher order in $\partial y/\partial x$. Only in this circumstance the net horizontal force on the differential element of string is zero and we can assume $ds \cong dx$ so that the displacement of each point is perpendicular to the x-axis and the tension T_0 remains unchanged in passing from point x to point $x + dx$. When the small slopes assumption ceases to be valid, the resulting differential equation is nonlinear.

Equation (8.2b) is the well-known one-dimensional differential wave equation which, by assigning the appropriate meanings to the quantities involved, represents a broad range of wave phenomena in many branches of physics and engineering (acoustics, electromagnetism, etc.).

Obviously, in our case, the motion ensues because the string has been disturbed from its equilibrium position ('plucked', for example) at some time. The tension then provides the restoring force but inertia delays the immediate return to the equilibrium position by overshooting the rest position. Note also that, for the time being, no consideration whatsoever is given to dissipative damping forces and to the effect of stiffness that, although generally negligible, occurs in real strings.

All books on basic physics show that the general solution of the one-dimensional wave equation is of the form

$$y(x, t) = f(x - ct) + g(x + ct) \tag{8.3}$$

which is called the d'Alembert solution of the wave equation. The functions f and g can be any two arbitrary and independent twice-differentiable functions whose forms depend on how the string has been started into motion, i.e. on the initial conditions. It is not difficult to see that $f(x - ct)$ represents a shape – or a profile – which moves without distortion in the positive x-direction with velocity c, while $g(x + ct)$ represents a similar wave (with a different shape if $f \neq g$) which moves in the negative x-direction with velocity c; the linearity of the wave equation implies that if both waveforms have a finite spatial duration, they can 'pass through' one another and reappear without distortion. Most of us will be familiar with these 'travelling wave' phenomena from childhood games with ropes.

8.2.1 The initial value problem

In the case of a string of indefinite length, we can gain some further insight by considering its motion due to some initial disturbance. This disturbance is specified by means of the initial conditions, i.e. the functions that determine the shape and velocity of the string at $t = 0$. Let these functions be

$$y(x, 0) = u(x)$$

$$\dot{y}(x, 0) \equiv \frac{\partial y}{\partial t}\bigg|_{t=0} = w(x) \tag{8.4}$$

From the general solution, at $t = 0$ we have

$$f(x) + g(x) = u(x)$$

$$-cf'(x) + cg'(x) = w(x) \tag{8.5}$$

where the primes represent here the derivatives of the functions f and g with respect to their arguments. The second of eqs (8.5) can be integrated to give

$$f(x) - g(x) = -\frac{1}{c} \int_0^x w(\xi)d\xi + C \qquad (8.6)$$

where the constant of integration C can be set to zero without loss of generality. From the first of eqs (8.5) and (8.6) it follows

$$f(x) = \frac{1}{2} u(x) - \frac{1}{2c} \int_0^x w(\xi)d\xi$$

$$g(x) = \frac{1}{2} u(x) + \frac{1}{2c} \int_0^x w(\xi)d\xi$$

which establish the initial values of the functions f and g. For $t \neq 0$ we replace the variable x with the appropriate arguments to get

$$y(x, t) = \frac{1}{2} [u(x - ct) + u(x + ct)] + \frac{1}{2c} \left(-\int_0^{x-ct} w(\xi)d\xi + \int_0^{x+ct} w(\xi)d\xi \right)$$

or, alternatively

$$y(x, t) = \frac{1}{2} [u(x - ct) + u(x + ct)] + \frac{1}{2c} \int_{x-ct}^{x+ct} w(\xi)d\xi \qquad (8.7a)$$

which, by integration of the second term on the r.h.s., i.e.

$$\int_{x-ct}^{x+ct} w(x)dx = \varphi(x + ct) - \varphi(x - ct)$$

can be written as

$$y(x, t) = \frac{1}{2} [u(x - ct) + u(x + ct)] + \frac{1}{2c} [\varphi(x - ct) - \varphi(x - ct)] \qquad (8.7b)$$

Equation (8.7b) – which has also been obtained by using transform techniques in Section 2.3 – physically represents identical leftward and rightward propagating disturbances containing separate contributions from the displacement and velocity initial conditions.

8.2.2 Sinusoidal waves, energy considerations and the presence of boundaries

Sinusoidal waves

Out of the infinite variety of functions permitted as solutions by the wave equation (as a matter of fact, any reasonable function of $x + ct$ or $x - ct$), it should be expected that sinusoidal waveforms deserve particular attention. This is because, besides their mathematical simplicity and the fact that many real-world sources of waves are nearly sinusoidal, we can represent as closely as desired any reasonable periodic and non-periodic function by the linear superposition of many sinusoidal functions (Fourier analysis – see Chapter 2 for more details). Mathematically, we can express an ideal sinusoidal wave of unit amplitude travelling along the string as

$$y(x, t) = \sin \frac{2\pi}{\lambda} (x - ct) = \sin(kx - \omega t) \tag{8.8}$$

where k is the so-called wavenumber: when the quantity kx increases by 2π the corresponding increment in x is the wavelength λ so that $k\lambda = 2\pi$ or, equivalently, $k = 2\pi/\lambda$.

Note that, in the light of preceding chapters, the symbols may be a bit misleading: here k is not a spring constant and λ is not an eigenvalue. However, these symbols for the wavenumber and the wavelength are so widely used that we adhere to the common usage: the meaning is generally clear from the context but precise statements will be made whenever some ambiguities may arise in the course if the discussion.

As far as time dependency is concerned, we already know that the period T is related to the frequency ν by $T = 1/\nu$ and that the angular frequency ω is given by $\omega = 2\pi\nu$, the fact that the wave moves to the right can be deduced by noting that, as time passes, increasing values of x are required to maintain the phase $\phi \equiv kx - \omega t$ constant. Two 'snapshots' of the waveform at times t_0 and $t_0 + T$ look exactly the same; this implies that the wave has travelled a distance λ in the time interval T so that

$$c = \frac{\lambda}{T} = \lambda\nu = \frac{\omega}{k} \tag{8.9}$$

Moreover, it is evident that also in this case the exponential representation (Chapter 1)

$$y(x, t) = e^{+i(kx - \omega t)} \tag{8.10}$$

is widely adopted and is often very convenient. The wave of eq (8.8) is obtained by taking the imaginary part of eq (8.10) but, as stated in previous chapters, the real-part convention may be adopted as well, and the difference is irrelevant as long as consistency is maintained. The general restriction of small amplitudes $\partial y / \partial x \ll 1$ translates for harmonic waves into $|iky| \ll 1$ or, in other words, into

$$y_{max} \ll \frac{\lambda}{2\pi} \tag{8.11}$$

which states that the maximum amplitude must be much smaller than the wavelength. One final word here to point out that the velocity of the propagation of the disturbance c must not be confused with the velocity of the individual particles of the string, i.e. with $\partial y / \partial t$; as a matter of fact, for a general waveform $y(x, t) = f(x - ct)$, since

$$\frac{\partial y}{\partial t} = -cf'(x - ct) = -c\frac{\partial y}{\partial x}$$

it follows that the small-amplitude approximation requires

$$\frac{\partial y}{\partial t} \ll c \tag{8.12}$$

where it must be understood that the string particles move in a transverse direction while the waveform propagates along the string. The word 'propagation' itself, as we shall see shortly, implies a transport of energy and momentum.

Energy considerations

From the previous discussion, it is apparent that the kinetic energy in a differential element of string is given by

$$dT = \frac{1}{2}\mu ds \left(\frac{\partial y}{\partial t}\right)^2 = \frac{1}{2}\mu ds \dot{y}^2$$

and the kinetic energy in a segment between x_1 and x_2 is then

$$T = \frac{1}{2}\mu \int_{x_1}^{x_2} \dot{y}^2 \left[1 + \left(\frac{\partial y}{\partial x}\right)^2\right]^{1/2} dx$$

which, for small deflections, can be approximated as

$$T = \frac{1}{2} \mu \int_{x_1}^{x_2} \dot{y}^2 dx \qquad (8.13)$$

The calculation of the potential energy is a bit more involved because second order terms come into play. The string must possess potential energy because some external work would have to be done to give it the deflected shape which, in turn, must locally stretch the string where the wave is present. This local stretching, however, must excite longitudinal waves that propagate along the string as well as the transverse waves. The coupling between longitudinal and transverse waves is expressed by nonlinear terms in the equation of motion, and precisely these terms is what we want to neglect. This difficulty can be circumvented by assuming a negligible Young's modulus (ideally $E = 0$, i.e. a string which is perfectly flexible). In this hypothesis we can consider the change in length of a portion of string of initial length dx: this is

$$ds - dx = \left[\sqrt{1 + \left(\frac{\partial y}{\partial x} \right)^2} - 1 \right] dx \cong \frac{1}{2} \left(\frac{\partial y}{\partial x} \right)^2$$

so that the potential energy between x_1 and x_2 is

$$V = \frac{1}{2} T_0 \int_{x_1}^{x_2} \left(\frac{\partial y}{\partial x} \right)^2 dx \qquad (8.14)$$

because the stretching takes place against a force of tension T_0. In the light of eqs (8.13) and (8.14), it is often convenient to speak of kinetic and potential energy densities

$$\mathcal{T}(x, t) = \frac{1}{2} \mu \left(\frac{\partial y}{\partial t} \right)^2$$

$$\qquad (8.15)$$

$$\mathcal{V}(x, t) = \frac{1}{2} T_0 \left(\frac{\partial y}{\partial x} \right)^2$$

although these definition have a certain degree of arbitrariness because it is often difficult – and sometimes meaningless – to keep track of the location in space and time of a given amount of energy.

Two points are worthy of notice at this point:

1. If we consider a general waveform $f(x - ct)$, the kinetic and potential energy densities are given by

    $$\mathcal{T} = \tfrac{1}{2}\mu c^2 f'^2$$

    $$\mathcal{V} = \tfrac{1}{2}T_0 f'^2$$

    respectively, and since $c^2 = T_0/\mu$, we see that the two expressions are equal. Moreover, we can consider a harmonic wave in the exponential form $Ae^{i(kx - \omega t)}$ – where A is just the amplitude which we assume now to be different from unity – and calculate the average kinetic and potential energy densities. Let us consider for example the potential energy density: we have from eq (8.15)

    $$\langle \mathcal{V} \rangle = \tfrac{1}{2}T_0 \langle (\partial y/\partial x)^2 \rangle$$

    where the bracket indicates the average over one period. If now we resort to the phasor convention of Section 1.3, we get

    $$\left\langle \left(\frac{\partial y}{\partial x} \right)^2 \right\rangle = \frac{1}{2}\,\mathrm{Re}\left\{ \left(\frac{\partial y}{\partial x} \right) \left(\frac{\partial y}{\partial x} \right)^* \right\} = \frac{1}{2}k^2 A^2$$

    from which it follows that

    $$\langle \mathcal{V} \rangle = \tfrac{1}{4}T_0 k^2 A^2 = \tfrac{1}{4}\mu\omega^2 A^2 \tag{8.16}$$

    By the same token, the reader is invited to calculate the average kinetic energy density, verify that $\langle \mathcal{T} \rangle = \langle \mathcal{V} \rangle$ and arrive at the same result by considering a harmonic wave in the form of eq (8.8).
2. The equation of motion (8.2) can be obtained by substituting the kinetic and potential energy densities in eq (3.109) where the Lagrangian density is given by $\mathcal{L} = \mathcal{T} - \mathcal{V}$.

In addition, we may be interested in the flux of energy past a given point x; this rate of energy transfer is just the instantaneous power flow from any piece of the string to its neighbour. Mathematically, it is obtained as the product of the vertical component of tension $-T_0(\partial y/\partial x)$ by the transverse velocity of the string at x (Fig. 8.1), i.e.

$$P = -T_0 \frac{\partial y}{\partial x}\frac{\partial y}{\partial t} \tag{8.17}$$

so that a positive value of P (watts) implies power flowing toward the positive

x-direction and a negative P means power flowing toward the negative x-direction. For a general travelling wave $f(x - ct)$ we have from eq (8.17)

$$P = cT_0 f'^2 \tag{8.18}$$

In the case of a sinusoidal wave, which again we take in the exponential form of the preceding paragraph, the average power transmitted by the wave can be obtained from

$$\langle P \rangle = -T_0 \left\langle \frac{\partial y}{\partial x} \frac{\partial y}{\partial t} \right\rangle \tag{8.19}$$

where, again, the phasor convention of Chapter 1 gives

$$\left\langle \frac{\partial y}{\partial x} \frac{\partial y}{\partial t} \right\rangle = \frac{1}{2} \operatorname{Re} \left\{ \left(\frac{\partial y}{\partial x} \right) \left(\frac{\partial y}{\partial t} \right)^* \right\} = -\frac{1}{2} k\omega A^2$$

so that eq (8.19) results in

$$\langle P \rangle = \tfrac{1}{2} T_0 k\omega A^2 = \tfrac{1}{2} c\mu\omega^2 A^2 \tag{8.20}$$

The calculation of the momentum in the x-direction associated with a transverse wave that obeys eq (8.2a or b) has to do with the small longitudinal motion that occurs when a transverse wave is present. We shall not perform such a calculation here but it can be shown (e.g. Morse and Ingard [1]; Elmore and Heald [2]) that the quantity

$$p(x, t) = -\mu \frac{\partial y}{\partial x} \frac{\partial y}{\partial t} \tag{8.21}$$

may be interpreted as a localized momentum density in the x-direction associated with a transverse wave. Note that

$$P(x, t) = c^2 p(x, t) \tag{8.22}$$

This is a general relationship connecting energy flow and momentum density for plane waves travelling in linear isotropic media.

The presence of boundaries

Real strings have a finite length and must be fastened somewhere. This circumstance affects the motion of the string by imposing appropriate boundary conditions which – as opposed to the initial conditions of Section

8.2.1 which are specified at a given time (usually $t = 0$) – must be satisfied at all times. Let us suppose that our string is attached to a rigid support at $x = 0$ and extends indefinitely in the positive x-direction (semi-infinite string). This is probably the simplest type of boundary condition and it is not difficult to see that such a 'fixed-end' situation mathematically translates into

$$y(0, t) = 0 \qquad (8.23)$$

for all values of t: a condition which must be imposed on the general solution $f(x - ct) + g(x + ct)$. The final result is that the incoming wave $g(x + ct)$ is reflected at the boundary and produces an outgoing wave $-g(x - ct)$ which is an exact replica of the original wave except for being upside down and travelling in the opposite direction. The fact that the original waveform has been reversed is characteristic of the fixed boundary.

Another simple boundary condition is the so-called free end which can be achieved, for example, when the end of the string is attached to a slip ring of negligible mass m which, in turn, slides along a frictionless vertical post (for a string this situation is quite artificial, but it is very important in many other cases). In physical terms, we can write Newton's second law stating that the net transverse force $F_y(0, t)$ (due to the string) acting on the ring is equal to $m(\partial^2 y / \partial t^2) \vert_{x = 0}$. Since $F_y = T_0(\partial y / \partial x)$ and m is negligible, the free-end condition is specified by

$$\frac{\partial y}{\partial x}(0, t) = 0 \qquad (8.24)$$

which asserts that the slope of string at the free end must be zero at all times.

By enforcing the condition (8.24) on the general d'Alembert solution, it is now easy to show that the only difference between the original and the reflected wave is that they travel in opposite directions: that is, the reflected wave has not been inverted as in the fixed-end case. Note that, as expected, in both cases – fixed and free end – the incoming and outgoing waves carry the same amount of energy because neither boundary conditions allow the string to do any work on the support. Other end conditions can be specified, for example, corresponding to an attached end mass, a spring or a dashpot or a combination thereof.

Mathematically, all these conditions can be analysed by equating the vertical component of the string tension to the forces on these elements. For instance, if the string has a non-negligible mass m attached at $x = 0$, the boundary condition reads

$$m \frac{\partial^2 y}{\partial t^2}(0, t) = T_0 \frac{\partial y}{\partial x}(0, t) \qquad (8.25)$$

or, say, for a spring with elastic constant k_0

$$k_0 y(0, t) = T_0 \frac{\partial y}{\partial x} (0, t) \qquad (8.26)$$

Enforcing such boundary conditions on the general d'Alembert solution makes the problem somewhat more complicated. However, on physical grounds, we can infer that the incident wave undergoes considerable distortion during the reflection process. More frequently, the reflection characteristics of boundaries are analysed by considering the incident wave as pure harmonic, thus obtaining a frequency-dependent relationship for the amplitude and phase of the reflected wave.

8.3 Free vibrations of a finite string: standing waves and normal modes

Consider now a string of finite length that extends from $x = 0$ to $x = L$, is fixed at both ends and is subjected to an initial disturbance somewhere along its length. When the string is released, waves will propagate both toward the left and toward the right end. At the boundaries, these waves will be reflected back into the domain $[0, L]$ and this process, if no energy dissipation occurs, will continually repeat itself. In principle, a description of the motion of the string in terms of travelling waves is still possible, but it is not the most helpful. In this circumstance it is more convenient to study standing waves, whose physical meaning can be shown by considering, for example, two sinusoidal waves of equal amplitude travelling in opposite directions, i.e. the waveform

$$y(x, t) = \frac{A}{2} \sin(kx - \omega t) + \frac{A}{2} \sin(kx + \omega t) \qquad (8.27a)$$

which, by means of familiar trigonometric identities, can be written as

$$y(x, t) = A \sin(kx)\cos(\omega t) \qquad (8.27b)$$

Two interesting characteristics of the waveform (8.27b) need to be pointed out:

1. All points x_j of the string for which $\sin(kx_j) = 0$ do not move at all times, i.e. $y(x_j, t) = 0$ for every t. These points are called **nodes** of the standing wave and in terms of the waveform (8.27a), we can say that whenever the crest of one travelling wave component arrives there, it is always cancelled out by a trough of the other travelling wave.

2.   At some specified instants of time that satisfy $\cos(\omega t) = 0$, all points x of the string for which $\sin(kx) \neq 0$ reach simultaneously the zero position and their velocity has its greatest value. At other instants of time, when $\cos(\omega t) = 1$, all the above points reach simultaneously their individual maximum amplitude value $A \sin(kx)$, and precisely at these times their velocity is zero. Among these points, the ones for which $\sin(kx) = 1$ are alternatively crests and troughs of the standing waveform and are called **antinodes**.

In order to progress further along this line of reasoning, we must investigate the possibility of motions satisfying the wave equation in which all parts of the string oscillate in phase with simple harmonic motion of the same frequency. From the discussions of previous chapters, we recognize this statement as the definition of normal modes.

The mathematical form of eq (8.27b) suggests that the widely adopted approach of separation of variables can be used in order to find standing-wave, or normal-mode, solutions of the one-dimensional wave equation.

So, let us assume that a solution exists in the form $y(x, t) = u(x)z(t)$, where u is a function of x alone and z is a function of t alone. On substituting this solution in the wave equation we arrive at

$$c^2 \frac{1}{u} \frac{d^2u}{dx^2} = -\frac{1}{z} \frac{d^2z}{dt^2}$$

which requires that a function of x be equal to a function of t for all x and t. This is possible only if both sides of the equation are equal to the same constant (the separation constant), which we call $-\omega^2$. Thus

$$\frac{d^2u}{dx^2} + \left(\frac{\omega}{c}\right)^2 u = 0$$

$$\frac{d^2z}{dt^2} + \omega^2 z = 0$$
(8.28)

The resulting solution for $y(x, t)$ is then

$$y(x, t) = (A \cos kx + B \sin kx)(C \cos \omega t + D \sin \omega t)$$
(8.29)

where $k = \omega/c$ and it is easy to verify that the product (8.29) results in a series of terms of the form (8.27b). The time dependent part of the solution represents a simple harmonic motion at the frequency ω, whereas for the space dependent part we must require that

$$u(0) = u(L) = 0$$
(8.30)

because we assumed the string fixed at both ends. Imposing the boundary conditions (8.30) poses a serious limitation to the possible harmonic motions because we get $A = 0$ and the **frequency equation**

$$\sin(kL) = 0 \qquad (8.31)$$

which implies $kL = n\pi$ (n integer) and is satisfied only by those values of frequency ω_n for which

$$\omega_n = \frac{n\pi c}{L} = \frac{n\pi}{L}\sqrt{\frac{T_0}{\mu}} \qquad (n = 1, 2, ...) \qquad (8.32)$$

These are the natural frequencies or eigenvalues of our system (the flexible string of length L with fixed ends) and, as for the MDOF case, represent the frequencies at which the system is capable of undergoing harmonic motion. Qualitatively, an educated guess about the effect of boundary conditions could have led us to argue that, when both ends of the string are fixed, only those wavelengths for which the 'matching condition' $n(\lambda/2) = L$ (where n is an integer) applies can satisfy the requirements of no motion at $x = 0$ and $x = L$. This is indeed the case and the allowed wavelengths satisfy $\lambda_1 = 2L$, $\lambda_2 = L$, $\lambda_3 = (2/3)L$, etc.

The first four patterns of motion (eigenfunctions) are shown in Fig 8.2: the motion for $n = 1, 3, 5, ...$ result in symmetrical (with respect to the point $x = L/2$) modes, while antisymmetrical modes are obtained for $n = 2, 4, 6, ...$ So, for a given value of n, we can write the solution as

$$y_n(x, t) = (A_n \cos \omega_n t + B_n \sin \omega_n t)\sin k_n x \qquad (8.33)$$

where, for convenience, the constant of the space part has been absorbed in the constants A_n and B_n. Then, given the linearity of the wave equation, the general solution is obtained by the superposition of modes:

$$y(x, t) = \sum_{n=1}^{\infty} (A_n \cos \omega_n t + B_n \sin \omega_n t)\sin k_n x \qquad (8.34)$$

where the (infinite) sets of constants A_n and B_n represent the amplitudes of the standing waves of frequency ω_n. The latter quantities, in turn, are related to the allowed wavenumbers by the equation $\omega_n = ck_n$. On physical grounds, since we observed that there is no motion at the nodes and hence no energy flow between neighbouring parts of the string, one could ask at this point how a standing wave gets established and how it is maintained. To answer this question we must remember that a standing wave represents a steady-state situation; during the previous transient state (which, broadly speaking, we may call the 'travelling wave era') the nodes move and allow the transmission of energy along the string. Moreover, it should also be noted that nodes are not perfect in real strings where friction is present; they are only points of minimum amplitude of vibration.

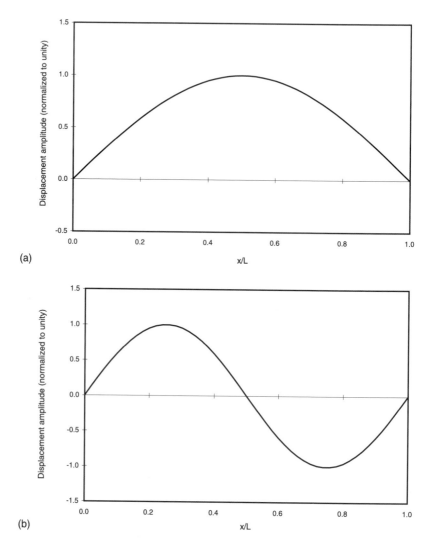

Fig. 8.2 Vibration of a string with fixed ends: (a) first, (b) second mode.

The values of the constants A_n and B_n can be obtained by imposing the initial conditions (8.4) on the general solution (8.34). We get

$$y(x, 0) = u(x) = \sum_{n=1}^{\infty} A_n \sin k_n x$$

$$\dot{y}(x, 0) = w(x) = \sum_{n=1}^{\infty} \omega_n B_n \sin k_n x$$

(8.35)

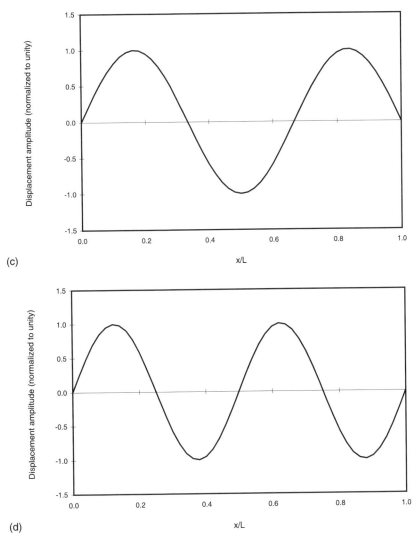

Fig. 8.2 Vibration of a string with fixed ends: (c) third and (d) fourth modes.

which we recognize as Fourier series with coefficients A_n and $\omega_n B_n$, respectively. Following the standard methods of Fourier analysis, we multiply both sides of eqs (8.35) by $\sin k_m x$ and integrate over the interval $[0, L]$ in order to obtain, by virtue of

$$\int_0^L \sin(k_n x)\sin(k_m x)dx = \frac{L}{2}\,\delta_{nm} \qquad (8.36)$$

the expressions

$$A_m = \frac{2}{L} \int_0^L u(x)\sin k_m x \, dx$$

$$B_m = \frac{2}{\omega_m L_0} \int_0^L w(x)\sin k_m x \, dx$$

(8.37)

which establish the motion of our system. Note that eq (8.34) emphasizes the fact that the string is a system with an infinite number of degrees of freedom, where, in the normal mode representation, each mode represents a single degree of freedom; furthermore, from the discussion above it is clear that the boundary conditions determine the mode shapes and the natural frequencies, while the initial conditions determine the contribution of each mode to the total response (or, in other words, the contribution of each mode to the total response depends on how the system has been started into motion). If, for example, we set the string into motion by pulling it aside at its centre and then letting it go, the ensuing free motion will comprise only the odd (symmetrical) modes; even modes, which have a node at the centre, will not contribute to the motion.

A final important result must be pointed out: when the motion is written as the summation of modes (8.34), the total energy E of the string – i.e. the integral of the energy density

$$\frac{1}{2}\mu\left(\frac{\partial y}{\partial t}\right)^2 + \frac{1}{2}T_0\left(\frac{\partial y}{\partial x}\right)^2$$

over the length of the string – is given by

$$E = \frac{L\mu}{4}\sum_{n=1}^{\infty}\omega_n^2(A_n^2 + B_n^2) = \frac{LT_0}{4}\sum_{n=1}^{\infty}k_n^2(A_n^2 + B_n^2)$$

(8.38)

where it is evident that each mode contributes independently to the total energy, without any interaction with other modes (recall Parseval's theorem stated in Chapter 2). The explicit calculation of (8.38), which exploits the relation (8.36) together with its cosine counterpart

$$\int_0^L \cos(k_n x)\cos(k_m x) \, dx = \frac{L}{2}\delta_{nm}$$

(8.39)

is left to the reader.

We close this section with a word of caution. Traditionally, cable vibration observations of natural frequencies and mode shapes are compared to those of the taut string model. However, a more rigorous approach must take into account the axial elasticity and the curvature of the cable (for example, power-line cables hang in a shape called 'catenary' and generally have a sag-to-span ratio between 0.02 and 0.05) and may show considerable discrepancies compared to the string model. In particular, the natural frequencies and mode shapes depend on a cable parameter $EA/\rho g A L_0$ (E = Young's modulus, A = cross-sectional area, ρg = cable weight per unit volume, L_0 = half-span length) and on the sag-to-span ratio. The interested reader may refer, for example, to Nariboli and McConnell [3] and Irvine [4].

8.4 Axial and torsional vibrations of rods

In the preceding sections we considered in some detail a simple case of continuous system – i.e. the flexible string. However, in the light of the fact that our interest lies mainly in natural frequencies and mode shapes, we note that we can explore the existence of solutions in which the system executes synchronous motions just by assuming a simple harmonic motion in time and asking what kind of shape the string has in this circumstance. This amounts to setting $y(x, t) = u(x)e^{i\omega t}$ and substituting it into the wave equation to arrive directly at the first of eqs (8.28) so that, by imposing the appropriate boundary conditions (fixed ends), we arrive at the eigenvalues (8.32) and the eigenfunctions

$$u_n(x) = C_n \sin\left(\frac{n\pi x}{L}\right) \qquad (n = 1, 2, 3, ...) \tag{8.40}$$

where C_n are arbitrary constants which, *a priori*, may depend on the index n.

If now we consider the axial vibration of a slender rod of uniform density ρ and cross-sectional area A in presence of a dynamically varying stress field $\sigma(x, t)$, we can isolate a rod element as in Fig. 8.3 and write Newton's second law as

$$-\sigma A + \left(\sigma + \frac{\partial \sigma}{\partial x} dx\right) A = \rho A dx \frac{\partial^2 y}{\partial t^2} \tag{8.41}$$

where $y(x, t)$ is the longitudinal displacement of the rod in the x-direction. If we assume the rod to behave elastically, Hooke's law requires that $\sigma = E(\partial y/\partial x)$, where $\varepsilon = \partial y/\partial x$ is the axial strain, and upon substitution in eq (8.41) we get

$$E \frac{\partial^2 y}{\partial x^2} = \rho \frac{\partial^2 y}{\partial t^2} \tag{8.42}$$

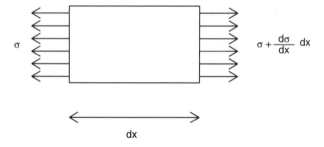

Fig. 8.3 Differential element of a thin rod.

which is the familiar wave equation. Now the wave velocity is given by

$$c = \sqrt{\frac{E}{\rho}} = \sqrt{\frac{EA}{\mu}} \qquad (8.43)$$

where $\mu = \rho A$ is the rod mass per unit length.

Alternatively, we could obtain eq (8.42) by writing the Lagrangian density (eq (3.119)) and by performing the appropriate derivatives required by eq (3.109). Note that this latter procedure leads also to the boundary conditions (3.121).

A completely similar line of reasoning leads to the equation of motion for the torsional vibration of rods which reads

$$C \frac{\partial^2 \theta}{\partial x^2} = \rho J \frac{\partial^2 \theta}{\partial t^2} \qquad (8.44)$$

where $\theta(x, t)$ is the angle of twist, C is the torsional rigidity of the rod (which, in turn, depends on the shear modulus G and on the type of cross-section) and J is the polar moment of inertia.

For a circular cross-section $C = JG$ and eq (8.44) becomes

$$G \frac{\partial^2 \theta}{\partial x^2} = \rho \frac{\partial^2 \theta}{\partial t^2} \qquad (8.45)$$

which is again in the form of a wave equation. As for the case of the string, it should be noted that some simplifying assumptions are implicit in the development of both eq (8.42) and eq (8.45): these assumptions are considered in any book on elementary strengths of materials and will not be repeated here. In the light of the formal analogy of the equations of motion stressed above, it is evident that, with the appropriate modifications for parameters and physical dimensions, the discussions of the preceding sections apply also for the cases above.

If now we separate the variables and assume a harmonic solution in time, we arrive at the ordinary differential equations

$$\frac{d^2u(x)}{dx^2} + \gamma^2 u(x) = 0$$

$$\frac{d^2u(x)}{dx^2} + \gamma^2 u(x) = 0$$

(8.46)

where, for convenience, we called $u(x)$ the spatial part of the solution in both cases (8.42) and (8.45) and the parameter γ^2 is equal to $\omega^2 \rho / E$ for the first of eqs (8.46) and to $\omega^2 \rho / G$ for the second.

Again, in order to obtain the natural frequencies and modes of vibrations we must enforce the boundary conditions on the spatial solution

$$u(x) = A \cos \gamma x + B \sin \gamma x \qquad (8.47)$$

One of the most common cases of boundary conditions is the clamped–free (cantilever) rod where we have

$$u(x) \big|_{x=0} = 0$$

$$\frac{du(x)}{dx}\bigg|_{x=L} = 0$$

(8.48)

so that substitution in (8.47) leads to

$$A = 0$$

$$\cos \gamma L = 0$$

which, in turn, translates into $\gamma_n = (2n-1)\pi/L$, meaning that the natural frequencies are given by

$$\omega_n = \frac{(2n-1)\pi}{2L} \sqrt{\frac{E}{\rho}}$$

$(n = 1, 2, 3, ...)$ $\qquad (8.49)$

$$\omega_n = \frac{(2n-1)\pi}{2L} \sqrt{\frac{G}{\rho}}$$

for the two cases of axial and torsional vibrations, respectively. Like the eigenvectors of a finite DOF system, the eigenfunctions are determined to within a constant. In our present situation

$$u_n(x) = B_n \sin \gamma_n x = B_n \sin\left(\frac{(2n-1)\pi x}{2L}\right) \tag{8.50}$$

where $u_n(x)$ must be interpreted as an axial displacement or an angle, depending on the case we are considering.

If, on the other hand, the rod is free at both ends, the boundary conditions

$$\frac{du}{dx} = 0 \qquad x = 0, L \tag{8.51}$$

lead to $B = 0$ and $\sin \gamma L = 0$, so that $\gamma L = n\pi$ and

$$\omega_n = \frac{n\pi}{L}\sqrt{\frac{E}{\rho}}$$

$$(n = 0, 1, 2, ...) \tag{8.52}$$

$$\omega_n = \frac{n\pi}{L}\sqrt{\frac{G}{\rho}}$$

are the eigenvalues for the two cases, respectively, while

$$u_n(x) = A_n \cos \gamma_n x \tag{8.53}$$

(where γ_n is as appropriate) are the eigenfunctions. Note that in the case of the free–free rod the solution with $n = 0$ is a perfectly acceptable root and does not correspond to no motion at all (see the taut string for comparison). In fact, for $n = 0$ we get $\gamma = \omega = 0$ and, from eq (8.46), $d^2u/dx^2 = 0$, so that $u_0(x) = C_1 x + C_2$ where C_1 and C_2 are two constants whose value is irrelevant for our present purposes. Enforcing the boundary conditions (8.51) gives

$$u_0 = C_2$$

which corresponds to a rigid body mode at zero frequency. As for the discrete case, rigid-body modes are characteristic of unrestrained systems.

For the time being, we do not consider other types of boundary conditions and we turn to the analysis of a more complex one-dimensional system – the beam in flexural vibration. This will help us generalize the discussion on continuous systems by arriving at a systematic approach in which the similarities with discrete (MDOF) systems will be more evident.

In fact, if now in eq (8.46) we define $\omega^2 = \lambda$, replace the differential operator with a stiffness matrix and the mass density ρ with a mass matrix, we may note an evident formal similarity with a matrix (finite-dimensional) eigenvalue problem. Moreover, it is not difficult to note that the same applies to the case of the flexible string.

8.5   Flexural (bending) vibrations of beams

Consider a slender beam of length L, bending stiffness $EI(x)$ and mass per unit length $\mu(x)$. We suppose further that no external forces are acting.

By invoking the Euler–Bernoulli theory of beams – namely that plane cross-sections initially perpendicular to the axis of the beam remain plane and perpendicular to the neutral axis during bending – and by deliberately neglecting the (generally minor) contribution of rotatory inertia to the kinetic energy, we can refer back to Example 3.2 to arrive at the governing equation of motion,

$$\frac{\partial^2}{\partial x^2}\left(EI\,\frac{\partial^2 y}{\partial x^2}\right) + \mu\,\frac{\partial^2 y}{\partial t^2} = 0 \qquad (8.54)$$

where the function $y(x, t)$ represents the transverse displacement of the beam. Equation (8.54) is a fourth-order differential equation to be satisfied at every point of the domain $(0, L)$ and it is not in the form of a wave equation. If, for simplicity, we also assume that the beam is homogeneous throughout its length, eq (8.54a) becomes

$$EI\,\frac{\partial^4 y}{\partial x^4} + \mu\frac{\partial^2 y}{\partial t^2} = 0 \qquad (8.55a)$$

or, alternatively

$$a^2\,\frac{\partial^4 y}{\partial x^4} + \frac{\partial^2 y}{\partial t^2} = 0 \qquad a = \sqrt{\frac{EI}{\mu}} = \sqrt{\frac{EI}{\rho A}} \qquad (8.55b)$$

where ρ is the mass density.

Note that a does not have the dimensions of velocity. We do not enter into the details of flexural wave propagation in beams, but is worth noting that substitution of a harmonic waveform $e^{i(kx - \omega t)}$ into eq (8.55) leads to the dispersion relation $\omega^2 = a^2 k^4$, and since the phase velocity of wave propagation is given by $c = \omega/k$ it follows that

$$c = ak = \frac{2\pi a}{\lambda} \qquad (8.56)$$

which shows that the phase velocity depends on wavelength and implies that, as opposed to the cases of the previous sections, a general nonharmonic flexural pulse will suffer distortion as it propagates along the beam. Energy, in this case, propagates along the beam at the **group velocity** $c_{group} = d\omega/dk$, which can be shown (e.g. Kolsky [5] and Meirovitch [6]) to be related to the phase velocity by

$$c_{group} = c_{phase} - \lambda \frac{dc_{phase}}{d\lambda}$$

Furthermore, eq (8.56) predicts that waves of very short wavelength (very high frequency) travel with almost infinite velocity. This unphysical result is due to our initial simplifying assumptions – i.e. the fact that we neglected rotatory inertia and shear deformation – and the price we pay is that the above treatment breaks down when the wavelength is comparable with the lateral dimensions of the beam. Such restrictions must be kept in mind also when we investigate the natural frequencies and normal bending modes of the beam unless, as it often happens, our interest lies in the first lower modes and/ or the beam cross-sectional dimensions are small compared to its length. When this is the case, we can assume a harmonic time-dependent solution $y(x, t) = u(x)e^{i\omega t}$, substitute it into eq. (8.55) and arrive at the fourth-order ordinary differential equation

$$\frac{d^4u}{dx^4} - \gamma^4 u = 0 \tag{8.57}$$

where we define

$$\gamma^4 = \frac{\omega^2}{a^2} = \frac{\omega^2 \mu}{EI}$$

We try a solution of the form $u(x) = e^{\alpha x}$ and solve the characteristic equation $\alpha^4 - \gamma^4 = 0$ which gives $\alpha = \pm\gamma$ and $\alpha = \pm i\gamma$ so that

$$u(x) = A_1 e^{\gamma x} + A_2 e^{-\gamma x} + A_3 e^{i\gamma x} + A_4 e^{-i\gamma x}$$

$$= C_1 \cosh(\gamma x) + C_2 \sinh(\gamma x) + C_3 \cos(\gamma x) + C_4 \sin(\gamma x) \tag{8.58}$$

where the arbitrary constants A_j or C_j ($j = 1, 2, 3, 4$) are determined from the boundary and initial conditions. The calculation of natural frequencies and eigenfunctions is just a matter of substituting the appropriate boundary conditions in eq (8.58); we consider now some simple and common cases.

Case 1. Both ends simply supported (pinned–pinned configuration)

The boundary conditions for this case require that the displacement $u(x)$ and bending moment $EI(d^2u/dx^2)$ vanish at both ends, i.e.

$$u(0) = u(L) = 0$$

$$\left.\frac{d^2u}{dx^2}\right|_{x=0} = \left.\frac{d^2u}{dx^2}\right|_{x=L} = 0 \tag{8.59}$$

where, in the light of the considerations of Section 5.5, we recognize that the first of eqs (8.59) are boundary conditions of geometric nature and hence represent geometric or essential boundary conditions. On the other hand, the second of eqs (8.59) results from a condition of force balance and hence represents natural or force boundary conditions.

Substitution of the four boundary conditions in eq (8.58) leads to $C_1 = C_2 = C_3 = 0$ and to the frequency equation

$$\sin \gamma L = 0 \tag{8.60}$$

which implies $\gamma L = n\pi$ and hence

$$\omega_n = \frac{n^2\pi^2}{L^2}\sqrt{\frac{EI}{\mu}} \qquad (n = 1, 2, 3, ...) \tag{8.61}$$

The eigenfunctions are then given by

$$u_n(x) = C_4 \sin \gamma_n x = C_4 \sin\left(\frac{n\pi x}{L}\right) \tag{8.62}$$

and have the same shape as the eigenfunctions of a fixed–fixed string.

Case 2. One end clamped and one end free (cantilever configuration)

Suppose that the end at $x = 0$ is rigidly fixed (clamped) and the end at $x = L$ is free; then the boundary conditions require that the displacement $u(x)$ and slope du/dx both vanish at the clamped end, i.e.

$$u(0) = 0$$

$$\left.\frac{du}{dx}\right|_{x=0} = 0 \tag{8.63a}$$

and that the bending moment and shear force $EI(d^3u/dx^3)$ both vanish at the free end, i.e.

$$\frac{d^2u}{dx^2}\bigg|_{x=L} = \frac{d^3u}{dx^3}\bigg|_{x=L} = 0 \tag{8.63b}$$

We recognize eqs (8.63a) as geometric boundary conditions and eqs.(8.63b) as natural boundary conditions. Substitution of eqs (8.63a and b) into (8.58) gives

$$C_1 + C_3 = 0$$

$$C_2 + C_4 = 0$$

$$C_1 \cosh \gamma L + C_2 \sinh \gamma L - C_3 \cos \gamma L - C_4 \sin \gamma L = 0 \tag{8.64a}$$

$$C_1 \sinh \gamma L + C_2 \cosh \gamma L + C_3 \sin \gamma L - C_4 \cos \gamma L = 0$$

which can be arranged in matrix form as

$$\begin{bmatrix} 1 & 0 & 1 & 0 \\ 0 & 1 & 0 & 1 \\ \cosh \gamma L & \sinh \gamma L & -\cos \gamma L & -\sin \gamma L \\ \sinh \gamma L & \cosh \gamma L & \sin \gamma L & -\cos \gamma L \end{bmatrix} \begin{bmatrix} C_1 \\ C_2 \\ C_3 \\ C_4 \end{bmatrix} = 0 \tag{8.64b}$$

and admits nontrivial solutions only if the determinant of the 4×4 matrix is zero, that is, if the frequency equation

$$1 + \cosh \gamma L \cos \gamma L = 0 \tag{8.65}$$

is satisfied. Equation (8.65) must be solved by some numerical method and the first few roots are given (in radians) by

$$\gamma_1 L = 1.875 \qquad \gamma_2 L = 4.694$$

$$\gamma_3 L = 7.855 \qquad \gamma_4 L = 10.996 \tag{8.66a}$$

$$\gamma_5 L = 14.137 \qquad \gamma_6 L = 17.279$$

so that

$$\omega_n = (\gamma_n L)^2 \sqrt{\frac{EI}{\mu L^4}} \tag{8.66b}$$

Note that for $n \geqslant 2$ the approximation $\gamma_n L \cong (n - \frac{1}{2})\pi$ is generally good. The eigenfunctions can be obtained from the first three of eqs (8.64a) which give

$$C_2 = -C_1 \left(\frac{\cosh \gamma_n L + \cos \gamma_n L}{\sinh \gamma_n L + \sin \gamma_n L} \right) \equiv -\kappa_n C_1$$

and, upon substitution into eq (8.58) lead to

$$u_n(x) = C_1[\cosh \gamma_n x - \cos \gamma_n x - \kappa_n(\sinh \gamma_n x - \sin \gamma_n x)] \tag{8.67}$$

where C_1 is arbitrary. One word of caution: because of the presence of hyperbolic functions, the frequency equation soon becomes rapidly divergent and oscillatory with zero crossings that are nearly perpendicular to the γL-axis. For this reason it may be very hard to obtain the higher eigenvalues numerically with an unsophisticated root-finding algorithm.

Case 3. Both ends clamped (clamped–clamped configuration)

All the boundary conditions are geometrical and read

$$u(0) = u(L) = 0$$

$$\left. \frac{du}{dx} \right|_{x=0} = \left. \frac{du}{dx} \right|_{x=L} = 0 \tag{8.68}$$

We can follow a procedure similar to the previous case to arrive at

$$\begin{bmatrix} 1 & 0 & 1 & 0 \\ 0 & 1 & 0 & 1 \\ \cosh \gamma L & \sinh \gamma L & \cos \gamma L & \sin \gamma L \\ \sinh \gamma L & \cosh \gamma L & -\sin \gamma L & \cos \gamma L \end{bmatrix} \begin{bmatrix} C_1 \\ C_2 \\ C_3 \\ C_4 \end{bmatrix} = 0$$

and to the frequency equation

$$1 - \cosh \gamma L \cos \gamma L = 0 \tag{8.69}$$

The first six roots of eq (8.69) are

$$\gamma_1 L = 4.730 \qquad \gamma_2 L = 7.853$$

$$\gamma_3 L = 10.996 \qquad \gamma_4 L = 14.137 \tag{8.70}$$

$$\gamma_5 L = 17.279 \qquad \gamma_6 L = 17.279$$

which, for $n \geq 2$, can be approximated by $\gamma_n L \cong (n + \frac{1}{2})\pi$. Note that the root $\gamma = 0$ of eq (8.69) implies no motion at all, as the reader can verify by solving eq (8.57) with $\gamma = 0$ and enforcing the boundary condition on the resulting solution.

As in the previous case, the eigenfunctions can be obtained from the relationships among the constants C_j and are given by

$$u_n(x) = C_1[\cosh \gamma_n x - \cos \gamma_n x - \kappa_n(\sinh \gamma_n x - \sin \gamma_n x)] \qquad (8.71)$$

where now

$$\kappa_n = \left(\frac{\cosh \gamma_n L - \cos \gamma_n L}{\sinh \gamma_n L - \sin \gamma_n L} \right) \qquad (8.72)$$

Case 4. Both ends free (free–free configuration)

The boundary conditions are now all of the force type, requiring that bending moment and shear force both vanish at $x = 0$ and $x = L$, i.e.

$$\frac{d^2u}{dx^2}\bigg|_{x=0} = \frac{d^3u}{dx^3}\bigg|_{x=0} = 0$$

$$\frac{d^2u}{dx^2}\bigg|_{x=L} = \frac{d^3u}{dx^3}\bigg|_{x=L} = 0 \qquad (8.73)$$

which, upon substitution into eq (8.58) give

$$\begin{bmatrix} 1 & 0 & -1 & 0 \\ 0 & 1 & 0 & -1 \\ \cosh \gamma L & \sinh \gamma L & -\cos \gamma L & -\sin \gamma L \\ \sinh \gamma L & \cosh \gamma L & \sin \gamma L & -\cos \gamma L \end{bmatrix} \begin{bmatrix} C_1 \\ C_2 \\ C_3 \\ C_4 \end{bmatrix} = 0$$

Equating the determinant of the 4×4 matrix to zero yields the frequency equation

$$1 - \cosh \gamma L \cos \gamma L = 0 \qquad (8.74)$$

which is the same as for the clamped–clamped case (eq (8.69)), so that the roots given by eq (8.70) are the values which lead to the first lower frequencies corresponding to the first lower elastic modes of the free–free beam. The elastic eigenfunctions can be obtained by following a similar procedure as in the previous cases. They are

$$u_n(x) = C_1[\cosh \gamma_n x + \cos \gamma_n x - \kappa_n(\sinh \gamma_n x + \sin \gamma_n x)] \qquad (8.75)$$

where κ_n is the same as in eq (8.72). In this case, however, the system is unrestrained and we expect rigid-body modes occurring at zero frequency, i.e. when $\gamma = 0$. On physical grounds, we are considering only lateral deflections and hence we expect two such modes: a rigid translation perpendicular to the beam's axis and a rotation about its centre of mass. This is, in fact, the case. Substitution of $\gamma = 0$ in eq (8.57) leads to

$$u_0(x) = Ax^3 + Bx^2 + Cx + D \tag{8.76a}$$

where A, B, C are constants. Imposing the boundary conditions (8.73) to the solution (8.76a) yields

$$u_0(x) = Cx + D \tag{8.76b}$$

which is a linear combination of the two functions

$$u_0^{(1)} = 1$$
$$u_0^{(2)} = x \tag{8.77}$$

where we omitted the constants because they are irrelevant for our purposes. It is not difficult to interpret the functions (8.77) on a mathematical and on a physical basis: mathematically they are two eigenfunctions belonging to the eigenvalue zero and, physically, they represent the two rigid-body modes considered above.

We leave to the reader the case of a beam which is clamped at one end and simply supported at the other end. The frequency equation for this case is

$$\tan \gamma L = \tanh \gamma L \tag{8.78}$$

and its first roots are

$$\gamma_1 L = 3.927 \qquad \gamma_2 L = 7.069$$

$$\gamma_3 L = 10.210 \qquad \gamma_4 L = 13.352 \tag{8.79}$$

$$\gamma_5 L = 16.493 \qquad \gamma_6 L = 19.635$$

Also, note that we can approximate $\gamma_n L \cong (n + \frac{1}{4})\pi$.

Finally, one more point is worthy of notice. For almost all of the configurations above, the first frequencies are irregularly spaced; however, as the mode number increases, the difference between the two frequency parameters $\gamma_{n+1}L$ and $\gamma_n L$ approaches the value π for all cases. This result is general and indicates, for higher frequencies, an insensitivity to the boundary conditions.

8.5.1   *Axial force effects on bending vibrations*

Let us consider now a beam which is subjected to a constant tensile force T parallel to its axis. This model can represent, for example, either a stiff string or a prestressed beam.

On physical grounds we may expect that the model of the beam with no axial force should be recovered when the beam stiffness is the dominant restoring force and the string model should be recovered when tension is by far the dominant restoring force. This is, in fact, the case. The governing equation of motion for the free motion of the system that we are considering now is

$$EI \frac{\partial^4 y}{\partial x^4} - T \frac{\partial^2 y}{\partial x^2} + \mu \frac{\partial^2 y}{\partial t^2} = 0 \qquad (8.80)$$

Equation (8.80) can be obtained, for example, by writing the two equilibrium equations (vertical forces and moments) in the free-body diagram of Fig. 8.4 and noting that, from elementary beam theory, $M = EI(\partial^2 y / \partial x^2)$. Alternatively, we can write the Lagrangian density

$$\mathcal{L} = \frac{1}{2} \left[\mu \left(\frac{\partial y}{\partial t} \right)^2 - EI \left(\frac{\partial^2 y}{\partial x^2} \right)^2 - T \left(\frac{\partial y}{\partial x} \right)^2 \right]$$

and arrive at eq (8.80) by performing the appropriate derivatives prescribed in eq (3.109). The usual procedure of separation of variables leads to a solution with a harmonically varying temporal part and to the ordinary differential equation

$$\frac{d^4 u}{dx^4} - \frac{T}{EI} \frac{d^2 u}{dx^2} - \frac{\mu \omega^2}{EI} u = 0 \qquad (8.81)$$

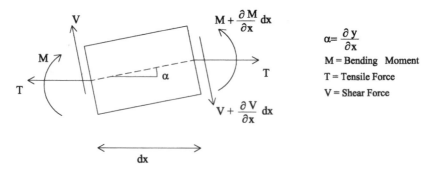

$$\alpha = \frac{\partial y}{\partial x}$$

M = Bending Moment
T = Tensile Force
V = Shear Force

Fig. 8.4 Beam element with tensile axial force (schematic free-body diagram).

where, as in the previous cases, we called $u(x)$ the spatial part of the solution. If now we let $u(x) = Ae^{\alpha x}$, eq (8.81) yields

$$\left.\begin{matrix} \alpha_1^2 \\ \alpha_2^2 \end{matrix}\right\} = \frac{T}{2EI} \pm \sqrt{\left(\frac{T}{2EI}\right)^2 + \frac{\mu\omega^2}{EI}}$$

where α_1^2 is positive and α_2^2 is negative. It follows that we have the four roots $\pm\eta$ and $\pm i\xi$ where we defined

$$\eta = \left[\sqrt{\left(\frac{T}{2EI}\right)^2 + \frac{\mu\omega^2}{EI}} + \frac{T}{2EI}\right]^{1/2}$$

$$\xi = \left[\sqrt{\left(\frac{T}{2EI}\right)^2 + \frac{\mu\omega^2}{EI}} - \frac{T}{2EI}\right]^{1/2}$$

(8.82)

The solution of eq (8.81) can then be written as

$$u(x) = C_1 \cosh(\eta x) + C_2 \sinh(\eta x) + C_3 \cos(\xi x) + C_4 \sin(\xi x) \tag{8.83}$$

which is formally similar to eq (8.58) but it must be noted that the hyperbolic functions and the trigonometric functions have different arguments. We can now consider different types of boundary conditions in order to determine how the axial force affects the natural frequencies. The simplest case is when both ends are simply supported; enforcing the boundary conditions (8.59) leads to $C_1 = C_3 = 0$ and to the frequency equation

$$\sinh(\eta L)\sin(\xi L) = 0 \tag{8.84}$$

which results in $\sin \xi L = 0$ because $\sinh \eta L \neq 0$ for any nonzero value of η. The allowed frequencies are obtained from $\xi_n L = n\pi$; this means

$$\omega_n = \frac{n^2\pi^2}{L^2}\sqrt{\frac{EI}{\mu} + \frac{TL^2}{n^2\pi^2\mu}} \tag{8.85a}$$

which can be rewritten as

$$\omega_n = \frac{n\pi}{L}\sqrt{\frac{T}{\mu}\left(1 + n^2\pi^2\frac{EI}{TL^2}\right)} \tag{8.85b}$$

where it is more evident that for small values of the nondimensional ratio $R \equiv EI/TL^2$ (i.e. when $EI \ll TL^2$) the tension is the most important restoring force and the beam behaves like a string. At the other extreme – when R is large – the stiffness is the most important restoring force and in the limit of $R \to \infty$ we return to the case of the beam with no axial force.

In addition to the observations above, note also that:

- In the intermediate range of values of R, higher modes are controlled by stiffness because of the n^2 factor under the square root in eq (8.85b).
- The eigenfunctions are given by (enforcement of the boundary conditions leads also to $C_2 = 0$)

$$u_n(x) = C_4 \sin(\xi x) \tag{8.86}$$

which have the same sinusoidal shape of the eigenfunctions of the beam with no tension (although here the sine function has a different argument).

The conclusion is that an axial force has little effect on the mode shapes but can significantly affect the natural frequencies of a beam by increasing their value in the case of a tensile force or by decreasing their value in the case of a compressive force. In fact, the effect of a compressive force is obtained by just reversing the sign of T. In this circumstance the natural frequencies can be conveniently written as

$$\omega_n = \frac{n^2\pi^2}{L^2}\sqrt{\frac{EI}{\mu}\left(1 - \frac{T}{n^2 T_{crit}}\right)} \tag{8.87}$$

where we recognize $T_{crit} = \pi^2 EI/L^2$ as the critical Euler buckling load. When $T = T_{crit}$ the lowest frequency goes to zero and we obtain transverse buckling.

In the case of other types of boundary conditions the calculations are, in general, more involved. For example, we can consider the clamped–clamped configuration and observe that placing the origin $x = 0$ halfway between the supports divides the eigenfunctions into even functions, which come from the combination

$$C_1 \cosh \eta x + C_3 \cos \xi x$$

and odd functions, which come from the combination

$$C_2 \sinh \eta x + C_4 \sin \xi x$$

In either case, if we fit the boundary conditions at $x = L/2$, they will also fit at $x = -L/2$. For the even functions the boundary conditions

$$u\left(\frac{L}{2}\right) = \frac{du}{dx}\bigg|_{x=L/2} = 0$$

lead to the equation

$$\tan\left(\frac{\xi L}{2}\right) = -\frac{\eta}{\xi} \tanh\left(\frac{\eta L}{2}\right) \tag{8.88a}$$

and for the odd functions we obtain

$$\frac{\eta}{\xi} \tan\left(\frac{\xi L}{2}\right) = \tanh\left(\frac{\eta L}{2}\right) \tag{8.88b}$$

Both eqs (8.88a and b) must be solved numerically: from eq (8.88a) we obtain the natural frequencies ω_1, ω_3, ω_5, ..., while the natural frequencies ω_2, ω_4, ω_6, ... are obtained from eq (8.88b).

8.5.2 The effects of shear deformation and rotatory inertia (Timoshenko beam)

It was stated in Section 8.5 that the Euler–Bernoulli theory of beams provides satisfactory results as long as the wavelength λ is large compared to the lateral dimensions of the beam which, in turn, may be identified by the radius of gyration r of the beam section. Two circumstances may arise when the above condition is no longer valid:

1. The beam is sufficiently slender (say, for example, $L/r \geqslant 10$) but we are interested in higher modes.
2. The beam is short and deep.

In both cases the kinematics of motion must take into account the effects of shear deformation and rotatory inertia which – from an energy point of view – result in the appearance of a supplementary term (due to shear deformation) in the potential energy expression and in a supplementary term (due to rotatory inertia) in the kinetic energy expression.

Let us consider the effect of shear deflection first. Shear forces τ result in an angular deflection θ which must be added to the deflection ψ due to bending alone. Hence, the slope of the elastic axis $\partial y/\partial x$ is now written as

$$\frac{\partial y}{\partial x} = \psi(x, t) + \theta(x, t) \tag{8.89}$$

and the relationship between bending moment and bending deflection (from elementary beam theory) now reads

$$M = EI \frac{\partial \psi}{\partial x} \tag{8.90}$$

Moreover, the shear force Q is related to the shear deformation θ by

$$Q = \kappa GA\theta \tag{8.91}$$

where G is the shear modulus, A is the cross-sectional area and κ is an adjustment coefficient (sometimes called the Timoshenko shear coefficient) whose value must generally be determined by stress analysis considerations and depends on the shape of the cross-section. In essence, this coefficient is introduced in order to satisfy the equivalence

$$\int_A \tau dA = \kappa GA\theta$$

and accounts for the fact that shear is not distributed uniformly across the section. For example, $\kappa \cong 0.833$ for a rectangular cross-section and other values can be found in Cowper [7].

In the light of these considerations, the potential energy density consists of two terms and is written as

$$\mathcal{V} = \frac{1}{2} EI \left(\frac{\partial \psi}{\partial x} \right)^2 + \frac{1}{2} \kappa GA\theta^2$$

$$= \frac{1}{2} EI \left(\frac{\partial y}{\partial x} \right)^2 + \frac{1}{2} \kappa GA \left(\frac{\partial y}{\partial x} - \psi \right)^2 \tag{8.92}$$

where in the last term we take into account eq (8.89).

On the other hand, the kinetic energy density must now incorporate a term that accounts for the fact that the beam rotates as well as bends. If we call J the beam mass moment of inertia density, the expression for the kinetic energy density is written as

$$\mathcal{T} = \frac{1}{2} \mu \left(\frac{\partial y}{\partial t} \right)^2 + \frac{1}{2} J \left(\frac{\partial \psi}{\partial t} \right)^2 \tag{8.93}$$

Moreover, J is related to the cross-section moment of inertia I by

$$J = \rho I = \frac{\mu}{A} I = \mu r^2 \tag{8.94}$$

where $\rho = \mu/A$ is the beam mass density and $r^2 = I/A$ is the radius of gyration of the cross-section. Taking eqs (8.92) and (8.93) into account, we are now in the position to write the explicit expression of the Lagrangian density

$$\mathcal{L} = \frac{1}{2}\,\mu\left[\left(\frac{\partial y}{\partial t}\right)^2 + r^2\left(\frac{\partial \psi}{\partial t}\right)^2\right]$$

$$-\frac{1}{2}\left[EI\left(\frac{\partial \psi}{\partial x}\right)^2 + \kappa GA\left(\frac{\partial y}{\partial x} - \psi\right)^2\right] \tag{8.95}$$

and perform the appropriate derivatives prescribed in eq (3.109) to arrive at the equation of motion. In this case, however, both y and ψ are independent variables and hence we obtain two equations of motion. As a function of y, the Lagrangian density is a function of the type $\mathcal{L} = \mathcal{L}(\dot{y}, y')$ where, following the notation of Chapter 3, the overdot indicates the derivative with respect to time and the prime indicates the derivative with respect to x.

So, we calculate the two terms

$$-\frac{\partial}{\partial t}\left(\frac{\partial \mathcal{L}}{\partial \dot{y}}\right) = -\mu\frac{\partial^2 y}{\partial t^2}$$

and

$$-\frac{\partial}{\partial x}\left(\frac{\partial \mathcal{L}}{\partial y'}\right) = \frac{\partial}{\partial x}\left[\kappa GA\left(\frac{\partial y}{\partial x} - \psi\right)\right]$$

to arrive at the first equation of motion

$$\frac{\partial}{\partial x}\left[\kappa GA\left(\frac{\partial y}{\partial x} - \psi\right)\right] - \mu\frac{\partial^2 y}{\partial t^2} = 0 \tag{8.96a}$$

and to the boundary conditions (eq (3.110))

$$\left[\kappa GA\left(\frac{\partial y}{\partial x} - \psi\right)\right]\delta y\,\Big|_0^L = 0 \tag{8.96b}$$

which take into account the possibility that either the term in brackets or δy can be zero at $x = 0$ and $x = L$.

A similar line of reasoning holds for the variable ψ; the Lagrangian function is of the type $\mathscr{L} = \mathscr{L}(\dot{\psi}, \psi', \psi)$ and the derivatives we must find are now given by

$$-\frac{\partial}{\partial t}\left(\frac{\partial \mathscr{L}}{\partial \dot{\psi}}\right) = -\mu r^2 \frac{\partial^2 \psi}{\partial t^2}$$

$$-\frac{\partial}{\partial x}\left(\frac{\partial \mathscr{L}}{\partial \psi'}\right) = \frac{\partial}{\partial x}\left(EI \frac{\partial \psi}{\partial x}\right)$$

$$\frac{\partial \mathscr{L}}{\partial \psi} = \kappa GA \left(\frac{\partial y}{\partial x} - \psi\right)$$

so that the second equation of motion is

$$\frac{\partial}{\partial x}\left(EI \frac{\partial \psi}{\partial x}\right) + \kappa GA \left(\frac{\partial y}{\partial x} - \psi\right) - \mu r^2 \frac{\partial^2 \psi}{\partial t^2} = 0 \qquad (8.97a)$$

with the boundary conditions (eq (3.110))

$$\left(EI \frac{\partial \psi}{\partial x}\right)\delta\psi \Bigg|_0^L = 0 \qquad (8.97b)$$

which take into account the possibility that either $EI(\partial \psi/\partial x)$ or $\delta\psi$ are zero at $x = 0$ and $x = L$.

Equations (8.96a) and (8.97a) govern the free vibration of a Timoshenko beam; we note that, in the above treatment, there are two 'modes of deformation' whose physical coupling translates mathematically into the coupling of the two equations.

If now, for simplicity, we assume that the beam properties are uniform throughout its length we can arrive at a single equation for the variable y. From eq (8.96) we get

$$\frac{\partial \psi}{\partial x} = \frac{\partial^2 y}{\partial x^2} - \frac{\mu}{\kappa GA} \frac{\partial^2 y}{\partial t^2} \qquad (8.98)$$

and by differentiating eq (8.97) we obtain

$$EI \frac{\partial^3 \psi}{\partial x^3} + \kappa GA \left(\frac{\partial^2 y}{\partial x^2} - \frac{\partial \psi}{\partial x}\right) - \mu r^2 \frac{\partial^3 \psi}{\partial x \partial t^2} = 0 \qquad (8.99)$$

Substitution of eq (8.98) into eq (8.99) yields the desired result, i.e.

$$EI \frac{\partial^4 y}{\partial x^4} + \mu \frac{\partial^2 y}{\partial t^2} - \left(\frac{EI\mu}{\kappa GA} + \mu r^2 \right) \frac{\partial^4 y}{\partial x^2 \partial t^2} + \frac{\mu^2 r^2}{\kappa GA} \frac{\partial^4 y}{\partial t^4} = 0 \qquad (8.100)$$

Now, a closer look at eq (8.100) shows that:

1. The term

$$-\frac{EI\mu}{\kappa GA} \left(\frac{\partial^4 y}{\partial x^2 \partial t^2} \right)$$

   arises from shear deformation and vanishes when the beam is very rigid in shear, i.e. when $GA \rightarrow \infty$.

2. The term

$$-\mu r^2 \frac{\partial^4 y}{\partial x^2 \partial t^2}$$

   is due to rotatory inertia and vanishes when $r \rightarrow 0$.

3. The term

$$\frac{\mu^2 r^2}{\kappa GA} \left(\frac{\partial^4 y}{\partial t^4} \right)$$

   results from a coupling between shear deformation and rotatory inertia. Note that this term vanishes when either $GA \rightarrow \infty$ or $r \rightarrow 0$.

4. When both shear deformation and rotatory inertia can be neglected we recover the Euler–Bernoulli case, which is represented by the first two terms.

With reference to the brief discussion on the velocity of wave propagation of Section 8.5, it may be interesting to note at this point that the unphysical result of infinite velocity as the wavelength $\lambda \rightarrow 0$ is removed by the introduction of the effects of shear deformation and rotatory inertia in the equations of motion. As a matter of fact, the introduction of rotatory inertia alone is sufficient to obtain finite velocities at any wavelength, but the results in the short-wavelength range are not in good agreement with the values of velocities calculated from the exact general elastic equations. A much better agreement is obtained by including the effect of shear deformation (e.g. Graff [8]).

From eq (8.100), if we assume a solution $u(x)e^{i\omega t}$ we arrive at the ordinary differential equation

$$\frac{d^4u}{dx^4} + \left(\frac{\mu\omega^2}{\kappa GA} + \frac{\mu r^2\omega^2}{EI}\right)\frac{d^2u}{dx^2} + \left(\frac{\mu^2 r^2\omega^4}{EI\kappa GA} - \frac{\mu\omega^2}{EI}\right)u = 0 \qquad (8.101)$$

which can be written in other forms (if the reader can find it convenient) by taking into account the relationships $\mu = \rho A$, $r^2 = I/A$, $E/G = 2(1+\nu)$ where ν is Poisson's ratio. From here, at the price of more cumbersome calculations, we can proceed as in the preceding cases to arrive at the values of the natural frequencies of the Timoshenko beam. We will not do so but we will investigate briefly the effects of shear deflection alone and of rotatory inertia alone on the eigenvalues of a pinned–pinned beam.

Case 1. Shear deflection alone

We neglect rotatory inertia in eq (8.101) and obtain

$$\frac{d^4u}{dx^4} + \frac{\mu\omega^2}{\kappa GA}\left(\frac{d^2u}{dx^2}\right) - \frac{\mu\omega^2}{EI}u = 0 \qquad (8.102)$$

The solution of eq (8.102) is formally analogous to the case of the beam with axial force: we obtain four roots $\pm\eta$ and $\pm i\xi$ where now we define

$$\eta = \left[\sqrt{\left(\frac{\mu\omega^2}{2\kappa GA}\right)^2 + \frac{\mu\omega^2}{EI}} - \frac{\mu\omega^2}{2\kappa GA}\right]^{1/2}$$

$$\xi = \left[\sqrt{\left(\frac{\mu\omega^2}{2\kappa GA}\right)^2 + \frac{\mu\omega^2}{EI}} + \frac{\mu\omega^2}{2\kappa GA}\right]^{1/2}$$

$$\qquad (8.103)$$

and the allowed frequencies are obtained from the condition $\xi_n x = n\pi$ as

$$\omega_n^{shear} = \omega_n^{(0)}\left[\frac{\kappa GAL^2}{\kappa GAL^2 + n^2\pi^2 EI}\right]^{1/2} \qquad (8.104a)$$

where we indicate with $\omega_n^{(0)}$ the nth eigenvalue of the pinned–pinned

Euler–Bernoulli beam. Equation (8.104a) can also be written as

$$\omega_n^{shear} = \omega_n^{(0)}\left[\frac{1}{1 + n^2\pi^2(E/\kappa G)(r/L)^2}\right]^{1/2} \tag{8.104b}$$

where the influence of the slenderness ratio is more evident: for small values of r/L and for low order modes we note that, as expected, $\omega_n^{shear}/\omega_n^{(0)} \rightarrow 1$.

Case 2. Rotatory inertia alone

We neglect now the effect of shear deformation in eq (8.101) and obtain the equation

$$\frac{d^4u}{dx^4} + \frac{\mu r^2\omega^2}{EI}\left(\frac{d^2u}{dx^2}\right) - \frac{\mu\omega^2}{EI}u = 0 \tag{8.105}$$

The procedure to be followed is exactly as in case 1 and it is not difficult to arrive at

$$\omega_n^{rot.} = \omega_n^{(0)}\left[\frac{1}{1 + n^2\pi^2(r/L)^2}\right]^{1/2} \tag{8.106}$$

where, again, $\omega_n^{(0)}$ refers to a pinned–pinned Euler–Bernoulli beam.

We note from eqs (8.104) and (8.106) that both shear deformation and rotatory inertia tend to decrease the beam eigenfrequencies and also that the shear deflection correction is more important that the rotatory inertia correction. These considerations are of general nature and retain their validity for types of boundary conditions other than the pinned–pinned configuration.

Finally, it may be worth pointing out that for hollow, thin-walled cross-sections the shear and rotatory effects tend to be more important because both the radius of gyration and the shear stresses are generally large.

8.6 A two-dimensional continuous system: the flexible membrane

The simplest two-dimensional continuous system is the flexible membrane which can be considered as the two-dimensional counterpart of the flexible string. As for strings, the assumption of flexibility implies that restoring forces in membranes arise from the in-plane tensile (or stretching) forces and that there is no resistance to bending and shear. In essence, it is the two-dimensional characteristics that distinguish this case from the case of the string.

In this regard, we note that now the tension at a point must be specified in terms of the pull across an elementary length of line drawn through the point:

this tension must equal the force tending to split the membrane along this line. This pull will be proportional to the length *ds* of the line element and the proportionality factor *T* is a force per unit length (in units of N/m) which may or may not be perpendicular to the line element. In other words, in the general case, *T* is a vector which is a function of the position and orientation of the line element. The common simplifying assumption is to consider membranes for which this tensile stress is the same at every point on the membrane and for every orientation of the line element: in this circumstance *T* is a constant and represents also the outward pull across each unit length of the membrane's boundary, i.e. its perimeter. With this in mind, we can picture a stretched membrane under the action of the in-plane stress *T* and consider a rectangular element of area *dxdy* (Figs 8.5(a), (b)).

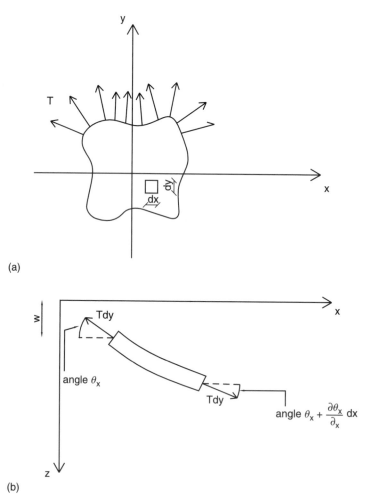

(a)

(b)

Fig. 8.5 Membrane area element: (a) plane view; (b) side view.

If we call $w = w(x, y, t)$ the coordinate used to measure the deflection of the membrane in the z direction it is easy to obtain from Fig 8.5(b) and from a similar figure in the $y - z$ plane

$$-Tdy\theta_x + Tdy\left(\theta_x + \frac{\partial\theta_x}{\partial x}\,dx\right) - Tdx\theta_y$$

$$+ Tdx\left(\theta_y + \frac{\partial\theta_y}{\partial y}\,dy\right) - \sigma dxdy\,\frac{\partial^2 w}{\partial t^2} = 0 \quad (8.107a)$$

where σ is the membrane mass per unit area and the angles θ_x and θ_y are given by $\partial w/\partial x$ and $\partial w/\partial y$ respectively. Note that small deflections have been assumed, so that $\sin\theta_x \cong \theta_x$, etc. and the area of the deflected element can still be written as $dxdy$. Equation (8.107a) results in the equation of motion

$$T\left(\frac{\partial^2 w}{\partial x^2} + \frac{\partial^2 w}{\partial y^2}\right) = \sigma\,\frac{\partial^2 w}{\partial t^2} \qquad (8.107b)$$

Obviously, we can arrive at eq (8.107b) by using Hamilton's principle and by considering that the Lagrangian density in this case is given by

$$\mathcal{L} \equiv \mathcal{T} - \mathcal{V} = \frac{1}{2}\sigma\left(\frac{\partial w}{\partial t}\right)^2 - \frac{1}{2}T\left[\left(\frac{\partial w}{\partial x}\right)^2 + \left(\frac{\partial w}{\partial y}\right)^2\right] \qquad (8.108)$$

We will not do it here, but we can follow an analogous line of reasoning as in Chapter 3 to arrive at the Euler–Lagrange equation for the membrane which reads (e.g. Morse and Ingard [1]; Meirovitch [9])

$$\frac{\partial}{\partial t}\left(\frac{\partial\mathcal{L}}{\partial w_t}\right) + \frac{\partial}{\partial x}\left(\frac{\partial\mathcal{L}}{\partial w_x}\right) + \frac{\partial}{\partial y}\left(\frac{\partial\mathcal{L}}{\partial w_y}\right) - \frac{\partial\mathcal{L}}{\partial w} = 0 \qquad (8.109)$$

where we write for simplicity $w_t = \partial w/\partial t$, $w_x = \partial w/\partial x$, etc.

Now, since the Laplacian operator ∇^2 in rectangular coordinates is expressed by

$$\nabla^2 = \frac{\partial^2}{\partial x^2} + \frac{\partial^2}{\partial y^2}$$

we can write eq (8.107b) as

$$\nabla^2 w = \frac{1}{c^2}\left(\frac{\partial^2 w}{\partial t^2}\right) \tag{8.110}$$

where $c = \sqrt{T/\sigma}$ has the dimension of a velocity. Equation (8.110) is the two-dimensional wave equation and has the advantage of being written in a form valid for all types of coordinates; the shape of the boundary (unless it is irregular) suggests which type of coordinates to use.

In order to obtain a solution of eq (8.110) we can once more adopt the method of separation of variables; we look for a solution in the form $w = ug$ where u is a function of the space variables alone and g is a function of time alone. We substitute this solution into eq (8.110), divide both members by ug and arrive at

$$\frac{\nabla^2 u}{u} = \frac{1}{c^2}\left(\frac{1}{g}\frac{d^2 g}{dt^2}\right)$$

where the left side is a function of the space variables only and the right side is a function of time only and therefore both members must be equal to a constant, which we call $-k^2$. The time equation becomes

$$\frac{d^2 g}{dt^2} + \omega^2 g = 0 \tag{8.111}$$

where $\omega = kc$ and the space equation is the so-called Helmholtz equation

$$\nabla^2 u + k^2 u = 0 \tag{8.112}$$

which assumes different explicit forms depending on the type of coordinates we decide to use. Note that $k = \omega/c$ is the familiar wavenumber. Equation (8.111) is well known and its solution is represented by a sinusoidal function of time; hence, we turn to eq (8.112).

As an example we will now consider the calculation of the natural frequencies and eigenfunctions of a circular membrane of radius R clamped at its outer edge. The geometry of the problem suggests the use of polar coordinates (r, θ) so that the Laplacian is written

$$\nabla^2 = \frac{1}{r}\frac{\partial}{\partial r}\left(r\frac{\partial}{\partial r}\right) + \frac{1}{r^2}\frac{\partial^2}{\partial\theta^2} = \frac{\partial^2}{\partial r^2} + \frac{1}{r}\frac{\partial}{\partial r} + \frac{1}{r^2}\frac{\partial^2}{\partial\theta^2}$$

and eq (8.112) reads explicitly

$$\frac{\partial^2 u}{\partial r^2} + \frac{1}{r}\frac{\partial u}{\partial r} + \frac{1}{r^2}\frac{\partial^2 u}{\partial\theta^2} + k^2 u = 0 \tag{8.113}$$

If now we assume a solution of the form $u(r, \theta) = f(r)\Theta(\theta)$ eq (8.113) leads to the two equations

$$\frac{d^2\Theta}{d\theta^2} = -\gamma^2\Theta$$

(8.114)

$$\frac{d^2f}{dr^2} + \frac{1}{r}\frac{df}{dr} + \left(k^2 - \frac{\gamma^2}{r^2}\right)f = 0$$

where we call γ^2 the separation constant. Now, by noting that the solution of the first equation is harmonic in θ, the continuity of membrane displacement requires that $u(r, \theta) = u(r, \theta + 2\pi)$. This periodicity condition can only be met if γ is an integer; hence the second of eq (8.114) becomes

$$\frac{d^2f}{dr^2} + \frac{1}{r}\frac{df}{dr} + \left(k^2 - \frac{n^2}{r^2}\right)f = 0$$

(8.115)

which is Bessel's equation of order n having the solution

$$f = A_1 J_n(kr) + A_2 Y_n(kr)$$

(8.116)

where J_n and Y_n are the Bessel functions of the first and second kind, respectively. The functions Y_n approach infinity as $r \to 0$ so, for a complete (i.e. without a hole in the centre) membrane, we must eliminate them and write $f = A_1 J_n(kr)$. If the membrane is clamped at its outer edge $r = R$, we must satisfy the boundary condition $f(R) = 0$, this means

$$J_n(kR) = 0$$

(8.117)

which is the frequency equation. The zeros of Bessel's function can be found in table form in various texts (e.g. Abramowitz and Stegun [10]) and some of the zeros are as follows:

$$J_0(x) = 0 \qquad x = 2.405;\ 5.520;\ 8.654;\ 11.792;\ ...$$

$$J_1(x) = 0 \qquad x = 3.832;\ 7.016;\ 10.173;\ 13.324;\ ...$$

at

$$J_2(x) = 0 \qquad x = 5.136;\ 8.417;\ 11.620;\ 14.796;\ ...$$

$$J_3(x) = 0 \qquad x = 6.380;\ 9.761;\ 13.015;\ 16.223;\ ...$$

It should be noted that $x = 0$ is also a root for all Bessel's functions of order $n \geq 1$, but this leads to trivial solutions for w and is therefore excluded.

The natural frequencies of our system can then be written as

$$\omega_{nm} = k_{nm}c \qquad (n = 0, 1, 2, ...; \quad m = 1, 2, 3, ...) \tag{8.118}$$

which means that for each value of n there exist a whole sequence of solutions labelled with the index m. For example, the frequencies $\omega_{01}, \omega_{02}, \omega_{03}, ...$ are the solutions of $J_0(kR) = 0$, the frequencies $\omega_{11}, \omega_{12}, \omega_{13}, ...$ are the solutions of $J_1(kR) = 0$ and so on; ω_{01} is the fundamental frequency and is given by

$$\omega_{01} = \frac{2.405}{R} c = \frac{2.405}{R} \sqrt{\frac{T}{\sigma}} \tag{8.119}$$

The overtones (which are not harmonics, i.e. are not integer multiples of the fundamental frequency) are, in increasing order

$$\omega_{11} = 1.593\omega_{01}$$

$$\omega_{21} = 2.136\omega_{01}$$

$$\omega_{02} = 2.295\omega_{01}$$

$$\omega_{31} = 2.653\omega_{01}$$

$$\vdots$$

The mode shapes are obtained by the product of the two spatial solutions: for the frequency ω_{0m} ($m = 1, 2, 3, ...$) we have the eigenfunction

$$u_{0m}(r) = A_{0m}J_0(k_{0m}r) \tag{8.120}$$

where A_{0m} is a constant and we note that u is a function of r alone. For each value of m, the corresponding mode has $(m - 1)$ nodal circles. A schematic representation for the first few modes can be given as in Fig 8.6.

For each one of the frequencies ω_{nm} ($n > 0$; $m = 1, 2, 3, ...$) we have the two eigenfunctions (A_{nm} and $\tilde{A}_{nm}$ are arbitrary constants)

$$u_{nm}(r, \theta) = A_{nm}J_n(k_{nm}r)\cos(n\theta)$$

$$\tilde{u}_{nm}(r, \theta) = \tilde{A}_{nm}J_n(k_{nm}r)\sin(n\theta) \tag{8.121}$$

which have the same shape and differ from one another only by an angular rotation of $90°$. This is an example of degeneracy (two or more eigenfunctions belonging to the same eigenvalue) which occurs frequently in two- and three-dimensional systems.

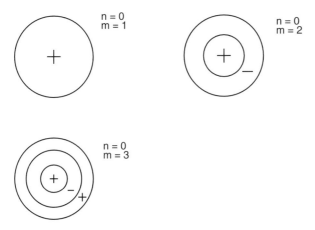

Fig. 8.6 First few modes for $n = 0$.

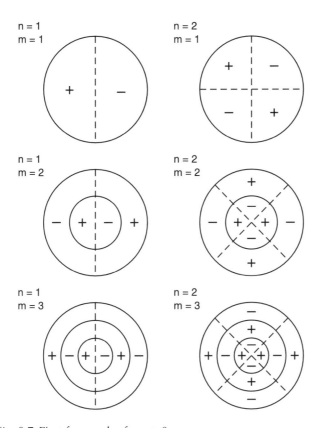

Fig. 8.7 First few modes for $n > 0$.

A schematic representation of a few of the functions (8.121) is shown in Fig 8.7 where we note that the (n, m)th mode has n nodal diameters and $(m - 1)$ circular nodes.

A final word for the curious reader. From the foregoing discussions it may seem that – provided that the mathematics is manageable – we can tackle any boundary value problem by the method of separation of variables which, in its own right, is a widely adopted method for finding solutions of boundary value problems for partial differential equations. However, strictly speaking, this is not so. In fact, although it is not our case, it is worth knowing that the same equation may allow a separation of variables in one system of coordinates but not in another; for example, the Helmholtz equation separates into ordinary differential equations in eleven different orthogonal coordinate systems [11], which, in fact, is sufficient to solve a large number of problems of physical significance.

8.7   The differential eigenvalue problem

So far in this chapter we limited ourselves to the discussion of some important continuous systems. These and many other distributed parameters systems, in spite of the significant differences in the order of the equations of motion and in the type of boundary conditions, can be analysed mathematically from a more general point of view which somehow broadens and extends the discussion and the formalism developed in the case of MDOF systems (Chapters 6 and 7).

In essence, the fundamental aspects of the analysis of n-DOF systems have their mathematical basis in the theory of finite-dimensional (with dimension n, depending on the number of degrees of freedom) vector spaces where the relevant quantities involved are vectors and matrices. In their own right, the finite-dimensional eigenvalue problems considered in Chapter 6 fit fully within this mathematical framework.

Matrices, in turn – besides being simply interpreted as arrays of numbers – can be seen as representations of linear operators in the vector space: a given operator is represented by different matrices when different bases (complete sets of linearly independent vectors) are chosen in the vector space and 'similar' matrices are just different basis representations of a single linear operator (recall that two matrices $\mathbf{A}$ and B are said to be similar when there exists an invertible matrix $\mathbf{S}$ such that $\mathbf{B} = \mathbf{S}^{-1}\mathbf{AS}$ and that similarity is an equivalence relation on the set of all $n \times n$ matrices).

Particularly important (Section 6.3.1) are Hermitian (or self-adjoint) matrices, i.e. matrices with complex entries for which $\mathbf{A} = \mathbf{A}^H$ or matrices with real entries for which $\mathbf{A} = \mathbf{A}^T$ (we speak of symmetrical matrices in this latter case). Among other desirable properties, recall that a Hermitian matrix has real eigenvalues and a complete set of orthonormal eigenvectors (Appendix A), where the concept of orthogonality implies that an inner product has been defined in the vector space.

With these considerations in mind, we can now go back to continuous systems and reconsider the cases of a finite flexible string and of the longitudinal and torsional vibrations of a bar of finite length (Sections 8.3 and 8.4). It is not difficult to show – for example, using Lagrange's equation (3.109) – that when the relevant physical parameters are not uniform throughout the length of the vibrating element, we arrive at the following equations of free motion:

$$\frac{\partial}{\partial x}\left(T_0\,\frac{\partial y}{\partial x}\right) - \mu\,\frac{\partial^2 y}{\partial t^2} = 0$$

$$\frac{\partial}{\partial x}\left(EA\,\frac{\partial y}{\partial x}\right) - \mu\,\frac{\partial^2 y}{\partial t^2} = 0 \qquad\qquad\qquad (8.122\text{a})$$

$$\frac{\partial}{\partial x}\left(GJ\,\frac{\partial \theta}{\partial x}\right) - \rho J\,\frac{\partial^2 \theta}{\partial t^2} = 0$$

which, upon separation of the time and space variables, lead to the ordinary differential equations for the space function $u = u(x)$:

$$-\frac{d}{dx}\left(T_0\,\frac{du}{dx}\right) = \lambda\mu u$$

$$-\frac{d}{dx}\left(EA\,\frac{du}{dx}\right) = \lambda\mu u \qquad\qquad\qquad (8.122\text{b})$$

$$-\frac{d}{dx}\left(GJ\,\frac{du}{dx}\right) = \lambda\rho J u$$

where, for reasons that will be clear shortly we call $-\lambda$ the separation constant. The formal structure of the above equations resembles the structure of an eigenvalue problem. In addition, it is worth recalling that what we did in Sections 8.3 and 8.4 was to look for the values of λ so that eqs (8.122b) admit nontrivial solutions which satisfy the appropriate boundary conditions dictated by the problem at hand.

If we note that in the case of finite-dimensional vector spaces we write the standard eigenvalue problem as $\mathbf{Au} = \lambda\mathbf{u}$, where A is an $n \times n$ matrix, we can leave aside mathematical rigour for the moment and observe that this concept can be generalized to the case of any linear operator, whether it be a matrix, a differential operator or an integral operator. The basic equation can again be written in the form

$$Au = \lambda u \tag{8.123}$$

where now u is whatever kind of object A can operate on: a column vector if A is a matrix, a (reasonably well-behaved) function if A is a differential operator and so forth. Furthermore, if we observe that the natural extension of finite-dimensional vector spaces with an inner product are the so-called Hilbert spaces (i.e. infinite-dimensional complete spaces with an inner product), we can introduce the concept of linear operators on a Hilbert space and, within this set, consider the fact that we often have to deal with Hermitian (symmetrical if we deal with spaces of real functions) operators which, in turn, have the following important properties:

1. Their eigenvalues are real.
2. Eigenfunctions belonging to different eigenvalues are orthogonal, where orthogonality means that their (appropriately defined) inner product is zero.

The point we are trying to make should be clear by now: despite the mathematical difficulties and intricacies – which are fundamental from a theoretical viewpoint – brought about by infinite dimensionality, for the cases of our interest there is a significant analogy with the finite-dimensional case. This analogy goes even further if we consider that, under conditions which are general enough for our purposes, there exists a counterpart of the expansion theorem stating that the eigenfunctions of a Hermitian operator form a complete set so that it is possible to expand any function belonging to an appropriate Hilbert space in a series of eigenfunctions.

Hence, with reference to eqs (8.122b), we note that the first member is just a linear differential operator acting on the function $u(x)$. In other words, when appropriate boundary condition have been specified, the above equations are particular cases of a differential eigenvalue problem. This differential eigenvalue problem is not precisely of the standard form (8.123) but is closer to the generalized problem where a 'mass operator' appears on the right-hand side. More specifically, eqs (8.122b) are just particular cases of the well-known Sturm–Liouville equation (Chapter 2), which is usually written in mathematics textbooks as

$$\frac{d}{dx}\left(p(x)\,\frac{du(x)}{dx}\right) - q(x)u(x) + \lambda\rho(x)u(x) = 0$$

where $p(x)$, $q(x)$ and $\rho(x)$ are real functions and $\rho(x)$, often called the density or weighting function, is assumed to be non-negative on the interval in question.

Now, the Sturm–Liouville problem is a second-order problem and, as such, deals with second-order systems. However, the foregoing general discussion goes further than this and covers a broad class of continuous systems, despite the fact that the mathematical difficulties of the calculations soon become formidable as the systems under investigation become more and more complex. In other words, although for a large number of complex systems no solutions in closed form are available, the above (and following) considerations retain their validity.

8.7.1 *The differential eigenvalue problem: some mathematical concepts*

This section is meant to serve the purposes of the present chapter and adapts to our present needs the mathematical concepts of Sections 2.5, 2.5.1 and 2.5.2. For this reason it will be kept short and results will be given without proof in the form of statements.

Since our interest lies mainly in one and two-dimensional continuous systems, let Ω be a two-dimensional finite domain with a smooth boundary S and let K and M be two differential operators of orders $2p$ and $2q$, respectively, where we assume $p > q$. Pursuing the analogy with MDOF systems, we call K and M the stiffness and mass operators, respectively.

The general differential eigenvalue problem is written symbolically as

$$Ku = \lambda Mu \tag{8.124}$$

and is supplemented by a set of p boundary conditions which we assume of the form

$$B_i u = 0 \qquad (i = 1, 2, ..., p) \tag{8.125}$$

where B_i are linear differential operators (boundary operators) of maximum order $2p - 1$. Equations (8.124) and (8.125) together define a homogeneous differential eigenvalue problem whose solution consists of determining the values of λ (eigenvalues) and the nontrivial functions (eigenfunctions) for which the differential equation (8.124) and the boundary conditions (8.125) are satisfied.

Now, let f and g be two real-valued square-integrable (hence finite-energy) functions in the domain Ω which, in addition, satisfy all the boundary conditions (8.125) (in mathematical terminology f, and g belong to an appropriate subspace of $L^2(\Omega)$). In other words, let f and g be two comparison functions. The integral

$$\int_\Omega fg\, d\Omega \equiv \langle f \mid g \rangle \tag{8.126}$$

defines an inner product in $L^2(\Omega)$ and the two functions are said to be orthogonal if $\langle f \mid g \rangle = 0$.

More generally, if A is a linear operator defined on a subspace of a Hilbert space, where we know from Chapter 2 that A is called symmetrical if

$$\langle f \mid Ag \rangle = \langle g \mid Af \rangle \tag{8.127a}$$

or, alternatively

$$\langle f \mid Ag \rangle = \langle Af \mid g \rangle \tag{8.127b}$$

and positive definite if $\langle f \mid Af \rangle > 0$ for every $f \neq 0$. If, on the other hand, the inner product $\langle f \mid Af \rangle$ can be zero for some $f \neq 0$, A is said to be positive semidefinite. Moreover, two functions f and g are said to be A-orthogonal if $\langle f \mid Ag \rangle = 0$.

Now, returning to the eigenvalue problem defined by eqs (8.124) and (8.125), we give without proof the following results:

1. The problem admits a countable infinite set of eigenvalues λ_i $(i = 1, 2, 3, ...)$.
2. The eigenvalues of a symmetrical eigenvalue problem (i.e. K and M symmetrical) are real and the eigenfunctions of a symmetrical eigenvalue problem can be chosen to be real.
3. If a symmetrical eigenvalue problem is positive definite (i.e. K and M symmetrical and positive definite) the eigenvalues are such that $\lambda_i > 0$ for every index i. If the problem is only positive semidefinite (i.e. K is positive semidefinite and M is positive definite) then $\lambda_i \geqslant 0$.
4. For a symmetrical eigenvalue problem, any two eigenfunctions u_i and u_j corresponding to different eigenvalues λ_i and λ_j $(i \neq j)$ are both K- and M-orthogonal, i.e. $\langle u_i \mid Ku_j \rangle = \langle u_i \mid Mu_j \rangle = 0$. This statement deserves an additional comment: in the case of m-fold degeneracy (and it can be proven that m is always finite) of an eigenvalue, say λ_k, there are m eigenfunctions associated with it. These functions, in general, are not mutually orthogonal, although they are orthogonal to the eigenfuctions belonging to λ_i with $i \neq k$. However, since it is always possible (by means of the Schmidt orthogonalization procedure) to construct a set of m mutually orthogonal eigenfunctions of λ_k by appropriate linear combinations of the original m eigenfunctions, all the eigenfunctions of a symmetrical problem can be considered as mutually orthogonal. In addition, their amplitude – which is arbitrary to within a multiplying constant because of the homogeneous nature of the problem – can be arranged in such a way as to make them mutually orthonormal, i.e. such that

$$\langle \phi_i \mid K\phi_j \rangle = \lambda_j \delta_{ij}$$

$$\langle \phi_i \mid M\phi_j \rangle = \delta_{ij} \tag{8.128}$$

where we indicate with the symbol ϕ_i the mass-orthonormal functions to distinguish them from the u_i in which the arbitrary amplitude constant is left undetermined.

5. The eigenfunctions u_i ($i = 1, 2, 3, ...$) of a symmetrical eigenvalue problem form a complete set, meaning that every function $f \in L^2(\Omega)$ satisfying the boundary conditions of the problem can be expanded in a series of eigenfunctions in the form

$$f = \sum_{j=1}^{\infty} c_j \phi_j \qquad (8.129)$$

where the c_i are the Fourier coefficients given by

$$
\begin{aligned}
c_i &= \langle \phi_i \mid Mf \rangle \\
\lambda_i c_i &= \langle \phi_i \mid Kf \rangle
\end{aligned}
\qquad (8.130)
$$

Note that eqs (8.130) are obtained by first making the operator K (or M for the first of eqs (8.130)) act on both members of eq (8.129) and then by taking the inner product with ϕ_i. The orthonormality conditions (8.128) lead immediately to eqs (8.130).

Given the results above, we can now return to Chapter 6 and consider the analysis of the free vibrations of undamped MDOF systems. We may note that it is sufficient to adopt the 'bracket notation' $\langle x \mid y \rangle$ to denote the inner product $x^T y$ between the two $n \times 1$ vectors x and y in order to have – for our purposes – a noteworthy analogy with the case of undamped continuous systems. The property of symmetry of the differential operators K and M parallels the symmetry of the matrices K and M, the othogononality conditions (6.44) and (6.42b) are the counterpart of eqs (8.128) and the expansion theorem is now an expansion in a series of eigenfunctions which reads explicitly

$$f = \sum_{j=1}^{\infty} \langle \phi_j \mid Mf \rangle \phi_j$$

and corresponds to the finite sum of eq (6.48b) where the p_j are the mass-orthonormal eigenvectors which satisfy the algebraic eigenvalue problem $Kp_j = \lambda_j Mp_j$.

We conclude this section by pointing out that the utility of the above discussion is threefold:

1. The class of symmetrical systems is quite large and includes all the continuous conservative systems considered in this chapter.
2. Regardless of the fact that a closed-form solution may not be available for a given system, all the above considerations hold true when it can be shown that the system is symmetrical.

3. We attain a significant economy of thought by considering that all the equations that describe the free vibration of a large number of conservative continuous systems are formally identical to the equations valid for MDOF systems. Moreover, for the former systems, the convergence of the series involved is mathematically guaranteed in all cases of practical interest.

8.7.2   *The differential eigenvalue problem: some examples and further considerations*

In the light of the formalism developed in the preceding section, we can now re-examine some of the systems encountered before. For one-dimensional systems (beams included) extending in the domain from $x = 0$ to $x = L$ the operators K, M and B_i can be expressed in the general forms

$$K = \sum_{j=0}^{p} \frac{d^j}{dx^j} \left[f_j(x) \frac{d^j}{dx^j} \right]$$

$$M = \sum_{j=0}^{q} \frac{d^j}{dx^j} \left[g_j(x) \frac{d^j}{dx^j} \right] \qquad (8.131)$$

$$B_i = \sum_{j=0}^{2p-1} \left[a_{ij} \left. \frac{d^j}{dx^j} \right|_{x=0} + b_{ij} \left. \frac{d^j}{dx^j} \right|_{x=L} \right]$$

where the $f_j(x)$ and $g_j(x)$ are continuous functions with continuous derivatives up to the order p and q, respectively, and the boundary conditions index i runs from $i = 0$ to $i = 2p$. More specifically, for the free vibrations of a fixed-fixed string with homogeneous properties (Section 8.3) we have $p = 1$, $q = 0$, $\lambda = \omega^2$ and

$$K = -T_0 \frac{d^2}{dx^2} \qquad (8.132)$$

$$M = \mu$$

Moreover, in the two boundary operators we have $a_{10} = b_{20} = 1$, $b_{10} = a_{11} = b_{11} = a_{20} = a_{21} = b_{21} = 0$ and the zeroth-order derivative is defined in such a way that $d^0 u(x)/dx^0 = u(x)$. Note that in eq (8.131) we expressed the boundary conditions in terms of $2p$ operators and not in terms of p operators as prescribed in (8.125). This is just a matter of our present convenience and it is immediate to see that in terms of (8.125) the boundary conditions are expressed as $B_1 = 1 = d^0/dx^0$ at $x = 0$ and $x = L$.

The operator M is clearly symmetrical because for any two comparison functions $f, g \neq 0$ we have

$$\langle f \mid Mg \rangle = \int_0^L f \mu g \, dx = \int_0^L g \mu f \, dx = \langle g \mid Mf \rangle \tag{8.133}$$

For the operator K we can write

$$\langle f \mid Kg \rangle = - \int_0^L f \left(T_0 \frac{d^2 g}{dx^2} \right) dx$$

$$= -f T_0 \frac{dg}{dx} \bigg|_0^L + \int_0^L T_0 \frac{df}{dx} \frac{dg}{dx} dx = \int_0^L T_0 \frac{df}{dx} \frac{dg}{dx} dx \tag{8.134a}$$

where we have integrated by parts and took into account the boundary conditions. The same procedure leads also to

$$\langle g \mid Kf \rangle = - \int_0^L g \left(T_0 \frac{d^2 f}{dx^2} \right) dx = \int_0^L T_0 \frac{df}{dx} \frac{dg}{dx} dx \tag{8.134b}$$

so that $\langle f \mid Kg \rangle = \langle g \mid Kf \rangle$ and the operator K is symmetrical. Also, we determined in Section 8.3 that the eigenfunctions of a fixed–fixed string are given by $u_n(x) = A_n \sin(n\pi x/L)$; since

$$\int_0^L \sin\left(\frac{n\pi x}{L} \right) \sin\left(\frac{m\pi x}{L} \right) dx = \frac{L}{2} \delta_{nm}$$

mass orthogonality is straightforward, i.e., $\langle u_n \mid Mu_m \rangle = 0$ for $n \neq m$, stiffness othogonality follows because $Ku_n = \lambda_n \mu u_n$ and the mass-orthonormal functions $\phi_n(x)$ are given by

$$\phi_n(x) = \sqrt{\frac{2}{\mu L}} \sin\left(\frac{n\pi x}{L} \right) \tag{8.135}$$

so that $\langle \phi_n \mid M\phi_n \rangle = 1$ and $\langle \phi_n \mid K\phi_n \rangle = (n\pi/L)^2 T_0/\mu = \omega_n^2$. Finally, if we let $f = g \neq 0$ in eqs (8.133) and (8.134a) it follows that $\langle f \mid Mf \rangle > 0$ and

$$\langle f \mid Kf \rangle = \int_0^L T_0 \left(\frac{df}{dx} \right)^2 dx > 0$$

This inequality holds because $f = const \neq 0$ is not a solution of the eigenvalue problem. Hence both M and K are positive definite.

On the other hand, for a uniform cantilevered Euler–Bernoulli beam (Section 8.5) we have $p = 2$, $q = 0$, $\lambda = \omega^2$; the stiffness and mass operators are

$$K = EI \, \frac{d^4}{dx^4}$$

(8.136)

$$M = \mu = \rho A$$

and for the boundary conditions we have now $a_{10} = a_{21} = b_{32} = b_{43} = 1$, with all other coefficients being zero. The property of symmetry of the operator K can be proven by a double integration by parts which takes into account the boundary conditions (8.63a) and (8.63b), M is trivially symmetrical and we leave it to the reader to show that K is positive definite. The proof of orthogonality and the determination of the normalizing constants involve now some cumbersome calculations which we do not pursue here.

By contrast, we can check the definite positiveness of the operator K in the two cases:

1. clamped-clamped beam
2. free-free beam.

In both cases, after a double integration by parts and due consideration to the boundary conditions, we get

$$\langle f \mid Kf \rangle = \int_0^L EI \left(\frac{d^2 f}{dx^2} \right)^2 dx$$

which – if $f \neq 0$ – can only be zero if a function f of the form $f = Ax + B$ is a solution of the differential eigenvalue problem. It is easy to verify that this is not possible for case 1 but is for case 2, meaning that, as expected, K is positive definite in case 1 and only positive semidefinite in case 2, where $\lambda = 0$ is a twofold degenerate eigenvalue whose eigenfunctions represent rigid-body motions.

We conclude this analysis on one-dimensional systems with a few additional considerations of particular interest. First, it is interesting to note that differential eigenvalue problems of the type (8.131) arise also in a number of problems related to elastic stability of one-dimensional members, where now the eigenvalue λ generally represents a critical load parameter beyond which the member's undeformed position of static equilibrium becomes unstable and leads to buckling. The eigenfunction then represents the buckled shape. Consider, for example, a clamped–pinned column centrally loaded by a compressive force F; the eigenvalue problem can be written as $EIu'''' = -Fu''$ (where the primes indicate space derivatives) and the boundary conditions read $u(0) = u''(0) = u(L) = u'(L) = 0$. In this circumstance (e.g. Bazant and Cedolin [12]) the lowest eigenvalue is the smallest critical load $\lambda_1 = F_{cr} \cong (\pi/0.7L)^2 EI$ and its eigenfunction u_1 represents the shape of the buckled column. Then, for $\lambda \gg \lambda_1$ the linear theory predicts a diverging amplitude but linearity soon ceases to be valid and a nonlinear description must be adopted for the postbuckling behaviour.

The second point we want to make concerns the perturbative method described in Section 6.4. If, for instance, we consider the first-order perturbation of the eigenvalues (eq (6.81)), we can write it in 'bracket notation' as

$$\lambda_n^{(1)} = \langle \mathbf{p}_n^{(0)} \mid (\mathbf{K}_1 - \lambda_n^{(0)} \mathbf{M}_1) \mathbf{p}_n^{(0)} \rangle$$

By analogy, for a differential eigenvalue problem this equation becomes

$$\lambda_n^{(1)} = \langle \phi_n^{(0)} \mid (\mathbf{K}_1 - \lambda_n^{(0)} \mathbf{M}_1) \phi_n^{(0)} \rangle \tag{8.137}$$

and fully retains its validity. As an example we can investigate the first-order effect of shear deformation on the eigenvalues of a pinned–pinned beam. The eigenvalue problem for this case is obtained from eq (8.102) and reads (all symbols have been introduced in Section 8.5)

$$EI \frac{d^4 u}{dx^4} = \lambda \left(\mu - \frac{E\mu r^2}{\kappa G} \frac{d^2}{dx^2} \right) u$$

where now we write λ in place of ω^2. The perturbative term is expressed by the operator

$$M_1 = - \left(\frac{E\mu r^2}{\kappa G} \right) \frac{d^2}{dx^2} \tag{8.138}$$

which is just the additional term representing shear in the equation of motion. Equation (8.62), in turn, yields the zeroth-order mass orthonormal eigenfunctions

$$\phi_n(x) = \sqrt{\frac{2}{\mu L}} \sin\left(\frac{n\pi x}{L} \right) \tag{8.139}$$

and since in the present case $K_1 = 0$, we can insert eqs (8.138) and (8.139) into eq (8.137) to get

$$\lambda_n^{(1)} = -\lambda_n^{(0)} \langle \phi_n^{(0)} \mid M_1 \phi_n^{(0)} \rangle$$

$$= -\lambda_n^{(0)} \frac{2}{\mu L} \frac{E\mu r^2}{\kappa G} \left(\frac{n\pi}{L} \right)^2 \int_0^L \sin^2\left(\frac{n\pi x}{L} \right) dx$$

Then, owing to eq (8.36), we obtain

$$\lambda_n^{(1)} = -\lambda_n^{(0)} n^2 \pi^2 \frac{E}{\kappa G} \left(\frac{r}{L}\right)^2$$

and finally, since the first-order perturbate eigenvalues are given by $\lambda_n = \lambda_n^{(0)} + \lambda_n^{(1)}$, our result is given by

$$\lambda_n = \lambda_n^{(0)} \left[1 - n^2 \pi^2 \frac{E}{\kappa G} \left(\frac{r}{L}\right)^2\right] \tag{8.140}$$

which is in perfect agreement with eq (8.104b). In fact, eq (8.140) is just the first-order series expansion of eq (8.104b) and is valid for small values of the ratio r/L and for low-order modes: precisely the hypotheses under which the shear effect can be considered as a small perturbation.

If now we turn our attention to a two-dimensional system, we can consider the clamped circular membrane of Section 8.6. The differential eigenvalue problem (8.112) can be written as

$$-T\nabla^2 u = \lambda \sigma u \tag{8.141}$$

which must be verified in a circular domain of radius R. We call this domain Ω and we call its boundary S. In order to prove the symmetry of the operator $K = -T\nabla^2$ we can resort to vector analysis and remember that $\nabla^2 = \text{div}(\text{grad})$ or, equivalently, using the operator 'del', $\nabla^2 = \nabla \cdot \nabla$. Now, since $\nabla \cdot (f \nabla g) = f \nabla^2 g + \nabla f \cdot \nabla g$ we have

$$\langle f \mid Kg \rangle = -\int_\Omega f(T\nabla^2 g) d\Omega$$

$$= \int_\Omega T(\nabla f \cdot \nabla g) d\Omega - \int_\Omega T\nabla \cdot (f \nabla g) d\Omega$$

Now we invoke the divergence theorem in the last integral, i.e.

$$\int_\Omega T\nabla \cdot (f \nabla g) d\Omega = \int_S Tf(\nabla g) \cdot \mathbf{n} dS = \int_S Tf \frac{dg}{dn} dS$$

where $\mathbf{n}$ is a unit vector perpendicular to S so that $\nabla g \cdot \mathbf{n} = dg/dn$ and dg/dn indicates the directional derivative of g along the normal to the boundary.

Substituting back into the expression of $\langle f \mid Kg \rangle$ we get

$$-\int_\Omega f(T\nabla^2 g)d\Omega = \int_\Omega T(\nabla f \cdot \nabla g)d\Omega - \int_S Tf \frac{dg}{dn} dS \tag{8.142}$$

and then, by exchanging f and g we arrive at a similar equation which, in turn, can be subtracted from eq (8.142) to give

$$\langle f \mid Kg \rangle - \langle g|Kf \rangle = -\int_S \left(f \frac{dg}{dn} - g \frac{df}{dn} \right) dS \tag{8.143}$$

Since the right-hand side of eq (8.143) is zero because the functions f and g obey the boundary condition of the problem it follows that the operator K is symmetrical. Also, we point out – without pursuing this point further here – that the right-hand side of eq (8.143) can be zero under more general boundary conditions (involving the directional derivatives of the comparison functions) than the present case.

If now we turn our attention to the eigenfunctions of the clamped circular membrane of radius R, we know in principle – owing to the symmetry of M and K – that they must be mutually orthogonal. This is indeed so and can be verified by considering the properties of Bessel's functions[1] of the first kind: one of these properties, provided that $J_n(kR) = 0$, can be written as

$$\int_0^R J_n(k_{nl}r)J_m(k_{ms}r)r dr = \delta_{nm}\delta_{ls} \frac{R^2}{2} [J_n'(k_{nl}R)]^2 \tag{8.144}$$

It is easy to see that eq (8.144) is immediately related to the property of mass orthogonality, and hence to stiffness orthogonality; moreover, when $n = m$ and $l = s$, eq (8.144) determines the constants that make the eigenfunctions mutually orthonormal.

8.8   Bending vibrations of thin plates

In many ways, a fundamental reference on the subject is Leissa [15]. This work – besides being a complete summary of all known results up to 1966 – contains a comprehensive set of results for the frequencies and mode shapes of free vibration of plates both according to the so-called 'classical theory' and with further complications such as anisotropy, variable thickness, in plane forces etc. For our part, since it is beyond the scope of this book to go into the

---

1 Bessel's functions have been extensively studied and tabulated (e.g. Abramowitz and Stegun [10], Jahnke *et al.* [13] and the tables published by Harvard University Press [14]).

details of this rich and mathematically delicate subject, we will limit ourselves to a discussion of some of its fundamental aspects within the framework of the classical theory.

In essence, plates are the two-dimensional counterpart of beams, much in the same way as membranes are the two-dimensional counterpart of strings. In other words, plates do have bending stiffness and the additional complications arise not only from the increased complexity of two-dimensional wave motion but also from the complex stresses that are set up when a plate is bent. In fact, when a plate element is bent, the material inside the bend becomes compressed and tends to expand laterally while the material outside the bend is stretched and tends to contract laterally, so that bending in one direction necessarily involves bending in a direction at right angles to it. It is well known that the ratio of the lateral extension (contraction) to compression (tension) is Poisson's ratio ν, which is approximately equal to 0.2–0.3 for most materials.

This 'sideways effect' was ignored in the case of beams because a beam element, in comparison with its length, is generally assumed to be thin enough as to make lateral bending negligible for most practical purposes.

Now, if we assume that our undisturbed plate lies in the $x - y$ plane, the basic assumptions of the classical theory of plates vibrations can be summarized as follows:

1. Only the transverse displacement $w = w(x, y, t)$ is considered.
2   The plate is thin, i.e., its thickness h is small compared to its lateral dimensions (say, $h \leqslant 0.1a$, where a is the smallest in plane dimension of the plate).
3. The stress in the transverse direction σ_z is zero. More specifically, since σ_z must vanish on the external layers at $z = \pm h/2$ and h is small, then σ_z is assumed to be zero for all values of z.
4. During bending, plane cross-sections remain plane and perpendicular to the midplane (just as in the Euler–Bernoulli beam development).
5. Only small deflections and slopes are considered (a maximum deflection of one-fifth of the thickness is generally considered the limit for small-deflection theory).

Given the above assumptions, it can be shown [see for example (e.g. Graff [8] or, for a variational approach, Meirovitch [9] and Gerardin and Rixen [16]) that the equation of motion for the free vibrations of a plate is given by

$$D\nabla^4 w + \rho h \frac{\partial^2 w}{\partial t^2} = 0 \tag{8.145}$$

where $\nabla^4 \equiv \nabla^2 \nabla^2$, the Laplacian of the Laplacian, is called the biharmonic operator, ρ is the mass density of the material (hence $\sigma \equiv \rho h$ is the plate mass density per unit area),

$$D = \frac{Eh^3}{12(1 - \nu^2)} \tag{8.146}$$

is the plate bending stiffness and E is the modulus of elasticity of the material.

Now, our interest is mainly in eigenvalues and eigenfunctions of finite plates, and we do not pursue the subject of waves propagation in infinite plates. However, in this regard it is worth noting that the plate – like the beam – is a dispersive medium, meaning that waves of different wavelength travel with different velocities. By contrast, in the case of finite plates, we note that eq (8.145) must be supplemented by appropriate boundary conditions in order to define a complete eigenvalue problem. The simplest types of boundary conditions are: simply supported (pinned), clamped and free. Without reference to any particular set of coordinates (rectangular, polar, etc.), we denote by s and n the coordinates in the tangential and normal directions to the contour and write the boundary conditions as:

- Simply supported edge,

$$w = M_n = 0 \tag{8.147a}$$

  where M_n is the normal bending moment per unit length.
- Clamped edge,

$$w = \frac{\partial w}{\partial n} = 0 \tag{8.147b}$$

- Free edge,

$$M_n = Q_n - \frac{\partial M_{ns}}{\partial s} = 0 \tag{8.147c}$$

  where Q_n is the shearing force per unit length and M_{ns} is the twisting moment per unit length about the direction n.

Note that the second boundary condition of eq (8.147c), i.e. $Q_n - (\partial M_{ns}/\partial s) = 0$, is by no means self-evident. In fact, at first glance, it would seem that the three stresses Q_n, M_n and M_{ns} should be independently equal to zero at a free edge; however, for a fourth-order equation, only two boundary conditions are required along each edge. It was left to Kirchhoff to show that Q_n and M_{ns} combine into a single edge condition as given above (e.g. Timoshenko and Woinowsky-Krieger [17] or Mansfield [18]).

For the above boundary conditions it can be shown with the aid of vector analysis that the biharmonic operator is symmetrical, so that the free vibration of plates also fits into the framework of symmetrical eigenvalue problems.

8.8.1   Circular plates

To be more specific, let consider the case of a uniform circular plate clamped at its outer edge $r = R$. As for the circular membrane of Section 8.6, the nature of the problem suggests the use of polar coordinates; assuming a harmonic time dependence we write the solution in the form $w(r, \theta, t) = u(r, \theta)e^{i\omega t}$ and obtain for the space part of the solution

$$\nabla^4 u - \gamma^4 u = 0 \tag{8.148a}$$

where $\gamma^4 = \omega^2 \sigma / D$. Equation (8.148a) can be rewritten in the form

$$(\nabla^2 + \gamma^2)(\nabla^2 - \gamma^2)u = 0 \tag{8.148b}$$

whose solution can be written as $u = u_1 + u_2$, u_1 being the solution of $(\nabla^2 + \gamma^2)u_1 = 0$ and u_2 being the solution of $(\nabla^2 - \gamma^2)u_2 = 0$. The equation for u_1 is formally equal to eq (8.112) for the circular membrane (note, however, that now the constant γ does not have the dimension of a wavenumber), so that we can separate the variables, write $u_1(r, \theta) = f_1(r)\Theta_1(\theta)$ and arrive at the solution (Section 8.6)

$$u_1(r, \theta) = [AJ_n(\gamma r) + BY_n(\gamma r)] \begin{Bmatrix} \sin(n\theta) \\ \cos(n\theta) \end{Bmatrix} \tag{8.149a}$$

were n is an integer because – as for the circular membrane – continuity considerations require that $u_1(r, \theta) = u_1(r, \theta + 2\pi)$.

The function u_2 can be explicitly obtained just by noting that its equation can be rewritten as $[\nabla^2 + (i\gamma)^2]u_2 = 0$; we separate the space variables and arrive at the so-called modified Bessel equation for the function $f_2(r)$ so that we obtain

$$u_2(r, \theta) = [CI_n(\gamma r) + DK_n(\gamma r)] \begin{Bmatrix} \sin(n\theta) \\ \cos(n\theta) \end{Bmatrix} \tag{8.149b}$$

where I_n and K_n are the modified Bessel functions of the first and second kind. They are related to J_n and Y_n by $i^n I_n(\gamma r) = J_n(i\gamma r)$ and $K_n(\gamma r) = Y_n(i\gamma r)$.

Putting eqs (8.149a) and (8.149b) together we have

$$u(r, \theta) = [AJ_n(\gamma r) + CI_n(\gamma r)] \begin{Bmatrix} \sin(n\theta) \\ \cos(n\theta) \end{Bmatrix} \qquad (8.150)$$

where the Bessel functions of the second kind Y_n and K_n have been eliminated, since they have singularities at $r = 0$. Note that for circular plates with a circular concentric hole of radius $r = a < R$ these functions must be retained in the solution because their singular behaviour at the origin no longer plays a role; in this case, however, two additional boundary conditions must be specified at $r = a$.

The boundary conditions for our case (plate clamped at its outer edge) read

$$u(R, \theta) = \frac{\partial u(r, \theta)}{\partial r}\bigg|_{r=R} = 0 \qquad (8.151)$$

so that we obtain the frequency equation

$$J_n(\gamma R)I'_n(\gamma R) - J'_n(\gamma R)I_n(\gamma R) = 0 \qquad (8.152a)$$

which, owing to the recursion relationships obeyed by Bessel's functions, can be equivalently written as

$$J_n(\gamma R)I_{n+1}(\gamma R) + I_n(\gamma R)J_{n+1}(\gamma R) = 0 \qquad (8.152b)$$

Equation (8.152a or b) must be solved numerically: for each value of n there are an infinite number of roots which we can identify with an index m, where $m = 1, 2, 3, \ldots$. If we now define the frequency parameter $\lambda_{nm} \equiv \gamma_{nm}R$ the natural frequencies can be written as

$$\omega_{nm} = \frac{\lambda^2_{nm}}{R^2}\sqrt{\frac{D}{\sigma}} \qquad (8.153)$$

where, for a given pair n, m there are two eigenmodes (except for $n = 0$) so that all modes with $n \neq 0$ are twofold degenerate. Furthermore, as in the case of the clamped circular membrane, we have n nodal diameters and $m - 1$ nodal circles. The first few values of λ^2_{nm} are as follows:

$$\lambda^2_{01} = 10.216 \qquad \lambda^2_{02} = 39.771 \qquad \lambda^2_{03} = 89.104$$

$$\lambda^2_{11} = 21.26 \qquad \lambda^2_{12} = 60.82 \qquad \lambda^2_{13} = 120.08$$

$$\lambda^2_{21} = 34.88 \qquad \lambda^2_{22} = 84.58 \qquad \lambda^2_{23} = 153.81$$

The eigenfuncions are written

$$u_{nm}(r, \theta) = A \left[J_n(\gamma_{nm} r) - \frac{J_n(\gamma_{nm} R)}{I_n(\gamma_{nm} R)} I_n(\gamma_{nm} r) \right] \left\{ \begin{array}{c} \sin(n\theta) \\ \cos(n\theta) \end{array} \right\} \qquad (8.154)$$

where the constant A – which, *a priori*, can depend on both n and m – can be fixed by means of normalization.

Different boundary conditions lead to more complicated calculations: for example, if our plate is simply supported at $r = R$ the boundary conditions to be imposed on the solution (8.150) are, from eq (8.147a)

$$w = M_r = 0$$

at $r = R$, and in polar coordinates the bending moment M_r is written explicitly

$$M_r = -D \left[\frac{\partial^2 w}{\partial r^2} + \nu \left(\frac{1}{r} \frac{\partial w}{\partial r} + \frac{1}{r^2} \frac{\partial^2 w}{\partial \theta^2} \right) \right]$$

Things are even worse for a completely free plate; in fact, in this case the boundary conditions read (eq (8.147c))

$$M_r = Q_r + \frac{1}{r} \frac{\partial M_{r\theta}}{\partial \theta} = 0$$

where M_r is as above and the transverse shearing force Q_r and the twisting moment $M_{r\theta}$ are given by

$$Q_r = -D \frac{\partial}{\partial r} (\nabla^2 w)$$

$$M_{r\theta} = -D(1 - \nu) \frac{\partial}{\partial r} \left(\frac{1}{r} \frac{\partial w}{\partial \theta} \right)$$

8.8.2   *Rectangular plates*

Due to its importance in many fields of applied engineering, let us now consider a uniform rectangular plate extending in the domain $0 \leqslant x \leqslant a$ and $0 \leqslant y \leqslant b$. The equation of motion of free vibrations is again (8.145) which, assuming a harmonic time dependence becomes eq (8.148a) for the function of the space variables $u = u(x, y)$. As in the preceding case, this equation can be written as

$$[\nabla^2 + \gamma^2] [\nabla^2 + (i\gamma)^2] u = 0$$

and we can express its solution as $u = u_1 + u_2$. Obviously, it is now convenient to adopt a system of rectangular coordinates so that the Laplacian and biharmonic operators are written explicitly as

$$\nabla^2 = \frac{\partial^2}{\partial x^2} + \frac{\partial^2}{\partial y^2}$$

$$\nabla^4 = \frac{\partial^4}{\partial x^4} + 2\frac{\partial^4}{\partial x^2 \partial y^2} + \frac{\partial^4}{\partial y^4}$$

The function u_1 satisfies the equation $(\nabla^2 + \gamma^2)u_1 = 0$: by separating the space variables and looking for a solution in the form $u_1(x, y) = f_1(x)g_1(y)$ we arrive at the two equations

$$\frac{d^2 f_1}{dx^2} + \alpha^2 f_1 = 0$$

$$\frac{d^2 g_1}{dy^2} + \beta^2 g_1 = 0$$

(8.155)

where $\alpha^2 + \beta^2 = \gamma^2$. Equations (8.155) have the solutions

$$f_1(x) = A \sin \alpha x + B \cos \alpha x$$

$$g_1(y) = C \sin \beta y + D \cos \beta y$$

so that

$$u_1(x, y) = A_1 \sin \alpha x \sin \beta y + A_2 \sin \alpha x \cos \beta y$$

$$+ A_3 \cos \alpha x \sin \beta y + A_4 \cos \alpha x \cos \beta y \qquad (8.156)$$

The equation satisfied by the function u_2 is $[\nabla^2 + (i\gamma)^2]u_2 = 0$, implying that its solution can be obtained from eq (8.156) by replacing the trigonometric functions by hyperbolic functions. This means that we can write the complete solution $u(x, y)$ as

$$u(x, y) = A_1 \sin \alpha x \sin \beta y + A_2 \sin \alpha x \cos \beta y$$

$$+ A_3 \cos \alpha x \sin \beta y + A_4 \cos \alpha x \cos \beta y + A_5 \sinh \alpha x \sinh \beta y$$

$$+ A_6 \sinh \alpha x \cosh \beta y + A_7 \cosh \alpha x \sinh \beta y + A_8 \cosh \alpha x \cosh \beta y$$

(8.157)

where the values of the constants A_j and parameters α and β depend on the boundary conditions. The simplest case is when all edges are simply supported and we must enforce the boundary conditions

$$u = \frac{\partial^2 u}{\partial x^2} = 0 \qquad \text{at } x = 0 \text{ and } x = a \qquad (8.158\text{a})$$

$$u = \frac{\partial^2 u}{\partial y^2} = 0 \qquad \text{at } y = 0 \text{ and } y = b \qquad (8.158\text{b})$$

where the conditions on the second derivative are obtained (eq (8.147a)) by noting that, in rectangular coordinates, the bending moments M_x and M_y are given by

$$M_x = -D\left(\frac{\partial^2 u}{\partial x^2} + \nu\,\frac{\partial^2 u}{\partial y^2}\right)$$

$$(8.159)$$

$$M_y = -D\left(\frac{\partial^2 u}{\partial y^2} + \nu\,\frac{\partial^2 u}{\partial x^2}\right)$$

By inserting the conditions of eqs (8.158a and b) into the solution (8.157) we obtain that only A_1 is different from zero so that

$$u(x, y) = A_1 \sin \alpha x \sin \beta y \qquad (8.160)$$

In addition we get the two characteristic equations

$$\sin \alpha a = 0$$

$$(8.161)$$

$$\sin \beta b = 0$$

which imply $\alpha = n\pi/a$ and $\beta = n\pi/b$ with $n, m = 1, 2, 3, \ldots$ and

$$\omega_{nm} = \pi^2 \left[\left(\frac{n}{a}\right)^2 + \left(\frac{m}{b}\right)^2\right]\sqrt{\frac{D}{\sigma}} \qquad (8.162)$$

The corresponding eigenfunctions are

$$u_{nm}(x, y) = A_{nm} \sin\left(\frac{n\pi x}{a}\right) \sin\left(\frac{m\pi y}{b}\right) \qquad (8.163\text{a})$$

where it is evident that $\langle u_{nm} \mid M u_{lk}\rangle = \sigma A_{nm}^2 \delta_{nl}\delta_{mk}$ and it is easy to see that the requirement $\langle u_{nm} \mid \sigma u_{nm} \rangle = 1$ yields the following mass-orthonormal eigenfunctions:

$$\phi_{nm}(x, y) = \frac{2}{\sqrt{\sigma ab}} \sin\left(\frac{n\pi x}{a}\right) \sin\left(\frac{m\pi y}{b}\right) \tag{8.163b}$$

The first few modes of a plate simply supported on all edges are shown in Fig. 8.8.

It is interesting to note at this point that trying to enforce different boundary conditions – say free or clamped – on the solution (8.157) is not at all an easy task. This has to do with the fact that in order to apply a separation of variables to the eigenvalue problem we must limit ourselves to the six combinations of boundary conditions where two opposite edges are simply supported.

Let us investigate this point a bit further. If we take a step back and write the solution in the form $u(x, y) = f(x)g(y)$, substitution into the eigenvalue problem $\nabla^4 u - \gamma^4 u = 0$ leads to

$$f''''g + 2f''g'' + fg'''' - \gamma^4 fg = 0 \tag{8.164}$$

and we can separate it into two independent equations if

$$f'' = -\alpha^2 f \tag{8.165a}$$

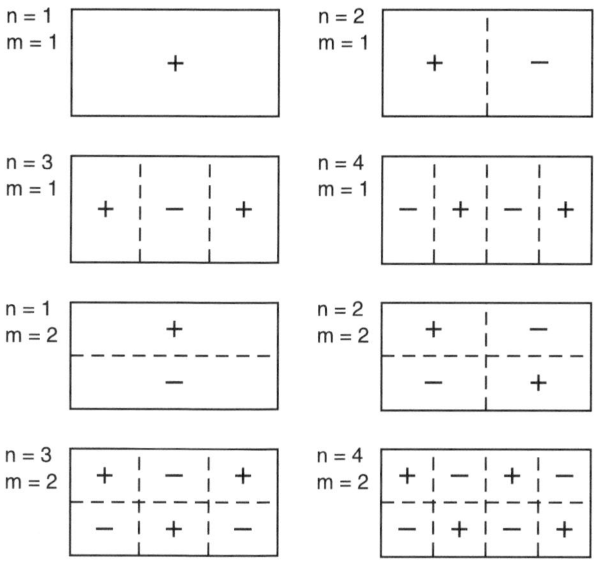

Fig. 8.8 A few lower-order modes for a rectangular plate simply supported on all edges.

or

$$g'' = -\beta^2 g \qquad (8.165b)$$

or both. Let us suppose that eq (8.165a) holds, this implies $f'''' = \alpha^4 f$ and

$$f(x) = \left\{ \begin{array}{c} \sin \alpha x \\ \cos \alpha x \end{array} \right\} \qquad (8.166)$$

If now we consider the boundary conditions of simply supported (SS), clamped (C) and free (F), along $x = 0$ we have

SS: $f(0)g(y) = f(0)[-\alpha^2 g(y) + \nu g''(y)] = 0$

C: $f(0)g(y) = f'(0)g(y) = 0$

F: $f(0)[-\alpha^2 g(y) + \nu g''(y)] = f'(0)[-\alpha^2 g(y) + (2-\nu)g''(y)] = 0$

which come from the expression of eqs (8.147a–c) in rectangular coordinates by noting that M_x is given in eq (8.159) and that the Kirchhoff condition reads

$$Q_x - \frac{\partial M_{xy}}{\partial y} = \frac{\partial^3 u}{\partial x^3} + (2-\nu)\frac{\partial^3 u}{\partial x \partial y^2} \qquad (8.167)$$

A set of similar conditions apply at $x = a$. Now, it is not difficult to show that only the SS conditions can be satisfied by a function of the form (8.166) and, more specifically, we need a sine function which satisfies $\sin \alpha x = 0$, i.e. $\alpha_n = n\pi/a$. If also eq (8.165b) holds, all sides are simply supported and an analogous line of reasoning yields $\beta_m = m\pi/b$. Moreover, substitution of eqs (8.165a and b) and of $f'''' = \alpha^4 f$, $g'''' = b^4 g$ into eq (8.164) yields

$$(\alpha^4 + 2\alpha^2\beta^2 + \beta^4 - \gamma^4)fg = 0 \qquad (8.168)$$

which can be solved for the frequency to give eq (8.162).

When the edges at $x = 0$ and $x = a$ are simply supported and we exclude the case of the other two edges simply supported, we are left with five possibilities for which we must solve the equation

$$g'''' - 2\alpha_n^2 g'' - (\gamma^4 - \alpha_n^2)g = 0 \qquad (8.169)$$

whose solution depends on whether $\gamma^4 > \alpha_n^4$, $\gamma^4 = \alpha_n^4$ or $\gamma^4 < \alpha_n^4$. However, even if separation of variables is possible in these latter cases, the information on natural frequencies and mode shapes is not easily obtained and the interested reader is urged to refer to the wide body of specific literature on the subject.

A final comment of general nature can be made on the orthogonality of the eigenfunctions. From our preceding discussion, we know that mass and stiffness orthogonality are guaranteed by the symmetry of the eigenvalue problem; however, it may be of interest to approach the problem from a different point of view. Let us consider two different eigenfunctions, say u_{nm} and u_{lk}: the equations

$$\nabla^4 u_{nm} = \gamma_{nm}^4 u_{nm}$$

$$\nabla^4 u_{lk} = \gamma_{lk}^4 u_{lk}$$

(8.170)

are identically satisfied. Now, since from static classical plate theory the differential equation of static deflection is written $\nabla^4 u = q/D$, we can interpret the first of eqs (8.170) as the equation of the static deflection of the plate under the action of the load $q_1 = D\gamma_{nm}^4 u_{nm}$ and, by the same token, we can say that our plate assumes the deflected shape u_{lk} when the load $q_2 = D\gamma_{lk}^4 u_{lk}$ is acting. In other words, the loads q_1 and q_2 represent two systems of generalized forces while u_{nm} and u_{lk} are the displacements caused by such forces.

We now invoke Betti's theorem which states that:

For a linearly elastic structure the work done by a system q_1 of forces under a distortion caused by a system q_2 of forces equals the work done by the system q_2 under a distortion caused by the system q_1.

In our case, this translates mathematically into

$$D\gamma_{nm}^4 \int_\Omega u_{nm} u_{lk} d\Omega = D\gamma_{lk}^4 \int_\Omega u_{nm} u_{lk} d\Omega$$

(8.171a)

where we had to integrate over the plate domain Ω in order to obtain the work expressions required by the theorem. Equation (8.171a) gives

$$(\gamma_{nm}^4 - \gamma_{lk}^4) \int_\Omega u_{nm} u_{lk} d\Omega = 0$$

(8.171b)

and hence, since we assumed $\gamma_{nm} \neq \gamma_{lk}$,

$$\int_\Omega u_{nm} u_{lk} d\Omega = \langle u_{nm} \mid u_{lk} \rangle = 0$$

(8.171c)

For our purposes, we can finish here our treatment on the free vibration of continuous systems referring the interested reader to the specific literature on the subject. In particular, an interesting discussion on one-dimensional eigenvalue problems in which boundary conditions contain the eigenvalue can be found in Humar [19] and Meirovitch [6, 9].

8.9 Forced vibrations and response analysis: the modal approach

The action of external time-varying loads on a continuous system leads to a nonhomogeneous partial differential equation. In general, in the light of the preceding sections, it is not difficult to obtain the governing differential equation also because – once the mass and stiffness operators have been introduced – its formal structure is similar to the matrix equation governing the forced vibrations of MDOF systems. However, now the boundary conditions come into play and we must pay due attention to them.

Two general methods of solutions can be identified:

1. integral transform methods, which are particularly well suited to systems with infinite or semi-infinite extension in space (note, in fact, that in these cases the concept of normal mode loses its meaning) and for problems with time-dependent boundary conditions;
2. the mode superposition method.

Here, we concentrate our attention on method 2, which we can call 'the modal approach'. The relevant equation of motion is written as

$$Kw + M\frac{\partial^2 w}{\partial t^2} = f \qquad (8.172)$$

where $w = w(x, t)$, $f = f(x, t)$ and, for brevity of notation, we indicate with x the set of space variables. Note that now f is a forcing function representing an external action (it is the counterpart of vector $\mathbf{f}$ of eq (7.1)) and must not be confused with the f functions of the preceding sections in this chapter, where this symbol was often adopted to identify a general function to be used for the specific needs of that part only. Equation (8.172) must be supplemented by the set of p boundary conditions

$$B_i w = 0 \qquad (i = 1, 2, ..., p) \qquad (8.173)$$

Now, by virtue of the results given in Section 8.7.1, the general response w can be expressed in terms of a superposition of eigenfunctions ϕ_i multiplied by a set of time-dependent generalized coordinates $y_i(t)$, i.e.

$$w = \sum_{i=1}^{\infty} \phi_i y_i \qquad (8.174)$$

which is the infinite-dimensional counterpart of eq (7.2). If we substitute the solution (8.174) into eq (8.172) and then take the inner product of the resulting equation by ϕ_j $(j \neq i)$ we get

$$\sum_i y_i \langle \phi_j | K\phi_i \rangle + \sum_i \ddot{y}_i \langle \phi_j | M\phi_i \rangle = \langle \phi_j | f \rangle$$

which, owing to the orthogonality relationships (8.128), reduces to the infinite set of 1-DOF uncoupled equations

$$\ddot{y}_j + \omega_j^2 y_j = \langle \phi_j | f \rangle \qquad (j = 1, 2, 3, \ldots) \tag{8.175}$$

whose solution is given by (Chapter 5 and eq (7.7))

$$y_j(t) = y_j(0)\cos \omega_j t + \frac{\dot{y}_j(0)}{\omega_j} + \frac{1}{\omega_j} \int_0^t \langle \phi_j | f \rangle \sin[\omega_j(t - \tau)]d\tau \tag{8.176}$$

where $y_j(0)$ and $\dot{y}_j(0)$ are the initial jth modal displacement and velocity, respectively. Moreover, it is not difficult to show that we can obtain these quantities from the inner products

$$y_j(0) = \langle \phi_j | Mw_0 \rangle$$
$$\dot{y}_j(0) = \langle \phi_j | M\dot{w}_0 \rangle \tag{8.177}$$

where $w(x, 0) = w_0$ and $\dot{w}_0 = \partial w / \partial t |_{t=0}$ are the initial conditions in physical coordinates. The analogy with eq (7.5b) is evident.

A few remarks can be made at this point.

1. First, it is apparent that the first step of the whole approach is the solution of the differential (symmetrical) eigenvalue problem $Ku = \lambda Mu$ satisfying the appropriate boundary conditions. The resulting set of eigenvalues and orthonormal eigenfunctions make it possible to express the general solution to the free vibrations problem (i.e. eq (8.172) with $f = 0$) as in eq (6.50), which reads

$$w(x, t) = \sum_{j=1}^{\infty} \left(\langle \phi_j | Mw_0 \rangle \cos \omega_j t + \frac{1}{\omega_j} \langle \phi_j | M\dot{w}_0 \rangle \sin \omega_j t \right) \phi_j \tag{8.178}$$

For example, suppose that we want to consider the longitudinal motion of a uniform clamped–free rod of length L and μ mass per unit length. Let the initial conditions be such that $w_0 = x/L$ and $\dot{w}_0 = 0$. Since the mass orthonormal functions are given by

$$\phi_j = \sqrt{\frac{2}{\mu L}}\, \sin\frac{(2j-1)\pi x}{2L}$$

we can calculate the rod free motion by means of eq (8.178). To this end we must first evaluate the inner product

$$\langle \phi_j \mid M w_0 \rangle = \frac{\mu}{L}\sqrt{\frac{2}{\mu L}} \int_0^L x \sin\frac{(2j-1)\pi x}{2L}\,dx$$

(the calculation is not difficult and is left to the reader) and then obtain the free motion as

$$w(x,t) = \sum_{j=1}^{\infty} \phi_j \langle \phi_j \mid M w_0 \rangle \cos \omega_j t$$

where we know from eq (8.49) that

$$\omega_j = \frac{(2j-1)\pi}{2L}\sqrt{\frac{EA}{\mu}}$$

2. When expressed in normal coordinates, the kinetic and potential energy of free vibration assume the particularly simple forms

$$T = \frac{1}{2}\sum_{j=1}^{\infty} \dot{y}_j^2$$

$$V = \frac{1}{2}\sum_{j=1}^{\infty} \omega_j^2 y_j^2$$

(8.179)

so that the Lagrangian has no coupling terms between the coordinates and is simply the Lagrangian function of an infinite number of independent harmonic oscillators.

3. The final remark has to do with the apparent discrepancy according to which our system may be equally well described by a continuous system of coordinates $w(x,t)$ or by a discrete one, i.e. $y_j(t)$. The general feeling is that this second set of coordinates cannot describe the same number of degrees of

freedom as the first one, although this number is infinite in both cases. However, it must be noted that, broadly speaking, the definition of normal coordinates itself somehow incorporates the boundary conditions from the outset, and this is not the case for $w(x, t)$. In fact, the coordinates $w(x, t)$ could be *a priori* chosen as completely independent of each other and this would allow us – in principle – to describe also discontinuous motions of our system. By contrast, the expansion in normal modes requires that the functions representing the point-by-point displacement of our system be reasonably well-behaved.

So, although the normal mode description is, as a matter of fact, mathematically more restrictive, it really does not put any significant restriction on the physical problem because the boundary conditions and the good behaviour of the displacement functions are dictated by the physical nature of our system and, ultimately, by the physics of the phenomena we are trying to describe. Also, note that for MDOF systems there was no need to make a similar remark because, whatever the coordinate system we choose to adopt, the elements of the stiffness matrix automatically take care of the boundary conditions.

We can now go back to the main discussion of this section and note that in Chapter 7 – where we dealt with the response of MDOF systems – we gave special emphasis to the modal impulse response functions and to the modal frequency response functions, also showing their relationships with impulse response functions and frequency response functions expressed in physical coordinates. We want now to extend those concepts to the case of distributed parameter systems.

Let us consider an undamped system: from the preceding chapters we know that the jth modal impulse response function $h_j(t)$ is the solution of the equation

$$\ddot{h}_j + \omega_j^2 h_j = \delta(t) \tag{8.180}$$

where δ is the Dirac's delta function. Moreover, we also know that (eqs (5.7a) and (5.7b))

$$h_j(t) = \frac{1}{\omega_j} \sin \omega_j t \tag{8.181}$$

where the term $\langle \phi_j \,|\, M\phi_j \rangle$ representing the jth generalized mass does not appear in the denominator of the right-hand side of eq (8.181) because we are considering mass-orthonormal eigenfunctions.

Now, if for simplicity we assume $y_j(0) = \dot{y}_j(0) = 0$, eq (8.176) reduces to

$$y_j(t) = \frac{1}{\omega_j} \int_0^t \langle \phi_j \,|\, f \rangle \sin[\omega_j(t - \tau)]d\tau = \int_0^t \langle \phi_j \,|\, f \rangle h_j(t - \tau)d\tau \tag{8.182}$$

If we further assume that the excitation is an impulse applied at position $x = x_k$ at time $\tau = 0$ we can write

$$f(x, \tau) = \delta(x - x_k)\delta(\tau)$$

so that

$$\langle \phi_j \, | \, f \rangle = \int_{\Omega} \phi_j(x)\delta(x - x_k)\delta(t)d\Omega = \phi_j(x_k)\delta(t) \tag{8.183}$$

which must be substituted in eq (8.182) to obtain $y_j(t)$.

Then, from eq (8.174) we can obtain the response in physical coordinates as

$$w(x, t) = \sum_{j=1}^{\infty} \phi_j(x)\phi_j(x_k) \int_0^t h_j(t - \tau)\delta(\tau)d\tau = \sum_{j=1}^{\infty} \phi_j(x)\phi_j(x_k)h_j(t) \tag{8.184}$$

More specifically, we can consider the response at the point $x = x_m$, i.e.

$$w(x_m, t) = \sum_{j=1}^{\infty} \phi_j(x_m)\phi_j(x_k)h_j(t) \tag{8.185a}$$

and note that this is just the displacement response of point x_m to a unit impulse applied at point x_k at the instant $t = 0$. In other words, eq (8.185a) represents the impulse response function $h(x_m, x_k, t)$ so that we can write

$$h(x_m, x_k, t) = \sum_{j=1}^{\infty} h_j(t)\phi_j(x_m)\phi_j(x_k) \tag{8.185b}$$

which is the continuous systems counterpart of eq (7.37b) where h_{jk} denoted the physical coordinates (displacement) impulse response function at the jth DOF due to an unit force impulse applied at the kth DOF at $t = 0$.

Alternatively, we can consider the frequency domain and note that the jth modal frequency response function (receptance in this case) can be obtained by assuming a forcing function in sinusoidal form. This means that $f = f_0(x)e^{i\omega t}$ so that eqs (8.175) become

$$\ddot{y}_j + \omega_j^2 y_j = \langle \phi_j \, | \, f_0 \rangle e^{i\omega t} \tag{8.186}$$

hence, assuming a response which is also in sinusoidal form, we have

$$y_j = \frac{1}{(\omega_j^2 - \omega^2)} \langle \phi_j \, | \, f_0 \rangle e^{i\omega t} \tag{8.187a}$$

so that the jth receptance FRF is

$$R_j(\omega) = \frac{1}{\omega_j^2 - \omega^2} \tag{8.187b}$$

If we now consider a harmonic forcing function of unit amplitude applied at the point $x = x_k$, i.e.

$$f = \delta(x - x_k)e^{i\omega t}$$

we have $\langle \phi_j \,|\, f \rangle = \phi_j(x_k)e^{i\omega t}$; the modal response is then

$$y_j = R_j\phi_j(x_k)e^{i\omega t} \tag{8.188}$$

and the (steady-state) solution in physical coordinates is given by

$$w(x, t) = \sum_{j=1}^{\infty} \phi_j(x)R_j\phi_j(x_k)e^{i\omega t} \tag{8.189a}$$

so that the response at point $x = x_m$ is

$$w(x_m, t) = \sum_{j=1}^{\infty} R_j\phi_j(x_m)\phi_j(x_k)e^{i\omega t} \tag{8.189b}$$

Finally, if we observe that the physical coordinate response at point x_m due to a harmonic excitation at x_k can be expressed in the general form

$$w(x_m, t) = R(x_m, x_k, \omega)e^{i\omega t}$$

where $R(x_m, x_k, \omega)$ is, by definition, the physical coordinates receptance function corresponding to points x_m and x_k, we get from eq (8.189b)

$$R(x_m, x_k, \omega) = \sum_{j=1}^{\infty} R_j(\omega)\phi_j(x_m)\phi_j(x_k) \tag{8.190}$$

which is the (undamped) continuous systems counterpart of eq (7.28).
Note that from eqs (8.185b) and (8.190) we get

$$h(x_m, x_k, t) = h(x_k, x_m, t)$$

$$R(x_m, x_k, \omega) = R(x_k, x_m, \omega) \tag{8.191}$$

which show that the reciprocity theorem holds. Moreover, the generalization of the second of eqs (8.191) to a general FRF function $H(\omega)$ other than receptance is straightforward and reads

$$H(x_m, x_k, \omega) = H(x_k, x_m, \omega) \tag{8.192}$$

Note that, for reasons outlined in Section 7.4.1, we only consider FRF functions in the forms of receptances, mobilities and accelerances; hence, for our purposes and unless otherwise stated, the symbol H will always be tacitly assumed to mean a FRF functions in one of these forms.

For its importance in modal testing, we refer to eqs (8.189b) and (8.190) to point out that a concentrated excitation force applied at a point x_k where $\phi_j(x_k) = 0$ for some index j results in no excitation of all these modes. In other words, for example, if we excite a beam by means of a concentrated load at mid-span, all even natural frequencies ($\omega_2, \omega_4, \ldots$) and modes ($\phi_2, \phi_4, \ldots$) will not contribute to the measured response. By the same token, if we pluck at mid-length a guitar string tuned to give the note A at 440 Hz, the second (A at 880 Hz), fourth (A at 1760 Hz) etc. harmonics will be missing and the sound we hear will be the superposition of the fundamental note A (440 Hz) plus the third (E at 1320 Hz), the fifth (C# at 2200 Hz) etc. harmonics. On the other hand, if we pluck the same string at one-third of its length we will still hear the same fundamental note A at 440 Hz, but the harmonic content of the sound will be different.

8.9.1 Forced response of continuous systems: some examples

Example 8.1 Consider a vertical clamped–free beam of length L, mass per unit length μ and flexural rigidity EI subjected to an excitation in the form of a lateral base displacement $u_{ground} = g(t)$. It is easy to realize that this situation, for example, can be used as a first approximation to model the response of a tall, slender building to an earthquake excitation. For our purposes, we ignore the fact that the definition of an appropriate $g(t)$ is very difficult in this case and it is one of the most uncertain steps of the analysis.

To solve this problem, we need to consider eq (8.55a); we write the beam displacement $y(x, t)$ as

$$y(x, y) = u(x, t) + g(t) \tag{8.193}$$

and substitute it into eq (8.55a). Note that $u(x, t)$ represents the displacement of the beam relative to the rigid-body translation of the ground. We get

$$EI \frac{\partial^4 u}{\partial x^4} - \mu \frac{\partial^2 u}{\partial t^2} = -\mu \frac{\partial^2 g}{\partial t^2} \tag{8.194}$$

where it is evident that the inertia forces depend on the total motion, whereas the stiffness (and damping, if it were included in the analysis) forces depend only on the relative motion. The r.h.s. of eq (8.194) is the effective earthquake force and it is usually indicated with the symbol f_{eff}. In the light of the discussion of preceding sections, it follows that – assuming the system initially at rest – we have (eq (8.182))

$$\langle \phi_j \,|\, f_{eff} \rangle = -\mu \frac{d^2 g}{dt^2} \int_0^L \phi_j(x) dx$$

and hence

$$y_j = -\frac{\mu}{\omega_j} \int_0^t \int_0^L \phi_j(x) \frac{d^2 g(\tau)}{d\tau^2} \sin \omega_j(t-\tau) dx d\tau$$

where ω_j are given by eq (8.66b) and the ϕ_j are the eigenfunctions (8.67), in which the constant C_1 has been chosen so as to make them mutually mass-orthonormal.

The relative response in physical coordinates is then obtained as the superposition

$$u(x,t) = -\mu \sum_{j=1}^{\infty} \frac{\phi_j(x)}{\omega_j} \int_0^t \int_0^L \phi_j(x) \frac{d^2 g}{d\tau^2} \sin \omega_j(t-\tau) dx d\tau \qquad (8.195)$$

where, in general, for seismic excitation the minus sign in (8.195) is irrelevant and – owing to the complicated expressions of the eigenfunctions and of the ground acceleration $\ddot{g}(t)$ – the integrations must be performed numerically. In general, only a few terms of the series must be considered in order to obtain a satisfactory representation of the actual response so that, in the end, we are brought back to the case of an n-DOF system, where only a finite number (n) of modal coordinates is needed to describe the response.

The total response $y(x,t)$ is finally obtained according to eq (8.193).

A simpler case arises if we consider the longitudinal motion of our system; the excitation $g(t)$ is now a vertical displacement and the relevant equation of motion is eq (8.42). The general form of the relative displacement is still given by eq (8.195) but the eigenvalues ω_j are given by the first of eq (8.49) and the eigenfunctions are given by eq (8.50), where $B_n = \sqrt{2/\mu L}$. In this circumstance, it is not difficult to show that the space integration in (8.195) results in

$$\langle \phi_j \,|\, f_{eff} \rangle = \frac{d^2 g}{dt^2} \sqrt{\frac{2\mu}{L}} \int_0^L \sin \frac{(2j-1)\pi x}{2L} dx = \frac{2\ddot{g}(t)}{(2j-1)\pi} \sqrt{2\mu L} \qquad (8.196)$$

and only the time integration needs to be performed numerically.

Example 8.2. Let us now consider a uniform clamped–free rod of length L and mass per unit length μ excited by a tip load at the free end, i.e. $f = p(t)\delta(x - L)$. If the rod is at rest before the excitation occurs the first two terms on the r.h.s. of eq (8.176) are zero and

$$\langle \phi_j \mid f \rangle = p(t)\phi_j(L) = p(t)\sqrt{\frac{2}{\mu L}}\sin\frac{(2j - 1)\pi}{2} \tag{8.197}$$

so that eq (8.176) reduces to

$$y_j(t) = \frac{(-1)^{j-1}}{\omega_j}\sqrt{\frac{2}{\mu L}}\int_0^t p(\tau)\sin[\omega_j(t - \tau)]d\tau \tag{8.198}$$

because $\sin[(2j - 1)\pi/2] = (-1)^{j-1}$. If we further assume that the explicit form of $p(t)$ is a unit step (Heaviside) function $\theta(t)$, i.e.

$$\theta(t) = \begin{cases} 1 & t \geqslant 0 \\ 0 & t < 0 \end{cases}$$

we can substitute $\theta(t)$ into eq (8.198), perform the integration by noting that

$$\int_0^t \theta(\tau)\sin[\omega_j(t - \tau)]d\tau = \frac{1}{\omega_j}(1 - \cos \omega_j t)$$

and obtain

$$y_j(t) = \frac{(-1)^{j-1}}{\omega_j^2}\sqrt{\frac{2}{\mu L}}(1 - \cos \omega_j t) \tag{8.199}$$

Finally the displacement in physical coordinates is obtained as the super-position

$$
\begin{aligned}
w(x, t) &= \frac{2}{\mu L}\sum_{j=1}^{\infty}\frac{(-1)^{j-1}}{\omega_j^2}(1 - \cos \omega_j t)\sin\frac{(2j - 1)\pi x}{2L} \\
&= \frac{8L}{\pi^2 EA}\sum_{j=1}^{\infty}\frac{(-1)^{j-1}}{(2j - 1)^2}(1 - \cos \omega_j t)\sin\frac{(2j - 1)\pi x}{2L}
\end{aligned}
\tag{8.200}
$$

where in the second expression we take into account the explicit form of ω_j^2. Also, it is worth pointing out that eq (8.200) is dimensionally correct because, since we have assumed a unit force, the dimensions of $w(x, t)$ are displacement per unit force (i.e. m/N).

At this point, it is interesting to note that the system above can be analysed either as:

1. an excitation-free system with a time-dependent boundary condition at $x = L$, or
2. a forced vibration problem with homogeneous boundary conditions.

As stated at the beginning of the preceding section, free-vibration problems with nonhomogeneous boundary conditions are often tackled by an integral transform (Laplace or Fourier) approach; however, the modal approach can also be adopted in consideration of the fact that a boundary value problem of type (1) can usually be transformed into a boundary value problem of type (2) (e.g. Courant and Hilbert [20] or Mathews and Walker [21]).

In general, it appears that in these cases a disadvantage of the modal approach – which is essentially a 'standing waves solution' – is that the resultant series converges quite slowly and many terms must be included in order to achieve a reasonable accuracy. By contrast, depending on how the inverse transformation is carried out, the Laplace transform method allows us the possibility to obtain a solution either in terms of standing waves or in terms of travelling waves (waves being reflected back and forth within the rod). This latter possibility – the travelling wave approach – leads to a solution in the form of a rapidly converging series, thus making this strategy more attractive. However, on physical grounds, we may argue that the time scale in which we are interested suggests the type of solution to adopt; in fact, the travelling wave solution converges rapidly when we consider the short-term response of our system whereas, if the long-term response is desired, more and more terms are needed. The situation is reversed for the modal solution: as time progresses, the terms corresponding to higher modes die out because of damping and we are left with a series in which, say, only the first two or three terms have a significant contribution.

We will not consider an integral transform strategy of solution here (the interested reader is referred to Meirovitch [22]) but, using the rod example above, we will show how a problem of type (1) can be transformed in a problem of type (2).

Our rod problem can be formulated as a type (1) problem in the following form:

$$EA \frac{\partial^2 w}{\partial x^2} - \mu \frac{\partial^2 w}{\partial t^2} = 0 \tag{8.201a}$$

$$w(0, t) = 0$$

$$EA \left. \frac{\partial w}{\partial x} \right|_{x = L} = p(t) \tag{8.201b}$$

where eq (8.201a) is the homogeneous equation of motion and eq (8.201b) are the boundary conditions. Since one boundary condition (the second) is nonhomogeneous we assume the solution of our problem in the form

$$w(x, t) = v(x, t) + \psi(x)p(t) \qquad (8.202)$$

where the term $\psi(x)p(t)$ – which we can define in compact notation as $u_{st}(x, t)$ – is a so-called 'pseudostatic' displacement brought about by the boundary motion and $v(x, t)$ is the displacement relative to the support displacement. Mathematically, the function u_{st} is chosen in such a way as to make the boundary conditions for $v(x, t)$ homogeneous. On physical grounds, the usual assumption made for the choice of u_{st} is that no inertia forces (i.e., no accelerations) are produced by the application of the support motion; hence the name 'pseudostatic'. For our case, this assumption implies that u_{st} obeys the equation

$$EA \frac{\partial^2 u_{st}}{\partial x^2} = 0 \qquad (8.203)$$

from which follows (provided that $p(t) \neq 0$)

$$\psi(x) = Cx + D \qquad (8.204)$$

Moreover, given the expression (8.202), the boundary conditions (8.201b) become

$$w(0, t) = v(0, t) + \psi(0)p(t) = 0$$

$$EA \left.\frac{\partial w}{\partial x}\right|_{x=L} = EA \left.\frac{\partial v}{\partial x}\right|_{x=L} + EAp(t) \left.\frac{\partial \psi}{\partial x}\right|_{x=L} = p(t)$$

from which – if we want homogeneous boundary conditions for $v(x, t)$ – it follows

$$\psi(0) = 0$$
$$\left.\frac{d\psi}{dx}\right|_{x=L} = \frac{1}{EA} \qquad (8.205)$$

Enforcing the boundary conditions (8.205) on the solution (8.204) leads to

$$\psi(x) = \frac{x}{EA} \qquad (8.206)$$

The transformation of the problem (8.201) into a type (2) problem is complete when we determine the nonhomogeneous equation of motion for the relative displacement $v(x, t)$: this is simply accomplished by substituting eq (8.202) into eq (8.201a) and results in

$$EA\frac{\partial^2 v}{\partial x^2} - \mu\frac{\partial^2 v}{\partial t^2} = \mu\psi(x)\frac{d^2 p}{dt^2} - EAp(t)\frac{d^2\psi}{dx^2} \tag{8.207a}$$

where the r.h.s. of eq (8.207) has clearly the dimensions of N/m and, for short, can be indicated with the symbol f_{eff} (effective force).

Equations (8.207a), (8.206) plus the homogeneous boundary conditions

$$v(0, t) = 0$$

$$EA\frac{\partial v}{\partial x}\bigg|_{x=L} = 0 \tag{8.207b}$$

constitute our type (2) boundary value problem which fits into the scheme of problems that can be more effectively tackled by the modal approach. In this light, we expand $v(x, t)$ in a series of eigenfunctions and calculate the normal coordinates as prescribed in eq (8.182) (note that, from eq (8.206), we get $d^2\psi/dx^2 = 0$), i.e.

$$y_j(t) = \int_0^t \langle\phi_j \mid f_{eff}\rangle h_j(t - \tau)d\tau$$

$$= \int_0^t \left\{\int_0^L \phi_j\mu\psi(x)\frac{d^2 p}{dt^2}\,dx\right\}h_j(t - \tau)d\tau \tag{8.208}$$

Upon substituting the explicit expressions of ϕ_j and ψ in eq (8.208), the space integral within braces gives

$$\frac{d^2 p}{dt^2}\frac{\mu}{EA}\sqrt{\frac{2}{\mu L}}\int_0^L x\sin\frac{(2j - 1)\pi x}{2L}\,dx = \frac{d^2 p}{dt^2}\frac{\mu}{EA}\sqrt{\frac{2}{\mu L}}\frac{4L^2(-1)^{j-1}}{(2j - 1)^2\pi^2}$$

so that eq (8.208) becomes

$$y_j = \frac{\mu}{EA}\sqrt{\frac{2}{\mu L}}\frac{4L^2(-1)^{j-1}}{(2j - 1)^2\pi^2}\int_0^t \frac{d^2 p(\tau)}{d\tau^2}h_j(t - \tau)d\tau \tag{8.209}$$

Now, in the problem we are considering, we assumed that $p(t)$ is the Heaviside function $\theta(t)$; since (eq (2.67a) or (2.84)) $d\theta(t)/dt = \delta(t)$, the time integral of eq (8.209) becomes

$$\int_0^t \frac{d^2 p}{d\tau^2} h_j(t - \tau)d\tau = \int_0^t \frac{d\delta(\tau)}{d\tau} h_j(t - \tau)d\tau = -\frac{dh_j(t)}{dt} = -\cos\omega_j t$$

where we take into account the properties of the Dirac's delta function (eq (2.69)) and the explicit expression of h_j. The final steps consist of substituting this result in eq (8.209), writing explicitly the series expansion of $v(x, t)$, i.e.

$$v(x, t) = \sum_{j=1}^\infty y_j(t)\phi_j(x)$$

and putting it all back together into the solution (8.202) which becomes

$$w(x, t) = \frac{x}{EA} - \frac{8L}{\pi^2 EA} \sum_{j=1}^\infty \frac{(-1)^{j-1}}{(2j-1)^2\pi^2} \sin\left[\frac{(2j-1)\pi x}{2L}\right]\cos\omega_j t \qquad (8.210)$$

This result must be compared with eq (8.200) and it is not difficult to show that they are equal. This is due to the fact that the function $(\pi^2 x/8L)$ can be expanded in a Fourier series as (the proof is left to the reader)

$$\frac{\pi^2 x}{8L} = \sum_{j=1}^\infty \frac{(-1)^{j-1}}{(2j-1)^2} \sin\frac{(2j-1)\pi x}{2L}$$

so that – after performing the product in eq (8.200) – the first term is exactly (x/EA), i.e. the function $\psi(x)$ of the pseudostatic displacement. The advantage of including explicitly the pseudostatic displacement from the outset lies in the more rapid convergence of the series (8.210) as compared to the series (8.200), the pseudostatic displacement representing the average position about which the vibration takes place.

Example 8.3. In modal testing, we are often concerned with the response of a given system to an impulse loading. So, consider the rod of Example 8.2 subjected to a unit impulse applied at $x = L$ at $t = 0$. The response in physical coordinates at $x = L$ is given by eqs (8.185) and reads

$$h(L, L, t) = \sum_{j=1}^\infty \phi_j^2(L)h_j(t) = \frac{2}{\mu L}\sum_{j=1}^\infty \frac{1}{\omega_j}\sin\omega_j t \qquad (8.211)$$

This result should be hardly surprising because we know from Chapter 5 (eq (5.42)) that the impulse response function is the time derivative of the Heaviside response function. So, in this circumstance we could have ignored eq (8.185) by simply noting that the result (8.211) can be obtained by calculating the time derivative of eq (8.200) and by substituting $x = L$ in it.

On the other hand, the receptance FRF can be obtained from eq (8.190): at $x = L$ this is

$$H(L, L, \omega) = \sum_{j=1}^{\infty} \frac{\phi_j^2(L)}{\omega_j^2 - \omega^2} = \frac{2}{\mu L} \sum_{j=1}^{\infty} \frac{1}{\omega_j^2 - \omega^2} \tag{8.212}$$

In the light of preceding chapters, we expect that $h(L, L, t)$ and $H(L, L, \omega)$ form a Fourier transform pair. However, the Fourier transform of eq (8.211) does not exist, but we may note that the Laplace transform of eq (8.211) does exist and is given by

$$H(L, L, s) = \frac{2}{\mu L} \sum_{j=1}^{\infty} \frac{1}{s^2 + \omega_j^2}$$

where s is the (complex) Laplace operator and can be expressed as $s = a + i\omega$. Hence, leaving aside mathematical rigour for a moment, we see that we can arrive at eq (8.212) by first taking the Laplace transform of eq (8.211) and then letting $a \to 0$. This mathematical trick is just for purposes of illustration and it would not be needed if the system had some amount of positive damping; as a matter of fact, this is always the case for real systems whose time response and FRFs (eq (8.211) and (8.212)) do not go to infinity when $\omega = \omega_j$.

Example 8.4.   Consider now the case of a constant force P moving at a constant velocity V along an Euler–Bernoulli beam simply supported at both ends. The engineering importance of this case is evident because this example can be used to model a number of common situations, the simplest one being a heavy vehicle travelling across a bridge deck. We also make the reasonable assumption that the mass of the vehicle is small in comparison with the beam mass (the bridge deck) and it does not alter appreciably its eigenvalues and eigenfunctions.

Mathematically, the moving load can be represented as

$$f(x, t) = \begin{cases} P\delta(x - Vt) & 0 \leqslant t \leqslant L/V \\ 0 & \text{otherwise} \end{cases} \tag{8.213}$$

and, with reference to eq (8.182), we obtain

$$\langle \phi_j \,|\, f \rangle = \sqrt{\frac{2}{\mu L}}\, P \sin\left(\frac{j\pi Vt}{L}\right) \qquad (j = 1, 2, 3, \ldots)$$

416 *Continuous or distributed parameter systems*

so that, assuming the beam at rest at $t = 0$, we get

$$y_j = \sqrt{\frac{2}{\mu L} \frac{P}{\omega_j}} \int_0^t \sin\left(\frac{j\pi V\tau}{L}\right) \sin \omega_j(t - \tau)d\tau \tag{8.214}$$

where now

$$\omega_j = \frac{j^2\pi^2}{L^2}\sqrt{\frac{EI}{\mu}}$$

are the eigenfrequencies of our pinned–pinned Euler–Bernoulli beam.
    A double integration by parts in eq (8.214) leads to

$$y_j = P\sqrt{\frac{2}{\mu L}}\left(\frac{L^2}{j^2\pi^2 V^2 - \omega_j^2 L^2}\right)\left[\frac{j\pi V}{\omega_j L}\sin\omega_j t - \sin\left(\frac{j\pi Vt}{L}\right)\right] \tag{8.215}$$

so that the displacement in physical coordinates is given by

$$w(x, t) = \frac{2P}{\mu L}\sum_{j=1}^{\infty}\frac{L^2 \sin(j\pi x/L)}{j^2\pi^2 V^2 - \omega_j^2 L^2}\left[\frac{j\pi V}{\omega_j L}\sin\omega_j t - \sin\left(\frac{j\pi Vt}{L}\right)\right] \tag{8.216}$$

    The solution (8.216) needs further comments. First, it is interesting to note that there are a series of values of velocity at which resonance may occur; they are

$$V_{j,\,crit} = \frac{L\omega_j}{j\pi} = \frac{j\pi}{L}\sqrt{\frac{EI}{\mu}} \tag{8.217a}$$

and the time of passage t_j at these values of speed is given by

$$t_j = \frac{L}{V_{j,\,crit}} = \frac{j\pi}{\omega_j} \tag{8.217b}$$

so that, calling $T_1 = 2\pi/\omega_1$ the fundamental period of vibration of the beam, we have $t_1 = (1/2)T_1$, $t_2 = T_1$, $t_3 = (3/2)T_1$, ... Furthermore, if we try to evaluate the response (8.216) at the critical speeds – i.e. when $j\pi V/L \to \omega_j$ – we run into an indeterminate $0/0$ situation. However, we can use L'Hospital's rule and obtain

$$\lim_{j\pi V/L \to \omega_j} w(x, t) = \frac{P}{\mu L}\sum_{j=1}^{\infty}\frac{\sin(j\pi x/L)}{\omega_j^2}\left[\sin\omega_j t - \omega_j t \cos\omega_j t\right] \tag{8.218}$$

which, owing to the finiteness of the time of passage, is a bounded quantity and does not grow indefinitely with time.

We will not consider an example of a two-dimensional system (say, a membrane or a plate) but it is evident that, besides the added mathematical complexities, the extension of the modal approach to these systems follows the same line of reasoning. Obviously, now the expansion (8.174) must include all modes and hence we must sum on all indexes, i.e. the mode indexes and the degeneracy indexes. For example, recalling the eigenfunctions of a circular membrane clamped at its outer edge (eqs (8.120) and (8.121)) the expansion in series of eigenmodes reads

$$w(r, \theta, t) = \sum_{m} y_{0m}(t)u_{0m}(r) + \sum_{n, m, \beta} y_{nm\beta}(t)u_{nm\beta}(r, \theta)$$

where the first series involves the Bessel functions of the first kind of order zero (no degeneracy), while the second series involves the Bessel functions of order n ($n \neq 0$) and the twofold degeneracy (expressed by eq (8.121)) which, in the expansion above, is taken into account by means of the summation index β ($\beta = 1, 2$), where we defined for convenience $u_{nm1} = u_{nm}$ and $u_{nm2} = \tilde{u}_{nm}$.

8.10 Final remarks: alternative forms of FRFs and the introduction of damping

For the sake of completeness, two final remarks are needed before closing this chapter. The first remark has to do with an alternative approach for finding a closed-form solution of the frequency response function of a continuous system. In essence, if we assume a harmonic excitation in the form

$$f(x, t) = F(x, \omega)e^{i\omega t} \tag{8.219}$$

we know that there will always exist a steady-state response of our system in the form

$$w(x, t) = W(x, \omega)e^{i\omega t} \tag{8.220}$$

so that, upon substituting eqs (8.219) and (8.220) into the appropriate equation of motion, the exponential terms cancel out and we obtain a linear differential equation in $W(x, \omega)$ together with the appropriate boundary conditions. Then, if we consider that the FRF $H(x_m, x_k, \omega)$ is, by definition, the multiplying coefficient of the harmonic solution when the response is measured at $x = x_m$ and the excitation is applied at the point $x = x_k$ – i.e. $F(x, \omega) = \delta(x - x_k)$, then we have

$$H(x_m, x_k, \omega) = W(x_m, \omega) \tag{8.221}$$

By this method, the solution is not obtained in the form of a series of eigenfunctions (for example, like eq (8.200)) and one of the advantages is that it can be profitably used in the case of support motion where, in general, a set of orthonormal functions cannot be obtained.

As an example of this method we can consider the longitudinal vibrations of a vertical rod (see end of Example 8.1) subjected to a support harmonic motion of unit amplitude. If, to be consistent with eq (8.220), we call $w(x, t)$ the longitudinal rod displacement, the relevant equation of motion is

$$EA\,\frac{\partial^2 w}{\partial x^2} - \mu\,\frac{\partial^2 w}{\partial t^2} = 0 \tag{8.222a}$$

with the boundary conditions

$$w(0, t) = e^{i\omega t}$$

$$\left.\frac{\partial w}{\partial x}\right|_{x=L} = 0 \tag{8.222b}$$

Assuming a solution in the form of eq (8.220) leads to

$$\frac{d^2 W}{dx^2} + \gamma^2 W = 0 \tag{8.223a}$$

where $\gamma^2 = \mu\omega^2/EA$. Moreover, the boundary conditions for W are obtained from eq (8.222b) as

$$W(0, \omega) = 1$$

$$\left.\frac{\partial W}{\partial x}\right|_{x=L} = 0 \tag{8.223b}$$

It is then easy to show that the solution of the problem (8.223) is

$$W(x, \omega) = H(x, 0, \omega) = \cos \gamma x + (\tan \gamma L)\sin \gamma x \tag{8.224}$$

which becomes unbounded (no damping has been considered) when $\cos \gamma L = 0$, i.e. in correspondence of the eigenvalues of a clamped–free rod.

The same line of reasoning applies if we reconsider the system of example 8.2 and we assume a harmonic excitation of unit amplitude at the free end $x = L$. Equations (8.222a) and (8.223a) still apply, but the boundary conditions for W are now

$$W(0, \omega) = 0$$

$$EA \left. \frac{\partial W}{\partial x} \right|_{x = L} = 1 \qquad (8.225)$$

which must be enforced on the solution $W(x, \omega) = B \cos \gamma x + C \sin \gamma x$ to give

$$W(x, \omega) = H(x, L, \omega) = \frac{1}{\gamma EA \cos \gamma L} \sin \gamma x \qquad (8.226)$$

Equation (8.226a) must be compared with the series solution obtained from eq (8.190), i.e. explicitly

$$H(x, L, \omega) = \frac{2}{\mu L} \sum_{j=1}^{\infty} \frac{(-1)^{j-1}}{\omega_j^2 - \omega^2} \sin \frac{(2j-1)\pi x}{2L} \qquad (8.227)$$

where the frequencies ω_j are given by the first of eqs (8.49). The fact that eqs (8.226) and (8.227) are the same is not obvious at first sight, but it left to the reader to prove that it is indeed so. (Hint: use the orthogonality property of the eigenfunctions ϕ_j of the clamped–free rod and calculate the inner product $\mu \langle \phi_j(x) \mid W(x, \omega) \rangle$.)

From eq (8.226) the FRF at $x = L$ is obtained as

$$H(L, L, \omega) = W(L, \omega) = \frac{1}{\omega \sqrt{\mu EA}} \tan \left(\omega \sqrt{\frac{\mu}{EA}} L \right) \qquad (8.228)$$

which must be compared with the series solution (8.212).

The second remark has to do with the fact that in none of the preceding sections have we taken into account the effect of energy dissipation. However, the inclusion of damping – both in free and forced vibration conditions – leads to results that parallel closely the MDOF case.

As stated on a number of occasions, damping is difficult to define and the general assumption of viscous damping is mostly a matter of mathematical convenience rather than an effective explanation of the physical phenomenon. In this light, if we call $w(x, t)$ the function that represents the displacement of our continuous system, the general equation of motion (8.172) can be written as

$$Kw + C \frac{\partial w}{\partial t} + M \frac{\partial^2 w}{\partial t^2} = f(x, t) \qquad (8.229)$$

where C is a linear homogeneous 'damping' operator which involves only space derivatives up to the order $2p$ (Section 8.7.1). Now, if we assume that we already solved the undamped free-vibration problem in terms of eigenvalues λ_j and mass orthonormal eigenfunctions ϕ_j, we can follow the modal approach and expand $w(x, t)$ as in eq (8.174). Then, taking the inner product

$$\sum_i y_i \langle \phi_j \mid K\phi_i \rangle + \sum_i \dot{y}_i \langle \phi_j \mid C\phi_i \rangle + \sum_i \ddot{y}_i \langle \phi_j \mid M\phi_i \rangle = \langle \phi_j \mid f \rangle$$

we arrive at

$$\ddot{y}_j + \sum_i \dot{y}_i \langle \phi_j \mid C\phi_i \rangle + \omega_j^2 y_j = \langle \phi_j \mid f \rangle \tag{8.230}$$

which, in general, are a set of coupled linear differential equations unless the damping operator is of the 'proportional' form

$$C = aM + bK \tag{8.231}$$

At this point the parallel with the MDOF case is evident: if eq (8.231) applies, eqs (8.230) become uncoupled and we can define the modal damping ratios ζ_j by means of

$$\langle \phi_j \mid C\phi_i \rangle = 2\omega_j \zeta_j \delta_{ij} \tag{8.232}$$

which is the infinite dimensional counterpart of eq (6.142). Also, it is understood that now the modal FRFs are given by eq (7.31).

As for the MDOF case, it may often be more convenient to introduce viscous damping at the modal level, without the need to specify a damping operator. In other words, we obtain the uncoupled set of undamped equations first, and only at this stage we introduce the terms $2\omega_j \zeta_j \dot{y}_j$, the values of ζ_j being chosen on the basis of experience and/or experimental measurements.

Alternatively, for a harmonic forcing excitation $f(x, t) = f_0(x)e^{i\omega t}$, the model of structural damping can be introduced. This is generally done by assuming a damping operator which is proportional to the stiffness operator so that the uncoupled equations read

$$\ddot{y}_j + (1 + i\gamma)\omega_j^2 y_j = \langle \phi_j \mid f_0 \rangle e^{i\omega t} \tag{8.233}$$

where here γ denotes the structural damping factor.

In the case of general damping, i.e. a damping operator which does not allow the uncoupling of the equation of motion, it may be interesting to note that, in principle, we can still adopt the approach described at the beginning of this section. In other words, if we want to obtain the frequency response of our (nonproportionally) damped system, we can assume a harmonic excitation and a harmonic response in the forms of eqs (8.219) and (8.220), substitute them in the equation of motion and arrive at a linear ordinary differential equation for the function $W(x, \omega)$. However, this differential equation has constant but complex coefficients and an analytical solution is often impossible to obtain.

8.11 Summary and comments

This chapter has dealt with the free and forced vibrations of continuous parameter systems. As a matter of fact, real-world vibrating systems are systems whose physical properties (mass, stiffness and damping) are continuously distributed although – very often – the modelling scheme one chooses to adopt is a discrete parameter model which lends itself more easily to the computer implementation of the necessary calculations. The rationale behind this choice is that continuous systems – which, mathematically speaking, have an infinite number of degrees of freedom – can be considered as the limit of finite DOF systems as the number of degrees of freedom tends to infinity. Then, provided that a sufficient number of DOFs is used in the appropriate modelling scheme, the finite DOF model can approximate the continuous system under investigation within an acceptable (and often good or very good) degree of accuracy.

In this light, it would seem that a specific analysis of continuous systems is not strictly necessary from a practical point of view. However, this analysis is of fundamental nature in its own right because it provides physical insight on the nature of the 'proper modes of vibration' of a structure (i.e. the eigenvectors in the finite DOF representation) which extends over a finite domain of space. In fact, these are specific motions of the system which arise from the superposition of travelling waves that propagate back and forth within the domain of space occupied by the structure. Ultimately, it is the presence of physical boundaries which is responsible for the onset on these 'standing waves'.

Our analysis starts with the study of one of the simplest continuous systems, the flexible string in transverse motion. Under some basic assumptions which are at the basis of the 'classical' treatment of the subject, we consider first the topic of transverse waves travelling along a string of infinite length, and only at a later stage (Section 8.3) do we turn our attention to strings of finite length. By adopting the well-known method of separation of variables, we arrive at the identification of the natural frequencies and proper modes of vibration of the system and obtain a solution of the relevant equation of motion in terms of these quantities. It is at this stage of the investigation that we point out the important role played by boundary conditions in the analysis of the vibrating characteristics of continuous systems.

Next, Section 8.4 deals with the free longitudinal (axial) and torsional vibrations of rods of finite length and the discussion develops along the same line of reasoning of the flexible string case. This is due to the fact that, mathematically speaking, the three cases are formally similar – that is, given the appropriate meaning to the mathematical symbols in each case, the equation of motion is always in the form of a one-dimensional wave equation. The same does not apply to the case of flexural vibrations of beams, a type of system whose equation of motion is in the form of a fourth-order differential equation (Section 8.5). Moreover, as far as travelling waves are concerned, the beam is a so-called 'dispersive' medium, where this term indicates the fact that waves of different frequency travel at different speed so that a flexural pulse (i.e. a given waveshape which is a superposition of a number of different sinusoidal components) will suffer distortion as it propagates along the beam. Restricting our attention to the analysis of a so-called 'Euler–Bernoulli' beam of finite length, we consider the natural frequencies (eigenvalues) and proper modes (eigenfunctions) of some of the most common configurations encountered in engineering practice, the various configurations differing in the nature of the boundary conditions. These are: (1) the pinned–pinned beam, (2) the cantilever beam, (3) the clamped–clamped beam and (4) the free–free beam. As expected from the developments of preceding chapters, we point out that the free–free (unrestrained) beam, besides the elastic modes of vibration, shows two rigid-body modes at zero frequency.

Then, with the above results in mind, Sections 8.5.1 and 8.5.2 deal with three types of complications: the first with the effect of an axial force (as, say, in the case of a prestressed beam) and the second effect of rotatory inertia and shear deformation. The situation in which the latter two types are taken into account (one generally speaks of a Timoshenko beam in this case) is of practical importance either when the beam is not sufficiently slender or we want to consider high-order modes of vibration. Moreover, the corrections introduced also eliminate the nonphysical result of infinite wave velocity at high frequencies encountered in the case of an Euler–Bernoulli beam.

Proceeding in order of increasing complexity, Section 8.6 deals with the two-dimensional counterpart of the flexible string – i.e. the flexible membrane – and investigates in detail the case of a circular membrane of radius R. The discussion then turns to more theoretical aspects of the analysis of continuous systems in general (Sections 8.7, 8.7.1 and 8.7.2), with the scope of focusing the attention on some unifying characteristics of the problem. Recalling some important results given in Chapter 2, we note that the ideas developed in Chapters 6 and 7 for finite DOF (discrete) systems can be extended, at the price of additional mathematical complexity, to the case of infinite DOF (continuous) systems. The generalized eigenvalue problem becomes now a differential eigenvalue problem where the system's matrices are replaced by appropriate linear symmetrical operators, the finite-dimensional vector space of the discrete case becomes an infinite-dimensional complete linear space with an inner product (a so-called Hilbert space and, more specifically, the Hilbert space $L^2(\Omega)$, where Ω is an appropriate finite domain of physical space) and the expansion theorem in terms of eigenvectors becomes a series expansion in terms of eigenfunctions whose convergence, in the general case, is understood in the L^2-sense.

These considerations retain their validity even when, for more complex systems, we are not able to obtain a solution in closed form because of the increasing mathematical complexity of the problem. As an example of these difficulties, we consider in Section 8.8 the free vibration of thin plates, a type of system which, qualitatively, represents the two-dimensional counterpart of the beam. Although it is still possible to obtain a closed-form solution for some types of boundary conditions – and we do so for a circular plate clamped at its outer edge and a rectangular plate simply supported on all sides – it is clear that the problem soon becomes impracticable and not amenable to an analytical solution. When this is the case, we have to resort to a discrete model with a finite number of degrees of freedom, which makes it possible to obtain an approximate solution and takes us back to the subject of MDOF systems.

Finally, also providing a number of worked-out examples, Sections 8.9 and 8.9.1 deal with the forced vibration of continuous systems, with particular attention to the modal approach, i.e. the strategy that allows the analyst to express the solution of the problem as a series expansion in terms of eigenmodes. Clearly, this is not the only method of attack (and sometimes it may not even be the best) but this choice is often preferred in the field of engineering vibrations because:

1. In the study of the response of a vibrating system to a given excitation there is the possibility of including only a limited number of modes and neglect all higher-order modes which do not significantly contribute to the response
2. The technique of experimental modal analysis (Chapter 10) allows the experimental measurement of the lowest-order eigenfrequencies and eigenmodes of a given vibrating system and makes it possible to compare experimental and theoretical results (these latter having been obtained, typically, from a finite-element model).

References

1. Morse, P.M. and Ingard, K.U., *Theoretical Acoustics*, Princeton University Press, 1986.
2. Elmore, W.C. and Heald, M.A., *Physics of Waves*, Dover, New York, 1985.
3. Nariboli, G.A. and McConnell, K.G., Curvature coupling of catenary cable equations, *International Journal of Analytical and Experimental Modal Analysis*, 3(2), 49–56, 1988.
4. Irvine, M., *Cable Structures*, Dover, New York, 1992.
5. Kolsky, H., *Stress Waves in Solids*, Dover, New York, 1963.

6. Meirovitch, L., *Analytical Methods in Vibrations*, Macmillan, New York, 1967.
7. Cowper, G.R., The shear coefficient in Timoshenko's beam theory, *ASME Journal of Applied Mechanics*, 33, 335–340, 1966.
8. Graff, K.F., *Wave Motion in Elastic Solids*, Dover, New York, 1991.
9. Meirovitch, L., *Principles and Techniques of Vibrations*, Prentice Hall, Englewood Cliffs, NJ, 1997.
10. Abramowitz, M. and Stegun, I., *Handbook of Mathematical Functions*, Dover, New York, 1965.
11. Page, C.H., *Physical Mathematics*, D. van Nostrand, Princeton, NJ, 1955.
12. Bazant, Z.P. and Cedolin L., *Stability of Structures*, Oxford University Press, 1991.
13. Jahnke, E., Emde, F. and Losch, F., *Tables of Higher Functions*, McGraw-Hill, New York, 1960.
14. Anon., *Tables of the Bessel Functions of the First Kind (Orders 0 to 135)*, Harvard University Press, Cambridge, Mass., 1947.
15. Leissa, A.W., *Vibration of Plates*, NASA SP-160, 1969.
16. Gerardin, M. and Rixen, D., *Mechanical Vibrations: Theory and Applications to Structural Dynamics*, John Wiley, New York, 1994.
17. Timoshenko, S. and Woinowsky-Krieger, S., *Theory of Plates and Shells*, McGraw-Hill, New York, 1959.
18. Mansfield, E.H., *The Bending and Stretching of Plates*, Pergamon Press, 1964.
19. Humar, J.L., *Dynamics of Structures*, Prentice Hall, Englewood Cliffs, NJ, 1990.
20. Courant, R. and Hilbert, D., *Methods of Mathematical Physics*, Vol. 1, Interscience, New York, 1961.
21. Mathews, J. and Walker, R.L., *Mathematical Methods of Physics*, 2nd edn, Addison-Wesley, Reading, Mass., 1970.
22. Meirovitch, L., *Elements of Vibration Analysis*, McGraw-Hill, New York, 2nd edn, 1986.

9 MDOF and continuous systems: approximate methods

9.1 Introduction

The reader is probably well aware of the fact that – in the last 30 years or so – the most successful approximate technique that is able to deal adequately with simple as well as with complex systems is the finite-element method (FEM). Moreover, since a number of finite-element codes are on the market at reasonable prices and more and more computationally sophisticated procedures are being developed, it is easy to predict that this current state of affairs is probably not going to change for many years to come. Finite-element codes for engineering problem solving were initially developed for structural mechanics applications, but their versatility soon led analysts to recognize that this same technique could be applied with profit to a larger number of problems covering almost the whole spectrum of engineering disciplines – statics, dynamics, heat transfer, fluid flow, etc. Since the essence of the finite-element approach is to establish and solve a (usually very large) set of algebraic equations, it is clear that the method is particularly well suited to computer implementation and that here, with little doubt, lies the key to its success.

However, since their advent, finite-element procedures have taken on a life of their own, so to speak, so that entire books are dedicated to the subject. This makes discussion here impractical for two reasons: first, it would divert us from the main topic of the book and, second, space limitations would necessarily imply that some important information had to be left out. So, although we will occasionally make some comments on FEMs in the course of the book, the interested reader is urged to refer to specific literature: for example, Bathe[1], Spyrakos[2] and Weaver and Johnston[3].

As a consequence of the considerations above, this chapter will be dedicated to more 'classical' approximation methods, basing our treatment on the fact that in common engineering practice it is often required, as a first approach to problems, to have an idea of only a few of the first natural frequencies – and eventually eigenfunctions – of a given vibrating system. In this light, discrete MDOF systems and continuous systems are considered together.

Finally, it must be noted that some of the concepts that will be discussed, despite the possibility to use them as computational tools, have important implications and far-reaching consequences that pervade all the field of engineering vibrations analysis.

9.2   The Rayleigh quotient

In Section 5.5.1 we first encountered the concept of Rayleigh's quotient. The line of reasoning is based on the consideration that for an undamped (or lightly damped) system vibrating harmonically at one of its natural frequencies the stiffness/mass ratio is equal to that particular frequency. To be more specific, consider a n-DOF system with symmetrical mass and stiffness matrices which is vibrating at its jth natural frequency ω_j. The motion of the system is harmonic in time so that the displacement vector is written as $\mathbf{u} = \mathbf{z}_j e^{i\omega_j t}$, where $\mathbf{z}_j$ is the jth eigenvector. The maximum potential and kinetic energies in this circumstance (since no energy is lost and no energy is fed into the system over one cycle) must be equal and are given by

$$E_{p,\,max} = \frac{1}{2}\,\mathbf{u}_{max}^T \mathbf{K}\mathbf{u}_{max} = \frac{1}{2}\,\mathbf{z}_j^T \mathbf{K}\mathbf{z}_j$$

$$\hat{E}_{k,\,max} = \frac{1}{2}\,\dot{\mathbf{u}}_{max}^T \mathbf{M}\dot{\mathbf{u}}_{max} = \frac{1}{2}\,\omega_j^2 \mathbf{z}_j^T \mathbf{M}\mathbf{z}_j$$

(9.1)

respectively. Hence, $E_{p,\,max} = E_{k,\,max}$ implies

$$\omega_j^2 \equiv \lambda_j = \frac{\mathbf{z}_j^T \mathbf{K}\mathbf{z}_j}{\mathbf{z}_j^T \mathbf{M}\mathbf{z}_j} = \mathbf{p}_j^T \mathbf{K}\mathbf{p}_j \qquad (9.2a)$$

where the $\mathbf{p}_j$s ($j = 1, 2, ..., n$) are the mass orthonormal eigenvectors. On the other hand, a symmetrical continuous system leads to the same result if we consider the parallel between MDOF and continuous systems outlined in Sections 8.7 and 8.7.1. The continuous systems counterpart of eq (9.2a) reads

$$\omega_j^2 = \frac{\langle u_j \mid K u_j \rangle}{\langle u_j \mid M u_j \rangle} = \langle \phi_j \mid K\phi_j \rangle \qquad (9.2b)$$

where the eigenfunctions ϕ_j ($j = 1, 2, 3, ...$) are chosen to satisfy the condition $\langle \phi_j \mid M\phi_j \rangle = 1$.

We will now consider a discrete n-DOF system and see what happens to the ratio (9.2a) when the vector entering the inner products at the numerator and denominator is not an eigenvector of the system under investigation.

Let $\mathbf{u}$ be a general vector and let $\mathbf{p}_j$, $j = 1, 2, ..., n$, be the set of mass orthonormal eigenvectors of our system. We define the Rayleigh quotient as

$$R(\mathbf{u}) \equiv \frac{\mathbf{u}^T \mathbf{K} \mathbf{u}}{\mathbf{u}^T \mathbf{M} \mathbf{u}} \tag{9.3}$$

By virtue of the expansion theorem, we can write

$$\mathbf{u} = \sum_{j=1}^{n} c_j \mathbf{p}_j \tag{9.4}$$

substitute it into eq (9.3) and obtain

$$R(\mathbf{u}) = \frac{\sum_{j=1}^{n} \lambda_j c_j^2}{\sum_{j=1}^{n} c_j^2} = \lambda_1 \frac{c_1^2 + \sum_{j=2}^{n} (\lambda_j/\lambda_1) c_j^2}{\sum_{j=1}^{n} c_j^2} \tag{9.5}$$

from which it follows, since $\lambda_j \geqslant \lambda_1$ for $j = 2, 3, ..., n$

$$R(\mathbf{u}) \geqslant \lambda_1 \tag{9.6}$$

meaning that the Rayleigh quotient for an arbitrary vector is always greater than the first eigenvalue; the equality holds only if $c_2 = c_3 = \cdots = c_n = 0$ or, in other words, when $\mathbf{u}$ coincides with the lowest eigenvector. Furthermore, if $\mathbf{u}$ is chosen in such a way as to be mass-orthogonal to the first $m - 1$ eigenfunctions, i.e. when $\mathbf{p}_j^T \mathbf{M} \mathbf{u} = 0$ for $j = 1, 2, ..., m - 1$ it follows that $c_1 = c_2 = \cdots = c_{m-1} = 0$ and

$$R(\mathbf{u}) = \lambda_m \frac{c_m^2 + \sum_{j=m+1}^{n} (\lambda_j/\lambda_m) c_j^2}{\sum_{j=m+1}^{n} c_j^2}$$

Hence

$$R(\mathbf{u}) \geqslant \lambda_m \tag{9.7}$$

and the equality holds when $\mathbf{u}$ coincides with the mth eigenvector. By the same token, we note that in writing the Rayleigh quotient we can factor out the highest eigenvalue to get

$$R(\mathbf{u}) = \lambda_n \frac{c_n^2 + \sum\limits_{j=1}^{n-1} (\lambda_j/\lambda_n)c_j^2}{\sum\limits_{j=1}^{n} c_j^2}$$

so that

$$R(\mathbf{u}) \leqslant \lambda_n \tag{9.8}$$

since $\lambda_j/\lambda_n \leqslant 1$ $(j = 1, 2, ..., n-1)$. Suppose now that the vector $\mathbf{u}$ is an approximation of the kth eigenvector $\mathbf{p}_k$, i.e. with ε small, we have

$$\mathbf{u} = \mathbf{p}_k + \varepsilon \mathbf{x} \tag{9.9}$$

where the term $\varepsilon \mathbf{x}$ takes into account the (small) contributions to $\mathbf{u}$ from all eigenvectors other than $\mathbf{p}_k$. Inserting eq (9.9) into eq (9.3) we get

$$R(\mathbf{u}) = \frac{\mathbf{p}_k^T \mathbf{K} \mathbf{p}_k + \varepsilon \mathbf{p}_k^T \mathbf{K} \mathbf{x} + \varepsilon \mathbf{x}^T \mathbf{K} \mathbf{p}_k + \varepsilon^2 \mathbf{x}^T \mathbf{K} \mathbf{x}}{\mathbf{p}_k^T \mathbf{M} \mathbf{p}_k + \varepsilon \mathbf{p}_k^T \mathbf{M} \mathbf{x} + \varepsilon \mathbf{x}^T \mathbf{M} \mathbf{p}_k + \varepsilon^2 \mathbf{x}^T \mathbf{M} \mathbf{x}} \tag{9.10}$$

and noting that we can expand the 'error' $\mathbf{x}$ as

$$\mathbf{x} = \sum_{j;j \neq k} a_j \mathbf{p}_j \tag{9.11}$$

so that, owing to the orthogonality properties of eigenvectors,

$$\mathbf{p}_k^T \mathbf{K} \mathbf{x} = \mathbf{x}^T \mathbf{K} \mathbf{p}_k = \mathbf{p}_k^T \mathbf{M} \mathbf{x} = \mathbf{x}^T \mathbf{M} \mathbf{p}_k = 0$$

$$\varepsilon^2 \mathbf{x}^T \mathbf{K} \mathbf{x} = \varepsilon^2 \sum_{j;j \neq k} \lambda_j a_j^2$$

$$\varepsilon^2 \mathbf{x}^T \mathbf{M} \mathbf{x} = \varepsilon^2 \sum_{j;j \neq k} a_j^2$$

eq (9.10) reduces to

$$R(\mathbf{u}) = \frac{\lambda_k + \varepsilon^2 \sum\limits_{j;j \neq k} \lambda_j a_j^2}{1 + \varepsilon^2 \sum\limits_{j;j \neq k} a_j^2}$$

where the denominator can be expanded according to the binomial approximation $(1 + \delta)^{-1} \cong 1 - \delta$ to give

$$R(\mathbf{u}) = \left(\lambda_k + \varepsilon^2 \sum_{j; j \neq k} \lambda_j a_j^2 \right) \left(1 - \varepsilon^2 \sum_{j; j \neq k} a_j^2 + \cdots \right)$$

$$= \lambda_k + \varepsilon^2 \left(\sum_{j; j \neq k} \lambda_j a_j^2 - \lambda_k \sum_{j; j \neq k} a_j^2 \right) + o(\varepsilon^3) \qquad (9.12)$$

The symbol $o(\varepsilon^3)$ means terms of order ε^3 or smaller. This result can be stated in words by saying that when the 'trial vector' $\mathbf{u}$ used in forming the Rayleigh quotient is an approximation of order ε of the kth eigenvector, then the Rayleigh quotient approximates the kth eigenvalue λ_k with an error of order ε^2. Alternatively, we can put it in more mathematical terms and say that the functional $R(\mathbf{u})$ has stationary values in the neighbourhood of eigenvectors: the stationary values are the eigenvalues, while the eigenvectors are the stationary points. To answer the question of whether the stationary points are maxima, minima or saddle points we must rely on some previous considerations and a few others that will follow.

9.2.1 Courant–Fisher minimax characterization of eigenvalues and the eigenvalue separation property

When no orthogonality constraints are imposed on the choice of $\mathbf{u}$ (such as in the discussion that leads to eq (9.5)) we may note that, as our trial vector ranges over the vector space, eqs (9.6) and (9.8) always hold. This leads to the important conclusions that Rayleigh quotient has a minimum when $\mathbf{u} = \mathbf{p}_1$ and a maximum when $\mathbf{u} = \mathbf{p}_n$, so that we can write

$$\lambda_1 = \min_{\mathbf{u} \neq 0} R(\mathbf{u})$$

$$\lambda_n = \max_{\mathbf{u} \neq 0} R(\mathbf{u}) \qquad (9.13)$$

and it is understood that $\mathbf{u}$ can be any arbitrary vector in the n-dimensional Euclidean space of the system's vibration shapes. On the other hand, the following heuristic argument can give us an idea of what happens at a stationary point other than λ_1 and λ_n, say at λ_m, when $\mathbf{u}$ is completely arbitrary. First, we write the obvious chain of inequalities

$$\ldots R(\mathbf{p}_{m-1}) = \lambda_{m-1} \leqslant R(\mathbf{p}_m) = \lambda_m \leqslant R(\mathbf{p}_{m+1}) = \lambda_{m+1} \ldots$$

and then we note that the Rayleigh quotient is a continuous functional of $\mathbf{u}$. Suppose now that, in ranging over the vector space, $\mathbf{u}$ finds itself in the vicinity of $\mathbf{p}_m$; continuity considerations imply that $R(\mathbf{u}) \cong \lambda_m$. Then, if our trial vector moves toward $\mathbf{p}_{m-1}$ the value of the Rayleigh's quotient will tend to decrease while it will tend to increase if $\mathbf{u}$ moves toward $\mathbf{p}_{m+1}$. The conclusion is that the stationary point at λ_m is a saddle point, i.e. the counterpart of a point of inflection with horizontal tangent when we look for the extremum points of a function $f(x)$ in ordinary calculus.

The situation is different if the trial vector is not completely arbitrary but satisfies a number of orthogonality constraints. In this case $\mathbf{u}$ is not free to range over the entire vector space and, referring back to the discussion that led to eq (9.7), we can write

$$\lambda_m = \min_{\mathbf{u} \perp \mathbf{p}_1, \ldots, \mathbf{p}_{m-1}} R(\mathbf{u}) \tag{9.14}$$

meaning that Rayleigh's quotient has a minimum value of λ_m (which occurs when $\mathbf{u} = \mathbf{p}_m$) for all trial vectors orthogonal to the first $m - 1$ eigenvectors $\mathbf{p}_1, \ldots, \mathbf{p}_{m-1}$.

If we turn now to the utility of the considerations above we note that Rayleigh's quotient may provide a method for estimating the eigenvalues of a given system. In practice, however, this possibility is often limited to the first eigenvalue because the calculation procedure (see also Section 5.5.1) must start with a reasonable guess of the eigenshape that corresponds to the eigenvalue we want to estimate. This means that – unless we are dealing with a very simple system, in which case we can attack the problem directly – only the first eigenshape can generally be guessed with an acceptable degree of confidence. Moreover, the deflection produced by a static (typically gravity) load often proves to be a good trial function for the estimate of λ_1, while no such intuitive hints exist for higher modes.

So, as far as the first eigenvalue is concerned, the method is very useful and can also be improved by forming a sequence of trial vectors designed to minimize the value of the functional $R(\mathbf{u})$ which, owing to eq (9.6), will tend to λ_1; this is exactly the procedure we followed in Section 5.5.1 and identified under the name of 'improved Rayleigh method'. It goes without saying that the lowest eigenvalue is the most important in a large number of applications.

By contrast, eq (9.14) is of little practical utility because we usually have no information on the lowest eigenvectors $\mathbf{p}_1, \ldots, \mathbf{p}_{m-1}$. At this point we could ask whether it is possible to obtain some information on the intermediate eigenvalues without any previous knowledge of the lower eigenvectors. This is precisely the result of the Courant–Fisher theorem. It must be pointed out that the importance of the theorem itself and of its consequences is not so much in the possibility of estimating eigenvalues independently, but in its fundamental nature; in fact, it provides a rigorous mathematical basis for a large number of developments in the solution of eigenvalue problems (e.g. Wilkinson [4], Bathe [1] and Meirovitch [5]).

The Courant–Fisher theorem, which we state here without proof, is generally given in terms of a single Hermitian (symmetrical, if all its entries are real) matrix in the following form:

Theorem 9.1. Let $\mathbf{A}$ be a Hermitian matrix with eigenvalues $\lambda_1 \leqslant \lambda_2 \leqslant \cdots \leqslant \lambda_n$ and let m be a given integer with $1 \leqslant m \leqslant n$. Then

$$\lambda_m = \max\left[\min_{\mathbf{u} \perp \mathbf{w}_1, \mathbf{w}_2, \ldots, \mathbf{w}_{m-1}} \frac{\mathbf{u}^T \mathbf{A} \mathbf{u}}{\mathbf{u}^T \mathbf{u}}\right] \tag{9.15}$$

and

$$\lambda_m = \min\left[\max_{\mathbf{u} \perp \mathbf{w}_1, \mathbf{w}_2, \ldots, \mathbf{w}_{n-m}} \frac{\mathbf{u}^T \mathbf{A} \mathbf{u}}{\mathbf{u}^T \mathbf{u}}\right] \tag{9.16}$$

where the $\mathbf{w}_i$s (in the appropriate number to satisfy eq (9.15) or (9.16)) are a set of (mutually independent) given vectors of the vector space.

A few comments are in order at this point.

First of all, we may note that the typical eigenvalue problem of vibration analysis involves two symmetrical matrices, while the theorem above is written for a single matrix alone. However, this is only a minor inconvenience, because we have shown in Chapter 6 (Section 6.8, eqs (6.165) and (6.166)) that the generalized eigenvalue problem can be transformed into a standard eigenvalue problem in terms of a single symmetrical matrix. Obviously, when we are dealing with this single matrix, which we call, $\mathbf{A}$ the Rayleigh quotient is defined as

$$R(\mathbf{u}) \equiv \frac{\mathbf{u}^T \mathbf{A} \mathbf{u}}{\mathbf{u}^T \mathbf{u}}$$

Second, when $m = 1$ or $m = n$, the theorem reduces to the statements $\lambda_1 = \min R(\mathbf{u})$ and $\lambda_n = \max R(\mathbf{u})$, which are the 'single matrix' counterpart of eqs (9.13).

In general, the statement of greatest interest to us is given by eq (9.15), because the attention is usually on lowest order eigenvalues. With this in mind, let us look more closely at this statement of the theorem. For example, suppose that we are trying to estimate λ_2; we can choose an arbitrary $n \times 1$ vector $\mathbf{w}$ and constrain our trial vector $\mathbf{u}$ to be orthogonal to $\mathbf{w}$, i.e. to satisfy the constraint equation

$$\mathbf{u}^T \mathbf{w} = 0 \tag{9.17}$$

Now, under the mathematical constraint expressed by eq (9.17), if $\mathbf{u}$ and $\mathbf{w}$ are allowed to vary within the vector space, the maximum value that can be obtained among the values $\min(\mathbf{u}^T\mathbf{A}\mathbf{u}/\mathbf{u}^T\mathbf{u})$ is exactly λ_2. If the eigenvalue we are trying to estimate is λ_3, two mathematical constraints are needed, meaning that we choose two vectors $\mathbf{w}_1, \mathbf{w}_2$ and our trial vector must satisfy both conditions $\mathbf{u}^T\mathbf{w}_1 = \mathbf{u}^T\mathbf{w}_2 = 0$. Therefore, as a matter of fact, the Courant–Fisher theorem can also be looked upon as an optimization procedure to estimate eigenvalues.

On more physical grounds, we may summarize the evaluation of, say, λ_2 by noting that enforcing the vibration shape $\mathbf{u}$ on our system – unless $\mathbf{u}$ coincides with one of the eigenshapes – necessarily increases the stiffness of our system, the mass being fixed. In practice, we are dealing with a new system whose first eigenvalue $\hat{\lambda}_1$ satisfies the obvious inequality $\lambda_1 \leqslant \hat{\lambda}_1$, but also, owing to the constraint (9.17), $\hat{\lambda}_1 \leqslant \lambda_2$ (this inequality is less obvious, but it is not difficult to prove; the proof is left to the reader). Then, the theorem states that the maximum value of $\hat{\lambda}_1$ that can be obtained under these conditions is λ_2. Likewise, the evaluation of λ_m implies $m - 1$ mathematical constraints of the form (9.17).

There are a number of important consequences of the Courant–Fisher theorem; for our purposes, one that deserves particular attention is the so-called separation property of the eigenvalues (or interlacing property), which we state here without proof in the form of the following theorem.

Theorem 9.2.   Let A be a given Hermitian $n \times n$ matrix with eigenvalues λ_j, $j = 1, 2, \ldots n$. If we consider the eigenproblems

$$\mathbf{A}^{(k)}\mathbf{u}^{(k)} = \lambda^{(k)}\mathbf{u}^{(k)} \tag{9.18}$$

where $\mathbf{A}^{(k)}$ is obtained by deleting the last k rows and columns of $\mathbf{A}$, we have the eigenvalue separation property

$$\lambda_1^{(k)} \leqslant \lambda_1^{(k+1)} \leqslant \lambda_2^{(k)} \leqslant \lambda_2^{(k+1)} \leqslant \cdots \leqslant \lambda_{n-k-1}^{(k)} \leqslant \lambda_{n-k-1}^{(k+1)} \leqslant \lambda_{n-k}^{(k)} \tag{9.19}$$

where the index k may range from 0 to $n - 2$.

In other words, if, for example, we turn our eigenvalue problem of order n into an eigenproblem of order $n - 1$ by deleting the last row and column from the original matrix, the eigenvalues of the $n - 1$ eigenproblem are 'bracketed' by the eigenvalues of the original problem. Conversely, if $\mathbf{A}$ is a $n \times n$ Hermitian matrix, $\mathbf{v}$ a given $n \times 1$ vector and b is a real number, the eigenvalues $\hat{\lambda}_j$ of the $(n+1) \times (n+1)$ matrix

$$\hat{\mathbf{A}} \equiv \begin{bmatrix} \mathbf{A} & \mathbf{v} \\ \mathbf{v}^T & b \end{bmatrix}$$

satisfy the inequalities $\hat{\lambda}_1 \leqslant \lambda_1 \leqslant \hat{\lambda}_2 \leqslant \lambda_2 \leqslant \cdots \leqslant \hat{\lambda}_n \leqslant \lambda_n \leqslant \hat{\lambda}_{n+1}$. The extension of Theorem 9.2 to the case of two real, positive definite $n \times n$ matrices is not difficult and it can be shown that the eigenvalues of the two eigenproblems

$$\mathbf{K}\mathbf{u} = \lambda \mathbf{M}\mathbf{u}$$

$$\hat{\mathbf{K}}\mathbf{u} = \hat{\lambda}\hat{\mathbf{M}}\mathbf{u}$$

in which the $n \times n$ matrices $\mathbf{K}$ and $\mathbf{M}$ are obtained by bordering $\hat{\mathbf{K}}$ and $\hat{\mathbf{M}}$ (of order $(n-1) \times (n-1)$) with the $(n-1) \times 1$ vectors $\mathbf{k}$ and $\mathbf{m}$ and the scalars k and m, respectively, satisfy the separation (interlacing) property.

9.2.2 Systems with lumped masses – Dunkerley's formula

In the preceding sections, we pointed out that, for a given system, the Rayleigh quotient provides an approximation of its lowest eigenvalue which satisfies the inequality $R(\mathbf{u}) \geqslant \lambda_1$. This means that, unless the choice of the trial vector is particularly lucky, $R(\mathbf{u})$ always overestimates the value of λ_1. For a limited (but not small) class of systems, we will now show that Dunkerley's formula provides a different method to estimate λ_1. Furthermore, the value that we obtain in this case is always an underestimate of λ_1.

Suppose that we are dealing with a positive definite n-DOF system in which the masses are localized (lumped) at n specific points. Then, if we choose the coordinates as the absolute displacements of the masses, the mass matrix is diagonal (Section 6.5).

The generalized eigenproblem for this system is written in the usual form as $\mathbf{K}\mathbf{u} = \lambda \mathbf{M}\mathbf{u}$, but it can also be expressed as a standard eigenproblem in terms of the flexibility matrix $\mathbf{A} = [a_{ij}] \equiv \mathbf{K}^{-1}$ (whose existence is guaranteed by positive definiteness), i.e.

$$\mathbf{A}\mathbf{M}\mathbf{u} = \frac{1}{\lambda}\mathbf{u} \qquad\qquad (9.20)$$

If the system has lumped masses and hence $\mathbf{M} = \text{diag}(m_j)$, the matrix $\mathbf{A}\mathbf{M}$ has the particularly simple form

$$\begin{bmatrix} a_{11}m_1 & a_{12}m_2 & \cdots & a_{1n}m_n \\ a_{21}m_1 & a_{22}m_2 & \cdots & a_{2n}m_n \\ \cdots & \cdots & \cdots & \cdots \\ a_{n1}m_1 & a_{n2}m_2 & \cdots & a_{nn}m_n \end{bmatrix}$$

so that – by virtue of a well known result of linear algebra stating that the trace (sum of its diagonal elements) of a matrix is equal to the sum of its eigenvalues – we can write

$$\text{trace}(\mathbf{AM}) = \sum_{j=1}^{n} a_{jj}m_j = \sum_{j=1}^{n} \frac{1}{\lambda_j} = \sum_{j=1}^{n} \frac{1}{\omega_j^2} \tag{9.21}$$

from which Dunkerley's formula follows, i.e.

$$\sum_{j=1}^{n} a_{jj}m_j > \frac{1}{\lambda_1} = \frac{1}{\omega_1^2} \tag{9.22a}$$

or, equivalently,

$$\left(\sum_{j=1}^{n} a_{jj}m_j \right)^{-1} < \lambda_1 \tag{9.22b}$$

The advantage of eq (9.22b) for lumped mass systems lies in the fact that the diagonal elements of the flexibility matrix are generally the easiest ones to evaluate and that, once the lumping of masses has been decided, the m_j are all known. As opposed to the Rayleigh quotient, the main drawbacks of Dunkerley's formula are that the method does not apply to unrestrained systems and that it is not possible to have an 'equals' sign in eqs (9.22a and b), meaning that, in other words, Dunkerley's formula always yields an approximate value.

9.3 The Rayleigh–Ritz method and the assumed modes method

The **Rayleigh–Ritz method** is an extension of the Rayleigh method suggested by Ritz. In essence, the Rayleigh method allows the analyst to calculate approximately the lowest eigenvalue of a given system by appropriately choosing a trial vector **u** (or a function for continuous systems) to insert in the Rayleigh quotient. The quality of the estimate obviously depends on this choice, but the stationarity of Rayleigh quotient – provided that the choice is reasonable – guarantees an acceptable result. Moreover, if the assumed shape contains one or more variable parameters, the estimate can be improved by differentiating with respect to this/these parameter(s) to seek the minimum value of $R(\mathbf{u})$. The Rayleigh–Ritz method depends on this idea and can be used to calculate approximately a certain number of undamped eigenvalues and eigenshapes of a given discrete or continuous system.

Consider for the moment a n-DOF system, where n is generally large. Our main interest may lie in the first m eigenvalues and eigenvectors, with $m \ll n$. In this light, we express the displacement shape of our system as the superposition of m independent Ritz trial vectors z_j, i.e.

$$\mathbf{u} = \sum_{j=1}^{m} c_j \mathbf{z}_j = \mathbf{Zc} \tag{9.23}$$

where the generalized coordinates c_j are, as yet, unknown and in the matrix expression we defined the $n \times m$ and $m \times 1$ matrices $\mathbf{Z} \equiv [\mathbf{z}_1, \mathbf{z}_2, ..., \mathbf{z}_m]$ and $\mathbf{c} \equiv [c_1, c_2, ..., c_m]^T$. Evidently, the closer the Ritz vectors are to the true vibration shapes, the better are the results.

The displacement shape (9.23) is then inserted in the Rayleigh quotient to give

$$R(\mathbf{u}) = \frac{\mathbf{c}^T \mathbf{Z}^T \mathbf{K} \mathbf{Z} \mathbf{c}}{\mathbf{c}^T \mathbf{Z}^T \mathbf{M} \mathbf{T} \mathbf{c}} = \frac{\mathbf{c}^T \hat{\mathbf{K}} \mathbf{c}}{\mathbf{c}^T \hat{\mathbf{M}} \mathbf{c}} \tag{9.24}$$

so that the coefficients c_j can be determined by making $R(\mathbf{u})$ stationary. The $m \times m$ matrices $\hat{\mathbf{K}}$ and $\hat{\mathbf{M}}$ in eq (9.24) are given in terms of the stiffness and mass matrix of the original system as

$$\hat{\mathbf{K}} = \mathbf{Z}^T \mathbf{K} \mathbf{Z}$$
$$\hat{\mathbf{M}} = \mathbf{Z}^T \mathbf{M} \mathbf{Z} \tag{9.25}$$

Before proceeding further, we may note that the assumption (9.23) consists of approximating our n-DOF by a m-DOF system, meaning that, in essence, we impose the constraints

$$c_{m+1} = c_{m+2} = \cdots = c_n = 0 \tag{9.26}$$

on the original system. Since constraints tend to increase the stiffness of a system, we may expect two consequences: the first is that the m eigenvalues obtained by this method will overestimate the lowest m 'true' eigenvalues and the second is that an increase of m will yield better estimates because, by doing so, we just eliminate some of the constraints (9.26).

The necessary conditions to make $R(\mathbf{u})$ stationary are

$$\frac{\partial R(\mathbf{u})}{\partial c_j} = 0 \qquad (j = 1, 2, ..., m) \tag{9.27}$$

which, taking eq (9.24) into account, become

$$\frac{\partial}{\partial c_j} (\mathbf{c}^T \hat{\mathbf{K}} \mathbf{c}) - R(\mathbf{u}) \frac{\partial}{\partial c_j} (\mathbf{c}^T \hat{\mathbf{M}} \mathbf{c}) = 0 \qquad (j = 1, 2, ..., m) \tag{9.28}$$

Now, owing to the symmetry of $\hat{\mathbf{K}}$ and $\hat{\mathbf{M}}$, the calculation of the derivatives in eq (9.28) leads to a set of equations that can be put together into the single matrix equation

$$\hat{\mathbf{K}}\mathbf{c} - R(\mathbf{u})\hat{\mathbf{M}}\mathbf{c} = 0 \tag{9.29}$$

which, by defining $\hat{\lambda} = \hat{\omega}^2 \equiv R(\mathbf{u})$, we recognize as a generalized eigenvalue problem of order m. This result shows that the effect of the Rayleigh–Ritz method is to reduce the number of degrees of freedom to a predetermined value m. In this regard, it is important to note that the number of eigenvalues and eigenvectors that can be obtained with acceptable accuracy is generally less than the number of Ritz vectors; in other words, if our interest is in the first m eigenpairs, it is advisable to include s Ritz shapes in the process, where, let us say, $s \cong 2m$.

The eigenproblem (9.29) can be solved by means of any standard eigensolver and the result will be a set of eigenvalues $\hat{\lambda}_1, ..., \hat{\lambda}_m$ with the corresponding eigenvectors $\mathbf{c}_1, ..., \mathbf{c}_m$; the eigenvalues are approximations of the true lower eigenvalues of the original system, while the eigenvectors are not the mode shapes of the original system. The $\mathbf{c}_j$s are orthogonal with respect to the matrices $\hat{\mathbf{K}}$ and $\hat{\mathbf{M}}$ and can be normalized by any appropriate normalization procedure. If we call these normalized eigenvectors $\hat{\mathbf{c}}_j$, we can obtain the approximations of the m mode shapes of the original system from eq (9.23), i.e. by writing

$$\hat{\mathbf{p}}_j = \mathbf{Z}\hat{\mathbf{c}}_j \qquad (j = 1, 2, ..., m) \tag{9.30}$$

and note that these approximate eigenvectors are orthogonal with respect to the matrices of the original system: that is, by virtue of eq (9.25), we have

$$\hat{\mathbf{p}}_i^T \mathbf{K} \hat{\mathbf{p}}_j = \hat{\mathbf{c}}_i^T \mathbf{Z}^T \mathbf{K} \mathbf{Z} \hat{\mathbf{c}}_j = \hat{\mathbf{c}}_i^T \hat{\mathbf{K}} \hat{\mathbf{c}}_j = \hat{K}_{jj} \delta_{ij}$$

$$\hat{\mathbf{p}}_i^T \mathbf{M} \hat{\mathbf{p}}_j = \hat{\mathbf{c}}_i^T \mathbf{Z}^T \mathbf{M} \mathbf{Z} \hat{\mathbf{c}}_j = \hat{\mathbf{c}}_i^T \hat{\mathbf{M}} \hat{\mathbf{c}}_j = \hat{M}_{jj} \delta_{ij} \tag{9.31a}$$

where we called $\hat{K}_{jj}$ and $\hat{M}_{jj}$ the jth generalized stiffness and mass of the reduced system, respectively (their values obviously depend on how we decide to normalize the vectors $\mathbf{c}_j$). The natural consequences of eq (9.31a) are that

$$\frac{\hat{\mathbf{p}}_j^T \mathbf{K} \hat{\mathbf{p}}_j}{\hat{\mathbf{p}}_j^T \mathbf{M} \hat{\mathbf{p}}_j} = \frac{\mathbf{c}_j^T \hat{\mathbf{K}} \mathbf{c}_j}{\mathbf{c}_j^T \hat{\mathbf{M}} \mathbf{c}_j} = \hat{\lambda}_j \qquad (j = 1, 2, ..., m) \tag{9.31b}$$

and that these approximate vectors can be used in the standard mode superposition procedure for dynamic analysis.

From the above considerations it appears that the choice of the Ritz shapes is probably the most difficult step of the whole method. In general, this is so; however, we may note that the line of reasoning adopted in the improved

Rayleigh method (Section 5.5.1) is still valid. Suppose, in fact, that we choose a set of m initial trial vectors arranged in the matrix $\mathbf{Z}^{(0)}$; on physical grounds we can argue that the deflected shapes originating from the action of the inertia forces due to $\mathbf{Z}^{(0)}$ represent a better set of Ritz vectors. These are given by $\tilde{\mathbf{Z}}^{(1)} = \lambda \mathbf{K}^{-1} \mathbf{M} \mathbf{Z}^{(0)}$ but cannot be calculated because of the unknown factor λ. So, we choose the vectors

$$\mathbf{Z}^{(1)} \equiv \mathbf{K}^{-1} \mathbf{M} \mathbf{Z}^{(0)} \tag{9.32}$$

and use them in eq (9.23) in order to arrive at the eigenvalue problem (9.29) where now, introducing the $n \times n$ flexibility matrix of the original system $\mathbf{A} = \mathbf{K}^{-1}$, we have

$$\hat{\mathbf{K}} \equiv \hat{\mathbf{K}}^{(1)} = (\mathbf{K}^{-1} \mathbf{M} \mathbf{Z}^{(0)})^T \mathbf{K} \mathbf{K}^{-1} \mathbf{M} \mathbf{Z}^{(0)} = \mathbf{Z}^{(0)^T} \mathbf{M} \mathbf{A} \mathbf{M} \mathbf{Z}^{(0)}$$
$$\hat{\mathbf{M}} \equiv \hat{\mathbf{M}}^{(1)} = \mathbf{Z}^{(0)^T} \mathbf{M} \mathbf{A} \mathbf{M} \mathbf{A} \mathbf{M} \mathbf{Z}^{(0)} \tag{9.33}$$

Again, note that the eigenvectors we obtain from this problem are not the eigenvectors of the original system but they must be transformed back by means of the matrix $\mathbf{Z}^{(1)}$. This procedure can also be seen as the first step of an iteration method which allows the analyst to obtain a good approximate 'reduced' solution even when the initial trial vectors do not represent what we might call 'a good guess' of the true vibration shapes. As a matter of fact, a robust numerical procedure based on the line of reasoning outlined above was developed by Bathe and it is called the 'subspace iteration method'. The interested reader may refer, for example, to Bathe[1] (Section 11.6) or Humar[6] (Section 11.3.4).

Also, it is worth noting that the eigenvalues that we obtain by solving the eigenproblem of order m are bracketed by those of the eigenproblem of order $m + 1$ because, in essence, we reduce by one the number of constraints of eq (9.26).

The Raleigh–Ritz method works equally well in the case of continuous systems. In this case, the initial choice consists of a set of m Ritz shape functions $z_j(x)$ and the deflected shape of the system is written as

$$u(x) = \sum_{j=1}^{m} c_j z_j = \mathbf{Z} \mathbf{c} \tag{9.34}$$

where $\mathbf{Z}$ is now the $1 \times m$ matrix $\mathbf{Z} = [z_1 \ldots z_m]$ and $\mathbf{c}$ is as in eq (9.23). Then, by forming the Rayleigh quotient

$$\hat{\lambda} \equiv R(u) = \frac{\langle u \mid Ku \rangle}{\langle u \mid Mu \rangle} = \frac{\displaystyle\sum_{i,j=1}^{m} c_i c_j \langle z_i \mid K z_j \rangle}{\displaystyle\sum_{i,j=1}^{m} c_i c_j \langle z_i \mid M z_j \rangle} = \frac{\displaystyle\sum_{i,j=1}^{m} k_{ij} c_i c_j}{\displaystyle\sum_{i,j=1}^{m} m_{ij} c_i c_j} \tag{9.35}$$

where K and M are, respectively, the symmetrical stiffness and mass operators of the system under investigation and we introduced the notation

$$k_{ij} \equiv \langle z_i \mid K z_j \rangle = \int_{\Omega} z_i K z_j \, d\Omega$$

$$(9.36)$$

$$m_{ij} \equiv \langle z_i \mid M z_j \rangle = \int_{\Omega} z_i M z_j \, d\Omega$$

By enforcing the conditions $\partial R(u)/\delta c_r = 0$ $(r = 1, 2, ..., m)$, we are led to the set of m algebraic equations

$$\sum_{j=1}^{m} (k_{rj} - \hat{\lambda} m_{rj}) c_j = 0 \qquad (r = 1, 2, ..., m)$$

$$(9.37)$$

which can be written in matrix form as the eigenproblem of order m

$$\hat{\mathbf{K}}\mathbf{c} = \hat{\lambda}\hat{\mathbf{M}}\mathbf{c}$$

$$(9.38)$$

where $\hat{\mathbf{K}}$ and $\hat{\mathbf{M}}$ are $m \times m$ symmetrical matrices whose entries are, respectively, k_{ij} and m_{ij}.

Also in this case, the quality of the result depends on the initial choice of the Ritz functions and better approximations are obtained when these functions resemble closely the eigenshapes of the system under investigation. In addition, for a given continuous system, we can intuitively expect that better approximations may be obtained by choosing a set of trial functions which satisfy as many boundary conditions as possible. This latter aspect, which has no counterpart in the discrete case, will be considered in a later section. For the moment it is interesting to note that the eigenfunctions of a simpler but similar system can be, in general, a good choice to represent the Ritz shapes of a more complex system; a typical example could be the use of the first eigenshapes of a beam with uniform properties as the Ritz functions of a beam with the same boundary conditions but a nonuniform mass and stiffness distribution along its length.

The **assumed modes method** is closely related to the Rayleigh–Ritz method and, as a matter of fact, leads to the same results (for this reason, some authors do not make a distinction between the two). In order to outline the assumed modes method, we may refer to a continuous system and note that, in this case, the solution is written in the form

$$u(x, t) = \sum_{j=1}^{m} q_j(t) z_j(x) = \mathbf{Z}\mathbf{q}$$

$$(9.39)$$

where the z_j, the assumed modes, are just a set of Ritz functions, whereas the generalized coordinates q_j depend now on the variable t. This means that, as opposed to Rayleigh–Ritz, the method starts before the elimination of the time-dependent part of the solution and it is used in conjunction with Lagrange's equations to obtain a finite number of ordinary differential equations that govern the time evolution of the q_j.

Given the approximate solution (9.39), the kinetic and potential energy of our system can be written as

$$T = \frac{1}{2} \sum_{i,j=1}^{m} m_{ij} \dot{q}_i \dot{q}_j = \frac{1}{2} \dot{\mathbf{q}}^T \hat{\mathbf{M}} \dot{\mathbf{q}}$$

$$V = \frac{1}{2} \sum_{i,j=1}^{m} k_{ij} q_i q_j = \frac{1}{2} \mathbf{q}^T \hat{\mathbf{K}} \mathbf{q}$$

(9.40)

where the matrices $\hat{\mathbf{M}} = [m_{ij}]$ and $\hat{\mathbf{K}} = [k_{ij}]$ are the symmetrical matrices of eqs (9.36) and (9.38). Next, by considering Lagrange's equations for a conservative holonomic system

$$\frac{d}{dt}\left(\frac{\partial T}{\partial \dot{q}_r}\right) - \frac{\partial T}{\partial q_r} + \frac{\partial V}{\partial q_r} = 0 \qquad (r = 1, 2, ..., m)$$

(9.41)

we can perform the prescribed derivatives to obtain

$$\sum_{j=1}^{m} m_{rj} \ddot{q}_j + \sum_{j=1}^{m} k_{rj} q_j = 0 \qquad (r = 1, 2, ..., m)$$

(9.42a)

which, in matrix form, reads

$$\hat{\mathbf{M}} \ddot{\mathbf{q}}(t) + \hat{\mathbf{K}} \mathbf{q}(t) = 0$$

(9.42b)

so that, assuming a harmonic time dependence for the generalized coordinates q_j, i.e. $\mathbf{q} = \mathbf{c} e^{i\omega t}$, we are led to the generalized eigenvalue problem of order m

$$\hat{\mathbf{K}} \mathbf{c} = \omega^2 \hat{\mathbf{M}} \mathbf{c}$$

(9.43)

which is identical to eq (9.38). Its solution consists of (1) a set of eigenvalues which represent the estimates of the first m eigenvalues of the original system and (2) a set of eigenvectors which represent the amplitudes of the time-dependent harmonic motion and can be used to obtain the first m eigenfunctions of the original system by means of eq (9.39).

Example 9.1. As a simple application of the Rayleigh–Ritz method which can be confronted with the closed form solution, we may consider the problem of approximating the first two eigenvalues of a clamped–clamped beam of length L, uniform flexural stiffness EI and uniform mass per unit length μ. For this example, we choose two Ritz functions which satisfy all the boundary conditions (8.68), i.e.

$$z_1(x) = x^4 - 2Lx^3 + L^2x^2$$

$$z_2(x) = x^5 - 3L^2x^3 + 2L^3x^2$$

(9.44)

then, we calculate the coefficients k_{ij} and m_{ij} as in eq (9.36)

$$k_{11} = \int_0^L z_1 \left(EI \frac{d^4z_1}{dx^4} \right) dx = 0.8EIL^5$$

$$k_{12} = k_{21} = \int_0^L z_1 \left(EI \frac{d^4z_2}{dx^4} \right) dx = 2EIL^6$$

$$k_{22} = \int_0^L z_2 \left(EI \frac{d^4z_2}{dx^4} \right) dx = 5.1428EIL^7$$

and

$$m_{11} = \int_0^L \mu z_1 z_1 dx = 0.001587\mu L^9$$

$$m_{12} = m_{21} = \int_0^L \mu z_1 z_2 dx = 0.003968\mu L^{10}$$

$$m_{22} = \int_0^L \mu z_2 z_2 dx = 0.0099567\mu L^{11}$$

and form the eigenvalue problem (9.38)

$$EIL^5 \begin{bmatrix} 0.8 & 2L \\ 2L & 5.1428L^2 \end{bmatrix} \mathbf{c} = \hat{\lambda}\mu L^9 \begin{bmatrix} 0.001587 & 0.003968L \\ 0.003968L & 0.0099567L^2 \end{bmatrix} \mathbf{c}$$

which admits nontrivial solutions only if

$$(0.8 - 0.001589\gamma)(5.1428 - 0.0099567\gamma) - (2 - 0.003968\gamma)^2 = 0 \qquad (9.45)$$

where we define $\gamma = \hat{\lambda}\mu L^4/EI$. Finally, from eq (9.45) we get

$$\hat{\omega}_1 = 22.45\sqrt{\frac{EI}{\mu L^4}}$$

$$\hat{\omega}_2 = 63.47\sqrt{\frac{EI}{\mu L^4}}$$

(9.46a)

These values must be compared to the exact eigenvalues (eq (8.70))

$$\omega_1 = 22.37\sqrt{\frac{EI}{\mu L^4}}$$

$$\omega_2 = 61.67\sqrt{\frac{EI}{\mu L^4}}$$

(9.46b)

showing that the relative error (with respect to the true eigenvalues) is 0.36% for $\hat{\omega}_1$ and 2.92% for $\hat{\omega}_2$. Moreover, as expected, both approximate frequencies are higher than the true values.

It is left to the reader to tackle the same problem by choosing as Ritz functions the first two eigenfunctions of a beam simply supported at both ends, i.e.

$$z_1(x) = \sin\left(\frac{\pi x}{L}\right)$$

$$z_2(x) = \sin\left(\frac{2\pi x}{L}\right)$$

(9.47)

which satisfy only two of the four boundary conditions of the clampled–clamped beam.

9.3.1 Continuous systems – a few comments on admissible and comparison functions

In forming the Rayleigh quotient – both in the Rayleigh and in the Raleigh–Ritz methods – we have pointed out more than once that a good choice of the trial function(s) translates into better approximations for the 'true' solution of the problem at hand: this means that, for continuous systems, the boundary conditions must be taken into account. In this regard we can refer back to Section 5.5 and recall, in the light of the developments of Chapter 8, the definitions of admissible and comparison functions.

Given a continuous system with stiffness and mass operators K and M of order $2p$ and $2q$ $(p > q)$, respectively:

- An **admissible function** is a function which is p times differentiable and satisfies only the geometric (or essential) boundary conditions of the problem.
- A **comparison function** is a function which is $2p$ times differentiable and satisfies all the boundary conditions of the problem.

It is evident that the eigenfunctions of the system constitute a subset of comparison functions (the comparison functions, in general, do not need to satisfy the differential equation of motion) and that, in turn, the comparison functions form a subset of admissible functions. So, on one hand, it would seem highly desirable to satisfy all the boundary conditions – thus limiting the choice to comparison functions – but, on the other hand, it is evident that the class of admissible functions allows more freedom of choice, particularly in view of the fact that force boundary conditions are often more difficult to satisfy than geometric ones.

If, for present convenience, we turn our attention to a specific case, we may consider, for example, the flexural vibrations of a beam simply supported at both ends, whose boundary conditions are given by eqs (8.59). Let us choose a set of (comparison) functions z_j which, by definition, satisfy all of eqs (8.59) and calculate the k_{ij} by means of the inner product $k_{ij} = \langle z_i \mid K z_j \rangle$. Explicitly, the stiffness operator is of order $p = 2$ and we have

$$k_{ij} = \int_0^L z_i \left(EI \, \frac{d^4 z_j}{dx^4} \right) dx$$

We can now integrate twice by parts to arrive at the expression

$$k_{ij} = \int_0^L EI \left(\frac{d^2 z_i}{dx^2} \right) \left(\frac{d^2 z_j}{dx^2} \right) dx \tag{9.48}$$

where the appropriate boundary conditions have been taken into account. The important point is that the eq (9.48) is defined for functions that are only p times differentiable, which is precisely the requirement for admissible functions.

In addition, we can consider other examples of continuous systems and note that we can form the Rayleigh quotient after having performed an appropriate number of integration by parts, so that some requirements on the Ritz functions can be relaxed and we can be free to choose from the larger class of admissible functions. Obviously, these considerations hold true for the Rayleigh method (only one function involved), the Rayleigh–Ritz method and the assumed modes method. With reference to the beam problem above, the consequence is that, say, in forming the Rayleigh quotient or in calculating the k_{ij} we either can adopt the inner-product expression in conjunction with comparison functions or adopt eq (9.48) in conjunction with admissible functions; when comparison functions are used in eq (9.48) the two forms are equivalent. It is left to the reader to show that the counterparts of eq (9.48) for a rod in longitudinal or torsional vibration are, respectively

$$k_{ij} = \int_0^L EA \frac{dz_i}{dx} \frac{dz_j}{dx} dx$$

$$k_{ij} = \int_0^L GJ \frac{dz_i}{dx} \frac{dz_j}{dx} dx$$

(9.49)

The discussion on the initial choice of a set of appropriate functions can be taken further by noting that, although convenient, the use of eq (9.48) (or (9.49), or the equivalent for the system under investigation) in conjunction with admissible functions obviously violates the natural boundary conditions. Hence, since comparison functions are often difficult to generate, the question arises whether we should abandon natural boundary conditions altogether. The answer is that yes, in most practical situations, this is the choice. However, it is interesting to note that a class of functions, called the quasi-comparison functions, has been devised in order to obviate this inconvenience; the interested reader is referred to Meirovitch and Kwak[7] or Meirovitch[5]. In general, the choice of such functions may not be easy and, owing to these difficulties, it is limited to one-dimensional systems.

In conclusion, there are two points we want to make in this section:

1. As far as the above methods are concerned, admissible functions are the most widely encountered choice. Nevertheless, when the problem formulation and physical insight permit, we may restrict our choice to comparison functions.
2. In forced-vibration problems – by taking a modal approach – we can obtain an approximate response by using the approximate m eigenvectors which result from the Rayleigh–Ritz method and it may happen that a particular response is better approximated by a set some judicious admissible functions rather than a set of comparison functions. This is because the forced response depends also on the spatial dependence of the forcing functions, and not only on the eigenfunctions of the free vibrating system.

Example 9.2.   As a second example in this chapter we consider a uniform beam of length L simply supported at both ends (pinned–pinned configuration); the flexural stiffness of the beam is EI and its mass per unit length is μ. We want to determine an approximate solution for the first two eigenvalues and the first two eigenfunctions. We begin by choosing the two Ritz functions

$$z_1 = \frac{x^2}{L^2} - \frac{x}{L}$$

$$z_2 = \frac{x^3}{L^3} - \frac{x^2}{L^2}$$
(9.50)

which we recognize as admissible functions because they do not satisfy the natural boundary conditions of the problem. We calculate the coefficients k_{ij} by means of eq (9.48) and the coefficients m_{ij} and we assemble them in the matrices

$$\hat{\mathbf{K}} = \frac{EI}{L^3} \begin{bmatrix} 4 & 2 \\ 2 & 4 \end{bmatrix}$$

$$\hat{\mathbf{M}} = \mu L \begin{bmatrix} \dfrac{1}{30} & \dfrac{1}{60} \\ \dfrac{1}{60} & \dfrac{1}{105} \end{bmatrix}$$

which, in turn, generate the eigenproblem

$$\begin{bmatrix} 4 & 2 \\ 2 & 4 \end{bmatrix} \mathbf{c} = \lambda \begin{bmatrix} \dfrac{1}{30} & \dfrac{1}{60} \\ \dfrac{1}{60} & \dfrac{1}{105} \end{bmatrix} \mathbf{c}$$
(9.51)

where we define $\gamma = \hat{\lambda}\mu L^4 / EI$. From eq (9.51) we obtain the two eigenvalues

$$\hat{\omega}_1 = 10.95 \sqrt{\frac{EI}{\mu L^4}}$$

$$\hat{\omega}_2 = 50.20 \sqrt{\frac{EI}{\mu L^4}}$$
(9.52a)

(which are, respectively, 10.9% and 27.1% higher with respect to the true eigenvalues) and the two mass-orthonormal eigenvectors

$$\hat{\mathbf{c}}_1^T = [5.477 \quad 0]$$

$$\hat{\mathbf{c}}_2^T = [-14.491 \quad 28.983] \tag{9.52b}$$

Then, the approximate eigenfunctions of the original problem can be recovered by means of eq (9.34), from which we obtain

$$u_1 = 5.447 z_1 = 5.447 \left(\frac{x^2}{L^2} - \frac{x}{L} \right)$$

$$u_2 = -14.491 z_1 + 28.983 z_2 \tag{9.53}$$

$$= -14.491 \left(\frac{x^2}{L^2} - \frac{x}{L} \right) + 28.983 \left(\frac{x^3}{L^3} - \frac{x^2}{L^2} \right)$$

These mode shapes are plotted in Fig. 9.1 with the exact eigenshapes of eq (8.62) as functions of the variable x/L. Note that the exact eigenshapes have been scaled to obtain the same maximum value as the approximate eigenfunctions.

Example 9.3. This last example is left to the reader and only a few comments will be made. Consider the longitudinal vibration of the rod shown in Fig. 9.2. The relevant parameters of the rod are as follows: axial stiffness

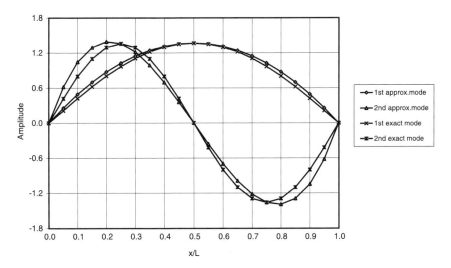

Fig. 9.1 Approximate and exact mode shapes (pinned–pinned beam).

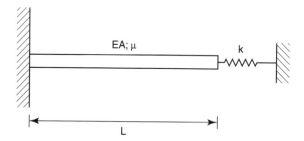

Fig. 9.2 Example 9.3: longitudinal vibration of a rod.

EA, length L and mass per unit length μ. In addition, k is the stiffness of the spring attached to the right end, and the idea is to estimate the first two eigenvalues of this system.

An easy and reasonable choice of two Ritz functions is represented by the polynomials

$$z_1 = \frac{x^3}{L^2} - 3x$$

$$z_2 = \frac{x^2}{L} - 2x$$

(9.54)

which satisfy all the boundary conditions – and hence are two comparison functions – for the fixed–free rod (eqs (8.48)). However, they are only admissible functions for the present case, whose boundary conditions read

$$u(x)\big|_{x=0} = 0$$

$$EA \frac{du}{dx}\bigg|_{x=L} = -ku(x)\big|_{x=L}$$

(9.55)

and it is evident that both functions eq (9.54) do not satisfy the natural boundary condition of axial force balance at $x = L$. A point worthy of notice is that, in this case, the coefficients k_{ij} are given by

$$k_{ij} = \int_0^L EA \frac{dz_i}{dx} \frac{dz_j}{dx}\,dx + kz_i(L)z_j(L)$$

(9.56)

In fact, if we consider two comparison functions f and g (i.e. two functions that satisfy eqs (9.55)), we can write

$$\langle f \mid Kg \rangle = -\int_0^L f\left(EA \frac{d^2g}{dx^2} \right) dx = -EA \left(f \frac{dg}{dx} \bigg|_0^L + \int_0^L EA \frac{df}{dx}\frac{dg}{dx} dx \right)$$

$$= -EA \left(f(L) \frac{dg(L)}{dx} + f(0) \frac{dg(0)}{dx} \right) + \int_0^L EA \frac{df}{dx}\frac{dg}{dx} dx$$

$$= kf(L)g(L) + \int_0^L EA \frac{df}{dx}\frac{dg}{dx} dx \tag{9.57}$$

where we have integrated by parts and taken into account the boundary conditions (9.55). The last expression is defined for admissible functions and is precisely the counterpart of the first of eq (9.49) for the case at hand. This result should not be surprising because the localized spring must contribute to the total potential energy of the system.

A final comment to note is that in the case of an elastic element – say, for example a beam in transverse vibration – with s localized springs and m localized masses, the coefficients k_{ij} and m_{ij} are obtained as

$$k_{ij} = \int_0^L EI \frac{d^2z_i}{dx^2}\frac{d^2z_j}{dx^2} dx + \sum_{l=1}^s k_l z_i(x_l)z_j(x_l)$$

$$m_{ij} = \int_0^L \mu z_i z_j dx + \sum_{r=1}^m m_r z_i(x_r)z_j(x_r) \tag{9.58}$$

where k_l ($l = 1, 2, ..., s$) are the stiffness coefficients of the springs acting at the locations $x = x_l$ and m_r ($r = 1, 2, ..., m$) are the localized masses at the locations $x = x_r$.

9.4 Summary and comments

On one hand, by means of increasingly sophisticated computational techniques, the power of modern computers allows the analysis and the solution of complex structural dynamics problems. On the other hand, this possibility may give the analyst a feeling of exactness and objectivity which is, to say the least, potentially dangerous. As a matter of fact, the user has limited control on the various steps of the computational procedures and sometimes – in the author's opinion – he does not even receive great help from the manuals that accompany the software packages. The numerical procedures themselves, in turn, are never 'fully tested' for two main reasons: first, because this is often an impossible task (furthermore, the software designer cannot be aware of the ways in which his software will be used) and, second, because of cost and time problems. So, it is always wise to look at the results of a complex numerical analysis with a critical eye. In this light, the importance of approximate methods cannot be overstated. This is why, even in the era of computers, a chapter on 'classical' approximate methods is never out of place. Here, the term 'classical' refers to methods that have been developed many years before the advent of digital computers (e.g. the fundamental text of Lord Rayleigh [8]) and whose 'only' requirements are a little patience, a good insight into the physics of the problem and, when necessary, a limited use of computer resources. Hence, discussion of the ubiquitous finite-element method – which is also an approximation method in its own right – is not included in this chapter.

Our attention is mainly focused on the Rayleigh and Rayleigh–Ritz methods, which are both based on the mathematical properties of the Rayleigh quotient (Sections 9.2 and 9.2.1) – a concept that pervades all branches of structural dynamics. For a given system, the Rayleigh method is used to obtain an approximate value for the first eigenvalue, while the Rayleigh–Ritz method is used to estimate the lowest eigenvalues and eigenvectors. Both methods start with an initial assumption on the vibration shape(s) of the system under study and their effectiveness is due to the stationarity property of the Raleigh quotient which guarantees that a reasonable guess of these trial shape(s) leads to acceptable results. Moreover, when the initial assumption seems too crude, both methods can be used iteratively in order to obtain better approximations of the 'true' values.

In the light of the fact that – unless the assumed shape coincides with the true eigenshape – the Rayleigh method always leads to an overestimate of the first eigenvalue, Section 9.2.2 considers Dunkerley's formula which, in turn, always leads to an underestimate of the first eigenvalue. Although its use is generally limited to positive definite systems with lumped masses, Dunkerley's formula can also be useful when we need to verify that the fundamental frequency of a given system is higher than a given prescribed value.

The Rayleigh and Rayleigh–Ritz methods apply equally well to both discrete and continuous systems, and so does the assumed modes method, which is closely related to the Rayleigh–Ritz method but uses a set of time dependent generalized coordinates in conjunction with Lagrange equations. However, for continuous systems the problem of boundary conditions must be considered when we choose the set of Ritz trial functions. Boundary conditions, in turn, can be classified as geometric (or essential) or as natural (or force). Geometric boundary conditions arise from constraints on the displacements and/or slopes at the boundary of a physical body, while natural boundary conditions arise from force balance at the boundary. Since the accuracy of the result depends on how well the chosen shapes approximate the real eigenfunctions, it may seem appropriate to choose a set of trial functions which satisfy all the boundary conditions of the problem at hand, i.e. a set of 'comparison functions'. However, natural boundary conditions are much more difficult to satisfy than geometric ones and the common practice is to choose a set of Ritz functions which satisfy only the geometric boundary conditions, meaning that the choice is made from the much broader class of 'admissible functions'. Again, this possibility ultimately relies on the stationarity property of the Rayleigh quotient and allows more freedom of choice to the analyst, often at the price of a negligible loss of accuracy for most practical purposes. Furthermore, when we adopt a modal approach to solve a forced vibration problem, a judicious choice of admissible Ritz functions may lead to an approximation of the true response which is just as good (or even better) as the approximation that we can obtain by choosing a set of comparison functions. This is because the response of the system depends both on the eigenfunctions of the system and on the spatial distribution of the forcing function(s).

References

1. Bathe, K.J., *Finite Element Procedures*, Prentice Hall, Englewood Cliffs, NJ, 1996.
2. Spyrakos, C., *Finite Element Modeling in Engineering Practice*, Algor Publishing Division, Pittsburgh, PA, 1996.
3. Weaver, W. and Johnston, P.R., *Structural Dynamics by Finite Elements*, Prentice Hall, Englewood Cliffs, NJ, 1987.
4. Wilkinson, J.H., *The Algebraic Eigenvalue Problem*, Oxford University Press, 1965.
5. Meirovich, L., *Principles and Techniques of Vibration*, Prentice Hall, Englewood Cliffs, NJ, 1997.
6. Humar, J.L., *Dynamics of Structures*, Prentice Hall, Englewood Cliffs, NJ, 1990.
7. Meirovitch, L. and Kwak, M.K., On the convergence of the classical Rayleigh–Ritz method and finite element method, *AIAA Journal*, **28**(8), 1509–1516, 1990.
8. Rayleigh, Lord J.W.S., *The Theory of Sound*, Vols 1 and 2, Dover, New York, 1945.

10 Experimental modal analysis

10.1 Introduction

In almost every branch of engineering, vibration phenomena have always been measured with two main objectives in mind: the first is to determine the vibration levels of a structure or a machine under 'operating' conditions, while the second is to validate theoretical models or predictions. Thanks to the developments and advances in electronic instrumentation and computer resources of recent decades, both types of measurements can now be performed effectively; one should also consider that the increasing need for accurate and sophisticated measurements has been brought about by the design of lighter, more flexible and less damped structures, which are increasingly susceptible to the action of dynamic forces.

Experimental modal analysis (EMA) is now a major tool in the field of vibration testing. As such, it was first applied in the 1940s in order to gain more insight in the dynamic behaviour of aircraft structures and, since then, it has evolved through various stages where the terms of 'resonance testing' or 'mechanical impedance' were used to define this general area of activity.

Modal testing is defined as the process of characterizing the dynamic behaviour of a structure in terms of its modes of vibration. More specifically, EMA aims at the development of a mathematical model which describes the vibration properties of a structure from experimental data rather than from theoretical analysis; in this light, it is important to understand that a correct approach to the experimental procedures can only be decided **after** the objectives of the investigation have been specified in detail. In other words, the right questions to ask are 'What do we need to know? What is the desired outcome of the experimental analysis?' and 'What are the steps that follow the experimental test and for what reason are they undertaken?'. As often happens in science and technology – and this easier said than done – posing the problem correctly generally results in considerable savings in terms of time and money. The necessity of stating the problem correctly is due to the fact that modal testing can be used to investigate a large class of problems – from finite-element model verification to troubleshooting, from component substructuring to integrity assessment, from evaluation of structural modifications to damage detection and so forth – and therefore the final goal has a significant influence on the practical aspects of what to do and how to do it. Obviously, the type and size of structure under test also play a major role in this regard.

Last but not least, it is worth noting that, on the experimenter's part, a correct approach to EMA requires a broad knowledge of many branches of engineering which, traditionally, have often been considered as separate areas of activity.

If we now refer back to the introduction of Chapter 7, we can once again adopt Ewins' definitions and note that in this chapter we will proceed along the 'experimental route' to vibration analysis which, schematically, goes through the following three stages:

1. the measurement of the response properties of a given system;
2. the extraction of its modal properties (eigenfrequencies, eigenvectors and modal damping ratios);
3. the definition of an appropriate mathematical model which, hopefully, describes within a certain degree of accuracy some essential characteristics of the original system and can be used for further analysis.

10.2   Experimental modal analysis – overview of the fundamentals

In essence, EMA is the process by which an appropriate set of measurements is performed on a given structure in order to extract information on its modal characteristics, i.e. natural frequencies of vibration, mode shapes and damping factors. Broadly speaking, the whole process can be divided into the three main phases as defined in the preceding section, which can be synthetically restated as:

1. data acquisition
2. modal parameters estimation
3. interpretation and presentation of results.

It is the author's opinion that the most delicate phase is the first one. In fact, no analysis can fix a set of poor experimental measurements, and it seldom happens that the experimenter is given a second chance. By contrast, a good set of experimental data can always be used more than once to go through phases 2 and 3.

A modal analysis test is performed under a controlled forced vibration condition, meaning that the structure is subject to a measurable force input and its vibratory response output is measured at a number of locations which identify the degrees of freedom of the structure. Three basic assumptions are made on the structure to be tested:

1. The structure is **linear**. This assumption means that the principle of superposition holds; it implies that the structure's response to a force

input is a linear combination of its modes and also that the structure's response to multiple input forces is the sum of the responses to the same forces applied separately. In general, a wide class of structures behave linearly if the input excitation is maintained within a limited amplitude range; hence, during the test, it is important to excite the structure within this range.

For completeness of information, It must be pointed out that there exists an area of activity called 'nonlinear modal analysis' whose main objective is the same as for the linear case, i.e. to establish a mathematical model of the structure under test from a set of experimental measurements. In this case, however, the principle of superposition cannot be invoked and the mathematical model becomes nonunique, being dependent on vibration amplitude.

2. The structure is **time invariant**. This assumption means that the parameters to be determined are constants and do not change with time. The simplest example is a mass–spring SDOF system whose mass m and spring stiffness k are assumed to be constant.

3. The structure is **observable**. This assumption means that the input–output measurements to be made contain enough information to adequately determine the system's dynamics. Examples of systems that are not observable would include structures or machines with loose components (that may rattle) or a tank partially filled with a fluid that would slosh during measurements: if possible, these complicated behaviours should be eliminated in order to obtain a reliable modal model.

In addition to the assumptions above, most structures encountered in vibration testing obey Maxwell's reciprocity relations provided that the inputs and outputs are not mixed. In other words, for linear holonomic–scleronomic systems reciprocity holds if, for example, all inputs are forces and all outputs are displacements (or velocities or accelerations); by contrast, reciprocity does not apply if, say, some inputs are forces and some are displacements and if some outputs are velocities and some are displacements. Unless otherwise stated, we will assume in the following that reciprocity holds; for our purposes, the main consequence of this assumption is that receptance, mobility, and acceleration and impulse response functions matrices are all symmetrical.

Given the assumptions above, a modal test can be performed by proceeding through phases 1–3. Since there is no such thing as 'the right way' valid for all circumstances, each phase poses a number of specific problems whose solutions depend, for the most part, on the final objectives of the investigation and on the desired results.

In phase 1 the problem to be tackled has to do with the experimental set-up and the questions to be answered are, for example: how many points (degrees of freedom) are needed to achieve the desired result? how do we excite the structure and how do we measure its response?

In phase 2, on the other hand, the focus is on the specific technique to be used in order to extract the modal parameters from the experimental measurements. This task is now accomplished by means of commercial software packages but the user, at a minimum, should at least have an idea of how the various methods work in order to decide which technique may be adopted for his/her specific application.

Finally, phase 3 has to do with the physical interpretation of results and with their presentation in form of numbers, graphs, animations of the modal shapes or whatever else is required for further theoretical analysis, if any is needed.

10.2.1   FRFs of SDOF systems

With the exception of the available electronic instrumentation and the basic concepts of digital signal analysis – which will be considered separately in the final chapters of this book – most of the theoretical concepts needed in EMA have been introduced and discussed in previous chapters (Chapters 4, 6 and 7) whose content is a prerequisite for the present developments. Nevertheless, in the light of the fact that the first step in a large number of experimental methods in modal analysis consists of acquiring an appropriate set of frequency response functions (FRFs) of the system under investigation, this section considers briefly some characteristics of these functions.

Consider, for example, the receptance function of an SDOF system whose physical parameters are mass m stiffness k and damping coefficient c. From eq (4.42) the magnitude of this FRF is given by

$$\left|\frac{X}{F_0}\right| \equiv |R(\omega)| = \frac{1}{\sqrt{(k-m\omega^2)^2 + (c\omega)^2}} \tag{10.1a}$$

or, alternatively (eq (4.44))

$$|R(\omega)| = \frac{1/k}{\sqrt{(1-\beta^2)^2 + (2\zeta\beta)^2}} \tag{10.1b}$$

where, as usual, $\omega_n = \sqrt{k/m}$, $\beta = \omega/\omega_n$, $\zeta = c/c_{cr}$ and $c_{cr} = 2\sqrt{km}$. When $\beta \to 0$ or $\beta \to \infty$ we have, respectively

$$|R(\omega)| \to 1/k \tag{10.2a}$$

and

$$|R(\omega)| \to 1/m\omega^2 \tag{10.3a}$$

Owing to the wide dynamic range of FRFs, it is often customary to plot the magnitude of FRF functions on log–log graphs or, more precisely, in dB (where the reference value, unless otherwise stated, is unity); this circumstance has also the additional advantage that data that plot as curves on linear scales become asymptotic to straight lines on log scales and provide a simple means for identifying the stiffness and mass of simple systems. In fact, eqs (10.2a) and (10.3a) become, respectively

$$\text{dB}\{\,|\,R(\omega)\,|\,\} = -20\,\log_{10}k \tag{10.2b}$$

and

$$\text{dB}\{\,|\,R(\omega)\,|\,\} = -20\,\log_{10}m - 40\,\log_{10}\omega \tag{10.3b}$$

so that in the low-frequency part of the graph we have a horizontal spring line and in the high-frequency part of the graph we have a mass line whose slope is -40 dB/decade (-12.04 dB/octave, or a downward slope of -2 on a log scale) and whose position is controlled by the value of m. The stiffness and mass lines intersect at a point whose abscissa is the resonant frequency of the system, i.e. when the spring and the inertia force cancel and only the damping force is left to counteract the external applied force.

As an example, a graph of this kind is plotted in Fig. 10.1 for a system with $k = 1 \times 10^{6}$ N/m, $m = 50$ kg and $c = 1200$ N s/m (implying $\omega_n = 141.1$ rad/s and $\zeta = 8.5 \times 10^{-2}$); note that, as expected, the stiffness line is at -120 dB, meaning that $k = 10^{120/20} = 1 \times 10^{6}$.

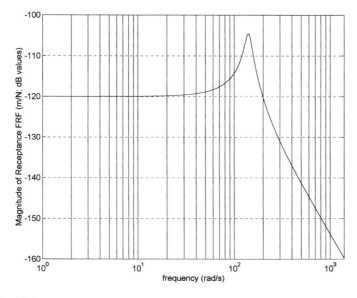

Fig. 10.1

A similar line of reasoning applies to mobility and accelerance FRFs; mobility graphs, for example, are symmetrical about the vertical axis at ω_n; in the low frequency range we note a stiffness line with an upward slope of +20 dB/decade (+6.02 dB/octave, or +1 on a log scale) while in the high-frequency range there is a mass line with a downward slope of −20 dB/decade (−1 on a log scale). Moreover, at resonance we get

$$| V(\omega) | \cong 1/c \tag{10.4}$$

implying that there is a horizontal line of viscous damping in the logarithmic representation (in this regard, the reader can verify that a horizontal line of hysteretic damping is obtained in receptance graphs).

By contrast, accelerance graphs display a stiffness line with an upward slope of +40 dB/decade in the low-frequency range and a horizontal mass line in the high frequency range. The graphs of mobility and accelerance for the SDOF system considered above are shown in Figs 10.2 and 10.3.

Equation (10.1a) (or (10.1b)), however, does not tell the whole story. Whether we consider an SDOF or an MDOF system, we know from previous chapters that FRFs are complex functions and cannot be completely represented on a standard $x - y$ graph. The consequence is that there are three widely adopted display formats:

- The **Bode diagram**. This consists of two graphs which plot, respectively, the FRF magnitude and phase as functions of frequency. The graph of

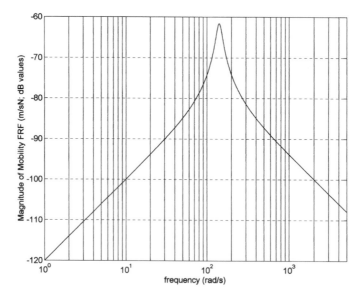

Fig. 10.2

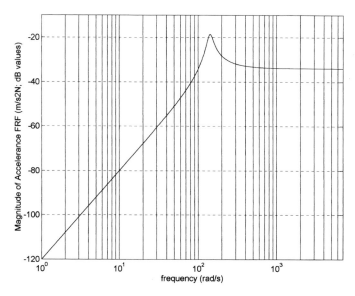

Fig. 10.3

magnitude versus frequency is usually displayed in $\log(y)-\log(x)$ scales, $dB(y)-\log(x)$ or $dB(y)-\text{linear}(x)$ scales (but linear–linear scales are sometimes used as well); in this regard it is worth noting that plotting the amplitude ratio in dB on a linear scale is equivalent to plotting the amplitude on a logarithmic scale.

- The **real and imaginary plots**. These display the FRF real and imaginary parts as functions of frequency.
- The **Nyquist diagram** (or polar graph). This is a single plot which displays the FRF imaginary part as a function of the real part (this format is particularly useful in many circumstances, but has the inconvenience of not showing explicitly the frequency information (Fig. 4.14); this information can be given by adding captions which indicate the values of frequency).

All of the above formats are generally available in commercial software packages.

In the Bode diagrams, the graphs of phase angles may sometimes be a source of confusion. If we adopt the phasor representation of rotating vectors (Chapter 1), we have stated on a few occasions that, provided that consistency is maintained, it is somewhat irrelevant to choose the convention $e^{-i\omega t}$ (clockwise rotating vector) or $e^{i\omega t}$ (counterclockwise rotating vector). However, if the forcing function is written as $f_0 e^{-i\omega t}$ it is customary (eq (4.41)) to write the displacement response as $x_0 e^{-i(\omega t - \phi)}$ so that ϕ, when

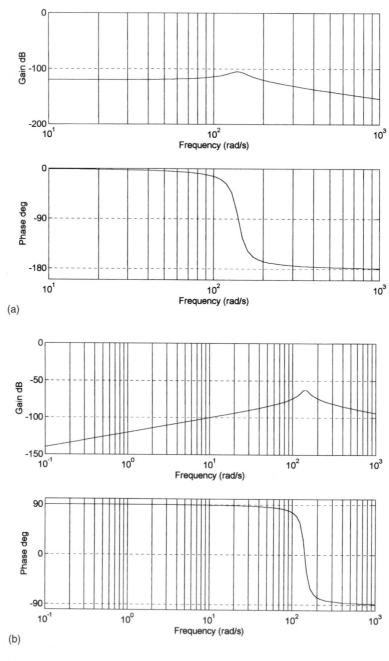

(a)

(b)

Fig. 10.4

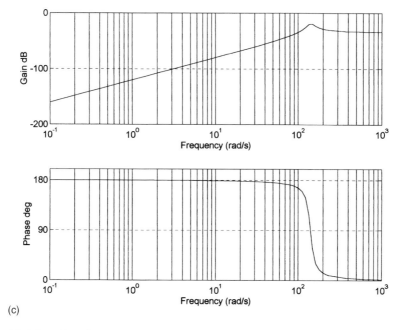

(c)

Fig. 10.4 (continued)

plotted as in Fig. 4.9, is to be understood as the angle of lag of displacement behind the external force, the two extreme situations being as follows:

- When $\beta \ll 1$ the displacement is in phase with the force and $\phi \cong 0$.
- When $\beta \gg 1$ the displacement lags behind the force of π radians and $\phi \cong 180°$.

By the same token, velocity is written as $\dot{x} = v_0 e^{-i(\omega t - \phi_v)}$ where ϕ_v is the angle of lag of velocity behind force and is given by $\phi_v = \phi - \pi/2$, since we know that velocity leads displacement by $\pi/2$ radians. When $\beta \ll 1$ velocity leads force by $\pi/2$ so that the velocity angle of lag behind force is $\phi_v = -\pi/2$; on the other hand, when $\beta \gg 1$, $\phi_v = \pi/2$. Similar considerations apply for the acceleration phase angle ϕ_a ($\phi_a = \phi - \pi = \phi_v - \pi/2$), which ranges from $-\pi$ ($\beta \ll 1$) to zero ($\beta \gg 1$) radians. In brief, in the negative exponential convention the phase angle is positive when it is an angle of lag, negative when it is an angle of lead and for an SDOF system all phase angles plotted as functions of frequency are monotonically increasing functions. The same situation arises if we adopt the positive exponential convention but we write displacement, velocity and acceleration as $x = x_0 e^{i(\omega t - \phi)}$, $\dot{x} = v_0 e^{i(\omega t - \phi_v)}$ and $\ddot{x} = a_0 e^{i(\omega t - \phi_a)}$, respectively.

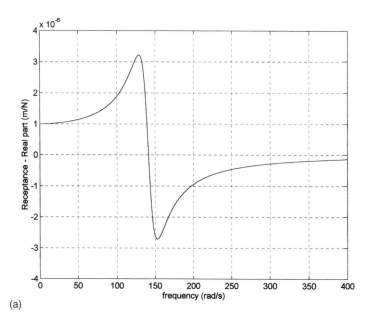

(a)

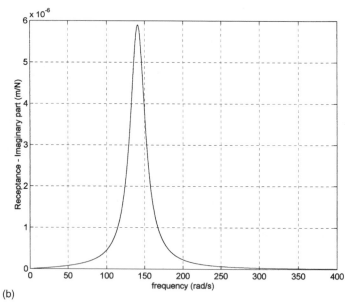

(b)

Fig. 10.5

By contrast, in the positive exponential convention, displacement, velocity and acceleration are sometimes written as $x = x_0 e^{i(\omega t + \phi)}$, $\dot{x} = v_0 e^{i(\omega t + \phi_v)}$ and $\ddot{x} = a_0 e^{i(\omega t + \phi_a)}$, respectively, so that the phase angles must be accompanied by a minus sign when they represent angles of lag. In this light, we have that ϕ ranges from zero ($\beta \ll 1$) to $-\pi$ ($\beta \gg 1$), $\phi_v = \phi + \pi/2$ ranges from $\pi/2$ to $-\pi/2$, $\phi_a = \phi_v + \pi/2 = \phi + \pi$ ranges from π to zero, and all phase angles plotted as functions of frequency are monotonically decreasing functions.

Obviously, whatever convention we choose, it must be consistent with the physical fact that – in steady-state conditions – displacement is in phase with force when $\beta \ll 1$, velocity is in phase with force at resonance, acceleration is in phase with force when $\beta \gg 1$ and that, in all cases, velocity leads displacement by π and acceleration leads velocity by $\pi/2$.

In order to illustrate this situation, Figs. 10.4(a), (b) and (c) show, respectively, the Bode diagrams of receptance, mobility and accelerance of the viscously damped SDOF system considered before. Those readers who are familiar with the MATLAB® environment have certainly noticed that these graphs have been drawn by using the 'Bode' command of MATLAB®. Magnitude graphs are the same as Figs. 10.1, 10.2 and 10.3 and the label 'Gain' on the y-axis comes from the terminology commonly adopted in the electrical engineering community.

The characteristic features of real and imaginary plots are that the real part of the receptance and accelerance has a zero crossing at the resonant frequency, while that of the mobility has a peak at resonance. On the other hand, the imaginary part of the receptance and accelerance has a peak at resonance, while that of the mobility has a zero crossing. Referring once again to the SDOF system considered before in this section, examples of such plots are shown in Figs. 10.5(a) and (b) (receptance), 10.6(a) and (b) (mobility) and 10.7(a) and (b) (accelerance).

Finally, the Nyquist plots for the same SDOF system are shown in Figs. 10.8–10.10. All these graphs have been drawn in the frequency range 0–400 rad/s with a frequency spacing of $\Delta \omega = 0.5$ rad/s, meaning that we have used 800 frequency lines to cover the whole range; the '+' markers on the curves identify these sampled frequency values. Since modern electronic instrumentation converts analogue signals into digital 'sampled' signals at an early stage of the measuring process (Chapters 13–15), the markers on the curves below might represent actual data from acquired FRFs.

Note that, on the curves, data points away from resonance are very close together (the markers overlap) while the arc spacing between markers becomes larger and larger as we approach the resonant region. This is an advantage and a disadvantage at the same time: the advantage is due to the fact that the resonant frequency can be identified on these graphs with good accuracy (i.e. better than other methods) by considering the maximum rate of change of arc length as a function of frequency, while the disadvantage is that for very lightly damped structures the typical circular shape may be lost if the number of frequency lines is insufficient. An example of this situation is

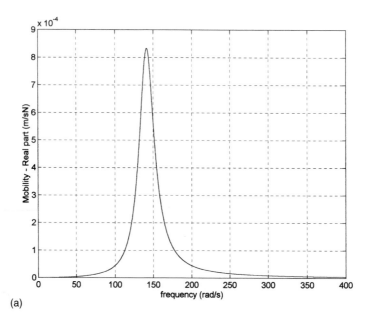

(a)

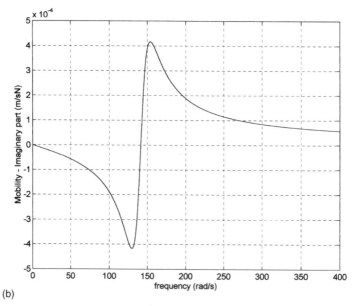

(b)

Fig. 10.6

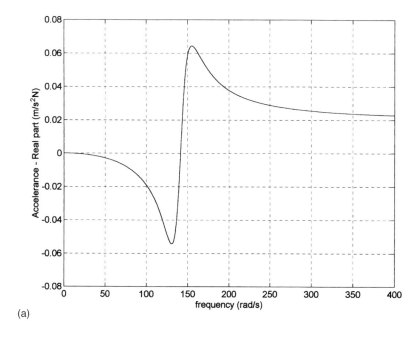

(a)

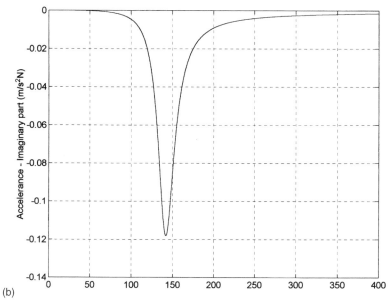

(b)

Fig. 10.7

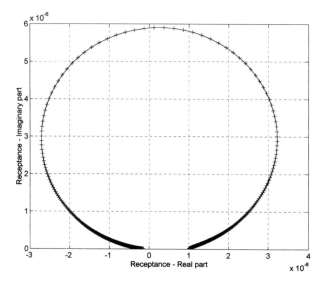

Fig. 10.8

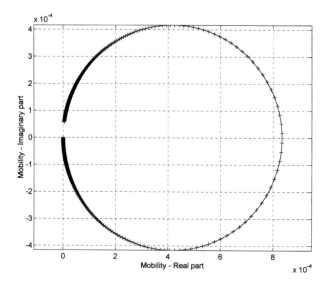

Fig. 10.9

shown in Fig. 10.11, where the SDOF system considered in this case has the same stiffness and mass as before, but it is much less damped ($c = 100$ N s/m, i.e. $\zeta = 0.0071$) and we have used 200 spectral lines to cover the range 0–400 rad/s (i.e. the markers are $\Delta\omega = 2$ rad/s apart).

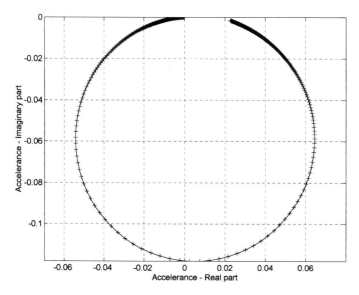

Fig. 10.10

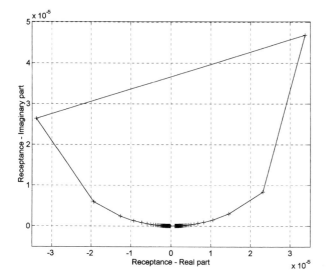

Fig. 10.11

From Figs 10.8–10.10, it is evident that the characteristic feature of Nyquist plots is to enhance the resonance region with a nearly circular shape which corresponds to the phase shift that the output undergoes with respect to the input. However, for a viscously damped system it must be noted that only mobility traces out an exact circle (see also eqs (4.95), (4.96) and (4.97)), while receptance and accelerance curves are distorted circles and tend to become more distorted as damping is increased. Figures 10.12–10.14 illustrate this situation: stiffness and mass are as before, but now $c = 4000$ N s/m, i.e. $\zeta = 0.28$. Also note that in this case the graphs have been drawn in the range 0–400 rad/s by using only 200 spectral lines and no information is lost on the shape of the curves.

Finally, from the graphs of mobility of Figs 10.9 and 10.13 we can easily obtain the value of viscous damping. In fact, eq (4.97) shows that the diameter D of the mobility circle is $1/c$, observing that $D = 8.33 \times 10^{-4}$ in Fig. 10.9 and $D = 2.5 \times 10^{-4}$ in Fig. 10.13 we get, as expected, $c = 1200$ N s/m in the first case and $c = 4000$ N s/m in the second case.

It is left to the reader to show that for a hysteretically damped system it is receptance that traces out an exact circle with centre at $(0, 1/(2k\gamma))$ and diameter $D = 1/k\gamma$. As a hint, define $U = \mathrm{Re}[R(\omega)]$, $V = \mathrm{Im}[R(\omega)] - 1/(2k\gamma)$ and note that

$$U^2 + V^2 = \left(\frac{1}{2k\gamma} \right)^2 \tag{10.5}$$

As an example, Fig. 10.15 shows the Nyquist plot of the FRF receptance

$$R(\omega) = \frac{1}{1 \times 10^6 - 50\omega^2 - 2 \times 10^5 i} \tag{10.6}$$

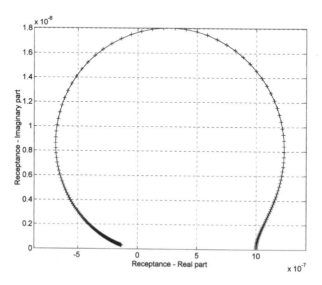

Fig. 10.12

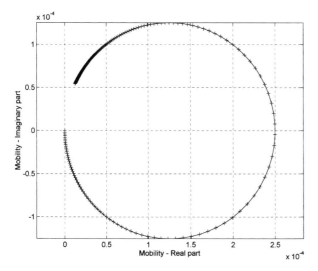

Fig. 10.13

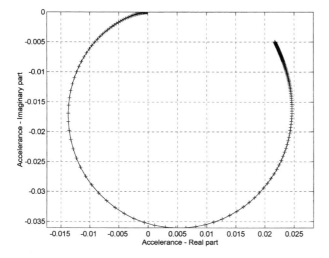

Fig. 10.14

which represents a hysteretically damped SDOF system with $k = 1 \times 10^6$ N/m, $m = 50$ kg, $k\gamma = 2 \times 10^5$ N/m (i.e. $\gamma = 0.2$); as expected, the diameter of the circle is $1/k\gamma = 5 \times 10^{-6}$. The graph covers the frequency range 0–400 rad/s using 400 frequencies and it is worth noting that we have adopted the negative exponential notation – i.e. the FRF is in the form of eq (4.72). The reader is also invited to verify that, in this case, the positive exponential notation leads to a Nyquist circle with centre at $(0, -1/(2k\gamma))$.

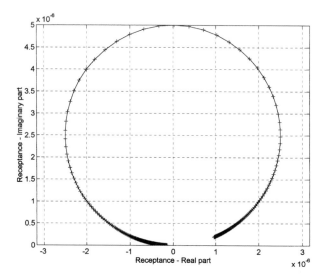

Fig. 10.15

10.2.2   *FRFs of MDOF systems*

Most of the considerations of the preceding section retain their validity when we turn our attention to FRFs of MDOF systems. However, some new features which have no counterpart in the SDOF case must be considered. The subject will be, for the most part, discussed qualitatively, with the intention of providing a general idea from an experimenter's point of view. In fact, during the data acquisition phase of a modal test – unless it is a laboratory test – there is generally not much time for detailed quantitative considerations; nonetheless, it is of fundamental importance to collect a 'good' set of data (typically FRFs) whose quality and consistency can often by be rapidly checked by a careful visual inspection.

For a *n*-DOF structure the main distinction that can be made is between **point** (or **driving**) **FRFs** and **transfer FRFs**: a point FRF is a function of the type $H_{jj}(\omega)$, meaning that the input and output are measured at the same point on the structure, while a transfer FRF is a function of the type $H_{jk}(\omega)$ with $j \neq k$. Whenever appropriate, both point and transfer FRFs can be further subdivided into **direct** and **cross** FRFs: the term direct meaning that both input and output are measured along the same direction and the term cross meaning that input and output are measured along different directions.

To make things clearer, suppose that we decided to test a given structure by taking measurements at two points – point 1 and point 2 – along both the *x* and y directions (the structure, for example, could be a beam with rectangular cross section, z being the longitudinal direction of the beam). Either:

1. we can perform two separate tests, one in the x direction and one in the y direction – and in both cases we would be dealing with a 2-DOF system, or
2. we can perform a single test in which the x and y directions are considered together, and in this case we would be dealing with a 4-DOF system.

If all FRFs are measured, each test of option 1 results in two direct point FRFs (H_{11} and H_{22}) and two direct transfer FRFs (H_{12} and H_{21}). Strictly speaking, no distinction between direct and cross FRFs is needed because no cross FRFs exist in this case. By contrast, in option 2 we would have four direct point FRFs (input and output measured in the same point along the same direction), four direct transfer FRFs (input and output at different points along the same direction), four cross point FRFs (input and output at the same point along different directions) and four cross transfer FRFs (input and output at different points along different directions). In this case, it is convenient to number the degrees of freedom from 1 to 4 referring, for example, to DOFs 1 and 2 for the measurements at point 1 and 2, respectively, along the x direction and to DOFs 3 and 4 for the measurements at points 1 and 2 along the y direction. With these definitions, the dynamic behaviour of the structure is described by the 4×4 matrix

$$\mathbf{H}(\omega) = \begin{bmatrix} H_{11} & H_{12} & H_{13} & H_{14} \\ H_{21} & H_{22} & H_{23} & H_{24} \\ H_{31} & H_{32} & H_{33} & H_{34} \\ H_{41} & H_{42} & H_{43} & H_{44} \end{bmatrix} \tag{10.7}$$

where the direct point FRFs are on the main diagonal, the cross transfer FRFs are on the secondary diagonal, the direct transfer FRFs are the elements H_{12}, H_{21}, H_{34} and H_{43}, and the remaining elements are the cross point FRFs.

In general, the most common situation in experimental tests is the case of MDOF systems in which input and output are measured in the same direction (i.e. a test of type 1, where no cross FRFs are acquired); for this reason, in this section we will focus our attention on such tests.

In order to examine the main characteristics of FRFs of MDOF systems, it will suffice for our purposes to consider the 2-DOF system of Section 7.9, because all the considerations that follow can be extended in a straightforward manner to systems with more than two degrees of freedom.

As expected, all graphs of magnitude versus frequency show two peaks which occur at the resonant frequencies of our system. However, we have already pointed out (Section 7.9) the appearance, between resonances, of an 'inverted peak' of antiresonance in the point FRFs $R_{11}(\omega)$ and $R_{22}(\omega)$. No such antiresonance exists in the magnitude graphs of the transfer FRFs $R_{12}(\omega)$ and $R_{21}(\omega)$. Moreover, it is interesting to note that a phase shift of $180°$ not only occurs at each resonance, but also at each (one in our case) antiresonance. As a rule, point FRFs must have antiresonances between resonances; by contrast, transfer FRFs may or may not have an antiresonance between two neighbouring resonances. In general, all that can be said in this latter case is that transfer FRFs corresponding to two points which are relatively close together on the structure will show more antiresonances than FRFs corresponding to points that are further apart on the structure. Let us investigate these statements in more detail.

For an undamped n-DOF system, it was shown in Chapter 7 that a general receptance FRF is written as

$$R_{jk}(\omega) = \sum_{m=1}^{n} \frac{p_{jm} p_{km}}{\omega_m^2 - \omega^2} \tag{10.8}$$

(see eq (7.28), where all $\zeta_m = 0$) where p_{jm} is the jth element of the mth eigenvector. For a point FRF $j = k$, implying that all the coefficients $p_{jm} p_{jm} = p_{jm}^2$ are positive. When $\omega < \omega_1$, all terms of the sum (10.8) are positive and the response is generally dominated by the first term (which has the smallest denominator). Right after the first resonance ($\omega_1 < \omega < \omega_2$), the first term of the sum still dominates but is now negative because its denominator is negative; hence R_{jj} becomes negative and this change of sign corresponds to the phase shift of $180°$. As we move towards ω_2, there will be a value of frequency at which the sum of all (positive) terms other than the first will exactly cancel out the contribution of the first term so that the magnitude at this point will be exactly zero. This is the antiresonance. As we pass this point and move towards values of increasing frequency, the sum (10.8) becomes positive again and this second change of sign at antiresonance corresponds to another $180°$ phase shift. Then – until the last resonance – the whole process repeats again and again as we keep moving in the direction of increasing frequencies.

If $j \neq k$ – depending on the type of structure and on the physical distance between point j and point k – the coefficients $p_{jm} p_{km}$ and $p_{jm+1} p_{km+1}$ do not necessarily have the same sign and no antiresonance may occur between any two neighbouring resonances; when point j and point k are close together on the structure it is more likely that the coefficients have the same sign and there will be an antiresonance. In our 2-DOF example ($R_{12}(\omega)$ and $R_{21}(\omega)$) the two neighbouring coefficients have different signs and there is no antiresonance between the two resonances.

Referring again to this 2-DOF system (Section 7.9), the graphs of mobilities M_{11}, M_{12} and accelerances A_{11}, A_{12} are shown in Figs. 10.16–10.19: part (a) of each figure plots the magnitude on dB(y)–linear(x) scales, while part (b) plots the magnitude on dB(y)–log(x) scales. The reader is invited to draw the

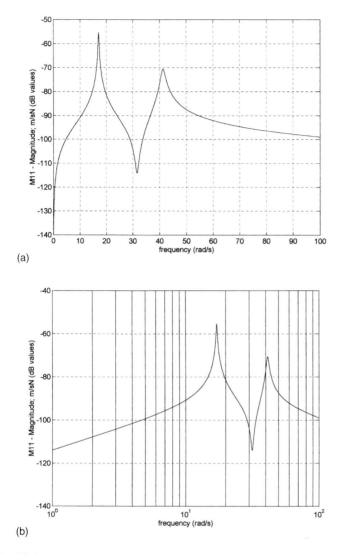

(a)

(b)

Fig. 10.16

graphs of M_{22} and A_{22} and the graphs of phase versus frequency; the phase information in mobility and accelerance FRFs is the same as in receptance FRFs, measuring velocity or acceleration rather than displacement merely introduces an offset of $90°$ or $180°$.

For completeness, we also show in Figs 10.20 and 10.21 the graphs of receptances R_{11} and R_{12} on dB(y)–log(x) scales (in Chapter 7 these graphs were drawn only on dB(y)–linear(x) scales).

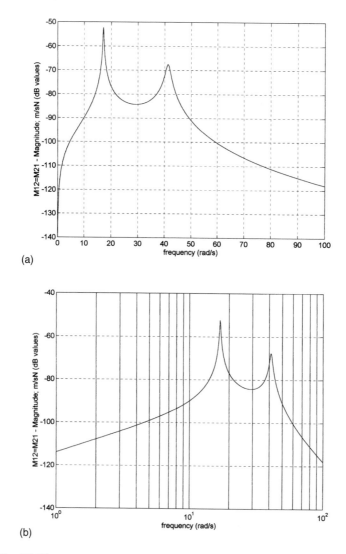

(a)

(b)

Fig. 10.17

Although this may not be immediately evident in our example, a visual comparison of the log–log graphs above also shows that resonances at higher frequencies tend to exhibit less displacement than lower frequency resonances. By contrast, the opposite situation occurs in acceleration graphs, which bias magnitude in proportion to the second power of frequency. This suggests that receptance may be the best choice if our attention is focused on low-frequency modes, while acceleration is better for high-frequency modes.

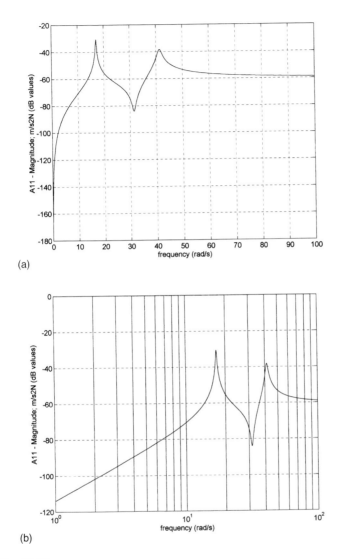

(a)

(b)

Fig. 10.18

However, when the frequency range of interest is relatively large, mobility graphs make the best use of the available dynamic range (i.e. the vertical space) because, broadly speaking, they give equal weight to all resonances in the frequency range. In this regard, the reader can find a more detailed explanation in Newland ([1], Chapter 3), where the 'skeleton' properties of logarithmic response graphs are considered, the 'skeleton' consisting of a sequence of straight line segments which change slope every time a resonance or an antiresonance is crossed.

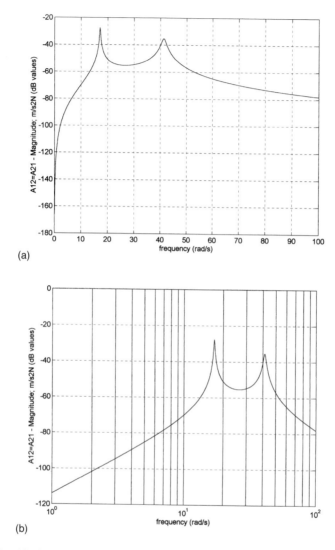

(a)

(b)

Fig. 10.19

Sometimes, the readability of the graphs may be problematic if linear scales are used. For example, higher-frequency resonances may hardly be noticed in FRF receptance graphs of real and imaginary part versus frequency or in a Nyquist plot, whereas low-frequency resonances may be difficult to see in FRF acceleration graphs. Although this is not the case for our 2-DOF example, the reader can have an idea by looking at Figs 10.22–10.24 which show,

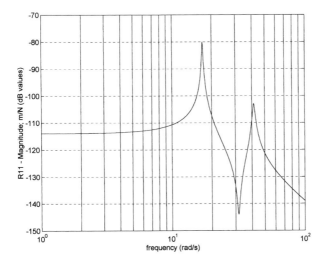

Fig. 10.20

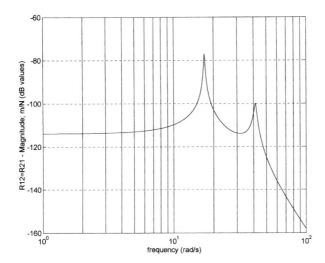

Fig. 10.21

respectively, different display formats of the transfer FRFs R_{12}, M_{12} and A_{12}. These figures also show the effect of different signs of the modal coefficients: in the Nyquist plot, for example, the two loops are not in the same half of the complex plane (as in point FRFs).

We close this section with two final observations. First, a careful inspection of the mobility (receptance if we had considered a hysteretically damped

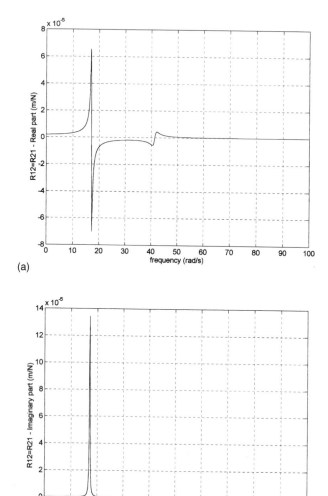

Fig. 10.22

system) Nyquist plot above shows that the loops are not exactly symmetrical with respect to the real axis. This is always the case for MDOF systems and it is due to the fact that each resonance loop 'feels' the presence of the other resonance loops. To be more specific, consider an n-DOF system with light hysteretic damping and well separated modes: in the vicinity, say, of the first natural frequency the response will be dominated by the first term of the sum

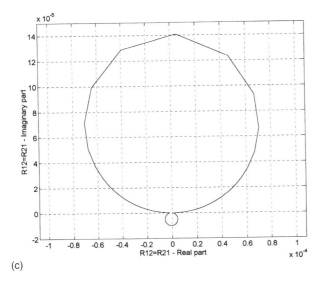

(c)

Fig. 10.22 (continued)

(eq (7.96b))

$$R_{jk}(\omega) = \sum_{m=1}^{n} \left(\frac{p_{jm}p_{km}}{\omega_m^2 - \omega^2 + i\gamma\omega_m^2} \right) \qquad (10.9)$$

so that, in this limited frequency range, we can write

$$R_{jk}(\omega) \cong \frac{p_{j1}p_{m1}}{\omega_1^2 - \omega^2 + i\gamma\omega_1^2} + a_{jk} \qquad (10.10)$$

where a_{jk} is a complex constant which is introduced to account for the effect of all the modes other than the first. So, even if the first term on the r.h.s. of eq (10.10) plots as a circle with centre on the imaginary axis, the presence of the second term displaces slightly this circle from its original position. The second observation is that this displacement is generally much more evident for non-proportionally damped systems. In fact, not only must we take into account the contribution of other modes by virtue of a complex constant as in eq (10.10), but we must consider that the modal coefficients themselves are now complex quantities with magnitude and phase. Hence, broadly speaking, the 'more complex' the modal coefficients of a resonant term, the more displaced the resonance loop from its 'symmetrical' position.

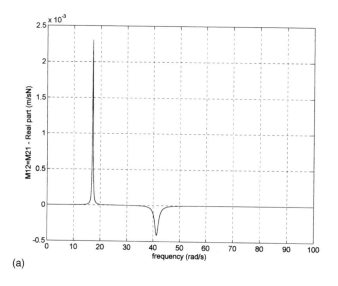

(a)

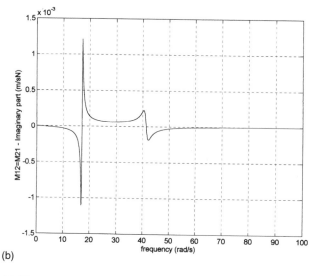

(b)

Fig. 10.23

10.3   Modal testing procedures

As often happens with experimental methods, there is no such thing as the 'right way' to perform a modal test. A good background theoretical knowledge is a fundamental prerequisite, but experience is invaluable in

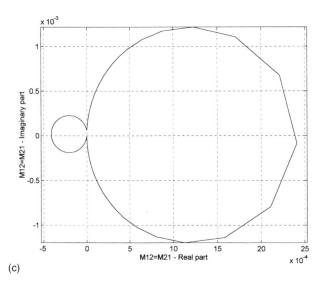

(c)

Fig. 10.23 (continued)

these circumstances. Basically, what we need to perform a modal test is:

- an excitation mechanism (shaker, impact hammer, etc.);
- a number of transducers (typically accelerometers) to measure the structure's response;
- an analyser with a minimum of two input channels (one for the excitation and one for a transducer).

The type, cost and level of sophistication of the instrumentation can vary but, in any case, a test planning phase is required in which the experimenter(s) must decide the test configuration, identify the frequency range of interest, the input and output locations (i.e. the points of the structure at which the excitation force is applied and at which the structure response is measured) and, in case, perform some preliminary measurements and check their quality. Furthermore, two other sources of error must be taken into account and should be minimized as much as possible: the first is due to the effects of the instrumentation on the structure (say, for example, exciter–structure interactions and 'mass loading' if the structure is very light), while the second has to do with improper use of the digital electronic instrumentation during the data acquisition phase (for example, with modern digital analysers, attention must be paid to frequency resolution, 'aliasing' effects, proper 'windowing' of the signal and 'leakage' errors, noise in both the excitation and response signal, etc.). A discussion of the electronic instrumentation is delayed to later chapters. Here, owing to the many ways and levels of sophistication in which a modal test can be performed,

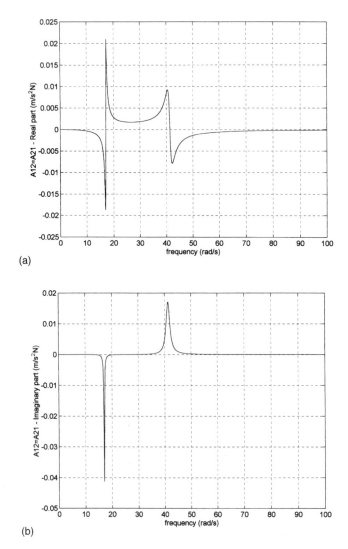

(a)

(b)

Fig. 10.24

we will limit ourselves to a general set of guidelines, also with the intention of making the reader aware of some common pitfalls and potential sources of error in the measurement phase.

10.3.1 Supporting the structure

If the experimenter has some control on this aspect of the test, the support condition of the structure under study must be considered. The two extremes

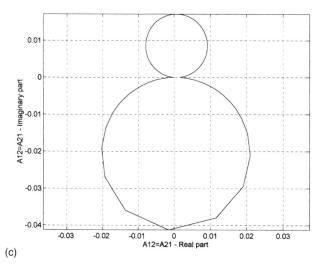

(c)

Fig. 10.24 (continued)

are the so-called 'free' configuration and the 'fixed' (or grounded) configuration; neither of the two conditions can be perfectly obtained in practice, but they can generally be approximated within a good degree of accuracy. The ideal free configuration implies that the structure is floating in the air without any support of any kind, which is obviously impossible. However, if we support the structure on very flexible springs or suspend it on light elastic rubber bands so that the highest rigid-body mode (note that, in this support condition, rigid-body modes no longer occur at $\omega = 0$) is well separated from the lowest elastic mode, we have simulated a good free configuration. A general rule of thumb in these circumstances is that the highest rigid-body mode should be at least 8–10 times smaller than the lowest elastic mode.

The 'fixed' condition, on the other end, can be simulated by grounding the structure with well-tightened bolts, clamps or other devices that prevent the movement of the structure at the supports. However, this is generally easier said than done because, even with massive and rigid foundations, it may not be so easy to provide sufficient grounding. In fact, strictly speaking, no supporting structure can be regarded as infinitely rigid; if in doubt, it may be desirable to perform a few preliminary FRF measurements on the supporting structure in the same frequency range of the test to verify that – near the 'grounding' points – the FRF levels at the base are much lower than the corresponding levels of the structure FRFs. However, even when this is the case, care should be exercised because rotational motion could be involved, and this is much more difficult to measure.

In any case, if we want to avoid as much as possible some undesired effects of the supports on the structure, we must assess the repeatability of the data, also in view of the fact that, in some circumstances, the absolute boundary conditions may not be as important as consistency between subsequent tests. Finally, due consideration must be given to the operating support conditions of the structure under investigation when – as often happens – the required results are the structure modes in operating conditions for comparison with an analytical model. Hence, for example, if the structure is practically grounded when operating, it would only increase the cost and time of subsequent analysis to test the structure in a free configuration.

10.3.2   *Excitation systems*

The excitation system is the device used to set the structure into motion, usually by the application of an appropriate driving force at the various preselected measurement locations. One of the primary decisions to be made during the phase of test planning is the type of force to be applied to the structure – i.e. the excitation function, which indirectly dictates the type of excitation system. Depending on the structure and on the desired results, this may vary from a stepped-sine to a periodic-random excitation, from transient (impact or chirp) to true random excitation, each method having its advantages and drawbacks in terms of measurement time, cost, signal-to-noise ratio etc. For our purposes, we can broadly classify excitation devices as attached and unattached: an attached device – typically a shaker – remains attached to the structure all through the duration of the test, while an unattached device – typically an impactor – is in contact with the structure only for short periods of time.

A shaker is an electromagnetic or electrohydraulic device which allows a considerable degree of control on the frequency and amplitude of the force applied to the structure. As compared to electroydraulic shakers, electromagnetic shakers are generally smaller and can operate at higher frequency ranges (from near zero up to 20–30 kHz). However, when larger force amplitudes are needed, it is more feasible to use electrohydraulic shakers which, by contrast, are generally more expensive and operate at frequency ranges from static (DC) to a maximum of 1 kHz or less. The exciter's moving element is attached to the structure under test via a force transducer (load cell), with the objective to transmit the desired excitation in a given direction while, at the same time, allowing the structure to move freely in all other directions. These latter movements are normal, but they can react back on the shaker and on the force transducer and result in a distortion of the force signal. To minimize this effect, it is customary to attach the shaker to the structure through a 'stinger', which consists of a short thin steel or nylon rod with large axial stiffness but low bending and shear stiffnesses. Shaker support is another factor that can produce unwanted consequences: in fact, any reaction force transmitted to the shaker should not be transmitted back into the structure. In other words the shaker should be isolated from the structure under test; this is generally done – when possible – by grounding the shaker to the floor and by suspending the structure on bungee cords, alternatively, one can mount the shaker on a mechanically isolated foundation or suspend the shaker. In this latter configuration, it may sometimes be necessary to mount an additional mass on the shaker in order to generate force amplitudes that can adequately excite the structure. Finally, another potential problem with shakers occurs when we try to excite the structure at one of its resonances, i.e. when a little force can produce large displacements of the structure. The result, especially for light damping, is a dip (force dropout) in the force spectrum which, in turn, can lead to significant measurement problems due to a low signal-to-noise ratio. One possible solution to these problems may be to use the so-called 'H_2 estimate' (which is not susceptible to noise in the input signal; see later sections) for the calculation of the FRF in the vicinity of resonances.

A very popular excitation method is an impact device, currently available on the market in the form of hammers of various sizes: from very small, for testing small and light structures, to sledgehammers, for testing heavy and more massive structures. The force input is measured by a force transducer mounted at the back of the hammer tip which, in turn, can be changed to have some degree of control on the frequency content of the force pulse. For a given test structure, a harder tip produces a pulse of shorter duration with a higher useful frequency range, while a softer tip – producing a pulse of longer duration – allows the impact energy to be concentrated in a lower frequency range. Typical frequency ranges may vary from zero to 4–5 kHz for smaller hammers and from zero up to 200–300 Hz for bigger and heavier hammers. Figure 10.25 shows two time histories of typical sledgehammer impacts obtained by the author by hitting the concrete floor of the laboratory. The impact represented by the thicker line has been obtained with a soft tip (the softer tip available with the hammer kit), while impact represented by the thinner line has been obtained with a harder tip (not the hardest available in the kit).

Basically, the maximum force amplitude is governed by the mass of the impactor and by the velocity of impact, while the duration of the pulse (and hence its frequency content) is determined by the stiffness of the contacting surfaces.

Figure 10.26 represents an indication of the useful part of the force spectrum of the above signals. It ranges from zero to a 'cutoff frequency' ν_c ($\nu_c \cong 190-200$ Hz in the figure) which is the largest value of frequency excited by the hammer blow. At higher frequencies the spectrum decays to zero force and the structure's resonances above ν_c would not receive enough energy to be excited adequately. As a rule of thumb, ν_c can be taken as the value where the frequency spectrum ceases to be reasonably flat and has fallen 10–20 dB from its maximum value. Although the figure stops at 195 Hz, it is evident that the value of ν_c for the softer tip is lower than the value of ν_c for the harder tip.

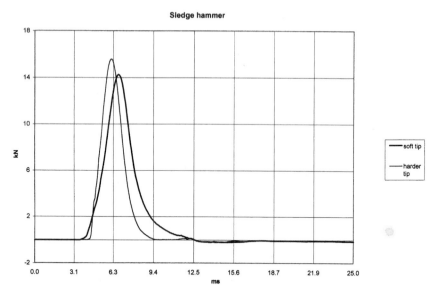

Fig. 10.25 Modally tuned sledgehammer blows with two different tips: time histories.

Also, while impacting a structure with a hammer, attention must be paid to the possibility of 'double strikes', because they cause difficulties in the signal processing stage. Double-impact measurements should be rejected because the force spectrum will have 'holes' of zero force at values of frequency which depend (inversely) on the time distance between the two impacts. A basic SDOF analysis shows that double impacts are more likely to occur when the ratio m/M – impactor mass to structure mass – is close to the classical coefficient of restitution e (ratio of relative velocities of the two bodies after and before impact). When this is the case, it is advisable to reduce the impactor mass until, approximately, $m/M < 0.2$.

The main advantages of the impact method are that it requires little hardware, it is relatively inexpensive and is fast. Its main drawbacks are that the energy imparted to the structure at any particular frequency is relatively small (low signal-to-noise ratio) and that the experimenter's control over amplitude and frequency content of the input force is very limited. In a typical impact test, the hammer input signal is fed to one channel of the analyser and the response output signal (say, from an accelerometer) is fed to the other channel. Two potential problems may arise in this situation: (1) the input signal is generally very short (from hundreds of µs to a few ms) compared to the time frame of the analyser; (2) for lightly damped structures, the response signal – an exponentially decaying oscillation – may not have decayed completely within the analyser time frame (whose length depends on the frequency range in which we are making the measurements). Problem (1)

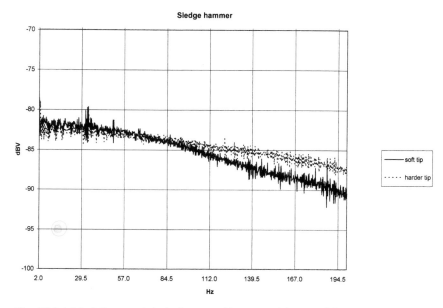

Fig. 10.26 Modally tuned sledgehammer blows: useful part of frequency spectrum.

implies that a significant part of the acquired input signal is practically noise, while problem (2) results in an undesired effect called, in digital signal processing terminology, 'leakage' (Part II).

Both of these problems can be addressed using a digital signal processing technique called 'windowing': the input signal is (digitally) multiplied by a 'force window' function which forces the signal to zero right after the end of the pulse while the response signal is forced to zero within the analyser's time frame through multiplication by an appropriate 'exponential window' function. If the structure is heavily damped, the response signal may also have a short duration when compared to the length of the analyser time frame, in which case application of an appropriate exponential window just eliminates unwanted noise and results in a higher signal-to-noise ratio. Both force and exponential 'windows' – together with other windows usually needed for other types of signals – are usually available on all commercial spectrum analysers. In this regard it must be noted that the exponential window adds artificial damping to the measurement and this effect must be taken into account after the signal processing phase.

10.3.3 Measurement of response

The vibration response is usually measured by means of acceleration transducers called accelerometers, the most common being the piezoelectric type. A large number of accelerometers are available on the market and this aspect of vibration transduction will be considered in more detail in Part II, which deals with electronic instrumentation. For the moment it suffices to say that – among others – some of the most important parameters to consider in choosing an accelerometer are: its sensitivity (usually given in mV/g; typical sensitivities for modal tests are either 100 mV/g or 1000 mV/g), its resonant frequency and, in some applications on light structures, its weight.

The sensitivity rates the voltage output for unit acceleration (1 $g = 9.81$ m/s^2); the higher the sensitivity, the higher the electrical signal for a given acceleration.

Given the transducer's damping ratio, the resonant frequency ν_{res} indirectly suggests the useful frequency range which – for a damping ratio of 0.65–0.70 – ranges approximately from near zero (piezoelectric accelerometers do not extend to DC, but other types of accelerometers do) to about $0.5\nu_{res}$. The weight (mass) of an accelerometer, in turn, may have undesired consequences on the measurements. In fact, mass loading of the structure can be a problem when the mass of the transducer is a significant fraction of the mass of the tested structure. A quick check can be made by making an FRF measurement with one accelerometer mounted on the structure and then repeating the same measurement with two accelerometers (the second accelerometer with the same mass as the first). If some frequency and amplitude shifts near the structure's resonant frequencies are noted, then mass loading can be a problem. For very light structures, it may be advisable to use a noncontacting transduction device.

In regard to mass loading, a word of caution is necessary. The general rule of thumb is that the mass of the transducer should be less than 1/10 of the mass of the measured structure. However, this statement is made more precise if we refer to the 'apparent dynamic mass' of a particular DOF associated with a specific mode of the structure. So, the mass of the transducer should be smaller that 1/10 of the 'apparent dynamic mass' of the structure at the attachment point. This latter quantity, in turn, may vary over a wide range depending on the point of the structure and the mode that we are considering; for practical applications it can be obtained by measuring a point (driving) FRF and calculating the modal coefficient $_mA_{jj}$ (eq (7.78b) and following). The reader can find an enlightening and practical discussion of this subject in Døssing [2].

Finally, great attention must be paid to the way the accelerometer is physically attached to the structure. Improper mounting can result in a significant reduction of the transducer's resonant frequency and hence to a reduction of the useful frequency range. The ideal mounting is by means of a threaded stud onto a flat, smooth surface. Other alternatives, in decreasing order of quality, are: a thin layer of beeswax, epoxy cement, thin double-sided adhesive, magnet and handheld.

10.3.4 Excitation functions

The last point we want to consider briefly in this section is the type of excitation function used to set the structure into motion. Impact testing is the simplest one. With a two-channel analyser and a n-DOF system (i.e. n locations geometrically identified on the structure under test), one generally acquires one row of the complete $n \times n$ FRF matrix by moving the excitation, i.e. the hammer, around the structure, while measuring the response at a single (fixed) location – preferably not close to the node of one or more of the structure's modes. Conversely, one column of the FRF matrix is acquired if we repeatedly hit the same point of the structure and move the accelerometer around. The two techniques are called, respectively, 'roving hammer' and 'roving accelerometer' techniques.

On the other hand, we can use a shaker and excite the structure sinusoidally at a single frequency, measure the response (or, more precisely the FRF value at that frequency) and then perform other measurements by varying the input frequency by small increments. This is the so called stepped-sine testing and care must be exercised because, at each frequency, the transient part of the response must be allowed to die out before a valid measurement can be taken (remember that, in these tests, only the steady state of the response is of interest to us). This technique generally has a high signal-to-noise ratio but, although the frequency step can be varied from smaller values near resonances to larger values away from resonances, it is lengthy.

One alternative is to excite the structure with a random signal (more about random signals can be found in Chapters 11 and 12). For the purposes of the test, a random signal is a signal with a continuous flat spectrum over the frequency range of interest. Typically, the excitation device is, once again, a shaker and the signal generator is often the analyser itself. Since the random excitation usually has a Gaussian probability distribution (Chapter 11), the response will also be random with a Gaussian probability distribution. From a digital signal processing point of view, such signals are sampled by the analyser over a finite period of time T (the time frame or time record) and the analyser FFT algorithm implicitly assumes that the acquired signals are periodic with period T. Then, since this is not generally the case, leakage occurs (unless the signal is transient and its entire duration fits completely within a time record). Once again, the remedy is to force the signal to zero at the beginning and end of the time frame by multiplying it by an appropriate window function, the so-called 'Hanning' window; windowing the signals does not eliminate leakage completely, but minimizes its effects. For purpose of illustration, consider Fig.10.27(a)–(c). Figure 10.27(a) shows a unit amplitude sinusoidal signal with frequency $\nu = 12$ Hz, which is clearly not periodic in the hypothetical analyser time frame of length $T = 1.7$ s. This occurrence would cause 'leakage' when the FFT is calculated. For this time frame, Fig.10.27(b), in turn, shows a typical Hanning window and Fig.10.27(c) shows the windowed signal (product of unwindowed signal times the window) which is used by the anlalyser to calculate the FFT.

Alternatively, the structure can be excited by a periodic–random signal, i.e. a signal which is random in character within the length of the time record T,

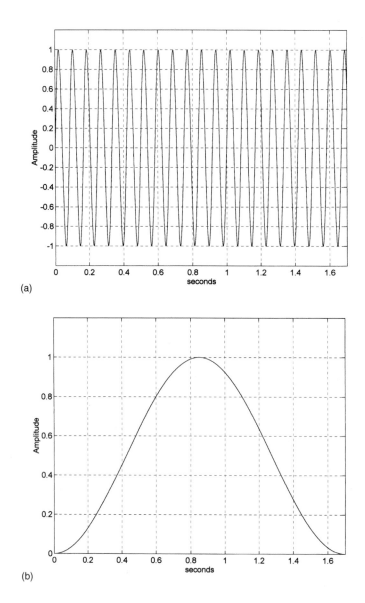

Fig. 10.27 (a) Unwindowed signal. (b) Hanning window.

but repeats itself with period T. Then, no window would be needed because, as a matter of fact, the signal is periodic within the analyser time frame.

Another problem with random excitation (as with all broadband signals) is that the signal-to-noise ratio may not be particularly favourable because the (limited) power supplied by the exciter is spread over the entire frequency

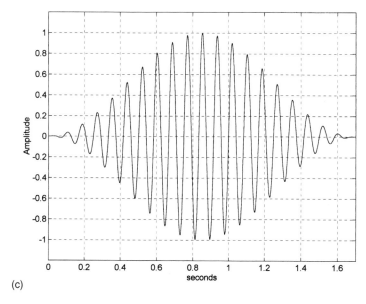

(c)

Fig. 10.27 (c) Windowed signal.

range. For each pair of input–output locations, this problem may be tackled by averaging a number (say, at least 50) of subsequent measurements, obviously at the expense of a longer measurement time.

Finally, mention should be made of two other facts. During a modal test with random or impact excitation and repeated averages, it is often advisable to inspect visually the so-called 'coherence' function (usually indicated with the symbol $\gamma(\omega)$, (Section 10.4.2)) which, ideally, should be unity over the whole frequency range of interest. For a single-input single-output system (e.g. Fig. 4.7) the coherence function represents a statistical level of confidence in the FRF estimate obtained from measurement and it is also an index the degree of causality between input an output (i.e. whether the measured response is entirely due to the measured input). Poor coherence may be due to a number of factors of random or of systematic nature, for example, random noise contamination of the measurements, leakage, insufficient frequency resolution or nonlinearities in the structure under test. However, random sources of error can generally be minimized by taking many averages, while this is not true for systematic (bias) sources of error.

The second fact worth mentioning is that reciprocity is another data quality check that should be performed. This is simply made by comparing two reciprocal measurements (i.e. the FRFs H_{jk} and H_{kj}) and observing any differences between them. Lack of reciprocity is usually indicative of a nonlinear system.

10.4   Selected topics in experimental modal analysis

The intention of the preceding section was to give the reader a general overview of the many problems that should be addressed during the measurement phase of a modal test. By contrast, this section considers some special topics that deserve particular attention. Owing to the available space and the large number of specific topics that could be considered, it is clear that some *a priori* choice had to be made. Therefore, there is no claim of completeness and the interested reader is urged to consult the wide body of specific literature on all the different aspects of experimental modal analysis [3].

First of all, as far as the testing method is concerned, it must be noted that there are many different approaches to modal testing. Traditionally, one can distinguish between **phase separation methods** and **phase resonance methods** (or tuned sinusoidal or force appropriation methods).

Phase separation methods owe their wide popularity to the advent of real-time FFT analysers; they are generally shorter in duration and less expensive. The basic underlying assumption is that the structure's response is a weighted linear summation of all its uncoupled modes of vibration; under this assumption, the excitation and the response are usually broadband signals (for example, random or impact) and the tests are often performed by using a two channel analyser and a single-input single-output configuration. The modal parameters are then extracted from the measured data by mathematical curve fitting (all modal analysis software packages offer a number of curve-fitting techniques) which, in turn, is usually performed on the acquired FRFs in the frequency domain. However, time-domain-fitting techniques are also becoming popular.

Phase resonance methods (often used, for example, in the aircraft industry) are generally more expensive and lengthy, require more hardware and considerable experimental skill. They are multishaker techniques based on the fact that a particular mode of vibration can be isolated by an appropriate sinusoidal excitation.

10.4.1   Characteristic phase-lag theory and Asher's method

Referring to the previous observation, let us consider an n-DOF system with viscous damping excited by a set of n forces of the same frequency ω and independently variable (as yet unknown) force amplitudes. The relevant equations of motions can be written as

$$\mathbf{M}\ddot{\mathbf{u}} + \mathbf{C}\dot{\mathbf{u}} + \mathbf{K}\mathbf{u} = \mathbf{f}_0 \sin \omega t \qquad (10.11)$$

and it is known that the response will be sinusoidal at the same frequency ω. However, let us look for a solution in which all the n degrees of freedom have the same phase lag θ with respect to the excitation, i.e. is of the form

$$\mathbf{u} = \mathbf{z} \sin(\omega t - \theta) \qquad (10.12)$$

Calculating the prescribed derivatives and substituting in eq (10.11), we get

$$[\mathbf{K} - \omega^2 \mathbf{M}]\mathbf{z} \sin(\omega t - \theta) + \omega \mathbf{C}\mathbf{z} \cos(\omega t - \theta) = \mathbf{f}_0 \sin \omega t \qquad (10.13)$$

so that, after use of well known trigonometric identities, we can separate the terms in $\sin \omega t$ and $\cos \omega t$ to obtain

$$[\mathbf{K} - \omega^2 \mathbf{M}]\mathbf{z} \cos \theta + \omega \mathbf{C}\mathbf{z} \sin \theta = \mathbf{f}_0$$
$$[\mathbf{K} - \omega^2 \mathbf{M}]\mathbf{z} \sin \theta - \omega \mathbf{C}\mathbf{z} \cos \theta = 0 \qquad (10.14)$$

where the unknowns are the vectors $\mathbf{z}, \mathbf{f}_0$ and the phase angle θ. Specifically, from the second of eqs (10.14) it is immediately evident that when $\theta = \pi/2$ we have

$$\mathbf{K}\mathbf{z} = \omega^2 \mathbf{M}\mathbf{z} \qquad (10.15a)$$

and, from the first of eq (10.14),

$$\mathbf{f}_0 = \omega \mathbf{C}\mathbf{z} \qquad (10.15b)$$

Equation (10.15a) is well known and is precisely the undamped eigenvalue problem for the system under investigation whose solution consists of n undamped eigenvalues ω_j and n real eigenvectors $\mathbf{z}_j$. More generally, if $\cos \theta \neq 0$, we can divide the second of eq (10.14) by $\cos \theta$ and obtain

$$[\mathbf{K} - \omega^2 \mathbf{M}]\mathbf{z} \tan \theta = \omega \mathbf{C}\mathbf{z} \qquad (10.16)$$

which can be interpreted as a generalized eigenvalue problem of order n and eigenvalue $\tan \theta$. This means that, for every value of excitation frequency ω, a nontrivial solution can be obtained when

$$\det[\tan \theta \mathbf{K} - \omega^2 \tan \theta \mathbf{M} - \omega \mathbf{C}] = 0 \qquad (10.17)$$

so that there are n values of θ and n deflection shapes (the so-called phase-lag modes which, strictly speaking, if $\omega \neq \omega_j$, should not be called modes but 'operating deflection shapes') $\mathbf{z}$ that satisfy eq (10.16). These modes can be interesting in themselves, but, as stated above, the most important case is when the excitation frequency equals one of the undamped eigenfrequencies ω_j. In this case the jth eigenvalue of problem (10.16) is such that $\theta_j \to \pi/2$ and the deflected shape equals the undamped real mode $\mathbf{z}_j$. Moreover, the force configuration needed to obtain this situation can be obtained from eq (10.15b), which can be rewritten as

$$\mathbf{f}_0^{(j)} = \omega_j \mathbf{C}\mathbf{z}_j \qquad (10.18)$$

and shows that the excitation forces must be in phase with the damping dissipation forces. Note that it is not important whether damping is proportional or not.

Let us summarize the above results in a few words. When the structure under test is excited (by means of an appropriate force configuration $\mathbf{f}_0^{(j)}$) at one of its undamped frequencies ω_j, all the displacement responses are in quadrature (90° phase shift) with the excitation forces and the structure's vibration shape coincides with the undamped real mode $\mathbf{z}_j$. So – in an experimental test with n shakers and n response transducers – once we have an idea of the frequency value ω_j and of the amplitude force configuration $\mathbf{f}_0^{(j)}$ we can 'isolate' (or 'tune') the undamped modal shape $\mathbf{z}_j$.

Example 10.1. As an example, let us once again consider the 2-DOF system of Section 7.9 with the intention of obtaining the set of forces needed to 'isolate' the two undamped modes. The undamped eigenvalues and eigenvectors are, respectively

$$\omega_1 = 17.114$$
$$\omega_2 = 41.317 \qquad (10.19)$$

and

$$\mathbf{z}_1 = [0.5774 \quad 0.8165]^T$$
$$\mathbf{z}_2 = [-0.5774 \quad 0.8165]^T \qquad (10.20)$$

(note that the eigenvectors have not been mass-orthonormalized; however, this is irrelevant for our purposes). Then, from eq (10.18) the force configuration needed to isolate the first mode is given by

$$\mathbf{f}_0^{(1)} \equiv \begin{bmatrix} f_{01}^{(1)} \\ f_{02}^{(1)} \end{bmatrix} = \omega_1 \mathbf{C} \mathbf{z}_1 = \omega_1 \begin{bmatrix} 1000 & -500 \\ -500 & 500 \end{bmatrix} \mathbf{z}_1 = 1 \times 10^3 \begin{bmatrix} 2.894 \\ 2.046 \end{bmatrix} \qquad (10.21)$$

The quantity of interest to us is the ratio $f_{01}^{(1)}/f_{02}^{(1)} = 1.41$. Similarly, for the second mode, we get $f_{01}^{(2)}/f_{02}^{(2)} = -1.41$, the minus sign indicating that the force applied at one of the two DOF must be in antiphase (180° phase shift) with the force applied at the other DOF.

From a practical point of view, the characteristic phase lag theory outlined above poses a number of problems. For example, we generally do not know the matrices $\mathbf{M}$, $\mathbf{K}$ and $\mathbf{C}$ and neither do we know the undamped frequencies ω_j. A method proposed by Asher addresses these problems. For a harmonic excitation $\mathbf{f} = \mathbf{f}_0\, e^{i\omega t}$ we can write

$$[\mathbf{K} - \omega^2 \mathbf{M} + i\omega \mathbf{C}]\mathbf{z} = \mathbf{f}_0$$

and hence, as in eq (7.24)

$$\mathbf{z} = [\mathbf{K} - \omega^2\mathbf{M} + i\omega\mathbf{C}]^{-1}\mathbf{f}_0 \cong \mathbf{R}(\omega)\mathbf{f}_0 \qquad (10.22\text{a})$$

where $\mathbf{R}(\omega)$ is the receptance matrix. Separating its real and imaginary parts, eq (10.22a) can be rewritten as

$$\mathbf{z} = \{\mathrm{Re}[\mathbf{R}(\omega)] + i\,\mathrm{Im}[\mathbf{R}(\omega)]\}\mathbf{f}_0 \qquad (10.22\text{b})$$

so that the condition of displacements in quadrature with forces implies

$$\{\mathrm{Re}[\mathbf{R}(\omega)]\}\mathbf{f}_0 = 0 \qquad (10.23)$$

This means that a preliminary measurement of the response by means of a sinusoidal sweep or broadband excitation can be used to extract the real part of the receptance matrix, calculate the undamped frequencies from the relation

$$\det\{\mathrm{Re}[\mathbf{R}(\omega)]\} = 0 \qquad (10.24)$$

and then the force configurations (eq (10.23)) at the frequency values $\omega = \omega_j$. Hence, the jth mode can be isolated (or 'tuned') by setting the shakers to produce the forces $\mathbf{f}_0^{(j)} \sin \omega t$ and measuring the response $\mathbf{u} = \mathbf{z}_j \sin(\omega t - \pi/2)$.

Two things should be noted at this point. First, the method is capable of extracting the undamped modes of the structure (most methods obtain the actual damped modes) and it has been found to produce very good results in a large number of circumstances. Second, we have made no mention of additional practical problems, for example, exciter location and rectangular receptance matrices (this is often the case, because tests are often performed using more response signals than excitation signals). However, a number of important contributions by many authors address these subjects, which are beyond the scope of this book.

10.4.2   Single-input single-output test configuration: FRF measurement

This section briefly considers a typical test situation in which we excite the structure (modelled as an n-DOF system) at one point and measure its response at the same or at another point. Little hardware is required in this circumstance and we can acquire a row or a column of the system's FRF matrix by, respectively, moving the excitation point and keeping the response point fixed or by exciting a specific point of the structure and moving the vibration transducer around. This type of test is usually performed by acquiring data with a dual-channel digital analyser: the excitation signal (say, from a shaker or from a instrumented hammer) is fed to channel 1, the response signal is fed to channel 2 and one FRF at a time is acquired and stored for further analysis. For each pair of point locations, i.e. excitation at point j and response at point k, the relevant input–output relationship in the frequency domain is (see eq (5.76))

$$X_k(\omega) = 2\pi H_{jk}(\omega)F_j(\omega) \tag{10.25a}$$

which, omitting the indexes j and k and switching to ordinary frequency ν ($\nu = \omega/2\pi$) for our present convenience, is written as

$$X(\nu) = H(\nu)F(\nu) \tag{10.25b}$$

The digital analyser can calculate the FRF by simultaneously sampling the excitation and the response signal. However, in order to deal with all types of signals (transient, periodic, random, etc.) and with the fact that real-world signals are always more or less contaminated by noise, it is not convenient to obtain the FRF by simply calculating the ratio $X(\nu)/F(\nu)$. A fundamental algorithm of any analyser (with two or more input channels) is based on the so-called 'trispectrum average' which can be implemented on two signals that have been simultaneously sampled. Basically, a number of repeated measurements are taken and averaged; in the averaging process the analyser calculates three spectra **estimates** from the two signals: the power autospectrum of each signal and the power cross-spectrum between the two signals. These three spectra are used to obtain the FRF and the coherence function (e.g. McConnell [4]). As is customary, let us indicate these three spectra with the symbols $G_{ff}(\nu)$ (force signal autospectrum), $G_{xx}(\nu)$ (response signal autospectrum) and $G_{fx}(\nu)$ (cross-spectrum between force and response signals). Also, it can be shown that the cross-spectrum $G_{xf}(\nu)$ between response and force signals is simply the complex conjugate of $G_{fx}(\nu)$, i.e. $G_{xf}(\nu) = G_{fx}^*(\nu)$.

Rigorous definitions of autospectra, cross-spectra and of other concepts in this section involve statistical considerations which will be made in the next two chapters. For the moment, let us just consider these spectra as convenient functions which serve the purpose of optimizing the calculation of the FRF.

Mathematically, we can multiply both sides of eq (10.25b) by $F^*(\nu)$ to obtain

$$F^*(\nu)X(\nu) = H(\nu)F^*(\nu)F(\nu) \tag{10.26}$$

which, upon averaging an infinite number of measurements of the two signals, leads to

$$G_{fx}(\nu) = H(\nu)G_{ff}(\nu) \tag{10.27}$$

so that

$$H(\nu) = \frac{G_{fx}(\nu)}{G_{ff}(\nu)} \tag{10.28a}$$

We must take into account, however, that in real situations we always average over a finite number of measurements so that what we really obtain are only estimates of these spectra. In this light, eq (10.28a) should be rigorously written as

$$\hat{H}(\nu) = \frac{\hat{G}_{fx}(\nu)}{\hat{G}_{ff}(\nu)} \equiv H_1(\nu) \tag{10.28b}$$

(where the symbol ' $\hat{}$ ' indicates that we are dealing with estimates and not with the 'true' quantities. With this in mind, eq (10.28b), defines the so-called 'H_1 estimate' of $H(\nu)$. Multiplying both sides of eq (10.25b) by $X^*(\nu)$, a similar line of reasoning leads to the definition of the H_2 estimate of $H(\nu)$, i.e.

$$H_2(\nu) \equiv \frac{G_{xx}(\nu)}{G_{xf}(\nu)} \tag{10.29}$$

(we abandon here the ' $\hat{}$ ' notation, but the statistical nature of the following discussions will be recalled whenever necessary).

Ideally (noise-free signals and an infinite number of measurements), H_1 and H_2 lead to the same result, namely, the function $H(\nu)$. However, this is generally not the case. This suggests that the ratio H_1/H_2 can be an indicator of the quality of our measurements; hence we define the ordinary coherence function as

$$\gamma^2(\nu) \equiv \frac{H_1(\nu)}{H_2(\nu)} = \frac{|G_{fx}(\nu)|^2}{G_{ff}(\nu)G_{xx}(\nu)} \tag{10.30}$$

which can be interpreted as a measure of how well the force signal is linearly related to the response signal and is the counterpart of the familiar correlation coefficient in ordinary regression analysis. The correlation function ranges from zero (no correlation) to unity (optimum correlation) and from eq (10.30) it can be anticipated that, when $\gamma < 1$, H_1 will tend to underestimate the actual FRF, while H_2 will tend to overestimate it. Needless to say, the coherence function calculated and displayed by all analysers – owing to the finite number of measurements – is only an estimate of the 'true' coherence function.

Before proceeding further we may note that there is a third possibility for estimating $H(\nu)$ which can be obtained by using only the magnitudes of both sides of eq (10.25b), i.e.

$$X^*(\nu)X(\nu) = H^*(\nu)H(\nu)F^*(\nu)F(\nu)$$

so that, upon averaging, we arrive at

$$|H(\nu)|^2 = \frac{G_{xx}(\nu)}{G_{ff}(\nu)} \equiv |H_a(\nu)|^2 \tag{10.31}$$

where the subscript a indicates that only autospectra are used in calculating the estimator H_a. Moreover, since we only consider magnitudes in this case, it is important to note that no phase information is given by the estimator H_a.

Now, suppose that some uncorrelated noise contamination exists in both the force and response signal. Even with a properly set and calibrated instrumentation chain, extraneous noise in a test environment may exist from a number of causes which, for practical purposes, we can classify as either of electrical or of mechanical nature. For example, the intrinsic noise of electronic circuitry is an electrical source of noise, while heavy construction activities in the vicinity of the test area results in unwanted mechanical excitation of the structure under test. Other sources of electrical noise such as random fluctuations of disturbances due to, say, electromagnetic and radio frequency interference, ground loops or insufficient shielding of long cables can be, at least in principle, eliminated by eliminating the disturbances themselves.

So, if we call the 'true' excitation and response signals $e(t)$ and $r(t)$, respectively, the measured signals will be

$$f(t) = e(t) + m(t)$$
$$x(t) = r(t) + n(t) \tag{10.32}$$

where $m(t)$ represents extraneous noise in the excitation signal and $n(t)$ represents extraneous noise in the response signal. Equations (10.32) imply that the measured auto- and cross-spectra will be

$$G_{ff}(\nu) = G_{ee}(\nu) + G_{mm}(\nu) + G_{em}(\nu) + G_{me}(\nu)$$
$$G_{xx}(\nu) = G_{rr}(\nu) + G_{nn}(\nu) + G_{rn}(\nu) + G_{nr}(\nu) \tag{10.33}$$
$$G_{fx}(\nu) = G_{er}(\nu) + G_{en}(\nu) + G_{mr}(\nu) + G_{mn}(\nu)$$

while the true FRF can be mathematically expressed by

$$H(\nu) = \frac{G_{er}(\nu)}{G_{ee}(\nu)} = \frac{G_{rr}(\nu)}{G_{re}(\nu)}$$

$$|H(\nu)|^2 = \frac{G_{rr}(\nu)}{G_{ee}(\nu)} \tag{10.34}$$

Also, from the first of eqs (10.34), we get

$$G_{er}(\nu)G_{re}(\nu) = G_{er}(\nu)G_{er}^*(\nu) = G_{rr}(\nu)G_{ee}(\nu) \tag{10.35}$$

Under the reasonable assumption that noise signals are uncorrelated with each other and with the excitation and response signals, i.e. that

$G_{em} = G_{me} = G_{rn} = G_{nr} = G_{en} = G_{mr} = G_{mn} = 0$, three cases of interest will be discussed in the following.

Case 1. Noise in excitation signal, no noise in response signal

Since $n(t) = 0$, eqs (10.33) become (omitting for brevity of notation the frequency dependence)

$$G_{ff} = G_{ee} + G_{mm}$$
$$G_{xx} = G_{rr} \qquad\qquad (10.36)$$
$$G_{fx} = G_{er}$$

Evaluation of H_1 according to eq (10.28) gives

$$H_1 = \frac{G_{er}}{G_{ee} + G_{mm}} = \frac{H}{1 + G_{mm}/G_{ee}} \qquad\qquad (10.37)$$

where the fist of eqs (10.34) has been taken into account. It follows that the true FRF is generally underestimated owing to the denominator being greater than unity. The phase information, however, is correct because it comes from the error-free cross-spectrum (autospectra contain magnitude information only). On the other hand, evaluation of H_2 according to eq (10.29) (taking (10.35) into account) leads to

$$H_2 = \frac{G_{rr}}{G_{er}^*} = \frac{G_{rr}G_{er}}{G_{rr}G_{ee}} = \frac{G_{er}}{G_{ee}} = H \qquad\qquad (10.38)$$

showing that, under the assumption of uncorrelated noise, H_2 is insensitive to noise in the input signal. At this point it is important to remember the intrinsic statistical nature of this whole discussion and note that eqs (10.37) and (10.38) are strictly true for an infinite number of measurements; in a real-world situation (i.e. finite number of averages) what these formula imply is that when noise is present in the excitation signal only, then $H_1(\nu)$ is a 'biased' estimator of the true FRF $H(\nu)$, while $H_2(\nu)$ is an 'unbiased' (and therefore preferable) estimator of $H(\nu)$.

The coherence function in this case is given by

$$\gamma^2 = \frac{1}{1 + G_{mm}/G_{ee}} \qquad\qquad (10.39)$$

and, as expected, depends on the signal-to-noise ratio G_{ee}/G_{mm}: the lower the signal-to-noise ratio, the closer the value of coherence to unity. Also, from eq (10.39) we get

$$S/N \equiv \frac{G_{ee}}{G_{mm}} = \frac{\gamma^2}{1 - \gamma^2} \tag{10.40}$$

In this regard, it is interesting to note that, since $G_{ee}H = G_{er}$ implies in this case $G_{ee} = G_{fx}/H_2$, from the definition of H_2 (eq (10.29)) we get

$$G_{ee} = \frac{|G_{xf}|^2}{G_{xx}} \tag{10.41}$$

and also, since $G_{mm} = G_{ff} - G_{ee}$, we have also

$$G_{mm} = G_{ff} - \frac{|G_{xf}|^2}{G_{xx}} \tag{10.42}$$

Equations (10.41) and (10.42) show that both G_{ee} and G_{mm}, although they cannot be directly measured, can be calculated (strictly speaking 'estimated') in terms of measured quantities.

Case 2. No noise in the excitation signal, noise in the response signal

Equations (10.33) now read

$$G_{ff} = G_{ee}$$
$$G_{xx} = G_{rr} + G_{nn} \tag{10.43}$$
$$G_{fx} = G_{er}$$

because $m(t) = 0$. Evaluation of H_1 gives

$$H_1 = \frac{G_{er}}{G_{ee}} = H \tag{10.44}$$

so that H_1 is insensitive to uncorrelated noise in the response signal in this case. On the other hand, by taking into account eqs (10.35) and the first of eqs (10.34)

$$H_2 = \frac{G_{rr} + G_{nn}}{G_{er}^*} = \frac{G_{er}(1 + G_{nn}/G_{rr})}{G_{ee}} = H\left(1 + \frac{G_{nn}}{G_{rr}}\right) \tag{10.45a}$$

which, by virtue of the second of eqs (10.34) can also be written as

$$H_2 = H\left(1 + \frac{G_{nn}}{|H|^2 G_{ee}}\right) \tag{10.45b}$$

Equations (10.45) show that H_2 generally overestimates the true FRF, because the term in parentheses is usually larger than unity. The phase information, however, is correct because it comes from the error-free cross-spectrum.

As for the previous case, the precise meaning of eqs (10.44) and (10.45) is as follows: when only the response signal is contaminated by noise, H_1 is an unbiased (and preferable) estimator of H, while H_2 is a biased estimator of H.

The coherence function is now

$$\gamma^2 = \frac{1}{1 + G_{nn}/G_{rr}} \tag{10.46}$$

and

$$S/N \equiv \frac{G_{rr}}{G_{nn}} = \frac{\gamma^2}{1 - \gamma^2} \tag{10.47}$$

Also, we can express G_{rr} and G_{nn} in terms of measured quantities because $G_{rr} = HG_{re} = H_1 G_{fx}^*$ and hence, from the definition of H_1,

$$G_{rr} = \frac{|G_{fx}|^2}{G_{ff}} \tag{10.48}$$

and

$$G_{nn} = G_{xx} - \frac{|G_{fx}|^2}{G_{ff}} \tag{10.49}$$

Case 3. Noise in both excitation and response signals

Equations (10.33) become now

$$G_{ff} = G_{ee} + G_{mm}$$
$$G_{xx} = G_{rr} + G_{nn} \tag{10.50}$$
$$G_{fx} = G_{er}$$

so that

$$H_1 = \frac{G_{er}}{G_{ee} + G_{mm}} = \frac{H}{1 + G_{mm}/G_{ee}} \tag{10.51}$$

and

$$H_2 = \frac{G_{rr} + G_{nn}}{G_{er}^*} = H\left(1 + \frac{G_{nn}}{G_{rr}}\right) \tag{10.52}$$

It is then evident that

$$H_1(\nu) < H(\nu) < H_2(\nu) \tag{10.53}$$

and that the coherence function

$$\gamma^2 = \frac{1}{(1 + G_{mm}/G_{ee})(1 + G_{nn}/G_{rr})} \tag{10.54}$$

can be less than unity because both excitation and response noises contribute together.

Since H_1 and H_2 are, respectively, a lower and an upper bound estimator for H, the geometric mean of these two quantities provides an estimator that lies in between the two. This is generally called H_ν and is given by

$$H_\nu(\nu) = \sqrt{H_1(\nu)H_2(\nu)} = H\sqrt{\frac{1 + G_{nn}/G_{rr}}{1 + G_{mm}/G_{ee}}} \tag{10.55}$$

Some conclusions can be drawn from the discussion above. If we consider that at the structure's resonances the excitation signal is particularly susceptible to noise because the structure is compliant at these values of frequency and little force is required to produce a significant displacement while, on the other hand (unless the response transducer is placed at a node for that particular mode) the response signal has generally a good signal-to-noise ratio, we can infer that better results can be expected by using the estimator H_2 near the resonances.

By contrast, away from resonances (and, specifically, around antiresonances) the structure is stiff and the response signal, rather than the excitation signal, may be more susceptible to noise contamination. Then, a better estimate of the actual FRF can be obtained by using the estimator H_1.

Furthermore, the coherence function is related to the variance on the estimate of H and is a statistical parameter based on averages in the quantities G_{fx}, G_{ff} and G_{xx}. The coherence for a single measurement is always unity (and is useless), even in the presence of noise. In a number of subsequent measurements, this function can be assumed to be a measure of the quality of our measurements with the following considerations in mind:

• With random excitation, low coherence does not necessarily imply a poor estimate of the FRF but it may just mean that more averages are needed for a reliable result (incidentally, we note that if we use a shaker with a 'stinger' which is too stiff in the transverse direction, the transverse shaker–structure interactions appear as noise in the excitation signal; however, this noise is not uncorrelated with the input signal and one of the assumptions of case 1 is not valid).

- With deterministic excitation (for example, impact or rapid 'chirp' sinusoidal sweeps), low coherence usually indicates bias errors such as nonlinearity, significant noise levels or, for improper windowing, leakage.

The considerations of this section can be extended to more general cases and, in particular to the most general case of a multiple-input multiple-output (MIMO) test configuration, i.e. where there is a number n of excitation points and a number m of measured responses and all signals are measured simultaneously. The main advantages of this procedure are an increase of accuracy and consistency in the estimates of the structure FRFs and a reduction in the testing time. Specifically, multiple input configurations allow the separation of closely spaced modes, a circumstance in which single-input configurations incur serious difficulties.

MIMO techniques, however, involve the inversion of a matrix (containing the input auto- and cross-spectrum information) which, in a number of practical situations, has been found to be singular. It has been shown that such an inverse exists and leads to a unique solution for FRFs when the inputs (excitations) are not correlated. In this light, the concepts of partial coherence and multiple coherence are introduced and partial coherence between the inputs is used for assessing whether the inputs are correlated. There are several sources to which the interested reader can refer for detailed discussion of this topic [5–10].

10.4.3 *Identification of modal parameters – curve fitting*

Once the experimental data (typically the FRFs) have been collected, the next task of interest is to extract from this information the modal parameters of the structure: natural frequencies, damping ratios and modal amplitudes associated with each natural frequency. For an SDOF system, this is an easy task; it has already been shown in Chapter 4 and in Section 10.2.1 how the FRF can be used to obtain the values of natural frequency, mass, stiffness and damping ratio. (For example, consider the simple SDOF system used to draw the FRF graphs of Section 10.2.1. Noting that for an SDOF system $|A(\omega)| \rightarrow 1/m$ as $\omega \rightarrow \infty$, the mass can be extracted from the $dB(y)$–$\log(x)$ graph of acceleration (Fig. 10.3) by noting that the mass line in the high-frequency range is approximately at -34 dB, so that $m = 10^{34/20} = 50.1$ kg.)

Similarly, suppose we have measured a column of the FRF matrix of an n-DOF system with widely spaced and lightly damped resonances, and suppose further that all n modes have been experimentally observed. Among other possibilities, we can determine the natural frequencies ω_j from:

- the peaks of the magnitude graphs;
- the peaks of the imaginary parts of the receptance or acceleration graphs;
- the zero crossings of the real part of the mobility graphs.

Then we can determine the modal damping ratios ζ_j from any one of the magnitude graphs by calculating

$$\zeta_j = \frac{\omega_a^{(j)} - \omega_b^{(j)}}{2\omega_j} \tag{10.56}$$

where $\omega_a^{(j)}$ and $\omega_b^{(j)}$ (the so-called half-power points, or -3 dB points) are obtained from (Section 4.4.1)

$$|H(\omega_a^{(j)})| = |H(\omega_b^{(j)})| = |H(\omega_j)|/\sqrt{2} \tag{10.57}$$

Alternatively, on the real part of receptance or accelerance graphs, $\omega_a^{(j)}$ and $\omega_b^{(j)}$ are those values of frequency at which the local maximum and minimum are attained (e.g. Fig. 10.22a).

When the ω_j and ζ_j are known we can calculate the magnitude of the modal coefficients as outlined in Section 7.4.1 for a 3-DOF system.

An alternative (generally leading to more accurate results) to this method is to consider only the imaginary part of receptance or accelerance FRFs and obtain directly the jth mode shape from the ratios:

$$\frac{\text{Im}[H_{km}(\omega_j)]}{\text{Im}[H_{mm}(\omega_j)]} \qquad (k = 1, 2, \ldots, n) \tag{10.58}$$

where in eq (10.58) we assumed that we have measured the mth column of the matrix $\mathbf{H}(\omega)$.

Let us clarify this point. Suppose we have a 3-DOF system with natural frequencies ω_1, ω_2 and ω_3 and that we have measured the first column of the receptance matrix $\mathbf{R}(\omega)$, i.e. the functions R_{11}, R_{21} and R_{31}. Since the real part of receptance is zero at each resonance, the response is purely imaginary and we can obtain the first mode shape (eigenvector) from

$$\mathbf{z}_1 = \begin{bmatrix} \text{Im}[R_{11}(\omega_1)]/\text{Im}[R_{11}(\omega_1)] \\ \text{Im}[R_{21}(\omega_1)]/\text{Im}[R_{11}(\omega_1)] \\ \text{Im}[R_{31}(\omega_1)]/\text{Im}[R_{11}(\omega_1)] \end{bmatrix} \tag{10.59a}$$

and, respectively, the second and third eigenvectors from

$$\mathbf{z}_2 = \begin{bmatrix} \text{Im}[R_{11}(\omega_2)]/\text{Im}[R_{11}(\omega_2)] \\ \text{Im}[R_{21}(\omega_2)]/\text{Im}[R_{11}(\omega_2)] \\ \text{Im}[R_{31}(\omega_2)]/\text{Im}[R_{11}(\omega_2)] \end{bmatrix} \tag{10.59b}$$

$$\mathbf{z}_3 = \begin{bmatrix} \text{Im}[R_{11}(\omega_3)]/\text{Im}[R_{11}(\omega_3)] \\ \text{Im}[R_{21}(\omega_3)]/\text{Im}[R_{11}(\omega_3)] \\ \text{Im}[R_{31}(\omega_3)]/\text{Im}[R_{11}(\omega_3)] \end{bmatrix} \tag{10.59c}$$

All three eigenvectors are normalized so that their first element is unity and the sign of each element is automatically taken into account. It is obvious that the same results are obtained by considering any column other than the first or any row of the matrix $\mathbf{R}(\omega)$ or, alternatively, any row or column of $\mathbf{A}(\omega)$.

Example 10.2. A simple numerical example can be given by considering once again the 2-DOF system of Section 7.9 whose mass-orthonormal eigenvectors are given in eq (7.110b). From the receptance functions R_{11} and R_{12}, the calculation the eigenvectors shown above leads to

$$
\mathbf{z}_1 = \begin{bmatrix} 1.00 \\ 1.41 \end{bmatrix} \qquad \mathbf{z}_2 = \begin{bmatrix} 1.00 \\ -1.41 \end{bmatrix} \tag{10.60}
$$

which are just the eigenvectors of eq (7.110b) with a different normalization; the minus sign in the second element of $\mathbf{z}_2$ comes from the fact that $\mathrm{Im}[R_{12}(\omega_2)]$ is negative (Fig. 10.22(b)) and indicates that, in the second mode, the two masses move in opposite phase. By contrast, all the peaks of the imaginary part of the point FRF ($R_{11}(\omega)$ in this case) must have the same sign because the response and excitation are measured at the same point.

The procedure shown above can lead to reliable results when its basic underlying assumption is verified, namely that in the vicinity of a resonance the response is dominated by only one term of the sum (10.9). This is generally reasonable for structures with widely spaced and lightly damped resonances, a case in which the contribution of off-resonant modes can be assumed to be negligible, and we treat each resonance as if the other resonances did not exist.

In practice, all modal analysis software packages incorporate some 'identification methods', that is, numerical procedures with the specific purpose of extracting the modal parameters from a set of experimental data. Some methods require the user to participate in various decisions throughout the analysis, while others – once the relevant data have been supplied – are completely automatic. From a user's point of view, however, the main concerns are not the detailed analytical aspects of the numerical procedures, but a general picture of the available possibilities in order to have an idea of which method may fit his/her particular needs. Then, the choice must be based on the available hardware (to perform the test) and computing resources, the scope of the investigation, the structure under test and the format of the experimental data, not necessarily in this order. In the following, we adopt this point of view, also because due consideration of the most popular methods would exceed the scope and boundaries of this chapter.

All the parameter identification methods are based on analytical curve fitting of the measured data.

The first and more general classification concerns the domain in which the data are treated numerically, that is, we can distinguish between **frequency-domain** and **time-domain** methods. The former methods operate on the system's response characteristics in the frequency domain, i.e. on the FRFs which are generally written as (eq (7.78b))

$$H_{jk}(\omega) = \sum_{m=1}^{n} \left(\frac{{}_mA_{jk}}{i\omega - \lambda_m} + \frac{{}_mA_{jk}^*}{i\omega - \lambda_m^*} \right) \tag{10.61}$$

where, in the usual modal analysis symbolism, λ_m is the mth eigenvalue (the term 'pole' is also common) and it is expressed as (eq (6.160))

$$\lambda_m = -\zeta_m\omega_m + i\omega_{m(damped)} = -\zeta_m\omega_m + i\omega_m\sqrt{1 - \zeta_m^2} \tag{10.62}$$

and ${}_mA_{jk} = z_{jm}z_{km}$ is the mth residue.

On the other hand, the latter methods perform the fitting in the time domain, i.e. on the impulse response functions (IRFs) and generally involve the calculation of the inverse FFT of the FRFs. The basic mathematical expression is now

$$h_{jk}(t) = \sum_{m=1}^{n} \left({}_mA_{jk}e^{\lambda_m t} + {}_mA_{jk}^*e^{\lambda_m^* t} \right) \tag{10.63}$$

which is the time-domain counterpart of eq (10.61).

In principle, no difference should exist between the two approaches but the numerical behaviour of the identification method and the fact that experimental measurements are always performed in a limited frequency band must be taken into account in practical applications.

Frequency-domain methods, in turn, can be further divided according to the number of modes that can be analysed; hence we have **SDOF methods** and **MDOF methods**. SDOF methods are based on the assumption that in the frequency region around a resonance the response is dominated by the resonant term corresponding to that mode only, so that the contribution of other modes can be either completely ignored or taken into account by means of a simple approximation term. In this regard, the strategy outlined for purposes of illustration at the beginning of this section is a typical SDOF method (as a matter of fact it is somewhat a mixture of the two simplest methods known as 'peak amplitude' and 'quadrature response').

Since the above assumption is not always justified and SDOF methods may lead to serious inaccuracies when the structure under test is not lightly damped and has closely spaced modes, a number of MDOF methods have been devised which fit a multiple-mode form analytical expression to the experimental FRFs. The distinction between SDOF and MDOF methods does not exist in the time-domain because all time domain methods are necessarily MDOF (from an IRF there is no way to make an *a priori* separation of the various modes).

Then, another classification is usually made for both frequency- and time-domain methods based on the number of functions that are analysed simultaneously. In fact, depending also on the set of experimental FRFs (or IRFs), we can decide to use a method which analyses one function at a time (SISO, i.e. single-input single-output methods) or more functions at the same time. This latter possibility comprises two cases: if the functions analysed simultaneously have been collected by exciting the structure at one fixed location and measuring its response at several different locations, the method is classified as **SIMO**: single-input multiple-output. On the other hand, if the experimental data have been collected by exciting the structure and measuring the response at a number of different locations, we can analyse all the available functions simultaneously by using a polyreference or **MIMO** (multiple-input multiple-output) method. Situations of multiple-input single-output (MISO) are also possible, but are used to a much lesser extent.

Finally, in both time and frequency domains, a general distinction exists between **indirect** and **direct** methods. Indirect methods base the identification procedure on the modal model, while direct methods work directly on the spatial model, i.e. the fundamental basic matrix equation from which all the treatment of MDOF systems is derived.

Without claim of completeness, some names may help the reader to find his/her way among the various possibilities. In the frequency domain some popular methods are:

- the peak amplitude method (SDOF, SISO)
- the quadrature response method (SDOF, SISO)
- the Kennedy–Pancu or circle-fitting method (SDOF, SISO)
- the inverse method (SDOF, SISO)
- Dobson's method (SDOF, SISO)
- the Ewins–Gleeson method (MDOF, SISO)
- the complex exponential frequency-domain, or CEFD method (MDOF, SISO)
- the rational fraction polynomial, or RFP method (MDOF, SISO)
- the global rational fraction polynomial, or GRFP method (MDOF, SIMO)
- the global Dobson method (MDOF, SIMO)
- the eigensystem realization algorithm in the frequency domain, or ERA-FD method (MDOF, MIMO).

On the other hand, in the time domain we have:

- the complex exponential, or CE method (SISO)
- the least squares complex exponential, or LSCE method (SIMO)
- the polyreference complex exponential, or PRCE method (MIMO)

- the Ibrahim time-domain, or ITD method (SIMO)
- the single station time-domain, or SSTD method (SISO)
- the eigensystem realization algorithm, or ERA method (MIMO).

All the above methods are classified as indirect; some direct methods in the frequency domain are:

- the Identification of Structural System Parameters, or ISSPA method (SIMO)
- the spectral method (MIMO)
- the simultaneous frequency-domain, or SFD method (SIMO)
- the multimatrix method (MIMO)
- the frequency-domain direct parameter identification, or FDPI method (MIMO).

Among the time-domain direct methods we find, for example:

- the autoregressive moving average, or ARMA method (SISO)
- the direct system parameter identification, or DSPI method (MIMO).

As may be expected, some of the above fitting procedures are just different versions or extensions of other methods, and new methods are continuously being developed. We leave the details to the specific literature cited in the references, but it is worth remembering here that the results of the fitting procedure, no matter how sophisticated, can be no better than the quality of the input data (FRFs or IRFs). In other words, the quality of the measurements is a necessary prerequisite for a reliable test, otherwise one will experience a classical case of 'garbage in, garbage out'.

In conclusion, given the appropriate computing capabilities, the choice is often a matter of personal preference and familiarity with a specific method. On a very general basis, it can be said that time-domain methods tend to provide the best results when a large frequency range or a large number of modes exist in the data, whereas frequency-domain methods tend to provide the best results when the frequency range of interest is limited and/or the number of modes is relatively small. Nonetheless, a major advantage of the frequency domain implementation is that we can take into account the effect of the modes outside the frequency range of interest – say, from ω_a to ω_b – by virtue of residual terms (not to be confused with the terms $_mA_{jk}$ of eq (10.61)) that are incorporated in the model by writing eq (10.61) as

$$H_{jk}(\omega) = P_{jk}(\omega) + \sum_{m=1}^{n} \left(\frac{_mA_{jk}}{i\omega - \lambda_m} + \frac{_mA_{jk}^*}{i\omega - \lambda_m^*} \right) + Q_{jk} \qquad (10.64)$$

where n is the number of modes in the frequency range of interest, $P_{jk}(\omega)$ is the lower residual (called inertia restraint or residual inertia) and is an inverse function of the frequency squared, and Q_{jk} is the upper residual (called residual flexibility), independent of frequency.

No such possibility exists in the time domain and no account is taken of the effect of modes outside the frequency range of analysis. However, it can be argued that the time-domain implementation is numerically better conditioned than the frequency-domain equivalent and is generally more suited to handle noisy measurements.

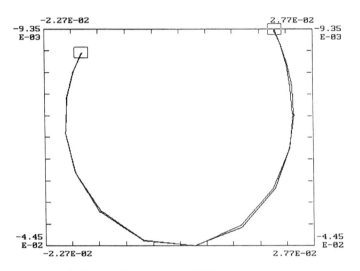

Fig. 10.28 Circle fitting of experimental FRF.

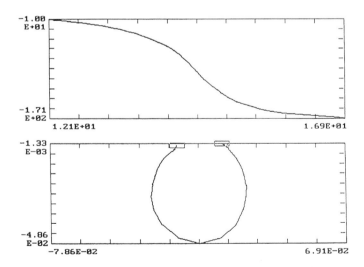

Fig. 10.29 Experimental FRF: phase versus frequency and Nyquist plot.

We close this chapter by giving some examples of actual measurements on engineering structures. Figure 10.28 shows the analytical fitting of a mode by means of the frequency-domain circle-fitting method. The excellent agreement of the two curves – i.e. the actual FRF and the analytical fit – is due to the fact that that mode (at about 14.4 Hz) was relatively isolated. Also note how the circle is well defined in the resonant region (from 13.50 to 15.58 Hz in this case) and encompasses approximately 250° of the circle. In this type of

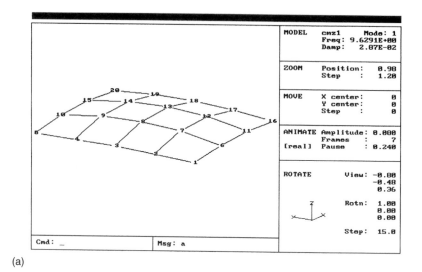

(a)

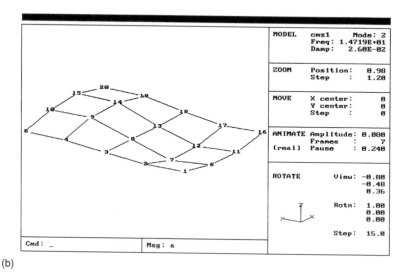

(b)

Fig. 10.30 Concrete floor: (a) first experimental mode; (b) second experimental mode. (Courtesy of Tecniter s.r.l., Milan, Italy.)

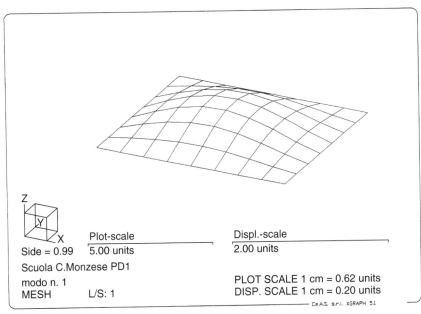

(a)

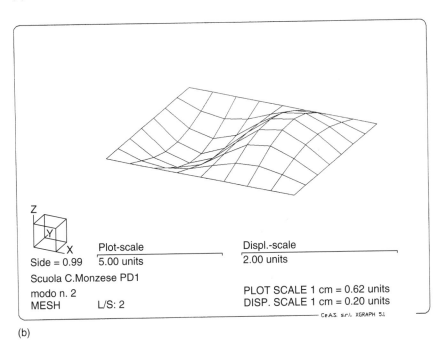

(b)

Fig. 10.31 Concrete floor: (a) first finite-element mode; (b) second finite-element mode. (Courtesy of Tecniter s.r.l., Milan, Italy.)

fitting, if possible, a span of at least $180°$ of the circle is advisable. The particular FRF of this example is a point FRF and was obtained by exciting the structure and measuring its response at location no. 13, i.e. $H_{13,13}$.

Figure 10.29, in turn, shows the unfitted curve of the same mode as seen from the FRF $H_{13,15}$, the upper curve being the phase and the lower curve being the Nyquist plot.

Finally, Figs 10.30(a) and (b) show the first (9.63 Hz) and second (14.72 Hz) mode of a light rectangular concrete floor (with dimensions of about 7×5 m) as obtained from a modal test. Figure 10.31(a) and (b) are the same modes as obtained from an independent finite-element analysis of the same structure. The experimental and theoretical (finite-element) modal frequencies were found to be in very good agreement.

10.5   Summary and comments

Experimental modal analysis (EMA) is a well-established field of activity in many branches of engineering. Basically, it is the process of characterizing the dynamic behaviour – or some aspects thereof – of a structure from a set of experimental measurements rather than from a theoretical (typically, finite-element) analysis. A modal test can be performed for a number of reasons, from troubleshooting and avoidance of vibration problems to finite-element model verification and updating, from nondestructive testing to evaluation of design modifications, from component substructuring to evaluation of structural integrity. The final scope of the investigation is the main factor that affects the quality and quantity of data to be collected. In fact, in some instances, a few data of good quality can do the job, resulting in substantial savings of time and money. On the other hand, other situations require a sophisticated analysis, which almost necessarily implies higher demands on hardware and computing capabilities. Although there exists an area of activity called 'nonlinear modal analysis' (which we do not consider), the basic assumption of EMA is that the structure under test is linear, time invariant, observable and obeys Maxwell's reciprocity relations.

We can distinguish three phases in a modal test: (1) data acquisition, (2) modal parameter estimation and (3) interpretation and presentation of results. Phase 1 is probably the most important because the quality of the final results cannot be any better that the quality of the experimental measurements. On the other hand, phase 2 – once the analysis has been completed – is generally a matter of good engineering judgement.

This chapter considers some fundamental aspects of phases 1 and 2, starting with the properties and formats of the data that are acquired in a modal test, i.e. the frequency response functions (FRFs). FRFs of SDOF and MDOF systems are considered in order to give the experimenter some guidelines for a quick check of the data during the test so that, if needed, he/she can repeat the measurements if there seems to be something wrong.

Then, a whole section deals with the experimental procedures, describing the instrumentation and potential problems associated with:

- the excitation mechanism (shaker, modally tuned hammer, etc.);
- the type of excitation signal (sinusoidal, impact, random, periodic–random, etc.);
- the measurement of the structure's response;
- the use of a multichannel FFT analyser.

Finally, the last section analyses three selected topics of EMA: the characteristic phase lag theory, the presence of noise in the input and/or the output signals in a single-input single-output test configuration and the curve-fitting methods, which are generally available on commercial software to extract and identify – starting from a set of measured FRFs or IRFs – the modal parameters of the structure under test.

Each part of this subject is interesting in its own right. The characteristic phase lag theory shows how it can be possible to obtain the undamped frequencies and mode shapes of a structure by an appropriate selection of the input force levels in a sinusoidal excitation type of test. This type of test is often performed on large structures, requires considerable hardware and experimental skill and one of its practical implementations is the so-called Asher method, whose line of reasoning is also shown.

The issue of (uncorrelated) noise in the input and/or at the output, in turn, is important to shed some light on the definition and properties of the most commonly adopted FRF estimators, i.e. $H_1(\omega)$, $H_2(\omega)$ and $H_v(\omega)$, and on the meaning of the ordinary coherence function $\gamma(\omega)$, which are all displayed by commercial FFT analysers.

Finally, a subsection on the curve-fitting procedures focuses the attention on the various possibilities in which the modal parameters (natural frequencies, damping factors and mode shapes) can be extracted from a set of measurements. In consideration of the fact that the user's main concern is not, in general, the specific numerical procedure, a general classification based on a set of basic criteria – frequency or time domain, number of functions processed simultaneously, etc. – is given in order to provide some information on the most popular methods.

References

1. Newland, D.E., *Mechanical Vibration Analysis and Computation*, Longman Scientific and Technical, 1989.
2. Døssing, O., The enigma of dynamic mass, *Sound and Vibration*, Nov., 16–21, 1990.
3. *Proceedings of the International Modal Analysis Conference, 1982–1997.* Available on CD-ROM; contact Mr Chris Tomko at SAVIAC (Shock and Vibration Information Analysis Center), Booz-Allen and Hamilton Inc., 3190 Fairview Park Drive (8th floor), Falls Church, VA 22042, USA.

4. McConnell, K.G., *Vibration Testing. Theory and Practice*, Wiley Interscience, 1995.

5. Allemang, R., Rost, B. and Brown, D., Multiple input estimation of frequency response functions: excitation consideration, ASME Paper no. 83-DET-73.

6. Bendat, J., Solution of the multiple input-output problem, *Journal of Sound and Vibration*, **44**(3), 311–325.

7. Leuridan, J., Multiple input estimation of frequency response functions for experimental modal analysis: currently used methods and some new developments, *Proceedings of the 9th International Seminar on Modal Analysis*, Vol. 2, 32 pp., Leuven, Belgium.

8. Leuridan, J., The use of principal inputs in multiple-input multi-output data analysis, *International Journal of Analytical and Experimental Modal Analysis*, July 1986, 1–8.

9. Bendat, J. and Piersol, S., *Random Data: Analysis and Measurement Procedures*, 2nd edn, John Wiley, New York, 1986.

10. Mendes Maia, N.M. and Montalvão e Silva, J.M. (eds), *Theoretical and Experimental Modal Analysis*, Research Studies Press, 1997.

11 Probability and statistics: preliminaries to random vibrations

11.1 Introduction

This chapter covers some fundamental aspects of probability theory and serves the purpose of providing the necessary tools for the treatment of random vibrations, which will be discussed in the next chapter. Probably, most of the readers have already some familiarity with the subject because probability theory and statistics – directly or indirectly – pervade almost all aspects of human activities, and, in particular, many branches of all scientific disciplines. Nonetheless, in the philosophy of this text (as in Chapters 2 and 3 and in the appendices), the idea is to introduce and discuss some basic concepts with the intention of following a continuous line of reasoning from simple to more complex topics and the hope of giving the reader a useful source of reference for a clear understanding of this text in the first place, but of other more specialized books as well.

11.2 The concept of probability

In everyday conversation, probability is a loosely defined term employed to indicate the measure of one's belief in the occurrence of a future event when this event may or may not occur. Moreover, we use this word by indirectly making some common assumptions: (1) probabilities near 1 (100%) indicate that the event is extremely likely to occur, (2) probabilities near zero indicate that the event is almost not likely to occur and (3) probabilities near 0.5 (50%) indicate a 'fair chance', i.e. that the event is just as likely to occur as not.

If we try to be more specific, we can consider the way in which we assign probabilities to events and note that, historically, three main approaches have developed through the centuries. We can call them the personal approach, the relative frequency approach and the classical approach. The personal approach reflects a personal opinion and, as such, is always applicable because anyone can have a personal opinion about anything. However, it is not very fruitful for our purposes. The relative frequency approach is more objective and pertains to cases in which an 'experiment' can be repeated many times and the results observed; $P[A]$, the probability of occurrence of event A is given as

$$P[A] = \frac{n_A}{n} \tag{11.1}$$

where n_A is the number of times that event A occurred and n is the total number of times that the experiment was run. This approach is surely useful in itself but, obviously, cannot deal with a one-shot situation and, in any case, is a definition of an *a posteriori* probability (i.e. we must perform the experiment to determine $P[A]$). The idea behind this definition is that the ratio on the r.h.s. of eq (11.1) is almost constant for sufficiently large values of n.

Finally, the classical approach can be used when it can be reasonably assumed that the possible outcomes of the experiment are equally likely; then

$$P[A] = \frac{n(A)}{n(S)} \tag{11.2}$$

where $n(A)$ is the number of ways in which outcome A can occur and $n(S)$ is the number of ways in which the experiment can proceed. Note that in this case we do not really need to perform the experiment because eq (11.2) defines an *a priori* probability. A typical example is the tossing of a fair coin; without an experiment we can say that $n(S) = 2$ (head or tail) and the probability of, say, a head is $P[A] \equiv P[\text{head}] = 1/2$. Pictorially (and also for historical reasons), we may view eq (11.2) as the 'gambler's definition' of probability.

However, consider the following simple and classical 'meeting problem': two people decide to meet at a given place anytime between noon and 1 p.m.. The one who arrives first is obliged to wait 20 min and then leave. If their arrival times are independent, what is the probability that they actually meet? The answer is 5/9 (as the reader is invited to verify) but the point is that this problem cannot be tackled with the definitions of probability given above.

We will not pursue the subject here, but it is evident that the definitions above cannot deal with a large number of problems of great interest. As a matter of fact, a detailed analysis of both definitions (11.1) and (11.2) – because of their intrinsic limitations, logical flaws and lack of stringency – shows that they are inadequate to form a solid basis for a more rigorous mathematical theory of probability. Also, the von Mises definition, which extends the relative frequency approach by writing

$$P[A] = \lim_{n \to \infty} \frac{n_A}{n} \tag{11.3}$$

suffers serious limitations and runs into insurmountable logical difficulties.

The solutions to these difficulties was given by the axiomatic theory of probability introduced by Kolmogorov. Before introducing this theory, however, it is worth considering some basic ideas which may be useful as guidelines for Kolmogorov's abstract formulation.

Let us consider eq (11.2), we note that, in order to determine what is 'probable', we must first determine what is 'possible'; this means that we have to make a list of possibilities for the experiment. Some common definitions are as follows: a possible outcome of our experiment is called an event and we can distinguish between simple events, which can happen only in one way, and compound events, which can happen in more than one distinct way. In the rolling of a die, for example, a simple event is the observation of a 6, whereas a compound event is the observation of an even number (2,4 or 6). In other words, simple events cannot be decomposed and are also called sample points. The set of all possible sample points is called a sample space.

Now, adopting the notation of elementary set theory, we view the sample space as a set W whose elements E_j are the sample points. If the sample space is discrete, i.e. contains a finite or countable number of sample points, any compound event A is a subset of W and can be viewed as a collection of two or more sample points, i.e. as the 'union' of two or more sample points. In the die-rolling experiment above, for example, we can write

$$A = E_2 \cup E_4 \cup E_6$$

where we call A the event 'observation of an even number', E_2 the sample point 'observation of a 2' and so on. In this case, it is evident that $P[E_2] = P[E_4] = P[E_6] = 1/6$ and, since E_2, E_4 and E_6 are mutually exclusive

$$P[A] = P[E_2 \cup E_4 \cup E_6] = P[E_2] + P[E_4] + P[E_6] = 1/2 \qquad (11.4a)$$

The natural extension of eq (11.4a) is

$$P[W] = P\left[\bigcup_{j=1}^{6} E_j\right] = \sum_{j=1}^{6} P[E_j] = 1 \qquad (11.4b)$$

Moreover, if we denote by $\overline{A}$ the complement of set A (i.e. $W = A \cup \overline{A}$), we have also

$$P[\overline{A}] = 1 - P[A] \qquad (11.4c)$$

and if we consider two events, say B and C, which are not mutually exclusive, then

$$P[B \cup C] = P[B] + P[C] - P[B \cap C] \qquad (11.4d)$$

where the intersection symbol $\cap$ is well known from set theory and $P[B \cap C]$ is often called the compound probability, i.e. the probability that events B and C occur simultaneously. (Note that one often finds also the symbols $A + B$ for $A \cup B$ and AB for $A \cap B$.) Again, in the rolling of a fair die, for example, let $B = E_2 \cup E_3$ and $C = E_1 \cup E_3 \cup E_6$, then $B \cap C = E_3$ and, as expected, $P[B \cup C] = 2/6 + 3/6 - 1/6 = 4/6$.

For three nonmutually exclusive sets, it is not difficult to extend eq (11.4d) to

$$P[B \cup C \cup D] = P[A] + P[B] + P[C]$$

$$- P[B \cap C] - P[B \cap D] - P[C \cap D] + P[B \cap C \cap D] \quad (11.4e)$$

as the reader is invited to verify.

Incidentally, it is evident that the method that we are following requires counting; for example, the counting of sample points and/or a complete itemization of equiprobable sets of sample points. For large sample spaces this may not be an easy task. Fortunately, aid comes from combinatorial analysis from which we know that the number of permutations (arrangements of objects in a definite order) of n distinct objects taken r at a time is given by

$$P_{n, r} = n(n - 1) \ldots (n - r + 1) = \frac{n!}{(n - r)!} \quad (11.5)$$

while the number of combinations (arrangements of objects without regard to order) of n distinct objects taken r at a time is

$$C_{n, r} = \binom{n}{r} \equiv \frac{n!}{r!(n - r)!} = \frac{P_{n, r}}{r!} \quad (11.6)$$

For example, if $n = 3$ (objects a, b and c) and $r = 2$, the fact that the number of combination is less than the number of permutations is evident if one thinks that in a permutation the arrangement of objects $\{a, b\}$ is considered different from the arrangement $\{b, a\}$, whereas in a combination they count as one single arrangement.

These tools simplify the counting considerably. For example, suppose that a big company has hired 15 new engineers for the same job in different plants. If a particular plant has four vacancies, in how many ways can they fill these positions? The answer is now straightforward and is given by $C_{15, 4} = 1365$. Moreover, note also that the calculations of factorials can be often made easier by using Stirling's formula, i.e. $n! \cong n^n e^{-n} \sqrt{2\pi n}$ which results in errors smaller that 1% for $n \geqslant 10$.

Returning now to our main discussion, we can make a final comment before introducing the axiomatic theory of probability: the fact that two events B and C are mutually exclusive is formalized in the language of sets as

$B \cap C = \emptyset$, where $\emptyset$ is the empty set. So, we need to include this event in the sample space and require that $P[\emptyset] = 0$. By so doing, we obtain the expected result that eq (11.4d) reduces to the sum $P[B] + P[C]$ whenever events B and C are mutually exclusive. In probability terminology, $\emptyset$ is called the 'impossible event'.

11.2.1 Probability – axiomatic formulation and some fundamental results

We define a **probability space** as a triplet $(W, \mathcal{B}, P)$ where:

1. W is a set whose elements are called **elementary events**.
2. $\mathcal{B}$ is a σ-algebra of subsets of W which are called **events**.
3. P is a **probability** function, i.e. a real-valued function with domain $\mathcal{B}$ and such that:

   (a) $0 \leqslant P[A] \leqslant 1$ for every $A \in \mathcal{B}$.
   (b) $P[W] = 1$ and $P[\emptyset] = 0$.
   (c) $P[\cup_{j=1}^{\infty} A] = \sum_{j=1}^{\infty} P[A_j]$ if the A_js are mutually disjoint events, i.e. $A_m \cap A_k = \emptyset$ when $m \neq k$ $(m, k = 1, 2, 3, \ldots)$.

For completeness, we recall here the definition of σ-algebra: a collection $\mathcal{B}$ of subsets of a given set W is a σ-algebra if

1. $W \in \mathcal{B}$.
2. If $A \in \mathcal{B}$ then $\overline{A} \in \mathcal{B}$.
3. If $A = \cup_{j=1}^{\infty} A_j$ and $A_j \in \mathcal{B}$ for every index $j = 1, 2, 3, \ldots$ then $A \in \mathcal{B}$.

Two observations can be made immediately. First – although it may not seem obvious – the axiomatic definition includes as particular cases both the classical and the relative frequency definitions of probability without suffering their limitations; second, this definition does not tell us what value of probability to assign to a given event $A \in W$. This is in no way a limitation of this definition but simply means that we will have to model our experiment in some way in order to obtain values for the probability of events. In fact, many problems of interest deal with sets of identical events which are not equally likely (for example, the rolling of a biased die).

Let us introduce now two other definitions of practical importance: conditional probability and the independence of events. Intuitively, we can argue that the probability of an event can vary depending upon the occurrence or nonoccurrence of one or more related events: in fact, it is different to ask in the die-rolling experiment 'What is the probability of a 6?' or 'What is the probability of a 6 given that an even number has fallen?'. The answer to the first question is $1/6$ while the answer to the second question is $1/3$. This is the concept of **conditional probability**, i.e. the probability of an event A given that an event B has already occurred. The symbol for conditional probability is $P[A \mid B]$ and its definition is

$$P[A \mid B] = \frac{P[A \cap B]}{P[B]} \tag{11.7}$$

provided that $P[B] \neq 0$. It is not difficult to see that, for a given probability space $(W, \mathscr{B}, P)$, $P[\,\cdot\mid B]$ satisfies the three axioms above and is a probability function in its own right. Equation (11.7) yields immediately the multiplication rule for probabilities, i.e.

$$P[A \cap B] = P[B]P[A \mid B] = P[A]P[B \mid A] \tag{11.8a}$$

which can be generalized to a number of events $A_1, A_2, \ldots, A_n$ as follows:

$$P\left[\bigcap_{j=1}^{n} A_j\right] = P[A_1]P[A_2 \mid A_1]P[A_3 \mid A_1 \cap A_2] \ldots P\left[A_n \,\middle|\, \bigcap_{j=1}^{n-1} A_j\right] \tag{11.8b}$$

If the occurrence of event B has no effect on the probability assigned to an event A, then A and B are said to be **independent** and we can express this fact in terms of conditional probability as

$$P[A \mid B] = P[A] \tag{11.9a}$$

or, equivalently

$$P[B \mid A] = P[B] \tag{11.9b}$$

Clearly, two mutually exclusive events are not independent because, from eq (11.7), we have $P[A \mid B] = 0$ when $A \cap B = \emptyset$. Also, if A and B are two independent events, we get from eq (11.7)

$$P[A \cap B] = P[A]P[B] \tag{11.10a}$$

which is referred to as the multiplication theorem for independent events. (Note that some authors give eq (11.10a) as the definition of independent events). For n mutually (or collectively) independent events eq (11.8b) yields

$$P\left[\bigcap_{j=1}^{n} A_j\right] = P[A_1]P[A_2] \ldots P[A_n] \equiv \prod_{j=1}^{n} p[A_j] \tag{11.10b}$$

A word of caution is necessary at this point: three (or more) random events can be independent in pairs without being mutually independent. This is illustrated by the example that follows.

Example 11.1. Consider a lottery with eight numbers (1–8) and let $E_1, E_2, \ldots, E_8$, respectively, be the simple events of extraction of 1, extraction of 2, etc. Let

$$A_1 = E_1 \cup E_2 \cup E_3 \cup E_4$$

$$A_2 = E_3 \cup E_4 \cup E_5 \cup E_8$$

$$A_3 = E_1 \cup E_2 \cup E_3 \cup E_5 \cup E_6 \cup E_8$$

Now, $P[A_1] = P[A_2] = 1/2$ and $P[A_3] = 3/4$. It is then easy to verify that $P[A_1 \cap A_2] = 1/4 = P[A_1]P[A_2]$, $P[A_2 \cap A_3] = 3/8 = P[A_2]P[A_3]$ and $P[A_3 \cap A_1] = 3/8 = P[A_3]P[A_1]$, which means that the events are pairwise independent. However, $P[A_1 \cap A_2 \cap A_3] = 1/8 \neq P[A_1]P[A_2]P[A_3] = 3/16$, meaning that the three events are not mutually, or collectively, independent.

Another important result is known as the total probability formula. Let $A_1, A_2, \ldots, A_n$ be n mutually exclusive events such that $\cup_{j=1}^{n} A_j = W$, where W is the sample space. Then, a generic event B can be expressed as

$$B = \bigcup_{j=1}^{n} (B \cap A_j) \tag{11.11}$$

where the n events $(B \cap A_j)$ are mutually exclusive. Owing to the third axiom of probability, this implies

$$P[B] = P\left[\bigcup_{j=1}^{n} A_j \right] = \sum_{j=1}^{n} P[B \cap A_j]$$

so that, by using the multiplication theorem, we get the total probability formula

$$P[B] = \sum_{j=1}^{n} P[A_j]P[B \mid A_j] \tag{11.12}$$

which remains true for $n \to \infty$.

With the same assumptions as above on the events A_j ($j = 1, 2, \ldots, n$), let us now consider a particular event A_k; the definition of conditional probability yields

$$P[A_k \mid B] = \frac{P[A_k \cap B]}{P[B]} = \frac{P[A_k \cap B]}{\sum_{j=1}^{n} P[A_j]P[B \mid A_j]} \tag{11.13}$$

where eq (11.12) has been taken into account. Also, by virtue of eq (11.8a) we can write $P[A_k \cap B] = P[A_k]P[B \mid A_k]$ so that substituting in eq (11.13) we get

$$P[A_k \mid B] = \frac{P[A_k]P[B \mid A_k]}{\displaystyle\sum_{j=1}^{n} P[A_j]P[B \mid A_j]} \tag{11.14}$$

which is known as **Bayes' formula** and deserves some comments. First, the formula is true if $n \rightarrow \infty$. Second, eq (11.14) is particularly useful for experiments consisting of stages. Typically, the A_js are events defined in terms of a first stage (or, otherwise, the $P[A_j]$ are known for some reason), while B is an event defined in terms of the whole experiment including a second stage; asking for $P[A_k \mid B]$ is then, in a sense, 'backward', we ask for the probability of an event defined at the first stage conditioned by what happens in a later stage. In Bayes' formula this probability is given in terms of the 'natural' conditioning, i.e. conditioning on what happens at the first stage of the experiment. This is why the $P[A_j]$ are called the *a priori* (or prior) probabilities, whereas $P[A_k \mid B]$ is called *a posteriori* (posterior or inverse) probability. The advantage of this approach is to be able to modify the original predictions by incorporating new data. Obviously, the initial hypotheses play an important role in this case; if the initial assumptions are based on an insufficient knowledge of the mechanism of the process, the prior probabilities are no better than reasonable guesses.

Example 11.2. Among voters in a certain area, 40% support party 1 and 60% support party 2. Additional research indicates that a certain election issue is favoured by 30% of supporters of party 1 and by 70% of supporters of party 2. One person at random from that area – when asked – says that he/she favours the issue in question. What is the probability that he/she is a supporter of party 2? Now, let

- A_1 be the event that a person supports party 1, so that $P[A_1] = 0.4$;
- A_2 be the event that a person supports party 2, so that $P[A_2] = 0.6$;
- B be the event that a person at random in the area favours the issue in question.

Prior knowledge (the results of the research) indicate that $P[B \mid A_1] = 0.3$ and $P[B \mid A_2] = 0.7$. The problem asks for the *a posteriori* probability $P[A_2 \mid B]$, i.e. the probability that the person who was asked supports party 2 given the fact that he/she favours that specific election issue. From Bayes' formula we get

$$P[A_2 \mid B] = \frac{P[A_2]P[B \mid A_2]}{P[A_1]P[B \mid A_1] + P[A_2]P[B \mid A_2]} = 0.778$$

Then, obviously, we can also infer that $P[A_1 \mid B] = 1 - P[A_2 \mid B] = 0.222$.

11.3   Random variables, probability distribution functions and probability density functions

Events of major interest in science and engineering are those identified by numbers. Moreover – since we assume that the reader is already familiar with the term 'variable' – we can state that a random variable is a real variable whose observed values are determined by chance or by a number of causes beyond our control which defy any attempt at a deterministic description. In this regard, it is important to note that the engineer's and applied scientist's approach is not so much to ask whether a certain quantity is a random variable or not (which is often debatable), but to ask whether that quantity can be modelled as a random variable and if this approach leads to meaningful results.

In mathematical terms, let x be any real number, then a **random variable** on the probability space $(W, \mathcal{B}, P)$ is a function $X : W \rightarrow \mathcal{R}$ ($\mathcal{R}$ is the set of real numbers) such that the sets

$$B_x \equiv \{w \in W : X(w) \leqslant x\}$$

are events, i.e. $B_x \in \mathcal{B}$. In words, let X be a real-valued function defined on W; given a real number x, we call B_x the set of all elementary events w for which $X(w) \leqslant x$. If, for every x the sets B_x belong to the σ-algebra $\mathcal{B}$, then X is a (one-dimensional) random variable.

The above definition may seem a bit intricate at first glance, but a little thought will show that it provides us precisely with what we need. In fact, we can now assign a definite meaning to expression $P[B_x]$, i.e. the probability that the random variable X corresponding to a given experiment will assume a value less than or equal to x. It is then straightforward, for a given random variable X, to define the function $F_X : \mathcal{R} \rightarrow [0, 1]$ as

$$F_X(x) \equiv P[B_x] = P[X \leqslant x] \tag{11.15}$$

which is called the **cumulative distribution function** (**cdf**, or the distribution function) of the random variable X. From the definition, the following properties can be easily proved:

$$F_X(x_2) - F_X(x_1) = P[x_1 < x \leqslant x_2]$$

$$F_X(x_1) \leqslant F_X(x_2)$$

$$F_X(-\infty) = 0 \tag{11.16}$$

$$F_X(+\infty) = 1$$

where x_1, and x_2 are any two real numbers such that $x_1 \leqslant x_2$. In other words, distribution functions are monotonically non-decreasing functions which start at zero for $x \to -\infty$ and increase to unity for $x \to \infty$. It should be noted that every random variable defines uniquely its distribution functions but a given distribution function corresponds to an arbitrary number of different random variables. Moreover, the probabilistic properties of a random variable can be completely characterized by its distribution function.

Among all possible random variables, an important distinction can be made between **discrete** and **continuous random variables**. The term discrete means that the random variable can assume only a finite or countably infinite number of distinct possible values $x_1, x_2, x_3, \ldots$. Then, a complete description can be obtained by knowing the probabilities $p_k = P[X = x_k]$ for $k = 1, 2, 3, \ldots$ by defining the distribution function as

$$F_X(x) = \sum_k p_k \theta(x - x_k) \tag{11.17}$$

where we use the symbol θ for the Heaviside function (which we already encountered in Chapters 2 and 5), i.e.

$$\theta(x) = \begin{cases} 0 & x < 0 \\ 1 & x \geqslant 0 \end{cases} \tag{11.18}$$

The distribution function of a discrete random variable is defined over the entire real line and is a 'step' function with a number of jumps or discontinuities occurring at any point x_k. A typical and simple example is provided by the die-rolling experiment where X is the numerical value observed in the rolling of the die. In this case, $x_1 = 1$, $x_2 = 2$, $x_3 = 3$ etc. and $p_k = 1/6$ for every $k = 1, 2, \ldots, 6$. Then

$$F_X(x) = \sum_{k=1}^{6} p_k \theta(x - x_k)$$

so that $F_X(x) = 0$ for $-\infty < x < 1$, $F_X(x) = 1/6$ for $1 \leqslant x < 2$, $F_X(x) = 2/6$ for $2 \leqslant x < 3$, $\ldots$, $F_X(x) = 1$ for $6 \leqslant x < \infty$.

A continuous random variable, on the other hand, can assume any value in some interval of the real line. For a large and important class of random variables there exist a certain non-negative function $p_X(x)$ which satisfies the relationship

$$F_X(x) = \int_{-\infty}^{x} p_X(\eta) d\eta \tag{11.19}$$

where $p_X(x)$ is called the **probability density function (pdf)** and η is a dummy variable of integration. The main properties of $p_X(x)$ can be summarized as follows:

$$p_X(x) \geqslant 0$$

$$\int_{-\infty}^{\infty} p_X(x)dx = 1 \tag{11.20}$$

$$p_X(x) = \frac{dF_X(x)}{dx}$$

The second property is often called the **normalization condition** and is equivalent to $F_X(\infty) = 1$. Also, it is important to notice a fundamental difference with respect to discrete random variables: the probability that the continuous random variable X assumes a specific value x is zero and probabilities must be defined over an interval. Specifically, if $p_X(x)$ is continuous at x we have

$$p_X(x)dx = P[x < X \leqslant x - dx] \tag{11.21a}$$

and, obviously

$$P[a < X \leqslant b] = \int_a^b p_X(x)dx \tag{11.21b}$$

Example 11.3. Discrete random variables – binomial, Poisson and geometric distributions. Let us consider a fixed number (n) of typical 'Bernoulli trials'. A 'Bernoulli trial' is an experiment with only two possible outcomes which are usually called 'success' and 'failure'. Furthermore, the probability of success is p and does not change from trial to trial, the probability of failure is $q \equiv 1 - p$ and the trials are independent. The discrete random variable of interest X is the number of successes during the n trials. It is shown in every book on statistics that the probability of having x successes is given by

$$p_X(x) = \binom{n}{x} p^x q^{1-x} = \frac{n!}{x!(n-x)!} p^x q^{1-x} \tag{11.22}$$

where $x = 1, 2, 3, \ldots, n$ and $0 < p < 1$. We say that a random variable has a binomial distribution with parameters n and p when its density function is given by eq (11.22).

Now, suppose that p is very small and suppose that n becomes very large in such a way that the product pn is equal to a constant λ. In mathematical terms, provided that $pn = \lambda$, we can let $p \to 0$ and $n \to \infty$; then

$$\lim_{n \to \infty} \binom{n}{x} p^x (1-p)^{n-x} = \lim_{n \to \infty} \frac{n(n-1) \ldots (n-x+1)}{x!} \left(\frac{\lambda}{n}\right)^x \left(1 - \frac{\lambda}{n}\right)^{n-x}$$

$$= \frac{\lambda^x}{x!} \lim_{n \to \infty} \left(1 - \frac{\lambda}{n}\right)^n \left(1 - \frac{\lambda}{n}\right)^{-x} \frac{n(n-1) \ldots (n-x+1)}{n^x}$$

$$= \frac{\lambda^x}{x!} \lim_{n \to \infty} \left(1 - \frac{\lambda}{n}\right)^n \left(1 - \frac{\lambda}{n}\right)^{-x} \left(1 - \frac{1}{n}\right) \left(1 - \frac{2}{n}\right) \ldots \left(1 - \frac{x-1}{n}\right)$$

$$= \frac{\lambda^x}{x!} e^{-\lambda}$$

because $\lim_{n \to \infty}(1 - \lambda/n)^n = e^{-\lambda}$. A random variable X with a pdf given by

$$p_X(x) = \frac{\lambda^x}{x!} e^{-\lambda} \tag{11.23}$$

is said to have a Poisson distribution with parameter λ. Equation (11.23) is a good approximation for the binomial equation (11.22) when either $n \geqslant 20$; $p \leqslant 0.05$ or $n \geqslant 100$; $pn \leqslant 10$. Poisson-distributed random variables arise in a number of situations, the most common of which concern 'rare' events, i.e. events with a small probability of occurrence. The parameter λ then represents the average number of occurrences of the event per measurement unit (i.e. a unit of time, length, area, space, etc.). For example, knowing that at a certain intersection we have on average 1.7 car accidents per month, the probability of zero accidents in a month is given by $(1.7^0/0!)e^{-1.7} = 0.18$. The fact that the number of accidents follows a Poisson distribution can be roughly established as follows. Divide a month into n intervals, each of which is so small that at most one accident can occur with a probability $p \neq 0$. Then, during each interval (if the occurrence of accidents can be considered as independent from interval to interval) we have a Bernoulli trial where the probability of 'success' p is relatively small if n is large and $pn = \lambda = 1.7$. Note that we do not need to know the values of n and/or p (which can be, to a certain extent, arbitrary), but it is sufficient to verify that the underlying assumptions of the Poisson distribution hold.

If now, in a series of Bernoulli trials we consider X to be the number of trials before the first success occurs we are, broadly speaking, dealing with the same problem as in the first case but we are asking a different question (the number of trials is not fixed in this case). It is not difficult to show that this circumstance leads to the geometric distribution, which is written

$$p_X(x) = pq^{x-1} = p(1-p)^{x-1} \tag{11.24}$$

where $x = 1, 2, 3, \ldots$ and $0 < p < 1$. Hence we will say that a random variable X has a geometric distribution with parameter p when its pdf is given by eq (11.24).

Example 11.4. Continuous random variable – the normal or Gaussian distribution. The most important and widely used continuous probability distribution was first described by de Moivre in the second half of the eighteenth century but was implemented as a useful practical tool only half a century later by Gauss and Laplace. Its importance is due to the central limit theorem which we will discuss in a later section. A random variable X is said to have a Gaussian (or normal) distribution with parameters μ and σ ($\sigma > 0$) if its pdf is given by

$$p_X(x) = \frac{1}{\sigma\sqrt{2\pi}} \exp\left(-\frac{(x-\mu)^2}{2\sigma^2}\right) \tag{11.25}$$

For practical use, it is convenient to cast eq (11.25) in a standardized format which can be more easily expressed in tabular form. It is not difficult to see that, by defining the new random variable Z as

$$Z \equiv \frac{X - \mu}{\sigma} \tag{11.26}$$

we obtain the standard form

$$p_Z(z) = \frac{1}{\sqrt{2\pi}} e^{-z^2/2} \tag{11.27}$$

whose cdf is given by

$$F_Z(z) = \frac{1}{\sqrt{2\pi}} \int_{-\infty}^{z} e^{-\eta^2/2} \, d\eta = \frac{1}{2} + \Phi(z) \tag{11.28}$$

since $\int_{-\infty}^{0} e^{-\eta^2/2} \, d\eta = 1/2$ and we defined $\Phi(z) = (2\pi)^{-1/2} \int_{0}^{z} e^{-\eta^2/2} \, d\eta$. Equation (11.28) has been given because either $F_Z(z)$ or $\Phi(z)$ are commonly found in statistical tables.

Also, it can be shown (local Laplace–de Moivre theorem, see for example Gnedenko [1]) that when n and np are both large – i.e. for $n \to \infty$ – we have

$$\binom{n}{x} p^x q^{n-x} \to \frac{1}{\sqrt{2\pi npq}} \exp\left(-\frac{(x-np)^2}{2npq}\right) \tag{11.29}$$

meaning that the binomial distribution can be approximated by a Gaussian distribution. The r.h.s. of eq (11.29) is called the Gaussian approximation to the binomial distribution.

Example 11.5. For purposes of illustration, let us take a probabilistic approach to a deterministic problem. Consider the sinusoidal deterministic signal $x(t) = x_0 \sin \omega t$. We ask, for any given value of amplitude $x < x_0$, what is the probability that the amplitude of our signal lies between x and $x + dx$?

From our previous discussion it is evident that we are asking for the pdf of the 'random' variable X, i.e. the amplitude of our signal. This can be obtained by calculating the time that the signal amplitude spends between x and $x + dx$ during an entire period $T = 2\pi/\omega$. Now, from $x(t) = x_0 \sin \omega t$ we get

$$dx = x_0\omega \cos(\omega t)dt = x_0\omega dt \sqrt{1 - \sin^2 \omega t}$$

$$= x_0\omega dt\sqrt{1 - (x/x_0)^2} = \omega dt\sqrt{x_0^2 - x^2}$$

which yields

$$dt = \frac{dx}{\omega\sqrt{x_0^2 - x^2}} \tag{11.30}$$

Within a period T the amplitude passes in the interval from x to $x + dx$ twice, so that the total amount of time that it spends in such an interval is $2dt$; hence

$$\frac{2dt}{T} = \frac{2dx}{\omega T\sqrt{x_0^2 - x^2}} = \frac{dx}{\pi\sqrt{x_0^2 - x^2}} \tag{11.31}$$

where the last expression holds because $T = 2\pi/\omega$. But, noting that $2dt/T$ is exactly $p_X(x)dx$, i.e. the probability that, within a period, the amplitude lies between x and $x + dx$, we get

$$p_X(x) = \frac{1}{\pi\sqrt{x_0^2 - x^2}} \tag{11.32}$$

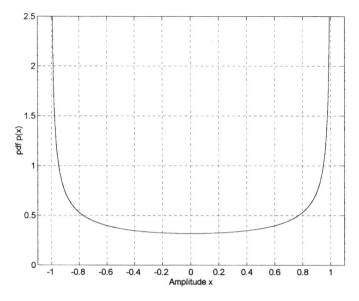

Fig. 11.1 Amplitude PDF of sinusoidal signal.

which is shown in Fig. 11.1 for $x_0 = 1$. From this graph it can be noted that a sinusoidal wave spends more time near its peak values than it does near its abscissa axis (i.e. its mean value).

11.4 Descriptors of random variable behaviour

From the discussion of preceding sections, it is evident that the complete description of the behaviour of a random variable is provided by its distribution function. However, a certain degree of information – although not complete in many cases – can be obtained by well-known descriptors such as the mean value, the standard deviation etc. These familiar concepts are special cases of a series of descriptors called **moments** of a random variable. For a continuous random variable X, we define the first moment of X, indicated by

$$E[X] \equiv \int_{-\infty}^{\infty} x p_X(x) dx \tag{11.33a}$$

or, for a discrete random variable

$$E[X] \equiv \sum_k x_k p_k \tag{11.33b}$$

528 *Probability and statistics*

Equations (11.33a) or (11.33b) define what is usually called in engineering terms the **mean** (or also the 'expected value') of X and is indicated by the symbol μ_X. Similarly, the second moment is the expected value of X^2 – i.e. $E[X^2]$ – and has a special name, the **mean squared value** of X, which for a continuous random variable is written as

$$E[X^2] = \int_{-\infty}^{\infty} x^2 p_X(x)dx \tag{11.34}$$

The square root of the second moment $\sqrt{E[X^2]}$ is called the root-mean-square value or **rms** of X.

By analogy, we define the mth moment $(m = 1, 2, 3, \ldots)$ of X as

$$E[X^m] = \int_{-\infty}^{\infty} x^m p_X(x)dx \tag{11.35a}$$

and in general, for a function $f(X)$ of the random variable we have

$$E[f(X)] = \int_{-\infty}^{\infty} f(x)p_X(x)dx \tag{11.35b}$$

Equations (11.35a) are just particular cases of eq (11.35b).

When we first subtract its mean from the random variable and then calculate the expected values, we speak of **central moments**, i.e. the mth central moment is given by

$$E[(X - \mu_X)^m] \equiv \int_{-\infty}^{\infty} (x - \mu_X)^m p_X(x)dx \tag{11.36}$$

In particular, the second central moment $E[(X - \mu_X)^2]$ is well known and has a special name: the **variance**, usually indicated with the symbols σ_X^2 or $\mathrm{Var}[X]$. Note that the variance can also be evaluated by

$$\sigma_X^2 \equiv E[(X - \mu_X)^2] = E[X^2 + \mu_X^2 - 2X\mu_X]$$
$$= E[X^2] + \mu_X^2 - 2\mu_X E[X] = E[X^2] - \mu_X^2 \tag{11.37}$$

which is just a particular case of the fact that central moments can be evaluated in terms of ordinary (noncentral) moments by virtue of the binomial theorem. In formulas we have

$$E[(X - \mu_X)^m] = E\left[\sum_{k=0}^{m} (-1)^k \binom{m}{k} X^{m-k} \mu_X^k\right]$$
$$= \sum_{k=0}^{m} \frac{(-1)^k m!}{k!(m-k)!} \mu_X^k E[X^{m-k}] \tag{11.38}$$

The square root of the variance, i.e. $\sqrt{E[(X - \mu_X)^2}$, is called the **standard deviation** and we commonly find the symbols σ_X or $SD[X]$.

Example 11.6. Let us consider some of the pdfs introduced in previous sections and calculate their mean and variance. For the binomial distribution, for example, we can show that

$$\mu_X \equiv E[X] = np$$

$$\sigma_X^2 \equiv E[X^2] = npq = np(1 - p) \tag{11.39}$$

The first of eqs (11.39) can be obtained as follows:

$$E[X] = \sum_{x=0}^{n} x \frac{n!}{x!(n-x)!} p^x q^{n-x} = np \sum_{x=1}^{n} \frac{(n-1)!}{(x-1)!(n-x)!} p^{x-1} q^{n-x}$$

$$= np \sum_{x=1}^{n} \frac{(n-1)!}{(x-1)![(n-1)-(x-1)]!} p^{x-1} q^{(n-1)-(x-1)} = np$$

where the last equality holds because the summation represents the sum of all the ordinates of the binomial distribution and must be equal to 1 for the normalization condition. For the second of eqs (11.39) we can use eq (11.37) so that we only need the term $E[X^2]$. This is given by

$$E[X^2] = \sum_x x^2 \frac{n!}{x!(n-x)!} p^x q^{n-x}$$

$$= np \sum_x (x-1+1) \frac{(n-1)!}{(x-1)!(n-x)!} p^{x-1} q^{n-x}$$

$$= np \left((n-1)p \sum_x \frac{(n-2)!}{(x-2)![(n-2)-(x-2)]!} p^{x-2} q^{(n-2)-(x-2)} \right.$$

$$\left. + \sum_x \frac{(n-1)!}{(x-1)![(n-1)-(x-1)]!} p^{x-1} q^{(n-1)-(x-1)} \right)$$

$$= np[(n-1)p + 1]$$

so that

$$\sigma_X^2 = E[X^2] - \mu_X^2 = E[X^2] - n^2 p^2 = np(1 - p)$$

We leave to the reader the proof that for a Poisson-distributed random variable we have

$$\mu_X = \sigma_X^2 = \lambda \qquad (11.40)$$

while for a geometric pdf we get

$$\mu_X = 1/p$$
$$\sigma_X^2 = q/p^2 \qquad (11.41)$$

If now we turn our attention to the continuous Gaussian pdf of eq (11.25) we can calculate the mean using eq (11.33a) with the change of variable $z = (x - \mu)/\sigma$ so that $x = \sigma z + \mu$, $dx = \sigma dz$ and we get

$$E[X] = \frac{1}{\sqrt{2\pi}} \int_{-\infty}^{\infty} z e^{-z^2/2} \, dz + \frac{\mu}{\sqrt{2\pi}} \int_{-\infty}^{\infty} e^{-z^2/2} dz = \mu \qquad (11.42)$$

because the first integral is equal to zero, while $\int_{-\infty}^{\infty} e^{-z^2/2} dz = \sqrt{2\pi}$. An analogous line of reasoning leads to

$$\sigma_X^2 = \sigma^2 \qquad (11.43)$$

so that the parameters μ and σ of the Gaussian pdf are precisely the mean and the standard deviation of the random variable X.

Often, it is convenient to use a standardized form of the random variable in question. Given a random variable X, we can define the random variable Z as $Z \equiv (X - \mu_X)/\sigma_X$ which is linearly related to X and always has zero mean and unit variance. The third and fourth moments of this standardized random variable are also given special names and are called **skewness** and **kurtosis**. Indicating these descriptors with the symbols α_3 and α_4, respectively, we have

$$\alpha_3 = E\left[\left(\frac{X - \mu_X}{\sigma_X}\right)^3\right] = \frac{E[(X - \mu_X)^3]}{\sigma_X^3}$$

$$\qquad (11.44)$$

$$\alpha_4 = E\left[\left(\frac{X - \mu_X}{\sigma_X}\right)^4\right] = \frac{E[(X - \mu_X)^4]}{\sigma_X^4}$$

The meanings of the quantities above are generally well known by every engineer or scientist. The mean (together with the median and the mode) is a measure of location, while the variance and the standard deviation are measures of scatter (dispersion) of the random variable about its mean. Skewness and kurtosis, in turn, have to do with the shape of the probability density function. More specifically, the skewness will be zero for a pdf symmetrical about the mean, positive for a pdf with a longer tail to the right and negative when the tail to the left is more prominent. Kurtosis, on the other hand describes the 'peakedness' of the pdf. For example, a Gaussian pdf has $\alpha_4 = 3$ and a pdf with, say, $\alpha_4 = 10$ has a high, narrow peak and fatter tails (with respect to a pdf with smaller kurtosis) far away from the mean.

For the measures of location or of central tendency, we can briefly say that the mean is the abscissa of the 'centre of gravity' of the area under the pdf curve, the median M_2 is the value of the random variable for which

$$F_X(M_2) = \int_{-\infty}^{M_2} p_X(\eta)d\eta = 1/2 \tag{11.45}$$

while the mode M_3 is the abscissa of the maximum of the pdf.

At this point, it is important to note that there exist many other distribution functions and we have only considered a few of the most frequently encountered. The interested reader is referred to specific texts for more information.

The usefulness of moment analysis is due to the fact that, in many problems, we do not know exactly the pdf but we only have an idea of which one it might be. Then, a knowledge of the first few moments – which, in turn, are estimated from the experimental data – can allow the evaluation of the parameters of the (unknown) pdf, in order to accept or reject our initial hypothesis. In other words, the lowest order moments of a random variable constitute a first step towards a description of the distribution function underlying a random process. The information they provide is incomplete (all the moments would be needed to have a complete description) but is very useful for many practical purposes.

11.4.1 *Characteristic function of a random variable*

We introduce here the concept of characteristic function of a random variable which, besides the cdf and the pdf, provides an alternative way of completely characterizing a random variable. The characteristic function of a random variable X is denoted $\varphi_X(\omega)$ and is defined as

$$\varphi_X(\omega) \equiv E[e^{-i\omega X}] = \int_{-\infty}^{\infty} p_X(x)e^{-i\omega x}dx \tag{11.46a}$$

where ω is a real variable. It is evident that $p_X(x)$ and $\varphi_X(\omega)$ bear a close resemblance to a Fourier transform pair. Furthermore, we know that the pdf has a Fourier transform because, owing to the normalization condition, the integral (11.46) verifies the Dirichlet condition $\int_{-\infty}^{\infty} |p_X(x)| \, dx < \infty$. Also

$$p_X(x) = \frac{1}{2\pi} \int_{-\infty}^{\infty} \varphi_X(\omega) e^{i\omega x} d\omega \qquad (11.46b)$$

A principal use of the characteristic function has to do with its moment-generating property. If we differentiate eq (11.46a) with respect to ω we obtain

$$\frac{d\varphi_X(\omega)}{d\omega} = -i \int_{-\infty}^{\infty} x p_X(x) e^{-i\omega x} dx$$

then, letting $\omega = 0$ in the above expression, we get

$$\left. \frac{d\varphi_X(\omega)}{d\omega} \right|_{\omega=0} = -i \int_{-\infty}^{\infty} x p_X(x) dx = -iE[X] \qquad (11.47)$$

Continuing this process and differentiating m times, if the mth moment of X is finite, we have

$$\left. \frac{d^m \varphi_X(\omega)}{d\omega^m} \right|_{\omega=0} = (-i)^m E[X^m] \qquad (11.48)$$

meaning that if we know the characteristic function we can find the moments of the random variable in question by simply differentiating that function and then evaluating the derivative at $\omega = 0$. Of course, if we know the pdf we always have to perform the integration of eq (11.46a), but if more than one moment is needed this is one integration only, rather than one for each moment to be calculated.

Thus, if all the moments of X exist we can expand in a Taylor series the function $\varphi_X(\omega)$ about the origin to get

$$\varphi_X(\omega) \cong 1 + (-i\omega)E[X] + \frac{(-i\omega)^2}{2!} E[X^2] + \frac{(-i\omega)^3}{3!} E[X^3] + \cdots \qquad (11.49)$$

For example, for a Gaussian distributed random variable X we can once again make use of the standardized random variable $Z = (X - \mu)/\sigma$ so that

$$\varphi_X(\omega) = E[e^{-i\omega X}] = E[e^{-i\omega\sigma Z} e^{-i\omega\mu}] = e^{-i\omega\mu} E[e^{-i\omega\sigma Z}]$$

Furthermore

$$E[e^{-i\omega\sigma Z}] = \frac{1}{\sqrt{2\pi}} \int_{-\infty}^{\infty} e^{-(i\omega\sigma z + z^2/2)}dz = \frac{1}{\sqrt{2\pi}} \int_{-\infty}^{\infty} e^{-(z+i\omega\sigma)^2/2}e^{-(\omega\sigma)^2/2}dz$$

$$= \frac{1}{\sqrt{2\pi}} e^{-(\omega\sigma)^2/2} \int_{-\infty}^{\infty} e^{-(z+i\omega\sigma)^2/2}dz = e^{-(\omega\sigma)^2/2}$$

and finally

$$\varphi_X(\omega) = e^{-(i\omega\mu + \omega^2\sigma^2/2)} \tag{11.50}$$

From eq (11.50) it is easy to determine that $d\varphi_X/d\omega\,|_{\omega=0} = -i\mu$ so that, as expected (eq (11.47)) $E[X] = \mu$. It is left to the reader to verify that $d^2\varphi_X/d\omega^2\,|_{\omega=0} = -\mu^2 - \sigma^2$. Then, by virtue of eqs (11.48) and (11.37) we get $\sigma_X^2 = \sigma$, which is the same result as eq (11.43). It must be noted that for a Gaussian distribution all moments are functions of the two parameters μ and σ only, meaning that the normal distribution is completely characterized by its mean and variance.

Finally, it may be worth mentioning the fact that the so called log-characteristic function is also convenient in some circumstances. This function is defined as the natural logarithm of $\varphi_X(\omega)$.

11.5   More than one random variable

All the concepts introduced in the previous sections can be extended to the case of two or more random variables. Consider a probability space $(W, \mathcal{B}, P)$ and let $X_j : W \rightarrow \mathfrak{R}$ $(j = 1, 2, 3, \ldots, n)$ be n random variables according to the definition of Section 11.3. Then we can consider n real numbers x_j and introduce the joint cumulative distribution function

$$F_{X_1 X_2 \ldots X_n}(x_1, x_2, \ldots, x_n) : \mathfrak{R}^n \rightarrow [0, 1]$$

as

$$F_{X_1 X_2 \ldots X_n}(x_1, x_2, \ldots, x_n) \equiv P \left[\bigcap_{j=1}^{n} (X_j \leqslant x_j) \right] \tag{11.51}$$

If and whenever convenient, both the X_js and the x_js can be written as column vectors, i.e. $\mathbf{X} = [X_1 \; X_2 \; \ldots \; X_n]^T$ and $\mathbf{x} = [x_1 \; x_2 \; \ldots \; x_n]^T$, so that the joint distribution function is written simply $F_{\mathbf{X}}(\mathbf{x})$. Equation (11.51) in words means that the joint cdf expresses the probability that all the inequalities $X_1 \leqslant x_1, \ldots, X_n \leqslant x_n$ take place simultaneously.

If now, for simplicity, we consider the case of two random variables X and Y (the 'bivariate' case) it is not difficult to see that the following properties hold:

$$F_{XY}(-\infty, -\infty) = F_{XY}(-\infty, y) = F_{XY}(x, -\infty) = 0$$
$$F_{XY}(x, \infty) = F_X(x)$$
$$F_{XY}(\infty, y) = F_Y(y)$$
$$F_{XY}(\infty, \infty) = 1$$

(11.52)

If there exists a function $p_{XY}(x, y)$ such that for every x and y

$$F_{XY}(x, y) = \int_{-\infty}^{x} \int_{-\infty}^{y} p_{XY}(\xi, \eta) d\xi d\eta$$

(11.53)

this function is called the joint probability density function of X and Y. This joint pdf can be obtained from $F_{XY}(x, y)$ by differentiation, i.e.

$$p_{XY}(x, y) = \frac{\partial^2 F_{XY}(x, y)}{\partial x \partial y}$$

(11.54)

The one-dimensional functions

$$F_X(x) = \int_{-\infty}^{x} \int_{-\infty}^{\infty} p_{XY}(\eta, y) d\eta dy$$
$$F_Y(y) = \int_{-\infty}^{\infty} \int_{-\infty}^{y} p_{XY}(x, \eta) dx d\eta$$

(11.55)

are called marginal distributions of the random variables X and Y, respectively. Also, we have the following properties for $p_{XY}(x, y)$:

$$p_{XY}(x, y) \geqslant 0$$
$$\int_{-\infty}^{\infty} \int_{-\infty}^{\infty} p_{XY}(x, y) dx dy = 1$$
$$\int_{-\infty}^{\infty} p_{XY}(x, y) dx = p_Y(y)$$
$$\int_{-\infty}^{\infty} p_{XY}(x, y) dy = p_X(x)$$

(11.56)

and the one-dimensional functions $p_X(x)$ and $p_Y(y)$ are called marginal density functions: $p_X(x)dx$ is the probability that $x \leqslant X < x + dx$ while Y can assume any value within its range of definition. Similarly, $p_Y(y)dy$ is the probability that $y \leqslant Y < y + dy$ when X can assume any value between $-\infty$ and $+\infty$. These concepts can be extended to the case of n random variables.

In Section 11.2.1 we introduced the concept of conditional probability. Following the definition given by eq (11.7) we can define the cdf $F_X(x \mid y)$ as

$$F_X(x \mid y) \equiv P[X \leqslant x \mid Y \leqslant y] = \frac{P[(X \leqslant x) \cap (Y \leqslant y)]}{P[Y \leqslant y]} = \frac{F_{XY}(x, y)}{F_Y(y)} \qquad (11.57)$$

and similarly for $F_Y(y \mid x)$. In terms of probability density functions, the conditional pdf that $x \leqslant X < x + dx$ given that $y \leqslant Y \leqslant y + dy$ can be expressed as

$$p(x \mid y)dx = \frac{p_{XY}dxdy}{dy \int_{-\infty}^{\infty} p_{XY}(x, y)dx} = \frac{p_{XY}dxdy}{p_Y(y)dy} \qquad (11.58)$$

provided that $p_Y(y) \neq 0$. From eq (11.58) it follows that

$$p_X(x \mid y) = \frac{p_{XY}(x, y)}{p_Y(y)} \qquad (11.59a)$$

where $p_Y(y)$ is the marginal pdf of Y. Similarly

$$p_Y(y \mid x) = \frac{p_{XY}(x, y)}{p_X(x)} \qquad (11.59b)$$

so that

$$p_{XY}(x, y) = p_X(x)p_Y(y \mid x) = p_Y(y)p_X(x \mid y) \qquad (11.60)$$

which is the multiplication rule for infinitesimal probabilities, i.e. the counterpart of eq (11.8a). The key idea in this case is that a conditional pdf is truly a probability density function, meaning that, for example, we can calculate the expected value of X given that $Y = y$ from the expression

$$E[X \mid Y = y] = E[X \mid y] = \int_{-\infty}^{\infty} x p_X(x \mid y)dx \qquad (11.61)$$

In this regard we may note that $E[X \mid y]$ is a function of y, i.e. different conditional expected values are obtained for different values of y. If now we let Y range over all its possible values we obtain a function of the random variable Y (i.e. $f(Y) \equiv E[X \mid Y]$) and we can calculate its expected value as (taking eqs (11.35b), (11.61) and (11.60) into account)

$$E[f(Y)] = \int_{-\infty}^{\infty} f(y) p_Y(y) dy = \int_{-\infty}^{\infty} E[X \mid Y] p_Y(y) dy$$

$$= \int_{-\infty}^{\infty} \int_{-\infty}^{\infty} x p_X(x \mid y) p_Y(y) dx dy$$

$$= \int_{-\infty}^{\infty} \int_{-\infty}^{\infty} x p_{XY}(x, y) dx dy = E[X]$$

which expresses the interesting result

$$E[E[X \mid Y]] = E[X] \tag{11.62}$$

Similarly $E[E[Y \mid X]] = E[Y]$. These formulas often provide a more efficient way for calculating the expected values $E[X]$ or $E[Y]$.

Proceeding in our discussion, we can now consider the important concept of independence. In terms of random variables, independence has to do with the fact that knowledge of, say, X gives no information whatsoever on Y and vice versa. This occurrence is expressed mathematically by the fact that the joint distribution function can be written as a product of the individual marginal distribution functions, i.e. the random variables X and Y are independent if and only if

$$F_{XY}(x, y) = F_X(x) F_Y(y) \tag{11.63}$$

or, equivalently

$$p_{XY}(x, y) = p_X(x) p_Y(y) \tag{11.64}$$

If now we consider the descriptors of two or more random variables we can define the joint moments of X and Y defined by the expression

$$m_{ij} = E[X^i Y^j] \equiv \int_{-\infty}^{\infty} \int_{-\infty}^{\infty} x^i y^j p_{XY}(x, y) dx dy \tag{11.65}$$

or the central moments

$$\mu_{ij} = E[(X - \mu_X)^i (Y - \mu_Y)^j] \tag{11.66}$$

where $\mu_X = E[X] \equiv m_{10}$ and $\mu_Y = E[Y] \equiv m_{01}$. Particularly important is the second-order central moment which is called the **covariance** ($\text{Cov}[X, Y]$, K_{XY} or Γ_{XY} are all widely adopted symbols) of X and Y, i.e.

$$K_{XY} \equiv \mu_{11} = \int_{-\infty}^{\infty} \int_{-\infty}^{\infty} (X - \mu_X)(Y - \mu_Y) p_{XY} dx dy$$

$$= m_{11} - \mu_X \mu_Y \qquad (11.67a)$$

which is often expressed in nondimensional form by introducing the **correlation coefficient** ρ_{XY}:

$$\rho_{XY} \equiv \frac{K_{XY}}{\sigma_X \sigma_Y} = \frac{E[(X - \mu_X)(Y - \mu_Y)]}{\sqrt{E[(X - \mu_X)^2]E[(Y - \mu_Y)^2]}} \qquad (11.67b)$$

For two independent variables eq (11.64) holds, this means $m_{11} \equiv E[XY] = E[X]E[Y] = \mu_X \mu_Y$ and $K_{XY} = 0$, so that if the two standard deviations σ_X and σ_Y are not equal to zero we have

$$\rho_{XY} = 0 \qquad (11.68)$$

Equation (11.68) expresses the fact that the two random variables are uncorrelated. It must be noted that two independent variables are uncorrelated but the reverse is not necessarily true, i.e. if eq (11.68) or $K_{XY} = 0$ holds, it does not necessarily mean that X and Y are independent. However, this statement is true for normally (Gaussian) distributed random variables.

The correlation coefficient satisfies the inequalities $-1 \leqslant \rho_{XY} \leqslant 1$ and is a measure of how closely the two random variables are linearly related. In the two extreme cases $\rho_{XY} = -1$ or $\rho_{XY} = 1$ there is a perfect linear relationship between X and Y.

In the case of n random variables $X_1, X_2, \cdots, X_n$ the matrix notation proves to be convenient and one can form the $n \times n$ matrix of products $\mathbf{XX}^T$ and introduce the covariance and the correlation matrices $\mathbf{K}$ and $\mathbf{r}$. This latter, for example, is given by

$$\mathbf{r} = \begin{bmatrix} 1 & \rho_{12} & \cdots & \rho_{1n} \\ \rho_{21} & 1 & \cdots & \rho_{2n} \\ \cdots & \cdots & \cdots & \cdots \\ \rho_{n1} & \rho_{n2} & \cdots & 1 \end{bmatrix} \qquad (11.69)$$

For n mutually independent random variables $\mathbf{r} = \mathbf{I}$ where $\mathbf{I}$ is the $n \times n$ identity matrix.

Among others, an example worth mentioning is the joint Gaussian pdf of two random variables X and Y. This is written

$$p_{XY}(x, y) = \frac{1}{2\pi\sigma_X\sigma_Y\sqrt{1-\rho^2}} \exp\left[-\frac{1}{2(1-\rho)}\left(\frac{(x-\mu_X)^2}{\sigma_X^2}\right.\right.$$

$$\left.\left. + \frac{(y-\mu_Y)^2}{\sigma_Y^2} - 2\rho\frac{(x-\mu_X)(y-\mu_Y)}{\sigma_X\sigma_Y}\right)\right] \tag{11.70}$$

where ρ is the correlation coefficient ρ_{XY}. The two-dimensional pdf (11.70) is often encountered in engineering practice. When the correlation coefficient is equal to zero it reduces to the product of two one-dimensional Gaussian pdfs, meaning that – as has already been mentioned – in the Gaussian case noncorrelation implies independence.

11.6   Some useful results: Chebyshev's inequality and the central limit theorem

Before considering two important aspects of probability theory, namely Chebychev's inequality and the central limit theorem, we will give some results that can often be useful in practical problems. Let X be a random variable with pdf $p_X(x)$. Since a deterministic relationship of the type $f(X)$ – where f is a reasonable function – defines another random variable $Y = f(X)$, we ask for its pdf. The simplest case is when f is a monotonic increasing function. Then, given a value y, $Y \leqslant y$ whenever $X \leqslant x$, where $y = f(x)$. Moreover, the function f^{-1} exists, is single valued and $x = f^{-1}(y)$. Hence

$$F_Y(y) = P[Y \leqslant y] = P[X \leqslant f^{-1}(y)] = F_X[f^{-1}(y)] \tag{11.71a}$$

and $p_Y(y)$ can be obtained by differentiation, i.e.

$$p_Y(y) = p_X[f^{-1}(y)]\frac{df^{-1}(y)}{dx} = \frac{p_X[f^{-1}(y)]}{[df(x)/dx]_{x=f^{-1}(y)}} \tag{11.72a}$$

If f is a monotonic decreasing function eq (11.71a) becomes

$$F_Y(y) = P[Y \leqslant y] = P[X > f^{-1}(y)]$$

$$= 1 - P[X \leqslant f^{-1}(y)] = 1 - F_X[f^{-1}(y)] \tag{11.71b}$$

and differentiating

$$p_Y(y) = -p_X[f^{-1}(y)]\frac{df^{-1}(y)}{dx} = -\frac{p_X[f^{-1}(y)]}{[df(x)/dx]_{x=f^{-1}(y)}} \tag{11.72b}$$

Then, noting that df/dx is positive if f is monotonically increasing and negative when f is monotonically decreasing, we can combine eq (11.72a) and (11.72b) into the single equation

$$p_Y(y) = \frac{p_X[f^{-1}(y)]}{|\,df(x)/dx\,|_{x=f^{-1}(y)}} \tag{11.73}$$

As a simple example, consider a random variable X with pdf $p_X(x) = e^{-x}$ ($x \geq 0$) and let $Y = f(X) = \sqrt{X}$. Then $x = f^{-1}(y) = y^2$, $p_X[f^{-1}(y)] = e^{-y^2}$ and $|\,df/dx\,|_{x=f^{-1}(y)} = (1/2\sqrt{x})_{x=y^2} = 1/2y$, so that

$$p_Y(y) = 2ye^{-y^2} \qquad (y \geq 0)$$

The reader is invited to sketch a graph of $p_X(x)$ and $p_Y(x)$ and note that the two curves are markedly different.

If f is not monotone, it can often be divided into monotone parts. The considerations above are then applied to each part and the sum taken.

The case of two or more random variables can also be considered. Suppose that we have a random $n \times 1$ vector $\mathbf{Y} = [Y_1 \ \ldots \ Y_n]^T$ which is a function of the basic random vector $\mathbf{X} = [X_1 \ \ldots \ X_n]^T$, i.e. $\mathbf{Y} = \mathbf{f}(\mathbf{X})$, this symbol meaning that $Y_1 = f_1(X_1, X_2, \ldots, X_n)$, $Y_2 = f_2(X_1, X_2, \ldots, X_n)$ etc. Suppose further that we know the joint pdf $p_X(\mathbf{x})$ and we ask for the joint pdf $p_Y(\mathbf{y})$. Then, if the inverse $\mathbf{f}^{-1}$ exists, we can obtain a result that resembles eq (11.73), i.e.

$$p_Y(\mathbf{y}) = \frac{p_X[\mathbf{f}^{-1}(\mathbf{y})]}{|\det(\mathbf{J})|_{\mathbf{x}=\mathbf{f}^{-1}(\mathbf{y})}} \tag{11.74}$$

where $\mathbf{J}$ is the Jacobian matrix

$$\mathbf{J} = \begin{bmatrix} \partial f_1/\partial x_1 & \partial f_1/\partial x_2 & \cdots & \partial f_1/\partial x_n \\ \partial f_2/\partial x_1 & \partial f_2/\partial x_2 & \cdots & \partial f_2/\partial x_n \\ \cdots & \cdots & \cdots & \cdots \\ \partial f_n/\partial x_1 & \partial f_n/\partial x_2 & \cdots & \partial f_n/\partial x_n \end{bmatrix} \tag{11.75}$$

Given two random variables X_1 and X_2 and their joint pdf, a problem of interest is to determine the pdf of their sum, i.e. of the random variable $Y_1 = X_1 + X_2$. Now, if we introduce the auxiliary variable $Y_2 = X_1$, we can adopt the vector notation that led to eq (11.74) and write the known pdf as $p_X(\mathbf{x}) = p_{X_1 X_2}(x_1, x_2)$. In this case, we have

$$\begin{aligned} Y_1 &= f_1(\mathbf{X}) = X_1 + X_2 \\ Y_2 &= f_2(\mathbf{X}) = X_1 \end{aligned} \tag{11.76}$$

and $|\det \mathbf{J}| = 1$. By noting that $x_1 = y_2$ and $x_2 = y_1 - x_1$ we can obtain the joint pdf $p_\mathbf{Y}(\mathbf{Y})$ from eq (11.74), i.e. $p_\mathbf{Y}(\mathbf{y}) = p_\mathbf{X}(x_1, y_1 - x_1)$, and then arrive at the desired result $p_{Y_1}(y_1)$ by calculating it as the marginal pdf of the random variable y_1, that is

$$p_{Y_1}(y_1) = \int_{-\infty}^{\infty} p_\mathbf{X}(x_1, y_1 - x_1)dx_1 \qquad (11.77a)$$

or – since the definition of the auxiliary variable is arbitrary – we can set $Y_2 = X_2$ and obtain the equivalent expression

$$p_{Y_1}(y_1) = \int_{-\infty}^{\infty} p_\mathbf{X}(y_1 - x_2, x_2)dx_2 \qquad (11.77b)$$

If the two original variables are independent, then $p_\mathbf{X}(x_1, x_2) = p_{X_1}(x_1)p_{X_2}(x_2)$ and we get

$$p_{Y_1}(y_1) = \int_{-\infty}^{\infty} p_{X_1}(x_1)p_{X_2}(y_1 - x_1)dx_1$$

$$= \int_{-\infty}^{\infty} p_{X_1}(y_1 - x_2)p_{X_2}(x_2)dx_2 \qquad (11.78)$$

which we recognize as the convolution integral of the two functions $p_{X_1}(x_1)$ and $p_{X_2}(x_2)$ (e.g. eq (5.24)). So, when the two original random variables are independent, we can recall the properties of Fourier transforms and infer from eq (11.78) that the characteristic function of the sum random variable Y_1 is given by the product of the two individual characteristic functions of X_1 and X_2, i.e.

$$\varphi_{Y_1}(\omega) = \varphi_{X_1}(\omega)\varphi_{X_2}(\omega)$$

where there is no 2π factor because no such factor appears in the definition of characteristic function (eq (11.46a)).

In this regard it is worth mentioning – and it is not difficult to prove – that if the two random variables X_1 and X_2 are individually normally distributed, then their sum $Y = X_1 + X_2$ is also normally distributed. Furthermore, the reverse statement is also true when the two variables are independent: if the pdf $p_Y(y)$ is Gaussian and the two random variables are independent, then X_1 and X_2 are individually normally distributed.

If, on the other hand, we now look for the pdf of the product of the two variables X_1 and X_2, we can set $Y_1 = X_1 X_2$ and $Y_2 = X_1$. Then $|\det \mathbf{J}| = |x_1|$ and since $x_2 = y_1/x_1$, we can obtain the desired result by integrating eq (11.74) in dx_1, that is

$$p_{Y_1}(y_1) = \int_{-\infty}^{\infty} \frac{1}{|x_1|} p_X\left(x_1, \frac{y_1}{x_1}\right) dx_1 \tag{11.79a}$$

or, equivalently,

$$p_{Y_1}(y_1) = \int_{-\infty}^{\infty} \frac{1}{|x_2|} p_X\left(\frac{y_1}{x_2}, x_2\right) dx_2 \tag{11.79b}$$

Finally, we can consider the ratio of the two original random variables X_1 and X_2. In this case it is convenient to set $Y_1 = X_1/X_2$ and $Y_2 = X_2$. Then $|\det J| = 1/|x_2|$ and from eq (11.74) we get

$$p_{Y_1}(y_1) = \int_{-\infty}^{\infty} |x_2| p_X(y_1 x_2, x_2) dx_2 \tag{11.80}$$

If we now turn to expected values, it is a common problem to consider a random variable Y which is a linear combination of n random variables $X_1, X_2, \cdots, X_n$, i.e.

$$Y = \sum_{j=1}^{n} a_j X_j \tag{11.81}$$

where the a_j are real coefficients. The expected value $E[Y]$ is easily obtained as

$$\mu_Y \equiv E[Y] = \sum_{j=1}^{n} a_j E[X_j] \tag{11.82}$$

while the variance σ_Y^2 can be calculated as follows:

$$\sigma_Y^2 \equiv E[(Y - \mu_Y)^2] = E\left[\left(\sum_j a_j X_j - \sum_j a_j E[X_j]\right)^2\right]$$

$$= E\left[\left(\sum_j a_j(X_j - E[X_j])\right)^2\right] = \sum_{j,k} a_j a_k \, \text{Cov}[X_j, X_k]$$

$$= \sum_j a_j^2 \, \text{Var}[X_j] + \sum_{j \neq k} a_j a_k \, \text{Cov}[X_j, X_k] \tag{11.83a}$$

meaning that, if the variables are pairwise uncorrelated,

$$\sigma_Y^2 = \sum_{j=1}^{n} a_j^2 \sigma_{X_j}^2 \tag{11.83b}$$

Obviously, eq (11.83b) holds also for the stronger condition of mutually independent X_js.

Chebychev's inequality

In practical circumstances, we often have to deal with random variables whose distribution function is not known. Although we lack important information, in these cases it would be nevertheless desirable to evaluate approximately the probability that the variable in question assumes a value in a given numerical range. An important result in this regard is given by Chebychev's inequality which can be stated as follows: let X be a random variable with a finite variance σ_X^2, then for any positive constant c

$$P[\,|X - \mu_X| \geqslant c] \leqslant \frac{\sigma_X^2}{c^2} \tag{11.84a}$$

Two remarks can be made immediately: first of all, it is important to note that eq (11.84) is valid for any probability distribution and, second, there is no requirement that the mean value μ_X is finite because it is not difficult to show that for a random variable with finite second-order moment – i.e. $E[X^2] < \infty$ – the first moment $\mu_X \equiv E[X]$ is also finite.

If now we take the constant c in the form $c = a\sigma_X$ (where a is a positive constant) and rearrange terms, eq (11.84a) can also be expressed as

$$P[(\mu_X - a\sigma_X) < X < (\mu_X + a\sigma_X)] \geqslant 1 - \frac{1}{a^2} \tag{11.84b}$$

or

$$P[\,|X - \mu_X| < a\sigma_X] \geqslant 1 - \frac{1}{a^2} \tag{11.84c}$$

A typical application of the Chebychev's inequality is illustrated in the following simple example.

Example 11.7. Suppose the steel rods from a given industrial process have a mean diameter of 20 mm and a standard deviation of 0.2 mm. Suppose further that these are the only available data about the process in question.

For the future, the management decides that the steel-rod production is considered satisfactory if at least 80% of the rods produced have diameters in the range 19.5–20.5 mm. Does the production process need to be changed?

Our random variable X is the rod diameter and the question is whether $P[19.5 < X < 20.5] \geqslant 0.8$. In this case we have $a = 0.5/\sigma_X = 2.5$ and Chebychev's inequality in the form of eq (11.84b) or (11.84c) leads to

$$P[19.5 < X < 20.5] \geqslant 1 - \frac{1}{(2.5)^2} = 0.84$$

so that, according to the management's standards, the process can be considered satisfactory and does not need to be changed.

In general, it must be noted that results of Chebychev's inequality are very conservative in the sense that the actual probability that X is in the range $\mu_X \pm a\sigma_X$ usually exceeds the lower bound $1 - 1/a^2$ by a significant amount. For example, if it was known that our random variable follows a Gaussian probability distribution we would have $P[19.5 < X < 20.5] = 0.988$.

The central limit theorem

In science and technology we often assume implicitly that the phenomenon or the process under investigation is affected by a large number of independent random factors and that each one of them contributes by a very small amount to the phenomenon or process as a whole. Furthermore, our interest lies in the final result and not in the individual effects of all these factors, which, in turn, cannot be known anyway in the majority of cases. In other words, the experimenter observes the phenomenon or process as a whole, which consists of the superposition of all these random effects. In these circumstances, the problem arises to consider the behaviour of the sum of a large number of independent random variables with unknown distribution functions assuming that each one of them has a small contribution on the total sum.

The fundamental result in this regard is the central limit theorem which can be stated as follows. Let $X_1, X_2, \ldots, X_n$ be a number of independent, identically distributed random variables with mean value μ and variance σ^2 and let

$$S_n = \sum_{j=1}^{n} X_j \tag{11.85}$$

Then, the standardized variable

$$Z_n = \frac{S_n - n\mu}{\sigma\sqrt{n}} \tag{11.86}$$

is asymptotically normally distributed, i.e. for any value of z

$$\lim_{n \to \infty} P[Z_n \leqslant z] = \frac{1}{\sqrt{2\pi}} \int_{-\infty}^{z} e^{-\eta^2/2} d\eta \tag{11.87}$$

meaning that, as n goes to infinity, the distribution function of Z_n approaches the standard normal distribution with zero mean and unit variance.

The proof of the theorem can be sketched as follows. We form the variables $Y_j = (X_j - \mu)/\sigma\sqrt{n}$ and expand their characteristic functions as in eq (11.49)

$$\varphi_{Y_j}(\omega) \cong 1 - i\omega E[Y_j] - \frac{\omega^2}{2} E[Y_j^2] + \cdots$$

$$= 1 - \frac{\omega^2}{2} \frac{E[(X_j - \mu)^2]}{n\sigma^2} + \cdots = 1 - \frac{\omega^2}{2n} + \cdots \tag{11.88}$$

Now, since the n variables are independent it follows that

$$\varphi_{Z_n}(\omega) = \prod_{j=1}^{n} \varphi_{Y_j}(\omega) \cong \left(1 - \frac{\omega^2/2}{n} + \cdots\right)^n$$

so that passing to the limit as $n \to \infty$ the higher-order terms go to zero and since $\lim_{n \to \infty}(1 - x/n)^n = e^{-x}$ we get

$$\lim_{n \to \infty} \varphi_{Z_n}(\omega) = e^{-\omega^2/2} \tag{11.89}$$

which we recognize as the characteristic function of the standard normal variable (eq (11.50) where $\mu = 0$ and $\sigma = 1$).

A few observations deserve to be made at this point:

1. As a matter of convenience, the conclusion of the central limit theorem is often replaced by the simpler statement that the variable

$$\hat{S}_n \equiv \frac{1}{n} \sum_{j=1}^{n} X_j \tag{11.90}$$

    is asymptotically normal with a mean of μ and a variance of σ^2/n.
2. The assumption of identically distributed variables can be relaxed and the theorem is still valid provided that the so-called Lindeberg condition is satisfied. This condition is very general and, in essence, requires that the contribution of each individual variable to the sum be sufficiently small.
3. There exists a multidimensional version of the central limit theorem.

Issues 2 and 3, however, are beyond our scope and the reader is referred to specific texts on probability theory (e.g. Gnedenko [1] Chistakov [2]).

In other words, the practical utility of the central limit theorem lies in the fact that, for large samples, it allows the use of a Gaussian distribution for overall measurements on effects of independently distributed causes, regardless of the probability distribution of the individual causes themselves.

11.7   A few final remarks

When dealing with random data in practical situations, we always perform a limited number of measurements which, in statistical terminology, form a 'sample' drawn from the entire 'population'. Broadly speaking, the population is the – finite or infinite, existent or conceptual – set of all possible observations that, in principle, could be performed for the statistical problem under investigation. By contrast, a sample is an existing entity, that is, a (finite) subset of the population which represents the available information that we collect when we perform our experiments.

Now, two types of problems arise; first, we are confronted with the fact that the idea of distribution functions of probability theory applies to populations, while all we have is one or more samples from the population and, second, the most common situations of science and engineering practice are those in which we either do not know at all or have a limited knowledge of the underlying distribution of the population from which our sample was taken. These problems bring us into the realm of mathematical statistics which, starting from the information available in form of samples, has developed methods and techniques to make inferences on the entire population. These inferences, in turn, can concern either the form of the underlying distribution function – if this is unknown – or some of its parameters (for example, mean, variance, etc.) if prior knowledge on the type of distribution function is available. As a matter of fact, finding answers to questions like, for example, 'What kind of distribution function can I use to describe these data?' or 'Can I consider these data to be observations from a population with such and such distribution function with such and such mean and variance?' is called **statistical hypotheses testing** and is a specific subject of mathematical statistics. For obvious reasons, it is well beyond the scope of this book to discuss these topics, and the interested reader is referred to the wide body of available literature. However, for our purposes it suffices to say that there exist a number of theorems which support our intuitive feeling that the statistical characteristics obtained from a (possibly large) sample can be used as reliable estimates for the corresponding characteristics of the entire population. In more mathematical terms, these theorems show that sampling characteristics (moments, just as an example) converge in probability to the corresponding population characteristics.

Moreover, it may be worthy of notice the fact that mathematical statistics also gives us the possibility to use distribution-free methods (the so-called 'nonparametric methods') in which no assumption whatsoever is made about the underlying probability distribution of the population when we make inferences about parameters or test specific statistical hypotheses. These methods can be particularly useful for small sample sizes.

The final remark that ends this chapter has to do with the central limit theorem which, although important for many reasons, is not a panacea. For some time in the past, because of this theorem, the approach often been followed has consisted of assuming that the underlying distribution of the data from an experiment was normal; then, all the subsequent statistical analysis revolved about this assumption. Care must be exercised, however, because the assumption of normality may often be reasonable but it is not always justified. Again, we fall back into the preceding discussion of this section in the sense that any assumption of normality (i.e. testing the hypothesis that our data come from a normally distributed population) should be checked by means of specific methods that belong to the discipline of mathematical statistics in their own right.

References

1.   Gnedenko, B.V., *Teoria della Probabilità*, Editori Riuniti, 1987.
2.   Chistakov, V.P., *A Course on Probability Theory*, Nauka, Moscow, 1987.

12 Stochastic processes and random vibrations

12.1 Introduction

A large number of phenomena in science and engineering either defy any attempt of a deterministic description or only lend themselves to a deterministic description at the price of enormous difficulties. Examples of such phenomena are not hard to find: the height of waves in a rough sea, the noise from a jet engine, the electrical noise of an electronic component or, if we remain within the field of vibrations, the vibrations of an aeroplane flying in a patch of atmospheric turbulence, the vibrations of a car travelling on a rough road or the response of a building to earthquake and wind loads. Without doubt, the question as to whether any of the above or similar phenomena is intrinsically deterministic and, because of their complexity, we are simply incapable of a deterministic description is legitimate, but the fact remains that we have no way to predict an exact value at a future instant of time, no matter how many records we take or observations we make. However, it is also a fact that repeated observations of these and similar phenomena show that they exhibit certain patterns and regularities that fit into a probabilistic description. This occurrence suggests taking a different and more pragmatic approach, which has turned out to be successful in a large number of practical situations: we simply leave open the question about the intrinsic nature of these phenomena and, for all practical purposes, tackle the problem by defining them as 'random' and adopting a description in terms of probabilistic statements and statistical averages.

In other words, we base the decision of whether a certain phenomenon is deterministic or random on the ability to reproduce the data by controlled experiments. If repeated runs of the same experiment produce identical results (within the limits of experimental error), then we regard the phenomenon in question as deterministic; if, on the other hand, different runs of the same experiment do not produce identical results but show patterns and regularities which allow a satisfactory description (and satisfactory predictions) in terms of probability laws, then we speak of random phenomenon.

12.2   The concept of stochastic process

First of all a note on terminology: although some authors distinguish between the terms, in what follows we will adopt the common usage in which 'stochastic' is synonymous with 'random' and the two terms can be used interchangeably.

Now, if we refer back to the preceding chapter, it can be noted that the concepts of event and random variable can be conveniently considered as forming two levels of a hierarchy in order of increasing complexity: the information about an event is given by a single number (its probability), whereas the information about a random variable requires the knowledge of the probability of many events. If we take a step further up in the hierarchy we run into the concept of stochastic or random process.

Broadly speaking, any process that develops in time or space and can be modelled according to probabilistic laws is a stochastic or random process. More specifically, a stochastic process $X(z)$ consists of a family of random variables indexed by a parameter z which, in turn, can be either discrete or continuous and varies within an index set Z, i.e. $z \in Z$. In the former case one speaks of a discrete parameter process, while in the latter case we speak of a continuous parameter process.

For our purposes, the interest will be focused on random processes $X(t)$ that develop in time so that the index parameter will be time t varying within a time interval T; such processes can also be generally indicated with the symbol $\{X(t), t \in T\}$. In general, the fact that the parameter t varies continuously does not imply that the set of possible values of $X(t)$ is continuous, although this is often the case. A typical example of a random time record with zero mean (velocity in this specific example, although this is not important for our present purposes) looks like Fig. 12.1, which was created by using a set of software-generated random numbers.

Also note that a random process can develop in both time and space: consider for example the vibration of a tall and slender structure under the action of wind during a windstorm. The effect of turbulence will be random not only in time but also with respect to the vertical space coordinate y along the structure.

The basic idea of stochastic process is that for any given value of t e.g. $t = t_0$, $X(t_0)$ is a random variable, meaning that we can consider its cumulative distribution function (cdf)

$$F_X(x, t_0) \equiv P[X(t_0) \leqslant x] \tag{12.1a}$$

or its probability density function (pdf)

$$p_X(x, t_0)dx = P[x < X(t_0) \leqslant x + dx] \tag{12.1b}$$

where we write $F_X(x, t_0)$ and $p_X(x, t_0)$ to point out the fact that, in general, these functions depend on the particular instant of time t_0. Note, however,

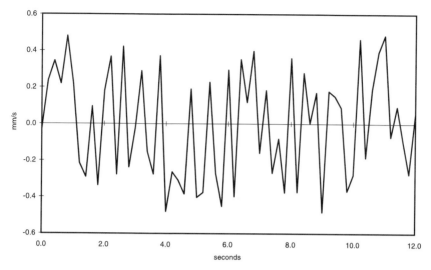

Fig. 12.1 Random (velocity) time record.

that if we adhere strictly to the notation of the preceding chapter we should write $F_{X(t_0)}(x)$ and $p_{X(t_0)}(x)$. By the same token, we can have information on the behaviour on the process $X(t)$ at two particular instants of time t_1 and t_2 by considering the joint cdf

$$F_{X_1 X_2}(x_1, x_2, t_1, t_2) \equiv P[(X \leqslant x_1) \cap (X \leqslant x_2)] \qquad (12.2a)$$

and the corresponding joint pdf

$$p_{X_1 X_2}(x_1, x_2, t_1, t_2) dx_1 dx_2$$
$$= P[(x_1 < X(t_1) \leqslant x_1 + dx_1) \cap (x_2 < X(t_2) \leqslant x_2 + dx_2)] \qquad (12.2b)$$

or, for any finite number of instants $t_1, t_2, \ldots, t_n$ we can consider the function

$$F_{X_1, X_2, \ldots, X_n}(x_1, x_2, \ldots, x_n, t_1, t_2, \ldots, t_n) = P\left[\bigcap_{j=1}^{n} (X(t_j) \leqslant x_j) \right] \qquad (12.3)$$

and its corresponding joint pdf so that, by increasing the value of n we can describe the probabilistic structure of the random process in finer and finer detail. Note that knowledge of the joint distribution function (12.3) gives information for any $m \leqslant n$ (e.g. the function of eq (12.2a) where $m = 2$), since these distribution functions are simply its marginal distribution functions. Similarly, we may extend the concepts above by considering more than one stochastic process, say $X(t)$ and $Y(t')$, and follow the discussion of Chapter 11 to define their joint pdfs for various possible sets of the index parameters t and t'.

Now, since we can characterize a random variable X by means of its moments and since, for a fixed instant of time $t \in T$, the stochastic process $X(t)$ defines a random variable, we can calculate its first moment (mean value) as

$$\mu_X(t) \equiv E[X(t)] = \int_{-\infty}^{\infty} x p_X(x, t) dx \tag{12.4}$$

or its mth order moment

$$E[X^m(t)] = \int_{-\infty}^{\infty} x^m p_X(x, t) dx \tag{12.5}$$

and the central moments as in eq (11.36). In the general case, all these quantities now obviously depend on t because they may vary for different instants of time; in other words if we fix for example two instants of time t_1 and t_2, we have $E[X^m(t_1)] \neq E[X^m(t_2)]$.

Similarly, for two instants of time we have the so-called autocorrelation function

$$R_{XX}(t_1, t_2) \equiv E[X(t_1)X(t_2)] = \int_{-\infty}^{\infty} x_1 x_2 p_{X_1 X_2}(x_1, x_2, t_1, t_2) dx_1 dx_2 \tag{12.6}$$

and the autocovariance

$$K_{XX}(t_1, t_2) \equiv E[(X(t_1) - \mu_X(t_1))(X(t_2) - \mu_X(t_2))] \tag{12.7}$$

which are related (eq (11.67a)) by the equation

$$K_{XX}(t_1, t_2) = R_{XX}(t_1, t_2) - \mu_X(t_1)\mu_X(t_2) \tag{12.8}$$

Particular cases of eqs (12.6) and (12.7) occur when $t_1 = t_2$ so that we obtain, respectively, the mean squared value and the variance

$$R_{XX}(t) = E[X^2(t)]$$
$$\sigma_X^2(t) = K_{XX}(t) \tag{12.9}$$

When two processes are studied simultaneously the counterpart of eq (12.6) is the cross-correlation function

$$R_{XY}(t_1, t_2) = E[X(t_1)Y(t_2)] \tag{12.10}$$

which is related to the cross-covariance

$$K_{XY}(t_1, t_2) \equiv E[(X(t_1) - \mu_X(t_1))(Y(t_2) - \mu_Y(t_2))] \tag{12.11}$$

by the equation

$$K_{XY}(t_1, t_2) = R_{XY}(t_1, t_2) - \mu_X(t_1)\mu_Y(t_2) \tag{12.12}$$

Consider now the idea of statistical sampling. With a random variable X we usually perform a series of independent observations and collect a number of samples, i.e. a set of possible values of X. Each observation x_j is a number and by collecting a sufficient number of observations we can get an idea of the underlying probability distribution of the random variable X. In the case of a stochastic process $X(t)$ each observation $x_j(t)$ is a time record similar to the one shown in Fig. 12.1 and our experiment consists of collecting a sufficient number of time records which can be used to estimate probabilities, expected values etc. A collection of a number – say n – of time records $x_1(t), x_2(t), \ldots, x_n(t)$ is the engineer's representation of the process and is called an **ensemble**. A typical ensemble of four time histories is shown in Fig. 12.2.

As an example, consider the vibrations of an aeroplane in a region of frequent atmospheric turbulence given the fact that the same plane flies through that region many times a year. During a specific flight we measure a vibration time history $x_1(t)$, during a second flight in similar conditions we measure $x_2(t)$ and so on, where, for instance, if the plane takes about 15 min

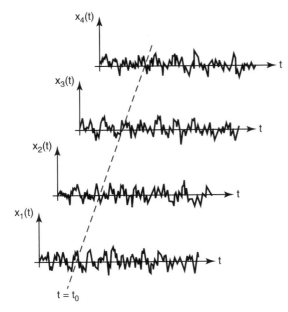

Fig. 12.2 Ensemble of four time histories for the stochastic process $X(t)$.

to fly through that region, $0 \leqslant t \leqslant 900$ s. The statistical population for this random process is the infinite set of time histories that, in principle, could be recorded in similar conditions.

We are thus led to a two-dimensional interpretation of the stochastic process which we can indicate, whenever convenient, with the symbol $X(j, t)$: for a specific value of t, say $t = t_0$, $X(j, t_0)$ is a random variable and $x_1(t_0), x_2(t_0), \ldots, x_n(t_0)$ are particular realizations, i.e. observed values, of $X(j, t_0)$; on the other hand, for a fixed j, say $j = j_0$, $X(j_0, t)$ is simply a function of time, i.e. a sample function $x_{j_0}(t)$.

With the data at our disposal, the quantities of eqs (12.4)–(12.9) must be understood as ensemble expected values, that is expected values calculated across the ensemble. However, it is not always possible to collect an ensemble of time records and the question could be asked if we can gain some information on a random process just by recording a sufficiently long time history and by calculating temporal expected values, i.e. expected value calculated along the sample function at our disposal. An example of such a quantity can be the temporal mean $\langle x \rangle$ obtained from a time history $x(t)$ as

$$\langle x \rangle = \lim_{T \to \infty} \frac{1}{T} \int_0^T x(t) dt \tag{12.13}$$

The answer to the question is that this is indeed possible in a number of cases and depends on some specific assumptions that can often (reasonably) be made about the characteristics of many stochastic processes of interest.

12.2.1 Stationary and ergodic processes

Strictly speaking, a stationary process is a process whose probabilistic structure does not change with time or, in more mathematical terms, is invariant under an arbitrary shift of the time axis. Stated this way, it is evident that no physically realizable process is stationary because all processes must begin and end at some time. Nevertheless the concept is very useful for sufficiently long time records, where by the expression 'sufficiently long' we mean here that the process has a duration which is long compared to the period of its lowest spectral components.

There are many kinds of stationarity, depending on what aspect of the process remains unchanged under a shift of the time axis. For example, a process is said to be **mean-value stationary** if

$$\mu_X(t + r) = \mu_X(t) \tag{12.14a}$$

for any value of the shift r. Equation (12.14a) implies that the mean value is the same for all times so that for a mean-value stationary process

$$\mu_X(t) = \mu_X \tag{12.14b}$$

Similarly, a process is **second-moment stationary** if

$$R_{XX}(t_1 + r, t_2 + r) = R_{XX}(t_1, t_2) \tag{12.15a}$$

for any value of the shift r. For eq (12.15a) to be true, it is not difficult to see that the autocorrelation and covariance functions must not depend on the individual values of t_1 and t_2 but only on their difference $\tau \equiv t_2 - t_1$ so that we can simply write

$$R_{XX}(t_1, t_2) = R_{XX}(t, t + \tau) = R_{XX}(\tau) \tag{12.15b}$$

By the same token, for two stochastic processes $X(t)$ and $Y(t)$ we can speak of joint second-moment stationarity when $R_{XY}(t_1, t_2) = R_{XY}(\tau)$. At this point it is easy to extend these concepts and define, for a given process, covariant stationarity and mth moment stationarity or, for two processes, joint covariant stationarity, etc. It must be noted that stationarity always reduces the number of necessary time arguments by one: i.e. in the general case the mean depends on one time argument, while for a stationary process it does not depend on time (zero time arguments); the autocorrelation depends on two time arguments in the general case and only on one time argument (τ) in the stationary case, and so on.

Other forms of stationarity are defined in terms of probability distributions rather than in terms of moments. A process is **first-order stationary** if

$$p_X(x, t + r) = p_X(x, t) \tag{12.16}$$

for all values of x, t and r; second-order stationary if

$$p_{X_1 X_2}(x_1, x_2, t_1 + r, t_2 + r) = p_{X_1 X_2}(x_1, x_2, t_1, t_2) \tag{12.17}$$

for all values of x_1, x_2, t_1, t_2 and r. Similarly, the concept can be extended to mth-order stationarity, although the most important types in practical situations are first- and second-order stationarities.

In general, a main distinction is made between strictly stationary processes and weakly stationary processes, strict stationarity meaning that the process is mth-order stationary for any value of m and weak stationarity meaning that the process is mean-value and covariant stationary (note that some authors define weak stationarity as stationarity up to order 2).

If we consider the interrelationships among the various types of stationarity, for our purposes it suffices to say that mth order stationarity implies all stationarities of lower order, while the same does not apply for mth moment stationarity. Furthermore, mth-order stationarity also implies mth moment stationarity so that, necessarily, an mth-order stationary process is also stationary up to the mth moment. Note, however, that it is not always possible to establish a hierarchy among different types of stationarities: for example it is not possible to say which is stronger between second-moment stationarity and first-order stationarity because they simply correspond to different behaviours. First-order stationarity certainly implies that all moments $E[X^m(t)]$ – which are calculated by using $p_X(x, t)$ – are invariant under a time shift, but it gives us no information about the relationship between $X(t_1)$ and $X(t_2)$ when $t_1 \neq t_2$.

Before turning to the issue of ergodicity, it is interesting to investigate some properties of the functions we have introduced above. The first property is the symmetry of autocorrelation and autocovariance functions, i.e.

$$R_{XX}(t_1, t_2) = R_{XX}(t_2, t_1)$$

$$K_{XX}(t_1, t_2) = K_{XX}(t_2, t_1)$$

(12.18)

which, whenever the appropriate stationarity applies, become

$$R_{XX}(-\tau) = R_{XX}(\tau)$$

$$K_{XX}(-\tau) = K_{XX}(\tau)$$

(12.19)

meaning that autocorrelation and autocovariance are even functions of τ. Also, if we note that

$$E[(X(t) \pm X(t + \tau))^2] \geq 0$$

we get $E[X^2(t) + X^2(t + \tau) \pm 2X(t)X(t + \tau)] = 2R_{XX}(0) \pm 2R_{XX}(\tau) \geq 0$, from which it follows that

$$R_{XX}(0) \geq |R_{XX}(\tau)|$$

(12.20)

for all τ. Similarly, for all τ

$$\sigma_X^2 = K_{XX}(0) \geq |K_{XX}(\tau)|$$

(12.21)

where the first equality is a direct consequence of the second of eqs (12.9) where stationarity applies. Moreover, it is not difficult to see that eq (12.8) now reads

$$K_{XX}(\tau) = R_{XX}(\tau) - \mu_X^2$$

(12.22a)

so that, as it often happens in vibrations, if the process is stationary with zero mean, then $K_{XX}(\tau) = R_{XX}(\tau)$. When $\tau = 0$, from eq (12.22a) it follows that

$$R_{XX}(0) = \sigma_X^2 + \mu_X^2$$

(12.22b)

Two things should be noted at this point: first (Chapter 11), Gaussian random processes are completely characterized by the first two moments, i.e.

by the mean value and the autocovariance or autocorrelation function. In particular, for a stationary Gaussian process all the information we need is the constant μ_X and one of the two functions $R_{XX}(\tau)$ or $K_{XX}(\tau)$. Second, for most random processes the autocovariance function rapidly decays to zero with increasing values of τ (i.e. $K_{XX}(|\tau| \to \infty) = 0$) because, as can be intuitively expected, at increasingly larger values of τ there is an increasing loss of correlation between the values of $X(t)$ and $X(t + \tau)$. Broadly speaking, the rapidity with which $K_{XX}(\tau)$ drops to zero as $|\tau|$ is increased can be interpreted as a measure of the 'degree of randomness' of the process.

If two weakly stationary processes are also cross-covariant stationary, it can be easily shown that the cross-correlation functions $R_{XY}(\tau)$ and $R_{YX}(\tau)$ are neither odd nor even; in general $R_{XY}(\tau) \neq R_{YX}(\tau)$ but, owing to the property of invariance under a time shift, they satisfy the relations

$$R_{XY}(\tau) = R_{YX}(-\tau)$$
$$R_{YX}(\tau) = R_{XY}(-\tau)$$

(12.23)

while eq (12.12) becomes

$$K_{XY}(\tau) = R_{XY}(\tau) - \mu_X \mu_Y$$

(12.24)

The final property of cross-correlation and cross-covariance functions of stationary processes is the so-called cross-correlation inequalities, which we state without proof:

$$|R_{XY}(\tau)|^2 \leqslant R_{XX}(0)R_{YY}(0)$$
$$|K_{XY}(\tau)|^2 \leqslant K_{XX}(0)K_{YY}(0) = \sigma_X^2 \sigma_Y^2$$

(12.25)

(We leave the proof to the reader; the starting point is the fact that $E[(aX(t) + Y(t + \tau))^2] \geqslant 0$, where a is a real number.)

Stated simply, a process is strictly **ergodic** if a single and sufficiently long time record can be assumed as representative of the whole process. In other words, if one assumes that a sample function $x(t)$ – in the course of a sufficiently long time T – passes through all the values accessible to it, then the process can be reasonably classified as ergodic. In fact, since T is large, we can subdivide our time record into a number n of long sections of time length Θ so that the behaviour of $x(t)$ in each section will be independent of its behaviour in any other section. These n sections then constitute as good a representative ensemble of the statistical behaviour of $x(t)$ as any ensemble that we could possibly collect. It follows that time averages should then be equivalent to ensemble averages.

Assuming that a process is ergodic simplifies both the data acquisition phase and the analysis phase. In fact, on one hand we do not need to collect an ensemble of time histories – which is often difficult in many practical situations – and, on the other hand, the single time history at our disposal can be used to calculate all the quantities of interest by replacing ensemble averages with time averages, i.e. by averaging along the sample rather than across the number of samples that form an ensemble. Ergodicity implies stationarity and hence, depending on the process characteristic we want to consider, we can define many types of ergodicity. For example, the process $X(t)$ is ergodic in mean value if the expression

$$\frac{1}{T} \int_0^T x(t)dt \tag{12.26}$$

where $x(t)$ is a realization of $X(t)$, tends to $E[X(t)]$ as $T \to \infty$. Mean value stationarity is obviously implied (incidentally, note that the reverse is not necessarily true, i.e. a mean-value stationary process may or may not be mean-value ergodic, and the same applies for other types of stationarities) because the limit of (12.26) cannot depend on time and hence (eq (12.13))

$$E[X(t)] = \mu_X = \langle x \rangle \tag{12.27}$$

Similarly, the process is second-moment ergodic if it is second-moment stationary and

$$R_{XX}(\tau) \equiv E[X(t)X(t+\tau)] = \lim_{T \to \infty} \frac{1}{T} \int_0^T x(t)x(t+\tau)dt \tag{12.28}$$

These ideas can be easily extended because, for any kind of stationarity, we can introduce a corresponding time average and an appropriate type of ergodicity.

There exist theorems which give necessary and sufficient (or simply necessary) conditions for ergodicity. We will not consider such mathematical details, which can be found in specialized texts on random processes but only consider the fact that in common practice – unless there are obvious physical reasons not to do so – ergodicity is often tacitly assumed whenever the process under study can be considered as stationary. Clearly, this is more an educated guess rather than a solid argument but we must always keep in mind that in real-world situations the data at our disposal are very seldom in the form of a numerous ensemble or in the form of an extremely long time history.

Stationarity, in turn – besides the fact that we can rely on engineering common sense in many cases of interest – can be checked by hypothesis testing noting that, in general, it is seldom possible to test for more than mean-value and covariance stationarity. This can be done, for example, by subdividing our sample into shorter sections, calculating sample averages for each section and then examining how these section averages compare with each other and with the corresponding average for the whole sample. On the basis of the amount of variation that we are willing to accept from one section to another in order to accept the assumption of stationarity, the statistical procedures of hypothesis testing provide us with the appropriate means to make a decision.

For instance, in common engineering practice, the vibration from continuous traffic is considered as a random stationary ergodic process and the length of the time record depends on the statistical error we are willing to accept. If, as generally happens, we accept a bias error of 4% and a variance error of 10%, the time record length is given by [1]

$$T = \frac{200}{\zeta \nu_n}$$

where ζ is the modal damping and ν_n is the natural frequency of the nth mode of the building. Also, as far as wind effects on structures are concerned, it should be noted that the vast majority of available results based on wind tunnel testing and/or analytical turbulence modelling are obtained under the assumption that the atmospheric flow is stationary. Hurricane flows, however, are highly nonstationary and some efforts to study nonstationary flow effects have been recently reported (e.g. Adhikari and Yamaguchi [2]). For the interested reader, it is worth mentioning that a technique which is becoming more and more popular for the study of nonstationary processes is called 'wavelet analysis', although in what follows we will be concerned with stationary processes (wide-sense stationary processes at least, unless otherwise stated) only.

12.3 Spectral representation of random processes

We noted in preceding chapters that the vibration analysis of linear systems can be performed either directly in the time domain or in the frequency domain via the classical tool of the Fourier transform. The two descriptions, in principle, are equivalent but the frequency domain is often preferred because it provides a perspective which lends itself more easily to engineering interpretation and synthesis of results. This is, indeed, the case also in the field of random vibrations.

However, if we consider a general stochastic process $X(t)$, two major difficulties arise. First, the expression

$$\frac{1}{2\pi} \int_{-\infty}^{\infty} X(t) e^{-i\omega t} dt$$

defines a new stochastic process on the index set of possible ω values, meaning that if we insert under the integral sign a particular realization $x(t)$ of $X(t)$ we do not obtain a frequency representation of the process but only of one member of it. Second, if the process is stationary (i.e. it goes on forever) the

Dirichlet condition

$$\int_{-\infty}^{\infty} |x(t)| \, dt < \infty \qquad (12.29)$$

is not satisfied and the sample function $x(t)$ is not Fourier transformable. These difficulties can be overcome by recalling the observation (Section 12.2.1) that for a large number of stationary random processes of engineering interest the autocorrelation tends to zero as the separation time τ tends to infinity (we assume, without loss of generality, processes with zero mean; when this is not the case, the following discussion applies to the covariance function).

More specifically, the autocorrelation function of many processes is of the form

$$R_{XX}(\tau) = e^{-\alpha|\tau|} f(\tau) \qquad (12.30)$$

where α is a positive constant and $f(\tau)$ is a well-behaved function of τ. Mathematically, this means that the autocorrelation function satisfies the Dirichlet condition and hence is Fourier transformable. This leads to the definition of the function $S_{XX}(\omega)$

$$S_{XX}(\omega) = \frac{1}{2\pi} \int_{-\infty}^{\infty} R_{XX}(\tau) e^{-i\omega\tau} \, d\tau \qquad (12.31a)$$

which is called the autospectral density, power spectral density (PSD, a term that comes from electrical engineering) or simply spectral density of the process $X(t)$. If $x(t)$ is a voltage signal, the units of the autocorrelation are volts squared and $S_{XX}(\omega)$ is expressed in volts squared per unit angular frequency; the relationship with the spectral density expressed in terms of ordinary frequency $\nu = \omega/2\pi$ is given by $2\pi S_{XX}(\omega) = S_{XX}(\nu)$ and the units of $S_{XX}(\nu)$ are V^2/Hz.

Inverse Fourier transform of eq (12.31a) yields

$$R_{XX}(\tau) = \int_{-\infty}^{\infty} S_{XX}(\omega) e^{+i\omega\tau} \, d\omega \qquad (12.31b)$$

and the result expressed by eqs (12.31a and b) are the so-called Wiener–Khintchine relations. Clearly, similar relations define the cross-spectral density $S_{XY}(\omega)$ between two stationary processes $X(t)$ and $Y(t)$ and we have

$$S_{XY}(\omega) = \frac{1}{2\pi} \int_{-\infty}^{\infty} R_{XY}(\tau) e^{-i\omega\tau} \, d\tau$$

$$R_{XY}(\tau) = \int_{-\infty}^{\infty} S_{XY}(\omega) e^{+i\omega\tau} \, d\omega$$

$$\qquad (12.32)$$

Before proceeding further, let us consider some properties of these spectral densities. First, the symmetry properties of the (real) autocorrelation and cross-correlation functions (see eqs (12.19) and (12.23)) lead to

$$S^*_{XX}(\omega) = S_{XX}(\omega) = S_{XX}(-\omega)$$

$$S^*_{XY}(\omega) = S_{YX}(\omega) = S_{XY}(-\omega)$$

(12.33)

where the first equation states that the autospectral density is a real, even function of ω, while the second equation tells us that, in general, the cross-spectral density is a complex-valued function that can be separated into its real and imaginary parts $\text{Re}[S_{XY}(\omega)]$ and $\text{Im}[S_{XY}(\omega)]$ which, in turn, are often called the co-spectrum and the quad-spectrum, respectively. Also, the symmetry property expressed by the first of eqs (12.33) implies that there is no loss of information if we only consider the frequency range $0 \leqslant \omega < \infty$. This has led to an alternative form of spectral density, the one-sided spectral density, which is usually denoted $G_{XX}(\omega)$ and is defined for positive frequencies only, as

$$G_{XX}(\omega) = 2S_{XX}(\omega)$$

(12.34)

The second consideration we want to make is that eq (12.31b) for $\tau = 0$ gives

$$\sigma^2_X = R_{XX}(0) = \int_{-\infty}^{\infty} S_{XX}(\omega)d\omega$$

(12.35)

This property is often used for calculations of variance values and shows that the variance of the stationary process can be obtained as the area under the autospectral density curve.

If now we proceed in our discussion, the question may arise as to whether, by Fourier transforming the correlation function, we are really considering the frequency content of the original process. The answer is yes and the following argument will provide some insight. Consider a stationary process $X(t)$ and a realization $x(t)$ of infinite duration. Let us define the Fourier transformable truncated version of $x(t)$ as

$$x_T(t) \equiv \begin{cases} x(t) & 0 \leqslant t \leqslant T \\ 0 & \text{otherwise} \end{cases}$$

(12.36)

we have $\lim_{T \to \infty} x_T(t) = x(t)$ and we can consider the truncated realization of the correlation function

$$R^{(T)}_{XX}(\tau) \equiv \frac{1}{T} \int_{-\infty}^{\infty} x_T(t)x_T(t+\tau)dt$$

(12.37)

Now, if we call $\tilde{X}_T(\omega)$ the Fourier transform of $x_T(t)$ it is not difficult to determine that

$$S_{XX}^{(T)}(\omega) \equiv \mathscr{F}\{R_{XX}^{(T)}(\tau)\} = \frac{2\pi}{T} \tilde{X}_T^*(\omega)\tilde{X}_T(\omega) = \frac{2\pi}{T} \mid \tilde{X}_T(\omega) \mid^2 \qquad (12.38)$$

where, as usual, $\mathscr{F}\{\,\cdot\,\}$ indicates the Fourier transform of the quantity within braces (recall the Fourier transform of a convolution product, (Chapter 2)).

In words, eq (12.38) states that the function $S_{XX}^{(T)}(\omega)$ – i.e. by definition the Fourier transform of the truncated autocorrelation – equals $2\pi/T$ the magnitude squared of the Fourier transform of the truncated process $x_T(t)$. The desired result can now be obtained from eq (12.38) by taking the ensemble average and passing to the limit as $T \to \infty$; under these operations it is not difficult to see that $S_{XX}^{(T)}(\omega) = \mathscr{F}\{R_{XX}^{(T)}(\tau)\} \to S_{XX}(\omega)$ so that

$$S_{XX}(\omega) = \lim_{T \to \infty} E\left[\frac{2\pi}{T} \mid \tilde{X}_T(\omega) \mid^2\right] \qquad (12.39a)$$

At this point, one might be tempted to argue that the ensemble average should not be needed if the process is ergodic. However, this is not so: the reason lies in the fact that the truncated function $S_{XX}^{(T)}(\omega)$, which is an estimator of the true spectral density, is not a 'consistent' estimator and its quality does not improve even for very large T. Hence, the version of eq (12.39a) without ensemble average, i.e.

$$S_{XX}(\omega) = \lim_{T \to \infty} \frac{2\pi}{T} \mid \tilde{X}_T(\omega) \mid^2 \qquad (12.39b)$$

applies to deterministic signals only.

This short argument, besides confirming our point that Fourier transforming the autocorrelation function preserves the frequency content of the original stationary signal, also shows that the spectral density obtained from a single sample is not a good estimator of the desired (and unknown) $S_{XX}(\omega)$. The typical approach to avoid this sampling difficulty is generally to replace $S_{XX}^{(T)}$ by a 'smoothed' version $\tilde{S}_{XX}^{(T)}$ whose variance tend to zero as $T \to \infty$. We will not go into more details here and refer the reader to specific literature (e.g. Papoulis [3], Bendat and Piersol [4]).

12.3.1 Spectral densities: some useful results

This section gives some general results which can be particularly useful when dealing with random processes. First of all, many transformations on random processes are in the form of linear, time-invariant operators and can be mathematically represented as an operator A which transforms a sample function $x(t)$ into another function $y(w)$, i.e. $A[x(t)] = y(w)$, where w may be time as well (for example if A is the derivative operator) or another variable $w \neq t$. Here, we give without proof the following results (more details will be given in subsequent sections):

- When the relevant quantities exist, the operator A and the operation of ensemble averaging can be exchanged, i.e. $E[A[x(t)]] = A[E[x(t)]]$.
- A weakly (strongly) stationary random process is transformed into a weakly (strongly) stationary random process.
- The linear operator A transforms a Gaussian process into a Gaussian process.

A second useful result can be obtained if we consider the meaning of the function $dR_{XX}(\tau)/d\tau$; we have

$$\frac{dR_{XX}(\tau)}{d\tau} = \frac{d}{d\tau} E[x(t)x(t+\tau)] = E[x(t)\dot{x}(t+\tau)] = R_{X\dot{X}}(\tau) \qquad (12.40a)$$

and also, since $E[x(t)x(t+\tau)] = E[x(t-\tau)x(t)]$,

$$\frac{dR_{XX}(\tau)}{d\tau} = -R_{\dot{X}X}(\tau) \qquad (12.40b)$$

so that eqs (12.40a and b) imply

$$\left.\frac{dR_{XX}(\tau)}{d\tau}\right|_{\tau=0} = R_{X\dot{X}}(0) = -R_{\dot{X}X}(0) = 0 \qquad (12.40c)$$

and only a little thought is needed to show that $dR_{XX}(\tau)/d\tau$ is an odd function of τ. The result of eq (12.40c) can also be obtained by noting that $E[X^2(t)]$ is a constant for a correlation covariant process; this implies

$$0 = \frac{d}{dt} E[X^2(t)] = 2E[X(t)\dot{X}(t)] = 2R_{X\dot{X}}(0)$$

In this regard, it is worth mentioning the often exploited fact that a maximum value for $R_{XX}(\tau)$ corresponds to a zero crossing for $dR_{XX}(\tau)/d\tau$, i.e. a zero crossing for the cross-correlation between the processes $X(t)$ and $\dot{X}(t)$. By a similar reasoning to the above we can show that

$$\frac{d^2R_{XX}(\tau)}{d\tau^2} = R_{X\ddot{X}}(\tau) = -R_{\dot{X}\dot{X}}(\tau)$$

$$\sigma_{\dot{X}}^2 \equiv E[\dot{x}^2(t)] = R_{\dot{X}\dot{X}}(0) = -\left.\frac{d^2R_{XX}(\tau)}{d\tau^2}\right|_{\tau=0} \qquad (12.41)$$

and that the second derivative of $R_{XX}(t)$ is an even function of τ. Similarly, we can obtain

$$\frac{d^4 R_{XX}(\tau)}{d\tau^4} = R_{\ddot{X}\ddot{X}}(\tau)$$

$$\sigma_{\ddot{X}}^2 \equiv E[\ddot{x}^2(t)] = \frac{d^4 R_{XX}(\tau)}{d\tau^4}\bigg|_{\tau=0}$$

(12.42)

Next, if we turn our attention to spectral densities we can start from the basic relation

$$R_{XX}(\tau) = \int_{-\infty}^{\infty} S_{XX}(\omega)e^{i\omega\tau} d\omega$$

and by noting that it is legitimate to take the derivative under the integral sign on the r.h.s., we can differentiate both sides to obtain

$$R_{X\dot{X}}(\tau) = \frac{dR_{XX}(\tau)}{d\tau} = \int_{-\infty}^{\infty} i\omega S_{XX}(\omega)e^{i\omega\tau} d\omega$$

$$R_{\dot{X}\dot{X}}(\tau) = \frac{d^2 R_{XX}(\tau)}{d\tau} = \int_{-\infty}^{\infty} \omega^2 S_{XX}(\omega)e^{i\omega\tau} d\omega$$

(12.43)

so that

$$S_{X\dot{X}}(\omega) = i\omega S_{XX}(\omega)$$
$$S_{\dot{X}\dot{X}}(\omega) = \omega^2 S_{XX}(\omega)$$

(12.44)

and also

$$S_{\ddot{X}\ddot{X}}(\omega) = \omega^4 S_{XX}(\omega)$$

(12.45)

Moreover

$$\sigma_{\dot{X}}^2 = \int_{-\infty}^{\infty} S_{\dot{X}\dot{X}}(\omega)d\omega = \int_{-\infty}^{\infty} \omega^2 S_{XX}(\omega)d\omega$$

$$\sigma_{\ddot{X}}^2 = \int_{-\infty}^{\infty} S_{\ddot{X}\ddot{X}}(\omega)d\omega = \int_{-\infty}^{\infty} \omega^4 S_{XX}(\omega)d\omega$$

(12.46)

showing that, if $x(t)$ is a displacement time history, we can calculate the mean square velocity and acceleration from knowledge of the spectral density $S_{XX}(\omega)$.

The final topic we want to consider in this section is the distinction that is usually made between narrow-band and wide-band random processes, these definitions having to do with the form of their spectral densities. Working, in a sense, backwards we can investigate what kind of time histories and autocorrelation functions result in narrow-band and wide-band processes.

Broadly speaking, a narrow-band process has a spectral density which is very small except within a narrow band of frequencies: i.e. $S_{XX}(\omega) \cong 0$ except in the neighbourhood of a frequency $\omega \cong \omega_0$. A typical example is given by the spectral density shown in Fig. 12.3, which is different from zero only in an interval of width $\omega_2 - \omega_1 = \Delta\omega$ centred at ω_0 where it has the constant value S_0.

In order to obtain the autocorrelation function we can simplify the calculations by noting that we are dealing with even functions of their arguments; then the inverse Fourier transform of $S_{XX}(\omega)$ can be written as a cosine Fourier transform and we get

$$R_{XX}(\tau) = \int_{-\infty}^{\infty} S_{XX}(\omega)\cos \omega\tau d\omega = 2S_0 \int_{\omega_1}^{\omega_2} \cos \omega\tau d\omega$$

$$= \frac{2S_0}{\tau}(\sin \omega_2\tau - \sin \omega_1\tau) = \frac{4S_0}{\tau}\sin\frac{\tau\Delta\omega}{2}\cos\frac{\tau}{2}(\omega_1 + \omega_2) \quad (12.47)$$

which is plotted in Figs. 12.4(a) and (b) for the values $\omega_0 = 50$ rad/s, $\Delta\omega = 4$ rad/s ($\Delta\omega/\omega_0 = 0.08$) and $S_0 = 1$. Figure 12.4(b) shows a detail of Fig. 12.4(a) in the vicinity of $\tau = 0$.

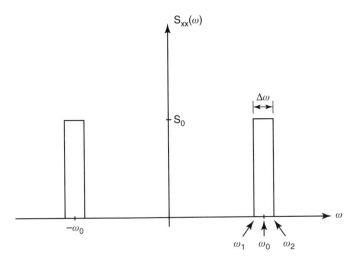

Fig. 12.3 Spectral density of narrow-band process.

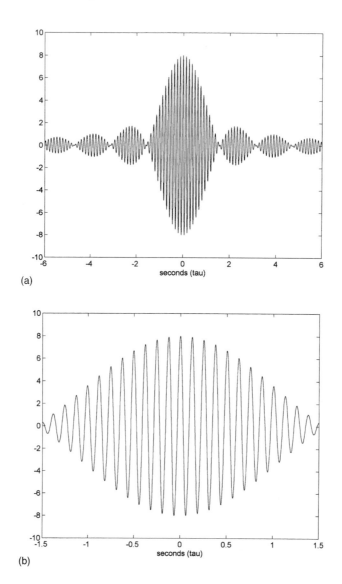

Fig. 12.4 Autocorrelation of narrow-band process.

In essence, for a typical narrow-band process $\Delta\omega/\omega_0 \ll 1$, so that the autocorrelation graph is a cosine oscillation at the frequency $(\omega_1 + \omega_2)/2$ enveloped by the slowly varying term

$$4S_0 \frac{\sin(\tau\Delta\omega/2)}{\tau}$$

which decays to zero for increasing values of $|\tau|$. In the limit of very small values of $\Delta\omega$, the spectral density becomes a Dirac delta 'function' at $\omega = \omega_0$ and

$$R_{XX}(\tau) = 2\int_{-\infty}^{\infty} S_0\delta(\omega - \omega_0)\cos \omega\tau d\tau = 2S_0 \cos \omega_0\tau \qquad (12.48)$$

so that the correlation function is a simple sinusoid. It is not difficult to show, for example, that such a correlation function can represent a process $X(t) = A \sin(\omega_0 t + \Theta)$, where A and ω_0 are deterministic quantities but the phase angle Θ is a random variable which can assume with equal probability any value between zero and 2π (or, in other words, has a pdf $p_\Theta(\theta) = 1/(2\pi)$ for $0 \leqslant \theta \leqslant 2\pi$ and zero otherwise). In fact

$$R_{XX}(t_1, t_2) = A^2 E[\sin(\omega_0 t_1 + \theta)\sin(\omega_0 t_2 + \theta)]$$

$$= \frac{A^2}{2} E[\cos[\omega_0(t_1 - t_2) - \cos[\omega_0(t_1 + t_2) + 2\theta]]$$

$$= \frac{A^2}{2} \cos \omega_0\tau - \frac{A^2}{2} \int_0^{2\pi} \frac{d\theta}{2\pi} \cos[\omega_0(t_1 + t_2) + 2\theta] = \frac{A^2}{2} \cos \omega_0\tau$$

$$(12.49)$$

By analogy, we can infer that a time history of the narrow-band process whose correlation function is given by eq (12.47) is surely not a sinusoidal function but, nonetheless, it may look 'quite sinusoidal' with a low degree of randomness.

At the other extreme we find the so-called wide-band processes, whose spectral densities are significantly different from zero over a broad band of frequencies. An example can be given by a process with a spectral density as in Fig. 12.3 but where now ω_1 and ω_2 are much more further away on the abscissa axis. For illustrative purposes we can set $\omega_0 = 50$ rad/s and $\Delta\omega = 80$ rad/s (i.e. $\omega_1 = 10$, $\omega_2 = 90$) and draw a graph of the autocorrelation function, which is still given by eq (12.47). This graph is shown in Fig. 12.5 where, again, we set $S_0 = 1$.

The fictitious process whose spectral density is equal to a constant S_0 over all values of frequencies represents a mathematical idealization called 'white noise' (by analogy with white light which has an approximately flat spectrum over the whole visible range of electromagnetic radiation). For this process it is evident that the spectral density is nonintegrable; however, we can once more use the Dirac delta function and note that the Fourier transform of the autocorrelation function

$$R_{XX}(\tau) = 2\pi S_0\delta(\tau) \qquad (12.50)$$

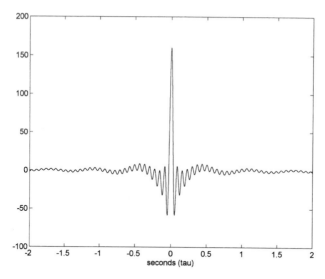

Fig. 12.5 Autocorrelation of wide-band process.

yields the desired spectral density $S_{XX}(\omega) = S_0$. A more realistic process, called 'band-limited white noise', has a constant spectral density from $\omega = 0$ up to a cutoff frequency $\omega = \omega_c$. In this case

$$R_{XX}(\tau) = 2S_0 \int_0^{\omega_c} \cos \omega\tau d\omega = \frac{2S_0}{\tau} \sin \omega_c\tau \qquad (12.51)$$

and ideal white noise is obtained by letting $\omega_c \to \infty$; if now we define the parameter $\varepsilon \equiv 1/\omega_c$ which tends to zero in the above limit, we get

$$\lim_{\varepsilon \to 0} \frac{2S_0}{\tau} \sin(\tau/\varepsilon) = 2\pi S_0 \delta(\tau) \qquad (12.52)$$

because one of the representations of the delta function as a limit (Chapter 2) is the Dirichlet or 'diffraction peak representation' which reads

$$\delta(x) = \lim_{\varepsilon \to 0} \frac{\sin(x/\varepsilon)}{\pi x}$$

The autocorrelation function of a band-limited white-noise signal (eq (12.51)) is shown in Fig. 12.6, where $\omega_c = 150$ rad/s and $S_0 = 1$.

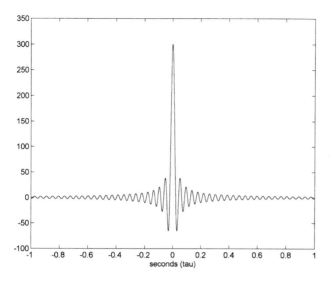

Fig. 12.6 Autocorrelation of band-limited white noise.

For obvious reasons, white-noise processes are also called 'delta-correlated', where this term focuses the attention on the time-domain correlation rather than on the flatness of the frequency-domain spectral density. At this point it is not difficult to figure out that the time histories of such processes are very erratic and show a high degree of randomness (e.g. Fig. 12.1), the reason being the fact that the random variables $X(t)$ and $X(t + \tau)$ are practically uncorrelated even for small values of τ. This confirms the qualitative statement of Section 12.2.1 that the rapidity with which the correlation function decays to zero is a measure of the degree of randomness of the process under investigation. Conversely, in the frequency domain some quantities have been devised in order to assign a numerical value to the concept of bandwidth of a random process. The interested reader is referred, for example, to Lutes and Sarkani [5], or Vanmarcke [6].

12.4 Random excitation and response of linear systems

We are now in a position to start the investigation of how linear vibrating systems respond to the action of one or more stochastic excitation inputs. The situations we are going to consider are those in which a random (and generally stationary, unless otherwise stated) input is fed into a deterministic linear system to produce a random output. For our purposes, the fact that the system is deterministic means that its physical characteristics – mass, stiffness and damping – are well-defined quantities independent of time. A higher level of sophistication is represented by the case in which these parameters are also considered as random variables and contribute to the randomness of the output in their own right. In this regard it may be interesting to mention the fact that the response of random parameters systems to deterministic initial conditions and under the action of deterministic loads is, as a matter of fact, a random quantity (e.g. Köylüoglu [7]). In our approach, however, the systems characteristics are fully represented by the impulse response functions $h(t)$ in the time domain or by frequency response functions $H(\omega)$ in the frequency domain.

The basic input–output relations can then be obtained as follows. Consider a linear physical system subjected to a forcing function in the form of a stationary random process $F(t)$ and let its response be the random output process $X(t)$. The mental picture we need is one of a large number of experiments where realizations $f(t)$ of the input force excite our deterministic system which, in turn, responds with realizations $x(t)$ of the output. If we refer back to Chapter 5 (eq (5.24)), the output of a typical sample experiment can be written as the Duhamel (or convolution) integral

$$x(t) = \int_{-\infty}^{\infty} f(t - \alpha)h(\alpha)d\alpha \tag{12.53}$$

so that, if the mean input level is given by $\mu_F \equiv E[F(t)]$, the first thing we can do is to calculate the mean output level $E[X(t)]$ by taking the ensemble average of both members of eq (12.53). Since it is legitimate to exchange the ensemble average operator with integration (this is always possible for stable systems subjected to random input provided that the mean square of the input is finite) we get

$$E[x(t)] = \int_{-\infty}^{\infty} E[f(t - \alpha)]h(\alpha)d\alpha = \mu_F \int_{-\infty}^{\infty} h(\alpha)d\alpha \tag{12.54}$$

Real and stable systems always possess some degree of damping which makes the function $h(t)$ decay to zero after some time. In these circumstances, eq (12.54) shows that a stationary input produces a stationary output. If, for example, our system is a simple damped SDOF system whose impulse response function is given by the second of eqs (5.7a), it is not difficult to determine that

$$E[x(t)] = \frac{\mu_F}{m\omega_d} \int_{-\infty}^{\infty} e^{-\zeta\omega_n\alpha} \sin \omega_d\alpha\, d\alpha$$

$$= \frac{\mu_F}{m\omega_d} \int_{0}^{\infty} e^{-\zeta\omega_n\alpha} \sin \omega_d\alpha\, d\alpha = \frac{\mu_F}{k} \tag{12.55}$$

showing that the mean input level is transmitted as any other static load.

Incidentally, we note that we do not even need to calculate the integral in eq (12.55); in fact, since $H(\omega) = 2\pi \mathcal{F}\{h(t)\}$, it follows that

$$H(0) = \int_{-\infty}^{\infty} h(t)dt \qquad (12.56)$$

and for an SDOF system (e.g. eq (4.42)) we have $H(0) = 1/k$, which leads precisely to the result of eq (12.55). More generally, eq (12.54) can also be written as

$$E[x(t)] = \mu_X = \mu_F H(0) \qquad (12.57)$$

Note that here and in what follows we represent the input as a force signal and the output as a displacement signal because this is the representation that we used for most part of the book. It is evident that this is merely a matter of convenience and it does not necessarily need to be so. The essence of the discussions remains the same and only a small effort is required to adjust to situations where different input and output quantities are considered.

If now we assume without loss of generality that the input process has zero mean value, we can turn our attention to the correlation function and write, by virtue of eq (12.53)

$$x(t)x(t+\tau) = \int_{-\infty}^{\infty} h(\alpha)f(t-\alpha)d\alpha \int_{-\infty}^{\infty} h(\gamma)f(t+\tau-\gamma)d\gamma$$

$$= \int_{-\infty}^{\infty} \int_{-\infty}^{\infty} h(\alpha)h(\gamma)f(t-\alpha)f(t+\tau-\gamma)d\alpha d\gamma$$

Taking the ensemble average on both sides we get

$$R_{XX}(\tau) = \int_{-\infty}^{\infty} \int_{-\infty}^{\infty} h(\alpha)h(\gamma)R_{FF}(\tau+\alpha-\gamma)d\alpha d\gamma \qquad (12.58)$$

because we assumed a covariant stationary input, meaning that its autocorrelation depends only on the time interval $(t+\tau-\gamma) - (t-\alpha) = \tau + \alpha - \gamma$. The immediate consequence is that the expected value on the l.h.s. – the output autocorrelation – is a function of τ only and the output process is also covariant stationary. In particular, if the input is a unit delta-correlated process

$$R_{FF}(\tau) = \delta(\tau) \qquad (12.59)$$

the response autocorrelation becomes a single integral, i.e.

$$R_{XX}(\tau) = \int_{-\infty}^{\infty} \int_{-\infty}^{\infty} h(\alpha)h(\gamma)\delta(\tau + \alpha - \gamma)d\alpha d\gamma$$

$$= \int_{-\infty}^{\infty} \left\{ \int_{-\infty}^{\infty} h(\alpha - (\gamma - \tau))h(\alpha)d\alpha \right\} h(\gamma)d\gamma$$

$$= \int_{\infty}^{\infty} h(\gamma - \tau)h(\gamma)d\gamma \tag{12.60a}$$

Furthermore, the response variance is given by

$$\sigma_X^2 = R_{XX}(0) = \int_{-\infty}^{\infty} h^2(\gamma)d\gamma \tag{12.60b}$$

In the frequency domain, the rather intimidating double integral of eq (12.58) turns into a simpler relationship. If we take the Fourier transform on both sides of eq (12.58) we get

$$S_{XX}(\omega) \equiv \mathcal{F}\{R_{XX}(\tau)\}$$

$$= \frac{1}{2\pi} \int_{-\infty}^{\infty} \left\{ \int_{-\infty}^{\infty} \int_{-\infty}^{\infty} h(\alpha)h(\gamma)R_{FF}(\tau + \alpha - \gamma)d\alpha d\gamma \right\} e^{-i\omega\tau}d\tau$$

$$= \int_{-\infty}^{\infty} \int_{-\infty}^{\infty} h(\alpha)h(\gamma) \left\{ \frac{1}{2\pi} \int_{-\infty}^{\infty} R_{FF}(\tau + \alpha - \gamma)e^{-i\omega\tau}d\tau \right\} d\alpha d\gamma$$

Then, in the integral within braces we make the change of variable $\tau + \alpha - \gamma = \theta$ so that $\tau = \theta + \gamma - \alpha$, $d\tau = d\theta$ and the equation above becomes

$$S_{XX}(\omega) = \int_{-\infty}^{\infty} \int_{-\infty}^{\infty} h(\alpha)h(\gamma)e^{-i\omega\gamma}e^{+i\omega\alpha} \left\{ \frac{1}{2\pi} \int_{-\infty}^{\infty} R_{FF}(\theta)e^{-i\omega\theta}d\theta \right\} d\alpha d\gamma$$

$$= S_{FF}(\omega) \int_{-\infty}^{\infty} \int_{-\infty}^{\infty} h(\alpha)e^{+i\omega\alpha}h(\gamma)e^{-i\omega\gamma}d\alpha d\gamma = S_{FF}(\omega)H^*(\omega)H(\omega)$$

$$\tag{12.61a}$$

which is the fundamental 'single-input single-output' relationship in the frequency domain for stationary random processes. Explicitly,

$$S_{XX}(\omega) = |H(\omega)|^2 S_{FF}(\omega) \tag{12.61b}$$

Also, by virtue of this last relationship we can obtain another expression for the variance σ_X^2 by writing the equation

$$R_{XX}(\tau) = \mathcal{F}^{-1}\{S_{XX}(\omega)\} = \int_{-\infty}^{\infty} S_{XX}(\omega)e^{+i\omega\tau}d\omega$$

from which it follows that

$$\sigma_X^2 = R_{XX}(0) = \int_{-\infty}^{\infty} |H(\omega)|^2 S_{FF}(\omega)d\omega \tag{12.62}$$

Other quantities of interest are the cross-relationships between input and output; from eq (12.53) we obtain

$$f(t)x(t+\tau) = \int_{-\infty}^{\infty} f(t)f(t+\tau-\alpha)h(\alpha)d\alpha$$

so that taking expectations on both sides and exploiting the covariance stationarity of the input yields

$$R_{FX}(\tau) = \int_{-\infty}^{\infty} h(\alpha)R_{FF}(\tau-\alpha)d\alpha \tag{12.63}$$

Equation (12.63) can then be Fourier transformed to give

$$S_{FX}(\omega) = H(\omega)S_{FF}(\omega) \tag{12.64a}$$

which expresses the input–output cross-spectral density in terms of the input autospectral density. Note that an important difference between the autospectral densities (eq (12.61b)) and the cross-spectral densities relations (eq (12.64)) is that the first is a real-valued relationship containing no phase information, while the second is a complex-valued relationship which can be broken down into a pair of equations to give both magnitude and phase information. This latter statement is of great practical importance because it means that the complete FRF of our system (i.e. magnitude and phase) can be obtained when both $S_{FX}(\omega)$ and $S_{FF}(\omega)$ are known, i.e.

$$H(\omega) = \frac{S_{FX}(\omega)}{S_{FF}(\omega)} \tag{12.64b}$$

thus justifying the H_1 FRF estimate of eq (10.28a), which was given in Chapter 10 without much explanation. (Note that eq (10.28a) is written in terms of one-sided spectral densities, the difference being only for practical purposes because these are the quantities displayed by spectrum analysers. Mathematically, the difference is irrelevant.)

By the same line of reasoning, it is now just a simple matter to obtain the output–input cross-relationships

$$R_{XF}(\tau) = \int_{-\infty}^{\infty} h(\alpha)R_{FF}(\tau + \alpha)d\alpha$$

(12.65)

$$S_{XF}(\omega) = H^*(\omega)S_{FF}(\omega)$$

from which we can obtain another expression for $H(\omega)$. In fact, putting together eq (12.61b) and the second of eqs (12.65) we have

$$S_{XX}(\omega) = H(\omega)H^*(\omega)S_{FF}(\omega) = H(\omega)H^*(\omega)\frac{S_{XF}(\omega)}{H^*(\omega)} = H(\omega)S_{XF}(\omega)$$

from which it follows that

$$H(\omega) = \frac{S_{XX}(\omega)}{S_{XF}(\omega)}$$

(12.66)

thus justifying the H_2 FRF estimate of eq (10.29).

Example 12.1. SDOF system subjected to broad-band excitation. From preceding chapters we know that the FRF of an SDOF system with parameters m, k and c is given by

$$H(\omega) = \frac{1}{k - m\omega^2 + i\omega c}$$

(12.67a)

so that

$$|H(\omega)|^2 = \frac{1}{(k - m\omega^2)^2 + \omega^2 c^2}$$

(12.67b)

Under the action of a random excitation with spectral density $S_{FF}(\omega)$, the system's response in the frequency domain is given by eq (12.61b), i.e.

$$S_{XX}(\omega) = \frac{S_{FF}(\omega)}{(k - m\omega^2)^2 + \omega^2 c^2} = \frac{S_{FF}(\omega)}{m^2[(\omega_n^2 - \omega^2)^2 + (2\zeta\omega\omega_n)^2]}$$

(12.68)

where, as usual, $\omega_n^2 = k/m$ is the system's natural frequency. If the excitation is in the form of a broad-band process whose spectral density is reasonably flat over a broad range of frequencies, we can approximate it as an 'equivalent' white noise by assuming $S_{FF}(\omega) \cong S_{FF}(\omega_n)$. The reason for this assumption comes from the fact that, for small damping, the function (12.67b) is sharply peaked in the vicinity of ω_n and small everywhere else – Fig. 12.7 being an example for $m = 10$, $k = 100$ and $\zeta = 0.05$. As a consequence, the product $|H(\omega)|^2 S_{FF}(\omega)$ will also show a similar behaviour, thus justifying the approximation above.

In physical terms, our system acts as a band-pass filter which significantly amplifies only the frequency components in the vicinity of its natural frequency and produces a narrow-band process at the output.

The variance of the output process can then be obtained from eq (12.62) as

$$\sigma_X^2 \cong S_{FF}(\omega_n) \int_{-\infty}^{\infty} |H(\omega)|^2 d\omega$$

$$= 2S_{FF}(\omega_n) \int_0^{\infty} |H(\omega)|^2 d\omega = \frac{\pi S_{FF}(\omega_n)}{kc} \qquad (12.69a)$$

where the last result can be obtained from tables of integrals. (Tables of integrals for $\int_{-\infty}^{\infty} |H(\omega)|^2 d\omega$, where the FRF is of the type

$$H(\omega) = \frac{B_0 + (i\omega)B_1 + (i\omega)^2 B_2 + \cdots + (i\omega)^{n-1} B_{n-1}}{A_0 + (i\omega)A_1 + (i\omega)^2 A_2 + \cdots + (i\omega)^n A_n}$$

and the A_j and B_j are real constants, are given, for example, in Newland [8].

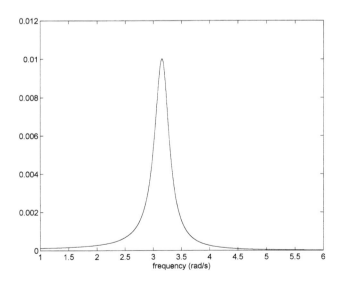

Fig. 12.7 FRF magnitude squared (SDOF).

In the light of this result, it is interesting to note that, since $c = 2m\zeta\omega_n$, we can write

$$\sigma_X^2 \cong \frac{\pi S_{FF}(\omega_n)}{2\zeta\omega_n km} = 2S_{FF}(\omega_n)\left(\frac{1}{4\zeta^2 k^2}\; \pi\zeta\omega_n\right) \qquad (12.70)$$

in which we note that $1/(4\zeta^2 k^2)$ is simply $|H(\omega_n)|^2$, while the quantity $\pi\zeta\omega_n$ can be interpreted as a 'mean-square' bandwidth. Both these quantities can be easily obtained from a graph of $|H(\omega)|$ without necessarily knowing the values of the parameters m, k and c. In essence, we obtain the area expressed by the integral $\int_0^\infty |H(\omega)|^2 d\omega$ by calculating the area of an 'equivalent' rectangle whose vertical and horizontal sides are given by the peak value of $|H(\omega)|^2$ and by the mean-square bandwidth, respectively.

The reader should not be deceived by the simplicity of this example. In practical situations it often happens that a specific mode of a structure – occurring, say, at a frequency ω_j – is subjected to a random forcing excitation which has an almost flat spectral density over the range of frequencies in the vicinity of ω_j. If the mode is lightly damped and well separated from other modes, we are in a situation which represents a good approximation of this SDOF example.

As a final comment in this section, it is worth pointing out that the expressions obtained above for the input–output relations refer to a steady-state condition, i.e. a situation in which the input has been acting for a while and the system has already had the time to adjust to its state of motion. As a matter of fact, however, real systems need some time to reach such a steady-state condition in which a stationary input produces a stationary output. In engineering terminology, we did not consider the transient part of the response, which occurs during a certain interval of time immediately following the onset of the input and dies away later because of the presence of damping.

During this period of time the response is obviously nonstationary. Now – resorting once again to the case of an SDOF system under the action of a white-noise excitation – we want to investigate how the variance of the response varies before reaching its steady-state value given by eq (12.69a).

Without loss of generality, we assume that the excitation is turned on at $t = 0$, that has a zero mean value and that its autocorrelation is in the form $R_{FF}(\tau) = R_0 \delta(\tau)$. This last condition implies that its white-noise spectral density is given by $2\pi S_0$, where $S_0 = R_0/2\pi$. There are different ways in which we can proceed; we will follow a time-domain approach by noting that the desired result can be obtained by rewriting eq (12.60b) as

$$\sigma_X^2(t) = 2\pi S_0 \int_0^t h^2(\gamma)d\gamma \qquad (12.71a)$$

or, explicitly,

$$\sigma_X^2(t) = \frac{2\pi S_0}{m^2 \omega_d^2} \int_0^t e^{-2\zeta\omega_n\gamma} \sin^2(\omega_d\gamma)d\gamma \tag{12.71b}$$

This integral can be slightly rearranged by making the change the variable $\omega_d\gamma = \varphi$ so that

$$\sigma_X^2(t) = \frac{2\pi S_0}{m^2 \omega_d^3} \int_0^{\omega_d t} e^{-2\zeta(\omega_n/\omega_d)\varphi} \sin^2 \varphi \, d\varphi \tag{12.71c}$$

Then, from a table of integrals we get

$$\int e^{ax} \sin^2 x \, dx = \frac{e^{ax} \sin x(a \sin x - 2 \cos x)}{a^2 + 4} + \frac{2}{a^2 + 4} \int e^{ax} \, dx$$

so that, being in our case $a = -2\zeta\omega_n/\omega_d$, it only takes a little patience to arrive at the final result

$$\sigma_X^2(t) = \frac{\pi S_0}{2m^2 \zeta\omega_n^3} \left\{ 1 - e^{-2\zeta\omega_n t} \left[1 + \frac{\zeta\omega_n}{\omega_d} \sin(2\omega_d t) + \frac{2\zeta^2\omega_n^2}{\omega_d^2} \sin^2(\omega_d t) \right] \right\} \tag{12.72}$$

where we note that the term outside the braces is exactly the steady-state value of eq (12.69a) and that

$$\lim_{t\to\infty} \sigma_X^2(t) = \frac{\pi S_0}{2m^2 \zeta\omega_n^3} = \frac{\pi S_0}{kc} = \sigma_X^2 \tag{12.73}$$

Considering an SDOF system with natural frequency $\omega_n = 10$ rad/s, Fig. 12.8 shows a plot of the ratio $W(t) \equiv \sigma_X^2(t)/\sigma_X^2$ (i.e. the term within braces of eq (12.72)) for two values of damping, namely $\zeta = 0.05$ and $\zeta = 0.10$. It is evident the effect of the damping ratio on the rapidity with which the steady-state value of variance is attained: the lower the damping ratio, the slower the convergence of $\sigma_X^2(t)$ to its asymptotic value σ_X^2.

12.4.1   One output and more than one random input

Consider first the case in which a linear system is subjected to two simultaneous random stationary inputs $f_1(t)$ and $f_2(t)$ (i.e. realizations of the random processes $F_1(t)$ and $F_2(t)$, respectively) and we measure one output $x(t)$. This may represent the common situation in which two excitations act at

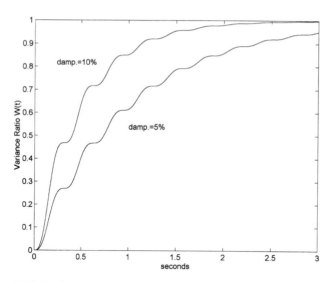

Fig. 12.8 Evolution towards steady state of the variance ratio $W(t)$ for two values of damping.

two different points of a structure (say, point A and point B) and we measure the resulting state of motion at a third point C. The time-domain response can obviously be written as

$$x(t) = \int_{-\infty}^{\infty} h_1(\alpha)f_1(t-\alpha)d\alpha + \int_{-\infty}^{\infty} h_2(\alpha)f_2(t-\alpha)d\alpha \qquad (12.74)$$

where $h_1(t)$ is the IRF between point A and point C and $h_2(t)$ is the IRF between point B and point C. The response mean value is readily obtained as

$$E[x(t)] = \mu_X = \mu_{F_1}\int_{-\infty}^{\infty} h_1(\alpha)d\alpha + \mu_{F_2}\int_{-\infty}^{\infty} h_2(\alpha)d\alpha \qquad (12.75a)$$

or also

$$\mu_X = \mu_{F_1}H_1(0) + \mu_{F_2}H_2(0) \qquad (12.75b)$$

since $H_1(\omega) = 2\pi\mathscr{F}\{h_1(t)\}$ and $H_2(\omega) = 2\pi\mathscr{F}\{h_2(t)\}$. The response autocorrelation can be obtained with some manipulations by calling α_1 the variable of integration in eq (12.74) and by writing

$$x(t+\tau) = \int_{-\infty}^{\infty} h_1(\alpha_2)f_1(t+\tau-\alpha_2)d\alpha_2 + \int_{-\infty}^{\infty} h_2(\alpha_2)f_2(t+\tau-\alpha_2)d\alpha_2$$

so that we get the expression

$$x(t)x(t+\tau) = \int_{-\infty}^{\infty} \int_{-\infty}^{\infty} h_1(\alpha_1)h_1(\alpha_2)f_1(t-\alpha_1)f_1(t+\tau-\alpha_2)d\alpha_1 d\alpha_2$$

$$+ \int_{-\infty}^{\infty} \int_{-\infty}^{\infty} h_1(\alpha_1)h_2(\alpha_2)f_1(t-\alpha_1)f_2(t+\tau-\alpha_2)d\alpha_1 d\alpha_2$$

$$+ \int_{-\infty}^{\infty} \int_{-\infty}^{\infty} h_1(\alpha_2)h_2(\alpha_1)f_2(t-\alpha_1)f_1(t+\tau-\alpha_2)d\alpha_1 d\alpha_2$$

$$+ \int_{-\infty}^{\infty} \int_{-\infty}^{\infty} h_2(\alpha_1)h_2(\alpha_2)f_2(t-\alpha_1)f_2(t+\tau-\alpha_2)d\alpha_1 d\alpha_2$$

$$(12.76)$$

which leads, after taking expectations on both sides, to

$$R_{XX}(\tau) = \int_{-\infty}^{\infty} \int_{-\infty}^{\infty} h_1(\alpha_1)h_1(\alpha_2)R_{F_1 F_1}(\tau+\alpha_1-\alpha_2)d\alpha_1 d\alpha_2$$

$$+ \int_{-\infty}^{\infty} \int_{-\infty}^{\infty} h_1(\alpha_1)h_2(\alpha_2)R_{F_1 F_2}(\tau+\alpha_1-\alpha_2)d\alpha_1 d\alpha_2$$

$$+ \int_{-\infty}^{\infty} \int_{-\infty}^{\infty} h_1(\alpha_2)h_2(\alpha_1)R_{F_2 F_1}(\tau+\alpha_1-\alpha_2)d\alpha_1 d\alpha_2$$

$$+ \int_{-\infty}^{\infty} \int_{-\infty}^{\infty} h_2(\alpha_1)h_2(\alpha_2)R_{F_2 F_2}(\tau+\alpha_1-\alpha_2)d\alpha_1 d\alpha_2 \qquad (12.77)$$

showing that the output process is also stationary because its autocorrelation depends on the variable τ only.

As for the single input case, in the frequency domain we obtain a simpler relationship. The autospectral density of the output process is obtained by Fourier transforming both sides of eq (12.77). For the first double integral we get

$$\frac{1}{2\pi} \int_{-\infty}^{\infty} \left\{ \int_{-\infty}^{\infty} \int_{-\infty}^{\infty} h_1(\alpha_1)h_1(\alpha_2)R_{F_1 F_1}(\tau+\alpha_1-\alpha_2)d\alpha_1 d\alpha_2 \right\} e^{-i\omega\tau} d\tau$$

$$= \int_{-\infty}^{\infty} \int_{-\infty}^{\infty} h_1(\alpha_1)h_1(\alpha_2) \left\{ \frac{1}{2\pi} \int_{-\infty}^{\infty} R_{F_1 F_1}(\tau+\alpha_1-\alpha_2)e^{-i\omega\tau} d\tau \right\} d\alpha_1 d\alpha_2$$

so that, by making the variable substitution $\varphi = \tau+\alpha_1-\alpha_2$ in the τ integral within braces, we have

$$\int_{-\infty}^{\infty}\int_{-\infty}^{\infty} h_1(\alpha_1)h_1(\alpha_2)\left\{\frac{1}{2\pi}\int_{-\infty}^{\infty} R_{F_1 F_1}(\varphi)e^{-i\omega\varphi}d\varphi\right\}e^{-i\omega(\alpha_2-\alpha_1)}d\alpha_1 d\alpha_2$$

$$= S_{F_1 F_1}(\omega)\int_{-\infty}^{\infty}\int_{-\infty}^{\infty} h_1(\alpha_1)e^{+i\omega\alpha_1}h_1(\alpha_2)e^{-i\omega\alpha_2}d\alpha_1 d\alpha_2$$

$$= S_{F_1 F_1}(\omega)H_1^*(\omega)H_1(\omega)$$

and similar results for the other three double integrals. Complete Fourier transformation yields

$$S_{XX}(\omega) = H_1^*(\omega)H_1(\omega)S_{F_1 F_1}(\omega) + H_1^*(\omega)H_2(\omega)S_{F_1 F_2}(\omega)$$

$$+ H_2^*(\omega)H_1(\omega)S_{F_2 F_1}(\omega) + H_2^*(\omega)H_2(\omega)S_{F_2 F_2}(\omega)$$

$$= \sum_{j=1}^{2}\sum_{k=1}^{2} H_j^*(\omega)H_k(\omega)S_{F_j F_k}(\omega) \tag{12.78a}$$

This important result can be readily extended to the case of N random inputs by writing

$$S_{XX}(\omega) = \sum_{j=1}^{N}\sum_{k=1}^{N} H_j^*(\omega)H_k(\omega)S_{F_j F_k}(\omega) \tag{12.78b}$$

which, as expected, becomes exactly eq (12.61b) in the case of one single input. (A word of caution about the notation: although the meaning is clear from the context, it must be pointed out that here the FRF functions $H_j(\omega)$ and $H_k(\omega)$ are **not** the jth and kth modal FRFs (which will come into play in a later section) but they simply represent the physical coordinate FRFs between the point of the system at which we measure the response and the points at which, respectively, the excitations $f_j(t)$ and $f_k(t)$ are acting. Obviously, the same applies to the IRFs $h_1(t)$ and $h_2(t)$ which appear in the equations above and are not modal IRFs.)

Furthermore, if the various inputs are uncorrelated with each other the cross-terms drop out and we get

$$S_{XX}(\omega) = \sum_{j=1}^{N} |H_j(\omega)|^2 S_{F_j F_j}(\omega) \tag{12.79}$$

The time-domain counterpart of eq (12.79) when $N=2$ is the somewhat more complicated expression

$$R_{XX}(\tau) = \int_{-\infty}^{\infty} \int_{-\infty}^{\infty} h_1(\alpha_1)h_1(\alpha_2)R_{F_1 F_1}(\tau + \alpha_1 - \alpha_2)d\alpha_1 d\alpha_2$$

$$+ \int_{-\infty}^{\infty} \int_{-\infty}^{\infty} h_2(\alpha_1)h_2(\alpha_2)R_{F_2 F_2}(\tau + \alpha_1 - \alpha_2)d\alpha_1 d\alpha_2 \qquad (12.80)$$

Knowledge of the output spectral density leads to the calculation of the output variance according to eq (12.35): in the case of mutually uncorrelated inputs, this reads

$$\sigma_X^2 = \int_{-\infty}^{\infty} S_{XX}(\omega)d\omega = \sum_{j=1}^{N} \int_{-\infty}^{\infty} |H_j(\omega)|^2 S_{F_j F_j}(\omega)d\omega \qquad (12.81)$$

The last quantities we want to consider in this section concern the input–output cross-relationships. For example, the cross-correlation between the first input and the output can be obtained by first writing explicitly the product $f_1(t)x(t + \tau)$ and then taking the ensemble average; these calculations lead to

$$R_{F_1 X}(\tau) = \int_{-\infty}^{\infty} h_1(\alpha)R_{F_1 F_1}(\tau - \alpha)d\alpha + \int_{-\infty}^{\infty} h_2(\alpha)R_{F_1 F_2}(\tau - \alpha)d\alpha \qquad (12.82)$$

If the inputs are uncorrelated and the excitation $f_1(t)$ is a white-noise process with autocorrelation given by $R_{F_1 F_1}(\tau - \alpha) = 2\pi S_0 \delta(\tau - \alpha)$, eq (12.82) becomes

$$R_{F_1 X}(\tau) = 2\pi S_0 \int_{-\infty}^{\infty} h_1(\alpha)\delta(\tau - \alpha)d\alpha = 2\pi S_0 h_1(\tau) \qquad (12.83)$$

showing that the cross-correlation between a white-noise input and the output is just $2\pi S_0$ times the IRF between the input point and the output point of the system. This result is also obviously true for the case of a single input (the first of eqs (12.65)) and is sometimes used to obtain the IRF experimentally.

For the cross-spectral density, Fourier transformation of both sides of eq (12.83) gives

$$S_{F_1 X}(\omega) = H_1(\omega)S_{F_1 F_1}(\omega) + H_2(\omega)S_{F_1 F_2}(\omega) \qquad (12.84)$$

which can immediately be extended to N inputs by writing the cross-spectral density between the kth input and the output as

$$S_{F_k X}(\omega) = \sum_{j=1}^{N} H_j(\omega)S_{F_k F_j}(\omega) \qquad (12.85)$$

For mutually uncorrelated inputs eq (12.85) becomes

$$S_{F_k X}(\omega) = H_k(\omega) S_{F_k F_k}(\omega) \tag{12.86}$$

with the consequence that a white-noise kth input results in a cross-spectral density $S_{F_k X}(\omega)$ which is proportional to the system's FRF $H_k(\omega)$. This last statement is just the frequency-domain counterpart of eq (12.83) and is the reason why uncorrelated broad-band processes ('nearly white noise' processes) are widely used as input excitations in many engineering applications (for example, the technique of experimental modal analysis described in Chapter 10).

One final comment worth making is that, in general, there is no simple relationship between the probability distribution of the input process and the probability distribution of the output process. One exception is given by Gaussian processes: if the input is Gaussian the output of a linear system is also Gaussian. Furthermore, in the case of multiple inputs and multiple outputs, it can be shown that if the input processes are jointly Gaussian then the output processes are jointly Gaussian as well. This property, together with the central limit theorem, and the fact that a Gaussian process is completely determined by its mean value and its autocorrelation (or its autocovariance) function are the reasons why the assumption of normality – unless there is strong evidence to the contrary – is so widely adopted and extensively used in many practical engineering applications.

12.5   MDOF and continuous systems: response to random excitation

The fundamental input–output relationships which have been obtained in the preceding sections form the basis of the stochastic analysis of linear systems. In order of increasing difficulty – although it is just a matter of extending those ideas – we can now turn our attention to the response of more complex systems.

If we consider a general system subjected to the action of n random inputs, the time response at the jth point of the structure is given by a straightforward extension of eq (12.74), i.e.

$$x_j(t) = \sum_{k=1}^{n} \int_{-\infty}^{\infty} h_{jk}(\alpha) f_k(t - \alpha) d\alpha \tag{12.87}$$

where we already know from preceding chapters that $h_{jk}(t)$ expresses, in physical coordinates, the response at point j due to a Dirac delta function excitation at point k. If, as is often the case, we model our system as an assemblage of n masses connected by appropriate springs and dampers, we are, in fact, dealing with a discrete n-DOF system whose response can be written in matrix form as

$$\mathbf{x}(t) = \int_{-\infty}^{\infty} \mathbf{h}(\alpha)\mathbf{f}(t-\alpha)d\alpha \tag{12.88}$$

where $\mathbf{h}(t)$ is the (symmetrical) IRF matrix in physical coordinates and it is evident that eq (12.87) is just the jth element of the $n \times 1$ vector $\mathbf{x}(t)$.

Now, before proceeding further, we make a short digression to extend the ideas of Section 11.5 to random processes. In fact, we note that n generic random processes $Z_1(t), Z_2(t), \ldots, Z_n(t)$ can be arranged in the $n \times 1$ matrix $\mathbf{Z} = [Z_1(t) \cdots Z_n(t)]^T$ so that the correlation matrix (see also eq (11.69)) can be defined as

$$\mathbf{R}_{ZZ}(\tau) \equiv E[\mathbf{Z}(t)\mathbf{Z}^T(t+\tau)] \tag{12.89}$$

where – since this is the case of most interest for our purposes – we assume that the n processes are stationary and with zero mean.

Then, returning to our main discussion we can write

$$\mathbf{x}(t+\tau) = \int_{-\infty}^{\infty} \mathbf{h}(\varphi)\mathbf{f}(t+\tau-\varphi)d\varphi \tag{12.90}$$

so that, putting together eqs (12.88) and (12.90) to form the product $\mathbf{x}(t)\mathbf{x}^T(t+\tau)$ and taking expectations on both sides we get the response correlation matrix as

$$\mathbf{R}_{XX}(\tau) = \int_{-\infty}^{\infty} \int_{-\infty}^{\infty} \mathbf{h}(\alpha)\mathbf{R}_{FF}(\tau+\alpha-\varphi)\mathbf{h}^T(\varphi)d\alpha d\varphi \tag{12.91}$$

where $\mathbf{R}_{FF}$ is the input correlation matrix. Explicitly, the jth diagonal element of eq (12.91) represents the autocorrelation of the jth response and reads

$$R_{X_j X_j}(\tau) = \sum_{r=1}^{n} \sum_{s=1}^{n} \int_{-\infty}^{\infty} \int_{-\infty}^{\infty} h_{jr}(\alpha)h_{js}(\varphi)R_{F_r F_s}(\tau+\alpha-\varphi)d\alpha d\varphi \tag{12.92}$$

Note that if we have two inputs and the jth response is the only output, then eq (12.92) becomes eq (12.77).

At this point we can turn to the frequency domain by noting that the spectral density matrix $\mathbf{S}_{XX}(\omega)$ is the Fourier transform of $\mathbf{R}_{XX}(\tau)$. Fourier transformation of both sides of eq (12.91) leads to

$$\mathbf{S}_{XX}(\omega) = \mathbf{H}^*(\omega)\mathbf{S}_{FF}(\omega)\mathbf{H}^T(\omega) \tag{12.93}$$

where $\mathbf{H}(\omega) = 2\pi\mathscr{F}\{\mathbf{h}(t)\}$. The diagonal elements of the matrix $\mathbf{S}_{XX}(\omega)$ are the response autospectral densities

$$S_{X_jX_j}(\omega) = \sum_{r=1}^{n} \sum_{s=1}^{n} H_{jr}^{*}(\omega)S_{F_rF_s}(\omega)H_{js}(\omega) \qquad (12.94a)$$

while the off-diagonal elements are the response cross-spectral densities

$$S_{X_jX_k}(\omega) = \sum_{r=1}^{n} \sum_{s=1}^{n} H_{jr}^{*}(\omega)S_{F_rF_s}(\omega)H_{ks}(\omega) \qquad (12.94b)$$

and again, in the case of n inputs and one output, eq (12.94b) becomes eq (12.78b). A word of caution is necessary to point out that the matrices $\mathbf{R_{XX}}(\tau)$ and $\mathbf{S_{XX}}(\omega)$ are sometimes called the autocorrelation matrix and the autospectral density matrix although most of their elements are, as a matter of fact, cross-correlation functions and cross-spectral density functions. Only their diagonal elements are autocorrelation functions.

By similar arguments as above, we can now obtain the cross-correlation matrix $\mathbf{R_{FX}}(\tau)$ between the inputs and the outputs, i.e.

$$\mathbf{R_{FX}}(\tau) = \int_{-\infty}^{\infty} \mathbf{R_{FF}}(\tau - \alpha)\mathbf{h}^{T}(\alpha)d\alpha \qquad (12.95)$$

whose diagonal elements are the cross-correlations $R_{F_jX_j}(\tau)$ while the off-diagonal elements are the cross-correlation functions $R_{F_jX_k}(\tau)$ with $j \neq k$.

Both sides of eq (12.95) can be Fourier transformed to obtain the cross-spectral density matrix

$$\mathbf{S_{FX}}(\omega) = \mathbf{S_{FF}}(\omega)\mathbf{H}^{T}(\omega) \qquad (12.96)$$

All the input–output relationships above have been obtained without taking into account the fact that the equations of motion of an n-DOF system can often be uncoupled into n SDOF equations. In Chapter 6 we determined how and when this can be done depending on the form of the damping matrix which, in common practice is sometimes neglected altogether – in which case uncoupling is always possible – or often assumed to be either 'proportional' (eq (6.141)) or in the 'Caughey' form of eq (6.146). This possibility leads to the concept of 'normal or modal coordinates' and, in the analysis of response to deterministic excitation (Chapter 7), to the concepts of modal IRFs $h_j(t)$ and modal FRFs $H_j(\omega)$. These latter functions, in turn, can be arranged in the form of diagonal matrices, i.e. the matrices $\text{diag}[h_j(t)] \equiv \hat{\mathbf{h}}(t)$ and $\text{diag}[H_j(\omega)] \equiv \hat{\mathbf{H}}(\omega)$ which are related to the IRF and FRF matrices in physical coordinates by the relationships (7.37a) and (7.34a), i.e.

$$\mathbf{h}(t) = \mathbf{P} \; \text{diag}[h_j(t)]\mathbf{P}^T = \mathbf{P}\hat{\mathbf{h}}(t)\mathbf{P}^T$$

$$\mathbf{H}(\omega) = \mathbf{P} \; \text{diag}[H_j(\omega)]\mathbf{P}^T = \mathbf{P}\hat{\mathbf{H}}(\omega)\mathbf{P}^T$$

(12.97)

where $\mathbf{P}$ is the $n \times n$ matrix of mass-orthonormal eigenvectors so that $\mathbf{P}^T\mathbf{M}\mathbf{P} = \mathbf{I}$ and $\mathbf{P}^T\mathbf{K}\mathbf{P} = \mathbf{L} \equiv \text{diag}(\lambda_j)$. In this light, we can now express the response quantities of eqs (12.91), (12.93), (12.95) and (12.96) in terms of modal characteristics. For example, eqs (12.91) and (12.93) become

$$\mathbf{R_{XX}}(\tau) = \int_{-\infty}^{\infty} \int_{-\infty}^{\infty} \mathbf{P}\hat{\mathbf{h}}(\alpha)\mathbf{P}^T\mathbf{R_{FF}}(\tau + \alpha - \varphi)\mathbf{P}^T\hat{\mathbf{h}}(\varphi)\mathbf{P}d\alpha d\varphi$$

(12.98)

and

$$\mathbf{S_{XX}}(\omega) = \mathbf{P}\hat{\mathbf{H}}^*(\omega)\mathbf{P}^T\mathbf{S_{FF}}(\omega)\mathbf{P}\hat{\mathbf{H}}(\omega)\mathbf{P}^T$$

(12.99)

where it should be noted that in eqs (12.98) and (12.99) we took into account the symmetries $\hat{\mathbf{h}}^T(t) = \hat{\mathbf{h}}(t)$, $\hat{\mathbf{H}}^T(\omega) = \hat{\mathbf{H}}(\omega)$ and the fact that $\mathbf{P}^* = \mathbf{P}$ because $\mathbf{P}$ is a matrix of real eigenvectors. Note that, explicitly, the (jk)th element of the spectral density matrix of eq (12.99) is written

$$[\mathbf{S_{XX}}(\omega)]_{jk} = \sum_{i=1}^{n} \sum_{l=1}^{n} \sum_{r=1}^{n} \sum_{s=1}^{n} p_{jl}p_{il}[\mathbf{S_{FF}}(\omega)]_{ir}p_{rs}p_{ks}H_l^*H_s$$

(12.100)

where $H_l(\omega)$ and $H_s(\omega)$ are the lth and sth modal FRFs, respectively. Suppose now that the excitation is in the form of 'nearly white noise' processes, i.e. $\mathbf{S_{FF}}(\omega) \cong \mathbf{S}_0$. The sum of eq (12.100) will contain some terms where the magnitudes squared of modal FRFs appear – say, for example $|H_s(\omega)|^2$ – and other terms where the cross-products $H_l^*(\omega)H_s(\omega)$ appear, with $l \neq s$. If – as often happens for the lowest-order modes of many structures – the modes of our system are well separated and lightly damped, then the magnitudes of the terms of the latter type will generally be much smaller than those of the former type and the contributions from cross-modal terms can be neglected without a significant loss of accuracy. In other words, the spectral densities on the l.h.s. of eq (12.100) will show peaks at the natural frequencies of the system so that, for all practical purposes, we can say that our system behaves as a selective filter which amplifies only the input contributions near its natural frequencies.

If now we turn our attention to continuous systems, two general remarks can be made before proceeding any further.

First of all, it is very common to model continuous systems as MDOFs systems by lumping masses at a number of locations. In general, a higher number of degrees of freedom corresponds to a better approximation for eigenvalues, eigenvectors and for the response characteristics of the real system under investigation. In any case, whenever we decide to adopt a similar approach it is evident that, in the case of random excitation, the above input–output relationships apply.

The second remark is that – when we do not follow this approach and we model our system as a truly continuous system – we must recall some general observations made in Chapter 8. In that chapter we noted that the behaviour of linear continuous systems bears noteworthy similarities to the behaviour of linear MDOF systems when the mathematical framework adopted to deal with these latter systems – i.e. finite-dimensional vector spaces – is extended to the idea of infinite-dimensional vector spaces with inner product, i.e. the so-called Hilbert spaces. The system's matrices of the MDOF case are replaced, in the continuous case, by appropriate differential symmetrical operators, the finite sets of eigenvalues and eigenvectors are replaced by countable infinite sets of eigenvalues and eigenfunctions and the expansion theorem is replaced by a series expansion in terms of eigenfunctions. Broadly speaking, we can say that the price we must pay is a higher level of mathematical difficulty in 'setting the stage' for our analysis.

However, if we look at the problem from a practical point of view and note that many fundamental concepts introduced in the case of MDOF systems retain their validity, we may observe that the dynamical behaviour of our system can still be described in terms of IRFs $h(\mathbf{r}, \mathbf{s}, t)$ or FRFs $H(\mathbf{r}, \mathbf{s}, \omega)$. The symbols used here are very general and indicate, respectively, the response in the direction of the unit vector $\mathbf{e}_r$ at the position identified by the vector $\mathbf{r}$ (with respect to a fixed origin) due to a Dirac delta excitation applied (at $t = 0$) in the direction of the unit vector $\mathbf{e}_s$ at the point identified by the vector $\mathbf{s}$ and the steady-state response at point $\mathbf{r}$ along the direction $\mathbf{e}_r$ due to a unit harmonic excitation applied at $\mathbf{s}$ in the direction $\mathbf{e}_s$. As usual, the following relations hold between these two functions:

$$H(\mathbf{r}, \mathbf{s}, \omega) = 2\pi\mathscr{F}\{h(\mathbf{r}, \mathbf{s}, t)\} = \int_{-\infty}^{\infty} h(\mathbf{r}, \mathbf{s}, t)e^{-i\omega t}dt$$

$$\hspace{10cm} (12.101)$$

$$h(\mathbf{r}, \mathbf{s}, t) = \frac{1}{2\pi}\mathscr{F}^{-1}\{H(\mathbf{r}, \mathbf{s}, \omega)\} = \frac{1}{2\pi}\int_{-\infty}^{\infty} H(\mathbf{r}, \mathbf{s}, \omega)e^{i\omega t}d\omega$$

Therefore, if we are dealing, for example, with a one-dimensional continuous system such as a string or a beam and let the realization $f(x_k, t)$ of a stationary random process $F(t)$ be the only excitation applied at point $x = x_k$, the autocorrelation and autospectral density functions of the response $w(x, t)$ (we are following the notation of Chapter 8: in this case, however, $w(x, t)$ represents a realization of the response process $W(t)$ at point x) at the specified point $x = x_m$ can be directly obtained from eqs (12.58) and (12.61b) as

$$R_{WW}(x_m, \tau) = \int_{-\infty}^{\infty} \int_{-\infty}^{\infty} h(x_m, x_k, \alpha)h(x_m, x_k, \gamma)R_{FF}(\tau + \alpha - \gamma)d\alpha d\gamma$$

$$S_{WW}(x_m, \omega) = |H(x_m, x_k, \omega)|^2 S_{FF}(x_k, \omega) \tag{12.102}$$

then, the mean squared displacement response at $x = x_m$ can be obtained (eq.12.35)), for example, from

$$E[w^2(x_m, t)] = \sigma_W^2(x_m) = \int_{-\infty}^{\infty} S_{WW}(x_m, \omega)d\omega \tag{12.103a}$$

and the mean squared velocity response can be obtained from the first of eqs (12.46), i.e.

$$\sigma_{dW/dt}^2(x_m) = \int_{-\infty}^{\infty} \omega^2 S_{WW}(x_m, \omega)d\omega \tag{12.103b}$$

If now we extend our reasoning to the case of multiple (say n) inputs and multiple outputs, we can be interested, for example, in the response characteristics at the point $x = x_m$. In the form of the output autospectral density, the desired result is given by (eq (12.94a))

$$S_{WW}(x_m, \omega) = \sum_{r=1}^{n} \sum_{s=1}^{n} H^*(x_m, x_r, \omega)H(x_m, x_s, \omega)S_{FF}(x_r, x_s, \omega) \tag{12.104}$$

where, for $r \neq s$, $S_{FF}(x_r, x_s, \omega)$ is the cross-spectral density between the two inputs applied at $x = x_r$ and $x = x_s$ while, on the other hand, for $r = s$ the function $S_{FF}(x_r, x_s, \omega) = S_{FF}(x_r, \omega)$ is the autospectral density of the input applied at $x = x_r$. If, on the other hand, we are interested in the cross-spectral density between the outputs at points $x = x_m$ and $x = x_k$, we get (eq (12.94b))

$$S_{WW}(x_m, x_k, \omega) = \sum_{r=1}^{n} \sum_{s=1}^{n} H^*(x_m, x_r, \omega)H(x_k, x_s, \omega)S_{FF}(x_r, x_s, \omega)$$

$$\tag{12.105}$$

At this point, provided that damping is either neglected or is 'proportional', we can recall from Chapter 8 that the IRFs and FRFs in physical coordinates can be expressed as series expansions in terms of the system mode shapes $\phi_j(x)$ and of modal IRFs or FRFs, respectively. These relations are given by eqs (8.185b) and (8.190), which we rewrite here for our present convenience:

$$h(x_m, x_k, t) = \sum_{j=1}^{\infty} h_j(t)\phi_j(x_m)\phi_j(x_k)$$

$$H(x_m, x_k, \omega) = \sum_{j=1}^{\infty} H_j(\omega)\phi_j(x_m)\phi_j(x_k)$$

(12.106)

When eqs (12.106) are substituted into the appropriate input–output relations we obtain the response characteristics in terms of modal contributions and, as a consequence, it is possible to include only the modes of interest and exclude those that are not.

We end this section here by noting that we have limited our discussion of continuous systems only to point excitations. The subject of distributed random loads has not been discussed and the interested reader is referred to specific literature on random vibrations, for example, Newland [8].

12.6   Analysis of narrow-band processes: a few selected topics

This section considers briefly some topics of general interest in many applications. The choice is subjective and it has been made only with the intention of introducing the reader to some concepts and ideas in the vast and specialized field of random vibrations.

12.6.1   *Stationary narrow-band processes: threshold crossing rates*

For isolated and lightly damped structural modes, we noted in preceding sections that the system output to a broad-band excitation is a narrow-band process $X(t)$ whose spectral density has a significant amplitude only in a limited range of frequencies in the vicinity of the mode natural frequency. Let us consider a time history $x(t)$ of such a process and ask if we can obtain some information on the number of times that our sample function crosses a given threshold level $x = a$ in a given time interval T. More specifically, we will be interested in the number of upward crossings – i.e. crossings with a positive slope – in time T. Let $n_a^+(T)$ be the number of such crossings in a typical sample function of duration T. Averaging over samples we obtain the number

$$N_a^+(T) \equiv E[n_a^+(T)]$$

(12.107)

and, since the process is stationary, we can easily expect that a sample twice as long will contain twice as many upwards crossings. This leads to the conclusion that $N_a^+(T)$ is directly proportional to T so that we can write

$$N_a^+(T) = v_a^+ T \tag{12.108}$$

where we interpret v_a^+ as the average frequency of upward crossings of the threshold $x = a$, i.e. the number of crossings per unit time. Now, by isolating a short (say, of length dt, between the instants t_0 and $t_0 + dt$) section of a sample time history, let us consider a typical situation in which an upward crossing is very likely to occur. The first condition to be met is that at the beginning of the interval – i.e. at time t_0 – we must have $x < a$. The second obvious condition is that the derivative dx/dt be positive. However, this is not enough: if we want an upward crossing to occur within the interval we must require that

$$\dot{x} \equiv \frac{dx}{dt} > \frac{a - x}{dt} \tag{12.109}$$

which, in essence, means that the slope must be steep enough to arrive at the threshold value within the time interval dt. Rearranging eq (12.109) we get the equivalent condition $x > a - \dot{x}dt$.

Since t_0 is arbitrary and the two conditions must be satisfied simultaneously, we can obtain the probability $v_a^+ dt$ of an upward crossing within the interval dt by expressing it as the double integral

$$v_a^+ dt = \int_0^\infty \int_{a - \dot{x}dt}^a p_{X\dot{X}}(x, \dot{x}) dx d\dot{x} \tag{12.110}$$

where $p_{X\dot{X}}(x, \dot{x})$ is the joint pdf for the process X and its time derivative $\dot{X}$. Now, since the interval dt is very small, it is reasonable to approximate the integral in dx as

$$\int_{a - \dot{x}dt}^a p_{X\dot{X}}(x, \dot{x}) dx = p_{X\dot{X}}(x = a, \dot{x})\dot{x}dt$$

so that eq (12.110) becomes

$$v_a^+ dt = dt \int_0^\infty p_{X\dot{X}}(x, \dot{x})\dot{x}d\dot{x}$$

and hence

$$v_a^+ = \int_0^\infty p_{X\dot{X}}(x, \dot{x})\dot{x}d\dot{x} \tag{12.111}$$

which is a general result valid for any probability distribution. A special case of eq (12.111) which deserves particular attention is the case of a Gaussian process, i.e. a process with distribution function

$$p_{X\dot{X}}(x, \dot{x}) = \frac{1}{2\pi\sigma_X\sigma_{\dot{X}}} \exp\left(-\frac{x^2}{2\sigma_X^2} - \frac{\dot{x}^2}{2\sigma_{\dot{X}}^2}\right) \tag{12.112}$$

Substitution of eq (12.112) into eq (12.111) leads to

$$\nu_a^+ = \frac{\sigma_{\dot{X}}}{2\pi\sigma_X} e^{-a^2/(2\sigma_X^2)} \tag{12.113}$$

which is a result obtained by Rice [9].

In this regard we can, for example, calculate the quantity ν_a^+ for the SDOF system of Example 12.1 when the input process is a stationary Gaussian white noise with spectral density S_0. From eq (12.69a) we have $\sigma_X^2 = \pi S_0/(kc)$. Then, the variance of the derived process $\dot{X}$ is obtained by combining the results of eq (12.62) and the first of eqs (12.46) to get

$$\sigma_{\dot{X}}^2 = \int_{-\infty}^{\infty} \omega^2 |H(\omega)|^2 S_0 d\omega = \frac{\pi S_0}{mc}$$

where $H(\omega)$ is given by eq (12.67a) and the last result on the r.h.s. has been obtained from tabulated integrals. Substitution in eq (12.113) gives

$$\nu_a^+ = \frac{\omega_n}{2\pi} e^{-a^2 kc/(2\pi S_0)} \tag{12.114}$$

where ω_n is the system natural frequency.

12.6.2   Stationary narrow-band processes: peak distributions

Consider again a sample function $x(t)$ of duration T of a stationary random process $X(t)$. If we call $p_p(\alpha)$ the peak probability density function, then the probability that a peak chosen at random has an amplitude that exceeds the amplitude a is given by

$$P[\text{peak} > a] = \int_a^{\infty} p_p(\alpha)d\alpha \tag{12.115}$$

Now, since we are considering a narrow-band process, any time history $x(t)$ is generally well behaved and not very dissimilar from a sinusoidal oscillation with varying amplitude. In this circumstance it is reasonable to assume that any upward crossing of the level $x = a$ will result in one peak with amplitude $>a$, so that the number of such peaks in the time interval T is given by $\nu_a^+ T$. Also, we can say that each upward crossing of the threshold $x = 0$ corresponds to one 'cycle' of our smoothly varying time history, so that there are, on average, $\nu_0^+ T$ 'cycles' in the time interval T. (Note that these assumptions are generally not true for a wide-band processes, which have highly erratic time histories. In this circumstance it cannot be assumed that each upcrossing of the threshold corresponds to one peak (or maximum) only.) Then, in the same interval, the favourable fraction of peaks greater than a can be expressed as the ratio ν_a^+/ν_0^+ and

$$\int_a^\infty p_p(\alpha)d\alpha = \frac{\nu_a^+}{\nu_0^+} \tag{12.116}$$

Differentiating both sides with respect to a gives the desired result, i.e. the probability density function for the occurrence of peaks

$$-p_p(a) = \frac{1}{\nu_0^+} \frac{d}{da}(\nu_a^+) \tag{12.117}$$

If, in particular, the narrow-band process has a Gaussian distribution, we can use eq (12.113) to obtain

$$p_p(a) = \frac{a}{\sigma_X^2} e^{-a^2/(2\sigma_X^2)} \tag{12.118}$$

where we took into account (eq (12.113)) that $\nu_0^+ = \sigma_{\dot X}/(2\pi\sigma_X)$. The distribution of eq (12.118) is well known in probability theory and is called the Rayleigh distribution. From this result it is easy to determine the probability that a peak chosen at random will exceed the level a: this is

$$P[\text{peak} > a] = e^{-a^2/(2\sigma_X^2)} \tag{12.119a}$$

or the probability that a peak chosen at random is less than level a, i.e. the Rayleigh cumulative probability distribution $P_p(a)$

$$P[\text{peak} \leqslant a] \equiv P_p(a) = 1 - e^{-a^2/(2\sigma_X^2)} \tag{12.119b}$$

Although the Rayleigh distribution is widely used in a large number of practical problems, it must be noted that the distribution of peaks may differ significantly from eq (12.118) if the underlying probability distribution of the original process is not Gaussian. In these cases, the Weibull distribution generally provides better results. This distribution in its general form is a two-parameter distribution and is often found in statistics books written as

$$F_X(x) = 1 - e^{-\beta x^\alpha} \tag{12.120a}$$

where α is a parameter which determines the shape of the distribution and β is a scale parameter which determines the spread of the values. From eq (12.120a) the Weibull probability density function can be obtained by differentiating with respect to x (see the third of eqs (11.20))

$$p_X(x) = \alpha \beta x^{\alpha-1} e^{-\beta x^\alpha} \tag{12.120b}$$

For our purposes, however, we can follow Newland and note that if we call a_0 the median (eq (11.45)) of the Rayleigh distribution (12.119b), we have

$$P_p(a_0) = 1 - e^{-a_0^2/(2\sigma_X^2)} = \tfrac{1}{2}$$

so that $a_0 = \sigma_X \sqrt{2 \ln 2}$, from which it follows $\sigma_X^2 = a_0^2/2 \ln 2$. Substitution of this result into eq (12.119b) gives the Rayleigh distribution in the form

$$P_p\left(\frac{a}{a_0}\right) = 1 - e^{-\ln 2(a/a_0)^2} \tag{12.121}$$

which, in turn, is a special case of the one-parameter Weibull distribution (eq (12.120b) with $x = a/a_0$, $\alpha = k$ and $\beta = \ln 2$), i.e.

$$P_p\left(\frac{a}{a_0}\right) = 1 - e^{-\ln 2(a/a_0)^k} \tag{12.122}$$

From eq (12.122) we obtain the Weibull pdf

$$p_p\left(\frac{a}{a_0}\right) = k \ln 2 \left(\frac{a}{a_0}\right)^{k-1} e^{-\ln 2(a/a_0)^k} \tag{12.123}$$

which is sketched in Fig. 12.9 for three different values of k, the case $k = 2$ representing the Rayleigh pdf.

(The reader is invited to sketch a graph of the Weibull cumulative probability distributions of eq (12.122) for the same values of k.)

At this point we may ask about the highest peak which can be expected within a time interval T. The average number of cycles in time T (and hence, for a narrow-band process, the average number of peaks) is given by $\nu_0^+ T$, where ν_0^+ has been introduced above in this section. Noting that there is no

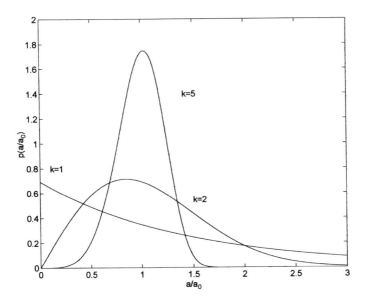

Fig. 12.9 Weibull pdf for different values of k.

loss of generality in considering the amplitude of peaks in median units, let us call A the (unknown) maximum peak amplitude expected, on average, in time T. In other words, we are putting ourselves in the situation in which the equation $\nu_A^+ T = 1$ applies, which in turn implies

$$\nu_A^+ = \frac{1}{T} \tag{12.124}$$

Furthermore, we know from eq (12.116) that

$$P[\text{peak} > A] = \frac{\nu_A^+}{\nu_0^+} \tag{12.125a}$$

and from eq (12.122) that

$$P[\text{peak} > A] = e^{-(\ln 2)A^k} \tag{12.125b}$$

so that, equating eqs (12.125a) and (12.125b) and taking (12.124) into account, we get

$$e^{-(\ln 2)A^k} = \frac{1}{\nu_0^+ T}$$

from which it follows that

$$A = \left[\frac{\ln(\nu_0^+ T)}{\ln 2} \right]^{1/k} \tag{12.126a}$$

Finally, noting that A expresses the maximum amplitude in median units and can therefore be written as $A = a_{max}/a_0$ where a_{max} is the maximum amplitude in its appropriate units, we get

$$\frac{a_{max}}{a_0} = \left[\frac{\ln(\nu_0^+ T)}{\ln 2} \right]^{1/k} \tag{12.126b}$$

Equation (12.126b) is a general expression for narrow-band processes when we can reasonably assume that any upcrossing of the zero level corresponds to a full cycle (and hence to a peak), so that the average number of cycles (peaks) in time T is given by $\nu_0^+ T$. It is left to the reader to sketch a graph of eq (12.126b) plotting a_{max}/a_0 as a function of the number of cycles $\nu_0^+ T$.

For example, if the peak distribution of our process is a Weibull distribution with $k = 1$, eq (12.126b) shows that, on average, a peak with an amplitude higher than four times the median can be expected every 16 cycles or, in other words, one peak out of 16 peaks will exceed, on average, four times the median. If, on the other hand, the peak distribution is a Rayleigh distribution, the average number of cycles needed to observe one peak higher than four times the median (i.e. $a_{max}/a_0 = 4$) is given by

$$\nu_0^+ T = e^{(4)^2 \ln 2} = 65\ 536$$

or, in other words, one peak out of approximately 65 500 peaks will exceed an amplitude of four times the median. Qualitatively, a similar result should be expected just by visual inspection of Fig. 12.9, where we note that higher values of k correspond to more and more strongly peaked probability density functions in the vicinity of the median, and hence to lower and lower probabilities for the occurrence of peak values significantly different from a_0. The interested reader can find further developments along this line of reasoning, for example, in Newland [8] or Sólnes [10].

12.6.3 *Notes on fatigue damage due to random excitation*

Fatigue is the process by which the strength of a structural member is degraded due to the cyclic application of load (stress) or strain so that the fatigue load that a structure can withstand is often significantly less than the load it would be capable of if the same load were applied only once. Broadly speaking, fatigue failure is caused by the gradual propagation of cracks in regions of high stress and the whole process can be divided into three main phases: crack initiation, crack growth and final failure. Although there exist some general guidelines, there is no clear distinction between the various phases and, as a matter of fact, traditional fatigue analysis makes no distinction between fatigue crack initiation and crack growth to failure. Moreover, it seems that none of the theories which have been developed to describe the actual mechanism of crack initiation is universally accepted.

Experimentally, the most important techniques of fatigue testing are performed by applying a constant amplitude and periodically varying load to a test specimen of the material to be tested, and estimating its 'fatigue life' by counting the number of cycles to failure (N_f). In this situation, the fatigue life depends significantly only on two characteristics of the stress time history: the stress range (maximum stress minus minimum stress) and the mean stress value, the effect of the former characteristic being generally more important than the effect of the latter. If, as it is often the case, we assume for the moment a zero mean stress value and consider only the stress range S and the number of cycles to failure N_f, then a typical experimental test leads to the so called S–N curve (or the Wöhler fatigue curve) which is essentially a plot of $\log S$ versus $\log N_f$. Analytical approximations of such graphs have generally the form

$$N_f(S) = bS^{-m} \tag{12.127}$$

where b and m are positive constants whose values depend on both the material and the geometry of the specimen. More specifically, eq (12.127) does not apply for all values of S and we can distinguish between two regimes of material behaviour: high-cycle fatigue, in which eq (12.127) applies and failure occurs in excess of approximately 10^3 cycles (e.g. ASTM Standard E468 [11]) and low-cycle fatigue in which failure occurs in relatively few cycles ($<10^3$). In the former case the deformations are small and within the material's elastic behaviour, while in the latter case crack growth is often accompanied by large-scale plastic deformation. Furthermore, in high-cycle fatigue testing, for many materials the value of N_f seems to go to infinity when S is smaller than some particular value S_e called the fatigue limit or endurance limit of the material.

Finally, two other preliminary considerations should be made before turning our attention to a stochastic approach to the problem. First, even for a given material tested in the deterministic conditions described above, fatigue tests often show appreciable scatter. This scatter is probably due to a number of factors such as specimen preparation (surface finish, heat treating etc.), specimen alignment, intrinsic material variability etc., and translates into the fact that statistical methods must be applied to fatigue test data. Second, the effect of mean stress value (when this is different from zero) is generally taken into account by means of empirical formulas such as the

Goodman or the Gerber formulas, just to name two of the most popular. The Goodman correction, for example, assumes that the fatigue damage done by a time-varying load $x(t)$ with mean value x_m and stress range S is the same as would be done by another loading with zero mean and stress range S' such that

$$\frac{S}{S'} + \frac{x_m}{x_u} = 1 \tag{12.128a}$$

where x_u represents the ultimate stress capacity of the material. Similarly, the Gerber formula reads

$$\frac{S}{S'} + \left(\frac{x_m}{x_u}\right)^2 = 1 \tag{12.128b}$$

and it has often been found to be in better agreement with experimental data. The utility of eqs (12.128a and b) is evident and lies in the fact that the experimenter can take into account both mean stress and stress range while at the same time maintaining the simplicity of a one-dimensional S–N relationship.

Now, despite the considerable practical utility of S–N plots in engineering practice, it should be remembered that most structures are subjected to loadings which are much more complicated than the deterministic periodic loadings of laboratory fatigue testing. As a matter of fact, a large number of these real-life loadings can be treated as random excitations which result in randomly varying stress amplitudes. In these circumstances, it is evident that the S–N curves cannot be used directly and some starting assumptions must be made to deal with the problem and compensate for our incomplete understanding of the basic mechanisms of fatigue.

In this light, we can introduce the concept of 'accumulated damage' described by a 'damage function' $D(t)$ which is used to denote the progress toward failure, whether this progress is physically observable or not. The function $D(t)$ is assumed to be a nondecreasing function of time that starts at zero (or very near zero) for a new structure ($t = 0$) and is normalized to unity when failure occurs; the instant of time t_f at which $D(t_f) = 1$ is obviously called the failure time. This definition is rather vague indeed, but it has nevertheless proven to be useful. In order to be more specific, let us call ΔD_j the damage increment produced by the jth cycle and let $N(t)$ be the number of stress cycles up to time t. Then

$$D(t) = \sum_{j=1}^{N(t)} \Delta D_j \tag{12.129}$$

and also, from the definition of damage function,

$$\sum_{j=1}^{N(t_f)} \Delta D_j = 1 \qquad (12.130)$$

where $N(t_f)$ is the number of cycles to failure. For a periodic loading at the constant amplitude of stress S, the upper limit of the sum (12.130) is clearly the number N_f obtained from the S–N curves and it is reasonable to assume (linear damage assumption) that all the ΔD_j values are the same, meaning that the damage increment per cycle is given by

$$\Delta D_j = \frac{1}{N_f(S)} \qquad (12.131)$$

and that the function $D(t)$ increases linearly with time. If we further assume that eq (12.129) retains its validity even in the case of more complex nonperiodic excitations, the problem is to devise a scheme to appropriately count the number of cycles and to determine an incremental damage ΔD_j for each one of these cycles. For complicated time histories, one of the most commonly adopted cycle identification scheme is called the 'rainflow' method and can be found in specific literature (e.g. Dowling [12], Fuchs and Stephens [13] or Downing and Socie [14]).

However, if for simplicity we limit ourselves to narrow-band processes, the issue of cycles identification becomes relatively simple, we can use the results of the preceding sections and we are only left with the problem of appropriately choosing the quantities ΔD_j to be assigned to each cycle. In order to tackle this problem, we adopt the Palmgren–Miner hypothesis which generalizes eq (12.131) to the case of stress cycles with variable amplitude. In essence, the Palmgren–Miner hypothesis can be formulated as follows: if the jth cycle occurs at the level of stress at which – according to S–N curves – N_j cycles cause failure, then the jth increment of damage is given by

$$\Delta D_j = \frac{1}{N_j} \qquad (12.132)$$

In other words, if we group cycles of approximately equal amplitude together, we will have a situation in which we can identify n_1 cycles at the stress level at which N_1 cycles would cause failure, n_2 cycles at the stress level at which N_2 cycles would cause failure, etc. Then, each one of these groups will produce an incremental damage of n_i/N_i so that the failure condition (12.130) now reads

$$\sum_i \frac{n_i}{N_i} = 1 \qquad (12.133)$$

For a narrow-band process (for example, a resonant system subjected to a broad-band excitation) we noted in Section 12.6.2 that, on average, there will be $\nu_0^+ T$ cycles in time T. Moreover, since $p_p(S)dS$ represents the probability that a cycle will have a stress peak amplitude in the range between S and $S + dS$, we will have $\nu_0^+ T p_p(S)dS$ cycles within this amplitude range. Then, if $N_f(S)$ represents the number of cycles to failure at the stress level S, the incremental damage produced by these cycles will be, according to the Palmgren–Miner hypothesis

$$\frac{\nu_0^+ T p_p(S)dS}{N_f(S)}$$

so that, summing on all possible stress peak levels, the total damage in time T is given by

$$D(T) = \nu_0^+ T \int_0^\infty \frac{p_p(S)}{N_f(S)} dS \tag{12.134}$$

We can now determine the failure time t_f by means of eq (12.130), i.e.

$$\nu_0^+ t_f \int_0^\infty \frac{p_p(S)}{N_f(S)} dS = 1$$

which gives

$$t_f = \left(\nu_0^+ \int_0^\infty \frac{p_p(S)}{N_f(S)} dS \right)^{-1} \tag{12.135}$$

Clearly, the failure time given by eq (12.135) is intrinsically a statistical quantity in which two types of error are typically involved. The first type – the error due to the random nature of the excitation and response time histories – can be reasonably reduced by analysing sufficiently long time histories and by keeping the experiment within the limits of the high-cycle fatigue regime for which the S–N curves apply (low to moderate stress amplitudes and a structure which is not too lightly damped), while the second type – the error due to our limited knowledge of materials fatigue – cannot, for the moment, be eliminated and is probably the result of the various assumptions made to arrive at the desired result. In fact, we must remember that these assumptions, although reasonable and generally in agreement with experimental results, form the basis of a 'useful' model of fatigue damage rather than identifying a 'true' model of the physical phenomenon of fatigue.

12.7   Summary and comments

In the light of the preliminary results on probability and statistics given in Chapter 11, this chapter has considered the subject of random vibrations. Random vibrations arise in a number of situations in engineering practice. More specifically, when it is not possible to give a deterministic description of the vibratory phenomenon under investigation but repeated observations show some underlying patterns and regularities, we resort to a description in terms of statistical quantities and we speak of a 'random (or stochastic) process'. This is precisely the subject of Section 12.2, where we also note that, in practical situations, the engineer's representation of a random process is an a so-called 'ensemble', i.e. a number of sufficiently long time histories (samples) which can be used, by averaging across the ensemble at specific instants of time, to calculate (or better 'estimate') all the quantities of interest. Luckily, a large number of natural vibratory phenomena have – or can be reasonably assumed to have – some properties that allow a noteworthy simplification of the analysis. These properties are stationarity and ergodicity (Section 12.2.1). There exist different levels of stationarity and ergodicity but, broadly speaking, the first property has to do with the fact that certain statistical descriptors of the process do not change with time, while the second property refers to the circumstance in which a sufficiently long time record can be considered as representative of the whole process. Furthermore, ergodicity implies stationarity and, in practice, when there is evidence that a given process is stationary, ergodicity is also tacitly assumed so that we can (1) record only one (sufficiently long) time history and (2) describe the process by taking time averages along this single sample rather than calculating ensemble averages across a number of different samples, the two types of averages being equal because of ergodicity. It should be noted, however, that the assumption of ergodicity is, more often than not, an educated guess rather than a proven fact.

Just as deterministic vibrations can be analysed in the time domain or in the frequency domain, there is the possibility of doing the same with random vibrations. However, some complications of mathematical nature do arise and the problem is tackled by Fourier-transforming correlation functions rather than the time signal itself (Wiener–Khintchine relations). This procedure leads to the concept of spectral density, whose definition and properties are the subject of Section 12.3, and to the notions of narrow-band and wide-band random processes.

Then, with all the above results at our disposal, we can consider the problem of determining the (random) response of a (deterministic) linear system to a random stationary source of excitation. Proceeding in order of increasing complexity – one input and one output, one output and more than one input, MDOF and continuous systems – we do so in Sections 12.4 and 12.5, where we establish the fundamental input–output relationships for linear systems and note that, once again, the system's characteristics are represented in terms of IRFs in the time-domain FRFs in the frequency domain. Moreover, also in this case, there is the possibility of expressing the output characteristics in terms of modal IRFs and FRFs.

Also, in the final part of Section 12.4, we pay due attention to the fact that the steady-state condition – in which a stationary input produces a stationary output – is not reached immediately after the onset of the input, but some time has to pass before the system, so to speak, adjusts to its new state of motion. During this time, the response is clearly nonstationary because its statistical characteristics (typically, its mean value if different from zero and its variance) vary from zero to their stationary value.

Finally, in order to give the reader an idea of the richness of the subject of random vibrations, which is now a specialized field of activity and research in its own right, Section 12.6 deals with specific topics of particular interest. Sections 12.6.1 and 12.6.2 are strictly related and consider, respectively, the threshold crossing rates and peak distributions of stationary narrow-band processes, while Section 12.6.3 introduces some basic concepts of fatigue damage of engineering materials and gives a brief account of how, based on our limited knowledge of the details of material fatigue, we can attack the frequently encountered problem of fatigue damage due to random excitation. In this circumstance, when this excitation is in the form of a narrow-band random process, it is also shown how we can use the results of Sections 12.6.1 and 12.6.2 to estimate the mean time to failure.

References

1. International Standard ISO 4866-1990, *Mechanical Vibration and Shock – Vibration of Buildings – Guidelines for the Measurement of Vibrations and Evaluation of Their Effects on Buildings*.
2. Adhikari, R. and Yamaguchi, H., A study on the nonstationarity in wind and wind-induced response of tall buildings for adaptive active control, *Wind Engineering, Proceedings of the 9th Wind Engineering Conference*, Vol. 3, pp. 1455–1466, Wiley Eastern Ltd., New Delhi, 1995.
3. Papoulis, A., *Signal Analysis*, McGraw-Hill, New York, 1981.
4. Bendat, J.S. and Piersol, A.G., *Random Data – Analysis and Measurement Procedures*, 2nd edn, John Wiley, New York, 1986.
5. Lutes, L.D. and Sarkani, S., *Stochastic Analysis of Structural and Mechanical Vibrations*, Prentice Hall, Englewood Cliffs, NJ, 1997.
6. Vanmarcke, E.H., Properties of spectral moments with applications to random vibration, *Journal of the Engineering Mechanics Division*, ASCE, 98(EM2), 425–446, 1972.
7. Köylüoglu, H.U., *Stochastic Response and Reliability Analyses of Structures with Random Properties Subject to Stationary Random Excitation*, Ph.D. Dissertation, Princeton University, Jan. 1995.
8. Newland, D.E., *An Introduction to Random Vibrations, Spectral and Wavelet Analysis*, 3rd edn, Longman Scientific and Technical, 1993.
9. Rice, S.O., Mathematical analysis of random noise, *Bell System Technical*

Journal, **23** (1944) and **24** (1945); reprinted in Wax, N. (ed.) *Selected Papers on Noise and Stochastic Processes*, Dover, New York, 1954.

10. Sólnes, J., *Stochastic Processes and Random Vibrations: Theory and Practice*, John Wiley, New York, 1997.

11. ASTM Standard E468, *American Society for Testing and Materials, Annual Book of ASTM Standards*, E468-2, Section 3, Vol. 03.01, ASTM, Philadelphia, 1983, pp. 577–587.

12. Dowling, N.E., Fatigue failure predictions for complicated stress–strain histories, *Journal of Materials*, **7**(1), 71–87, 1972.

13. Fuchs, H.O. and Stephens, R.I., *Metal Fatigue in Engineering*, John Wiley, New York, 1980.

14. Downing, S.D. and Socie, D.F., Simple rainflow counting algorithms, *International Journal of Fatigue*, **4**(1), 31–40, 1982.

Further reading to Part I

Ainsworth, M., Levesley, J., Light, W.A. and Marletta, M. (eds) *Wavelets, Multilevel Methods and Elliptic PDEs*, Oxford Science Publications, Clarendon Press, Oxford, 1997.

Barton, G. *Elements of Green's Functions and Propagation: Potentials, Diffusion and Waves*, Clarendon Press, Oxford, 1989.

Boas, M.L. *Mathematical Methods in the Physical Sciences*, 2nd edn, John Wiley, New York, 1983.

Boswell, L.F. and D'Mello, C. *The Dynamics of Structural Systems*, Blackwell Scientific, Oxford, 1993.

Broman, A. *Introduction to Partial Differential Equations: from Fourier Series to Boundary-Value Problems*, Dover, New York, 1970.

Cercignani, C. *Spazio, Tempo, Movimento. Introduzione alla Meccanica Razionale*, (in Italian), Zanichelli, Bologna, 1976.

Champeney, D.C. *Fourier Transforms in Physics*, Adam Hilger, Bristol, 1985.

Chisnell, R.F. *Vibrating Systems*, Routledge & Kegan Paul, London, 1960.

Clough, R.W. and Penzien, J. *Dynamics of Structures*, McGraw-Hill, New York, 1975.

Diana, G. and Cheli F. *Dinamica e Vibrazioni dei Sistemi Meccanici*, Vols 1 and 2, Utet, Torino, 1993.

Genta, G. *Vibrazioni delle Strutture e delle Macchine*, Levrotto & Bella, Torino, 1996. (Also available in English, *Vibration of Structures and Machines*, Springer-Verlag, New York, 1993–5.)

Griffin, M.J. *Handbook of Human Vibration*, Academic Press, London, 1990.

Hartog, J.P.D. *Mechanical Vibrations*, Dover, New York, 1984.

Hewlett-Packard Application Note 243-3. *The Fundamentals of Modal Testing*, Hewlett-Packard Co., 1986.

IOtech, Inc. *Signal Conditioning and PC-based Data Acquisition Handbook*, 1997.

Ivchenko, G. and Medvedev, Yu.I. *Mathematical Statistics*, Mir Publishers, Moscow, 1990.

Kolmogorov, A.N. and Fomin, S.V. *Introductory Real Analysis*, Dover, New York, 1970.

Köylüoglu, H.U. *Theory and Applications of Structural Vibrations*, CIV 362 – Lecture Notes, Princeton University, 1995.

Landau, L.D. and Lifshitz, E.M. *Meccanica*, Editori Riuniti, Rome, 1982. (Also available in English: Landau and Lifshitz – *Course of Theoretical Physics*, Vol. 1, *Mechanics*, Pergamon Press.)

Lebedev, N.N., Skalskaya, I.P. and Uflyand, Y.S. *Worked Problems in Applied Mathematics*, Dover, New York, 1965.

Lembgrets, F. Parameter estimation in modal analysis, *L.M.S. Seminar on Modal Analysis*, Milan, 25–26 May, 1992.

Milton, J.S. and Arnold J.C. *Introduction to Probability and Statistics*, 2nd edn, McGraw-Hill, New York, 1990.

Mitchell, L.D. Modal test methods – quality, quantity and unobtainable, *Sound and Vibration*, Nov., 10–17, 1994.

Norton, M.P. *Fundamentals of Noise and Vibration Analysis for Engineers*, Cambridge University Press, Cambridge, 1989.

Ohayon, R. and Soize, C. *Structural Acoustics and Vibration*, Academic Press, London, 1998.

Pettofrezzo, A.J. *Matrices and Transformations*, Dover, New York, 1966.

Petyt, M. *Introduction to Finite Element Vibration Analysis*, Cambridge University Press, Cambridge, 1990.

Piersol, A.G. Optimum resolution bandwidth for spectral analysis of stationary random vibration data, *Shock and Vibration*, 1(1), 33–43, 1993–4.

Przemieniecki, J.S. *Theory of Matrix Structural Analysis*, Dover, New York, 1968.

Reddy, B.D. *Introductory Functional Analysis with Applications to Boundary Value Problems and Finite Elements*, Springer-Verlag, New York, 1998.

Richardson, M.H. Is it a mode shape, or an operating deflection shape? *Sound and Vibration*, Jan., 54–61, 1997.

Rudin, W. *Real and Complex Analysis*, McGraw-Hill, New York, 1966.

Rudin W. *Functional Analysis*, McGraw-Hill, New York, 1973.

Scavuzzo R.J. and Pusey H.C. *Principles and Techniques of Shock Data Analysis*, edited and produced by the Shock and Vibration Information Analysis Center (SAVIAC, Arlington, Virginia), SVM-16, 2nd edn, 1996.

Shephard, G.C. *Spazi Vettoriali di Dimensioni Finite*, Cremonese, Rome, 1969. (Also available in English: *Vector Spaces of Finite Dimension*, Oliver & Boyd Ltd.)

Smith, J.D. *Vibration Measurement and Analysis*, Butterworths, London, 1989.

Suhir, E. *Applied Probability for Engineers and Scientists*, McGraw-Hill, New York, 1997.

Thomsen, J.J. *Vibrations and Stability. Order and Chaos*, McGraw-Hill, New York, 1997.

Thomson, W.T. *Theory of Vibration with Applications*, 4th edn, Chapman & Hall, London, 1993.

Timoshenko, S., Young, D.H. and Weaver, W. Jr. *Vibration Problems in Engineering*, 4th edn, John Wiley, New York, 1974.

Towne, D.H. *Wave Phenomena*, Dover, New York, 1967.

Ventsel, E.S. *Teoria delle probabilitá*, Edizioni MIR, 1983.

Vu, H.V. and Esfandiari, R.S. *Dynamic Systems. Modeling and Analysis*, McGraw-Hill, New York, 1998.

Zaveri, K. *Modal Analysis of Large Structures – Multiple Exciter Systems*, Bruel & Kjaer, 1985.

Part II

Measuring instrumentation

Vittorio Ferrari

13 Basic concepts of measurement and measuring instruments

13.1 Introduction

The importance of making good measurements is readily understood when considering that the effectiveness of any analysis is strongly determined by the quality of the input data, which are typically obtained by measurement. Since analysis and processing methods cannot add information to the measurement data but can only help in extracting it, no final result can be any better than such data originally are.

With the intention of highlighting correct measurement practice, this chapter presents the fundamental concepts involved with measurement and measuring instruments. The first two sections on the measurement process and uncertainty form a general introduction. Then three sections follow which describe the functional model of measuring instruments and their static and dynamic behaviour. Afterwards, a comprehensive treatment of the loading effect caused by the measuring instrument on the measured system is presented, which makes use of the two-port models and of the electro-mechanical analogy. Worked out examples are included. Finally, a survey of the terminology used for specifying the characteristics of measuring instruments is given.

This chapter is intended to be propaedeutic and not essential to the next two chapters; the reader more interested in the technical aspects can skip to Chapters 14 and 15 regarding transducers and the electronic instrumentation.

13.2 The measurement process and the measuring instrument

Measurement is the experimental procedure by which we can obtain quantitative knowledge on a component, system or process in order to describe, analyse and/or exert control over it. This requires that one or more quantities or properties which are descriptive of the measurement object, called the **measurands**, are individuated. The measurement process then basically consists of assigning numerical values to such quantities or, more formally stated, of yielding measures of the measurands. This should be accomplished in both an empirical and objective way, i.e. based on experimental procedures and following rules which are independent of the observer. As a relevant consequence of the numerical nature of the measure of a quantity, measures can be used to express facts and relationships involving quantities through the formal language of mathematics.

The practical execution of measurements requires the availability and proper use of measuring instruments. A measuring instrument has the ultimate and essential role of extending the capability of the human senses by performing a comparison of the measurand against a reference and providing the result expressed in a suitable measuring unit. The output of a measuring instrument represents the measurement signal, which in today's instruments is most frequently presented in electrical form.

The process of comparison against a reference may be direct or, more often, indirect. In the former case, the instrument provides the capability of comparing the unknown measurand against reference samples of variable magnitude and detecting the occurrence of the equality condition (e.g. the arm-scale with sample masses, or the graduated length ruler). In the latter case, the instrument's functioning is based on one or more physical laws and phenomena embodied in its construction, which produce an observable effect that is related to the measurand in a quantitatively known fashion (e.g. the spring dynamometer).

The indirect comparison method is often the more convenient and practicable one; think, for instance to the case of measurement of an intensive quantity such as temperature. Motion and vibration measuring instruments most frequently rely on an indirect measuring method.

Regardless of whether the measuring method is direct or indirect, it is fundamental for achieving objective and universally valid measures that the adopted references are in an accurately known relationship with some conventionally agreed standard. Given a measuring instrument and a standard, the process of determination and maintenance of this relationship is called **calibration**. A calibrated and properly used instrument ensures that the measures are traceable to the adopted standard, and they are therefore assumed to be comparable to the measures obtained by different instruments and operators, provided that calibration and proper use is in turn guaranteed.

If we refer back to the definition of measurement, it can be recognized that measurement is intrinsically connected with the concept of information. In fact, measuring instruments can be thought of as information-acquiring machines which are required to provide and maintain a prescribed functional relationship between the measurand and their output [1]. However, measurement should not be considered merely as the collection of information from the real world, but rather as the extraction of information which requires understanding, skill and attention from the experimenter. In particular, it should be noted that even the most powerful signal postprocessing techniques and data treatment methods can only help in retrieving the information embedded in the raw measurement data, but have no capability of increasing the information content. As such, they should not be misleadingly regarded as substitutive to good measurements, nor a fix for poor measurement data. Therefore, carrying out good measurements is of primary importance and should be considered as an unavoidable need and prerequisite to any further analysis. A fundamental limit to the achievable knowledge on the measurement object is posed at this stage, and there is no way to overcome such a limit in subsequent steps other than by performing better measurements.

13.3 Measurement errors and uncertainty

After realizing the importance of making good measurements as a necessary first step, we may want to be able to determine when measurements are good or, at least, satisfying to our needs. In other words, we become concerned with the problem of qualifying measurement results on the basis of some quantifiable parameter which characterizes them and allows us to assess their reliability. We are essentially interested in knowing how well the result of the measurement represents the value of the quantity being measured. Traditionally, this issue has been addressed by making reference to the concept of measuring **error**, and error analysis has long been considered an essential part of measurement science.

The concept of error is based on the reasonable assumption that a measurement result only approximates the value of the measurand but is unavoidably different from it, i.e. it is in error, due to imperfections inherent to the operation in nonideal conditions. Blunders coming from gross defects or malfunctioning in the instrumentation, or improper actions by the operator are not considered as measuring errors and of course should be carefully avoided.

In general, errors are viewed to have two components, namely, a **random** and a **systematic** component. Random errors are considered to arise from unpredictable variations of influence effects and factors which affect the measurement process, producing fluctuations in the results of repeated observation of the measurand. These fluctuations cancel the ideal one-to-one relationship between the measurand and its measured value. Random errors cannot be compensated for but only treated statistically. By increasing the number of repetitions, the average effect of random errors approaches zero or, more formally stated, their expectation or expected value is zero.

Systematic errors are considered to arise from effects which influence the measurement results in a systematic way, i.e. always in the same direction and amount. They can originate from known imperfections in the instrumentation or in the procedure, as well as from unknown or overlooked effects. The latter sources in principle always exist due to the incompleteness of our knowledge and can only be hopefully reduced to a negligible level. Conversely, the former sources, as they are known, can be compensated for by applying a proper correction factor to the measurement results. After the correction, the expected value of systematic errors is zero.

Although followed for a long time, the approach based on the concept of measurement error has an intrinsic inconsistency due to the impossibility of determining the value of a quantity with absolute certainty. In fact, the **true value** of a quantity is unknown and ultimately unknowable, since it could only be determined by measurement which, in turn, is recognizably imperfect and can only provide approximate results. As a consequence, the measurement error is unknowable as well, since it represents the deviation of the measurement result from the unknowable true value. As such, the concept of error can not provide a quantitative and consistent mean to qualify measurement results on a theoretically sound basis.

As a solution to the problem, a different approach has been developed in the last few decades and is currently adopted and recommended by the international metrological and standardization institutions [2]. It is based on recognizing that when performing a measurement we obtain only an estimate of the value of the measurand and we are uncertain on its correctness to some extent. This degree of uncertainty is, however, quantifiable, though we do not know precisely how much we are in error since we do not know the true value. The term measurement **uncertainty** can be therefore introduced and defined as the parameter that characterizes the dispersions of the values that could be attributed to the measurand. In other words, the uncertainty is an estimate of the range of values within which the true value of a measurand lies according to our presently available knowledge. Therefore uncertainty is a measure of the 'possible error' in the estimated value of a measurand as obtained by measurement.

It is worth noting that the result of a measurement can unknowably be very close to the value of the measurand, hence having a small error, nonetheless it may have a large uncertainty. On the other hand, even when the uncertainty is small there is no absolute guarantee that the error is small, since some systematic effect may have been overlooked because it is unknown or not recognized and, as such, not corrected for in the measurement result.

From this standpoint, a different meaning can be attributed to the term true value in which the adjective 'true' loses its connotation of uniqueness and becomes formally unnecessary. The true value, or simply the value, of a measurand can be conventionally considered as the value obtained when the measurement with lowest possible uncertainty according to the presently available knowledge is performed, i.e. when an exemplar measuring method which minimizes and corrects for every recognized influencing effect is used.

In practical cases, the idea of an exemplar method should be commensurate with the accuracy needed for the particular application; for instance, when we measure the length of a table with a ruler we consciously disregard the influence of temperature on both the table and the ruler, since we consider this effect to be negligible for our present measuring needs. We simply acknowledge that our result has an uncertainty which is higher than the best obtainable, but is suitable for our purposes.

However, we may be in the situation of negligible uncertainty of the instrument (the ruler in this case) compared to that caused by temperature on the measurement object (the table), for which we are therefore able to detect and measure the thermal expansion. The converse situation is that of negligible uncertainty of the measurement object compared to that of the measuring instrument and procedure. This is the case encountered when testing an instrument by using a reference or standard of low enough uncertainty to be ignored. Thus the value of the reference or standard can be conventionally assumed as the true value, and the test thought of as a mean to determine the errors of the measuring instrument and procedure. Quantifying such errors and correcting those due to systematic effects is actually no different from performing a calibration of the measuring instrument under test.

Summarizing, the introduction of the concept of uncertainty removes the inconsistency of the theory of errors, and directly provides an operational mean for characterizing the validity of measurement results. In practice, there are many possible sources of uncertainty that, in general, are not independent, for example: incomplete definition of the measurand, effect of interfering environmental conditions and noise, inexact calibration and finite discrimination capability of measuring instruments and variations in their readings in repeated observations under apparently identical observations, unconscious personal bias in the operation of the experimenter.

In principle, the influence of each conceivable source of uncertainty could be evaluated by the statistics of repeated observations. In the practical cases this is essentially impossible and, therefore, many source of uncertainty can be more conveniently quantified *a priori* by analysing with scientific judgment the pool of available information, such as tabulated data, previous measurement results, instrument specifications. The results of the two evaluation methods are called respectively type A and type B uncertainties, which are classified as different according to their derivation but do not differ in nature and, therefore, are directly comparable. A detailed treatment of the methods used to evaluate uncertainty can be found in [2] and [3].

13.4 Measuring instrument functional model

Irrespective of the measured variable and the operating principle involved, a measuring instrument can be represented by the block diagram of Fig. 13.1. This is a simplified and general model which focuses on the very fundamental features that, with various degrees of sophistication in the implementation, are typical of every measuring instrument.

The measuring instrument can be seen as composed of three cascaded blocks, which provide an information transfer path from the measurand quantity to the observer. The first block, named the sensing element, is the

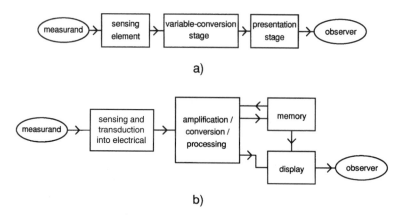

Fig. 13.1 Functional model of (a) a measuring instrument and (b) an electronic measuring system.

stage being in contact with the measurand and interacting with it in order to sense its value. This interaction should be perturbing as little as possible so that negligible load is produced by the instrument on the measured object, as discussed in Section 13.7. The output of the sensing element is in the form of some physical variable which is in a known relationship with the measurand. If we take, for example, a mercury glass thermometer, than the sensing element is constituted by the mercury, and its output is the thermal expansion of the fluid volume in the bulb. As we shall see, in electronic instruments and systems the sensing element function is performed by sensors and transducers.

The second block, named the variable-conversion stage, accepts the output of the sensing element and converts it into another variable and/or manipulates it with the general aim of obtaining a representation of the signal more suitable to its presentation, yet preserving the original information content. In our example of the glass thermometer the variable-conversion stage is the capillary tube that transduces the volume expansion into the elongation of the fluid column.

The third block, named the presentation stage, undertakes the final translation of the measurement signal into a form which is perceived and understood by humans, once again preserving the original information content. The role of this stage is straightforward, but its importance should not be overlooked. In fact, the degree of discrimination between closely spaced values of the measurand that an instrument allows, i.e. the resolution, is strongly related, among other factors, to the design and construction of its presentation stage. This can be readily recognized if we think at our glass thermometer for which the presentation stage is the gridmark pattern on the capillary tube. Although the mercury expansion is a continuous function of temperature, the discrete spacing of the gridmarks enables discrimination no better than 0.1 °C to the naked eye, which is, nevertheless, all that is needed in many applications.

It should be observed that the distinction between functional blocks does not necessarily reflect a physical separation of such blocks in the real instruments. On the contrary, there are many cases in which several functions are somewhat distributed among different pieces of hardware so that it is difficult, besides essentially useless, to distinguish and parse them.

Nowadays, most of the measurement tasks in any field are performed by instruments and systems which measure physical quantities by electronic means. Basically, the use of electronics in measuring instrumentation offers higher performance, improved functionality and reduced cost compared to purely mechanical systems. A very general block representation of an electronic measuring instrument or system is given in Fig. 13.1(b), which is fairly similar to that of Fig. 13.1(a) with some important differences. In this case the sensing function is performed by sensors, or transducers, which respond to the physical stimulus caused by the measurand with a corresponding electrical signal. Such a signal is then amplified, possibly converted into a digital format and processed in order to extract the information of interest contained in the sensor signal, and filter out the unwanted spurious components. All such processing operations are carried out in the electrical domain irrespective of the nature of the measurand, and therefore they may take advantage of the high capabilities of modern electronic elaboration circuitry.

The obtained results can then be presented to the observer through a display stage, and/or possibly stored into some form of memory device, most typically electronic or magnetic. The memory storage capability offered by many electronic measuring instruments is of fundamental importance, as it enables analysis, processing and comparisons on measurement data to be performed offline, that is, arbitrarily later than the time when the data are captured. Some instruments are optimized for extremely fast cycles of data storage–retrieval–processing so that they can perform specialized functions, such as filtering, correlation or frequency transforms, in real time, i.e. with a delay inessential for the particular application.

Transducers and electronic signal amplification and processing will be treated in Chapters 14 and 15 respectively.

A fundamental fact resulting from both block diagrams of Fig. 13.1 is that the measuring instrument occupies the position at the interface between the observer and the measurand. Moreover, all of them are under the global influence of the surrounding environment. This influence is generally a cause of interference on the information transfer path from the measurand to the observer, producing a perturbing action which ultimately worsens the measurement uncertainty. This fact may be represented by considering the output y of a measuring instrument being a function not only of the measurand x, as we ideally would like to happen, but also of a number of further quantities q_i related to the effects of the boundary conditions. Such quantities are named the **influencing** or **interfering quantities**. Typical influencing quantities may be of an environmental nature, such as temperature, barometric pressure and humidity, or related to the instrument operation, such as posture, loading conditions and power supply.

Besides observing that y, x and the quantities q_i are actually functions of time $y(t)$, $x(t)$ and $q_i(t)$, we may even consider time itself as an influencing quantity, since in the most general case the output of a real measuring instrument depends to some extent on the time t at which the measurement is performed. This means that the same input combination of measurand and influencing quantities applied at different time instants of the instrument's operating life may, in general, produce different output values due to instrument ageing and drift. Considered as an influencing quantity, time has a peculiar nature due to the fact that, unlike what theoretically can be done for the q_is, the observer cannot exert any kind of control over it.

Developing a formal description of measuring instruments which globally takes into account all the involved variables as functions of time with the aim of deriving the time evolution of the output is a difficult task. Usually, a more practicable approach is followed which, besides, provides a better understanding of the instrument performances and a deeper insight into its operation. It consists of distinguishing between static and dynamic behaviour, each of which can be analysed separately. Operation under static conditions can be analysed by neglecting the time dependence of the measurand and the influencing quantities, therefore avoiding the solution of complicated partial-derivative differential equations. The consequent reduction in complexity enables a detailed description of the output-to-measurand relationship and the evaluation of the impact due to influencing quantities.

On the other hand, the analysis of dynamic operation is essentially performed by taking into account the time evolution of the measurand only and the resultant time dependence of the instrument output, thereby requiring only ordinary differential equations. The effect of the influencing quantities on dynamic behaviour is generally evaluated by a semiquantitative extension of the results obtained for the static analysis. Though this approach it is not strictly rigorous, it offers a viable solution to an otherwise unmanageable problem and, as such, it is of great practical utility.

13.5   Static behaviour of measuring instruments

Let us assume that the measurand x and the influencing quantities q_is are constant and independent of time. It should be noted that this assumption is not in contradiction with regarding x and the q_is as variables. In fact, we consider that the x and the q_is are subject to variations over a range of values, but we do not take into account the time needed by such variations to take place. In other words, we consider only the static combinations of constant inputs once the transients have died out. Under such an assumption, the relationship between the instrument output y and the measurand x, the q_is and the time t at which the measurement is performed is given by the following expression:

$$y = f_g(x, q_1, \ldots, q_i, \ldots, t) \tag{13.1}$$

where f_g is a function which defines the global conversion characteristic of the measuring instrument.

The differential of y is given by

$$dy = \frac{\partial f_g}{\partial x} dx + \frac{\partial f_g}{\partial q_1} dq_1 + \cdots + \frac{\partial f_g}{\partial q_i} dq_i + \cdots + \frac{\partial f_g}{\partial t} dt \tag{13.2}$$

The quantities $\partial f_g/\partial x$, $\partial f_g/\partial q_i$ and $\partial f_g/\partial t$ represent the **sensitivities** of the measuring instrument in response to the measurand x, the ith influence quantity q_i and the time t. The term $\partial f_g/\partial t$ is responsible for the time stability of the conversion characteristic or, better, of its instability. Higher values of $\partial f_g/\partial t$ imply a more pronounced ageing effect on the instrument and require a more frequent calibration. An instrument for which $\partial f_g/\partial t = 0$ is called **time-invariant**.

The instrument is the more selective for x the lower the value of the terms $\partial f_g/\partial q_i$ are compared to $\partial f_g/\partial x$, so that their effect on the output is negligible with respect to the measurand. If all the terms $\partial f_g/\partial q_i$ were ideally zero, the instrument would respond to the measurand only and would be called **specific** for x. In the real cases, given the desired level of accuracy and estimated the ranges of variability of x and the q_is, the comparison between $\partial f_g/\partial x$ and the $\partial f_g/\partial q_i$ allows us to determine the influence quantities which actually play a role and need to be taken into account in the case at hand.

In principle, the contribution of the significant influence quantities could be experimentally evaluated by varying each of them in turn over a given interval, while keeping the measurand and the other q_is constant and monitoring the instrument output. In practice, this is hardly possible and usually the contribution is estimated partly from experimental data and partly from theoretical predictions.

Of course, it is expected that the instrument is mostly responsive to the measurand x, and, therefore, the above procedure is primarily applied to the experimental determination of the measurand-to-output relationship. The curve obtained in this way is the static **calibration** or **conversion characteristic** of the instrument under given conditions of the influencing quantities. Under varying conditions, a family of calibration characteristics is obtained, which contain information on the impact of the considered q_is.

Assuming a reference condition for which the influencing quantities are kept constant at their nominal or average values q_{oi}, and ageing effects are neglected, it follows that the output y depends on the measurand only and eq (13.1) reduces to

$$y = f(x) \tag{13.3}$$

The function f represents the instrument's static conversion characteristic in the reference condition. For the instrument to be of practical utility, $f(x)$ should be monotonic so that its inverse function, which relates the instrument reading with the measurand, is single-valued.

The term $\partial f / \partial x = S$ is called the **sensitivity** of the instrument with respect to the measurand x. In general, the sensitivity is not constant throughout the measurand range but is itself a function of x, i.e. $S = S(x)$. In most cases, however, the instrument is built to ensure a relationship of proportionality between y and x of the type $y = f(x) = kx + y_0$. In these cases the instrument is said to be **linear** if $y_0 = 0$ and **incrementally linear** if $y_0 \neq 0$, and the sensitivity S becomes a constant given by the coefficient k, which is typically called the instrument **scale factor, calibration factor** or **conversion coefficient**. The term y_0 is called the instrument offset and represents the output at zero applied measurand.

Figure 13.2 shows the conversion characteristics for both an incrementally linear and a nonlinear instrument. For incrementally linear instruments, the variations in the coefficients k and y_0 about their reference values induced by the influencing quantities are generally adopted to specify their effect. Taking temperature as an example, we may therefore find widespread usage of the terms temperature coefficient of the scale factor and of the offset, meaning the temperature-induced variations in k and y_0 respectively.

It is very important to point out that nonlinear instruments may be linearized by considering a small interval of the input x about an average value x_0, and approximating dy with $S(x_0)dx$ in such an interval. For suitably small variations around x_0 the sensitivity can therefore assumed to be constant equal to $S(x_0)$ and the instrument considered as locally linear. This procedure is the so called small-signal linearization.

The property of linearity is extremely important for measuring instruments, as it is for every system, since it implies the validity of the superposition principle. Essentially, this means that a linear system responds to the sum of two inputs with an output which is the sum of the two single responses caused by each input when applied alone. As a consequence, linear systems, and linear instruments in particular, produce an output which is a scaled replica of the input, i.e. the readings of the instrument provide an undistorted image of the measurand variations.

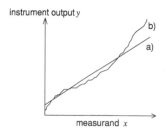

Fig. 13.2 Examples of (a) incrementally linear and (b) nonlinear conversion characteristics.

It is worth noting that for an instrument to have a linear conversion characteristic there is no need that each of the blocks of Fig. 13.1 is linear. In fact, this is only a sufficient condition for overall linearity, and we may as well have several blocks with nonlinear behaviours which mutually cancel, giving rise to a globally linear instrument. This property is very often exploited when input or intermediate stages are intrinsically nonlinear, and such a nonlinearity is compensated for within an additional conversion stage or even within the presentation stage. As an example, you may think of an instrument that, to correct a nonlinearity of some intermediate stages, uses a needle indicator whose reading scale has unequally spaced marks, as happens in logarithmic paper. Of course, an unfortunate drawback of this expedient is the possible reduction of the indicator readability in some parts of its range.

13.6 Dynamic behaviour of measuring instruments

Let us consider that the measurand x is actually a function of time $x(t)$ and assume that the effect of the influencing quantities is negligible. Besides, suppose that time ageing and drift phenomena are absent or, as generally happens, very slow compared to the time evolution of the measurand signal, so that they can be overlooked and the instrument considered as time-invariant.

Then we may rewrite eq (13.3) which now takes the form

$$y(t) = F[x(t)] \qquad (13.4)$$

F is conceptually different from f in eq (13.3), since F is an operator, in the sense that it represents a correspondence between entities which are themselves functions of time and not scalar values as for f. In eq (13.3) F defines the dynamic conversion characteristic of the measuring instrument and generally contains time derivatives and integrals of both $x(t)$ and $y(t)$, giving rise to integrodifferential nonlinear equations.

We restrict the field of the many mathematical forms that F can take, by assuming that it has the property of linearity and, therefore, we limit ourselves to considering linear instruments.

Briefly, a linear dynamic system, and an instrument in particular, is one for which the superposition principle is valid when input and output, respectively considered as cause and effect, are regarded as functions of time.

It is worth pointing out that the linearity of F, which could be indicated as dynamic linearity, is not equivalent to the linearity of f, that is the static linearity described in the preceding section. In fact, they refer to two different ideas of the concept of linearity, namely operational in the former case and functional in the latter. Indeed, the dynamic linearity is a more restrictive condition than static linearity. That is, we may have a system for which the superposition principle holds for constant values of the input, and, on the contrary, does not apply when the input is considered as a function of time.

For example, a system for which the input–output relationship is given by $y(t) = F[x(t)] = ax^2 dx/dt + bx$ is not linear in the dynamical sense, though it is statically linear, since for x independent of time the output becomes $y = bx$. Conversely, dynamic linearity implies static linearity.

For a time-invariant dynamically linear instrument for which input and output are real functions of time, eq (13.4) takes the form of a linear ordinary differential equation with constant coefficients, which can be generally written as

$$a_n \frac{d^n y(t)}{dt^n} + a_{n-1} \frac{d^{n-1} y(t)}{dt^{n-1}} + \cdots + a_0 y(t)$$

$$= b_m \frac{d^m x(t)}{dt^m} + b_{m-1} \frac{d^{m-1} x(t)}{dt^{m-1}} + \cdots + b_0 x(t) \quad (13.5)$$

The coefficients a_i and b_i are a combination of instrument parameters assumed to be independent of time, and are therefore real and constant numbers.

Equations of the form of eq (13.5) are encountered in a wide number of fields of engineering and science, and standard methods have been developed for their solution. We will not go into details about this aspect, on which the interested reader can find many exhaustive references, such as [4]. We would rather like to point out the main lines of reasoning that can be followed to approach the problem, and illustrate the modelling of measuring instruments as dynamic systems [5, 6].

The first approach is that of directly solving eq (13.5) in the time domain. It is well known that the general form of the solution $y(t)$ is

$$y(t) = y_f(t) + y_i(t) \tag{13.6}$$

where $y_f(t)$ is the forced response, and $y_i(t)$ is the free response determined by the initial conditions. In turn, $y_f(t)$ is the sum of a steady-state term $y_{fS}(t)$ and a transient term $y_{fT}(t)$.

The time-domain approach becomes rather complex unless low-order systems with simple input functions are considered, and is therefore of limited practical utility. Instead, it is very fruitful to take advantage of the property of linearity and the consequent validity of superposition principle. The generic input $x(t)$ can be decomposed as a finite or infinite sum of elementary functions for which eq (13.5) simplifies to a set of readily solvable algebraical equations. The solutions of such equations are then summed to produce the overall response $y(t)$ to the original stimulus $x(t)$. Depending on the type of the elementary functions used as a decomposition basis, either complex exponentials $e^{i\omega t}$ or damped complex exponentials $e^{(\alpha + i\omega)t}$ with α real, the above procedure leads to the methods of Fourier and Laplace transform respectively.

In the Fourier transform method, solving eq (13.5) in the time domain becomes equivalent to solving the following complex algebraical equation in the frequency domain

$$Y(\omega) = \frac{b_m(i\omega)^m + b_{m-1}(i\omega)^{m-1} + \cdots + b_0}{a_n(i\omega)^n + a_{n-1}(i\omega)^{n-1} + \cdots + a_0} X(\omega) = T(\omega)X(\omega) \qquad (13.7)$$

where $X(\omega)$ and $Y(\omega)$ are complex functions of the angular frequency ω called the Fourier (or $\mathcal{F}$-) transform of $x(t)$ and $y(t)$, given by

$$X(\omega) = \int_{-\infty}^{+\infty} x(t)e^{-i\omega t} dt \qquad (13.8a)$$

$$Y(\omega) = \int_{-\infty}^{+\infty} y(t)e^{-i\omega t} dt \qquad (13.8b)$$

and $T(\omega)$ is called the **frequency**, or sinusoidal, **response function** of the system. For a given angular frequency ω, $T(\omega)$ is a complex number whose magnitude and argument respectively represent the gain and phase shift between the sinusoidal input of angular frequency ω and the corresponding sinusoidal output.

The Fourier transform method can be applied to the class of functions of time for which the $\mathcal{F}$-transform exists, i.e. the integral given in eq (13.8) converges. In the most general case, such functions are suitably regular nonperiodic functions with their $\mathcal{F}$-transform being nonzero over a continuous spectrum of frequencies. A subset of such functions is represented by the periodic functions, for which the integral of eq (13.8) becomes a summation over a discrete spectrum of frequencies and the $\mathcal{F}$-transform becomes the series of Fourier coefficients. The method of analysis based on the expression of periodic functions of time as Fourier series is called the harmonic analysis.

In the Laplace transform method, damped complex exponentials are used as the elementary functions constituting the decomposition basis, thereby extending the transform method to functions which are not $\mathcal{F}$-transformable but, nevertheless, have great practical importance, such as the linear ramp and exponential functions. Again, solving eq (13.5) in the time domain becomes equivalent to solving the following complex algebraical equation in the domain of the complex angular frequency $s = \alpha + i\omega$:

$$Y(s) = \frac{b_m s^m + b_{m-1}s^{m-1} + \cdots + b_0}{a_n s^n + a_{n-1}s^{n-1} + \cdots + a_0} X(s) = T(s)X(s) \qquad (13.9)$$

where $X(s)$ and $Y(s)$ are complex functions being the Laplace (or $\mathcal{L}$-) transforms of $x(t)$ and $y(t)$ given by

$$X(s) = \int_{-\infty}^{+\infty} x(t)e^{-st}dt \qquad (13.10a)$$

$$Y(s) = \int_{-\infty}^{+\infty} y(t)e^{-st}dt \qquad (13.10b)$$

and the complex function $T(s)$ is called the **transfer function** of the system. As the $\mathcal{L}$-transform of the impulse function $\delta(t)$ is unity, $T(s)$ is the $\mathcal{L}$-transform of the system response when subject to an impulsive stimulus, sometimes called a ballistic excitation.

As can be seen by comparing eqs (13.8) with eqs (13.10), the $\mathcal{L}$-transform is a generalization of the $\mathcal{F}$-transform based on substituting ω with $s = \alpha + i\omega$, where α is such that the integral converges. The Laplace transform method offers the desirable advantage that it takes into account the initial conditions in a consistent way, thereby being a powerful tool for dealing with transient problems.

The use of the transform method enables us to describe the dynamic behaviour of a linear instrument by simply analysing its frequency response function or its transfer function. In turn, they may be derived by defining elementary blocks which compose the instrument and properly combining the respective $T(\omega)$ or $T(s)$ of such blocks. As a rule, the frequency response or transfer function of cascaded blocks is the product of the individual $T(\omega)$ or $T(s)$. As a result a block representation of the instrument can be obtained, which is shown in Fig. 13.3 for both the $\mathcal{F}$- and $\mathcal{L}$-transforms.

It is important to point out that several relevant features of the system under consideration can be analysed directly in the frequency domain by using $T(\omega)$ and $T(s)$, without the need to formulate the problem in the time domain, thereby avoiding the related difficulties. $T(\omega)$ and $T(s)$ can be experimentally measured by monitoring the outputs generated by swept-sine and impulse inputs respectively. In practice, it is sometimes preferable to use a step excitation in place of the impulse, which may be more difficult to generate. Since the unitary step function $1(t)$ is the integral of the impulse $\delta(t)$, if the

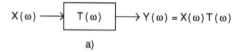

X(ω) $\longrightarrow$ T(ω) $\longrightarrow$ Y(ω) = X(ω) T(ω)

a)

X(s) $\longrightarrow$ T(s) $\longrightarrow$ Y(s) = X(s) T(s)

b)

Fig. 13.3 Block-diagram representation of a measuring instrument in the (a) Fourier and in (b) Laplace domains.

system is linear the step response can be derived to obtain the impulse response and, thereafter, the transfer function $T(s)$.

Three main models of measuring instruments can be distinguished, which differ in their dynamic behaviour according to the degree n of the denominator of their respective transfer functions, alternatively seen as the order of the differential equation in the time domain. They are the **zeroth-**, **first-** and **second-order** instrument models.

For a zeroth-order instrument the input–output relationship in the time domain is given by

$$a_0 y(t) = b_0 x(t) \tag{13.11}$$

which in the s-domain becomes

$$Y(s) = \frac{b_0}{a_0} X(s) \tag{13.12}$$

We can observe that, both in time- and frequency-domain representations, the output is proportional to the input. This means that $y(t)$ instantly follows $x(t)$ whatever its time evolution is, differing from it only by the scale factor $k = b_0/a_0$ which represents the instrument sensitivity. In particular, the step response of a zeroth-order instrument is a step function itself as shown in Fig. 13.4, and the sinusoidal frequency response $T(\omega)$ is flat throughout the frequency axis (Fig. 13.5). An example of a zeroth-order instrument is a resistive potentiometer displacement transducer.

For a first-order instrument the input-output relationship in the time domain is given by

$$a_1 \frac{dy(t)}{dt} + a_0 y(t) = b_0 x(t) \tag{13.13}$$

which in the s-domain becomes

$$Y(s) = \frac{b_0}{a_1 s + a_0} X(s) = \frac{b_0}{a_0} \frac{1}{\frac{a_1}{a_0} s + 1} X(s) \tag{13.14}$$

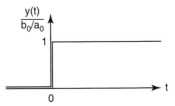

Fig. 13.4 Step response of a zeroth-order instrument.

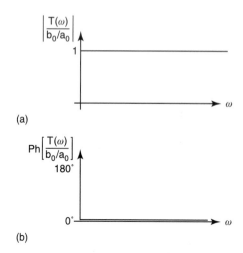

(a)

(b)

Fig. 13.5 Frequency response of a zeroth-order instrument: (a) magnitude; (b) phase.

The static sensitivity is given by b_0/a_0. Considering the expression of the transfer function which has a pole at $s = -a_0/a_1$ and applying the Laplace method, the step response can be determined. As shown in Fig. 13.6, the output $y(t)$ is a rising exponential function with a time constant τ given by a_1/a_0, and a steady-state value given by b_0/a_0. Therefore, the output lags the input when abrupt variations take place, while it reaches the steady-state value after a characteristic response time (after 5τ the output is at 99% of its final value).

The frequency response function $T(\omega)$ is plotted in Fig. 13.7, showing the existence of a cutoff or corner frequency $\omega_c = 1/\tau$ at which the magnitude is attenuated of $1/\sqrt{2} = -3$ dB compared to the low-frequency value $|b_0/a_0|$, and after which it becomes to decrease with an asymptotic log–log slope of -20 dB/decade. For $b_0/a_0 > 0$ the phase shift is zero at $\omega = 0$, $-45° = -\pi/4$

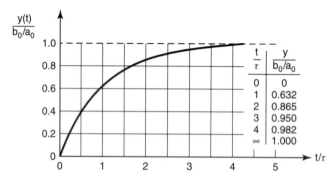

Fig. 13.6 Step response of a first-order instrument ([6, p. 177], reproduced with permission).

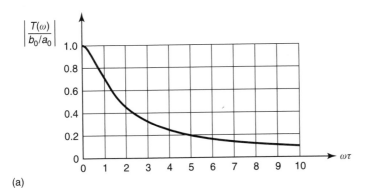

(a)

(b)

Fig. 13.7 Frequency response of a first-order instrument: (a) magnitude; (b) phase ([6, p. 122], reproduced with permission).

at $\omega = \omega_c$ and it asymptotically reaches $-90° = -\pi/2$ for $\omega \to \infty$. The system then behaves as a first-order low-pass filter with a -3 dB bandwidth extending from DC, i.e. zero frequency, to ω_c. An example of a first-order instrument is a thermometer with a finite thermal resistance and capacitance.

For a second-order instrument the input–output relationship in the time domain is given by

$$a_2 \frac{d^2y(t)}{dt^2} + a_1 \frac{dy(t)}{dt} + a_0 y(t) = b_0 x(t) \tag{13.15}$$

which in the s-domain becomes

$$Y(s) = \frac{b_0}{a_2 s^2 + a_1 s + a_0} X(s) = \frac{b_0}{a_0} \frac{1}{\dfrac{a_2}{a_0} s^2 + \dfrac{a_1}{a_0} s + 1} X(s) \tag{13.16}$$

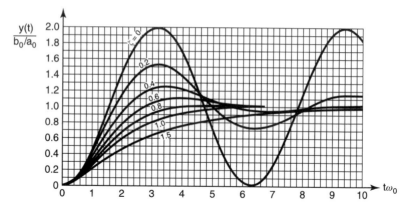

Fig. 13.8 Step response of a second-order instrument ([6, p. 129], reproduced with permission).

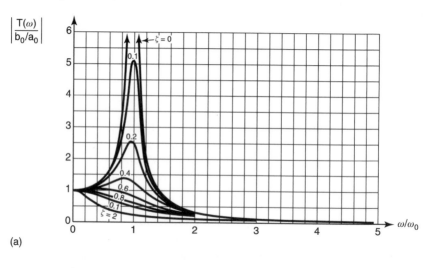

(a)

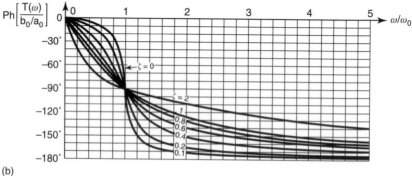

(b)

Fig. 13.9 Frequency response of a second-order instrument: (a) magnitude. (b) phase ([6, p. 135], reproduced with permission).

Second-order dynamic systems were extensively treated in Chapter 4, and therefore we simply limit ourselves to few considerations here.

The static sensitivity is given by b_0/a_0. Figure 13.8 shows the step response which, depending on the damping ratio $\zeta = a_1/2\sqrt{a_0 a_2}$, follows a monotonic or oscillatory trend toward the steady-state value b_0/a_0. The ringing approaches a constant amplitude oscillation at $\omega_0 = \sqrt{a_0/a_2}$ when the damping ratio ζ tends to zero. ω_0 is called the undamped natural frequency.

The frequency response function $T(\omega)$ is plotted in Fig. 13.9. For ω around ω_0 the magnitude of $T(\omega)$ starts decreasing from its low-frequency value $|b_0/a_0|$ with a log–log slope of -40 dB/decade. The damping ratio ζ determines the amount of peaking and the steepness of the phase curve in the region around ω_0. For $\zeta = 1$ the magnitude attenuation at $\omega = \omega_0$ is equal to $\frac{1}{2} = -6$ dB.

For $b_0/a_0 > 0$ the phase shift is zero at $\omega = 0$, $-90° = -\pi/2$ at $\omega = \omega_0$ and it asymptotically reaches $-\pi$ for $\omega \to \infty$. The system then behaves as a second-order low-pass filter with a bandwidth extending from DC to ω_0, with the extension of the flatness region, which depends on the amount of damping. A typical example of a second-order instrument is a seismic accelerometer.

13.7 Loading effect

13.7.1 *General considerations*

Measurement is a transfer of information which implies an exchange of energy. In fact, the measuring instrument interacts with the measurement object and unavoidably perturbs it by altering its energetic equilibrium. As a consequence of this perturbation, we are fundamentally unable to determine the value of a quantity without simultaneously modifying it due to the action of measuring. This effect, which is very important to understand and take into account, is named the **loading effect**, as the measuring instruments loads the measurement object. As will be illustrated shortly, if all relevant quantities were exactly known the loading effect could be evaluated and therefore compensated for as a systematic effect. In practice this is not generally feasible, and it is a better practice to reduce loading as much as possible, in order to minimize its quantitative impact.

In general, the perturbing actions caused by the instrument on the measurement object can be divided into two categories. The first category comprises those interactions which alter the measurand value and globally modify the conditions of the system under measurement, which thereby assumes a different configuration. For example, if a flowmeter is inserted into a tube to measure fluid velocity, the presence of the meter not only alters the local value of the flow but also distorts the overall flow field producing a global effect.

The case is different for the second category of interactions which can be exemplified, for instance, by a mass–spring–damper vibrating system to which an accelerometer is attached. It can be readily realized that the accelerometer adds its mass to the vibrating mass, causing a loading action which affects both the amplitude and the frequency of vibration.

The difference between the two categories of interactions lies in the fact that the former comprises situations typical of distributed-parameter systems, whose analysis requires the solution of partial-derivative differential equations and is therefore rather complicated, whereas the latter example represents interactions that can be analysed by making reference to simplified lumped-parameter systems and, as such, more easily treated and understood. However, the second category is actually a particular subset of the first one. In fact, if we suppose that the mass–spring–damper system represents the lumped-parameter model of a continuous structure, then the real effect caused by the accelerometer on such a structure might need to be considered in more detail. Possibly, it may happen that the dimensions and/or the location of the accelerometer on the structure are such that the mode shapes are appreciably modified and the analysis based on a lumped-parameter model is no longer appropriate. Eventually, an approach based on a distributed-parameter representation of the structure–accelerometer system could be required.

13.7.2   *The two-port representation*

As far as the action of the instrument on the measurement object can be represented as the interaction between two blocks modelled as lumped-parameter systems, the correspondent loading effect can be analysed by making use of the concept of two-port devices. A two-port device is a system seen as a black box which exchanges energy with the external world through two connections, named ports, positioned at its input and output. Irrespective of the domain to which the input and output of a two-port device belong, i.e. mechanical, electrical, thermal or other, the energy transfer across each port is realized by means of a couple of variables, namely a **flow** or **through** variable f, and an **effort** or **across** variable e. Flow and effort variables are functions of time and are characterized by the feature that their product $P(t) = f(t)e(t)$ is the instantaneous power transferred across the port [1, 6]. Examples of flow variables are given by velocity, current and heat flux, and the respective effort variables are force, voltage and temperature drop. A block diagram of a two-port device is shown in Fig. 13.10.

The representation of a system as a two-port device is strongly related to the physical behaviour of the system and to the configuration of its input and

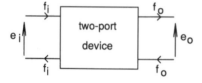

Fig. 13.10 Two-port representation of a measuring instrument.

output connections in their effect on the flow of energy. As such, it is conceptually different and more descriptive then the functional representation of Fig. 13.3, which is actually limited to the description of the flow of signals, without direct reference to their physical nature.

Of the four variables involved in a two-port device, namely e_i, f_i, e_o and f_o, only two can be independent. The remaining two variables are necessarily dependent. The choice of which couple among the four variables should be considered as independent is arbitrary from a formal point of view, but, depending on the case at hand, it may be more convenient to choose one representation or another among the six combinations available. One possible choice which is convenient for analysing the loading effect is that of considering e_i and f_o as independent variables, and e_o and f_i as dependent variables.

We will hereafter assume that the two-port is linear and time-invariant, and indicate with $E_i(s)$, $F_i(s)$, $E_o(s)$ and $F_o(s)$ the $\mathscr{L}$-transforms of e_i, f_i, e_o and f_o. In place of the $\mathscr{L}$-transforms, the $\mathscr{F}$-transforms could be used as well, but since the $\mathscr{L}$-transform method is actually a generalization of the $\mathscr{F}$-transform method, we will use $\mathscr{L}$-transforms in the following. Therefore, it can be written that

$$\left.\begin{array}{l} E_o(s) = A(s)E_i(s) + B(s)F_o(s) \\ F_i(s) = C(s)E_i(s) + D(s)F_o(s) \end{array}\right\} \tag{13.17}$$

If the two-port is not linear we can linearize it locally, and the above equations remain valid with each variable substituted by its incremental value and with $A(s)$, $B(s)$, $C(s)$ and $D(s)$ becoming dependent on the point around which the linearization is performed. Without loss of generality we may therefore make reference to the eqs (13.17) for both linear and locally linear two-port devices.

The terms $A(s)$, $B(s)$, $C(s)$ and $D(s)$ are complex functions of s which have the following meaning:

$$A(s) = \left.\frac{E_o(s)}{E_i(s)}\right|_{F_o(s)=0} = K_e^F(s)$$

is the forward effort-transfer function under no load, i.e. with no power drawn from the output port since $F_o(s) = 0$.

$$B(s) = \left.\frac{E_o(s)}{F_o(s)}\right|_{E_i(s)=0} = -Z_o(s) \quad \text{where} \quad Z_o(s) = \frac{1}{Y_o(s)}$$

is the generalized output impedance and $Y_o(s)$ is the generalized output admittance. The minus sign comes from the choice of taking f_o positive in the outward direction, so that power is considered as positive when transferred by the device to its output. Note that Z_o and Y_o are defined under the condition of zero input effort, i.e. $E_i(s) = 0$.

$$C(s) = \frac{F_i(s)}{E_i(s)}\bigg|_{F_o(s)=0} = Y_i(s) = \frac{1}{Z_i(s)}$$

is the generalized input admittance and $Z_i(s)$ is the generalized input impedance. Note that Y_i and Z_i are defined under the condition of zero output flow, i.e. $F_o(s) = 0$.

$$D(s) = \frac{F_i(s)}{F_o(s)}\bigg|_{E_i(s)=0} = K_f^R(s)$$

is the reverse flow-transfer function under no load, i.e. with no power drawn from the input port, since $E_i(s) = 0$.

It should be carefully noted that in the definition of $K_e^F(s)$ and $K_f^R(s)$ the no-load condition is identified in the first case with $F_o(s) = 0$, and in the second case with $E_i(s) = 0$. This apparent contradiction is indeed consistent with the variable taken as the output in the two cases, $E_o(s)$ and $F_i(s)$ respectively, and with the requirement of zero power flow across the port which defines the no-load situation. To differentiate between these two occurrences of the no-load condition it may be helpful to borrow the terminology of electrical circuits and call $K_e^F(s)$ the **open-circuit** forward effort-transfer function, and $K_f^R(s)$ the **short-circuit** reverse flow-transfer function.

In the following we further assume that the reverse transfer is equal to zero, i.e. in this case $K_f^R(s) = 0$, therefore considering the two-port to be **unilateral**. This assumption is quite reasonable for measuring instruments, since the application of the measurand at the input produces an output, but the opposite is obviously not true. It should be noticed that here we are not referring to the behaviour of the instrument components, which may often be bilateral when taken singularly (e.g. the piezoelectric or electrodynamic elements), but to that of the whole instrument, which is generally designed to be unilateral.

To simplify the notation, we will then drop the superscript F from $K_e^F(s)$ by implicitly assuming that we hereafter refer to forward transfer functions only. Therefore eqs (13.17) become

$$\left. \begin{array}{l} E_o(s) = K_e(s)E_i(s) - Z_o(s)F_o(s) = K_e(s)E_i(s) - \dfrac{1}{Y_o(s)}F_o(s) \\[4mm] F_i(s) = Y_i(s)E_i(s) = \dfrac{1}{Z_i(s)}E_i(s) \end{array} \right\} \qquad (13.18)$$

The same line of reasoning can be applied to the cases in which other couples of variables, different from E_i and F_o, are taken as independent. Since for analysing the loading effect it is convenient that the couple of independent variables is formed by an input and an output variable, the following four systems of equations resume all the possibilities for linear unilateral two-port devices:

$$
\left.
\begin{aligned}
E_o(s) &= K_e(s)E_i(s) - Z_o(s)F_o(s) \\
F_i(s) &= Y_i(s)E_i(s)
\end{aligned}
\right\}
\tag{13.19a}
$$

$$
\left.
\begin{aligned}
F_o(s) &= K_f(s)F_i(s) - Y_o(s)E_o(s) \\
E_i(s) &= Z_i(s)F_i(s)
\end{aligned}
\right\}
\tag{13.19b}
$$

$$
\left.
\begin{aligned}
F_o(s) &= K_{ef}(s)E_i(s) - Y_o(s)E_o(s) \\
F_i(s) &= Y_i(s)E_i(s)
\end{aligned}
\right\}
\tag{13.19c}
$$

$$
\left.
\begin{aligned}
E_o(s) &= K_{fe}(s)F_i(s) - Z_o(s)F_o(s) \\
E_i(s) &= Z_i(s)F_i(s)
\end{aligned}
\right\}
\tag{13.19d}
$$

The terms $K_e(s)$, $K_f(s)$, $K_{ef}(s)$ and $K_{fe}(s)$ are the open-circuit effort-, short-circuit flow-, short-circuit effort-to-flow- and open-circuit flow-to-effort-transfer functions respectively. For a given device they are not mutually independent, consistent with the fact that eqs (13.19a) to (13.19d) are just different representations of the same unique system, and the following relationships hold:

$$
\left.
\begin{aligned}
K_f(s) &= K_e(s)\frac{Z_i(s)}{Z_o(s)} \\[2mm]
K_{fe}(s) &= K_{ef}(s)Z_i(s)Z_o(s) \\[2mm]
K_{ef}(s) &= K_e(s)\frac{1}{Z_o(s)} = K_f(s)\frac{1}{Z_i(s)}
\end{aligned}
\right\}
\tag{13.20}
$$

Incidentally, it can be noticed that, while K_e and K_f are dimensionless numbers, K_{ef} and K_{fe} do have dimensions, and they are therefore sometimes called hybrid transfer functions.

The representations of a two-port device can be particularized to the case of a device having only two connecting terminals, called a one-port device. As shown in Fig. 13.11, a one-port can be thought as a two-port with the output terminals only, and therefore two variables, i.e. E_o and F_o, are sufficient to describe it. They are linked by the one-port constitutive equations which, in

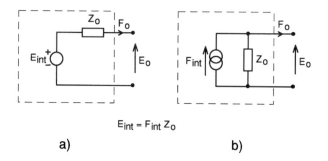

$$E_{int} = F_{int} Z_0$$

a)   b)

Fig. 13.11 Equivalent representations of one-ports: (a) Thevenin; (b) Norton.

the hypothesis of linearity, can take either one of the two following forms:

$$E_o(s) = E_{int}(s) - Z_o(s)F_o(s) \tag{13.21a}$$

$$F_o(s) = F_{int}(s) - \frac{E_o(s)}{Z_o(s)} \tag{13.21b}$$

Equation (13.21a), called the Thevenin representation, expresses the output effort E_o as a function of the flow F_o. Equation (13.21b), called the Norton representation, expresses the output flow F_o as a function of the effort E_o. The terms E_{int} and F_{int} represent an internal effort and flow source respectively, which may be absent if the one-port is purely passive.

Since both eqs (13.21a) and (13.21b) are nothing but two different representations of the same device they must be equivalent. Therefore, E_{int} and F_{int} must be related by

$$E_{int}(s) = F_{int}(s)Z_o(s) \tag{13.22}$$

where Z_o is the generalized output, or internal, impedance.

In contrast to the two-ports, which behave as energy converters, one-ports are energy sources or sinks, depending on the sign of E_{int} (or F_{int}), or energy dissipating and/or storage elements if $E_{int} = F_{int} = 0$ with $Z_o \neq 0$.

Two-port devices can be cascaded, as schematized in Fig. 13.12 provided that each input quantity is homogeneous with the output quantity of the preceding device and, as a special case, the first device of the chain may as well be a one-port. As far as the interconnection with the following block is concerned, this circumstance is indistinguishable from the more general situation in which both blocks are two-ports and, as such, a single analysis approach can be followed.

We can firstly assume that devices 1 and 2 are both represented by the form of eqs (13.19a), or of eqs (13.21a) in the case of a one-port, i.e. they are thought

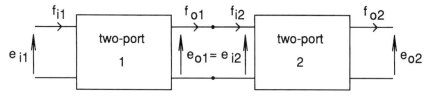

Fig. 13.12 Cascade connection of two-port blocks.

as effort devices. Considering that the cascade connection implies $E_{o1}(s) = E_{i2}(s)$ and $F_{o1}(s) = F_{i2}(s)$, after some simple algebra it can be concluded:

$$
\left.
\begin{aligned}
E_{o2}(s) &= K_{e1}(s)K_{e2}(s) \frac{1}{1 + Z_{o1}(s)/Z_{i2}(s)} E_{i1}(s) - Z_{o2}(s)F_{o2}(s) \\
&= K_{e12}(s)E_{i1}(s) - Z_{o2}(s)F_{o2}(s) \\
F_{i1}(s) &= \frac{1}{Z_{i1}(s)} E_{i1}(s)
\end{aligned}
\right\} \tag{13.23}
$$

It can be observed that the overall effort-transfer function $K_{e12}(s)$ is not simply the product of $K_{e1}(s)$ and $K_{e2}(s)$ but it is scaled by the term $1/[1 + Z_{o1}(s)/Z_{i2}(s)]$ which represents the loading exerted by the input port of the device 2 on the output port of the device 1. This can be understood by realizing that the output of device 1 is not working in an open-circuit condition and, accordingly, its effective transfer function is different from the open-circuit transfer function $K_{e1}(s)$.

The term $Z_{o1}(s)/Z_{i2}(s)$, which is in general a complex function of s, determines an effort-divider action at the interconnection between the two devices and can be called the effort loading factor $L_e(s)$. The effort loading effect is minimized when $Z_{o1}(s)$ is negligible compared to $Z_{i2}(s)$, being ideally zero either for $Z_{o1}(s) = 0$ or $Z_{i2}(s) = \infty$. In such cases $L_e(s)$ is equal to zero and $K_{e12} = K_{e1}K_{e2}$.

A similar situation is encountered in the case where devices 1 and 2 are both represented by the form of eqs (13.19b), i.e. they are thought as flow devices. By taking into consideration that the cascade connection again implies $E_{o1}(s) = E_{i2}(s)$ and $F_{o1}(s) = F_{i2}(s)$ we can write the equivalent of eqs (13.23):

$$
\left.
\begin{aligned}
F_{o2}(s) &= K_{f1}(s)K_{f2}(s) \frac{1}{1 + Y_{o1}(s)/Y_{i2}(s)} F_{i1}(s) - Y_{o2}(s)E_{o2}(s) \\
&= K_{f12}(s)F_{i1}(s) - Y_{o2}(s)E_{o2}(s) \\
E_{i1}(s) &= Z_{i1}(s)F_{i1}(s)
\end{aligned}
\right\} \tag{13.24}
$$

Again the overall flow-transfer function $K_{f12}(s)$ is not simply the product of $K_{f1}(s)$ and $K_{f2}(s)$ but it is scaled by the term $1/[1 + Y_{o1}(s)/Y_{i2}(s)]$, which represents the loading exerted by the input port of the device 2 on the output port of the device 1. In this case the output of device 1 is not working in a short-circuit condition and, accordingly, its effective transfer function is different from the short-circuit transfer function $K_{f1}(s)$.

The term $Y_{o1}(s)/Y_{i2}(s) = Z_{i2}(s)/Z_{o1}(s)$, which is in general a complex function of s, determines a flow-divider action at the interconnection between the two devices and can be called the flow loading factor $L_f(s)$. The flow loading effect is minimized when $Y_{o1}(s)$ is negligible compared to $Y_{i2}(s)$, being ideally zero either for $Y_{o1}(s) = 0$ or $Y_{i2}(s) = \infty$. In such cases $L_f(s)$ is equal to zero and $K_{f12} = K_{f1}K_{f2}$.

The above analysis of the loading effect for the effort and the flow devices is directly extendible to the case of effort-to-flow and flow-to-effort devices. In fact, irrespective of the overall input and output quantities of the cascaded devices, i.e. effort or flow, it can be realized that the loading effect of device 2 on device 1 will always be represented by one of the two cases already discussed. Therefore the analysis necessarily reduces to taking into consideration either the effort or the flow loading factor:

- The effort loading factor $L_e(s)$ needs to be as close to zero as possible in each of the following three cascade configurations: effort/effort, effort/effort-to-flow, flow-to-effort/effort.
- Conversely, the flow loading factor $L_f(s)$ needs to be as close to zero as possible in each of the remaining three cascade configurations: flow/flow, flow/flow-to-effort, effort-to-flow/flow.

We therefore can conclude that, for a given pair of systems 1 and 2 represented by linear unilateral two-port devices, the effort and flow loading factors L_e and L_f are parameters which enable to quantify the loading effect produced by system 2 on system 1. Furthermore, it is worthwhile observing that L_e and L_f are not independent but are linked by the relationship $L_e = 1/L_f$, consistent with the fact that they simply refer to different representations of the same phenomenon of interaction between the two systems.

The foregoing conclusions can be directly applied to the analysis of the measurement loading error by considering systems 1 and 2 as the measurement object and the measuring instrument respectively. We will shortly show some examples of application of this concept in the measurement of mechanical dynamic quantities, such as displacement, velocity, acceleration and force. To do this we first need to introduce a simplification in the description of the two-port devices consisting of exploiting the so-called **electromechanical analogy**.

13.7.3   *The electromechanical analogy*

The electromechanical (EM) analogy is based on the fact that the linear differential equations describing mechanical systems and electrical circuits are formally identical. Therefore, a correspondence can be established between lumped mechanical components and lumped electrical elements, and the formalism of electrical circuits can be used to describe mechanical systems [7, 8]. Such a correspondence is not unique and, without defining any requirement that it should satisfy, none of the choices is preferable with respect to the others.

A consistent approach is firstly to agree on which quantities are taken as the effort and flow variables in the mechanical and electrical domains. The choice of mating force F with voltage V as the effort variables, and velocity u with current I as the flow variables is that which renders the usual definitions of both the mechanical impedance $Z_M = F/u$ and the electrical impedance $Z_E = V/I$ consistent with that of the generalized impedance given by the effort/flow ratio.

Afterwards, in both domains the **series** connection can be defined as that in which all elements are subjected to the same value of the flow variable, while the **parallel** connection is that for which all elements see the same value of the effort variable.

Therefore, in a **series electrical** circuit the same current I flows through each element, while the total voltage drop V is the sum of the individual voltage drops. In a **parallel electrical** circuit the voltage drop V is the same across each element, while the total current I is the sum of the individual currents.

Similarly, in a **series mechanical** circuit each element undergoes the same velocity u, while the total force F is the sum of the forces acting on each individual element. In a **parallel mechanical** circuit the same force F acts on each element, while the total velocity u is the sum of the individual velocities.

Now, a correspondence criterion between series electrical and series mechanical circuits can be established, giving rise to the **direct** EM analogy. This is equivalent to considering the mechanical effort and flow variables analogous to the electrical effort and flow variables. Thus V is analogous to F, and I is analogous to u. It follows that the electrical resistance R, inductance L and capacitance C are analogous to mechanical resistance R_m, mass m and compliance $1/K$ respectively. Conversely, a correspondence between series electrical and parallel mechanical circuits (and *vice versa*) can be established, giving rise to the **inverse** EM analogy. This is equivalent to considering the mechanical effort and flow variables analogous to the electrical flow and effort variables in a cross-linked correspondence. Thus V is now analogous to u, and I is analogous to F, leading to electrical impedance V/I being analogous to mechanical **mobility** u/F. It follows the electrical resistance R, inductance L and capacitance C are now analogous to $1/R_m$, $1/K$ and m respectively. Both EM analogies are summarized in Fig. 13.13.

The main drawback of the inverse EM analogy is that the mechanical and electrical impedances of the analogous components have opposite behaviours versus the frequency, as opposed to the case of the direct analogy. For example, a spring element has $Z_M = K/s$ but its inverse analogous inductor

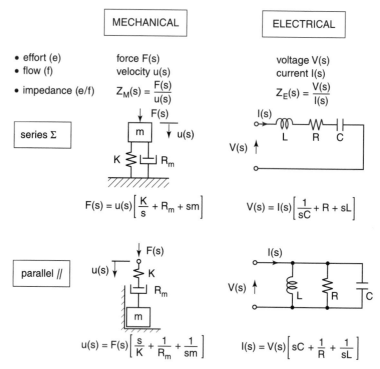

Fig. 13.13 Summary of the electromechanical analogies.

has $Z_E = sL$, while its direct analogous capacitor has $Z_E = 1/(sC)$. For this reason, the direct EM analogy is more often preferred and will be adopted hereafter.

13.7.4 Examples

Example 13.1. Consider a series-connected mass–spring–damper system excited by a force F, as shown in Fig. 13.14. Attention should be paid to the fact that, though the elements are in a 'side-by-side' configuration suggesting the parallel connection, they are actually connected in series, since they undergo the same velocity.

Suppose first that we want to measure the acceleration a, which is the same for all the elements. To this purpose, an accelerometer of mass m_a is attached to the mass m as shown in Fig. 13.15(a).

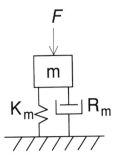

Fig. 13.14 Series mass–spring–damper system excited by a force F treated in Examples 13.1 and 13.2.

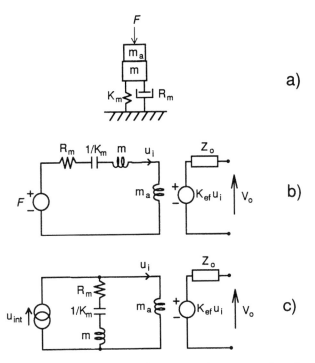

Fig. 13.15 Measurement of acceleration on the mass–spring–damper system of Fig. 13.14 by an attached accelerometer of mass m_a. (a) mechanical representation; (b) analogous electrical circuit; (c) Norton repesentation.

By applying the direct EM analogy the system can be converted into the electrical circuit of Fig. 13.15(b). The accelerometer is represented as a two-port device with a mechanical velocity input u_i and an electrical voltage output V_o, as described by the following equations:

$$\left.\begin{array}{l} V_o(s) = K_{fe}(s)u_i(s) - Z_o(s)I_o(s) = S_a(s)a_i(s) - Z_o(s)I_o(s) \\[2mm] F_i(s) = Z_i(s)u_i(s) \end{array}\right\} \tag{13.25}$$

where $S_a = K_{ef}/s$ is the acceleration sensitivity in $V/(m\ s^{-2})$, Z_o is the electrical output impedance, and $Z_i = sm_a$ is the mechanical input impedance.

The system under measurement behaves as a one-port with the flow variable, i.e. the velocity, being the integral of the measurand, the acceleration. It is therefore more convenient to pass to the Norton representation shown in Fig. 13.15(c), where the velocity $u_{int} = F/Z_{int}$ is the internal source variable, and $Z_{int} = (sm + R_m + K_m/s)$ is the internal impedance.

At this point it is very simple to derive the measurand-to-output transfer function in the hypothesis of no electrical loading of the accelerometer, i.e. $I_o = 0$:

$$V_o(s) = S_a(s)a_i(s) = S_a(s)su_i(s) = S_a(s)s\,\frac{Z_{int}(s)}{Z_{int}(s) + Z_i(s)}\,u_{int}(s)$$

$$= S_a(s)\,\frac{sm + R_m + \dfrac{K_m}{s}}{\left(sm + R_m + \dfrac{K_m}{s}\right) + sm_a}\,a \tag{13.26}$$

The term $Z_i(s)/Z_{int}(s)$ represents the flow loading factor L_f, which in this case is a velocity loading factor L_u. In the ideal case of $m_a = 0$ it results that $L_u = 0$, no loading error occurs and the output provides the measurand acceleration $a = su_{int}$.

The presence of the accelerometer mass m_a causes $L_u \neq 0$ and brings about two effects. The first effect is that at each fixed frequency s_o, the measured acceleration differs from the real one by a factor $L_u(s_o)$, and the resulting error is of the order of m_a/m. The second effect is that the resonant frequency of the system is diminished, changing from $2\pi\sqrt{K_m/m}$ in the unperturbed case to $2\pi\sqrt{K_m/(m + m_a)}$ in the loaded case. Both these loading effects are of critical importance in vibration measurements in general and modal analysis in particular, and should be carefully minimized by choosing an accelerometer with a mass m_a being as small as possible compared to the system mass m. In other words, the measuring instrument, due to its connection in series with the measuring system, should have a negligible impedance compared to Z_{int} of the measured structure.

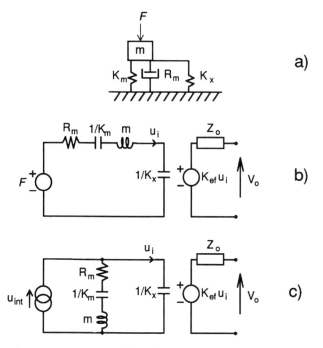

Fig. 13.16 Measurement of displacement on the mass–spring–damper system of Fig. 13.14 by an attached displacement transducer of negligible mass and damping, and stiffness K_x: (a) mechanical representation; (b) analogous electrical circuit; (c) Norton representation.

Example 13.2. Suppose now that we want to measure the displacement x of the same system. For this purpose, we can use a spring-loaded transducer modelled by a stiffness element K_x with negligible mass and damping, as schematized in Fig. 13.16(a). The resulting analogue electrical circuit is shown in Fig. 13.16(b). The system under measurement is the same one-port as before, except that now the measurand, i.e. the displacement, is the integral of the output flow variable, i.e. the velocity.

The transducer is modelled as a two-port governed by the following equations:

$$\left. \begin{array}{l} V_o(s) = K_{fe}(s)u_i(s) - Z_o(s)I_o(s) = S_x(s)x_i(s) - Z_o(s)I_o(s) \\ F_i(s) = Z_i(s)u_i(s) \end{array} \right\} \tag{13.27}$$

where $S_x = K_{fe}s$ is the displacement sensitivity in V/m, Z_o is the electrical output impedance, and $Z_i = K_x/s$ is the mechanical input impedance.

Again, we can pass to the Norton equivalent of the one-port as shown in Fig. 13.16(c) and obtain the measurand-to-output transfer function under the hypothesis of no electrical loading of the displacement transducer, i.e. $I_o = 0$:

$$V_o(s) = S_x(s)x_i(s) = S_x(s)\frac{u_i(s)}{s} = S_x(s)\frac{Z_{int}(s)}{Z_{int}(s) + Z_i(s)}\frac{u_{int}(s)}{s}$$

$$= S_x(s)\frac{sm + R_m + \dfrac{K_m}{s}}{\left(sm + R_m + \dfrac{K_m}{s}\right) + \dfrac{K_x}{s}}x \qquad (13.28)$$

The term $Z_i(s)/Z_{int}(s)$ represents the flow loading factor L_f, which is now determined by the stiffness of the measuring instrument. In case of static displacement, i.e. for $s = 0$, it reduces to K_x/K_m which shows that the loading error increases the higher is the transducer stiffness K_x. The dynamic behaviour is also affected. In particular, the resonant frequency of the system is augmented by the measuring instrument stiffening action, changing from $2\pi\sqrt{K_m/m}$ in the unperturbed case to $2\pi\sqrt{(K_m + K_x)/m}$ in the loaded case. To reduce both these errors it is necessary to use a transducer as compliant as possible, i.e. with a very low K_x, in order to ensure a negligible impedance compared to Z_{int}. Often, especially in the cases of small-sized and lightweight systems, the only way to obtain such a condition is by adopting a noncontact measuring method which ensures a K_x equal to zero.

Example 13.3. As a third example, consider the mass–spring–damper system of Fig. 13.17 held in motion at a velocity u with respect to the reference frame. It should be noted that, though the system elements are in a 'stacked' arrangement suggesting the series connection, they are connected in parallel, since they are all subject to the same force F.

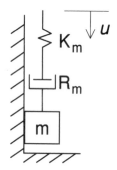

Fig. 13.17 Parallel mass–spring–damper system held in motion at a velocity u treated in Example 13.3.

Suppose that we are interested in measuring such a force and, to this purpose, an elastic-element force transducer is adopted and attached to the system as shown in Fig. 13.18(a). By applying the direct EM analogy the system can be converted into the electric circuit of Fig. 13.18(b).

The force transducer is represented as a two-port device with a mechanical force input F_i and an electrical voltage output V_o, as described by the following equations:

$$\left.\begin{array}{l} V_o(s) = K_e(s)F_i(s) - Z_o(s)I_o(s) = S_F(s)F_i(s) - Z_o(s)I_o(s) \\ u_i(s) = Y_i(s)F_i(s) \end{array}\right\} \qquad (13.29)$$

where $S_F = K_e$ is the force sensitivity in V/N, Z_o is the electrical output impedance, and $(Y_i)^{-1} = Z_i = K_F/s$ is the mechanical input impedance.

The system under measurement again behaves as a one-port but now the effort variable, i.e. the force, is the measurand quantity. In this circumstance it is advisable to pass to the Thevenin representation as shown in Fig. 13.18(c),

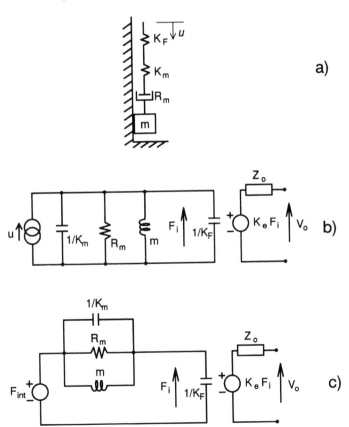

Fig. 13.18 Measurement of force on the mass–spring–damper system of Fig. 13.17 by a force transducer of stiffness K_F: (a) mechanical representation; (b) analogous electrical circuit; (c) Thevenin representation.

where the force $F_{int} = uZ_{int} = F$ is the internal source variable, and $Z_{int} = sm \parallel R_m \parallel K_m/s$, is the internal impedance.

Again assuming no electrical load for the transducer, i.e. $I_o = 0$, the measurand-to-output transfer function can be determined:

$$V_o(s) = S_F(s)F_i(s) = S_F(s) \frac{Z_i(s)}{Z_{int}(s) + Z_i(s)} F_{int}(s)$$

$$= S_F(s) \frac{\dfrac{K_F}{s}}{\left(sm \parallel R_m \parallel \dfrac{K_m}{s} \right) + \dfrac{K_F}{s}} F \qquad (13.30)$$

This time, the stiffness K_F of the measuring instrument determines the effort loading factor L_e represented by the term $Z_{int}(s)/Z_i(s)$. The loading error is now smaller the more rigid the transducer is, consistent with the fact that it is connected in parallel with the system and, therefore, its mechanical impedance should be much greater than Z_{int}.

As a concluding comment, it should now be clear what we qualitatively anticipated at the beginning of this section by saying that it is theoretically possible to compute the loading error and compensate for it. In fact, it appears to be sufficient to analyse the system with the method outlined above to determine, depending on the case at hand, either L_e or L_f. However, this would require the generalized impedances of all the involved elements to be exactly known, which is a situation that rarely occurs in practice. The preferred and most used approach is that of properly choosing the instrument in order to minimize its impact by maintaining L_e (or L_f) as small as possible.

13.8   Performance specifications of measuring instruments

The quality of measurements taken and their suitability for our purposes depend on the measuring instrument characteristics which, therefore, need to be properly specified. Measuring instruments are complex systems, and specifying their performance in a clear and consistent way is not a trivial task. It is of fundamental importance that a suitable terminology is both understood and adopted [9]. The reason is not merely a matter of formal rigorousness, but rather lies in the substantial need to express concepts in a concise and objective way, avoiding incompleteness and misinterpretations.

The following is a glossary of recommended terms which apply to measuring instruments in general, irrespective of their operating principle and construction. Though it is not exhaustive, it is intended to help the reader in understanding the basic characteristics of an instrument to properly chose and operate it.

Range (input/output)

The input range identifies the limits between which the measurand can be applied with the instrument operating properly and in compliance with its specifications. Such limits have the dimensions of the measurand.

The output range identifies the limits between which the instrument output lies when the measurand is within the input range. Such limits have the dimensions of the instrument output.

Span

The span is the difference between the limits which define the range (either input or output).

Full-scale value

The full-scale value identifies the upper limit of the range (either input or output). When the lower limit is zero, the full-scale value equals the span. When this is not the case, full-scale value and span differ and, strictly speaking, they should not be confused. However, it is common practice to pay little attention to this subtle matter and in many occasions, such as when specifying linearity or other parameters, full-scale value and span are used as synonyms.

Sensitivity

The sensitivity S is the rate of change of the instrument output y with respect to input measurand x, i.e. $S = dy/dx$. In the general case, the sensitivity is a function of the measurand value x, i.e. $S = S(x)$. For linear instruments the sensitivity is constant throughout the range and takes the name of scale factor, calibration factor or conversion coefficient.

When the measurand and the instrument output are functions of time and linearity holds, they can be decomposed into a superposition of sinusoids and the sensitivity becomes in general a complex function of frequency representing the instrument frequency response function.

Linearity

Linearity is the ability of the instrument to follow a prescribed linear relationship between input and output, either in the case they are static quantities or functions of time. The nonlinearity, or nonlinearity error, is the maximum deviation of the instrument output, at any measurand value, from that calculated assuming such a linear relationship. To properly quantify the nonlinearity it is essential to specify which line is assumed as the reference. The following two alternatives are generally adopted:

- End-point linearity: the reference line is the one passing through the limiting points, or end points, of the instrument calibration characteristic.
- Least-squares linearity: the reference line is the one which best fits the experimental points of the instrument calibration characteristic by the method of the least squares.

Usually, the nonlinearity is expressed as a percentage of the full-scale output (%FSO or %FS) and it is, therefore, a dimensionless number.

Resolution

Resolution identifies the minimum variation in the measurand which produces a detectable variation at the instrument output. Such a minimum variation is called resolution threshold or discrimination threshold. Therefore, the resolution describes the ability of the instrument of discriminating small changes in the measurand. An ideal instrument with unlimited discrimination capability is said to have continuous or infinite resolution.

The resolution is specified by quoting the resolution threshold and may be referred to the input or to the output. In the former case it is usually expressed as an absolute value having the dimensions of the measurand quantity; in the latter case it is mostly expressed as a percentage of full-scale output (%FSO or %FS). Notice that the improper use of the term sensitivity in place of resolution is sometimes encountered. Since sensitivity has a markedly different significance, this misuse should be avoided.

Dynamic range

Dynamic range is defined as the range of measurand values contained between the minimum detectable level, given by the resolution, and the full-scale value. The dynamic range DR is often expressed in decibel, that is $DR \text{ (dB)} = 20 \log_{10} DR$. For a given full-scale value, the instrument with the highest resolution (i.e. the highest discrimination capability) has the widest dynamic range.

Repeatability

Repeatability is the ability of an instrument of providing the same output reading when the same value of the measurand is repeatedly applied in the same conditions. It expresses the amount of random fluctuations that affect the instrument output in the ideal condition of a perfectly defined and stable measurand value.

Reproducibility

Reproducibility is the ability of an instrument of providing the same output reading when the same value of the measurand is subsequently applied in different conditions, such as different time, observer, location.

Hysteresis

Hysteresis is the effect by which an instrument provides, for each given value of the measurand, two different outputs when such a value has been approached by increasing or by decreasing the measurand, resulting in an upward and a downward calibration characteristic. The hysteresis error is the maximum difference between such upward and downward calibration characteristics.

The hysteresis error is usually expressed as a percentage of the full-scale output (%FSO or %FS) and it is, therefore, a dimensionless number.

Stability
Stability is the ability of an instrument of providing the same output reading over time when a value of the measurand is applied and maintained constant. To properly express the stability it is advisable to specify the duration of the time interval to which it refers. To provide a qualitative reference, the terms short-, medium- or long-time stability are often used.

Drift
The drift of an instrument is its tendency to provide an output reading which gradually changes over time without any relationship with the measurand. Drift can be caused by lack of stability, by instrument ageing or by the action of influencing quantities, e.g. the thermal drift related to temperature fluctuations.

Temperature influence
Temperature influence is the effect in which the instrument output for a constant applied value of the measurand changes when the environmental temperature changes. It is usually expressed as a temperature coefficient defined as the output variation, absolute or fractional, for a unitary variation of the temperature.

Bandwidth
The bandwidth of a measuring instrument is the range of frequencies of the input measurand for which the magnitude of the instrument frequency response function is no lower than 3 dB of its peak value. A decrease of 3 dB is equivalent to an attenuation of $1/\sqrt{2}$.

Response time
The response time is the length of time required for the instrument output to reach a specified percentage of its final value (typically 95% or 98%) when subject to a step change in the measurand.

Time constant
The time constant is the length of time required for the instrument output to reach the 63.2% of the final value when subject to a step change in the measurand. For a first-order instrument the time constant coincides with the reciprocal of the transfer function -3 dB cutoff angular frequency ω_c.

Accuracy
The accuracy of an instrument is the extent to which its output may differ from the true value of the measurand, with the significance given to the term

'true value' discussed in Section 13.2. Therefore, high accuracy means low measurement uncertainty. Accuracy may be distinguished as static and dynamic. The static accuracy, or better the inaccuracy, is determined by such effects as nonlinearity, nonrepeatability, hysteresis, drift, instability and environmental influences. The dynamic accuracy is influenced also by the effects of bandwidth limitation and finite response time.

Accuracy may be expressed by quoting the associated uncertainty (for a given confidence level) as an absolute interval, as a percentage of the reading (%rdg) or as a percentage of the full-scale output (%FSO or %FS). In the first case it has the dimensions of the measurand, while in the remaining two it is a dimensionless number.

Precision

The precision of an instrument has been long identified with its ability to provide the same output reading when the same measurand value is applied in the same conditions. A precise instrument therefore ensures a low scattering of the readings. The term repeatability is currently preferred to define this property, and the use of the term precision is discouraged by the international metrological organizations. The term precision is definitely not synonymous with accuracy, and should not be confused with it.

13.9   Summary

Measurement is the experimental procedure to assign numerical values to real-world quantities called measurands, in order to describe them quantitatively. To this purpose, measuring instruments are used. To allow comparison between measurement results of the same quantity obtained by different instruments they need to be calibrated against a reference which is traceable to a conventionally agreed standard.

The extent to which the result of a measurement approximates the value of the measurand has traditionally been evaluated with reference to the concepts of random and systematic errors. This approach has the problem that errors are defined with respect to the true value of the measurand which, however, is an unknown and unknowable quantity. The inconsistency is removed by introducing the concept of measurement uncertainty, which represents a quantitative estimate of the range of values within which the value of the measurand lies with a given confidence level when using a particular instrument in specified conditions. The lower the measurement uncertainty, the higher the accuracy.

Many traditional measuring instruments are nonelectronic, but most of the measurement tasks are today performed by electronic measuring instruments and systems. Compared to purely mechanical systems, they offer higher performance, improved functionality and reduced cost.

Irrespective of its construction, the measuring instrument occupies the position at the interface between the observer and the measurand and is under the influence of the environment. This influence generally produces a perturbing action due to several influencing quantities which need to be recognized and kept under control to avoid a detrimental effect on the measurement accuracy.

The behaviour of a measuring instrument can be analysed both statically and dynamically, depending on considering the measurand as a constant or as a function of time. For linear instruments, the superposition principle holds and the dynamic behaviour can be analysed making use of the Fourier- and Laplace-transform methods.

During the measurement process, the measuring instrument interacts with the measured system and unavoidably perturbs it by altering its energetic equilibrium. As a consequence, the measurement result necessarily differs from the unperturbed measurand value. This is called the loading effect caused by the measuring instrument on the measured system.

To analyse the loading effect, linear instruments can be represented as two-port devices with a proper choice of the effort, or across, and flow, or through, variables. When mechanical systems are involved which lend themselves to a lumped-element representation, it may be convenient to make use of the electromechanical analogy to analyse the interaction between instrument and system with the formalism of electrical circuits. This method allows us to determine which parameters are responsible for the loading effect and what can be done to reduce it to a minimum.

To properly choose and operate measuring instruments, and to effectively exchange information with other people on the subject of measurement, it is important to know and make use of the correct terminology. Striving to do so as a habit helps to avoid incompleteness and misinterpretations, and provides a deeper insight into the fundamental concepts related to measurement and measuring instruments.

References

1. Finkelstein, L, and Watts, R.D., Mathematical models of instruments – fundamental principles, in B.E. Jones (ed.), *Instrument Science and Technology*, Adam Hilger, Bristol, 1982, Ch. 2.
2. *Guide to the Expression of Uncertainty in Measurement*, International Organization for Standardization (ISO), Geneva, 1993.
3. *Calibration: Philosophy in Practice*, 2nd edn, Fluke Corporation, Everett, WA, 1994.
4. Dettmann, J.W., *Mathematical Methods in Physics and Engineering*, Dover Publications, Mineola, NY, 1988.
5. Oppenheim, A.V., Willsky, A.S. and Young, I.T., *Signals and Systems*, Prentice Hall, Englewood Cliffs, NJ, 1983.
6. Dobelin, E.O., *Measurement Systems Application and Design*, McGraw-Hill, New York, 1975.
7. Raven, F.H., *Automatic Control Engineering*, McGraw-Hill, New York, 1968.

8. D'Azzo, J.J. and Houpis, C.H., *Linear Control System Analysis and Design,* McGraw-Hill, New York, 1975.
9. *International Vocabulary of Basic and General Terms in Metrology (VIM),* International Standardization Organization (ISO), Geneva, 1993.

14 Motion and vibration transducers

14.1 Introduction

Transducers are the interface elements between the physical world and the electronic measurement and/or control system [1–4]. Input transducers convert physical quantities into electrical signals. Conversely, output transducers produce a physical action in response to an electrical driving stimulus. Typical examples of input and output transducers in the vibration field are accelerometers and electromagnetic shakers, respectively. Input transducers, or simply transducers, are also called sensors, even if, historically, this term was preferred for indicating the primary sensing element, whereas the term transducer was used for a composite measuring device incorporating one or more sensors. Output transducers are also named actuators.

Sensors may be either active (self-generating), or passive (modulating). Self-generating sensors, such as the piezoelectric accelerometer, provide an electrical signal without the need for an electrical power supply, unless for additional amplification or processing. Conversely, modulating sensors, such as the resistive potentiometer, require an external source of energy in the form of an excitation which is modulated by the measurand quantity and produces the output signal.

In the present chapter, the main types of transducers which find application in the measurement of structural and mechanical vibrations are illustrated. The treatment is not intended to be exhaustive, nor deeply involved in the analysis of the functioning principles for which the interested reader can refer to the references given and to the manufacturers' technical literature. Instead, the approach is chiefly illustrative, basically oriented towards pointing out the main features of the presented devices as an hopefully helpful aid to their proper choice and use.

14.2 Relative- and absolute-motion measurement

The position of an object is defined by its coordinates in space with respect to an assigned reference. The displacement indicates the difference of the object coordinates from one position to another. If the initial coordinates of the still object are taken as the reference, then position and displacement coincide and the two terms become interchangeable, although in the most general case they are not. The variation of the object position means that the displacement is a function of time, indicating the presence of motion. Displacement may be linear or angular, i.e. the motion may be translational or rotational, or a combination of the two. For both the linear and the angular cases, the first and second time derivatives of displacement are the object velocity and acceleration respectively.

There are two methods for measuring the motion of an object with reference to a fixed point in space.

- **Relative-motion method**: the transducer is attached between the object and the reference point fixed in space, then the motion is referred to such a point. The kind of attachment may be of contact type, i.e. mechanical, as well as of noncontact type, such as electromagnetic or optical.
- **Absolute-motion method**: a seismic transducer made by a mass–spring–damper system is attached to the object and the motion of the object is derived from the motion of the mass relative to the transducer base. In this case, the fixed reference point is represented by the position in space of the seismic mass and, consequently, of the object to which it is attached, under the condition of absence of motion.

14.3   Contact and noncontact transducers

Contact transducers function by establishing a mechanical link between the measured object and the fixed reference point. This fact unavoidably produces a loading error which is more or less pronounced depending on the mechanical output impedance of the object compared to the input impedance of the transducer, as expressed by the generalized effort- and flow-loading factor parameters $L_e(s)$ and $L_f(s)$ illustrated in Chapter 13.

In high-precision prolonged measurements under variable temperature conditions, the mechanical contact link may be an additional source of problems due to thermal expansions in it and in the transducer mounting fixture. On the other hand, in dynamic measurements of relatively short duration these effects are usually not an issue, and the main point of concern remains the loading effect with the possible alteration in the system frequency response that it may cause. Moreover, it must be ensured that the reference point is securely motionless, though in the practical cases this is hardly completely achievable.

Noncontact transducers avoid the use of any mechanical link between the measurement object and the fixed reference point. This fact has two positive consequences that represent the main advantages of noncontact transducers. Firstly, they generate no loading on the measured object and are therefore best suited for small and lightweight structures. Secondly, the absence of mechanical links allows their use for measurements on rotating parts such as shafts, motors and bearings.

Most noncontact transducers measure over an area of the target object, hence their response can be somewhat dependent on the geometric properties of the object surface, such as shape, roughness or presence of cracks, as well as on its electrical characteristics. This is not the case with optical transducers, which are capable of spot measurements on both insulating and conducting target materials, but on the other hand they are affected by the surface optical properties.

14.4 Relative-displacement measurement

14.4.1 *Resistive potentiometers*

A potentiometer is a three-terminal resistor with two fixed contacts and one movable contact, named a wiper, connected to an accessible mechanical termination called a cursor or slider. The potentiometer body is fixed at the reference point, while the cursor normally terminates in a shaft attached to the measured object. In place of the shaft a spring-tensioned wire is used in the draw-wire models.

The position of the cursor along the stroke of the potentiometer determines the ratio of the two resistances into which the overall resistance is divided, and therefore the position can be indirectly determined by the measurement of this resistance ratio. Potentiometers can be either linear or rotary, and thus can be used for the measurement of both translation and rotation.

Potentiometers are passive transducers for which the measurement of resistance is usually done indirectly. A common, but not the only, method is shown in Fig. 14.1, where a voltage excitation V_E is applied and the correspondent voltage V_o is read between the cursor and one the other

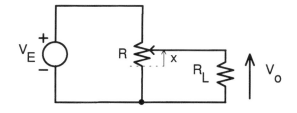

$$R_L \gg R \quad \rightarrow \quad V_o = V_E \frac{xR}{xR + (1-x)R} = xV_E$$

Fig. 14.1 Readout scheme commonly adopted for resistive potentiometers. To ensure linearity, the load resistance R_L, representing the input resistance of the readout instrument, should be much greater than R.

terminals. If the resistance per unit length is constant, the output voltage V_o is linearly related to the cursor position x.

Linearity is lost if the potentiometer is connected to a load resistor of comparable or lower value than the transducer resistance R which, in order to keep the power dissipation and the self-heating within tolerable limits, is generally in the range of 100 Ω to 10 kΩ. Therefore, it is necessary to use high-input-impedance reading instruments or amplifiers, which in turn may cause problems of noise and interference pick-up when connected remotely from the transducer by means of long wires.

Since V_o is proportional to V_E, the excitation voltage determines the transducer sensitivity and therefore it must guarantee a stable and precisely known value. This is somewhat costly to obtain and it is often preferable to use the ratiometric method, consisting of measuring both V_o and V_E and then electronically computing their ratio which is solely dependent on position x.

The mechanical sliding contact of the wiper along the resistor length and the associated friction produce a gradual worsening of the electrical contact through wear, possibly becoming intermittent, and the generation of spurious signals in the form of noise. These unwanted effects are usually enhanced by environmental factors, such as dirt, humidity and vibrations.

If the transducer resistor element is made of wirewound metal, the resistance produced is a step-like function of cursor position as the wiper slides from one wire turn to the next. This effect ultimately produces a limit on the achievable resolution. The problem is much reduced with transducers using continuous conductive tracks, such as carbon films or conductive plastics, but generally at the expense of a higher track-to-wiper contact resistance and a poorer temperature stability.

The typical overall accuracy obtainable with potentiometers is $\pm 0.5\%$ FS. When higher accuracy is required, different devices, such as linear variable differential transformers (LVDT) or noncontact transducers described in the following are to be preferred.

The potentiometer is essentially a zeroth-order instrument throughout the frequency region in which its electrical impedance can be considered as purely resistive. However, the transducer mechanical impedance can produce a significant loading effect on the measured body which may appreciably alter its dynamic behaviour, especially at high frequency. This is particularly true when spring-loaded shaft potentiometers are used to ensure stable and bounceless mechanical contact with the measured body. In this case, the mechanical impedance includes the terms due to friction and mass plus a relevant spring term. Therefore, potentiometers are not generally suitable for use in dynamic measurements on lightweight structures, and in any case not beyond a frequency of the order of 100 Hz.

14.4.2 *Resistance strain gauges*

Resistance strain gauges, or simply strain gauges, are small and lightweight resistive elements that are cemented on a structure of which they measure the local deformation through the variation in resistance caused by elongation or contraction.

Strain gauges function on a principle based on the expression $R = \rho L/A$, which gives the resistance of a uniform conductor of resistivity ρ, length L and cross-section area A.

The fractional change in resistance is then given by

$$\frac{\Delta R}{R} = \frac{\Delta L}{L}(1 + 2\nu) + \frac{\Delta \rho}{\rho} = \left(1 + 2\nu + \frac{\Delta\rho/\rho}{\Delta L/L}\right)\frac{\Delta L}{L} = K\frac{\Delta L}{L} \tag{14.1}$$

where ν is Poisson's ratio of the conductor material, and $\Delta L/L = \varepsilon$ is the strain.

According to eq (14.1), strain gauges do not actually measure displacement but strain, i.e. the average gauge elongation or contraction divided by the gauge length. The parameter K is called the gauge factor, which accounts for the resistance variations due to dimensional changes, represented by the term $(1 + 2\nu)$, and for those caused by the strain-induced resistivity variations $(\Delta\rho/\rho)/(\Delta L/L)$. This latter effect is called the piezoresistive effect.

Depending on the material of which the strain gauge is made, the gauge factor assumes different values, ranging from close to 2 for nickel–copper (constantan) and 2.1 nickel–chromium (karma) alloys, to about 3.5 for isoelastic, to above 100 for semiconductors.

Metal alloy strain gauges are the most widely used and, as shown in Fig. 14.2, they typically have the form of grid foils of various dimensions and geometry supported by an insulating backing carrier which allows them to be bonded to the body under test. The backing carrier performs the fundamental function of transferring the strain from the specimen to the gauge with maximum fidelity.

The nominal values of resistance are normally 120, 350, 700 or 1000 Ω, with strain-induced variations that are usually quite small, as low as few parts per million (ppm), and therefore require special care in their measurement. Moreover, the temperature appreciably influences both the gauge resistance and the gauge factor, producing the so-called thermal output, which is due to the temperature coefficient of resistance (TCR) and of gauge factor (TCGF) combined with the thermal expansion of the specimen.

A typical solution is given by the use of the Wheatstone bridge configuration with voltage or current excitation of either DC or AC type (Chapter 15). Special arrangements including multiple active and/or dummy gauges are used to maximize linearity and compensate for the thermal effects, and proper wiring techniques allow for lead wire resistance cancellation in distant connections between the bridge and the excitation and amplification circuitry.

With properly employed good-quality signal conditioning circuits, strain levels in the microstrain range can be ordinarily detected and values lower than 1 $\mu\varepsilon$ are possible. Such figures enable the use of strain gauges for stress

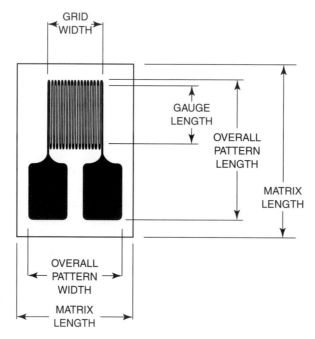

Fig. 14.2 Etched metal foil resistance strain gauge (by courtesy of Measurements Group, Inc.).

analysis, i.e. determination of the stress in a structure by knowing the strain and the elastic modulus, and for the measurement of microdisplacements occurring in rigid structures.

Thanks to the fact that they are zeroth-order elements and they cause negligible mechanical loading due to their reduced size and weight, strain gauges have great potential for use in dynamic and shock measurements. In particular, isoelastic gauges are typically preferred in these cases due to their high gauge factor. In conditions of steep strain gradient the averaging effect over the gauge length must be considered for proper selection of the gauge dimensions. However, the process of instrumenting the structure under test by bonding the gauges and connecting the wiring is time consuming, requires special precautions, such as dirt removal and surface treatment, and it is therefore rather impractical in spot tests. In some cases, strain gauges are permanently attached to structures or machinery in order to allow for continuous or periodical monitoring, both for functionality assessment and maintenance purposes.

Besides being used as transducers themselves, strain gauges are also widely used as primary sensing elements in a large number of transducers for the measurement of strain-related quantities, such as displacement, acceleration, force, torque or pressure. Such transducers are called strain-gauge-based or piezoresistive. In these cases, the strain gauges are not exclusively made by metal foils bonded to the transducer structure, but may be of different materials and construction. For instance, they may be conductors deposited in thin- or thick-film technology, or integrated semiconductors, such as in silicon micromachined sensors.

14.4.3 *Linear variable differential transformers*

The linear variable differential transformer (LVDT) is a transducer based on the magnetic induction principle. It is made by a transformer with one primary coil and two secondary coils with a movable core of ferromagnetic material which is placed coaxially in the coils without touching them, as shown in Fig. 14.3. The core terminates in a shaft or plunger which is attached to the target object either by threading or spring loading. The transducer body containing the primary and secondary coils is mounted in the reference position. As the core moves, it produces a variation of the magnetic coupling depending on its position along the coil axis. When the primary coil is excited with a sinusoidal voltage of amplitude V_E an induced voltage V_o is collected across the secondary coils, which is linearly related to the core position through the mutual inductance coefficient M. Since the secondary coils are connected in series opposition, M equals zero when the core is centered and it changes sign according to the sign of the core off-centre position. For the rotary variable differential transformer (RVDT) the operating principle is the same, with the difference that the rotary core movement allows the measurement of angular rather than linear displacement.

A measure of the core displacement can then be obtained by rectifying the voltage V_o while taking into account of its phase relative to V_E. This readout operation is typically carried out by dedicated electronic circuitry, and is usually obtained by employing an oscillator and a phase-sensitive demodulation stage followed by low-pass filtering (Chapter 15).

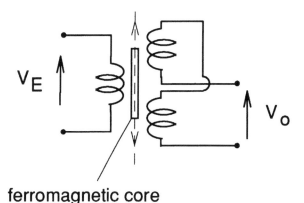

ferromagnetic core

Fig. 14.3 Schematic diagram of linear variable differential transformer (LVDT).

The oscillator signal for the primary coil excitation is preferably sinusoidal to avoid the generation of harmonics, and often a particular frequency is recommended by the transducer manufacturer for obtaining maximum compensation of residual phase shift at null core position. In the case of dynamic measurements, the oscillator frequency should be set at a value up to ten times higher than the highest motion frequency, to avoid frequency overlapping. Since the maximum recommended excitation frequencies of typical LVDTs are in the order of 20 kHz, the useful measurement bandwidth is generally limited to several kilohertz.

The mechanical input impedance of LVDTs is mainly given by the mass associated with the core inertia in threaded-core types, with an added spring effect when spring-loaded plunger-type cores are used. Friction is generally almost absent, since there is no contact during the core motion within the transducer. This fact offers an ideally unlimited resolution and virtually no hysteresis, which represent fundamental advantages of LVDTs compared, for instance, with resistive potentiometers. Practical devices can reach resolutions better than 0.01% of the range, therefore submicron displacements may be appreciated with transducers with a stroke of few millimetres.

These features, joined to a typically rugged construction and good immunity to environmental factors and electromagnetic interference, makes LVDTs the first choice for transducers in many precision measurement applications. They are often considered the electrical equivalent of the mechanical dial gauge, or micrometer.

As a drawback, LVDTs are not generally cheap and require dedicated signal conditioning electronics, which is comparatively costly. In this respect, some devices which include the excitation and amplification electronics within the transducer case are particularly advantageous, providing a DC voltage output signal ready to be acquired.

As a precaution, nonmagnetic materials such as aluminium or plastic should be used for the mounting fixture in order not to alter the sensitivity.

Similar in shape to the LVDTs is a type of variable-inductance transducer which includes only two series-connected coils in an autotransformer configuration, i.e. analogous to the secondary windings of the LVDTs with the primary absent. When the core is halfway between the coils their respective inductances are equal, while they become different according to the amount and sign of the core off-centre displacement. The transducer then essentially works as an inductive potentiometer whose cursor is represented by the core, with the advantage of no internal electrical contact. The inductance imbalance can be measured by connecting the transducer in an AC-excited bridge and reading the correspondent bridge output.

14.4.4 *Inductive transducers*

The functioning principle is based on the variable inductance of a coil wound on a core caused by the changes in the magnetic flux reluctance when the distance from a ferromagnetic target varies. If the measured object is ferrous it can act as the target, otherwise a ferromagnetic target must be attached to the object. The inductance changes are usually measured in an AC bridge circuit, or by making it part of a resonant circuit and detecting the resonant frequency shift. Two geometries may be used, namely the closed-loop and open-loop magnetic system, as shown in Fig. 14.4. In both cases, provided that the magnetic permeability of both the core and the target are much greater than that of air practically equal to that of vacuum $\mu_0 = 1.26 \times 10^{-6}\,\mathrm{Hm^{-1}}$ (hence the requirement for a ferromagnetic target) the coil inductance L may be approximated by

$$L \cong \frac{N^2 A \mu_0}{2d} \tag{14.2}$$

where A is the area of the core facing the target, N is the number of coil turns and d is the distance from the target.

It can be observed that the relationship between L and d is nonlinear, therefore these transducers are best suited for use as proximity sensors rather than distance measuring devices. To obtain linear operation, electronic linearization circuitry is generally added, often within the sensor housing, and typical residual linearity errors are in the order of $\pm 1\%$ FS.

This type of transducer is inherently sensitive to stray magnetic fields, and to ferromagnetic materials in proximity of the sensing coil (especially in the open-loop magnetic geometry), therefore attention should be paid to this aspect in its positioning and mounting.

14.4.5 Eddy-current transducers

When a coil is driven by an alternating voltage it generates an electromagnetic field. If an electroconductive object is placed in proximity of the coil, the

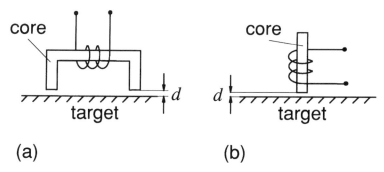

Fig. 14.4 Schematic diagrams of inductive displacement transducers based on variable reluctances: (a) closed magnetic loop design, (b) open magentic loop design.

electromagnetic field induces **eddy currents** (the name comes from the circular nature of their flow within the object). Such eddy currents in turn produce an electromagnetic field which opposes to the original field, and has the ultimate effect of changing both the inductance and the quality factor, i.e. the losses, of the coil. Therefore, from the measurement of the coil impedance, the distance from the object can be derived.

The principle is generally applied by making use of a two-coil arrangement, which includes a driving coil and a sensing coil both oriented with their axis perpendicular to the target, as shown in Fig. 14.5. Both shielded and unshielded constructions exist, which differ in that the former provides a more directional field that ensures a higher immunity from stray effects caused by metallic objects near the sides of the transducer. When two or more transducers need to be mounted in close proximity, the shielded construction is preferred to minimize mutual interference.

Eddy-current inductive sensors are responsive to both the magnetic permeability μ and the electrical conductivity σ of the target material but, as opposed to the variable-reluctance type, they do not require that the magnetic permeability is high. Therefore, they operate properly even with non-ferromagnetic yet conductive target materials, or at temperatures higher than the Curie temperature of ferromagnetic materials.

As a drawback, different target materials give different sensitivities and, therefore, require different calibrations. Moreover, the thickness of the target is also influential on the sensitivity, since the eddy currents have a finite penetration depth δ in the material which depends on its μ and σ and on the field frequency f through the relationship

$$\delta = 1/\sqrt{\pi f \mu \sigma} \qquad\qquad\qquad (14.3)$$

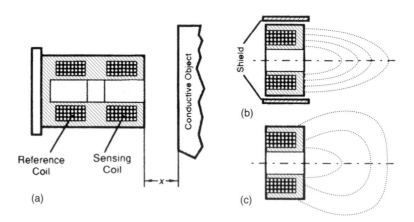

Fig. 14.5 Inductive displacement transducer based on eddy currents: (a) schematic diagram of the two-coil configuration; (b) shielded design; (c) unshielded design ([4, p. 279], reproduced with permission).

On the basis of this formula it is possible to use eddy-current probes to measure the thickness of metal foils or coatings and to detect material cracks.

Eddy-current sensors are almost always provided with built-in or external signal-conditioning electronic circuits which drive the coil, amplify the signal and linearize it to a typical value of $\pm 1\%$ FS.

A resolution as high as $\pm 0.1\%$ FS can be obtained. Typically, with small-size short-range devices with 5 mm of probe diameter and 0–1 mm measuring distance, submicron resolutions can be achieved. The time and temperature stability can be very high making eddy-current sensors very suitable for long-term operation even in harsh and dirty environments.

A frequency response typically ranging from zero to several kilohertz or a few tens of kilohertz and the absence of mechanical loading because of their noncontact operation make eddy-current sensors ideal for the measurement of vibration. For instance, they are well suited and widely used for measuring the vibrations and the eccentricity of rotating shafts, or the looseness of bearings.

As far as the mounting is concerned, attention should be paid to ensuring that the lateral dimensions of the target are at least two to three times the probe diameter and the target surface is as flat as possible. Especially for the unshielded version, the side-mounting of more transducers or the use of metallic fixtures may perturb the sensitivity and, therefore, the recommendations of the manufacturer should be followed to keep distances at safe values.

14.4.6 *Capacitive transducers*

Capacitive transducers are based on the principle that the capacitance of two electrical conductive bodies (armatures) separated by a dielectric medium varies if either the dielectric constant of the medium or the system geometry vary. The change in the dielectric properties of the separating medium is exploited, for instance, in liquid level or air humidity sensors. The change of geometry is well suited to use in dimensional measurements, such as linear and rotational displacement sensing. For instance, the principle can be applied in devices where an armature terminating in a shaft is guided to move between fixed armatures in a cylindrical geometry, giving rise to a capacitive linear potentiometer.

The capacitive effect is well suited to noncontact displacement measurements according to the expression of the capacitance C of two parallel plates of area A separated by an air gap d given by

$$C = \frac{\varepsilon A}{d} \tag{14.4}$$

where ε is the dielectric constant of air practically equal to that of vacuum $\varepsilon_0 = 8.85 \times 10^{-12}\,\text{Fm}^{-1}$.

With reference to the above formula, two alternative methods may be used, as illustrated in Fig. 14.6, namely the variation of the distance d between the armatures and the variation of the area of overlap A.

The noncontact capacitive probes are based on the former principle, i.e. they use capacitance variations to measure the air-gap d between two parallel conductive plates, of which one is the fixed reference and the second is attached to the moving object. If the object material is conductive it may act as the electrode, otherwise it may be equipped with a metal target or made conductive with the aid of conductive paint or rubbed graphite. As opposed to the eddy-current probes, the conductivity value of the target is not influential on the sensitivity. Capacitive transducers for nonconductive targets are also on the market, but their sensitivity typically depends on the target material and are mostly suited to proximity detection.

Equation (14.4) shows that the capacitance between two conductor plates varies nonlinearly with the plate spacing d. The problem may be partially overcome by operating the transducer over a reduced portion of its usable range to approach linearity. A better and elegant solution is given by employing an electronic readout scheme which provides an output signal proportional to the modulus of the transducer impedance $|Z|$. Since $|Z| = 1/\omega C = d/\omega \varepsilon A$, the output signal is proportional to d at a fixed frequency ω.

A further limitation to linearity comes from the fringing effect caused by the electric field lines diverging from parallel at the border of the plates due to their finite extension. As shown in Fig. 14.7, this problem may be solved with the help of the so-called **guard electrode** which encircles the moving armature and, by an electronic active driving circuitry, is kept at its same potential without, however, establishing any physical short-circuit between the two. In this way, the fringing effect is moved to the external border of the guard electrode, while the inner field lines in the region facing the sensitive electrode are steered to be perfectly parallel.

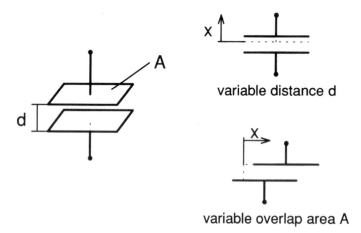

Fig. 14.6 Variable-distance and variable-area methods for measuring displacement along the direction x making use of a parallel-plate capacitive transducer.

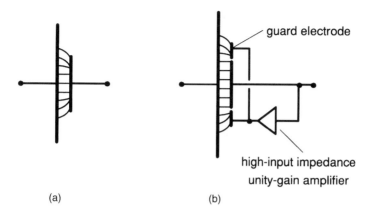

Fig. 14.7 Electric field lines in a capacitive displacement transducer: (a) without guarding electrode; (b) with guarding electrode.

Capacitive transducers with air-gap dielectric are passive transducers with high internal impedance, hence they are prone to interference pick-up and therefore require special shielding precautions. Since the capacitance values are typically small (few tens of picofarads or less), they could be swamped by the capacitance of coaxial cables of even moderate length. For these reasons, the signal-conditioning electronics needs to employ special readout techniques and is usually located in close proximity with the sensing head or often integrated in the sensor housing.

Extremely high resolutions can be obtained, essentially limited by the signal-to-noise ratio in the electronics (Chapter 15), and high temperature and time stability is ensured by the stable dielectric properties of air. However, air humidity can be significantly influential and should be as low as possible to avoid the occurrence of condensation. With dust-free electrodes made of high-quality corrosion-protected materials, such as invar steel, having low surface roughness, displacements of some hundredths of a micron may be detected.

Because the capacitive transducer is a zeroth-order instrument, the measurement frequency range is virtually limited only by the conditioning electronics. The market offers devices with a bandwidth exceeding 40 kHz which is a value well suited to vibration measurements. Nevertheless, capacitive probes are essentially special-purpose devices for high-end applications, and, even for reasons of cost, their use is still not as widespread as their high potential might suggest.

14.4.7 Optical transducers

Displacement transducers based on optical detection methods have the advantage of not requiring any particular electrical or dimensional characteristic in the target object, and of being essentially immune to electromagnetic interference. On the other hand, they are typically sensitive to the optical properties of the target object, such as colour, reflectivity, surface roughness, presence of dirt or dust, and of the optical path between the sensor and the object. Therefore, their use is generally confined to clean environments.

A very simple principle makes use of a light source, such as a light-emitting diode (LED), coupled to a light detector in a side-by-side arrangement contained in a single unit which is positioned in front of the target object. The amount of reflected light collected by the photosensor depends on the target distance. The method is simple and cheap but gives a limited range of linearity, and suffers from a significant dependence on the optical properties of the target.

Transducers based on the triangulation method employ the configuration depicted in Fig. 14.8. The light beam emitted by a visible or infrared light source, such as an LED or a semiconductor laser diode, is reflected by the target object and reaches a linear position-sensitive detector (PSD) at a particular point x of its length. Such a point is related to the target distance by trigonometric relationships, therefore the properly processed signal from the PSD gives a measure of the distance. Triangulation transducers typically have a working range of few millimetres around a stand-off distance that can be as high as several centimetres. The resolution is in the micron range with a frequency response no wider than few hundred hertz which is generally inversely dependent on the resolution.

The method with the highest performance and cost is that based on the laser interferometer. A Michelson configuration is generally adopted in which the laser light beam is split into two beams which travel along different paths. One path has a fixed length and works as the reference, while the other one comprises the distance from the light source to the measurand object usually equipped with a mirrored reflecting target. The two beams recombine in a photodetector and, due to the high coherence of the laser light, produce a neat interference pattern whose number of fringes can be counted and related to the target distance. The achievable resolution can be as high as 1 nm, and the frequency response extends from DC to tens of kilohertz or more. Such

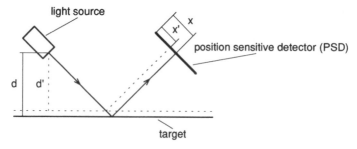

Fig. 14.8 The optical triangular method for noncontact displacement measurement.

performance makes the laser interferometer the preferred instrument where high-quality displacement and vibration measurements are needed, such as for laboratory calibration purposes or for highly demanding applications.

14.5 Relative-velocity measurement

14.5.1 *Differentiation of displacement*

In principle, the output signal from any displacement sensor can be differentiated with respect to time to obtain a velocity signal. This may be done either electronically, by making use of differentiation circuits cascaded to the transducer output, or as a postprocessing step on the recorded data. This indirect approach, however, has possible problems with the fact that the process of differentiation inherently enhances the high-frequency components in a signal, since the amplitude of each sinusoidal component at a frequency ω results multiplied by a factor ω. Therefore, any spurious high-frequency component in the displacement signal is amplified to a level which may impair the detectability of the true velocity signal. The situation may be critical, for instance, with wire wound potentiometers, due to their staircase characteristic which causes stepping output signals under rapid cursor movement, as well as with AC excited transducers, such as the LVDTs, due to possible residual ripple in the output signal.

Although the differentiating method may prove satisfying in several noncritical situations, the choice which provides a more general applicability and is therefore often preferred is to make use of velocity measuring transducers.

14.5.2 *Electrodynamic transducers*

The operating principle is based on the Faraday–Lenz law of magnetic induction for which the electromotive force (EMF), i.e. the voltage E, generated in a closed circuit is equal to the time derivative of the magnetic flux Φ linked with such a circuit. That is, $E = -d\Phi/dt$, with the minus sign representing the fact that the magnetic flux generated by the induced current caused by E opposes to the original flux Φ. On this principle, it is possible to develop self-generating sensors where velocity is converted into variations of the magnetic flux concatenated with a coil, and therefore produces a proportional output voltage signal.

A simple and effective method is that of making use of a permanent bar magnet positioned inside a coil and free to move relative to it. There are two alternatives, called the moving-coil and the moving-magnet designs, differing in that the former has the element fixed to the reference point, while the latter has it attached to the moving object.

The moving-magnet geometry is widely used for linear velocity transducers which generally base on the two-coil configuration shown in Fig. 14.9. The

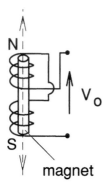

V_o

magnet

Fig. 14.9 Schematic diagram of an electrodynamic relative-velocity transducer.

two coils are mounted in series opposition and are needed to generate a net output voltage related to the magnitude and sign of the magnet velocity. Otherwise, with a single coil, the voltage induced by the movement of one end of the magnet would cancel that generated by the opposite end, with a zero net effect. It can be observed that the design of Fig. 14.9 is similar to that of an LVDT or a variable-inductance transducer (Section 14.4.3), but in this case a permanent magnet is used instead of the ferromagnetic core.

Electrodynamic velocimeters have the advantage of a noncontact and thus frictionless movement of the magnet with respect to the coils, and provide a high-level output with a typical sensitivity in the order of 10 mV/(mm/s). A method to increase sensitivity is to increase the number of coil turns, but this raises the coil resistance, thereby requiring more care in the design of the readout electronics and possibly limiting the high-frequency response. Typically, they can operate up to several kilohertz. The minimum detectable velocity, i.e. the resolution threshold, generally depends on the noise floor, especially for electromagnetic noise transmitted by nearby AC high-current equipment. Moreover, the presence of magnetic fields is another important source of interference which can perturb the sensitivity.

The electrodynamic principle, either in the moving-coil or moving-magnet designs, is widely and successfully adopted within the absolute velocity transducers based on the seismic instrument, as described in Section 14.7.3.

As a further application of the electrodynamic principle, it is possible that both the magnet and the coil are fixed, while the target object moves in front of them. In this case, the magnetic flux variation is produced by the velocity of the object, which needs to be ferromagnetic. Typical examples of such a configuration are given by the pick-ups of the electric guitars which sense the vibrations of the nickel-wound strings and convert them into a voltage signal, or by the toothed-rotor tachometers for the measurement of velocity of rotating members.

14.5.3 Laser velocimeters

The laser velocimeters are based on the Doppler effect, for which a light wave of frequency f reflected by a target moving at a velocity v relative to the light source becomes shifted to a frequency $f + \Delta f$, where Δf depends on the ratio v/c, and $c = 2.998 \times 10^8$ m/s is the speed of light in vacuum. The magnitude and sign of v can then be determined by measuring the difference between the frequencies of the emitted and reflected laser beams. Laser velocimeters are specialized and costly instruments which have the advantage of providing a noncontact measuring method. They are used for instance to measure vibrations in rotating blades or acoustic emitting surfaces, such as loudspeakers.

14.6 Relative-acceleration measurement

Ideally, acceleration may be derived by time-differentiating velocity or by doubly differentiating displacement, either electronically on the signals or numerically on the recorded data. However, it should be noted that the double-differentiation process is still more sensitive to high-frequency spurious components than the single differentiation is. Therefore, obtaining reliable acceleration data from displacement readings is hardly feasible except for very smooth signals. Starting from velocity signals may be less problematic depending on the particular case.

In general, given the availability of reliable and high-performance transducers which directly measure acceleration, as illustrated in the following section, it is common practice to make use of such devices, and the application of the differentiation techniques can be considered as an exception.

14.7 Absolute-motion measurement

14.7.1 The seismic instrument

Consider a single-degree-of-freedom mass–spring–damper system mounted within a case, as shown in Fig. 14.10, which is subject to motion when rigidly attached to a vibrating structure. Motion is assumed to be directed along the instrument sensitive axis, which is oriented vertically in the figure. The coordinate x_0 gives the position of the transducer base with respect to a motionless absolute reference, indicated as ground, while z_0 and y_0 give the position of the mass with respect to the same reference and to the base, respectively. The variations of x_0, z_0 and y_0 from their initial unperturbed values are indicated by x, z and y, which therefore represent the displacements referenced to ground, for x and z, and to the transducer base, for y.

The mass m is called the **proof** or **seismic** mass. The term seismic comes from the fact that vibrating the transducer base puts in motion the mass, in the same way as buildings are acted on through their foundations by earthquakes.

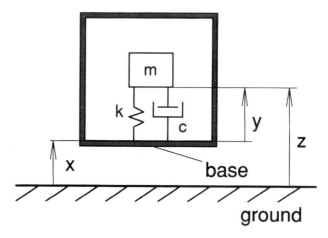

Fig. 14.10 Mechanical model of the seismic instrument.

The force equilibrium condition for the proof mass m requires that the elastic force f_k, the damping force f_c and the force f_g due to gravity balance the inertial force $m\ddot{z}$. Considering that $z = x + y$, this condition is expressed by the relationship

$$m\ddot{x} = -(m\ddot{y} + c\dot{y} + ky) - f_g \tag{14.5}$$

Passing to the $\mathscr{F}$-transforms X, Y and F_g, eq (14.5) becomes:

$$Y(\omega) = -\frac{\left(\dfrac{i\omega}{\omega_0}\right)^2}{\left(\dfrac{i\omega}{\omega_0}\right)^2 + 2\zeta\dfrac{i\omega}{\omega_0} + 1} X(\omega) - \frac{\dfrac{F_g}{k}}{\left(\dfrac{i\omega}{\omega_0}\right)^2 + 2\zeta\dfrac{i\omega}{\omega_0} + 1}$$

$$= -Y_x(\omega) - Y_g(\omega) \tag{14.6}$$

where $\omega_0 = \sqrt{k/m}$ is the transducer natural frequency, and $\zeta = c/(2\sqrt{km})$ is the damping ratio.

If the masses of the case and of the structure to which the transducer is attached were taken into account, the value of ω_0 would increase to the so-called mounted resonant frequency. The mounted resonant frequency tends to ω_0 for an infinite mass of the measuring structure. Therefore, ω_0 provides a useful conservative estimate of the mounted resonant frequency and approximates it well when, as generally happens, the transducer seismic mass m is negligible compared to that of the measuring structure.

Equation (14.6) shows that the transducer absolute displacement X causes a mass–base relative displacement Y_x which depends on frequency and on the transducer mechanical parameters.

Additionally, Y includes the term Y_g caused by the force due to gravity. If, as assumed in Fig. 14.10, the sensitive axis is oriented vertically and the transducer spans a limited altitude, F_g is nothing but the $\mathscr{F}$-transform of the constant proof mass weight mg, where $g = 0.981 \text{ m/s}^2$ is the gravity acceleration, i.e. $F_g = mg\delta(\omega)$. Therefore, the resulting term Y_g seen in the time domain produces only a static displacement $y_g = mg/k$. (For the mathematical details on the fact that the $\mathscr{F}$-transform of a constant k is given by $k\delta(\omega)$ the reader can refer to Chapter 2.) If, however, the transducer's sensitive axis has a different orientation, f_g and in turn y_g change accordingly both in magnitude and sign, being zero for a perfectly horizontal orientation. This implies that if the transducer happens to experience a motion with a nonzero rotational component in a vertical plane, then y_g is no longer a constant but depends on time and, as such, it becomes a signal source indistinguishable from that due to the absolute displacement $x(t)$. Failing to recognize such an orientation-dependent effect may cause significant errors in measurement using seismic transducers.

The transducers working on the seismic principle employ an internal method to measure the relative displacement $y(t)$, or some related quantity, and infer the absolute motion represented by $x(t)$ or its time derivatives on the basis of eq (14.6). The internal relative displacement is measured by a secondary sensor, the type of which can vary depending on the construction technology, and which provides the electrical output of the whole transducer.

14.7.2 Seismic displacement transducers

Assume that the contribution due to gravity is constant and directed in the negative x_0 direction (Fig. 14.10), so that the antitransform of the term Y_g in eq (14.6) reduces to the static displacement $y_g = mg/k$.

With reference to eq (14.6), the expression of Y_x is $Y_x(\omega) = T_d(\omega)X(\omega)$ where $T_d(\omega)$ is given by

$$T_d(\omega) = \frac{\left(\dfrac{i\omega}{\omega_0}\right)^2}{\left(\dfrac{i\omega}{\omega_0}\right)^2 + 2\zeta\dfrac{i\omega}{\omega_0} + 1}. \tag{14.7}$$

$T_d(\omega)$ can be called the displacement frequency response function. The magnitude and phase curves of $T_d(\omega)$ are shown in Fig. 14.11. The damping

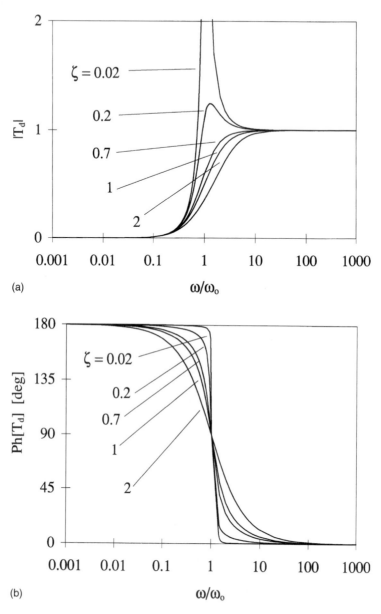

Fig. 14.11 (a) Magnitude and (b) phase of the displacement frequency response
function of a seismic instrument.

ratio ζ controls the peaking of the magnitude and the steepness of the phase around ω_0. For frequencies higher than ω_0, $T_d(\omega)$ approaches unity and therefore the relative displacement y_x becomes equal to the displacement x. Since $z = x + y$, this means that above ω_0 the mass remains still at $z = -y_g$.

Considering $x(t)$ as a pure sinusoidal displacement represented by the complex exponential $x(t) = x_M e^{i\omega t}$, then $y_x(t) = T_d(\omega)x_M e^{i\omega t}$.

The sinusoidal transfer function or sensitivity function $-k_e(\omega)$ of the displacement secondary sensor can now be introduced, where the minus sign is taken for notation convenience. According to eq (14.6) the electrical output $e(t)$ of the transducer is then given by

$$e(t) = -k_e(\omega)y(t) = k_e(\omega)y_x(t) + k_e(0)y_g$$

$$= k_e(\omega)T_d(\omega)x_M e^{i\omega t} + k_e(0)\,\frac{mg}{k} \tag{14.8}$$

Provided that $k_e(\omega)$ is constant in the region $\omega \gg \omega_0$ where $T_d(\omega)$ is unitary, the first term results in an output time signal which, in such a frequency range, is proportional to $x(t)$. Since $T_d(0) = 0$, the system is not sensitive to static displacements, therefore it is only capable of motion measurements. The second term represents the gravity-induced electrical signal. Since for nonvertical orientations such a signal would vary, in the absence of motion the transducer behaves as an **inclinometer** referenced to the gravity axis. It is worth noting that if $k_e(0) = 0$, i.e. if the secondary transducer is not sensitive to static displacement (as for instance in piezoelectric elements), then the gravity term is cancelled. This occurrence has no detrimental consequences on the sensitivity to $x(t)$, which is already zero at DC due to $T_d(0) = 0$.

The displacement sensitivity function $S_d(\omega)$ of the overall transducer is therefore given by

$$S_d(\omega) = k_e(\omega)T_d(\omega) \tag{14.9}$$

In practical transducer designs, the secondary sensing element which converts displacement into an electrical signal may be based on resistive potentiometers, LVDTs or noncontact displacement sensors. Alternatively, strain gauges bonded to a flexible member which behaves as the spring element, may be used.

To extend the useful bandwidth towards the low frequencies, the natural frequency ω_0 should be as low as possible. This requires low stiffness and high mass, which however imply reduced robustness and possibly higher cross-sensitivity, and unavoidably increase size and loading effect.

14.7.3 Seismic velocimeters

One possible way to measure velocity is by electronically differentiating the signal from a seismic displacement transducer. The possible problems related to the time differentiation operation are essentially the same as those already discussed for the case of the relative-displacement transducers in Section 14.5.1.

A more efficient and widely used solution is that of equipping the seismic system of Fig. 14.10 with a secondary sensor inherently sensitive to relative velocity, rather than displacement.

To analyse this solution, assume a purely sinusoidal displacement represented by the complex exponential $x(t) = x_M e^{i\omega t}$, so that $y_x(t) = T_d(\omega) x_M e^{i\omega t}$, and a fixed orientation in space, so that the gravity term y_g is constant. Therefore, the relative velocity is given by $\dot{y}(t) = T_d(\omega) x_M i\omega e^{i\omega t} = T_d(\omega)\dot{x}(t)$.

Introducing the sensitivity function $-k_e(\omega)$ of the velocity sensor, the electrical output $e(t)$ of the transducer is given by

$$e(t) = -k_e(\omega)\dot{y}(t) = k_e(\omega)T_d(\omega)x_M i\omega e^{i\omega t} = k_e(\omega)T_d(\omega)\dot{x}(t) \tag{14.10}$$

Therefore, the velocity sensitivity function $S_v(\omega)$ of the overall transducer results

$$S_v(\omega) = k_e(\omega)T_d(\omega) \tag{14.11}$$

It can be observed that as long as the secondary sensor sensitivity k_e is independent of frequency, the frequency response of the system as a velocity transducer is again given by the displacement response function $T_d(\omega)$, and therefore it may be analysed by referring back to Fig. 14.11. In particular, even in the case of k_e extending to DC, that is the secondary sensor being responsive to constant velocity, the overall sensitivity $S_v(\omega)$ has a low-frequency cutoff given by the natural frequency ω_0, which poses an inferior limit to the usable measurement bandwidth.

The most widely used method to convert relative velocity into an electrical signal is the electrodynamic principle, either in the moving coil or moving magnet variant. Seismic velocimeters of this kind are often called **vibrometers** or, in the case of transducers with very low resonant frequency (typically below 1 Hz), **seismometers**. Electrodynamic transducers have their main advantages in their self-generating nature with a high-level output, significant sensitivity (typical values range from 1 to 10 V/(m/s) and reliability of operation due to the absence of electrical contact and friction in the secondary sensor.

The drawbacks are that they are susceptible to magnetic fields, even if magnetically shielded versions are on the market, and tend to be sensitive to their orientation, since the damping and stiffness of the seismic system are somewhat influenced by the gravity force.

Regarding the frequency response, extending the bandwidth towards the low frequencies requires ω_0 to be as low as possible, but, as discussed for the seismic displacement transducers, this is usually obtained at the expense of an increased mass loading and reduced robustness. By setting the damping factor ζ typically close to 0.7, with oil filling or some other means, the usable bandwidth can be extended slightly below the resonant frequency ω_0 without appreciable phase distortion. The high-frequency limit is often posed by the first contact resonance between the transducer case and the structure and it is therefore dependent on the mounting method. Overall, the typical extension of the usable bandwidth is from few hertz to few kilohertz.

Seismic velocimeters are becoming less widely used due to the availability of reliable and high-performance accelerometers, but still find their way into those applications where low-frequency signals of interest are mixed with extraneous high-frequency components which would possibly cause an overrange condition in an accelerometer.

14.7.4 Seismic accelerometers

Suppose we are now interested in measuring the absolute acceleration $\ddot{x}(t)$, whose $\mathcal{F}$-transform is given by $\ddot{X}(\omega) = (i\omega)^2 X(\omega)$. With reference to eq (14.6), the term Y_x can be written as $Y_x(\omega) = T_a(\omega)\ddot{X}(\omega)$ where $T_a(\omega)$ is given by

$$T_a(\omega) = \frac{T_d(\omega)}{(i\omega)^2} = \frac{\left(\dfrac{1}{\omega_0}\right)^2}{\left(\dfrac{i\omega}{\omega_0}\right)^2 + 2\zeta\dfrac{i\omega}{\omega_0} + 1}. \tag{14.12}$$

$T_a(\omega)$ can be called the acceleration frequency response function, which can be recognized as the characteristic response function of a second-order system.

The magnitude and phase curves of $T_a(\omega)$ for various values of the damping ratio ζ are shown in Fig. 14.12. For frequencies from zero to around ω_0, $T_a(\omega)$ is equal to $(1/\omega_0)^2$ and, therefore, in this frequency region the relative displacement y_x, except for the sign reversal, is proportional to the absolute acceleration $\ddot{x}$.

Assume a purely sinusoidal acceleration $\ddot{x}(t)$ represented by the complex exponential $\ddot{x}(t) = \ddot{x}_M e^{i\omega t}$, so that $y_x(t) = T_a(\omega)\ddot{x}_M e^{i\omega t}$, and that the contribution due to gravity is constant, so that the antitransform of the term Y_g in eq (14.6) reduces to the static displacement $y_g = mg/k$.

The secondary sensor is chosen to be responsive to displacement, with $-k_e(\omega)$ being its sensitivity function. In this way, the electrical output $e(t)$ of the transducer results to be given by

$$e(t) = -k_e(\omega)\ddot{y}(t) = k_e(\omega)y_x(t) + k_e(0)y_g$$
$$= k_e(\omega)T_a(\omega)\ddot{x}_M e^{i\omega t} + k_e(0)\frac{mg}{k} \tag{14.13}$$

Provided that $k_e(\omega)$ is constant in the region $0 < \omega \ll \omega_0$ where $T_a(\omega)$ is constant, the first term results in an output signal which in such a frequency

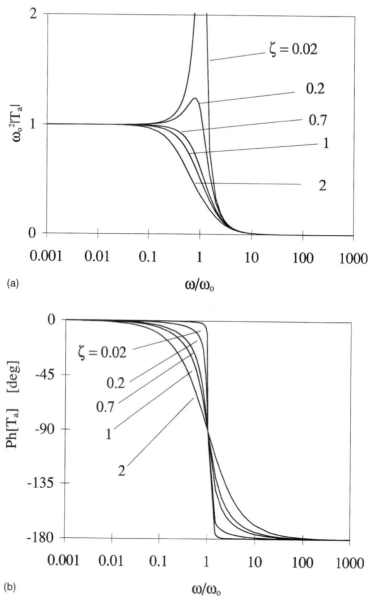

Fig. 14.12 (a) Magnitude (b) phase of the displacement frequency response function of a seismic instrument.

range is proportional to the measurand acceleration $\ddot{x}(t)$. The second term is due to the effect of gravity on the relative displacement y. If $k_e(0) = 0$ the transducer is not responsive to gravity and static accelerations. If on the contrary $k_e(0) \neq 0$, in the absence of motion it works as an inclinometer.

The acceleration sensitivity function $S_a(\omega)$ of the overall transducer is given by

$$S_a(\omega) = k_e(\omega)T_a(\omega) \tag{14.14}$$

Due to the fact that the low-frequency value of $T_a(\omega)$ decreases with increasing ω_0, a trade-off between sensitivity and high-frequency response is required. Depending on the damping value, the flat-band region, i.e. the range of frequencies where the sensitivity is constant and therefore the shape of the input signal spectrum is not distorted, extends more or less towards ω_0. Since the damping ratio depends on the particular accelerometer construction, the manufacturer's specification must be consulted for the case at hand. Typically, operation below $0.2\omega_0$ would ensure negligible sensitivity variations even for lightly damped transducers.

Usually, the unit used for acceleration is not the m/s^2 SI unit, but the gravitational acceleration $g = 9.81$ m/s^2. Therefore, in the case of an accelerometer with voltage output, the sensitivity may be expressed in volts per g.

Seismic accelerometers are probably the most widely used transducer for the measurement of vibrations, and they are well suited to the measurement of both continuous vibrations and transients, or shocks. Their functioning principle requires a high resonant frequency to obtain a wide measuring bandwidth, therefore they tend to have high stiffness and small mass. This makes them typically rugged and small size devices, especially in the high-frequency versions, thereby ensuring a reduced loading effect.

The demand for a high-sensitivity secondary sensor to detect the correspondingly small internal relative displacements is satisfied by several sensing methods, which result in diverse accelerometer types and construction technologies, as discussed in the following section. Depending on such sensing methods, the readout of the electrical output may require some precautions and special circuitry, but generally high-level signals with good linearity over a wide dynamic range may be obtained at moderate price. This, in turn, makes affordable the ever-increasing use of several devices in a multipoint test on the same structure.

In many cases, accelerometers prove useful in providing displacement and velocity signals by means of time integration, which is a technique intrinsically insensitive to noise and high-frequency disturbances due to its averaging nature.

14.8   Accelerometer types and technologies

14.8.1   *Piezoelectric*

Piezoelectric elements are insulators which become electrically polarized when subject to mechanical stress (direct piezoelectric effect), and conversely elongate or contract when electrically excited by the application of a voltage (reverse piezoelectric effect) [5].

Materials exhibiting the piezoelectric effect fall in two categories. The natural piezoelectrics, such as quartz (SiO_2), are intrinsically piezoelectric due to their structure. The artificial piezoelectrics are ferroelectric materials in which the piezoelectric effect is permanently induced by a poling process at the manufacturing stage. Poling consists of applying a high-intensity electric field to align the electric dipoles.

The reverse piezoelectric effect can be used for actuating purposes, such as in the case of ultrasound generation transducers. The direct piezoelectric effect can be exploited for sensing all those mechanical quantities which ultimately produce a stress on the piezoelectric element, such as force, torque or pressure. Piezoelectric sensors are self-generating. In piezoelectric accelerometers, the piezoelectric material has the role of the relative displacement secondary sensor. It behaves as a continuous elastic element generating an output charge Q proportional to the strain induced in the element itself by the inertial force of an overlying seismic mass. In some designs the seismic mass may be missing, leaving only the distributed mass of the piezoelectric element itself, which therefore embodies both the seismic system and the secondary sensor.

The proportionality factor between relative displacement and output charge is given by the charge sensitivity function $k_Q(\omega)$, which depends on the elastic properties and dimensions of the piezoelectric element, and by its intrinsic charge piezoelectric coefficient d, usually expressed in picocoulombs per newton. As will be discussed shortly, different piezoelectric materials have different values of d. Generally, in the region of interest, the charge sensitivity function $k_Q(\omega)$ can be considered as independent of frequency. Therefore, according to eq (14.14), the overall charge sensitivity $S_{Qa}(\omega)$ is given by

$$S_{Qa}(\omega) = k_Q T_a(\omega) \qquad (14.15)$$

The charge sensitivity is usually expressed in picocoulombs per g (pC/g) and its frequency behaviour is determined by $T_a(\omega)$.

As shown in the equivalent circuit of Fig. 14.13, the charge $S_{Qa}(\omega)\ddot{X}(\omega)$ generated by an acceleration $\ddot{X}(\omega)$ is developed across the capacitor C, which is formed by the portion of the piezoelectric material delimited by the two electrode faces. Piezoelectrically generated charges do not last indefinitely but

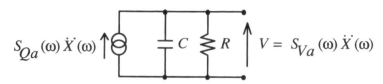

Fig. 14.13 Equivalent circuit of a piezoelectric accelerometer. $\ddot{X}(\omega)$ is the acceleration, $S_{Qa}(\omega)$ and $S_{Va}(\omega)$ indicate the charge and voltage sensitivity respectively.

tend to neutralize due to the fact that the material is an imperfect insulator, with losses represented by the resistance R in the equivalent circuit. These losses are responsible for an intrinsic discharging effect of the capacitor C. As a consequence, if the voltage V is taken as the electrical output quantity and the voltage sensitivity function $k_V(\omega)$ is introduced, it follows that $k_V(\omega) = k_Q i\omega R/(1 + i\omega RC)$. Similarly to $S_{Qa}(\omega)$, the overall voltage sensitivity $S_{Va}(\omega)$ can then be defined, expressed in volt per g, which obeys the expression

$$S_{Va}(\omega) = k_V(\omega)T_a(\omega) = k_Q \frac{i\omega R}{1 + i\omega RC} T_a(\omega) \tag{14.16}$$

From the comparison of the expressions for $S_{Qa}(\omega)$ and $S_{Va}(\omega)$ follows a very important fact regarding piezoelectric accelerometers. A low-frequency limit exists given by $\omega_L = 1/RC$ under which the voltage sensitivity $S_{Va}(\omega)$ drops, and at zero frequency $S_{Va}(0) = 0$. That is, piezoelectric accelerometers are not able to respond with their output voltage to DC acceleration. In particular, they are not sensitive to gravity acceleration and orientation. Seen in the time domain, this implies that the response to a step-changing acceleration is a decreasing exponential with a discharge time constant (DTC) given by $\tau_L = RC$.

The problem is virtually absent if the output charge is considered. However, in practice, the charge has to be extracted from the piezoelectric material in some way to be measured, and this in turn necessarily involves the presence of a time constant which is nothing but the product of the equivalent R_{eq} and C_{eq} of the electronic circuit used for the charge readout (Section 15.4.2). By properly choosing R_{eq} and C_{eq} the low-frequency cutoff limit can be varied with respect to $\omega_L = 1/RC$.

Charge- and voltage-output readout schemes also differ when the presence of the connecting cable is considered, which contributes with its equivalent capacitance C_S in parallel to the sensor capacitance C. As will be discussed in more detail in Chapter 15, the presence of C_S has essentially no consequences on the sensitivity $S_{Qa}(\omega)$ when charge amplification is employed. On the other hand, when voltage readout is adopted, the shunting action of C_S determines a diminution in the voltage sensitivity from its open-circuit expression $S_{Va}(\omega)$ of a factor proportional to C_S/C, and a corresponding variation in the low-frequency limit. In order to keep the value of C_S constant and as low as possible, the voltage readout is generally accomplished by inserting the amplifier within the transducer case, giving rise to the so-called low-impedance voltage-output transducers. Both charge- and low-impedance voltage-output piezoelectric accelerometers are on the market, with the former indicated more for laboratory use or where high operating temperatures would damage the built-in electronics, and the latter commonly used more for general purpose field applications.

At the high-frequency end, both $S_{Qa}(\omega)$ and $S_{Va}(\omega)$ are limited in the same way by the behaviour of $T_a(\omega)$, i.e. by the natural frequency $f_0 = \omega_0/2\pi$. The damping is typically low, resulting in a narrow region of resonance and in a steep phase change. Usually, the flat-band region is individuated by the upper and lower limiting frequencies where the voltage sensitivity is within 5% of its midband value given by $S_{Va} = k_Q/C\omega_0^2$. When proper mounting is adopted, operation up to $f_0/3$ and $f_0/5$ typically ensures a deviation from the midband sensitivity of 12% (1 dB) and 6% (0.5 dB) respectively.

Piezoelectric materials typically used in acceleration sensors are either quartz, or poled ceramics mainly of the lead zirconate-titanate or barium titanate family. For high temperatures tourmaline or lithium niobate are used. Quartz has a crystalline structure, is highly stable both thermally and over time, and it offers good repeatability. It represents, therefore, the best solution for transducers used for continuous monitoring over prolonged durations at temperatures between −190 and 240°C and for calibration reference standards. It has a piezoelectric coefficient d of 2.3 pC/N, which is a rather low value and, as such, the charge sensitivity S_{Qa} is not very high. On the other hand, since its dielectric constant is comparatively low, such a charge produces a rather high value of the open-circuit voltage sensitivity S_{Va}.

Poled ceramics have a piezoelectric coefficient which is much higher than quartz, reaching a typical value of 350 pC/N for lead zirconate-titanate (PZT). This causes a very high charge sensitivity, even if, by properly tailoring the material composition to keep its dielectric constant low, significant values of the voltage sensitivity can also be obtained. However, poled ceramics are polycrystalline materials and, as such, they are typically less stable then quartz. They tend to depolarize over time in a process called time depoling, and typically suffer from a significant sensitivity to temperature. Moreover, they generally suffer from a significant pyroelectric effect, consisting of the generation of charge under temperature variations which adds unwantedly to the piezoelectric signal of interest. In general, they lack stability when exposed to extreme mechanical or thermal shocks, even though accelerometers based on ceramics can operate between −190 and 400°C.

Irrespective of the material employed, the piezoelectric accelerometer construction basically conforms to one of the three configurations shown in Fig. 14.14. The compression mounting is the traditional and simplest construction where the piezoelement operates with the electrodes placed perpendicular to the transducer sensitivity axis. Generally, it includes a preloading spring and offers a moderately high sensitivity-to-mass ratio. However, it is rather sensitive to base bending and thermally induced inputs. Its typical use is in shock tests, where the signals are high, or in controlled laboratory environments for calibration purposes.

In the shear mounting the piezoelement operates with the electrodes placed parallel to the transducer sensitivity axis. This configuration generally offers a high sensitivity-to-mass ratio and a good thermal stability. Moreover, it gives a reduced sensitivity to base strain and a minimal cross-sensitivity, which makes the shear configuration the best choice as a general purpose accelerometer.

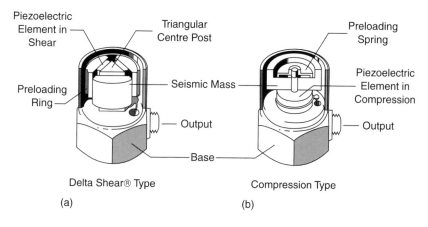

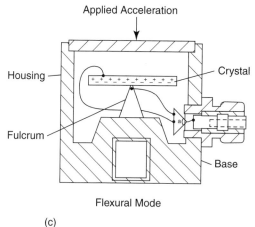

Fig. 14.14 Piezoelectric accelerometer designs: (a) shear, (b) compression ([6, p. 100], reproduced with permission) and (c) flexural beam (by courtesy of PCB Piezotronics, Inc.) configurations.

In the flexural beam configuration the piezoelement is shaped as a clamped beam which bends under the action of the acceleration. This configuration offers low cost with high output signal in a small size. Additionally, it has low sensitivity to base strain, thermal transients and transverse motion. However, the natural frequency is typically rather low, and it is therefore best suited to applications such as structural testing.

Independently of their configuration, piezoelectric accelerometers are reliable and rugged transducers which are durable, since they do not have moving parts. They present high stiffness and low mass, and therefore have high natural frequency and enhanced survivability to overloads up to a limit as high as 10^5 g. The sensitivity can be very high with low cross-sensitivity, and dynamic ranges as wide as 1 : 1000 are attainable, which are especially helpful around structural antiresonances where the motion is virtually absent and the signal is low compared to background noise. As a limitation, they are not sensitive to static acceleration and have a low frequency cutoff, but this in turn may be an advantage in those applications where it is required to have independence from sensor spatial orientation.

General purpose accelerometers may have a frequency range from 1 to 5 kHz or more, with a voltage sensitivity of up to 100 mV/g. High-frequency and shock types are available that work well beyond 10 kHz; however, these yield few millivolts per g. Low-frequency and low-acceleration level applications such as in vibration of bridges and large structures are addressed by the seismic-range accelerometers, which have large size and weight but can provide typically 1 V/g from a tenth of a hertz up to several hundred hertz.

Some manufacturers sell multifunction transducers which incorporate a piezoelectric accelerometer plus an additional sensing function, such as temperature or velocity/displacement obtained through electronic integration, in a single unit.

When talking about piezoelectric materials it is worth mentioning the piezofilms. They are polymers mostly derived from polyvinylidene fluoride (PVDF) shaped in thin flexible layers coated with metal electrodes. The piezoelectric coefficient d of PVDF is about 20 pC/N which is ten times greater than quartz and an order of magnitude lower than PZT. Piezofilms are very light and can be bent and deformed a great amount, which makes it possible to directly bond them even to small and irregularly shaped structures to detect vibrations. However, since the deformation pattern can be very complex, it is not very easy to determine exactly the motion direction. Though piezofilms have great potential in specialized applications as embedded sensors and have even been used as sensing elements in lightweight low-cost accelerometers, they are hardly suitable to be employed in precision vibration measurements.

14.8.2 *Combined linear-angular*

The flexural bending of piezoelectric beams can be exploited for the simultaneous measurement of both translational and rotational accelerations with a single device. The Translational-Angular PiezoBEAM® (TAP) sensor[1] accomplishes this task by making use of two clamped–free beams [7]. Each beam is actually a flexure bimorph made by two bonded piezoelectric layers electrically connected. The signals from the two beams are both summed and subtracted, providing two separate outputs. The beams are mounted in such a geometry that a translational acceleration produces a signal at the summing output, while a rotational acceleration determines a signal at the difference output, thereby producing separate information on both components without any appreciable cross-sensitivity.

---

[1] PiezoBEAM® is a trademark of Kistler Instrumentation.

These are very compact devices basically suited to low-frequency low-level applications such as modal testing. When three sensors are attached to a point with orthogonal orientations of the sensitive axes, the motion at such a point can be measured with respect to all six of its degrees of freedom.

14.8.3 *Piezoresistive*

Piezoresistive accelerometers are based on bending elements sustaining a seismic mass, whose relative displacement is detected by strain gauges properly located on the bending element. The more simple configurations use a seismic mass at the end of a clamped-free cantilever beam, equipped with strain gauges close to the clamping where the strain is maximum, as shown in Fig. 14.15. More sophisticated designs employ suspended bridge or diaphragm geometries which increase robustness and overload survivability, and reduce cross-sensitivity.

Piezoresistive sensors are passive, or modulating, sensors and therefore they require a source of excitation to provide an electrical output. They typically employ multiple strain gauges connected in a Wheatstone bridge configuration, which will be discussed in more detail in the following chapter on electronics. This allows us to maximize sensitivity by taking advantage of more than a single active gauge, and to virtually cancel the errors due to temperature-induced resistance variations.

In the case of a raw sensor bridge, the sensitivity $S_a(\omega)$ is usually given as fractional bridge imbalance, i.e. bridge imbalance voltage divided by the bridge excitation voltage, per unit of acceleration, that is $(\mathrm{mV/V})/g$. There are, however, many sensors on the market with built-in signal conditioning electronics which offer high-level outputs in several voltage or current ranges and, in such cases, the sensitivity is expressed accordingly.

A fundamental property of strain gauges is sensitivity to static strain. Therefore, according to eq (14.14), $k_e(\omega)$ and in turn, the sensitivity $S_a(\omega)$,

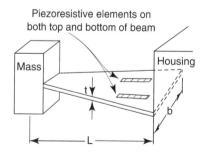

Fig. 14.15 Simplified structure of a piezoresistive accelerometer adopting strain gauges bonded to an elastic bending element which sustains the seismic mass ([8, p. 12.20], reproduced with permission).

include DC. That is, piezoresistive accelerometers are capable of responding to static accelerations and, in particular, to gravity.

Traditionally, piezoresistive accelerometers have been manufactured using bonded metal foil strain gauges on properly machined elastic elements. As usual, high sensitivity requires massive and compliant structures which end up having a low natural frequency and are prone to breakage. To avoid this occurrence, they often employ some method for increasing damping, such as filling the case with oil, and include mechanical stops. As an alternative method to improve the sensitivity, semiconductor strain gauges are used, since they provide a higher gauge factor. Overall, the traditional manufacturing method is labour intensive, costly and demands special precautions in the critical process of bonding the strain gauges; therefore it tends to be confined to special applications and its use is likely to decrease in the future.

The state-of-the-art piezoresistive accelerometers, and piezoresistive sensors in general, are instead made in silicon by the so-called silicon micromachining technology, which now represents a great portion of the market. This technology is based on the mature semiconductor processes used for the fabrication of microelectronic integrated circuits, and currently is extending to the development of so-called microelectromechanical systems (MEMS), which include electronic circuitry as well as sensing and actuating capabilities on the same chip. By making use of photomasking and chemical etching, silicon micromachining allows for the fabrication of three-dimensional suspended structures, such as bridges and diaphragms, with typical dimensions of few micrometres. Silicon has excellent elastic behaviour and mechanical properties and, by adding dopant impurities, it allows the creation of strain gauges, or piezoresistors, embedded in the structure bulk without bonding. Properly dimensioned air-filled spacings surrounding the elastic microstructure provide a controlled amount of damping by the so-called 'squeeze film' effect. This technological opportunity leads to batch manufacturing of low-cost miniaturized sensors which typically also integrate the signal-conditioning electronics in the same package or even on the same silicon substrate. In some cases, both filtered and unfiltered outputs are available so that the user can take advantage of either wide bandwidth or reduced noise according to the application.

Silicon integrated piezoresistive accelerometers are tiny, lightweight, have a generally high sensitivity because they include amplification and the gauge factor of semiconductor strain gauge is intrinsically high, and are capable of good high-frequency response. As a drawback, they generally cannot operate at temperatures higher than 100–120 °C and, within the operating range, they tend to be sensitive to temperature. Compared to the piezoelectric accelerometers, the piezoresistive ones have a generally smaller dynamic range and resolution. They are generally used in applications where extreme high-frequency response is not demanded and, conversely, DC sensitivity is required, such as in vehicle dynamics and biomedical motion analysis.

14.8.4 Capacitive

Capacitive seismic accelerometers measure the internal relative displacement y by detecting the capacitance change between a stationary armature and an armature attached to the seismic mass. The principle is that of the noncontact capacitive displacement sensors already illustrated in this chapter.

Most often, a differential configuration is employed by incorporating two capacitances which, under the motion of the seismic mass, undergo changes of the same magnitude but of opposite sign in a push–pull fashion. By measuring the difference of the two capacitances, the acceleration-induced displacements can be detected, while offset and drift are virtually cancelled.

Due to the fact that both dynamic and static displacements cause a capacitance change, capacitive accelerometers are capable of DC measurements, i.e. their sensitivity extends to zero frequency.

Despite their apparently simplistic structure, capacitive accelerometers require high-precision and low-tolerance manufacturing methods, in order to ensure high sensitivity, suitable ruggedness and overload survivability at a reasonable size and weight.

As in the case of piezoresistive sensors, also for capacitive accelerometers the technology of silicon micromachining offers the way to obtain high-performance and low-cost integrated sensors, and represents the leading technology in the field. Silicon capacitive accelerometers are based on suspended elastic elements incorporating electrodes, closely faced to fixed plates at typical distances of few microns. The geometry of the suspended element may be planar, giving rise to a variable-gap capacitor, or employ more elaborate comb-like configurations to increase sensitivity-to-size ratio.

Typically, silicon capacitive accelerometers also include the signal conditioning and amplification electronics, resulting in tiny and lightweight integrated sensors with a high-level output signal. Some manufacturers offer the bare sensor without the electronics, which comes in an external unit to be located close to the sensing head. This is due to the fact that capacitive sensing elements may operate at comparatively high temperatures (in the order of 150 °C), unless limited by the working range of the electronics (generally no more than 120 °C).

Silicon capacitive accelerometers with integrated microelectronics have typically a voltage output with sensitivity in the range of 100 mV/g to 1 V/g accompanied by low noise and high resolution and a mass of few grams. The bandwidth is a few kilohertz or less, and its upper limit is governed by the natural frequency of the seismic structure but is sometimes electronically reduced by internal filtering stages. As a very important and convenient feature, some devices incorporate a self-test feature, consisting of internally applying a fixed voltage to the sensing element which produces an electrostatic deflection of the elastic element corresponding to a preselected value of DC acceleration.

Typical applications of integrated capacitive accelerometers are in aircraft and automobile testing where they can withstand very high shocks, and they are rapidly replacing purely mechanical devices in car airbag deployment systems. Besides, their choice must be considered in all those cases where they are competitors of piezoresistive accelerometers in providing DC response with a high-level output signal.

14.8.5   *Force-balance or servo*

Force-balance accelerometers use the null method for quantifying the relative displacement of the seismic mass, as opposed to the sensors we have described so far, which rely on the deflection method. Force-balance accelerometers are feedback, or servo, systems which incorporate an electromagnetic or electrostatic actuator that continuously acts on the seismic mass in such a way to impede its relative motion as sensed by a position detector. Therefore, instead of measuring the seismic mass deflection caused by the acceleration, the motion is nulled by the proportional restoring force generated by the actuator which essentially works as an electrically controlled spring. The electrical input of the actuator is a measure of such a balancing force, and is therefore taken as the transducer output signal.

Since the mass remains essentially still, the influence of friction and mechanical hysteresis is virtually eliminated. Moreover, both the damping and, to some extent, the resonant frequency can be electrically set within the feedback loop, rendering them more controllable versus temperature and influence factors in general. The requirements set on the position detector are not particularly stringent, since the seismic mass experiences extremely reduced excursions around the neutral position.

For all these reasons, servo accelerometers are generally precision-class and costly devices which can reach a 0.1% FS overall accuracy. Their sensitivity and resolution are typically very good with a frequency response including DC. Therefore, they are often used to measure tilt in special versions called servo inclinometers, or in inertial navigation systems for avionics and military industry.

14.8.6   *Multiaxial accelerometers*

There are some applications which require the simultaneous measurement at a single point of the acceleration components in an orthogonal coordinate system. The market offers two kinds of solution to the problem, namely the multiaxial arrangement of monoaxial accelerometers, and the intrinsically multiaxial accelerometers.

In the former case, several manufacturers offer tridimensional mounting fixtures and slotted blocks to which up to three accelerometers are fitted in an orthogonal configuration. This solution is convenient and cost-effective, since it uses ordinary monoaxial sensors and gives the opportunity of mounting less than three devices on a block when sensor availability or cost are a factor. As the drawbacks, it increases the mass load on the measured structure due to the mass of the mounting fixture, and is somewhat critical with respect to mounting, since it typically tends to worsen the high-frequency response of the sensors due to the presence of an intermediate body between the accelerometers and the test object. Moreover, the mutliple output connectors and cables may cause problems when space is tight.

Essentially all of the above limitations are overcome by the intrinsically multiaxial accelerometers, which incorporate up to three orthogonal sensors sealed in a single housing. They are optimized for minimum mass load, reduced cross-axis sensitivity and usable bandwidth, with the output provided on a single multipolar wire or multipin connector. The cost of this solution should be carefully considered according to the application and budget, since a triaxial unit is probably cheaper than a monoaxial triplet plus mounting block, but the second approach offers more flexibility and reusability.

Multiaxial accelerometers generally are either piezoelectric, piezoresisitive or capacitive, with the respective features already illustrated for monoaxial sensors. The miniaturization capability of the silicon micromachining technology has recently brought onto the market a $\pm 50\ g$ biaxial capacitive accelerometer with electronics all integrated in tiny package like those ordinarily used for integrated circuits [9]. Such devices are factory calibrated and temperature compensated, and due to their low cost and size can be used for instance in permanent monitoring systems for machinery and struc- tures [10].

14.9 Accelerometer choice, calibration and mounting

14.9.1 *Choice*

As can be expected, none of the above illustrated principles and technologies for accelerometers is generally the best choice for every application. On the contrary, the preferred solution should be individuated by taking into account the accelerometer characteristic features, and how they globally match with the requirements of the particular application. The main factors influencing the choice are listed and briefly commented upon in the following.

Measurement range

For measuring high-level accelerations up to several thousand g, such as in shock tests, piezoelectric or piezoresisitive sensors are generally used. The second type may be less sensitive, but in some cases is more lightweight.

For medium-level accelerations, such as those encountered in most vibration measurements and in modal tests on average-sized machinery and structures, the choice widens to essentially all the sensor types illustrated, and generally involves the consideration of several concurring factors, such as sensitivity, bandwidth and operating conditions.

For low-level accelerations in the sub-*g* range, such as in vibration tests on large structures or buildings, piezoelectric seismic-range types or force-balance transducers are typically used.

Sensitivity, resolution and dynamic range

These three parameters have been grouped together, since they are actually mutually correlated, though they represent different concepts, as discussed in Chapter 13. High sensitivity means a large output amplitude for a given measurand level, but the minimum detectable level is set by the resolution limit. The dynamic range is the range of input levels that the transducer can measure, usually expressed as the maximum-to-minimum ratio. Accelerometers manufactured using different technologies may provide the same sensitivity value, which is typically higher in devices with lower bandwidth. However, not all the accelerometers are the same as far as the dynamic range and resolution are concerned. This fact may become an issue when there is the need of detecting low-level signals riding on a high-level background, or when both structural resonances and antiresonances must be detected neatly as in modal testing.

Piezoelectric accelerometers have typically the widest dynamic range compared to piezoresistive and capacitive types.

Frequency response

As a general rule for seismic accelerometers, stiff and lightweight sensors have high natural frequency and moderate sensitivity, while compliant and massive sensors increase sensitivity but have lower bandwidth.

When no DC response is required, piezoelectric types are generally preferred. Several manufacturers provide different configurations and sensitivity values in the frequency range of interest for modal testing, i.e. below 10 kHz, and even for shock measurements at frequency of several tens of kilohertz or beyond. Piezoelectric accelerometers can extend their low frequency limit below 1 Hz by properly tailoring the discharge time constant within the electronic preamplifier (Chapter 15). However, this practice can prove detrimental in the presence of thermal transients which generate low-frequency signals due to the pyroelectric effect.

When true DC response is needed, piezoresistive or capacitive accelerometers are generally adopted, though this does not mean that high frequency cannot be measured with these types of sensors. Indeed, piezoresistive silicon accelerometers are on the market with specified frequency limits as high as 150 kHz. When extremely low frequencies are involved, the sensors are typically required to have a significant sensitivity, since the displacements, and hence the acceleration levels, are necessarily limited. In these cases, force-balance sensors may be used, or highly amplified and noise-filtered piezoresistive and capacitive integrated sensors when size and weight is at a premium.

Environmental factors

The environmental factor of main concern is normally the temperature. In general, silicon microeletronic circuits cannot operate above 100–120 °C or below −25 °C (or, in special cases, −50 °C). Therefore, those sensors which incorporate electronics have more stringent temperature limitations compared to those without internal amplification.

Charge-output piezoelectric accelerometers are those which can reach the widest temperature operating range (from −269 to 750°C), while piezo-resisitive types tend to be the most temperature sensitive with respect to both sensitivity and damping characteristics. In some cases, thermal jackets can be fitted to the sensor to shield thermal transients. In cases of operation in a humid or wet environment, hermetically sealed sensors should be used, and attention should be paid to the perfect insulation of electrical connectors. If any leakage exists, significant variations in the low-frequency response may result, especially with charge-output piezoelectric accelerometers.

A factor which is often overlooked is the acoustic sensitivity of accelerometers, i.e. their unwanted responsivity to sound pressure fields which often accompany vibration phenomena. Piezoelectric sensors based on the shear configuration are generally less prone to acoustic interference than compression or flexure types.

The choice of sensors with metal housings for electromagnetic shielding purposes is suggested as often as possible, but it becomes mandatory in the presence of high electromagnetic interference (EMI), such as in contact or proximity of electric motors or machinery switching high-level currents.

Special working conditions, such as operation in high magnetic or ionizing radiation fields, are addressed by special purpose transducers supplied by some manufacturers.

Mass loading effect

Mass loading may become a problem in those structures with low effective mass of vibration, and in these cases lightweight accelerometers must be used. As a rule of thumb, the mass of the accelerometer should be no greater than one tenth of the structure equivalent mass to reduce the shift in both sensitivity and resonant frequency to negligible levels.

Unfortunately, it is not always easy to know the structural mass which is actually contributing to a particular vibration mode. Therefore, of practical utility is the method of inspecting the signal from an accelerometer located at a point of the structure, and compare it with the signal obtained when a second identical sensor is positioned in close proximity or, when possible, firmly attached to the first one. If significant differences are evident between the two responses, it means that the structure at that point is particularly sensitive to mass loading and the measurements will be affected by errors.

ming modal analysis on lightweight structures by subsequently ometers around the structure, inconsistencies in data may arise able mass loading resulting. Moreover, during the time needed to easurements, the modal parameters may vary due to change in the al factors determining the failing of the time invariance assump- problems are partially overcome by taking simultaneous measure- ments w.... several lightweight transducers placed on the structure. In this way test times shorten, and the mass load is minimal and evenly distributed.

Cost

When the application requires a large number of transducers to carry out the test, cost becomes an important factor. In particular, for modal analysis of complex structures with multichannel systems the low-cost piezoelectric modal array transducers are a good solution. The future trend is the use of smart accelerometers compliant with the IEEE 1451 standard, which have on-board memory to store calibration and identification parameters in a transducer electronic data sheet (TEDS).

Silicon micromachined sensors offer cost reduction as one of the major benefits, and for this reason they are increasingly installed permanently on machinery for routine monitoring and online fault diagnosis.

14.9.2 Calibration

The calibration of an accelerometer consists of accurately determining by experimental methods its sensitivity within the frequency range of interest. Calibration methods may be basically divided into comparison and absolute methods. Comparison methods involve checking the response of the accelerometer to be calibrated against that of a reference transducer, which is usually traceable to the primary standards maintained by the national metrological laboratories. Absolute methods are those in which the acceleration levels applied to the transducer under test are accurately determined through derivation from basic units.

The two most common comparison methods are based on harmonic excitation, and on gravimetric transient (or shock) excitation. The harmonic excitation method is also known as the back-to-back method, since it consists of mounting the accelerometer under test on the top side of the reference accelerometer whose opposite side is firmly attached to an electromagnetic vibration shaker. The shaker driving signal is swept in frequency and magnitude within the range of interest, and the output signals from both transducers are recorded. The sensitivity of the accelerometer to be calibrated is then derived with respect to that of the reference. Attention must be paid to ensuring a rigid mechanical coupling between the accelerometers, which is usually made easier by the provision of mounting holes and flattened surface on the reference transducer. Moreover, if the mass of the sensor to be calibrated is not negligible compared to the equivalent mass of the shaker plus the reference transducer, a significant loading error may result which is generally frequency dependent.

The typical obtainable accuracy with the back-to-back method, taking into account a 1–5% uncertainty in the calibration of the reference, is 5–10%. Electromagnetic shakers can usually operate from few hertz to few tens of kilohertz. For extremely low frequencies where calibration along linear strokes would imply inpractically large displacements, the modulation of the centrifugal acceleration of a rotating platform can be exploited [11].

As a variant of the harmonic excitation method, a broad-band noise signal can be applied to the shaker, resulting in a random vibration excitation. The output signals from the two transducers are then fed to a dual-channel FFT analyser which calculates the frequency response function. This gives the relative magnitude and phase of the sensitivity of the unknown transducer versus frequency.

A simplified version of the harmonic excitation method is implicitly used also in handheld calibration exciters. They are battery-powered units which provide fixed and precalibrated vibration levels to test accelerometers in a rapid and convenient way and are particularly useful in the field.

The gravimetric transient (or shock) method, sometimes called the drop-test method, consists of mounting the accelerometer to be calibrated on a steel mass which experiences a free-fall motion guided by a vertical tube. At the bottom of the tube is mounted a force sensor which detects the impulsive signal caused by accelerometer-plus-mass system impacting the sensor and stopping its fall. The signals from both the force transducer and the accelerometer are recorded with a digital-storage oscilloscope.

By Newton's law of motion, the impact force is the impact acceleration multiplied by the carriage mass inclusive of that of the accelerometer. Therefore, the unknown accelerometer sensitivity can be readily derived with reference to the values of the mass and of the force transducer sensitivity. Alternatively, the output signal from the force transducer with the static weight of accelerometer-plus-mass system applied can be previously measured and used as a scaling factor. In this way, the unknown accelerometer sensitivity becomes referred only to the local value of gravity acceleration g, but depends on the accuracy of the measurement of the static weight output.

To avoid exciting resonances in the transducers, the surface exposed to the impact is generally covered with a damping material working as a cushion. The gravimetric method is relatively simple and permits shock calibration from a few hundred to several thousand g, over a frequency range which depends on the particular system used.

Coming to the absolute calibration methods, probably the most simple one is that which uses the gravity field static acceleration. It simply consists of measuring the output signal of the accelerometer to be calibrated before and after rotating it by 180° along the vertical direction, i.e. perpendicularly to the earth's surface. The sensitivity is readily derived by taking the difference of the signals in the upward and downward orientations and dividing by $2g$.

The method offers the advantage that gravity acceleration is available everywhere and its value $g = 9.8062$ m/s^2 at sea level and 45° lat. changes by only 0.4% over the entire planet. As a drawback, it is only suited to accelerometers having true DC response, therefore it cannot normally be applied to piezoelectric sensors. Moreover, it gives no information on transducer linearity and frequency response. When attention is paid to ensuring a turnover rotation of precisely 180°, the typically obtainable accuracy is 1–2%. Alternatively, the 1g-step obtained by sudden free fall can be exploited.

The most sophisticated and high-quality absolute calibration method is that based on laser interferometry. Normally it uses a Michelson interferometer configuration to measure the displacement of the vibration shaker to which the accelerometer to be calibrated is mounted. The frequency of the sinusoidal signal exciting the shaker is measured by an electronic counter. Therefore, the generated acceleration can be calculated by the values of the motion displacement and frequency. This method offers very high accuracy and significant flexibility in varying the level and frequency of the test vibration. On the other hand, the cost is high and the typical use is confined to the laboratory calibration of primary standard transducers, or of transfer standard transducers of the highest quality and stability.

Though sensitivity is normally the most important parameter to be considered in accelerometer calibration, other relevant quantities may need to be checked. The transverse sensitivity may be determined by mounting the accelerometer on top of the reference transducer on the shaker with an inclination angle of 90°, and then monitoring the corresponding output.

The dynamic linearity is checked by exciting the accelerometer at various levels and verifying the amount of sensitivity variations. It can be also done indirectly by measuring the harmonic distortion.

The frequency response can be typically determined by sweeping the excitation frequency and monitoring the variation in sensitivity with respect to a reference frequency. Such frequency is chosen to be adequately distant from power line frequency and its harmonics to avoid interference, and it is normally 100 Hz in the USA and 160 Hz in Europe. The latter value has the advantage of being equal to 1000 rad/s, which simplifies calculations when integrations are involved. An estimate of the transducer natural frequency and damping can be obtained by suspending the accelerometer by the signal wire and tapping the case while recording the output signal. The frequency of the resulting damped oscillation is approximately equal to the transducer natural frequency, and the decay time of the envelope gives an indication of the damping.

The discharge time constant of piezoelectric accelerometers can be measured by turning the transducer upside-down while recording the output with a digital storage oscilloscope. The resulting signal has a stepping rising edge caused by the sudden application of a 2g input, followed by an exponential decay. The time required to the signal to fall to 37% of its initial amplitude is the discharge time constant.

14.9.3 Mounting

The importance of properly mounting the accelerometer to the structure under test should not be overlooked. Ideal mounting should allow vibrations to be transferred with maximum fidelity from the structure to the transducer. To this purpose it must be ensured that neither the frequency bandwidth nor the dynamic range of the accelerometer are limited by its poor mounting. Additionally, the attachment of the accelerometer to the structure must not appreciably alter the vibration characteristics compared to the unloaded conditions.

Insufficiently tight coupling due to poor mounting results in an equivalent spring interposed between the transducer and the structure, which produces mounting resonances and alters the frequency response and the sensitivity by an unknown amount.

The preferred mounting method is to use a threaded stud which is screwed into the structure and provides a rigid support for the accelerometer. The surfaces must be as flat as possible and before fastening the accelerometer it is advisable to add a thin film of silicone grease to improve high-frequency coupling. The stud method cannot be applied to those structures where drilling a hole is not possible or allowed, or when the transducer is not suitable for stud mounting. In such cases the transducer can be glued with fast-setting cyanoacrylate adhesives (ensuring thin glue lines), epoxy adhesives or dental cement for highest coupling stiffness. These bonding methods are permanent; once installed, the transducer cannot easily be removed and reused.

When temporary mounting is required, double-sided adhesive tape or self-adhesive discs can be employed. Mounting wax is also a possibility, but, due to the soft, unpredictable and temperature-dependent coupling action provided, its use should be confined to cases where no viable alternatives exist. When the test surface is made of ferromagnetic metal, the use of magnetic mounting adapters is convenient, especially when the accelerometer needs to be repeatedly moved to different places on the structure. For applications where measurement rapidity and convenience are more important than accuracy, the accelerometer can be mounted on handheld probes whose tip is kept by the operator in contact with the test structure. This quick-test method can give only indicative results, since the frequency response is heavily affected and the repeatability is poor.

In every case, the contacting surfaces of both the structure and the accelerometer should be smooth, and must be cleaned to eliminate every trace of dirt or grease. The location of the accelerometer on the structure should be carefully chosen, preferable on flat surfaces and avoiding nodal points when their position is known in advance. Attention should be paid to selecting positions with minimum transverse motion, which otherwise would cause measurement errors through the transducer cross-axis sensitivity.

When high-level shocks and transients are present which can mask weaker vibrations of interest at lower frequency, mechanical low-pass filters can be used. They consist of suitably dimensioned metal bodies which are stud-mounted between the accelerometer base and the test structure and provide a controlled damping action. The accelerometer becomes protected against high-frequency overloading, so that a transducer with smaller range and higher sensitivity can be used, which provides a better resolution in the useful frequency range.

Since the metal housing of most piezoelectric accelerometers is tied to the common side of the output signal circuit, problems of ground loops may arise when the transducer is attached to a point at nonzero potential or is providing a 'dirty' ground, such as on electrical machines (Section 15.5). In these cases, the use of mica washers or electrical isolation mounting bases may prove useful or necessary. For maximum accuracy, the effect of the isolation base on the accelerometer sensitivity can be determined by performing a previous calibration of the transducer with the isolation base fitted.

The connecting cable should be taped or clamped to the test structure in the vicinity of the accelerometer to relieve the transducer of cable-induced stresses and to avoid spurious signals caused by cable motion. In moist or wet environments it is advisable to arrange the cable in a recessed drip loop to prevent water from draining towards the connector side.

14.10   General considerations about motion measurements

When considering vibration problems in theory, it appears to be irrelevant which of the kinematic parameters displacement, velocity and acceleration is actually measured, since they are all related through subsequent time differentiation and in principle they all contain the same information. However, each vibration phenomenon has its own energy distribution along the frequency spectrum. Therefore, to maximize resolution and reduce the dynamic range requirements of the instrumentation, it is preferred in practice to measure the parameter which gives the most flat response over the frequency band of interest for the case at hand.

In vibration phenomena, appreciable displacement generally occurs only at low frequency, since increasingly high energies are required to excite high amplitudes at rising frequencies. Normally, displacement is measured in large structures vibrating at low frequency, such as buildings, ships or bridges. Another typical situation where displacement is measured is in testing of rotating elements such as shafts. In such cases a noncontact transducer, for instance an eddy-current probe, is generally a good choice. When a fixed reference point is not available to perform a relative measurement, a seismic vibration sensor can be mounted to the measured structure and its output electronically integrated to yield displacement.

Velocity is directly related to vibration energy and therefore the measurement of this parameter is indicated for vibration severity assessment. Additionally, velocity measurements are useful, for instance, in balancing rotating machinery. If the frequencies are not too high and the structure can tolerate the mass loading, a moving-coil seismic velocimeter is the traditionally preferred transducer.

As an alternative, a piezovelocity sensor can be used, consisting of a piezoelectric accelerometer with built-in or external integration circuitry. Compared to moving-coil velocimeters the piezovelocity sensors are lighter and thus provide a lower loading error, have possibly a higher resolution and a typically wider frequency response. In addition, they offer lower cross-sensitivity and are not influenced by mounting orientation.

As a further solution, a noncontact electrical or optical method can be adopted depending on operative conditions and budget.

The measurement of acceleration is privileged in terms of sensitivity when high-frequency components are of interest, such as in modal testing of comparatively small structures, in machinery monitoring, or in transient and pyroshock tests. Besides, acceleration is probably the most versatile parameter to be measured, since accelerometers can provide high dynamic and frequency ranges which permit velocity and displacement to be derived by signal integration. However, integration may be a problem with transient and shock signals, for which its use should be avoided, and when small offsets are present which would determine a drift in the integrated signal.

In some special cases involving failure monitoring where information is at very high frequency it can be useful to measure jerk, which is the time derivative of acceleration. Though dedicated jerk meters can be developed and exist, jerk is almost always obtained by electronically differentiating the signal from an accelerometer having a good high-frequency response.

14.11   Force transducers

14.11.1   *Functioning principle*

In vibration testing, force transducers, sometimes called load cells, are typically employed to measure the force exerted on a structure which is necessary to put it in a particular state of motion. By the simultaneous measurement of some characteristic parameter of the motion, for instance the acceleration, the frequency response function (FRF) of the structure between the excitation and detection points can be determined. If these two points are differently located on the structure a **transfer FRF** is derived, whereas if they coincide a **driving-point FRF** results.

A force transducer can be represented by the two-degree-of-freedom system of Fig. 14.16(a). The masses m_1 and m_2 represent the front, i.e. the face in contact with the structure under test at side 1, and the back mass at side 2. The front and back masses are separated by an elastic element of stiffness k and damping c. The applied forces f_1 and f_2 determine the elongation or contraction $x_1 + x_2 = x_0$ of the elastic element which are proportionally converted into an electrical output by some internal sensing element.

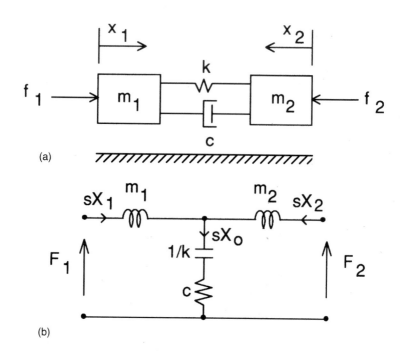

Fig. 14.16 Schematic representation of a force transducer: (a) mechanical model; (b) electromechanical analogue circuit.

Force transducers used in vibration measurements are generally operated in one of three ways: attached to a fixed base in a rigidly backed configuration, fitted to an impact hammer, or mounted between an exciter and the structure under test. Each of the three configurations has different boundary conditions at sides 1 and 2, and therefore is characterized by different measurement transfer functions. The three configurations can be analysed by deriving the corresponding mechanical equations, as is done for instance in [12]. We propose here a different approach based on the direct (voltage–force) electromechanical analogy discussed in Chapter 13. The electrical circuit analogue of the force transducer is shown in Fig. 14.16(b), where all the quantities are expressed in the Laplace domain as functions of the complex angular frequency s.

For the rigidly backed configuration, the displacement x_2 at side 2 is zero. Therefore, the equivalent circuit becomes that shown in Fig. 14.17(a), and solving the circuit for $X_0(s)$ as a function of $F_1(s)$ yields

$$X_0(s) = \frac{F_1(s)}{s} \frac{1}{sm_1 + k/s + c} = \frac{F_1(s)}{k} \frac{1}{s^2 m_1/k + sc/k + 1} \qquad (14.17)$$

It can be noticed that the static behaviour is determined by spring constant k and that the system resonates at $\omega_1 = \sqrt{k/m_1}$. Often, the parameter $f_1 = \omega_1/(2\pi)$ is specified by the manufacturer and called the rigidly mounted resonant frequency.

For the hammer mounting configuration, m_1 is the mass of the hammer impact head and tip, while m_2 is that of the hammer body including optional extenders. At side 2 it is present only the inertia of m_2 and no external force is applied, hence F_2 is zero. Therefore, the equivalent circuit is that shown in Fig. 14.17(b), and it follows that

$$
\begin{aligned}
X_0(s) &= \frac{F_1(s)}{s} \frac{m_2}{m_1 + m_2} \frac{1}{sm_1 m_2/(m_1 + m_2) + k/s + c} \\
&= \frac{F_1(s)}{k} \frac{m_2}{m_1 + m_2} \frac{1}{s^2 m/k + sc/k + 1}
\end{aligned}
\tag{14.18}
$$

where $m = m_1 m_2/(m_1 + m_2)$ is the system effective mass. It can be noticed that now the displacement is scaled by the factor $m_2/(m_1 + m_2)$ at low frequency, and the system resonates at $\omega = \sqrt{k/m}$ which is higher than ω_1.

For the case of the transducer mounted on an exciter, a force F_2 is applied on the exciter (side 2), producing a reaction force F_1 which depends on the mechanical impedance $Z(s)$ of the structure at the excitation point. Therefore, with reference to the equivalent circuit of Fig. 14.17(c), we have

$$
X_0(s) = \frac{F_1(s)}{k} \frac{1}{1 + sc/k} \left(1 + \frac{sm_1}{Z(s)}\right) = \frac{F_1(s)}{k} \frac{1}{1 + sc/k} (1 + A(s)m_1) \tag{14.19}
$$

where $A(s)$ is the structure **accelerance**, i.e. acceleration divided by force, at the measurement point. Two important facts can be observed in eq (14.19). Firstly, there is no resonance due to the transducer but only a high-frequency filtering action. Secondly, the term $(1 + A(s)m_1)$ introduces an error dependent on the unknown characteristics of the measured structure which becomes particularly large in the vicinity of the structural resonances. Methods for compensation of such an error are described for instance in [12].

14.11.2 *Piezoelectric force transducers*

Though several principles, such as the piezoresisitive and capacitive, are adopted for measuring the elastic element elongation in the construction of general purpose load cells, the most widely used force transducers in vibration tests are of the piezoelectric type. Piezoelectric force transducers offer essentially the same advantages of piezoelectric accelerometers. They have high resolution over a wide frequency range, high stiffness, good time and

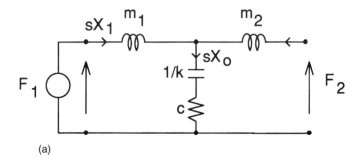

(a)

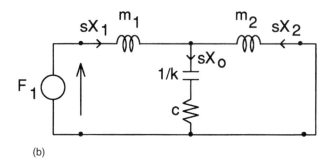

(b)

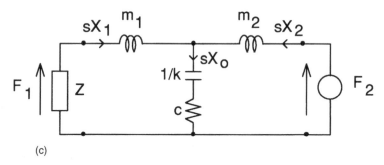

(c)

Fig. 14.17 Electromechanical analogue circuit for a force transducer under different operating conditions. (a) Rigidly backed configuration at side 2 with force F_1 applied at side 1. (b) Hammer mounting at side 2 with F_1 being the impact force at side 1. (c) Exciter mounting at side 2 with F_2 being the force generated by the exciter, F_1 the force transmitted to the test structure and Z its mechanical impedance.

temperature stability, rugged construction and reliable operation. Quartz is often used as the piezoelectric material due to its high stiffness which is comparable to that of steel. High resolution and dynamic range are important requisites that piezoelectric transducers satisfy, since the testing of a structure around its resonances, which is often a situation of interest, involves the detection of weak excitation forces.

Piezoelectric force transducers are supplied in different design geometries, such as rings or washers, and low-profile disc and axial links, depending on application, mounting requirements and range. Multiaxial units are also available. In Fig. 14.18 is shown the structure of a general purpose piezoelectric force transducer for tension–compression–impact use. Typically, some form of preloading of the piezoelement is internally provided in order to allow linear operation in both compression and tension.

The electrical output primarily generated by the piezoelectric element is a charge Q proportional to the experienced displacement x_0. The proportionality factor k_Q depends on the material piezoelectric coefficient and geometry, and in the region of interest can be considered independent of frequency. When charge output is considered, the force sensitivity S_{Qf} is generally expressed in picocoulombs per newton (pC/N) and in the Laplace domain is given by

$$S_{Qf}(s) = k_Q \frac{X_0(s)}{F_1(s)} \qquad (14.20)$$

where the mechanical transfer function $X_0(s)/F_1(s)$ has the expressions given in eqs (14.17), (14.18) and (14.19) respectively for the rigidly backed, hammer and exciter mounting.

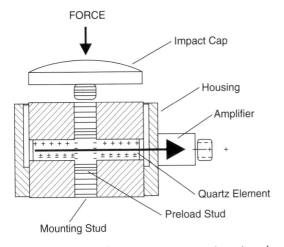

Fig. 14.18 Structure of a tension–compression piezoelectric force transducer with internal electronic amplification (by courtesy of PCB Piezotronics, Inc.).

The generated charge Q injected across the parallel of the piezoelement capacitance C and resistance R develops an open circuit voltage V that can be taken as the output signal. When voltage output is considered, the force sensitivity S_{Vf} in volts per newton (V/N) is given by

$$S_{Vf}(s) = k_Q \frac{sR}{1 + sRC} \frac{X_0(s)}{F_1(s)} \tag{14.21}$$

where again $X_o(s)/F_1(s)$ depends on the mounting configuration. Both eqs (14.20) and (14.21) can be expressed in the Fourier domain by replacing s with $i\omega$. It can then be observed that the voltage sensitivity $S_{Vf}(\omega)$ is zero at DC, therefore piezoelectric force transducers are not suitable for the measurement of static forces such as weight.

Similarly to what happens for accelerometers, the voltage sensitivity can significantly decrease from its open-circuit value when the signal cable is connected, due to the cable capacitance C_S which adds in parallel to C (Section 14.8.1). To overcome this problem, several manufacturers integrate a miniaturized voltage amplifier within the transducer housing which enables one to drive long cables without signal degradation (Section 15.4.3). These internally amplified transducers are generally defined as a low-impedance voltage-output type and are widely used in most applications. Alternatively, charge-output transducers are on the market which need to be coupled to charge amplifiers, but they are more suited to use in a laboratory environment.

14.11.3 *Impedance heads*

As an all-in-one device for driving-point FRF measurements, in particular for mechanical impedance testing, the impedance head transducer is available. It consists of a mechanical element incorporating a piezoelectric force transducer and an accelerometer mounted in such a way that their sensitivity axes are collinear. Impedance heads are useful and convenient because they combine both the measurement parameters of interest in the same instrument and referred to a single point. They tend, however, to have a comparatively large mass, and therefore attention should be paid with light structures to the possible loading error.

14.12 Summary

Input transducers, or sensors, convert physical quantities into electrical signals. Conversely, output transducers, or actuators, produce a physical action in response to an electrical driving stimulus. Self-generating sensors, such as the piezoelectric accelerometer, provide an electrical signal without the need for electrical power supply, unless for additional amplification or processing. Modulating sensors, such as the resistive potentiometer, require an external excitation which is modulated by the measurand quantity to give rise to the output signal.

When considering the means to measure the motion of an object, there are two methods that can be followed. In the relative-motion method the transducer is attached between the object and a reference point fixed in space, then the motion is referred to such a point. In the absolute-motion method a seismic transducer made by a mass–spring–damper system is attached to the object and the motion of the object is derived from the motion of the mass relative to the transducer base. In this case, the reference point is represented by the position in space of the system in the absence of motion.

In relative-motion measurement the attachment between the transducer and the measured object can be either of a contact or a noncontact type.

The most widely used contact displacement transducers are the resistive potentiometer and the linear variable differential transformer (LVDT). The resistance strain gauge can be also considered in the same category, even if it is somewhat different in that it actually measures strain and it generates a typically negligible loading error.

Noncontact transducers avoid the use of any mechanical link and, therefore, they generate no loading on the measured object and are suitable for measurements on rotating parts.

Noncontact displacement transducers can be inductive, capacitive or optical. The first two kinds require particular electromagnetic properties in the target material, such as high magnetic permeability or conductivity. The third kind is typically sensitive to the optical properties of the target object, such as colour, reflectivity, surface roughness, presence of dirt or dust, and of the optical path between the sensor and the object.

Velocity may be in principle derived indirectly by time differentiating a displacement signal. This is, however, a noise-enhancing process of limited practical applicability. Typically the direct measurement of relative velocity is carried out making use of electrodynamic transducers in either the moving-magnet or moving-coil configuration. For high-end applications demanding high performance irrespective of cost, laser velocimeters can be used.

There is no widely used method dedicated to the measurement of relative acceleration, given the availability of reliable absolute accelerometers.

Absolute-motion transducers are based on the seismic instrument, which comprises a mass–spring–damper system mounted within a case rigidly attached to the vibrating structure. The motion of the seismic mass relative to the case is measured by an internal secondary sensor which provides the output electrical signal. With a displacement secondary sensor, the transducer output is proportional to absolute displacement in the range of frequencies higher than the natural frequency of the seismic system. In the same frequency range, the transducer output is proportional to absolute velocity if a velocity secondary sensor is adopted. Again with a displacement secondary sensor but in the range of frequencies lower than the natural frequency of the seismic system, the transducer output is proportional to absolute acceleration.

Accelerometers are probably the most widely used seismic transducers due to their typically high sensitivity over a wide frequency range, small size and moderate cost. Piezoelectric accelerometers have a wide dynamic range and bandwidth, are well suited to both the measurement of vibration and shock, whereas they cannot measure constant accelerations. They are available either as charge-output or as low-impedance voltage-output transducers.

Piezoresistive and capacitive accelerometers are capable of DC response, can be manufactured by silicon micromachining and often have built-in amplification electronics. Sensitivity and bandwidth are very much dependent on the fabrication technology and are subject to wide differences. Force-balance accelerometers are capable of high accuracy and are suited to low-frequency and DC measurements.

Several vendors supply mutiaxial accelerometers which provide the measurement of acceleration at a point along three orthogonal directions.

Accelerometer calibration can be performed by either a comparison or an absolute method. Comparison methods involve checking the response of the accelerometer to be calibrated against that of a reference transducer, such as in back-to-back or in gravimetric transient calibration. In absolute methods the acceleration levels applied to the transducer under test are accurately determined through derivation from basic units, such as when gravity field static acceleration or a laser interferometer is used.

To obtain the maximum performance from an accelerometer it is necessary to properly mount it on the structure under test to ensure that vibrations are transferred with the highest fidelity to the transducer. Several mounting methods are available and, when possible, the one providing the highest mounting rigidity should be used.

Though displacement, velocity and acceleration are all related through subsequent time differentiation and in principle they all contain the same information, it is preferred in practice to try and measure the parameter which gives the most flat response over the frequency band of interest for the case at hand. In the case of motion of large structures or with low-frequency vibrations, the measurement of displacement can the best choice, in either the contact or noncontact mode. For medium-frequency analysis, for instance in vibration severity assessment or balancing of rotating machinery, the measurement of velocity can prove the most useful. The measurement of acceleration is privileged in terms of sensitivity when high-frequency components are of interest, such as in modal testing of comparatively small structures, in machinery monitoring, or in transient and pyroshock tests. Moreover, the measurement of acceleration has the advantage that in many cases it allows velocity and displacement to be derived by signal integration.

Force transducers used in vibration measurements are generally operated in one of three ways: attached to a fixed base in a rigidly backed configuration, fitted to an impact hammer, or mounted between an exciter and the structure under test. In all cases, they are employed to measure the force exerted on a structure which is necessary to put it in a particular state of motion. For this purpose piezoelectric force transducers are used almost invariably, and can be built in several shapes to satisfy different mounting and range requirements. As with piezoelectric accelerometers, both charge-output and low-impedance voltage-output force transducers are on the market, the latter ones being generally preferred for field applications. For driving-point measurements, the impedance head transducer can be used, which incorporates a piezoelectric force transducer and an accelerometer mounted with their sensitivity axes aligned.

References

1. Neubert, H.K.P., *Instrument Transducers*, 2nd edn, Clarendon Press, Oxford, 1975.
2. Norton, H.N., *Handbook of Transducers*, Prentice Hall, Englewood Cliffs, NJ, 1989.
3. Khazan, A.D., *Transducers and Their Elements*, Prentice Hall, Englewood Cliffs, NJ, 1994.
4. Fraden, J., *AIP Handbook of Modern Sensors*, AIP Press, New York, 1995.
5. Cady, W.G., *Piezoelectricity*, Dover Publications, New York, 1964.
6. Broch, J.T., *Mechanical Vibration and Shock Measurement*, Bruel and Kjaer, 1980.
7. Kubler, J.M. and Bill, B., *PiezoBEAM® Accelerometers: A Proven and Reliable Design for Modal Analysis*, Application Note, Kistler Instruments.
8. Eller, E.E. and Whittier, R.M., Piezoelectric and piezoresistive transducers, in C.M. Harris (ed.), *Shock and Vibration Handbook*, 3rd edn, McGraw-Hill, New York, 1988, Ch. 12.
9. Doscher, J. and Kitchin, C., Monitoring machine vibration with micromachined accelerometers, *Sensors Magazine*, May, 33–38, 1997.
10. Robinson, J.C. *et al.*, Using accelerometers to monitor complex machinery vibration, *Sensors Magazine*, June, 36–42, 1997.
11. Marioli, D., Sardini, E. and Taroni, A., A system for the generation of static and very low-frequency reference acclerations, *IEEE Transactions on Instrumentation and Measurement*, **46**(1), 27–30, 1997.
12. McConnell, K.G., *Vibration Testing*, John Wiley, New York, 1995.

15 Signal conditioning and data acquisition

15.1 Introduction

The role of the electronic chain starting at the transducers' output and ending at the data acquisition and analysis instruments is that of collecting the often weak and barely detectable measurement signals from sensors and enhancing the useful information content that they carry, while discarding the background components of no interest. This is primarily carried out in the signal-conditioning stage, which is often erroneously regarded as a piece of electronic circuitry which essentially increases the measurement sensitivity by signal amplification. This is only partly true, since the role of the signal-conditioning circuits is not merely that of amplifying the signal, but rather that of augmenting the signal magnitude over the background noise.

As an example, imagine you are sitting in the audience of a theatre and are tape-recording the music played by an orchestra on the stage. If your neighbours are speaking loud enough that their voices are picked up by the recorder's microphone and obscure the music, you do not gain any advantage in merely turning up the recording level. In fact, this operation would increase both the desired music and the unwanted background voices by the same amount, with no net improvement in the music intelligibility. To change the situation and solve the problem you may either get closer to the stage, i.e. increase the signal level, or ask people around you to be quieter, i.e. reduce the noise, or both. This is essentially what the signal-conditioning stage is designed to do, that is to provide selective and specifically tailored amplification to improve the signal-to-noise ratio. When dealing with measurement signals, this is equivalent to increasing the achievable resolution and, ultimately, the amount of information that can be extracted by the measurement process. Such information then needs to be carefully acquired, processed and made available and understandable to the human operator by further stages in order to make it useful for the purpose of interest.

Following this outline, this chapter is devoted to the electronic chain from transducers to readout instruments and is intended to provide the reader with some basic information on its typical functionality, capability and use. The coverage is principally aimed at signals and systems encountered in vibration measurements, but the approach is rather general and several of the concepts introduced are suitable to be extended to cases different to those explicitly treated.

The concept of signal-to-noise ratio is firstly illustrated, then some examples on how it can be improved by both signal amplification and noise reduction are described. Then the subject of analogue-to-digital conversion is introduced, and its main features are presented.

Finally, the instruments and systems for data acquisition and signal analysis are briefly illustrated as far as their functioning and basic use are concerned. No emphasis is given to the signal-analysis techniques and data-processing methods that such systems and instruments enable to perform, since they are outside the scope of this book. The interested reader is invited to consult the references on the topic listed in the further reading section.

15.2   Signals and noise

The term noise in electronic systems is used, in analogy with sound, to indicate spurious fluctuations of a signal around its average value due to various interfering causes which obscure the information of interest in the signal [1, 2]. It can be distinguished as an intrinsic noise, called electronic noise, which is caused by phenomena occurring in electronic components and amplifiers and is inherent to their operation and construction. Electronic noise can be minimized but not completely cancelled, since it depends on fundamental laws of nature governing the operation of electronic components.

In addition to electronic noise, there is generally present an amount of interference noise caused by external sources of disturbance, such as nearby power electrical machines, radiowave transmitters, or cables carrying significant amount of time-variant current. Therefore, interference noise results from nonideal experimental conditions and, in contrast to electronic noise, can be virtually eliminated if all the external sources of disturbance are identified and neutralized.

Noise may be of a random or deterministic nature depending on the phenomena which cause it. Electronic noise is typically random, while interference noise may often show up as deterministic to some degree. Deterministic interference noise can be caused by external sources generating a disturbing action with a somewhat regular and predictable behaviour, such as for fluorescent lamps or mains transformers which generate noise at the mains frequency and its harmonics. After these introductory considerations, we will simply use the term noise, as is customarily done in practice, to include both the electronic noise and the interference, differentiating between the two con-tributors only when required by the specific context in which they are treated.

Focusing attention on random noise, the fluctuations which are super-imposed on the average signal and constitute noise cannot be represented by a definite function of time, since the instantaneous values are unknown and cannot be predicted. Random noise is in fact a stochastic process that can only be described in terms of its statistical properties, as discussed in Chapter 12.

Usually, it is assumed that the noise amplitude probability distribution is Gaussian with zero mean, and that the stochastic process is stationary and ergodic, so that the ensemble averages are equivalent to time averages of any particular process realization.

Therefore, indicating with $x_S(t)$ a signal, such as a voltage or a current, and with $x_N(t)$ the amplitude of the superimposed noise fluctuations so that $x(t) = x_S(t) + x_N(t)$, it follows that the noise average value $\overline{x_N(t)}$ is

$$\overline{x_N(t)} = 0 \qquad (15.1)$$

and that the noise mean-square value $\overline{x_N^2(t)}$ is given by Parseval's theorem and is equal to

$$\overline{x_N^2(t)} = \int_0^{+\infty} S_N(f)\, df \qquad (15.2)$$

where $S_N(f)$ is the monolateral (i.e. considering the frequency f varying from 0 to $+\infty$) power spectral density of the noise.

If $S_N(f)$ is a constant independent of frequency, the noise is called **white noise**, in analogy with white light which is composed of an even mixture of all the frequencies. Examples of electronic noise which are white over a large frequency range are the thermal, also called Nyquist or Johnson, noise of resistors and the shot, or Schottky, noise of semiconductors. A kind of noise which is encountered in a wide variety of systems, from electronic, to mechanical, thermal and biological, is one for which $S_N(f)$ varies with frequency as $|1/f|^\alpha$ with α usually very close to unity. This kind of noise is normally called **1/f noise**, but other popular terms are low-frequency, flicker or pink noise. The $1/f$ noise is very important in measurement systems of slowly variable quantities, because it mainly affects the low-frequency region where the signal of interest is located.

We are now in a position to introduce the **signal-to-noise (S/N) ratio**, which can be defined as the ratio between the mean square values of the signal and the noise. To make this definition consistent, it is important that both the signal and the noise are considered at the same point in the system. Usually, all the noise contributions present in the system are divided by the appropriate gain factors and referred to the system input. The referred-to-input (RTI) noise and the input signal are then directly comparable and undergo the same amplification toward the system output. Assuming that $x_N(t)$ is the RTI noise, the S/N ratio, which is usually expressed in decibels (dB), is given by

$$\frac{S}{N}[\text{dB}] = 10 \log\left(\frac{\overline{x_S^2(t)}}{\overline{x_N^2(t)}}\right) = \frac{\displaystyle\int_0^{+\infty} S_S(f)\,df}{\displaystyle\int_0^{+\infty} S_N(f)\,df} \qquad (15.3)$$

where $S_S(f)$ is the signal monolateral power spectral density.

In practical cases, eq (15.3), which has a general theoretical validity, modifies for two aspects. Firstly, real signals are necessary band-limited between, say, f_{min} and f_{max}, with $S_S(f) \cong 0$ outside such a frequency range. Secondly, every real system has a finite bandwidth extending from f_1 to f_2, with $f_1 = 0$ in the case of a DC-responsive system. Of course, f_1 and f_2 must be chosen so that $f_1 \leqslant f_{min}$ and $f_2 \geqslant f_{max}$ to include the signal into the system bandwidth. Therefore, eq (15.3) in practice becomes

$$\frac{S}{N} \, [\text{dB}] = 10 \log \left(\frac{\overline{x_S^2(t)}}{\overline{x_N^2(t)}} \right) = \frac{\displaystyle\int_{f_1}^{f_2} S_S(f)df}{\displaystyle\int_{f_1}^{f_2} S_N(f)df} = \frac{\displaystyle\int_{f_{min}}^{f_{max}} S_S(f)df}{\displaystyle\int_{f_1}^{f_2} S_N(f)df} \qquad (15.4)$$

This result points out the importance of properly tailoring the system bandwidth according to both the signal and the noise characteristics.

If the noise is white or has significant components outside the signal bandwidth, it is desirable to reduce the system bandwidth $[f_1, f_2]$ as close as possible to $[f_{min}, f_{max}]$ by proper filtering, since this operation has the effect of maximizing the S/N ratio. On the other hand, keeping the system bandwidth much wider than the signal bandwidth is useless and has the only detrimental effect of collecting more noise. Unfortunately, the portion of the noise which resides within the signal bandwidth cannot be directly removed without affecting the signal as well. Special techniques can be used in these cases, such as the modulation which will be briefly presented later in this chapter.

15.3   Signal DC and AC amplification

15.3.1   *The Wheatstone bridge*

The Wheatstone bridge represents a classical and very widespread method for measuring a small resistance variation ΔR superimposed on a much higher average value R. This situation represents a rather typical occurrence in transducers, and is for instance encountered in strain-gauge-based sensors, where $\Delta R/R$ can be as low as 1 part per million (ppm), and other resistive sensors such as resistive temperature detectors (RTD).

The Wheatstone bridge consists of four resistors arranged as two resistive dividers connected in parallel to the same excitation source, as shown in Fig. 15.1. Such a source can be either constant or a function of time, and either made by a current or a voltage generator. In the following, we shall consider a constant voltage excitation V_E, which is the most frequently used in practice.

The bridge output voltage V_o is given by:

$$V_o = V_E \left(\frac{R_1}{R_1 + R_2} - \frac{R_3}{R_3 + R_4} \right) \qquad (15.5)$$

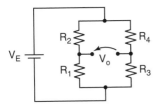

Fig. 15.1 The Wheatstone bridge with a DC voltage excitation.

When the condition $R_1/R_2 = R_3/R_4$ is satisfied, it follows that $V_o = 0$ and the bridge is said to be balanced. It should be noted that the balance condition is independent of the excitation voltage V_E.

The bridge can be operated in two modes, namely balance and deflection operation. In balance operation, one of the bridge resistors, say R_1, is the unknown resistor, and R_2 and R_4 are constant while R_3 is adjusted, either manually or automatically, until the bridge is balanced. At that point, R_1 can be calculated from the balance condition and the known values of the remaining three resistors. Deflection operation is more often used in transducer design and consists of letting the bridge work in the off-balance condition. The imbalance voltage V_o is then measured and related to the resistance variations of one or more resistors in the bridge.

Suppose that $R_1 = R + \Delta R$ and $R_2 = R_3 = R_4 = R$, with $\Delta R \ll R$. This condition is named the quarter-bridge configuration; R_1 is the active resistor and R_2, R_3 and R_4 are the bridge completion resistors. In this condition the bridge output voltage is given by

$$V_o = V_E \left(\frac{R + \Delta R}{2R + \Delta R} - \frac{1}{2} \right) \cong V_E \frac{\Delta R}{4R} \tag{15.6}$$

That is, the voltage output is proportional to the fractional resistance variation $\Delta R/R$ (provided it is sufficiently small) which can be determined by measuring V_o and knowing V_E. Equation (15.6) contains the essence of the bridge deflection approach to the measurement of small resistance variations. Instead of measuring $R + \Delta R$ and then requiring the subtraction of the offset R to retrieve the value of ΔR, the bridge intrinsically performs the subtraction and directly outputs the variation ΔR.

In piezoresistive sensors, the active resistor R_1 is a strain gauge. Almost always, multiple strain gauges are used and connected in pairs properly located on the elastic structure, so that one element in the pair elongates while the other one contracts by an equal or proportional amount. If one or two tension–compression pairs are used, the corresponding configurations are named the half- or full-bridge configuration respectively.

A survey of the possible configurations is given in Fig. 15.2. The bridge imbalance voltage can be generally expressed as

$$V_o = V_E \gamma \tag{15.7}$$

where γ is the bridge fractional imbalance which is approximately equal to $\Delta R/(4R)$, and exactly equal to $\Delta R/(2R)$ and $\Delta R/R$ in the quarter-, half- and full-bridge respectively.

It can be observed that the use of tension–compression pairs increases the sensitivity over the quarter-bridge. Moreover, the nonlinearity inherent in the quarter-bridge configuration is removed since the current in each arm is constant. Another advantage of making use of the configurations incorporating multiple piezoresistors is the intrinsic temperature compensation provided. In fact, if all the strain gauges have the same characteristics and are located closely so that they experience the same temperature, their thermally induced resistance variations are equal and, as such, they do not contribute any net imbalance voltage. The same result can hardly be obtained in the quarter-bridge configuration, because the strain gauge and the

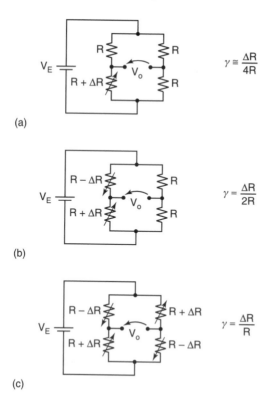

(a)

(b)

(c)

Fig. 15.2 Wheatstone bridge configurations for resistive measurements: (a) quarter bridge; (b) half bridge; (c) full bridge.

completion resistors normally have different thermal coefficients of resistance (TCR) and, moreover, are subject to different temperatures.

In practical cases, the excitation voltage V_E is in the range of few volts and the bridge imbalance voltage V_o can be as low as few microvolts, and therefore it requires amplification. This is generally accomplished by a differential voltage amplifier, called an instrumentation amplifier (IA), with an accurately set gain typically ranging from 100 to 2000, and a very high input impedance in order not to load the bridge output by drawing any appreciable current.

Since V_o is proportional to V_E, any fluctuation in V_E directly reflects on V_o causing an apparent signal. To overcome this problem, a ratiometric readout scheme is sometimes used in which the ratio V_o/V_E is electronically produced within the signal conditioning unit, thereby providing a result which is only dependent on γ. In turn, γ is related to the input mechanical quantity to be measured through the gauge factor and the material and geometrical parameters of the elastic structure.

The Wheatstone bridge can be also used with resistance potentiometers. In this case, with reference to Fig. 15.1, one side of the bridge, say the left, is made by the potentiometer so that R_1 and R_2 represents the two resistances into which the total potentiometer resistance R_P is divided according to the fractional position x of the cursor. That is $R_1 = xR_P$ and $R_2 = (1 - x)R_P$ with $0 \leqslant x \leqslant 1$. Then, the system works in the half-bridge configuration and, assuming $R_3 = R_4$, according to eqs (15.5) and (15.7) the bridge fractional imbalance is given by $\gamma = (x - \frac{1}{2})$.

The Wheatstone bridge with DC excitation may be critical in terms of S/N ratio when the signal γ is in the low-frequency region. In fact, in this case the bandwidth of the bridge output voltage V_o becomes superimposed with that of the system low-frequency noise, which is typically the largest noise component in real systems.

Moreover, an additional spurious effect comes from the DC electromotive forces (EMF) arising across the junctions between different conductors present in the bridge circuit, and from their slow variation due to temperature called the **thermoelectric effect**. This causes a low-frequency fluctuation of the bridge imbalance indistinguishable from the signal of interest.

Both problems may be greatly reduced by adopting an AC carrier modulation technique, as illustrated in the following section.

15.3.2 AC bridges and carrier modulation

If reactive components have to be measured instead of resistors, such as for capacitive or inductive transducers, the bridge configuration of Fig. 15.1 can again be adopted with the resistors now substituted by the impedances Z_1, Z_2, Z_3 and Z_4.

Since the impedance of inductors and capacitors at DC is either zero or infinite, the bridge now requires an AC excitation, which we can assume to be a sinusoidal voltage expressed in complex exponential notation as $V_E(t) = V_{Em}e^{i\omega_E t}$. An expression equivalent to eq (15.5) can then be written for the bridge output $V_o(t)$, leading to:

$$V_o(t) = \left(\frac{Z_1}{Z_1 + Z_2} - \frac{Z_3}{Z_3 + Z_4} \right) V_{Em}e^{i\omega_E t} \tag{15.8}$$

Similarly to the resistive bridge, the balance condition is given by $Z_1/Z_2 = Z_3/Z_4$, which, however, involves complex impedances and hence actually implies two balance requirements, one for the magnitude and one for the phase. The balance condition is independent of the excitation amplitude V_E but, in general, does depend on the frequency ω_E.

Equation (15.8) also describes the bridge deflection operation, with the term

$$\left(\frac{Z_1}{Z_1 + Z_2} - \frac{Z_3}{Z_3 + Z_4} \right)$$

representing the bridge fractional imbalance γ introduced in eq (15.7) which is now a complex function of the excitation frequency. In general, both the amplitude and the phase of $V_o(t)$ depend on γ and, as such, they may vary with frequency. Therefore, the determination of γ from $V_o(t)$ for a given known excitation $V_E(t)$ can be rather involved.

Fortunately, there are several cases of practical interest where the situation simplifies considerably. Suppose, for instance, that Z_1 and Z_2 represent the impedances of the two coils of an autotransformer inductive displacement transducer as described at the end of Section 14.4.3, or alternatively, the impedances of the two capacitors of a differential (push–pull) configuration used for the measurement of the seismic mass displacement in capacitive accelerometers, as mentioned in Section 14.8.4. In both cases, it can be readily shown that $Z_1 = Z(1 + x)$ and $Z_2 = Z(1 - x)$, where x is the fractional variation of impedance induced by the measurand around the average value Z. If the completion impedances Z_3 and Z_4 are chosen so that $Z_3 = Z_4 = Z$, which is most typically accomplished by using equal resistors $Z_3 = Z_4 = R_3 = R_4$, then γ reduces to a real number which equals $x/2$.

In this circumstance, eq (15.8) may be rewritten avoiding the complex exponential notation with $V_E(t) = V_{Em} \cos(\omega_E t)$, yielding

$$V_o(t) = \gamma V_{Em} \cos(\omega_E t) = \frac{x}{2} V_{Em} \cos(\omega_E t) \tag{15.9}$$

which is equivalent to the resistive half-bridge configuration. It can be noticed that the output voltage $V_o(t)$ becomes a cosinusoidal signal synchronous with the excitation voltage with an amplitude controlled by the bridge fractional imbalance γ. Hence, $V_E(t)$ behaves as the carrier waveform over which γ exerts an amplitude modulation.

The process of extracting γ from $V_o(t)$ is called demodulation. To properly retain the sign of γ, i.e. to preserve its phase, it is necessary to make use of a so-called **phase-sensitive** (or coherent, or synchronous) **demodulation** method. In fact, if pure rectification of $V_o(t)$ were adopted then both $+\gamma$ and $-\gamma$ would result in the same rectified signal, thereby losing any information on the measurand sign.

A typically adopted method to implement phase-sensitive demodulation employs a multiplier circuit. Such a component accepts two input voltages $V_{M1}(t)$ and $V_{M2}(t)$ and provides an output given by $V_{Mo}(t) = K_M V_{M1}(t) V_{M2}(t)$, where K_M is the multiplier gain factor.

With reference to the block diagram of Fig. 15.3(a), the bridge output voltage is first amplified by a factor A, then is band-pass filtered around ω_E, for a reason that will be shortly illustrated, and then fed to one of the multiplier inputs, while the other one is connected to the excitation voltage $V_E(t)$. The multiplier output $V_{Mo}(t)$ is then given by

$$V_{Mo}(t) = \gamma A K_M (V_{Em})^2 \cos^2(\omega_E t)$$

$$= \frac{x}{2} \frac{A K_M (V_{Em})^2}{2} [1 + \cos(2\omega_E t)] \qquad (15.10)$$

In eq (15.10) can be observed the fundamental fact that, due to the nonlinearity of the operation of multiplication, $V_{Mo}(t)$ includes a constant component proportional to the input signal x. The oscillating component at $2\omega_E$ can be easily removed by low-pass filtering, and the overall output $V_{out}(t)$ becomes a DC voltage proportional to x given by:

$$V_{out}(t) = \frac{A K_M (V_{Em})^2}{2} \gamma = \frac{A K_M (V_{Em})^2}{4} x \qquad (15.11)$$

To maximize accuracy, both the excitation amplitude V_{Em} and the gains A and K_M need to be kept at constant and stable values. The excitation frequency ω_E is instead not critical, since it does not appear in eq (15.11).

The configuration schematized in Fig. 15.3(a) for either inductive or capacitive transducers can also be adopted for resistive sensors connected in any variant of the Wheatstone bridge.

Moreover, the method of AC excitation followed by phase-sensitive demodulation also represents a typical readout scheme used for LVDTs (Section 14.4.3), as illustrated in Fig. 15.3(b). In this case, for the particular transducer used, ω_E is usually chosen equal to the value which zeroes the parasitic phase-shift between the voltages at the primary and the secondary at null core position.

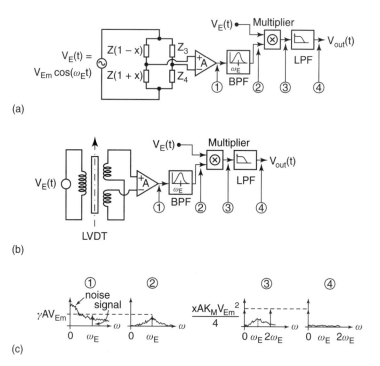

(a)

(b)

(c)

Fig. 15.3 The amplification method based on amplitude carrier modulation followed by phase-sensitive detection. (a) Block diagram in case of an AC excited bridge formed by either inductive, capacitive or resistive transducers. (b) Block diagram for the case of an LVDT. (c) Qualitative shape of the signal and noise spectra in relevant positions of the above systems.

It is worth pointing out that the main advantage of the AC amplification method followed by synchronous demodulation lies in the fact that a constant input signal is displaced in frequency from DC to ω_E. Conversely, most of the noise and interference contributions as well as the main sources of errors of the input stage, such as contact EMFs and the amplifier offset voltages, are located in the low frequency region. Therefore, they can be efficiently filtered out without affecting the signal which is 'safely' positioned at ω_E. This is exactly what is done by the aforementioned band-pass filter inserted after the input amplifier in Fig. 15.3(a) and (b). By means of the following multiplication and low-pass filtering, the signal is then brought back to DC which is now a 'cleaner and quieter' region after most of the noise and disturbances have been removed.

This same line of reasoning can be applied without significant differences to the most general case when the input signal x is not constant but has a certain frequency spectrum, as shown for instance in [3] and [4]. If the carrier frequency ω_E is chosen adequately higher than the maximum frequency of the signal, usually ten times greater, and the bandwidths of the band-pass and low-pass filters are properly set, then the output $V_{out}(t)$ reproduces the input signal without frequency distortions.

15.4 Piezoelectric transducer amplifiers

15.4.1 Voltage amplifiers

In Section 14.8.1 piezoelectric accelerometers were discussed, and the equivalent electrical circuit of Fig. 14.13 was derived in which the sensor is modelled as a charge generator proportional to acceleration in parallel with the internal resistance R and capacitance C. This model generally applies to all piezoelectric transducers, such as accelerometers, force or pressure transducers, and accounts for the fact that piezoelectric sensors are self-generating.

Depending on the strength of the mechanical input signal and the value of C, the voltage developed across the sensor terminals may sometimes be directly detectable by a recording instrument, such as an oscilloscope or a spectrum analyser, without any amplification. However, due to the finite internal impedance of the transducer $Z_{int} = R \mid\mid 1/(sC)$, the input impedances of the readout instrument and of the connecting cable itself generally cause significant loading of the transducer output in the case of direct connection. Therefore, the measured voltage can be considerably reduced compared to the open-circuit voltage, and the sensitivity is diminished by a factor which is neither constant nor controllable. Moreover, the direct connection is prone to interference pick-up which may significantly degrade the signal.

Avoiding these effects requires voltage amplification to raise the signal level, and impedance conversion to decrease the loading by the cable and the readout instrument. This may be accomplished by making use of a voltage amplifier, whose ideal features are infinite input impedance, zero output impedance and gain G independent of frequency. Figure 15.4(a) shows the circuit diagram inclusive of the equivalent capacitances and resistances of the sensor (C, R), the sensor-to-amplifier cable (C_S, R_S), the input stage of a real voltage amplifier (C_i, R_i), and the amplifier-to-instrument cable plus the instrument input (C_o, R_o).

The voltage amplifier may be as simple as a single operational amplifier (OA) in the noninverting configuration as shown in Fig. 15.4(b) for which the gain G is equal to $(1 + R_2/R_1)$ [5]. If G is made equal to one, it becomes a unity-gain or buffer amplifier, also called a voltage follower, since the output follows the input signal without any gain added.

The voltage at the readout instrument input, in the Laplace domain, is given by

$$V_o = Q \frac{sR_T}{1 + sC_T R_T} G \tag{15.12}$$

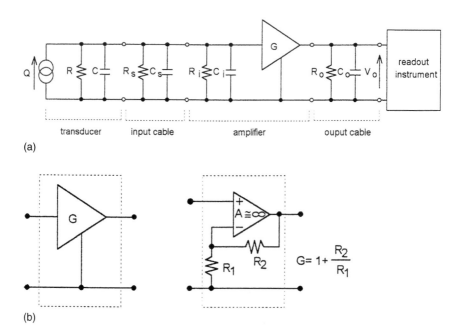

(a)

(b)

Fig. 15.4 (a) Voltage amplifier configuration. (b) Voltage amplifier implemented with an operational amplifier.

where Q is the generated charge, and $C_T = C + C_S + C_i$ and $R_T = R \parallel R_S \parallel R_i$ are the total capacitance and resistance seen in parallel with the transducer charge generator. Even if we are dealing with a voltage amplifier, it makes sense to regard the charge Q as the quantity actually sensed because, assuming that we are in the frequency region below the transducer natural frequency ω_0, the charge is proportional to the mechanical input signal. For a piezoelectric accelerometer, in particular, we had already shown in Section 14.8.1 how the output charge is given by $Q(\omega) = S_{Qa}(\omega)\ddot{X}(\omega)$, where $\ddot{X}(\omega)$ is the acceleration and $S_{Qa}(\omega)$ is the charge sensitivity.

In Fig. 15.5 is plotted the magnitude in decibel of the charge-to-voltage transfer function V_o/Q versus frequency in logarithmic scale called the Bode plot of the amplifier. The gain curve has a low-frequency cutoff limit at $\omega_L = 1/(R_T C_T)$, where $R_T C_T$ is the effective discharge time constant (DTC) of the transducer–cable–amplifier system. For angular frequencies higher than $1/(R_T C_T)$ the gain curve is flat and eq (15.12) simplifies to

$$V_o = \frac{Q}{C + C_S + C_i} G \qquad (15.13)$$

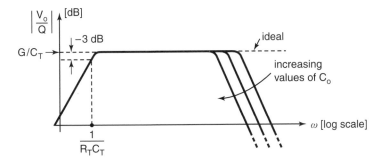

Fig. 15.5 Gain magnitude versus frequency for the voltage amplifier configuration.

The ratio $V_o/Q = G/C_T$ gives the midband gain or amplification, usually expressed in volts per picocoulomb (V/pC). It should be noticed that the amplification is critically dependent on the capacitances C_S and C_i. For ordinary coaxial cables C_S is typically of the order of 100 pF per metre and in most cases it dominates C_i. Therefore, for a given amplifier, cable type or length cannot be changed without affecting the calibration constant. For transducers based on piezoceramics this effect is less evident than with quartz, due to the fact that the internal capacitance C is generally much higher in the former case and may eventually dominate C_S. The cable should be of the low-noise type, that is it must be coaxial with the outer shield devoted to blocking the radio-frequency and electromagnetic interference (RFI and EMI) and it must not suffer from the **triboelectric effect**. This effect consists in charge generation across the cable inner insulator due to friction when the cable is bent or twisted. Such a spurious charge appears across the cable capacitance and is directly added to the signal charge Q, therefore it may impair its detectability. The triboelectric effect can be minimized by choosing a cable of noise-free construction incorporating a lubricant layer between the insulator and the shield and, anyway, preventing cable movement by securing it in a fixed position by cable clamps or adhesive tape.

The connection of an extra capacitor, sometimes called a ranging capacitor, in parallel with the amplifier input increases C_T and produces a decrease in amplification that may be adjusted to scale down the sensitivity to the desired level without acting on the amplifier gain G.

For a good low-frequency response the discharge time constant (DTC) $R_T C_T$ must be high. A possible method would seem that of making C_T very high, but this is not a good choice since it decreases the midband gain according to eq (15.13). It is better to increase R_T as much as possible by choosing a high input resistance amplifier and by paying attention to any possible cause of loss of insulation in cabling and connectors, such as dirt or humidity. In the ideal case of a perfect cable and amplifier, the DTC would reduce to that intrinsic of the transducer given by RC.

As voltage amplifiers, and OAs in particular, have virtually zero output impedance, to first order the presence of C_o and R_o causes no loading effect, as demonstrated by the fact that they do not appear in eqs (15.12) and (15.13). In practice, the output of a voltage amplifier can typically drive sufficiently long cables; however the high-frequency response drops the higher the capacitive load and, therefore, the longer the cable as qualitatively shown in Fig. 15.5. As a significant cost advantage over the use of costly low-noise cable, ordinary coaxial cable can be used at the output. In fact, the virtually zero output impedance of the amplifier shunts the cable impedance and the input impedance of the readout instrument, therefore it prevents the tribolectric charge developing a spurious voltage at the instrument terminals.

Voltage amplifiers are most usually sold as in-line units that must be connected as near as possible to the transducer and, occasionally, can fit on top of its case. In the former case, it should be remembered that the length of the input cable must be kept fixed to preserve calibration.

15.4.2 Charge amplifiers

The role of a charge amplifier is not that of augmenting the charge generated by the sensor, which is impossible to attain since such a charge is fixed by the strength of the mechanical input. Instead, charge amplifiers behave as charge converters which are able to transform the input charge into a voltage output through a gain factor that is virtually independent of both the sensor and the cable impedance.

The circuit diagram of a charge amplifier is shown in Fig. 15.6(a). It can be noticed the presence of a voltage amplifier having a negative voltage gain $-A$, which is usually very high and assumed to be ideally infinite, and the parallel connection of the capacitor C_f and the resistance R_f which provide a feedback path from the output to the input. Again, the equivalent resistances and capacitances of the sensor, of the cables and of the input stage of the real amplifier are taken into account by inserting the corresponding lumped elements in the circuit diagram. This scheme is most often implemented in practice by making use of an OA in the inverting configuration [5], as shown in Fig. 15.6(b).

By applying Kirchhoff's current law at the amplifier input node and remembering that the current entering an ideal voltage amplifier is zero due its infinite input impedance, it can be written that

$$V_{in} = sQ(1/(sC_T)) \,||\, R_T + \frac{(V_o - V_{in})}{(1/(sC_f)) \,||\, R_f}\, (1/(sC_T)) \,||\, R_T \qquad (15.14)$$

with $C_T = C + C_S + C_i$ and $R_T = R \,||\, R_S \,||\, R_i$. By considering that $V_o = -AV_{in}$,

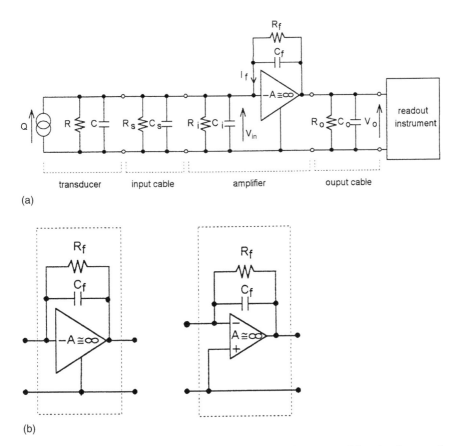

Fig. 15.6 (a) Charge amplifier configuration. (b) Charge amplifier implemented with an operational amplifier.

it can be shown that

$$V_{in} = \frac{sQ(1/(sC_T)) \,||\, R_T}{1 + (1 + A)\dfrac{(1/(sC_T)) \,||\, R_T}{(1/(sC_f)) \,||\, R_f}}$$

(15.15)

from where the voltage output V_o becomes

$$V_o = -AV_{in} = -\frac{AsQ(1/(sC_T)) \,||\, R_T}{1 + (1 + A)\dfrac{(1/(sC_T)) \,||\, R_T}{(1/(sC_f)) \,||\, R_f}}$$

(15.16)

If A is made sufficiently high so that $(1 + A)(1/(sC_T)) \,||\, R_T \gg (1/(sC_f)) \,||\, R_f$ which, neglecting the resistances which are usually very high, reduces to $(1 + A)C_f \gg C_T$, it follows that eq (15.16) simplifies to

$$V_o = -sQ(1/(sC_f)) \,||\, R_f = -Q\,\frac{sR_f}{1 + sC_f R_f} \qquad (15.17)$$

It can be observed that eq (15.17) is equivalent to eq (15.12) valid for a voltage amplifier. The differences are that R_f and C_f now replace R_T and C_T, the voltage gain G is absent, and the presence of the minus sign determines an inversion of the output voltage with respect to the input charge.

It is important to notice that, as long as A is sufficiently high so that eq (15.16) can be replaced by eq (15.17), the voltage output is now insensitive to the sensor internal impedance, the cable impedance, and the amplifier voltage gain and input impedance. The charge-to-voltage transfer function, whose magnitude Bode plot is shown in Fig. 15.7, is only dependent on R_f and C_f, which are external components that may be properly chosen to set both the low-frequency limit $\omega_L = 1/(R_f C_f)$, or equivalently the DTC given by $R_f C_f$, and the midband amplification $-1/C_f$ expressed in volts per picocoulomb (V/pC).

The sometimes-encountered statement that charge amplifiers have a high input impedance is not correct. In fact, it is the voltage amplifier around which the charge amplifier is built that has a high input impedance. On the contrary, owing to the negative feedback, the charge amplifier actually works as a virtual short-circuit to ground, which presents an ideally zero input impedance to the transducer. In fact, $V_{in} \to 0$ for $A \to \infty$. It is for this reason that a charge amplifier has the fundamental capability of bypassing the transducer and cable impedances and drawing all the generated charge Q. For signal frequencies beyond ω_L such a charge is then conveyed into C_f, developing a proportional output voltage V_o.

The condition of a high value of the voltage gain A is usually well satisfied with OAs, which typically provide a voltage gain in the order of 10^5 at low

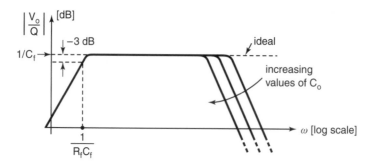

Fig. 15.7 Gain magnitude versus frequency for the charge amplifier configuration.

frequency. This figure is so high that OAs are usually said to have virtually infinite gain. However, A usually drops in real amplifiers for increasing frequencies, hence for accurate prediction of the output voltage in the high-frequency region the exact expression of eq (15.16) needs to be considered rather than the simplification of eq (15.17). This is still more true when C_f is chosen low to obtain a high gain.

To extend the low-frequency response, R_f can be made very large. However, R_f cannot be infinite since, in such a case, the input voltage and current offsets of the real amplifier would charge C_f causing the output V_o to steadily drift toward a saturation level determined by the circuit power supply.

The DTC can be made virtually infinite only momentarily by using a switch in place of R_f. With the switch open, the circuit works without R_f and, as such, no low-frequency limit exists and a DC response is obtained. However, the circuit must be periodically reset by closing the switch to discharge C_f, to bring V_o back to zero and prevent output saturation. Amplifiers employing this method are sometimes denoted electrostatic amplifiers. They can provide a quasistatic response, which enables the measurement of phenomena lasting up to several minutes. Their most typical use is for quasi-DC calibration of piezoelectric transducers (generally made with thermally stable quartz or shear-geometry ceramics to minimize thermal drift), but they are not suitable for continuous amplification of time-variable signals owing to the need for a periodical reset.

A fundamental feature of charge amplifiers is that the sensitivity is, to first order, unaffected by changing the sensor-to-amplifier cable type or length, since neither R_S nor C_S enters the expression of eq (15.17). However, the longer the cable and the higher its capacitance the worse the system high-frequency response, as can be understood if the exact expression of eq (15.16) is taken into consideration, remembering that for real amplifiers A tends to decrease with frequency.

Moreover, it could be demonstrated that the intrinsic electronic noise of the amplifier appears at the output amplified by a factor proportional to the cable capacitance C_S. Therefore, augmenting the cable capacitance has the overall effect of decreasing the S/N ratio. The situation is rather similar for a voltage amplifier, since rising C_S does not influence the noise; however, it decreases the signal amplification (eq (15.13)) and, as a consequence, the S/N ratio again worsens. To avoid introducing further disturbances in the measurement chain, the input cable needs to be of the low-noise kind as for the case of voltage amplifiers, i.e. free from the triboelectric effect and well shielded against RFI and EMI, and should be prevented from moving during the measurement.

On the output side, since charge amplifiers have a voltage output with ideally zero output impedance, C_o and R_o cause no loading effect to first order. In practice, the loading effect is mostly due to C_o and is more evident at high frequency, causing the gain to drop with increasing the output cable capacitance and length, as happens for voltage amplifiers. Generally, the high-frequency gain roll-off due to the capacitive load tends to be more pronounced in charge amplifiers than in voltage amplifiers, therefore attention should be paid to consulting the manufacturer's specifications for the maximum bandwidth available when the amplifier needs to be positioned at some distance from the readout system. Ordinary coaxial cable can be used at the output, since the low output impedance of the amplifier swamps the triboelectric charge possibly generated in the cable.

In general, the main advantages offered by charge amplifiers over voltage amplifiers are that both the sensitivity and the low-frequency limit can be set within the amplifier independently from the sensor and cable impedance. This is particularly valuable for laboratory use, where it is generally advantageous to use a single unit capable of adjusting its amplification and dynamic range to interface with transducers having different sensitivity, providing standardization of the system output.

Charge amplifiers are well suited to ceramic piezoelectric transducers, which generally have a high charge sensitivity but a significant internal capacitance that would cause considerable signal attenuation if voltage amplification were adopted. They are also useful for remote connection to transducers operating at high temperatures, since the electronics can be positioned at some distance in a less hostile environment without signal degradation due to the connecting cable. In humid and dirty environments, attention should be paid to adequately sealing the cable and connectors to prevent any loss of insulation, which would cause low-frequency drifts.

Charge amplifiers are typically sold either as rack-mounted instruments or as in-line units. Rack-mounted charge amplifiers are designed for laboratory use and are very versatile since they generally include in a single unit several signal treatment options, such as coarse and fine adjustment of the amplification to accurately match with the transducer sensitivity (the so-called 'dial-in sensitivity' feature), setting of the bandwidth, additional gain and filtering stages, integration for velocity and displacement, peak hold capability, overload indication, and optional remote control by personal computer through RS-232 or IEEE-488 interfaces.

In-line units are compact and rugged devices which are connected relatively close to the transducer and are suited to field operation. In most cases they have fixed amplification and bandwidth, but some models have trimmable gain, giving the provision for adjusting to the characteristics of different transducers for the standardization of system sensitivity. As an advantage, they are less costly than rack units. Additionally, since they are generally battery powered, they may sometimes offer a higher resolution as they do not suffer from power-line-induced noise.

15.4.3 Built-in amplifiers

As seen in the two preceding sections, to reduce the influence of the input cable on sensitivity and noise it is necessary to keep its length to a minimum by bringing the amplifier maximally close to the piezoelectric sensing element. In-line amplifiers serve this purpose by being able to drive the possibly long cable to the readout instrument by means of a low-impedance voltage output, while the distance travelled by the weak and high-impedance signal of the transducer is minimized. As a limiting case, such a distance can be reduced to zero by enclosing a microelectronic amplifying circuit directly within the transducer case. This operation advantageously turns a raw high-impedance piezoelectric transducer into an amplified low-impedance voltage-output sensing unit. Moreover, it strongly enhances immunity to interference, since the metal housing of the transducer provides an effective shielding action.

The problem of supplying the power to the built-in amplifier and extracting the output signal in an effective way can be solved by adopting a constant-current loop in which the voltage is modulated by the signal, as shown in the symbolic representation of Fig. 15.8. This approach enables both the power supply and the signal to be carried on the same two wires, which most often are the conductor and shield of an ordinary coaxial cable.

The external power supply unit provides the transducer with a constant current I_B to bias the internal amplifier. As a consequence, the output voltage at zero mechanical input settles at a bias level V_B that depends on the transducer and the value of I_B. The piezoelectric charge is converted into a voltage signal V_{oQ} that superimposes on V_B, producing an overall voltage output V_o given by $V_o = V_B + V_{oQ}$.

The readout instrument, represented by its input resistance R, can be connected to V_o either by DC coupling or AC coupling. In the former case, the instrument input voltage V_o' is equal to V_o and therefore the piezoelectric signal of interest rides on the bias voltage V_B. In the latter case, the decoupling

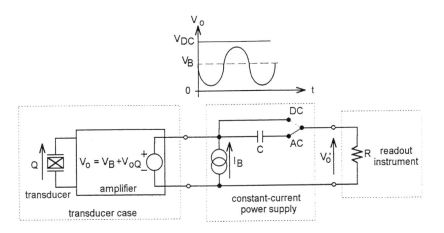

Fig. 15.8 Symbolic diagram of the built-in amplification scheme based on constant supply current and variable output voltage (ICP® concept).

capacitor C removes the offset V_B and causes V_o' to be equal to V_{oQ}, therefore referencing the piezoelectric signal to ground.

Based on the above-illustrated concept for the built-in amplification of piezoelectric transducers, there are many products from different manufacturers which are essentially identical in operation, such as ICP® (by PCB Piezotronics Inc.), ISOTRON® (by Endevco Co.), PIEZOTRON® (by Kistler Instruments), DeltaTron® (by Bruel & Kjaer), LIVM® (by Dytran Instruments Inc.) to name a few [6–8]. Presumably for market reasons, the ICP has become an industry standard so that, currently, many vibration equipment manufacturers and users simply employ the term ICP as a short form for generally indicating a built-in amplification scheme based on constant current and variable voltage.

Coming to the practical implementation of the internal amplifier, there are two possibilities, namely voltage amplifier or charge amplifier. The simplified circuit diagrams of both versions are shown respectively in Fig. 15.9(a) and (b). The voltage amplifier makes use of metal-oxide-semiconductor field-effect transistor (MOSFET) working in the source follower configuration,

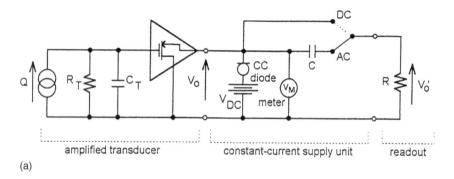

(a)

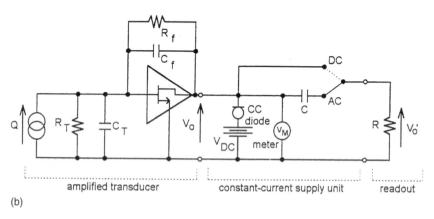

(b)

Fig. 15.9 Different implementations of built-in amplification schemes: (a) MOSFET-based voltage amplifier; (b) JFET-based charge amplifier.

which provides an almost unitary voltage gain G (this is why this configuration is often indicated as a voltage follower) and a low-output impedance. R_T and C_T include the impedance of the transducer, of the amplifier input and of the ranging capacitor if present. The product $R_T C_T$ gives the system DTC, and $\omega_L = 1/(R_T C_T)$ sets the low-frequency limit. With reference to eqs (15.12) and (15.13) with G now equal to one, for frequencies higher than ω_L the output voltage V_o is given by

$$V_o = V_B + V_{oQ} = V_B + Q/C_T \tag{15.18}$$

The charge amplifier is based on a junction-field-effect transistor (JFET) with R_f and C_f forming the negative feedback network. The system DTC and the low-frequency limit are given by $R_f C_f$ and $\omega_L = 1/(R_f C_f)$. According to eq (15.17), for frequencies higher than ω_L the output voltage V_o is then given by

$$V_o = V_B + V_{oQ} = V_B - Q/C_f \tag{15.19}$$

The voltage-sensing scheme is mostly used for low-capacitance quartz elements, while charge sensing is best suited to high-charge-output piezo-ceramic transducers. In both cases, the amplification V_{oQ}/Q rated in volts per picocoulomb (V/pC) is fixed internally and cannot be modified unless by adding following amplification (or attenuation) stages. The voltage-sensing method generally allows for a higher frequency response than the charge amplifiers at parity of operating conditions. Irrespective of the amplification method, the DTC may range from few seconds in most cases, to several thousand seconds in extended low-frequency response transducers. Both circuits have a low output impedance (in the order of 100 Ω) and can then drive a considerable length of ordinary coaxial cable without appreciable signal degradation. The output connectors commonly adopted by the majority of the transducer manufacturers are either the standard 10-32 threaded male microdot coaxial connector, or the two-contact MIL-C-5015 socket.

The power unit generally consists of a DC voltage supply, coming either from a battery pack (usually two or three PP3 9 V cells) or from rectified mains, in series with a constant-current diode which fixes the current in the loop at I_B. The value of the DC voltage supply V_{DC} determines the upper limit of the output dynamic range, while the lower one is set by the value of the bias voltage V_B. Typically, V_B is between 8 and 14 V, and V_{DC} is between 18 and 30 V, while the commonly adopted nominal output ranges are ± 3 V, ± 5 V or ± 10 V. The bias current I_B may range from 2 to 20 mA depending on the application. Generally, higher values of I_B are needed to preserve high-frequency response when driving longer cables at significant voltage levels. This is caused by a nonlinear phenomenon occurring in the amplifier, called slew-rate limiting. The manufacturer's specifications should be consulted to determine the maximum allowed frequency for the case at hand. As typical values, a current $I_B = 5$ mA allows for a $f_{max} = 150$ kHz with about 300 m of a 100 pF/m coaxial cable and a ± 1 V signal swing. It is not advisable to use high I_B values unless necessary, since this causes overheating of the amplifier which increases thermal drifts and the electronic noise level, reducing the resolution.

The voltmeter V_M is often included in the power unit to continuously monitor V_B and allow the detection of a short in cables or connectors (the reading is zero), a cable-open (the reading is about V_{DC}) or a low-battery condition (with no transducer connected the reading is lower than the nominal V_{DC} value). In some cases, further voltage amplification may be provided inside the power supply unit.

When the readout instrument is AC coupled, the decoupling capacitor C and the instrument resistance R form a high-pass filtering network at the output which adds to that due to the DTC of the transducer plus amplifier, hence the overall circuit becomes a dual time-constant system. As will be discussed in the following section, the presence of the output time constant RC may result in a bandwidth limitation on the low-frequency side.

For this reason, when the maximum low-frequency response allowed by the transducer DTC needs to be exploited, DC coupling is to be adopted at the expense of having a nonzero-referenced output signal. Alternatively, some power units incorporate a level shifting circuit based on the use of a difference amplifier to subtract the bias voltage V_B from V_o, therefore providing a DC-coupled zero-referenced output without the insertion of a second time constant.

Built-in amplification is commonly adopted for all types of piezoelectric transducers, such as accelerometers, force and pressure sensors. The general advantages include good resolution independent of cable length (up to several hundred metres) or type (no low-noise cable required), sensitivity and bandwidth set at the manufacturing stage, rugged and sealed construction, low per-channel cost. The fundamental limitations come from the limited temperature operating range and shock survivability compared to the charge-output sensors, owing to the presence of the internal electronics, which cannot withstand temperatures more than typically 120°C, or extreme mechanical shock.

15.4.4   Frequency response of amplified piezoelectric accelerometers

Making reference to Section 14.8.1, and considering a piezoelectric accelerometer followed by either a voltage or a charge amplifier, the general expression of the output voltage as a function of the angular frequency is

$$V_o(\omega) = G_Q(\omega)Q(\omega) = G_Q(\omega)S_Q(\omega)\ddot{X}(\omega) = G_{Qo} \frac{i\omega\tau_1}{1 + i\omega\tau_1} k_Q T_a(\omega)\ddot{X}(\omega)$$

$$(15.20)$$

where $Q(\omega)$ is the charge, $G_Q(\omega)$ is the electrical gain function, with G_{Qo} indicating the midband gain, $\ddot{X}(\omega)$ is the acceleration and $S_Q(\omega) = k_Q T_a(\omega)$ is the transducer charge sensitivity, with k_Q being the charge sensitivity coefficient and $T_a(\omega)$ the acceleration frequency response function of the seismic system. For a voltage amplifier (Section 15.4.1) $G_{Qo} = G/C_T$, with G being the amplifier gain and C_T the total capacitance at the input, and $\tau_1 = R_T C_T$ is the DTC. The product

$$\frac{1}{C_T} \frac{i\omega\tau_1}{1 + i\omega\tau_1} S_Q(\omega)$$

reduces to the transducer open-circuit voltage sensitivity $S_V(\omega)$ in the ideal case of infinite cable and amplifier impedance. For a charge amplifier (Section 15.4.2) $G_{Qo} = -1/C_f$, and $\tau_1 = R_f C_f$ is the DTC.

For constant-current internally-amplified transducers (Section 15.4.3) with DC output coupling, eq (15.20) is again valid, with the only difference that V_o now includes the bias voltage V_B instead of being ground-referenced. In the case of AC output coupling, two time constants are involved and the eq (15.20) becomes

$$V_o(\omega) = G_{Qo} \frac{i\omega\tau_1}{1 + i\omega\tau_1} \frac{i\omega\tau_2}{1 + i\omega\tau_2} k_Q T_a(\omega)\ddot{X}(\omega) \tag{15.21}$$

with $\tau_2 = RC$ being the output time constant caused by the decoupling capacitor C and the input resistance R of the readout instrument, as shown in Fig. 15.18.

Both eqs (15.20) and (15.21) show that on the high-frequency side the signal from an amplified accelerometer reflects the behaviour of $T_a(\omega)$ (Section 14.7.4) with its resonance peak at the transducer natural frequency ω_0. Nonidealities in the amplifiers, such as nonzero output impedance or the influence of the output cable, or poor transducer mounting also affect the high-frequency response (as discussed in the preceding sections) in addition to the fundamental limitation posed by $T_a(\omega)$.

The low-frequency response is determined by the time constant τ_1, representing the DTC of the transducer, and by τ_2 if present. Such time constants introduce a high-pass filtering action and the system is not responsive to DC acceleration. If only the DTC τ_1 is present, at $\omega_1 = 1/\tau_1$ the overall gain is attenuated by -3 dB with respect to its midband value, and it decreases at a 20 dB/decade (or 6 dB/octave) rate for $\omega < \omega_1$. The phase shift is $\pi/2$ at low frequency ($\omega \ll \omega_1$), becomes $\pi/4$ at ω_1, and tends to zero for $\omega > \omega_1$.

If both τ_1 and τ_2 are present owing to AC output coupling, it is important to consider their relative magnitude. If $\tau_1 = \tau_2 = \tau_{12}$ then at $\omega_{12} = 1/\tau$ the gain attenuation is -6 dB and the roll-off rate is -40 dB/decade (or -12 dB/octave) for $\omega < \omega_{12}$. The phase shift is π at low frequency, equals $\pi/2$ at ω_{12}, and tends to zero for $\omega > \omega_{12}$. If τ_1 and τ_2 are not equal the exact calculations are rather involved. However, it can be shown that the dual time-constant behaviour can be approximated by that due to a single effective time constant τ_{eff} given by $\tau_{eff} = \tau_1\tau_2/(\tau_1 + \tau_2)$. This is to say that the low-frequency response is essentially dominated by the lowest between the $DTC = \tau_1$ and the output time constant $RC = \tau_2$. Typically, C is of the order of 10 μF and R can range from 10 kΩ to 1 MΩ, yielding to a time constant between 0.1 and 10 s. By properly choosing the value of C for a given instrument resistance R, RC can be made smaller than the transducer DTC, resulting in $\tau_{eff} \cong \tau_2 = RC$ when it is desired to filter out unwanted low-frequency components, such as thermal drifts. On the other hand, when, as often happens, it is not desired that the output time constant should limit the transducer intrinsic low-frequency response, τ_2 is chosen, say, ten times greater then τ_1 resulting in $\tau_{eff} \cong \tau_1 = DTC$.

To summarize, the generalized transfer function valid for amplified accelerometers is plotted in Fig. 15.10, where the low-frequency behaviour is assumed to be due to a single time constant τ_{LF}. For DC coupling this assumption is exact with $\tau_{LF} = \tau_1$. For AC coupling it represents a convenient approximation which is valid for $\tau_{LF} = \tau_{eff}$.

15.4.5 Time response of amplified piezoelectric accelerometers

The time behaviour of the output voltage $V_o(t)$ caused by a transient input acceleration can be in principle calculated by expressing the eqs (15.20) and (15.21) in the Laplace domain and then antitransforming the resulting output voltage $V_o(s)$. However, considerable insight is gained in trying to analyse and predict the time response to elementary excitation waveforms by starting from the system frequency response.

We have seen that the high-frequency response is affected by the combination of $T_a(\omega)$ and the possible amplifier and mounting nonidealities, while at low frequency the system behaves as (or can be approximated by) a high-pass network with a single time constant τ_{LF}. Therefore, fast time signals with sharp edges involving high-frequency components will be ultimately

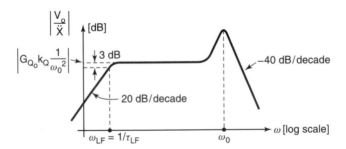

Fig. 15.10 Magnitude of the generalized transfer function of amplified piezoelectric accelerometers.

limited by the time response of T_a assuming that nonidealities are absent, whereas slowly varying and static signals will be attenuated or blocked according to the combined action of the transducer DTC and of the output time constant if present.

These considerations can well be applied to the analysis of the voltage output caused by a step input acceleration. The initial abrupt change in the input excites the high frequencies and, as such, it involves T_a. For a duration of the order of $1/\omega_0$, where ω_0 is the transducer natural frequency, the output voltage follows the general behaviour of Fig. 13.8 with the rise time and amount of ringing determined by ω_0 and the damping factor ζ. As time t elapses, the initial transient dies out and only the static excitation remains active. When t becomes comparable to τ_{LF} the system low-frequency response becomes involved.

The normalized output voltage then behaves as plotted in Fig. 15.11(a). After an initial step whose finite rise time is not distinguishable in the figure due to the abscissa scale factor, it follows a decreasing exponential that will diminish to essentially zero after $5\tau_{LF}$. Therefore, to accurately measure the step amplitude, $V_o(t)$ needs to be read before it droops appreciably and causes a significant error. Considering that the exponential decay is approximately linear to about $0.1\tau_{LF}$, then to obtain a 1% accuracy the reading should be taken within 1% of τ_{LF}. This explains the importance of having a very long time constant when quasistatic measurements need to be performed accurately.

When the input acceleration is a square pulse of duration T the normalized output voltage takes the form plotted in Fig. 15.11(b). The amplitudes of the rising and falling steps are equal since they depend on the high-frequency response. As a consequence, in correspondence to the downward transition at T, V_o undergoes a negative undershoot equal to the voltage loss accumulated during the discharge time T, then it finally approaches zero by following a rising exponential trend. This behaviour is justified by the fact that a system with no DC response, such as a piezoelectric transducer, excited by an input of finite duration responds with an output whose time average, i.e. the DC value, is equal to zero. In other words, the area subtended by the positive and negative portions of the function $V_o(t)$ are equal.

The qualitative behaviour described for the square pulse is observed also for other pulse shapes of interest in vibration measurements, such the triangular and half-sine pulse. In general, the amount of undershoot depends on the relative magnitude of the pulse duration T and the system time constant τ_{LF}, becoming increasingly accentuated the longer T is compared to τ_{LF}. As a conservative rule of thumb, the percentage relationship can be used for undershoot estimation for any pulse shape, leading to an undershoot value of $x\%$ for an $x\%$ value of the ratio T/τ_{LF} (with x lower than 10).

The pulsed input can be generalized to a pulse-train excitation where pulses are assumed to repeat at intervals of T_P. If T_P is of the same order of magnitude of τ_{LF} the corresponding output signal is shown in Fig. 15.11(c). Due to the lack of DC response, $V_o(t)$ shows a decaying trend with

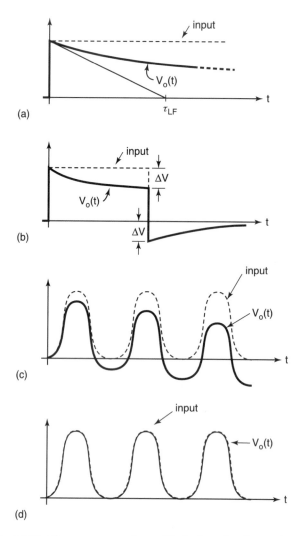

Fig. 15.11 Time-response of amplified piezoelectric accelerometers with the low-frequency behaviour of overall transfer function assumed to be determined by a single time constant: (a) step response; (b) square-pulse response; (c) pulse-train response; (d) pulse-train response after a DC restorer circuit.

exponential envelope which causes a baseline shift and ultimately produces a zero-average output for a nonzero-average input. To avoid the baseline shift phenomenon, which is particularly detrimental in slow transient measurements, some signal conditioners incorporate special nonlinear circuits called zero-clamps or DC restorers that are capable of recovering the average value of the input, as shown in Fig. 15.11(d).

15.4.6 *Electronic integrating networks*

Accelerometers provide high-resolution signals over a wide frequency range, making it possible to electronically integrate such signals to obtain velocity and displacement measurements with a dynamic range and bandwidth often unattainable with dedicated transducers, as discussed in Section 14.10. Time integration corresponds in the Fourier and Laplace domains to division for the terms $i\omega$ and s respectively. Therefore, an integrating network should have a sinusoidal transfer function of the form $1/i\omega$. A simple network which is generally exploited for the purpose and is inserted in many amplifiers and conditioning units consists of the RC circuit shown in Fig. 15.12(a). Its transfer function has the expression

$$\frac{V_o(\omega)}{V_i(\omega)} = \frac{1}{1 + i\omega RC} \tag{15.22}$$

which has the same form as a first-order system (Section 13.6). For high frequencies, so that $\omega RC \gg 1$, eq (15.22) becomes

$$\frac{V_o(\omega)}{V_i(\omega)} \cong \frac{1}{i\omega RC} \tag{15.23}$$

which exactly corresponds to integration, apart from the scaling factor $1/(RC)$ that can be accounted for in calibration.

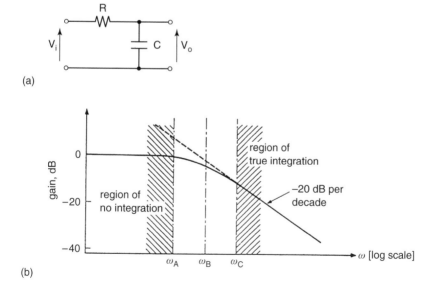

(a)

(b)

Fig. 15.12 The RC low-pass filter circuit as an approximate integrator: (a) circuit diagram; (b) magnitude of the frequency response characteristic ([9, p. 13.4], reproduced with permission).

Therefore, the *RC* circuit is an integrator which is called approximate, since it approximates true integration for sufficiently high frequencies, whereas at low frequency and DC it simply passes the input signal without attenuation. This is shown in Fig. 15.12(b), which shows the magnitude characteristic of the circuit which, seen in the frequency domain, behaves as a first-order low-pass filter with cutoff angular frequency $\omega_c = 1/(RC)$.

For signal frequencies below ω_A no integration takes place, while above ω_B correct integration is performed. For normal accuracy requirements on repetitive signals, the limiting frequencies ω_A and ω_B can be considered to be

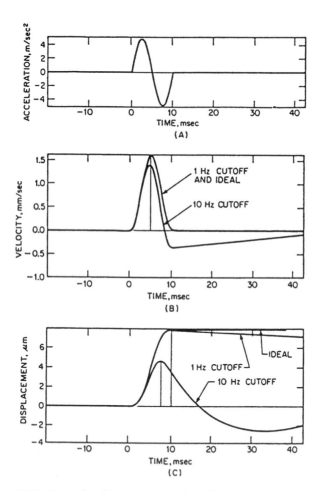

Fig. 15.13 Example of integration and double integration of an acceleration pulse made by one period of a 100 Hz sinusoid, for comparison between approximate integration with different values of the limiting frequency $f_B = \omega_B/(2\pi)$ and ideal integration: (a) input acceleration signal; (b) velocity signal after single integration; (c) displacement signal after double integration ([9, p. 13.5], reproduced with permission).

distant from ω_c by a factor of 3 to 5. Therefore for accurate integration of a signal with minimum frequency component ω_i it must be ensured that $\omega_B = 5\omega_c \leqslant \omega_i$.

Extremely low values of ω_c are difficult to obtain with this circuit and DC integration is not performed. This may be a source of significant inaccuracies when dealing with transient inputs, which deserve great attention when integration is carried out. In fact, a transient is a nonperiodic function of time and, as such, it has a continuous frequency spectrum which can extend to considerably low frequencies. In general, the shorter the transient duration the wider the involved frequency range. The portion of the range which falls below ω_B is not correctly integrated and therefore causes an output signal that departs from the ideal behaviour. When double integration is performed by connecting two *RC* stages in cascade to obtain displacement from acceleration, this problem becomes still more evident due to error accumulation.

As shown in the example of Fig. 15.13 dealing with an acceleration pulse made by a single-period sinusoid, it is of fundamental importance to choose the integration limiting frequency ω_B properly to determine the peak values of velocity and displacement with suitable accuracy. As a practical guideline, it should be ensured that $f_B = \omega_B/(2\pi) < 1/(30t_p)$ for single integration and $f_B = \omega_B/(2\pi) < 1/(50t_p)$ for double integration, where t_p is the time taken by the input pulse to reach its peak (2.5 ms in the example).

15.5   Noise and interference reduction

15.5.1   *Ground noise and ground loops*

Consider the case encountered in many practical situations where the signal from a transducer needs to be sent to a receiving unit located at some distance. The following sections will deal with the aspects concerning the different wiring connections that may be used for this purpose, and their effect on the system immunity to electrical interference [10, 11].

In general, the signal source may be either an unamplified self-generating sensor or a sensor unit incorporating amplification. In both cases, it is assumed the output signal to be voltage, as it most frequently happens, hence a Thevenin equivalent representation can be used. The receiving unit, in turn, is a voltage-input device that might be a signal amplifier or a readout instrument. The input resistance R_i of the receiver is ideally infinite and in practice very high (in the range of hundreds or thousands of kilohms).

The most simple kind of connection is the one shown in Fig. 15.14, where a single wire is used from the transducer to the receiver. The wire resistance R_c is included in the scheme for completeness, though it is usually very low (in the range of few tenths of an ohm for 10 m of signal wire approximately). The signal return path is provided by connecting the low terminals of both the transducer and the receiver to the common ground. In the most general sense,

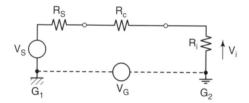

Fig. 15.14 Voltage transmission between transducer and readout unit with the signal return path made by the ground conductor.

the ground need not be the earth ground but it might be any structural or mechanical part made of electrically conductive material, such as the metallic frame of a machine or the chassis of a car, that has the function of closing the electrical circuit.

This solution of Fig. 15.14 is of utmost simplicity but, however, it may cause severe problems of interference pickup. In fact, the material making the ground is not an ideal conductor and, in practice, it has low but finite resistance and inductance values. In addition the ground path generally carries significant current flow coming for instance from power return paths of equipment or machinery which share the same physical part of the ground conductor.

The combination of these two facts causes an important consequence that has to be remembered as a fundamental property of real grounds. As opposed to ideal zero-resistance paths which behave as short circuits, a ground connection cannot be considered as perfectly equipotential. On the contrary, across different points of the ground conductor there will be present in general spurious voltages of unpredictable amplitude and time behaviour reflecting the random nature of the ground currents. This is called the **ground noise,** which is of higher intensity the more the ground is electrically 'dirty' and the longer the distance separating the connection points.

In the scheme of Fig. 15.14 can be observed the presence of the ground noise voltage V_G that appears across the ground connections G1 and G2 at the transducer and receiver sides, which are also shown graphically different to emphasize that they are not truly equipotential. It becomes evident that V_G is directly summed to the signal voltage V_S in contributing to the received voltage V_i. Under the usually satisfied condition $R_i \gg R_S + R_c$, the result is that $V_i = V_S + V_G$ and the S/N ratio at the receiver input is thus severely degraded. In the extreme case of low-level signal, particularly dirty ground and transducer and receiver very distant from each other, V_G can completely swamp V_S causing the signal information in V_i to be totally obscured by noise. Therefore, the wiring scheme of Fig. 15.14 should be in general avoided unless the signal is of high level, the travelled distance is short, the ground is 'clean' and the measurement accuracy is not of fundamental importance.

Now consider the case of Fig. 15.15(a) in which the signal return path is provided by a second wire, again of resistance R_c, connected between G1 and G2. This is a very common situation in practice, and it is often thought that grounding both the source and the receiver at the respective locations is a good practice to generally avoid interferference pick-up. Actually, this is far from true. In fact, it can be readily realized from Fig. 15.15(a) that in this case again the ground noise V_G is directly summed to the signal V_S in developing the received voltage V_i. As the return wire resistance R_c is low but cannot be made exactly zero, the ground noise V_G is not shunted and prevented to appear across the receiver input terminals.

As a matter of fact, the effective solution to the ground noise problem is that of avoiding the voltage V_G being concatenated within the same loop of the signal V_S and the receiver input terminals. That is, no more than one ground connection should exist within the signal circuit, and loops including multiple ground points, called **ground loops,** must be broken.

This can be obtained in the circuit of Fig. 15.15(a) by disconnecting from ground either the transducer or the receiver, depending on which solution is more practical in each application. It should be noticed that the same operation cannot be made on the circuit of Fig. 15.14, since the ground conductor is an integral part of the signal loop and breaking it would interrupt the return path.

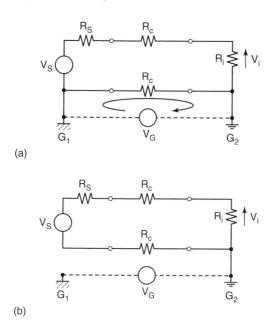

(a)

(b)

Fig. 15.15 Two-wire voltage transmission between transducer and readout unit. (a) Ground noise V_G affects the readout voltage V_i due to ground-loop problem. (b) Ground loop elimination by one-point connection to ground.

Then, starting from Fig. 15.15(a) and assuming that the transducer side is the one that can be lifted off ground, the resulting situation is depicted in Fig. 15.15(b). Now the ground noise V_G, though still present, no longer influences the signal loop and, for $R_i \gg R_S + 2R_c$, the readout signal V_i equals V_S.

15.5.2 Inductive coupling

The preceding section has pointed out that a two-wire connection with a single-sided ground is advantageous to avoid ground loops. However, the existence of a signal loop of finite area automatically implies a degree of susceptibility to inductively coupled interference. In fact, consider a magnetic field which couples into the circuit as happens, for instance, when a current carrying cable runs parallel to the signal wires, as depicted in Fig. 15.16(a). The wire resistances R_c have been omitted in the circuit diagram because they are inessential to the following discussion.

If I_{Ext} is the current in the external cable, the magnetic field induction B generated at a distance h from the cable is $B = KI_{Ext}/h$, where K is a proportionality constant. Then a magnetic flux $\Phi = AB = AKI_{Ext}/h_o$ is coupled with the circuit, where h_o is the average distance between the cable and the signal loop and A is the loop area. Now for the Faraday–Lenz law of magnetic induction a spurious electromotive force, i.e. an interference voltage V_M, is generated into the signal loop given by

$$V_M = -\frac{d\Phi}{dt} = -\frac{d\left(AKI_{Ext}/h_o\right)}{dt}$$

Therefore, if A, h_o or I_{Ext} varies with time then a noise voltage is coupled into the signal circuit adding to V_i a noise term given by $V_M R_i/(R_S + R_i)$. For a given geometry, external current I_{Ext} and receiver resistance R_i, the inductively coupled noise decreases the higher is the source resistance R_S. Striving to keep constant the parameters determining the flux Φ is not a viable way to eliminate V_M, since at least the current I_{Ext} is unpredictable and out of control.

One thing that can be effective is the use of magnetic shields made of materials with high magnetic permittivity, but due to cost this practice is mostly limited to neutralization of particularly strong magnetic sources such as transformers. In general, positioning cables carrying time-varying currents of high intensity close to signal wires should be avoided. In particular, avoid running them parallel; if power cables and signal wires have to cross it is preferred that they do it at right angles. In fact, the flux concatenated in the loop is orientation dependent and, when possible, it is desirable to vary the cable position until the condition of minimum pickup is achieved. This operation, besides, is helpful to assess if the nature of the interference

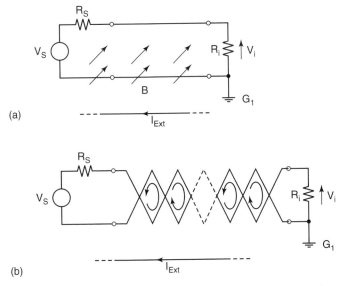

Fig. 15.16 Two-wire voltage transmission between transducer and readout unit: (a) inductive coupling of interference; (b) reduction of inductive coupling by use of a twisted-pair cable.

phenomenon is magnetic induction, in which case it should be significantly orientation dependent, or not. A rule of more general applicability is that of keeping the area A of the loop as small as possible by avoiding loose wiring and preferring to run the two signal cables as close as possible to each other.

The best results in terms of immunity from inductively coupled noise are obtained by using the trick of twisting the two signal conductors as shown in Fig. 15.16(b) to form a multitude of tight loops. Voltages induced in adjacent loops have opposite signs (since the loop axis switches by 180°) and virtually equal magnitude, hence they cancel. Therefore, the overall effect on V_M is ideally null, as if the effective loop area were zero.

Twisted-pair cables are then highly recommended in environments subject to potentially high inductively coupled interference, such as close to electrical power lines, motors and machinery switching high currents, particularly if the signal of interest is of low level.

15.5.3 *Capacitive coupling*

Let us now consider what happens in the two-wire configuration with single-sided ground when an external voltage source V_{Ext} is present near to the circuit, so that the two stray capacitances C_H and C_L exist between the source and the signal wires, as shown in Fig. 15.17(a). V_{Ext} can be due to a number

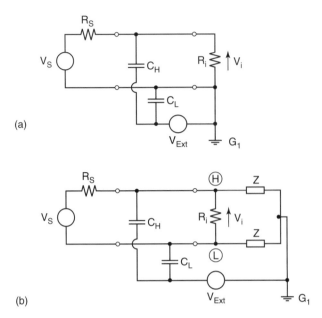

Fig. 15.17 Two-wire voltage transmission between transducer and readout unit: (a) capacitive coupling of interference; (b) reduction of capacitive coupling by use of a differential-input receiver.

of causes such as power lines, electrical machines or even parts of the same electronic circuit or system that we are considering such as, for instance, adjacent conductors at varying voltage or the power supply transformer.

It can be readily recognized that the interfering voltage V_{Ext} is coupled into the signal circuit via the capacitive effect solely due to C_H. Conversely, C_L is virtually uninfluential, since it shunts V_{Ext} directly to ground without affecting the receiver input signal V_i. The unwanted contribution of V_{Ext} to V_i is given by

$$V_{Ext}(\omega) \frac{i\omega C_H(R_S \,||\, R_i)}{1 + i\omega C_H(R_S \,||\, R_i)}$$

which shows that V_{Ext} is coupled to the circuit through a term decreasing with frequency and vanishing at DC, consistent with the fact that we are dealing with a capacitive effect. For a given geometry (determining the value of C_H), external voltage V_{Ext} and receiver resistance R_i, the capacitively coupled noise decreases the lower the source resistance R_S.

The simpler way to minimize the capacitive pickup is again given by rearranging the circuit geometry in order to minimize the stray capacitances, in particular C_H. This can be done by keeping sensitive signal circuits, especially if involving high impedance sources, far from regions where high time-varying voltages are present.

A different approach of high effectiveness is that of passing from a receiver with a **single-ended input** as in Fig. 15.17(a) where the signal V_i is ground-referenced, i.e. the low terminal is tied to ground, to the **differential input** configuration of Fig. 15.17(b). The single-ended and differential configuration will be compared in Section 15.5.5. The impedances Z in Fig. 15.17(b) are typically very high so that in most cases they can be neglected and the input considered floating. Including them in the diagram in the present case, however, helps to better visualize the circuit operation.

It can be observed that now C_L is no longer uninfluential but it intervenes in coupling V_{Ext} towards the low input terminal L, while C_H does the same thing towards the high input terminal H. The idea is that if C_H and C_L are made equal the circuit becomes symmetrical, therefore the amounts of noise coupled to H and L are equal and, as such, they are cancelled by the differential input. In such a case it is said that the noise is transformed into a purely **common-mode** contribution, i.e. equal at the H and L inputs, while the **differential-mode**, or **normal-mode**, noise is virtually zero. In general, the stray capacitances C_H and C_L can be made as equal as possible by keeping the signal wires very close to each other and possibly twisting them to increase symmetry.

Another approach that can be successfully followed even without making use of a differential input at the receiver, though this would be of additional advantage, is that of using an electrostatic shield.

15.5.4 Electrostatic shielding

The charge distribution and relative potentials of bodies encircled by a closed surface made of electrically conductive material are not influenced by charges and electric potentials present outside such a closed surface. This is the Faraday cage concept and the conductive shell is called an electrostatic shield or screen.

The principle can be advantageously applied to reducing capacitive coupling from external voltages by using a two-conductor shielded cable and by enclosing both the source and the receiver within metal housings, as shown in Fig. 15.18(a). It can be noticed how the connection of the cable shield to both the source and the receiver housings creates a unique screen which completely surrounds the signal path. To be maximally effective, the electrostatic shield has to be connected to the point representing the zero reference of the signal, the terminal L in our circuit, which in turn is tied to ground. In this case, the external voltage source V_{Ext} can only couple its interference into the cable shield through the stray capacitance C_S which routes the spurious current to ground, without any perturbation induced into the signal circuit.

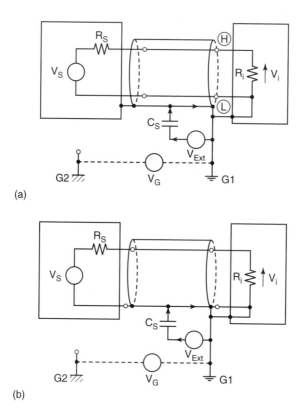

(a)

(b)

Fig. 15.18 Electrostatic shielding: use of (a) twinaxial and (b) coaxial shielded cable.

It should be noticed that the shield must be tied to ground at one point only and this must be at the signal zero-reference ground, which is at the receiver side G1 in the example of Fig. 15.18(a). Otherwise, a significant fraction of the ground noise V_G can be coupled into the signal circuit by means of the cable shield itself. Therefore, the recommended practice in this case is to leave the transducer housing electrically floating as shown.

To save cable cost, the two-conductor-plus-shield cable, often called twin axial, can be replaced by a one-conductor-plus-shield, or coaxial cable, as shown in Fig. 15.18(b). Again the shield drives the stray currents due to V_{Ext} to ground but, since now the shield coincides with the return conductor of the signal, the noise immunity of this unbalanced configuration is typically less than that of the balanced configuration of Fig. 15.18(a). In this case it is imperative that the transducer case is electrically isolated from its local ground G2, otherwise the ground noise V_G becomes directly hooked into the signal circuit, causing severe noise problems. This is a point of fundamental importance which is a direct consequence of the unbalanced link using the coaxial cable, and is often a cause of ground-loop problems in practical installations of metal-case transducers.

To help avoid the problem, a double shield is sometimes provided within the transducer, as for instance in many piezoelectric accelerometers. An internal shield screens the sensing element and is shorted to the signal return terminal, while it is isolated from the outer shield made by the external housing which can be thus safely connected to the local ground.

Figure 15.19(a) shows how the double shielding concept is generally applied with internally amplified piezoelectric accelerometers. The receiver in this case can be either the readout instrument or the constant-current supply unit. The use of a two-conductor twisted and shielded cable gives the system an industrial grade protection, and makes it most noise immune even in factory environments where a high level of interference is typically present. Note that the cable shield is left floating at the source side to avoid ground loops. In less demanding applications, the general purpose solution of using a standard coaxial cable as shown in Fig. 15.19(b) is typically adequate and less costly.

A point that is worth remembering about shielding is that unless the shield is made of material with high magnetic permeability, which is rarely the case due to cost, shielded cables do not screen inductively coupled noise. Cable twisting is needed for that purpose.

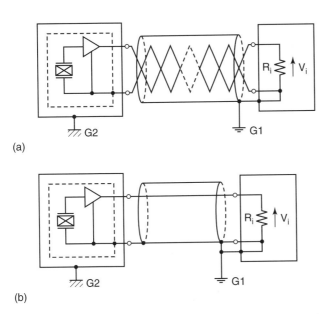

(a)

(b)

Fig. 15.19 Typical shielding schemes used for piezoelectric transducers with built-in amplifiers: (a) double-shielding with twisted-shielded cable; (b) shielding with coaxial cable.

However, shielded cable and metal enclosures give effective protection against electromagnetic and radiofrequency interference (EMI and RFI). Both phenomena may arise when the frequency of the interfering sources is sufficiently high so that electromagnetic waves are generated which can couple into some part of the signal circuit that, due to its dimensions and location, happens to work as a receiving antenna. Fortunately, electromagnetic waves are screened by conducting surfaces. Therefore, a well designed and connected screen is also effective in reducing EMI and RFI.

In practice, the screen surface need not be perfectly closed but can include flaws and openings, such as required for instance with ventilation holes in instrument metal cabinets. However, in order to guarantee the full effectiveness of the shielding action, the dimensions of such openings should be made less than the wavelength of the incoming interference.

15.5.5 *Single-ended and differential connection*

So far, the problem of possible ground loops has been tackled by suggesting that the source reference terminal be disconnected from its local ground and the receiver side be grounded. There are cases, however, where the transducer is intrinsically ground-referenced and cannot be floated. A solution is then float the receiver side by using a receiver with a differential input as shown in Fig. 15.20. The input signal V_i is taken as the voltage difference between the input H and L terminals respectively, none of which is ground connected. Strictly speaking, there are very high impedances Z connecting H and L to ground G1, as previously shown in Fig. 15.17(b), but for the present discussion they can be neglected. Then we have passed from a single-ended ground-referenced input to a differential floating input.

It can be observed from Fig. 15.20 that now the ground loop is broken at the receiver side. As a consequence, the ground noise V_G appears as a common-mode input which is therefore rejected by the differential input receiver.

The capability to respond to the difference of the input signals only, irrespective of their common-mode value is expressed by the common-mode

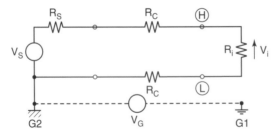

Fig. 15.20 Readout unit with a differential input: the ground noise V_G does not affect the readout voltage V_i.

rejection ratio (CMRR) of the receiver, defined as the ratio of the differential gain over the common mode gain. High CMRR means close-to-ideal differential behaviour.

For real differential receivers or amplifiers the range of input values over which they can operate correctly and without damage is limited both in the difference and in the common mode. The differential input range (DIR) defines the maximum acceptable difference between the input voltages, while the common-mode input range (CMIR) sets the maximum allowed value of their average. Typically, the CMIR is larger than the DIR and is limited by the power supply voltages.

As a general rule, the differential configuration is more noise immune than the single ended, since every disturbance which is induced equally in the two signal wires appears as a common-mode input at the receiver and is then rejected. With a differential receiver, the strategy for reduction of noise pickup should then be aimed at obtaining the maximum symmetry in the circuit, in order to leave the signal of interest as the only differential contribution.

Besides the noise immunity aspects, the use of a differential configuration at the receiver is necessary when interfacing to differential output transducers, such as for Wheatstone bridges excited with a ground-referenced supply. The typical situation is depicted in Fig. 15.21 where the differential input of the receiver, with the input resistance R_i considered infinite, is capable of reading the bridge imbalance voltage as a floating voltmeter. Conversely, using a single-ended input would essentially make the bridge configuration vanish by short-circuiting R_3.

Another case where the differential input is very advantageous is in cancelling cable resistance effects with remotely powered transducers. Consider, for instance, the example of Fig. 15.22(a) illustrating the three-wire connection of a resistive potentiometer excited by the constant current I_E coming from the receiver site, where the signal is read in a single-ended way. The cable resistances R_{c1} and R_{c2} are uninfluential on the voltage output due to current excitation and the infinite input impedance of the receiver. Conversely, R_{c3} causes a voltage drop $I_E R_{c3}$ which might be neither small nor constant and is unwantedly read as a signal. Instead, with the four-wire

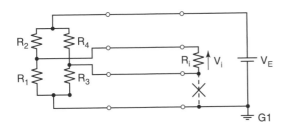

Fig. 15.21 Benefits of using a differential-input receiver with Wheatstone bridge transducers.

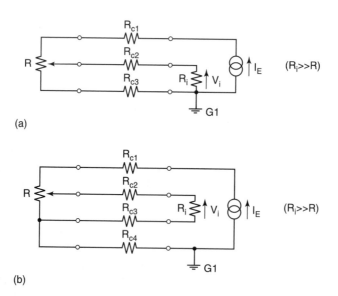

(a)

(b)

Fig. 15.22 Connection of a resistive potentiometer to a readout unit having (a) a single-ended input and (b) a differential input.

differential configuration of Fig. 15.22(b) the voltage drop $I_E R_{c4}$ is transformed into a common-mode signal which is then rejected.

15.5.6   *Optical, magnetic and capacitive isolation*

There are situations where it is impossible to float either the source or the receiver, therefore in order to avoid ground-loop problems it is necessary to create an intermediate break in the circuit. The single circuit comprising the source and the receiver referred to their respective local grounds should be replaced by two circuits, between which some sort of coupling has to exist in order to pass the signal but any direct current flow must be impeded. That is, the two-half circuits should be signal coupled but galvanically isolated. Such an isolation should have a low ohmic leakage and a high dielectric breakdown voltage, i.e. should behave as close as possible to an open circuit at DC over an extended voltage range. There are essentially three methods to accomplish this result which make use of magnetic, capacitive or optical isolation.

Magnetic isolation makes use of a transformer with the primary coil connected to the source and the secondary to the receiver. No electrical contact exists between the coils, hence the galvanic isolation is ensured. The signal is transferred from the source to the receiver scaled according to the transformer turn ratio. Since transformers do not transfer DC signals, the method cannot be applied if the source signal is constant or varying at low frequency. The solution in this case is to of use modulation to shift the signal bandwidth to a frequency region where the magnetic transmission is effective, and then demodulate the transformer output to recover the original signal. Magnetic isolation can be very effective, but transformers tend to be bulky and costly.

Capacitive isolation makes use of one or more high-quality capacitors to couple the signal from the source to the receiver circuit. Since capacitors do not pass DC currents, the galvanic isolation is straightforward but, again, constant and slowly-varying signals cannot be transmitted without resorting to modulation. Capacitive isolation is compatible with a significant degree of miniaturization, which enables this technique to be implemented in an integrated circuit, including the capacitors and the modulation/demodulation stages in the same package.

In optical isolators the source signal is transformed into a light intensity by means of a light-emitting diode (LED) which is mounted facing a photodiode, or a phototransistor, that generates an output current proportional to the incoming light and feeds it to the receiver. Therefore, the signal is transferred by means of light, without any electrical contact required. Optical coupling in principle allows DC signals to be transmitted, since they are converted into a constant light intensity and in turn transformed into a constant signal at the receiver. However, due to unavoidable offset currents in the photodetectors, the operation at DC can be in practice problematic except for simple on/off signals, such as in switches, or digital signals. For analogue signals it is again typically preferred to use modulation. Generally the LED, the photodetector and the driver circuits are all included in the same package sold as a single unit called an optoisolator or optocoupler.

All the three methods can achieve very high degrees of isolation which, depending on size and technology, may be up to several thousand volts. This makes isolators useful not only for eliminating problems with multiple grounds, but also as a means to acquire signals of moderate magnitude riding on much higher average levels without damaging the readout unit or amplifier. For instance, signals of a fraction of a volt superimposed on a background of hundreds of volts can be directly extracted and referenced to the receiver ground. In other words, isolators allow the common-mode input range of the readout unit to be extended significantly beyond the level which, in the case of direct connection, would cause its failure or destruction.

Isolation amplifiers also find application in medical instrumentation, such as electrocardiographs, where for safety reasons it is necessary that no galvanic continuity exists between the electrodes on the patient and the equipment connected to the power lines.

Somewhat related to isolators, though different in intended applications, are the noncontact transmission links. When transducer and receiver must operate in a condition of relative motion incompatible with cable connection, such as with sensors mounted on rotating shafts, there is the need for methods for extracting the signal without interfering with the motion. Traditionally, sliding contacts called sliprings have been adopted, but they suffer from poor contact quality and wear due to friction. Superior performance is obtained by noncontact methods, typically based on radio frequency, inductive or optical coupling, which transfer the signal from the moving transducer to the steady receiver. Since such noncontact methods imply galvanic isolation, they are insensitive to ground-loop problems.

15.5.7 Current signal transmission

Some significant advantages in signal integrity and system cost can be obtained by using current transmission instead of voltage for connecting transducers to remote receiving units. As shown in the general scheme of Fig. 15.23, the signal voltage V_S is converted into a current I_S by the V/I converter circuit placed close to or within the transducer. The current I_S is sent to the receiver where it is made to flow in a drop resistance R_d and converted into the readout voltage V_i. The commonly adopted standard range for the current I_S is from 4 to 20 mA, and the connection scheme is called the 4–20 mA current loop. R_d is usually 200–500 Ω to give a voltage V_i in the range of volts.

The current transmitter is powered by the unregulated DC voltage V_{supply} which is usually located at the receiver site. The current I_{supply} is the sum of the signal current I_S and the current I_P required to power the converter and, on occurrence, to excite the sensor, such as for strain-gauge bridges. The return path for I_P is provided by wire number 3 in the figure, and the resulting scheme is called a three-wire current loop. Often I_P is suitably low, so that the 4 mA offset in I_S is sufficient to cover the power current demand. In these cases, wire number 3 is dropped and the resultant scheme is called a two-wire current loop.

The current transmission is not affected by the cable resistances R_c, since the resultant voltage drops are not reflected in V_i, which is in any case given by $I_S R_d$. Therefore, distant transmission up to some kilometres can be accomplished without signal degradation.

As opposed to voltage-transmitting circuits, current loops are low-impedance circuits due to the small value of R_d compared to R_i. As a consequence, they are very immune to electrostatic crosstalk coming through

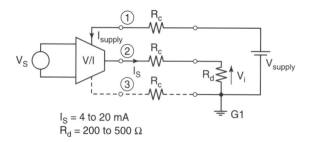

Fig. 15.23 Block diagram of a 4–20 mA current transmitter.

the capacitive coupling, so no expensive shielded cables are generally required. On the other hand, they tend to be slightly sensitive to inductively coupled noise, but this can be typically suppressed easily by inexpensive twisted-pair cables. With the 4–20 mA standard, a fault condition caused by a broken wire (0 mA) can be readily distinguished from a zero signal (4 mA), which is a benefit not offered by voltage transmission. Current transmission is also minimally affected by ground noise problems and, though not shown in Fig. 15.23, the transducer can also be tied to its local ground without typically compromising the system accuracy. However, it should be considered that, particularly when transmission distances are long, the voltages across different ground points can be very high, especially on a transient basis, therefore it is often preferable to include an isolation stage.

The current value inferred by the measurement of the voltage across the drop resistor R_d can only be as accurate as the resistor itself. Therefore, it is always preferable for this function to use high-stability and low-tolerance metal-film resistors as opposed to less stable carbon ones.

Current transmission is not very widespread in vibration-measuring transducers, though it is not totally absent, while it is more typically used for slowly varying quantities, such as temperature, pressure and weight, for process control in plants and factory environments.

15.5.8 Basics of low-noise amplification

So far we have been concerned with the reduction of interference, i.e. of that kind of noise which is not intrinsic in the operation of the circuit components but comes from the surrounding environment and, as such, could be virtually eliminated under ideal experimental conditions.

We now briefly consider intrinsic electronic noise, which is related to the active and passive electronic components making the measurement chain. Part of this noise, called the excess noise, is due to deficiencies and imperfections in the components and could in principle be eliminated by using better devices, though this may not be possible practically. However, most of the intrinsic electronic noise is unavoidably present due to fundamental laws of nature, hence all we can do is understand its features in order to minimize its unwanted effects.

A real amplifier of gain A with its intrinsic noise sources can be modelled as in Fig. 15.24(a) [1, 2, 4]. All the noise contributions can be referred to the amplifier input (RTI) and condensed into the equivalent voltage and current noise generators V_N and I_N which are assumed to be uncorrelated. The respective unilateral voltage- and current-power spectral densities can be indicated with $S_{VN}(f)$ and $S_{IN}(f)$. The noise generators are followed by a noiseless amplifier having a voltage gain A. The noise spectra $S_{VN}(f)$ and $S_{IN}(f)$ depend on the particular amplifier both in their magnitude and in their frequency dependence. Most typically $S_{IN}(f)$ is almost white, while $S_{VN}(f)$ may have a more or less pronounced $1/f$ component superimposed on a white background.

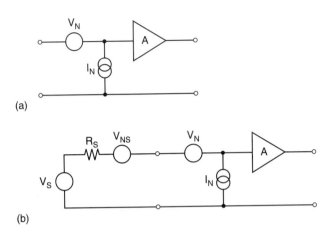

(a)

(b)

Fig. 15.24 Electronic noise in amplifiers: (a) equivalent representation of a noisy amplifier with voltage and current noise sources at the input of a noiseless amplifier; (b) noisy amplifier connected to a transducer for which the thermal noise V_{NS} due to the internal resistance R_S is considered.

Such an equivalent model can be in principle applied to each amplifier in the measurement chain. However, it is generally sufficient to consider only the noise of the input amplifier, since it handles the signal at its lowest level. The noise contributions of all the subsequent stages are of decreasing importance as the signal becomes increasingly amplified. Therefore the input amplifier needs to have the best low-noise characteristics in the chain, as it ultimately determines the noise performance of the whole system.

It is important to point out that the overall noise contribution of an amplifier is not only determined by its own generators V_N and I_N but also depends on the signal source for two reasons. Firstly, the signal source may add its own noise contribution. Secondly, its internal impedance affects the way in which V_N and I_N combine to give rise to the overall output noise.

As an example, consider a noisy amplifier connected to a real voltage source as shown in Fig. 15.24(b). The source resistance R_S has its own thermal noise represented by an equivalent voltage generator V_{NS} with unilateral power spectral density given by $4kTR_S$, where $k = 1.38 \times 10^{-23}\,\mathrm{JK}^{-1}$ is the Boltzmann constant and T the absolute temperature.

By making the simplifying assumption that the amplifier has a rectangular-shaped bandwidth of width Δf, for each noise source of power density $S_x(f)$ it can be calculated the corresponding mean-square value $\overline{x^2}$ as

$$\overline{x^2} = \int_{\Delta f} S_x(f)df \qquad (15.24)$$

The output noise voltage $\overline{v_{Nout}^2}$ is then obtained by quadratically adding the terms, leading to

$$\overline{v_{Nout}^2} = A^2(\overline{v_N^2} + \overline{i_N^2}R_S^2 + 4kTR_S^2\Delta f) \qquad (15.25)$$

Referring $\overline{v_{Nout}^2}$ to the input and taking the root gives the RTI root-mean-square voltage noise

$$v_{Nin,\,rms} = \sqrt{\frac{\overline{v_{Nout}^2}}{A}}$$

which represents the minimum value of the signal V_S which can be distinguished from noise, that is the value for which the S/N ratio is one. This sets the minimum detection limit or the resolution of the system.

Equation (15.25) shows that, for a given amplifier, the voltage noise is minimum for $R_S = 0$. This consideration is, however, not of great help, since R_S is typically fixed by the sensing element, being for instance the leakage resistance of a piezoelectric accelerometer. On the other hand, for a given value of R_S the amplifier which adds the minimum possible noise to the unavoidable thermal noise of R_S can be found. This is equivalent to calculating the S/N ratio at the input and at the output and taking their ratio, called the **noise figure** (NF), and searching for the minimum of NF. From equation (15.25) it can be easily derived that this minimum condition happens for an amplifier for which $\overline{v_N^2}/\overline{i_N^2} = R_S^2$, that is said to be noise-matched to the source.

15.5.9 *Filtering*

Filters are mostly linear processing blocks which provide a frequency-selective behaviour. This property can be exploited for attenuating unwanted frequency components in a signal while passing those of interest. In particular, filtering is of fundamental importance to reduce the residual amount of noise and interference which, for technical or economical reasons, it has not been possible to prevent entering the measurement system. It is important to realize that filtering should not be erroneously thought capable of totally suppressing noise by repairing any possible deficiency and imperfection of the experimental equipment and method. Indeed, the use of filtering for noise reduction can only be effective if done in conjunction with the noise prevention techniques described in the preceding sections, not in substitution.

A filter may be either analogue or digital. The distinction between the analogue and digital domains will be illustrated in the Section 15.6. Plainly speaking, analogue filters are characterized by the property that they handle signals represented by continuous functions of time. Conversely, digital filters are processing blocks that handle data resulting from the conversion of the signal to be filtered into a stream of numbers which represent its values at discrete time instants. We will limit our discussion to analogue filters, while the topic of digital filtering is beyond our scope.

An analogue filter is characterized by its frequency response function $H(\omega)$. Assuming that we consider voltage as the processed quantity, calling $V_i(\omega)$ and $V_o(\omega)$ the $\mathcal{F}$-transform of the filter input and output respectively, it holds that

$$V_o(\omega) = H(\omega)V_i(\omega) \tag{15.26}$$

The filtering action works on the basis that, depending on the shape of $H(\omega)$, some frequencies are enhanced or passed without alteration, and others are attenuated. Passing to the input and output power spectral densities $S_{Vi}(\omega)$ and $S_{Vo}(\omega)$ which can represent either signal or noise, it follows that

$$S_{Vo}(\omega) = |H(\omega)|^2 S_{Vi}(\omega) \tag{15.27}$$

Therefore, the knowledge of the gain function $|H(\omega)|$ of the filter and the power spectral density of the signal or noise at the input allows us to determine the mean-square value of the corresponding quantity at the filter output.

The term equivalent noise bandwidth (ENBW) is typically used to define the rectangular bandwidth $\Delta\omega_N$ whose area is the same of that subtended by $|H(\omega)|^2$, so that for white noise at the input the rms output noise is proportional to $\sqrt{\Delta\omega_N}$.

Four categories of filters can be distinguished according to the shape of their frequency response, namely the **low-pass** (LP), **high-pass** (HP), **band-pass** (BP) and **band-reject** (BR) or **notch** filters. The gain functions $|H(\omega)|$ of each of them are graphically shown in Fig. 15.25. A simple example of a LP filter is the RC network discussed in Section 15.4.6 with regard to its time-integration capability.

In all the plots of Fig. 15.25 can be distinguished a frequency region called the passband where $|H(\omega)|$ does not vary and the signal is simply multiplied by a constant gain $|H_B|$, a region where attenuation takes place called the stopband, and a boundary region between the two called the transition band. If $|H_B|$ is unity the filter can be built with passive components and is termed passive, while if a gain greater than one is needed then active stages providing amplification must be included and the filter is named active.

The limiting frequency separating the passband from the transition region is called the cutoff or breakpoint frequency ω_c of the filter and is typically defined on a conventional basis. Typically, it is assumed to be the frequency where the attenuation reaches -3 dB.

Filters can be fixed or tunable. The latter ones provide the capability to vary the cutoff frequency (for LP and HP types), or the centre frequency and possibly the bandwidth (for BP and BR types). Some BP tunable filters have an auxiliary input for a synchronization signal, such as a trigger pulse from rotating speed sensor on a rotating part, which is directly used to control the filter centre frequency. This feature makes sure that the filter bandwidth is

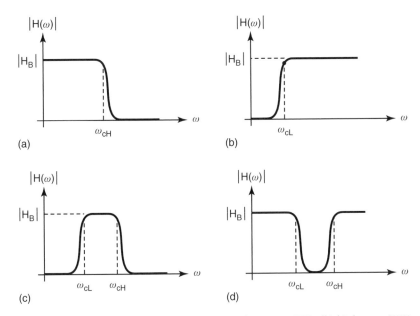

Fig. 15.25 Frequency response of filter types: (a) low-pass (LP); (b) high-pass (HP); (c) band-pass (BP); (d) band-reject (BR) or notch.

constantly tracking the frequency of the signal of interest even if it varies, such as in machine run-up or coast-down.

An ideal filter would have perfectly flat passband and complete attenuation outside, with infinitely sharp boundaries separating the two regions. This rectangular shape is not realizable in practical filters but can only be approximated. The roll-off slope of $|H(\omega)|$ in the transition regions is related to the filter order n, which is the number of poles of $H(\omega)$, and amounts to $n20$ dB per decade of frequency or, equivalently, to $n6$ dB per octave. Filters of high order, for instance 5 or more, are available providing in many cases a satisfactory approximation of infinitely sharp edges. However, the higher the filter order the higher the tendency of $|H(\omega)|$ to ripple within the passband causing an uneven transmission of the signal frequencies. Moreover, the phase response due to $Ph[H(\omega)]$ is also a point to be considered, since it may cause significant signal distortion by adding different delays to each frequency component and causing a variation of the signal shape. It can be shown that in order to ensure no distortion the filter phase shift should vary linearly with the frequency. Sometimes, a so-called **all-pass** (AP) filter can be added which gives a unitary $|H(\omega)|$ but a specific phase response tailored to equalize the overall phase behaviour and minimize distortion. Another point worth remembering is that, in general, the smaller the filter passband, i.e. the more the filter is selective, the longer the settling time, defined as the time needed by the output to return to zero after the input has been removed.

The potential of filters to reduce the noise lies in the fact that if signal and noise are at distinct frequencies then the noise can be selectively attenuated without affecting the signal. To this purpose the filter should be centred on the signal and dimensioned in order to leave out the noise, as shown in Fig. 15.26(a). The higher the filter order the better its capability to reduce the noise while retaining the maximum of the signal bandwidth.

If not filtered out, both wideband noise and interference at a discrete frequency which reside within the bandwidth of the measuring system would be uselessly amplified with the only detrimental effect of reducing the S/N ratio (Section 15.2). Therefore, the system bandwidth should be narrowed as much as possible around the signal bandwidth.

As an example, if you are interested in measuring the vibration amplitude of a light part oscillating at say 500 Hz, it is advisable to insert a BP filter centred on such a frequency with a suitable bandwidth to suppress wideband noise and possible interference components such as mains frequency (50 Hz in Europe, 60 Hz in the USA) or stray pickup. The point of the measuring chain where such a filtering stage is positioned is not in practice uninfluential, though it is theoretically irrelevant under the assumption of

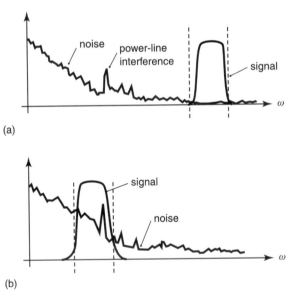

(a)

(b)

Fig. 15.26 Filtering as a mean for passing the signal and rejecting noise and interference. (a) Negligible overlap of signal and noise spectra, therefore a BP filter centred on the signal spectrum removes most of the noise and retains the signal. (b) Substantial overlap of signal and noise spectra, therefore a BP filter centred on the signal spectrum would pass a significant amount of noise.

perfect linearity. As a guideline, it is recommended to filter as close as possible to the signal source, i.e. in the first stages of the receiving unit, in order to maximize the S/N ratio from the beginning and avoid the possibility that particularly large amounts of interference, if not removed, cause saturation of the amplifiers.

When the signal and noise bandwidths overlap as shown in Fig. 15.26(b), the simple frequency discrimination achieved by filtering loses effectiveness, since a significant amount of noise cannot be removed without at the same time attenuating the signal. This situation is typical in the measurement of low-frequency and DC signals which have to compete with the ubiquitous and often intense $1/f$ noise. A possible solution is to make use of modulation as illustrated in Section 15.3.2. Other methods can provide the equivalent of a bandwidth narrowing effect by taking advantage of the fact that the signal repeats itself over time, so that some sort of averaging can be carried out.

15.5.10 *Averaging*

Filtering is most immediately regarded as a process carried out in the frequency domain; however, due to the frequency–time duality it has a direct correspondence in the time domain. The fact that in the frequency domain the output of a linear filter (as of any linear system in general) is given by the multiplication of the input for the frequency response, as expressed by eq (15.26), has its counterpart in the convolution operation.

Indicating with $v_i(t)$ and $v_o(t)$ the filter input and output voltages now expressed as functions of time, it holds that

$$v_o(t) = \int_{-\infty}^{\infty} v_i(\tau)h(t-\tau)d\tau = v_i(t) * h(t) \tag{15.28}$$

where $h(t)$ is the filter impulse response, equal to the Fourier antitransform of the frequency response function $H(\omega)$. The symbol $*$ is used to indicate the convolution operation. Equation (15.28) states that the output v_o at time t can be obtained by the multiplication of the input v_i by a time-reversed and shifted version of h, and the product be integrated over the entire time axis.

Equivalently, the output v_o can be regarded as the time average of the input v_i weighted by a function $w(t, \tau) = h(t - \tau)$ which depends on the difference $(t - \tau)$ between the time instants when the output is observed and when the input is considered. Since in causal systems the effect cannot occur before the cause, $h(t - \tau)$ is zero for $\tau > t$ and the upper integration limit in eq (15.28) could be as well substituted by t.

This latter perspective helps to realize the important fact that filtering in the frequency domain can be associated with the process of a running, or moving, average in the time domain. The shape and nature of the **weighting function** $w(t, \tau)$ determines the effective length of time over which the input is averaged to produce the output at present time t, and the amount of importance given to the past history of the input in each instant $\tau < t$.

For example, consider again the simple first-order low-pass filter made with the RC network of Fig. 15.27(a). Since the pulse response is $h(t) = e^{-t/(RC)}$, the output is given by an exponentially weighted running average of the past input values. The longer the time constant RC, the wider the moving time window over which averaging is performed. This translates into an increased capability of the circuit to smooth sudden input variations, resulting in a low-pass filtering action as illustrated in Fig. 15.27(b). It is important to observe how a high value of RC necessarily implies a slow response with a long output tail, meaning that in fact the system keeps a memory of the input for a time of the order of $5RC$. If it is desired to preserve the signal shape, the positioning of the cutoff frequency, and hence the amount of the filtered out noise, reflects a necessary trade-off on the achievable S/N ratio.

If $h(t) = 1(t)$ then the output becomes the time integral of the input, i.e. the time average of all its past history with each instant equally weighted. This is the limiting case of the exponential weighting obtained for RC going to infinity.

If we now consider applying to the integrator a constant input $v_i(t) = v_{iK}$ to which a white noise v_{iN} is superimposed, the following fundamental result can be derived. The signal v_{iK} steadily accumulates over time leading to an output v_{oK} increasing proportionally to t. On the other hand, the noise v_{iN} randomly adds and subtracts from its immediately previous value resulting in an mean output amplitude equal to zero. Only the mean-square value $\overline{v_{iN}^2}$ actually accumulates, leading to a mean-square output noise $\overline{v_{oN}^2}$ which increases proportionally to t. As a positive consequence, the S/N ratio at the output given by $v_{oK}/(\overline{v_{oN}^2})^{1/2}$ increases as $\sqrt{T_{int}}$, where T_{int} is the integration time.

Therefore, it can be concluded that time averaging by signal integration gives basically the same advantage gained by performing repeated measurements of the same quantity, with the only difference that measurements are taken over a continuous time interval. As a matter of fact the input signal accumulates at the output, while the noise on average cancels.

In the case of an input noise which is not completely uncorrelated, i.e. has some amount of periodicity, the S/N ratio improvement with increasing the integration time is accordingly less than $\sqrt{T_{int}}$. If the input noise is periodic with a frequency f_N, such as for power-line interference, it can be entirely suppressed by making an integer number of noise periods fall within the integration time, i.e. by choosing T_{int} equal to or multiple of $1/f_N$.

In practice, a circuit capable of providing a continuous output which instantaneously represents the average of the input over the past duration T_{int} is rather difficult to implement. Commonly, a somewhat different processing scheme is used where the output is made available only at the end of the integration time T_{int}, after which the system may either pause in a hold condition or restart another integration cycle.

When the input signal is not constant, the extension of the concept of S/N ratio improvement due to averaging deserves some attention. The fundamental

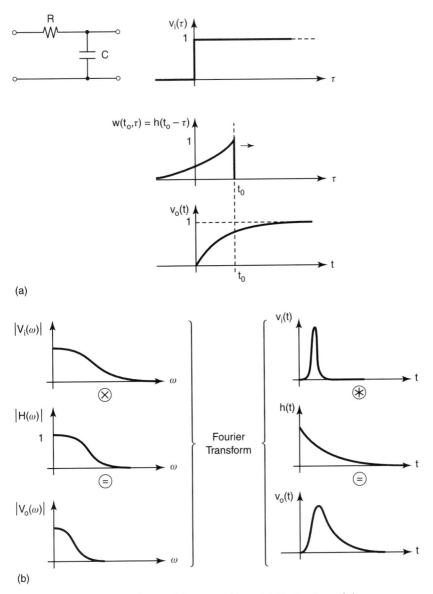

Fig. 15.27 Example on the *R*–*C* low-pass filter. (a) Derivation of the step response as the convolution of the input signal with the filter impulse response *h*(*t*). (b) Time convolution and frequency multiplication are related by the Fourier transform.

point in all the above discussion is that a constant input repeats itself exactly so it experiences a cumulative effect, while the noise is random. Therefore, when considering the theoretical situation in which the input signal is made by a succession of infinitesimally short pulses of equal height immersed in noise as shown in Fig. 15.28(a), the integration has to be performed in such a way to include just the signal pulses and leave out the time intervals where only the noise is present. This is equivalent to saying that the integration can no longer be free-running, but must be gated synchronously with the input signal. As a result, the integration time T_{int} is no longer made up of contiguous time instants, but is a sum of gating intervals relative to subsequent signal pulses. The signal adds synchronously, while the uncorrelated noise adds randomly as it would do over a continuous duration and hence it is again averaged out.

The **gated integrator** is an example of time-variant filter, that is a filter whose response is dependent on the time of arrival of the input. Time-invariant filters,

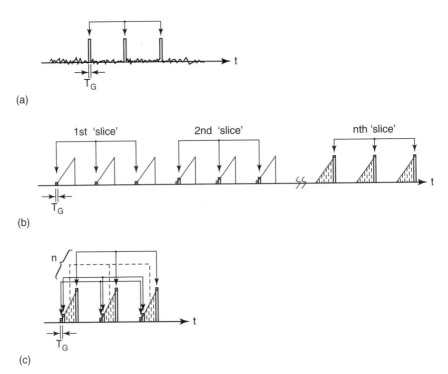

Fig. 15.28 Averaging over subsequent signal repetitions. The linked arrows indicate the signal portions which are averaged together. (a) Averaging of pulses immersed in noise (gated integration). (b) Averaging of sliding portions of a signal (boxcar integration). To obtain three averages of the complete triangular waveform, $3n$ repetitions are required, where n is the signal duration divided by the gate time T_G. (c) All n signal portions are averaged in parallel (waveform averaging). To obtain three averages, only three signal repetitions are sufficient.

such as the RC circuit, can be entirely represented by the impulse response h dependent only on the difference $(t - \tau)$ and equivalently analysed either in the time domain or in the frequency domain by means of the frequency response $H(\omega)$. Conversely, time-variant filters have h dependent on both $(t - \tau)$ and τ which poses problems in determining $H(\omega)$ and makes preferable the description in the time domain in terms of the weighting function $w(t, \tau)$.

The concept of the gated integrator can be readily extended from the particular case of a succession of infinitesimally short pulses to any repetitive waveform, which in fact can be regarded as a series of adjacent pulses of different height. The averaging can be carried out as shown in Fig. 15.28(b), where the gate is first positioned at the beginning of the signal and then the average is taken over a number of signal repetitions (three in the figure). Afterwards, the gate is slid further by a time T_G equal to the gate width and the average is taken over the next signal repetitions. The process then goes on by 'slicing' the input waveform into adjacent portions which are averaged with the correspondent ones of subsequent signal repetitions. This method, which is sometimes called the **boxcar integrator**, since the rectangular gate pulse sliding along the signal waveform resembles a car moving on a road, is effective but is lengthy. In fact, many repetitions are needed to collect a reasonable number of averages on signals of significant duration compared to the gate width T_G. At the same time, T_G cannot be made long, otherwise the signal would be averaged with its adjacent portions, resulting in an unwanted low-pass filtering effect.

The solution comes from the **waveform averaging** method illustrated in Fig. 15.28(c). Now all the slices of the input signal are acquired simultaneously from each signal repetition and averaged with the corresponding ones of the next incoming signal. Waveform averaging can be thought of as a set of gated integrators working in parallel, each one making a processing channel devoted to a particular time portion of the input signal. The improvement in speed is evident, since m signal repetitions are now sufficient to collect m averages of the entire waveform.

All the above considerations are based on the assumptions that the input signal is repetitive and the examples of Fig. 15.28 assume in particular that it is periodic. The repetitiveness is a necessary condition; the periodicity, however, is not a fundamental requirement. All that is needed is a method to synchronize the averaging process with the incoming signal, in order to avoid that corresponding signal slices from subsequent repetitions scroll over different gating intervals causing an error in the determined average.

If the signal is intrinsically repetitive, or periodic in particular, this can be achieved by triggering the start of the averaging at a fixed point of the signal repetition interval or, if the noise level is relatively low, at the instant when the input reaches a given threshold value. Attention should be paid to the fact that any interfering component which is exactly time-correlated with the signal, i.e. shares the same periodicity, cannot be reduced by averaging but it is actually enhanced as it adds in phase over different repetitions. In such a case, if allowed, it is advisable either to change the rate of the signal repetitions or force them to happen at irregular instants, or to filter the interfering frequency if it falls out of the signal bandwidth.

If the signal is not repetitive, it can in most cases be made repetitive by repeatedly exciting the phenomenon that causes it, such as when several impact responses from an accelerometer are obtained by subsequent hammer blows and are waveform-averaged together. As long as the noise can be considered uncorrelated, the time of occurrence of the blows and their time separation are uninfluential and only their total number, i.e. the number of averages, determines the S/N ratio at the output.

As a concluding observation, it should be realized that synchronous averaging is actually a bandwidth-narrowing technique. The measurement bandwidth increasingly reduces at a rate proportional to the integration time T_{int} or, equivalently, to the number of averages m and shrinks on the signal while leaving out the noise. As opposed to filtering in the frequency domain, the signal harmonic components need not be adjacent for the method to be effective but may as well be separated. The averaging process is able to detect them and enhance their amplitude over wideband noise by selectively concentrating the energy accumulated from signal repetitions.

15.6   Analogue-to-digital conversion

Analogue signals take their name from the fact that their time behaviour is analogous to, i.e. an exact replica of, that of the real-world quantity that they represent. Analogue signals, therefore, are continuous functions of time and can assume an infinite number of values within a range. Conversely, digital signals, also called numerical signals, have defined values only at discrete time instants and can assume only a finite number of stepping values within a range.

For instance, if we measure the temperature of a room with an electronic thermometer and continuously plot the results on a strip-chart recorder, we obtain an example of an analogue signal or, better stated, an analogue representation of the temperature as a function of time. In contrast, if we decide to take the measurement once every hour and to round-off the readings to a resolution of say $1\,°C$, we obtain a sequence of data pairs, i.e. the time of measurement and the corresponding reading, which represent the temperature over time in a digital form. It is important to notice that, as the above example illustrates, the digital representation of a signal implies both a time discretization, called **sampling**, and an amplitude value round-off, called **quantization**. Sampling and quantizing are fundamental steps in passing from analogue to digital signals, i.e. in performing an analogue-to-digital conversion.

It is important to note that a time-discrete signal is not necessarily a digital signal unless amplitude quantization also occurs. On the other hand, it is essentially impossible to quantize a signal without acquiring its value for a time interval, however short. Therefore, practical amplitude quantization involves time sampling. In digital signals the time sampling most invariably occurs at regularly spaced instants; the interval between successive samples is called the sampling interval and the number of samples per unit time is called the **sample rate**.

The quantized values of digital signals are usually coded into binary format, that is they are expressed as numbers in base 2. The binary numeration system makes use of two symbols, 0 and 1, which are called **bits** from the contraction of 'BInary digiT'. The choice of the base-2 coding is motivated by the fact that it is particularly easy and convenient to obtain electronic circuits which use two voltage or current levels to represent the equivalent of binary digits 1 and 0. In contrast, it would be rather difficult and inefficient to obtain ten different voltage or current levels necessary to implement the decimal coding. The numeration in base 10 is well suited to humans, but rather problematic for adoption by machines.

Binary-coded signals obey the formal rules of the Boolean logic, which is based on the two states 'false' and 'true' that can be made to correspond to binary levels 0 and 1. For this reason, binary-coded digital systems are usually called logic systems. Most often digital signals are represented by voltage levels. When the absence of signal, i.e. a voltage level low, is coded as 0 and the signal presence, i.e. a voltage level high, is coded as 1 the logic is said to be positive. When the inverse correspondence applies, the logic is called negative.

Most of today's electronic instrumentation makes an extensive use of digital circuitry and processing techniques to manipulate signals. In fact, due to the advent and widespread diffusion of microprocessors, microcontrollers, and digital signal processors (DSP) this is accomplished in an efficient an convenient way. Moreover, thanks to the availability of memory circuits and devices, digital signals are more easily stored and retrieved without degradation.

On the other hand, the real world variables are typically analogue in nature. Therefore, the need is always present for devices capable of converting signals from the analogue to the digital domain and vice versa. They are respectively called analogue-to-digital converters (ADC), and digital-to-analogue converters (DAC). In the present section we will mainly concentrate on ADCs, even if many presented concepts apply equally well to DACs.

15.6.1   *Quantization: resolution, number of bits, conversion time*

We will consider the input of an ADC as an analogue voltage signal $v_i(t)$ which, for sake of simplicity, is supposed to be always greater than zero, i.e. unipolar and positive. Each ADC has an input range represented by the full-scale value V_{FS} which specifies the maximum input level acceptable for conversion. When $v_i = V_{FS}$ the ADC output code is the maximum possible.

The number of intervals into which V_{FS} is divided is 2^n, where n is the number of bits used to represent the output in digital format. The AD conversion is performed by assigning the value of the input signal amplitude to the corresponding interval identified by a digital code. The resolution of an ADC is the smallest increment in the input v_i which causes the output code to change by a unitary step. For example, an 8-bit ADC divides V_{FS} into $2^8 = 256$ intervals numbered as 00000000, 00000001, ..., 11111111. Thus the resolution of an 8-bit ADC is one part in 256, equivalent to 0.39% of V_{FS}, which for $V_{FS} = 10$ V corresponds to 39 mV. A set of eight bits grouped to represent a single number is called a **byte**. The rightmost and leftmost bits are respectively called the least-significant bit (LSB) and the most-significant bit (MSB).

Therefore, resolution and number of bits are equivalent terms defining the same concept, i.e. the width of the discretization interval referred to the full scale. Related to the resolution is the dynamic range, usually expressed in dB as $20 \log_{10} 2^n$ where n is the number of bits. This leads to approximately 6 dB per bit, hence an 8-bit ADC has a dynamic range of 48 dB.

Depending on the conversion method and the technology, ADCs with markedly different resolutions are available. For general purpose applications 12–14 bits are typical, and for slowly varying signals 16 bits are achievable with 20 bits and beyond encountered in top-end instrumentation.

For illustration purposes, the ideal conversion characteristic of a 3-bit ADC is shown in Fig. 15.29. The staircase output resulting from amplitude discretization is responsible for the **quantization error**, representing the intrinsically unavoidable difference between the converted output and the corresponding input. The quantization error has a typical sawtooth shape with maximum amplitude of ± 0.5 LSB. This can be treated as a random noise, called quantization noise, superimposed on the input with a resulting rms value equal to $\pm 1/\sqrt{12}$ LSB. If a sinusoidal signal of peak amplitude $V_{FS}/2$ is taken as a reference, the S/N ratio can be calculated as the ratio between the rms signal and rms quantization noise, yielding

$$S/N[dB] = 20 \log \frac{2^{n-1} LSB/\sqrt{2}}{LSB/\sqrt{12}} = 6.02n + 1.76 \qquad (15.29)$$

An ideal ADC would be limited only by the associated quantization error, that is by the resolution, which is, however, more a design parameter rather than a performance specification. Real ADCs are affected by additional non-idealities, such as offset and scale errors, nonlinearity errors, possible missing codes and temperature-induced errors, which overall combine in worsening the actual conversion accuracy and decrease the S/N ratio below the ideal limit set by the quantization error given by eq (15.29).

The quantization process is not instantaneous but takes some time to be carried out. This time is called quantization or conversion time and usually depends on the type of the ADC and sometimes also on the signal amplitude. The reciprocal of the conversion time is called conversion rate.

For the AD conversion to be carried out accurately it is important that the input signal be constant within the conversion time. Some ADCs are

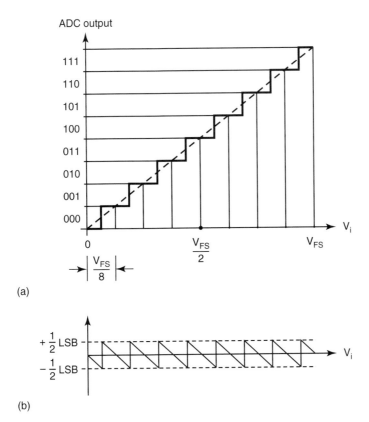

Fig. 15.29 (a) Ideal characteristic and (b) associated quantization noise of a 3-bit ADC.

particularly sensitive to this potential problem, hence they are often preceded by a stage called **sample and hold** (SH), which has the function of taking a snapshot of the input signal and holding the value for a time sufficiently long for the quantization to be performed.

15.6.2 *Sampling: sampling theorem, aliasing*

Sampling consists of the time discretization of a continuous signal. When performing repeated AD conversions on a signal we at least have to allocate the conversion time between successive quantizing operations, hence we cannot expect to have a continuous flow-rate of digitized values. The theoretical superior limit of such a flow-rate is the conversion rate, even if in practice it is always less. As a consequence, sampling is inherently present in a real quantization process, although it should be realized that it does not necessarily imply that quantization occurs.

In this section we are primarily concerned with sampling itself for its effect on the processing of time-varying signals. The fact that the sampled signal is subsequently quantized to perform an AD conversion is not important to the following considerations.

Taking again as an example the measurement of the temperature in a room, imagine that we have monitored the temperature during a period of 24 h. Then we are unable to ascertain if there have been temperature variations between daytime and nighttime if we have not taken at least two readings at a 12 h distance. Similarly, we cannot determine possible temperature fluctuation during a single 12 h daylight period if we do not take at least two readings at a 6 h interval. That is, to catch the presence of a periodicity in a continuous signal we need to sample it at a rate which is at least twice such a periodicity. The principle intuitively suggested by this example is formalized in the sampling theorem by Shannon (previously implicitly formulated by Nyquist), which states that to reconstruct a continuous signal having its highest frequency component at f_M from its sampled version, the sampling frequency f_S must be at least two times f_M, that is it must be ensured that

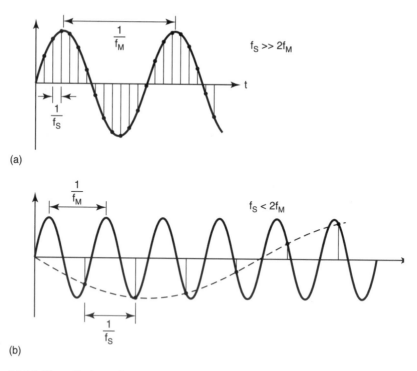

(a)

(b)

Fig. 15.30 The aliasing phenomenon seen in the time domain: (a) absence of aliasing; (b) presence of aliasing.

$f_S \geqslant 2f_M$. The frequency f_M is sometimes called the Nyquist frequency of the signal. Thus, the minimum allowed sampling rate, called the **Nyquist rate**, is twice the Nyquist frequency.

The concept is illustrated in Fig. 15.30 showing the sampling of a sinusoidal signal. If f_S satisfies the Nyquist condition $f_S \geqslant 2f_M$, the sampled signal is a faithful representation of the continuous signal with no information lost in the sampling, since the original waveform can be readily recovered by interpolating the sampled values. Of course, the higher f_S the more the sampled waveform resembles the continuous signal, but practical limits necessarily impede an arbitrary increase of f_S. On the other hand, if $f_S < 2f_M$ the sampled values are no longer uniquely representative of the original signal. In particular, it can be observed how they may as well be attributed to the dashed waveform, which is completely different from the original signal and actually nonexistent at the input. Such a spurious waveform resulting from undersampling (i.e. insufficient sampling rate) is called an 'alias' and the phenomenon is named '**aliasing**'.

Aliasing can be better understood if the sampling process is analysed in the frequency domain. The sampling operation is actually the multiplication of the continuous time signal by a series of pulses equally spaced by $1/f_S$, where f_S is the sampling frequency [12]. This, seen in the frequency domain, corresponds to the fact that the spectrum of the sampled signal is a periodic repetition of that of the underlying continuous waveform at a regularly spaced distance given by f_S, as shown in Fig. 15.31. This follows from the fact that sampling is basically equivalent to amplitude modulation.

If $f_S \geqslant 2f_M$ as in Fig. 15.31(b) the frequency bands of adjacent spectrum repetitions are separated and the original signal can be reconstructed by low-pass filtering the sampled signal. Conversely, if $f_S < 2f_M$ as in Fig. 15.31(c) the frequency bands of adjacent repetitions overlap, since each component at a frequency $f > f_S/2$ is folded back at a frequency $f - f_S$ superimposing on the spectrum of the original signal. This is the aliasing condition and no linear filtering can recover the original signal from the sampled version.

The aliasing phenomenon finds practical applications for instance in the stroboscope, where a pulsed light illuminates a rotating or vibrating object. If the frequency f_S of the light pulses is made equal to that of the moving target f_M, the latter appears still. Furthermore, if f_S is slightly greater than f_M a negative frequency alias is produced which manifests as an apparent inversion of the target motion. Stroboscopes can then be used to determine the unknown frequency f_M in a noncontact way by tuning f_S until the motion apparently stops.

Aliasing can only be avoided by sampling fast enough. In practical cases, the bandwidth of the input signal is not always known in advance to properly choose the sampling frequency. In addition, high-frequency interference and wide bandwidth noise can unpredictably enter the system and appear at the sampler input. All these circumstances may harmfully cause aliasing, which is generally very difficult to detect when the actual input signal is unknown. To

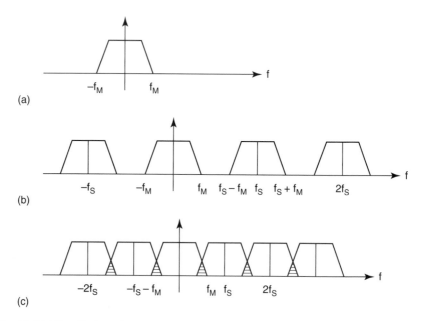

Fig. 15.31 The aliasing phenomenon seen in the frequency domain: (a) spectrum of the input signal; (b) absence of aliasing; (c) presence of aliasing as shown by spectrum overlapping.

avoid the problem it is typically desirable to deliberately limit the bandwidth at the sampler input by inserting a filter in front of it, called an **antialiasing filter**.

An ideal antialiasing filter would have a rectangular shape with flat passband, infinitely sharp roll-off and no transmission in the stopband. Practical antialiasing filters are high-order analogue filters (6–8 poles is typical) with a cutoff frequency f_c set to include the expected signal bandwidth, and they have a finite transition region. For these reasons, a sampling frequency higher than the theoretical Nyquist limit given by $2f_c$ is usually adopted. Typical oversampling factors may range from 3 to 5.

After the signal has been properly sampled, amplitude quantization can be performed to carry out a complete AD conversion of a time-varying analogue input. In practice, a sample cannot be acquired instantaneously but it takes a finite time, often called the aperture time of the sampler. Therefore the digitization time of an ADC, i.e. the time from the start of the sampling of an input value to the availability of the correspondent digital number at the output, is determined by the combination of the sampler aperture time and the quantizer conversion time. Actual ADC circuits incorporate both the sampler and the quantizer functions, which, however, often may not be quite distinct depending on the functioning principle of each particular device.

The antialiasing filter, if present, is a separate circuit. Its performance specification can be somewhat relaxed compared to the ideal requirements by taking advantage of the finite quantization resolution. In fact, all the residual aliasing components falling below the rms quantization noise level are of no concern, since they are not converted.

15.6.3 Main ADC types

The functioning principles of the principal ADCs types are briefly illustrated and their main characteristics are collected in Table 15.1.

Parallel or flash

The analogue input is applied simultaneously to a set of voltage comparators with equally spaced thresholds derived by a voltage reference at V_{FS} and a multiple resistive divider. The output levels from all the comparators are then processed by an encoding block which yields a quantized representation of the input in binary format.

This technique is the fastest available, since all the bits are determined in parallel at the same time instant. For this reason, flash ADCs may reach conversion rates of several hundred megahertz and find typical application in transient digitizers and digital oscilloscopes. On the other hand, the method leads to rather complex and expensive hardware, since for n bits of resolution $(2^n - 1)$ comparators are required. This is why flash ADCs are typically available with a maximum of 8 bits, corresponding to 255 comparators.

Successive approximation

The analogue input signal is applied to a single comparator which confronts it with the output from an internal digital-to-analogue converter (DAC). At the start of conversion the DAC begins a strategy of binomial search by operating subsequent bisections of its output from the initial values $V_{FS}/2$ guided by the comparator output levels. At the end of the search, which lasts n clock pulses,

Table 15.1 Typical values of speed range and resolution for most common ADC types

Type	Speed	Resolution (bits)
Parallel (flash)	100 kHz to 100 MHz	4–8
Successive approximation	10 kHz to 1 MHz	10–16
Integrating	50 (60) Hz*	12–18
Voltage-to-frequency	50 (60) Hz*	14–24

*For line frequency rejection.

where n is the number of bits of the ADC, the input of the DAC represents the analogue input in digital form and is then taken as the output of the ADC.

Successive approximation ADCs are relatively fast, since they only need n comparisons to produce a n-bit output, enabling conversion rates up to 1 MHz. This fact, coupled with moderate cost, makes them general purpose devices extensively used in most data acquisition (DA) boards. As a drawback, they tend to be very sensitive to input sudden changes or spikes and then typically require a sample-and-hold stage to freeze the input during the n clock cycles needed for the conversion.

Integrating

The input voltage is converted into a current which is used to charge an internal capacitor at a reference voltage. The time interval necessary to complete the charging is measured by a digital counter which provides a quantized representation of the input averaged over the integration time. The most popular version is the dual-slope ADC which actually charges the capacitor with the input signal for a fixed amount of time, and then measures the variable time required to discharge the capacitor at a constant reference current.

Dual-slope ADCs are able to provide resolutions as high as 20 bits and more; however, they are slow due to their inherent integrating nature. Most often the integration time is set equal to, or to a multiple of, the power-line period (20 ms at 50 Hz, and 16.66 ms at 60 Hz) in order to average out possible interference and increase to overall noise immunity. As a consequence, the highest conversion rate is 50 or 60 Hz, and even less if multiple cycle integration is adopted.

They tend to be more expensive than successive approximation ADCs, and their typical use is in digital voltmeters, or in DA boards dedicated to the measurement of slowly-varying signals such as temperature, static pressure or weight.

Voltage/frequency conversion

The analogue input signal is converted into a pulse train with frequency proportional to the input voltage. The frequency is then measured by a digital counter, which counts the number of pulses within a fixed time interval. Such a pulse number is then taken as the ADC output.

ADCs based on V/f conversion can reach resolutions as high as 24 bits, and are very immune to noise since the input is actually integrated over the counting time. On the other hand, they are slow since, as in dual slope ADCs, the quantization scheme inherently requires the input signal to be acquired for a significant time duration. As a consequence, V/f ADCs are not suitable for dynamic signals and especially find application in remote sensing of slowly-varying quantities. In such cases, the V/f conversion can be done at the remote sensor location and the frequency signal transmitted to the counter, in this way offering a markedly higher noise immunity than is achievable when sending analogue amplitude signals over long distances.

In this regard, it is often affirmed that the frequency conversion provides a digital representation of a signal. This is incorrect, since the frequency of a signal is a continuous function of time and no quantization actually takes place until such a frequency is converted into a number by counting. The fact that a frequency signal often has the form of squarewave should not be misleadingly regarded as indicating a digital nature. It simply means that information is carried analogically in the time scale rather than in the signal amplitude, which is the exact reason why frequency signals are particularly insensitive to amplitude fluctuations due to noise.

15.7   Data acquisition systems and analysis instruments

15.7.1   *Vibration meters*

Vibration meters are portable instruments which connect to accelerometers or handheld probes and provide the measurement and display of one or more vibration parameters. Some units are pocket-sized for on-the-spot tests. Often they measure velocity, but most frequently they measure acceleration and extract velocity and displacement by integration. The result is usually displayed on an analogue needle indicator or on a digital liquid crystal display (LCD), or frequently on both.

In general, vibration meters measure the amplitude of the vibration parameter of interest over a range of frequencies, therefore giving an integral result related to the measurement bandwidth, which is generally user-selectable. By inserting a tunable narrow band-pass filter (also called a resonant filter) at the input, a selective frequency analysis can be performed by sweeping the filter frequency and taking the corresponding readings. Some units have the tunable filter internally. Typically the displayed reading is related to the rms value of the measured quantity, but almost always the instrument may also indicate the peak value or the crest factor, i.e. the ratio of peak-to-rms value.

Depending on the model, some additional features may be present, such as input charge-mode or constant-current-mode amplifiers for piezoelectric accelerometers, an interface to a personal computer or printer, relay contacts to activate external controls or alarms on occurrence of threshold trespassing.

Vibration meters are suitable for the measurement of continuous vibration levels, but not for transients. They are most typically used for machinery inspection and maintenance, often coupled to handheld probes. In particular, they find wide application in tests on rotating machines in a frequency range which is generally between 10 Hz and 10 kHz. Several models can be directly used to perform vibration severity and exposure measurements in accordance to ISO 2954, 2631 and 8041. Some manufacturers offer special versions usable as human hand–arm vibration meters in compliance with ISO 5349.

15.7.2   Tape recorders

Tape recorders enable the acquisition, storage and playback of electrical signals coming from transducers by converting them into the magnetization of a ferromagnetic tape. There are two types of tape recorder, differing in the format in which the signals are transferred and stored, namely the analogue and the digital recorders.

Analogue

The input signal is recorded by modulating the tape magnetization as a continuous function of time. To this purpose, two alternative methods are adopted, which are the direct recording (DR) and the frequency modulation (FM) methods. In the DR mode the input signal amplitude as a function of time directly modulates the degree of magnetization of the tape along its length. In the FM mode, the input amplitude is converted into a frequency signal which is used to magnetize the tape at the saturation levels, resulting in the information being contained in the number of magnetization inversions for unit tape length.

The two methods have similarities and differences. They are similar in the fact that they can use the same tape, standard audio or VHS cassettes, and, as such, several recorder models use both DR and FM and provide the option to choose between the two techniques. For both methods, the frequency response increases with the tape speed, which can also be different between recording and playback. As a consequence, for a given tape length, higher frequency response implies shorter available recording times.

As for the differences, the DR mode cannot record and reproduce DC signals while, on the other hand, its upper frequency limit can be considerably high. The typical frequency response obtainable with VHS cassette recorders is from 20 Hz to well above 100 kHz. Since the degree of magnetization of the tape can change over time due to tape deterioration and ambient conditions, the DR mode provides poor preservation of the recorded information.

The FM mode has the advantage that it can record DC signals, as they correspond to a magnetization at a constant frequency, but it has an upper frequency limit generally around 50 kHz which is typically lower than achievable with DR. Moreover, for a given upper frequency limit it requires a faster tape speed than DR, hence the available recording time is consequently less. The preservation of information on tapes recorded with FM is good.

An important characteristic of tape recorders is the dynamic range or, equivalently, the S/N ratio. In this regard, the FM method tends to be superior to the DR on a wide-frequency-range basis. However, since the former method has typically a smaller bandwidth than the latter, the actual comparison on the same narrow band can provide somewhat different results. Anyway, the average S/N ratio achievable is around 50–60 dB. In general, for both the RD and the FM recording methods the accuracy in the tape transport mechanism is a fundamental limiting factor of the achievable performances.

Analogue multichannel tape recorders are on the market that can acquire up to 24 signals and have an auxiliary channel connected to a microphone for memo recording. In tests on rotating machines one channel is usually dedicated to the signal from a one-per-revolution sensor, in order to provide a reliable synchronizing signal for averaging on playback.

An important feature which characterizes analogue recording and differentiates it from digital is that the bandwidth of a multichannel unit is only dependent on the tape speed irrespective of the number of channels. Additionally, the intrinsic low-pass characteristic during recording related to the adopted tape speed can be exploited as an effective antialiasing filter for signal to be subsequently converted into digital form.

Digital

In digital recording the analogue input signal is converted into a digital data stream by an ADC and then transferred into the tape magnetization by means of a pulse-coded modulation (PCM) technique [13]. Digital audio tape (DAT) recorders are very versatile and powerful instruments, and with all the features offered by the digital techniques they are increasingly replacing analogue units.

They have no problem from the tape transport mechanism, since the synchronization of recording and playback depends on the accurately set sampling frequency. The dynamic range is very high and, depending on the number of bits of the ADC used, which is typically 14 or 16, the S/N ratio is of the order of 75–80 dB. DATs allow the simultaneous recording on many channels with a phase difference among different channels typically as low as 1°. As a typical property of digital sampling instruments, the bandwidth available on each channel depends, for a given unit, on the number of channels activated. The total bandwidth given by the individual channel bandwidth multiplied by the number of channels is a constant for a given tape speed, hence doubling the channel number halves the individual channel bandwidth.

Most units employ a multiple speed technique for different recording and playback times for optimizing tape usage, and give the user the opportunity to select the channel-frequency configuration most suitable to the application, with often the possibility of assigning different bandwidths to different channels. In general, the upper frequency limit of DATs can be satisfactorily high but tends to be lower than achievable with DR analogue recorders at parity of channel number. As an example of the achievable performance, a 16-channel unit, expandable to 32, can typically acquire all the channels with a 16-bit resolution and a frequency response from 0 to 20 kHz on each channel.

The problem of the large quantity of data to be stored has been one of the limiting factors of DATs, but its relevance continuously decreases as technology progresses. Currently, there are units on the market which provide a tape storage capability of 25 Gbytes, corresponding to an available recording time that, even at the highest sample rates, is generally sufficient for most applications. Limiting the bandwidth from DC to 5 kHz on 16 channels, the available recording time can be several hours.

DATs invariantly come with an interface for connection to digital computers for data analysis and automatic operation control, and in most cases are portable units which can be used conveniently in the field.

15.7.3   *Computer-based data acquisition boards and systems*

In a large number of applications requiring the measurement, processing and storage of signals from multiple transducers the use of data acquisition systems (DASs) employing an analogue-to-digital (AD) board interfaced to a personal computer (PC) represent the preferred solution, from both technical and cost points of view. Traditionally, such systems used to be limited to quasi-static or low-frequency signals. However, due to the significant improvements in computer performance, including computational power, speed and memory storage capability, and the availability of microcontrollers and digital signal processor (DSP) chips of ever increasing performance and reduced price, they have currently become well suited to dynamic signals as well, such as are encountered in vibration tests. As a matter of fact, the vibration measurement applications where a good multichannel DAS based on a fast AD board interfaced to a PC proves unsatisfactory are increasingly few. Moreover, the presence of the PC represents a great advantage, especially for field operation, since it incorporates in a single unit the functions of measuring instrument, and data analysis and storage system. Some manufacturers, for instance, use this kind of architecture comprising a dynamic DAS and a notebook PC with dedicated software to implement a completely portable modal testing system. In general, state-of-the-art PC-based dynamic DAS, also called waveform digitizers, can offer better resolutions than the typical 8 bits of a digital storage oscilloscope (DSO), and longer recording times than the usually more expensive transient recorders.

The simplified block diagram of a multichannel dynamic DAS is shown in Fig. 15.32. Each input signal firstly enters a dedicated antialiasing filter to remove the frequency components beyond half the sampling rate. In low-frequency AD systems the antialiasing filters are typically absent, but dynamic boards most invariantly have them either internally or as add-on modules. Antialiasing filters generally have a very sharp roll-off and a variable cutoff frequency related to the selected sampling frequency. They may optionally be bypassed entirely to provide visualization of the signal without any possible distortion and delay introduced by the filter, but in such cases, of course,

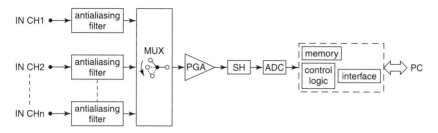

Fig. 15.32 Block diagram of a multichannel dynamic data acquisition system.

possible aliasing problems may occur. It should be noted that antialiasing filters are necessarily analogue as they must come before the sampling stage, while digital filters cannot be used for the purpose even if they may be useful after conversion to smooth the data or reduce noise.

Since using an ADC for each channel would be very costly, the solution usually adopted is to use a single ADC to serve multiple channels sequentially connected at its input by means of a multiplexer (MUX). A multiplexer is a circuit capable of connecting upon command one among several inputs to a single output, and is essentially equivalent to an electronically controlled rotary switch.

Multiplexers can be either made using multiple relay switches or transistor-based electronic switches. The former have extremely low resistance (0.01 Ω) in the closed position, negligible parasitic capacitive effects and high survivability to overvoltages, but they are costly and, most of all, very slow with a maximum switching rate lower than 1 kHz. Therefore, dynamic systems invariably use electronic multiplexers as their switching rate can be several orders of magnitude higher. On the other hand, they have higher resistance (10 Ω), are more sensitive to voltage overloads and tend to suffer from charge injection and parasitic capacitive effects which can cause settling problems and crosstalk between adjacent channels. To minimize settling time, some systems use voltage buffers in front of the MUX.

The MUX output is then fed to a programmable-gain amplifier (PGA) which provides different gain values according to a digital control code, hence allowing optimal matching of signal levels from different transducers to the ADC input range. At slow rates such a control code could be generated by software from the PC and varied between successive commutations of the MUX in order to set a different gain and range for each scanned channel. This procedure is, however, unsuitable for dynamic signals because of its slowness and time unpredictability due to software delays. In particular, it is not compatible with a subsequent data analysis using the fast-Fourier-transform (FFT) algorithm which instead requires equidistant samples. The commonly used solution employs a dedicated sequencer on the board which can be programmed with the number of channels to be scanned by the MUX, the scan order and the relative gain settings of the PGA, thereby ensuring maximum speed and time accuracy. This feature is called programmable channel-gain list or queue. Following the PGA there is the sample-and-hold (SH) stage followed by the ADC. Most often the ADC is of the successive approximation type for its good speed compared to the number of bits, which is typically from 12 to 16.

The SH and ADC are properly synchronized by the controlling logic on the board to operate at the selected sampling rate. Generally, several options are possible to trigger the AD conversion, including hardware (preferred) and software triggering. Most often the system incorporates a ring memory buffer where data are stored continuously but retained and visualized only in relation to the triggering event, hence allowing for pre-, post- and about-trigger acquisition.

It is of fundamental importance to realize that the use of a single ADC working at a sampling rate f_S multiplexed across n channels limits the rate at which the signal from each individual channel can be sampled and converted into digital form. In fact, as the channels are scanned sequentially each of them is actually sampled at rate equal to f_S/n. The quantity f_S is called the **aggregate sampling rate** (or frequency), and the manufacturer specifies its highest value, expressed in samples/s or hertz, as an indication of the maximum conversion speed achievable while using a single channel. For example, a DAS with a maximum aggregate sampling rate of 200 ksample/s can digitize the signals from eight multiplexed channels at no more than 25 ksample/s per channel. The aggregate sampling rate specification should not be confused with the system bandwidth, which refers to a different concept related to the analogue domain and defines the highest signal frequency which can be passed into the channel without being attenuated.

A fundamental limitation of the multiplexed ADC connection is that it introduces time skews between different channels due to the readings being not taken at the same instants but sequentially. This is particularly detrimental with fast signals, especially when preserving the phase relationship among different channels is required, as typically happens in vibration analysis. A possible solution is that of using a dedicated ADC for each channel but this is very costly and then rarely adopted. Alternatively, there exist methods for time skew correction by intervening on the digitized data, but they are of limited applicability, especially with transients.

The preferred approach consists of performing simultaneous sampling on all the channels by employing multiple sample-and-hold blocks, as shown in Fig. 15.33. In this way, the samples from all the channels that are sequentially converted by the ADC are always relative to the same instants, therefore the corresponding digitized signals become synchronized. Simultaneous-sampling DASs should be generally preferred for dynamic applications, and become essential for performing high-quality vibration measurements, such as in modal testing.

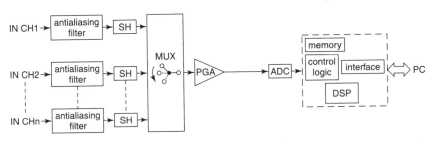

Fig. 15.33 Block diagram of a multichannel dynamic data acquisition system with simultaneous sample-and-hold on each channel.

After AD conversion the digital data are temporarily memorized and then transferred to the PC under the control of dedicated logic circuitry which supervises and coordinates all the system functions. Ever more often, dynamic DASs incorporate DSP chips, as shown in Fig. 15.33, which enormously extend the onboard computational power and enable real-time data processing and analysis without burdening the PC.

In relation to the type of connection to the PC, DASs can be classified as plug-in boards or external systems. Plug-in boards are mounted inside the computer cabinet and directly connect to the PC bus, which ultimately determines the maximum allowed data transfer speed to the computer memory. External systems are more convenient but their transfer speed might be limited by the interface used for connection to the PC. However, both the PCMCIA and the enhanced parallel port (EPP) interfaces can currently ensure significantly high transfer rates and hence they are increasingly adopted as connection links to external DAS, which consequently are becoming more popular.

The characteristics which specify the performances of a DAS can be divided into static (DC) and dynamic specifications. Static specifications include ADC resolution (number of conversion bits) and DC accuracy. It should be kept in mind that resolution and accuracy are different, with the former being the theoretically achievable limit of the latter when no errors other than the conversion error are present. High-quality systems have a DC accuracy very close to their resolution limit: on the other hand, beware systems which claim an attractively high resolution without specifying accuracy. In such cases what often happens is that the resolution is used to measure 'accurately' the system errors, such as amplifier gain and offset errors or thermal drift.

DC accuracy can be specified as a number of bits, as a percentage of the reading (%rdg) plus a number of bits, or as a percentage of the conversion range. Resolution and DC accuracy are insufficient to describe the DAS performances under dynamic operation, such as the errors resulting from multiplexer settling time, signal distortion caused by the antialiasing filter, amplifier bandwidth limitations, or sample-and-hold and ADC nonidealities especially influential when multiple channels are scanned with different gains. Moreover, a high DC accuracy does not necessarily imply good dynamic performance.

A global figure of merit of DAS performance under dynamic operation which is often taken as the parameter to specify the overall dynamic accuracy is the equivalent (or effective) number of bits (ENOB). The ENOB is the number of bits n which satisfies eq (15.29) when the S/N ratio is not the ideal one resulting from quantization noise only, but is the one determined from actual measurements on the systems under dynamic conditions. According to this definition, the ENOB is given by

$$\text{ENOB} = \frac{\text{S/N[dB]} - 1.76}{6.02} \tag{15.30}$$

For example, a hypothetical 12-bit system with a ENOB of 11 can be 'trusted' under dynamic operation to one part over $2048 = 2^{11}$, and not to one over $4096 = 2^{12}$.

When generically referring to the speed of a DAS the term throughput is often used. The throughput actually specifies the rate at which a signal can be converted and the resultant data transferred to the computer memory. Therefore, it takes into account both the digitization time, depending on the selected sampling frequency and number of channels, and the data transfer time. For high-speed systems the latter factor can be as important as the former or even dominant.

The data transfer method can be based on programmed input/output (PIO), either software-controlled or interrupt-driven, or make use of direct memory access (DMA). PIO is too slow to support the typical requirements of dynamic applications, while DMA, as it is hardware-controlled, can be very fast and is therefore the generally adopted method.

It is generally advisable to ascertain if the specified throughput refers to burst or continuous transfer rates, which may be significantly different in value. The fastest systems have onboard memory for temporary storage of the data when they are acquired faster than transferred to the computer, so that no data are lost and the DAS performance is not limited by the speed of the computer bus.

When dealing with dynamic signals it is not only important how fast the data can be acquired, but also for how long. Long recording times require the computer to have enough random access memory (RAM) and fast access routines to a high-capacity hard-disk for continuous data streaming.

DASs generally come with several optional features, such as onboard counters and digital I/Os, or the capability to connect to expansion boards to increase the channel count, usually, however, at the expense of speed. One of the most important features present in high-quality dynamic DASs is an internal DAC to output a signal usable for driving a vibration shaker or actuator for excitation purposes. Typically, several DAC signal options are provided including sine, random, user-defined and playback of acquired data.

15.7.4 *Frequency and dynamic signal analysers*

The analysis of signals in the frequency domain is an extremely powerful tool to investigate the nature of dynamic phenomena and mechanical vibrations in particular. The evaluation of the frequency content of a complex signal may often reveal signal features and details otherwise undetectable with an analysis in the time domain. Moreover, the majority of the signal characteristics observable in the time domain become more clearly identifiable and quantifiable when seen in the frequency domain, for instance with resonances.

When processing the signals analogically, as done in the past, different instruments need to be used for analysis in the time domain and in the frequency domain. Today, a single instrument can convert the incoming signals into digital form and then perform both types of analysis thanks to the progress in electronics and digital signal processing.

In the following paragraphs we will give a brief description of analogue and digital frequency analysers, with an emphasis on the latter due to their higher capabilities for vibration measurements and widespread usage in this field.

Analogue frequency analyzers

Analogue frequency analysers are also called analogue spectrum analysers. The basic functioning principle consists of passing the input signal through a bank of selective band-pass (BP) filters centred at adjacent frequencies and measuring the power at the output of each filter to determine the signal component at the corresponding frequency. To obtain good frequency resolution the filters must be highly selective, i.e. have a narrow passband, therefore to cover a suitably wide measuring span a large number are required and the consequent cost is excessive. An alternative could be that of using a tunable filter which can be swept in frequency across the signal bandwidth to successively measure the power level at each frequency component. However, tunable filters of suitably high quality are difficult to obtain.

The preferred and commonly adopted solution is that of using a single BP filter of high selectivity at a fixed frequency f_F, and then sliding the signal along the frequency axis to intersect the filter passband with different portions of the translated signal bandwidth.

This process of translation of the signal bandwidth is called heterodyning and is commonly used, for instance, in radio receivers. Heterodyning is carried out in practice by multiplying the input signal with a sinusoidal signal of fixed amplitude coming from a local oscillator of frequency f_L. This is no different from the amplitude modulation concept described in Section 15.3.2 where amplitude multiplication in time corresponds to frequency translation, therefore each signal component at a frequency f_i becomes shifted at $f_L \pm f_i$. By properly sweeping the frequency f_L of the local oscillator, the translated signal frequency crosses the filter frequency f_F and then for every frequency f_i contained in the signal the corresponding power can be measured. The results are then presented on an XY display such as that of an oscilloscope.

Analogue spectrum analysers only provide the measurement of the signal amplitude spectrum with no phase information, since each frequency composing the signal is actually measured at different times because of the frequency sweeping. Moreover, as the readings only refer to the frequency components present in the signal at the corresponding measuring instants along the sweep time, they are not suitable for nonstationary signals such as transients.

Analogue spectrum analysers have been traditionally widely employed in acoustics for fraction-of-octave analysis over a limited frequency range, and currently find common application for very-high-frequency signals (up to the gigahertz range) such as encountered in telecommunications.

Digital frequency analysers

Digital frequency analyzers work in a completely different way with respect to their analogue counterpart. The fundamental difference is that in this case the analogue input signal is first converted into digital form and memorized, then all the analysis work is actually carried out on the data representing the sampled and quantized signal, rather than on the original signal itself.

This conversion step brings about many significant advantages basically connected with the opportunity of processing and examining the signal from different points of view to better extract the desired information. In fact, once a signal has been acquired it can be subject to either time or frequency analysis, and very often also octave and order analysis are available in a single instrument. Due to this flexibility, digital frequency analysers have earned the more general name of **dynamic signal analysers** (DSAs).

To perform the analysis in the frequency domain a DSA starts from the input signal in the time domain and calculates its Fourier transform which, as the signal is sampled, is actually a discrete-Fourier transform (DFT). The DFT is, however, very computation-intensive, as a time record of N samples requires N^2 calculations. The solution comes from the fast-Fourier transform (FFT) algorithm proposed in 1965 by Cooley and Tukey [14] which has revolutionized the application of Fourier techniques in instrumentation. The FFT enables us to calculate the transform in $N\log_2 N$ steps, thereby gaining a considerable reduction in computation time as N increases. As a consequence, the FFT is universally adopted in dynamic signal analysers which, for this reason, are also named **FFT analysers**.

The simplified block diagram of an FFT analyser is shown in Fig. 15.34. The input signal $x(t)$ is firstly antialiasing-filtered and then converted into digital form, resulting in a sequence of data separated in time by a constant interval $\Delta t = 1/f_S$, where f_S is the sampling frequency. The data sequence is broken into blocks corresponding to time records made by a given number of samples N, which is usually 1024 or 2048. For the moment we ignore in Fig. 15.34 parts 1 and 2 enclosed within the dotted lines, which will be illustrated along the way, and consider the ADC output as if it were directly connected to the FFT processor.

Since the signal $x(t)$ is sampled in time, its spectrum $X(f)$ is periodic in frequency with a period equal to f_S. Therefore, the FFT calculated on N samples gives N points in the frequency domain within the band $[-f_S/2, +f_S/2]$. As the signal $x(t)$ is real, its spectrum $X(f)$ is even in magnitude and odd in phase, i.e. is a conjugate even function of frequency. Therefore, it is sufficient to retain $N/2$ points in frequency, the remaining ones adding no further information, and as such the resulting frequency bandwidth is $[0, f_S/2]$.

As the antialiasing filter cannot have an ideal brick-wall shape, at frequencies close to $f_S/2$ there is some distortion, and therefore it is usually chosen to visualize a number of frequency points N_V lower than $N/2$ corresponding to a visualized bandwidth accordingly narrower than $[0, f_S/2]$. For example, for $N = 1024$ and $f_S = 128$ kHz the value of N_V can be 400 and the visualized bandwidth equal to [0, 51.2 kHz].

It is important to point out that the data output by the FFT algorithm are complex numbers which retain information on both amplitude and phase, the latter being generally referred to the start of the time record. Hence they represent a complex spectrum which can be visualized on an XY display in various forms, such as magnitude and phase or real and imaginary parts versus frequency. The frequency values in which the FFT is calculated are called bins, and the distance between adjacent bins is equal to f_S/N which gives the resolution in frequency achieved in the spectrum estimation. This has the fundamental consequence that, for a given value of the sampling frequency f_S, and hence of the visualized bandwidth, the FFT resolution is inversely proportional to the length N of the time record and, therefore, to the measurement time. Thus an arbitrary high resolution can be in principle obtained by acquiring the signal for a sufficiently long time, but this in turn requires an exceedingly large memory, which is not practically feasible.

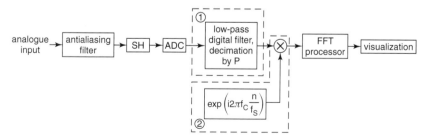

Fig. 15.34 Block-diagram of a FFT dynamic signal analyser.

Moreover, special attention must be paid to nonstationary signals whose features may change significantly within the measurement time.

When the analysis can be restricted to only the lower part of the bandwidth, this can be done at an increased resolution by diminishing the value of f_S. This operation, however, must be accompanied by a corresponding reduction in the cutoff frequency of the antialiasing filter. To achieve this purpose modern FFT analysers use a clever trick called the fixed sample rate method, consisting of operating the ADC at its maximum sampling rate and setting the antialiasing filter accordingly. Then any reduction of the effective sampling rate is obtained by a digital low-pass filter at the ADC output in which one sample out of P is retained, as shown in the dotted part 1 of Fig. 15.34. This process is called decimation and P is named the filter decimation factor. The result is an effective sampling rate of f_S/P with no aliasing, and a corresponding frequency resolution of f_S/PN. In practice, the frequency span has been decreased and the record length augmented by the same factor P. This method works fine but is limited by the fact that the lower end of the frequency span is constrained to be zero, i.e. DC.

To translate the frequency span at other than DC the heterodyning method is adopted, as already encountered in the analogue spectrum analyser. The difference is that now the modulation operation is performed digitally by multiplying the acquired data by a complex exponential sequence of the form $e^{i2\pi f_C n/f_S}$, where f_C is the centre frequency at which the bandwidth will be translated and the integer n spans the record length. This is shown in the dotted part 2 of Fig. 15.34.

The combination of bandwidth narrowing by sample decimation and centre frequency translation by heterodyning is usually referred as a **zoom** operation, since the displayed frequency window can be expanded around the region of interest.

The strength of modern FFT analysers is that for input signals of considerably high frequency all the computations involved are done in less time than necessary to acquire the data record. In this condition there is no dead time, hence no data is lost and the analyser is said to work in real time. The **real-time bandwidth** (RTBW) is the maximum bandwidth of the input signal that the analyser can process in real time. Typical values of RTBW are of some tens of kilohertz but in excess 100 kHz is possible, depending on the instrument and also on the kind and amount of processing that it performs on the data.

Most analysers have an overlap feature which consists of calculating the next FFT spectrum without actually waiting for a complete data record to be acquired but using some data of the previous one, hence gaining in speed, particularly in narrow-span analysis.

Concerning the usage of the FFT analyser, it should be firstly pointed out that the amount of information gained by the visualization of the signal already in the time domain can be considerable. For example, the presence of backlash or free play in mechanical parts can be readily detected by the presence of harmonics which, due to their phase relationship, give a particular shape to the time signal. The frequency spectrum of the same signal would clearly make evident the presence of harmonics, but the evaluation of their phase to figure out if malfunctioning is present is much less immediate. In time-analysis mode, most FFT analysers can be externally triggered and provide pre-, post- and about-trigger visualization.

Sometimes, when inspecting a signal in the time domain the antialiasing filter can be turned off to eliminate its associated distortion, which typically manifests on the phase of the highest frequency components. In such cases, it should be ensured that no aliasing takes place to avoid misinterpreting the results.

The presence of aliasing can be possibly identified by an analysis in the frequency domain in one of the following ways:

- Increase the sampling rate, if possible, and see what happens to the spectrum. If some frequency components appear to change position along the frequency axis while the input signal does not change, they are most likely due to aliasing.
- Alternatively, when possible, the frequency of the input signal can be changed, for instance increased. Then the nonaliased components will move to the right along the frequency axis, and the aliased ones, if present, will move to the left.
- If the signal has sharp edges this determines high-frequency harmonics with an amplitude typically decreasing with frequency. Conversely, if aliasing occurs, the folded-back harmonics appear with an increasing amplitude trend.

Most FFT analysers used for vibration measurements are two-channel instruments. In particular, the two-channel FFT analyser is a fundamental tool for modal testing, as both the signal from the excitation source, hammer or shaker, and that from the accelerometer can be acquired simultaneously. The analyser then allows the complex frequency response function to be determined, i.e. including magnitude and phase, by computing the auto- and cross-spectra of the signals, and provides visualization of the coherence function representing the proportion of the output signal actually due to the input excitation (Chapter 10). The usage of FFT-analysers for modal testing is illustrated in detail elsewhere (the interested reader should consult the further reading list), and will not be covered here.

One of the points where the attention and understanding of the experimenter is mostly required when performing frequency analysis with an FFT instrument is that related to the problem of **spectral leakage**. The problem arises from the fact that the input data record is limited in length as it refers to a finite duration during which acquisition is performed. But an intrinsic characteristic of the FFT algorithm is to assume that the input signal is periodic, with the period given by the data record length. If such a periodicity does not exist, as often happens when the data record is a limited portion of a signal initiated before the acquisition start and continuing after the acquisition stop, then the FFT algorithm attributes the signal discontinuities between the first and last samples in the record to the presence of frequency components outside the signal bandwidth. Such extraneous frequencies are displayed in the spectrum, which then smears along the frequency axis attenuating or even totally obscuring the components due to the signal. The spectral leakage, or smearing, caused by the time record truncation can be prevented by three methods:

- When possible make the signal periodic within the record length, that is change the signal frequency in order to fit the time record with an integer number of signal periods.
- For transient signals increase the record length by zooming, so that the signal has the time to extinguish and return to zero without suffering any truncation.
- Deliberately force the signal to be zero at both extremes of the time record by multiplying it by appropriate weighting functions centred in the middle of the record length having a bell shape and tapered ends. Such functions are called **windows** and the method is named windowing.

Time windows act like filters in the frequency domain. The more the window is tapered in the time domain the more the filter side-lobes are low, hence the spectral leakage is reduced and the amplitude accuracy of the spectrum is increased. At the same time, however, lower filter side-lobes imply a wider centre-lobe which results in a loss of frequency resolution.

The choice of the window is then always a matter of trade-off between amplitude accuracy gained by leakage reduction, and frequency resolution lost due to the enlargement of the window centre-lobe. The most commonly used windows in order of decreasing resolution and increasing amplitude accuracy are the following:

- Rectangular, or uniform, or boxcar window (meaning that all the data in the records are multiplied by unity)
- Hamming
- Hanning
- Blackman/Harris

As a practical rule, when using tapered windows the signal portion of interest should be in the middle of the time record for the leakage suppression to be maximally effective without appreciably distorting the signal shape.

Another aspect where the proper choice of a window can be useful is in reducing the scallop or **picket-fence effect**. These terms indicate the amplitude error occurring for those frequencies of the signal which do not fall exactly on a frequency bin, but lie between two adjacent bins. In this case, it is as if the signal spectrum were observed through the openings of a picket-fence which may possibly screen the true signal peak. In this circumstance, the use of a window of the flat-top family offers a good trade-off between leakage suppression and sufficiently wide centre-lobe to reduce the picket-fence effect.

As we illustrated in Section 15.5.10, an improvement in the measurement S/N ratio can be obtained by averaging. FFT analysers generally offer two averaging options, namely time-domain or frequency-domain averaging. In time-domain averaging, the corresponding sample points of repeated time records are summed together and then divided by the number of repetitions. For this process to be effective, the repetitions must be triggered so that they are synchronized to one another and the time records exactly overlap. In this way the signal is enhanced, while the noise and the interfering components uncorrelated with the signal average out eventually to zero.

This method has its exact counterpart in frequency-domain vectorial averaging, in which the spectra of signal repetitions are summed as complex functions, i.e. by taking into account of both the magnitude and the phase. Again, noise can be averaged out if the signal is properly triggered.

In frequency-domain rms averaging, the rms spectra of signal repetitions are summed hence ignoring the phase information. This method smooths fluctuations in the signal, which does not need to be triggered any more, but does not reduce the noise. Indeed, rms averaging can be used to obtain a good estimation of the random noise floor present in the measurement bandwidth.

Usually, for both time- and frequency-domain averaging, we can choose between three modes of calculating the average. Let us indicate with X_i and Y_i respectively the ith input and ith output of the averaging process performed on n repetitions. In the linear or additive averaging mode, the averaged output Y_n is given by

$$Y_n = \frac{1}{n} \sum_i X_i \tag{15.31}$$

Most often used is the following recursive formula which allows us to update the displayed data along with the averaging calculation, without having to wait until the last repetition is acquired:

$$Y_i = \frac{1}{i} [X_i + (i-1)Y_{i-1}] \tag{15.32}$$

Linear averaging works well for stationary signals. For tracking the trend of nonstationary signals, exponential averaging is more suitable, as given by the following expression:

$$Y_i = (1-k)X_i + kY_{i-1} \tag{15.33}$$

The term k with $0 \leqslant k \leqslant 1$ is a weighting factor expressing how much of the past signal history enters the average computation.

The last is the peak-hold mode, which is not really an averaging mode. It simply means that each signal repetition is compared with the previous one on a point-to-point basis and the highest value is retained and memorized for comparison with the subsequent repetition. The data available at the end of the process represent the envelope of the occurred peaks and may be particularly useful for determining infrequent events which otherwise would be averaged out.

15.8 Summary

This chapter has been devoted to the description of the electronic measuring chain starting at the transducers' output and ending at the data acquisition and analysis instrument.

Section 15.1 introduces the role of the signal conditioning stage as a mean to selectively amplify the signal of interest over the unwanted disturbing components called noise, and of the acquisition instrument for collecting, processing and displaying the measurement signal.

Section 15.2 discusses how the term noise can be generally used to indicate both the intrinsic random fluctuations which are unavoidably present in the signal due to fundamental laws of nature, and the interfering disturbances which result from nonideal experimental conditions and could be virtually eliminated in a 'perfect' environment. The concept of signal-to-noise (S/N) ratio is introduced, and it is anticipated how the S/N ratio can be enhanced by properly tailoring the measuring system bandwidth in order to amplify the signal and reduce the noise.

Section 15.3 deals with the basic methods for the amplification of DC and AC signals. It is shown how the Wheatstone bridge allows the measurement of very small resistance variations superimposed on high stationary values, such as encountered in strain-gauge based transducers. The AC excitation of bridges is then illustrated to be adopted with reactive transducers, as LVDTs and capacitive elements. The important concepts of amplitude modulation and phase-sensitive detection are then presented as fundamental methods to extract the signal from AC bridges and achieve a high S/N ratio.

Section 15.4 is dedicated to the amplifier options for piezoelectric transducers. The voltage and charge amplifiers are presented both as stand-alone units and in their built-in versions, and it is shown how the built-in amplifiers generally offer many advantages especially for field use. Both the time and the frequency responses of amplified piezoelectric transducers are then discussed, and the simple and widely used RC integrating network is presented, pointing out how its region of true integration has a low frequency limitation.

Section 15.5 is dedicated to the basic methods and techniques for the reduction of noise, of both of the interference and intrinsic types. Firstly the possible interference problems related to the connection of a transducer to a readout unit are explained and discussed, including those deriving by ground loops, inductive and capacitive couplings. Then some remedies are presented including the use of electrostatic shielding, differential versus single-ended connection scheme, galvanic isolation and current signal transmission. Afterwards, there is a brief treatment of the problem of the reduction of intrinsic noise by low-noise amplification, showing how the overall noise depends on both the contribution of the amplifier and the transducer internal resistance.

Then the basics of analogue filters are illustrated with reference to their use in removing noise without affecting the signal. A general discussion of the concept of averaging then follows, especially oriented towards pointing out analogies and differences with respect to filtering and giving indications on when to use the former or the latter method to enhance the measurement S/N ratio. While filtering is a powerful tool for improving the detectability of the signal over the noise if the respective frequency bandwidths are separated, synchronous averaging applied to a repetitive signal is a powerful bandwidth-narrowing technique virtually capable of extracting a signal from any noise or disturbance, provided that they are uncorrelated with the signal and that enough averages are taken.

Section 15.6 is dedicated to analogue-to-digital conversion and explains how it implies both a time discretization called sampling, and an amplitude quantization called quantizing. The resolution of an analogue-to-digital converter (ADC) is expressed by the number of bits n of the ADC and is equal to one part over 2^n of the conversion range. The AD conversion error has an associated noise called the quantization noise due to the finite number of intervals used to represent a signal which actually spans a continuous range. The quantization noise is the ideal accuracy limit achievable by an ADC; real ADCs have additional sources of errors which worsen the accuracy.

Sampling must satisfy the theorem by Shannon, which states that to reconstruct a continuous signal having its highest frequency component at f_M from its sampled version, the sampling frequency f_S must be $f_S \geqslant 2f_M$. The frequency $2f_M$ is the minimum allowed sampling rate and is called the Nyquist rate.

If a signal is undersampled, i.e. sampled at a frequency below the Nyquist rate, the phenomenon of aliasing occurs, where the samples are not able to uniquely represent the original analogue signal. Aliasing can be prevented by an antialiasing low-pass filter in front of the ADC to remove any frequency component greater than half the sampling frequency f_S.

Section 15.7 deals with the main instruments and systems for the acquisition and analysis of dynamic signals. Vibration meters are briefly described, then analogue and digital tape recorders are illustrated and compared.

Afterwards, attention is focused on computer-based data acquisition systems and boards whose characteristics are described in some detail. Such systems can currently perform the majority of the measurement usually required in vibration testing, with the additional advantage that the associated PC is exploitable for data storage, processing and analysis.

Finally, the importance of the analysis in the frequency domain is introduced and the frequency and dynamic signal analysers are presented. Firstly, the analogue spectrum analysers are briefly described. Secondly, the digital dynamic signal analyser (DSA) is presented which enables frequency analysis to be performed in real time by implementing the fast-Fourier-transform (FFT) algorithm. Since FFT analysers represent the most universally used instruments for measuring dynamic signals and vibrations in particular, their principle of operation and capabilities are illustrated in some detail and some hints are given for their practical usage.

References

1. Van der Ziel, A., *Noise: Sources, Characterization, Measurement*, Prentice-Hall, Englewood Cliffs, NJ, 1970.
2. Motchenbacher, C.D. and Fitchen, F.C., *Low Noise Electronic Design*, John Wiley & Sons, New York, 1973.
3. Malmstadt, H.V., Enke, C.G. and Crouch, S.R., *Electronic Measurements for Scientists*, W.A. Benjamin, Menlo Park, CA, 1974.
4. Pallas Areny, R. and Webster, J.G., *Sensors and Signal Conditioning*, Wiley Interscience, New York, 1991.
5. Horowitz, P. and Hill, W., *The Art of Electronics*, 2nd edn, Cambridge University Press, 1989.
6. *General Signal Conditioning Guide to Piezoelectric ICP® and Charge Output Sensor Instrumentation*, Application note, PCB Piezotronics Inc.
7. Kistler, W.P., *The Piezotron Concept as a Practical Approach to Vibration Measurement*, Application Note, Kistler Instruments.
8. Change, N.D., Application notes collection by Dytran Instruments, Inc., 1988.
9. Randall, R.B., Vibration measurement equipment and signal analyzers, in C.M. Harris (ed.), *Shock and Vibration Handbook*, 3rd edn, McGraw-Hill, New York, 1988, Ch. 13.
10. Morrison, R., *Grounding and Shielding Techniques in Instrumentation*, 2nd edn, John Wiley, New York, 1975.
11. Ott, H.W., *Noise Reduction Techniques in Electronic Systems*, Wiley Interscience, New York, 1976.
12. Oppenheim, A.V., Willsky, A.S. and Young, I.T., *Signals and Systems*, Prentice-Hall, Englewood Cliffs, NJ, 1983.
13. Jorgensen, F., *The Complete Handbook of Magnetic Recording*, McGraw-Hill, New York, 1996.
14. Cooley, J.W. and Tukey, J.W., An algorithm for the machine calculation of complex Fourier series, *Math. Comput.*, **19**, 297–301, 1965.

Further reading to Part II

Analog Devices ADXL Accelerometer Family Data Sheets.

Beckwith, T.G. and Marangoni, R.D., *Mechanical Measurements*, 4th edn, Addison-Wesley, Reading, Mass., 1990.

Bell, D.A., *Electronic Instrumentation and Measurements*, 2nd edn, Prentice-Hall, Englewood Cliffs, 1994.

Bertolaccini, M., Bussolati, C. and Manfredi, P.F., *Elettronica per Misure Industriali*, CLUP Politecnico di Milano, 1984.

Bolton, W., *Instrumentation and Measurement*, Newnes, Oxford, 1991.

Buckingham, M.J., *Noise in Electronic Devices and Systems*, Ellis Horwood, 1983.

Dally, J.W., Riley, W.F. and McConnell, K.G., *Instrumentation for Engineering Measurements*, 2nd edn, John Wiley, New York, 1993.

Harris, C.M. (ed.), *Shock and Vibration Handbook*, 3rd edn, McGraw-Hill, New York, 1988.

Herceg, E.E., *Handbook of Measurement and Control*, Schaevitz Engineering, Pennsauken, NJ, 1986.

Jones, B.K., *Electronics for Experimentation and Research*, Prentice-Hall, Hemel Hempstead, 1986.

Kail, R. and Mahr, W., *Piezoelectric Measuring Instruments and their Applications*, Application Note, Kistler Instruments.

Kovacs, G.T.A., *Micromachined Transducers Sourcebook*, WCB-McGraw-Hill, New York, 1998.

Lally, D.M., Jing, L. and Lally, R.W., *Low Frequency Calibration with the Structural Gravimetric Technique*, Application Note, PCB Piezotronics Inc.

Lynn, P.A., *An Introduction to the Analysis and Processing of Signals*, Macmillan, London, 1973.

Mandel, J., *The Statistical Analysis of Experimental Data*, Dover Publications, New York, 1984.

Marven, C. and Ewers, G., *A Simple Approach to Digital Signal Processing*, Texas Instruments, 1993.

Methods for the Calibration of Vibration and Shock Transducers – Part I: Basic Concepts, ISO 16063-1, Geneva, 1998.

Oppenheim, A.V. and Schafer, R.W., *Digital Signal Processing*, Prentice Hall, Englewood Cliffs, NJ, 1975.

Ovaska, S.J. and Väliviita, S., Angular acceleration measurement: a review, *Proc. IEEE Instrumentation and Measurement Techn. Conf.*, St Paul, USA, 875-880, May 18–21, 1998.

Scavuzzo, R.J. and Pusey, H.C., *Principles and Techniques of Shock Data Analysis*, 2nd edn, SAVIAC, Arlington, VA, 1996.

Serridge, M. and Licht, T.R., *Piezoelectric Accelerometers and Vibration Preamplifiers*, Bruel and Kjaer, 1987.

Smith, J.D., *Vibration Measurement and Analysis*, Butterworths, London, 1989.

Standard for a Smart Transducer Interface for Sensors and Actuators, IEEE 1451.2, 1997.

Sydenham, P.H. (ed.), *Handbook of Measurement Science – Theoretical Fundamentals*, Vol. 1, John Wiley, New York, 1982.

Tompkins, W.J. and Webster, J.G., *Interfacing Sensors to the IBM PC*, Prentice Hall, Englewood Cliffs, NJ, 1988.

Waanders, J.W., *Piezoelectric Ceramics: Properties and Applications*, Philips, Eindhoven, The Netherlands, 1991.

Wowk, V., *Machinery Vibration Measurement and Analysis*, McGraw-Hill, New York, 1991.

Appendices

Paolo L. Gatti

A Finite-dimensional vector spaces and elements of matrix analysis

A.1 The notion of finite-dimensional vector space

The fact that finite-dimensional vector (or linear) spaces are the fundamental setting of matrix theory is probably known to the reader. On the other hand, it is also true that – by simply considering a matrix as an array of numbers – many aspects of matrix theory and manipulation can be examined without ever even mentioning the notion of linear space. In the writer's opinion, this kind of approach tends to conceal the important interplay between matrices and linear transformations (or operators) defined on vector spaces – typically $\mathcal{R}^n$ or $\mathscr{C}^n$ – in which a matrix is just a particular representation of a linear transformation and different matrices may represent the same linear operator. It is in this light that we approach the subject by giving some basic concepts and definitions of finite-dimensional vector spaces.

Underlying a vector space is a **field** $\mathscr{F}$, or set of scalars, which for our purposes will always be the field of real numbers $\mathcal{R}$ or the field of complex numbers $\mathscr{C}$ with the usual operations of addition and multiplication.

A **vector** (or **linear**) **space** V over a field $\mathscr{F}$ is a set of objects (called vectors) where two operations are defined: addition between elements of the set and multiplication by elements of the field $\mathscr{F}$. The vector space is closed under these two operations which, in turn, must satisfy the following properties.

1. Addition:

    (a) $\mathbf{x} + \mathbf{y} = \mathbf{y} + \mathbf{x}$ for every $\mathbf{x}, \mathbf{y} \in V$.
    (b) $(\mathbf{x} + \mathbf{y}) + \mathbf{z} = \mathbf{x} + (\mathbf{y} + \mathbf{z})$ for every $\mathbf{x}, \mathbf{y}, \mathbf{z} \in V$.
    (c) There exists a unique null vector $\mathbf{0} \in V$ such that $\mathbf{x} + \mathbf{0} = \mathbf{x}$ for every $\mathbf{x} \in V$.
    (d) For every $\mathbf{x} \in V$ there exists a unique element $-\mathbf{x}$ such that $\mathbf{x} + (-\mathbf{x}) = \mathbf{0}$.

2. Multiplication by a scalar:

    (a) $\alpha(\mathbf{x} + \mathbf{y}) = \alpha\mathbf{x} + \alpha\mathbf{y}$ for every $\mathbf{x}, \mathbf{y} \in V$, $\alpha \in \mathscr{F}$.
    (b) $(\alpha + \beta)\mathbf{x} = \alpha\mathbf{x} + \beta\mathbf{x}$ for every $\mathbf{x} \in V$, $\alpha, \beta \in \mathscr{F}$.

(c) $(\alpha\beta)\mathbf{x} = \alpha(\beta\mathbf{x})$ for every $\mathbf{x} \in V$, $\alpha, \beta \in \mathscr{F}$.
(d) $1\mathbf{x} = \mathbf{x}$ for every $\mathbf{x} \in V$.

In 2(d) it is understood that 1 is the multiplicative unity of the scalar field $\mathscr{F}$. Also, from the above properties it is evident that we will indicate vectors with boldface letters while scalars will be indicated either by Greek or Latin letters.

Example A.1.   For a fixed positive integer n, the set of all n-tuples of real (or complex) numbers $\mathbf{x} = (x_1, x_2, ..., x_n)$ forms a vector space on the real (complex) field when we define the addition of vectors and multiplication by a scalar as

1. $\mathbf{x} + \mathbf{y} = (x_1, x_2, ..., x_n) + (y_1, y_2, ..., y_n) = (x_1 + y_1, x_2 + y_2, ..., x_n + y_n)$
2. $\alpha\mathbf{x} = \alpha(x_1, x_2, ..., x_n) = (\alpha x_1, \alpha x_2, ..., \alpha x_n)$.

Depending on whether $\mathscr{F} = \mathscr{R}$ or $\mathscr{F} = \mathscr{C}$ we obtain the vector spaces $\mathscr{R}^n$ or $\mathscr{C}^n$, which are the basic vector spaces in a large number of applications in physics and engineering.

Example A.2.   The reader is invited to verify that the set $M_{m \times n}(\mathscr{F})$ of all $m \times n$ matrices with elements in the field $\mathscr{F}$ is a vector space where addition is defined as ordinary element-wise addition and multiplication by a scalar is ordinary multiplication of a matrix by a real or a complex number.

Moreover, we also leave to the reader the proof of the following theorem.

Theorem A.1.   let V be a vector space on the field $\mathscr{F}$, then

1. $0\mathbf{x} = \mathbf{0}$
2. $(-1)\mathbf{x} = -\mathbf{x}$
3. $\alpha\mathbf{0} = \mathbf{0}$

where 0 is the null element of the scalar field.

Other basic definitions are as follows:
We call a **subspace** U of a vector space V a subset of V that is, by itself, a vector space over the same scalar field with respect to the same operations (addition and multiplication by a scalar) defined in V.

A set of vectors $\mathbf{u}_1, \mathbf{u}_2, ..., \mathbf{u}_k \in V$ is said to be **linearly dependent** if there exists a set of scalars $\alpha_1, \alpha_2, ..., \alpha_k \in \mathscr{F}$ (not all zero) such that

$$\alpha_1\mathbf{u}_1 + \alpha_2\mathbf{u}_2 + \cdots + \alpha_k\mathbf{u}_k = 0 \tag{A.1}$$

where this definition implies that at least one of the vectors $\mathbf{u}_i$ can be written as a linear combination of the other vectors. A set of vectors which is not linearly dependent is called **linearly independent** and in this case the relationship

$$\sum_{i=1}^{k} \alpha_i \mathbf{u}_i = 0$$

implies $\alpha_1 = \alpha_2 = \cdots = \alpha_k = 0$.

A finite set $\mathbf{u}_1, \mathbf{u}_2, ..., \mathbf{u}_n$ of elements of a vector space V is said to **span** V if every $\mathbf{x} \in V$ can be written in the form

$$\mathbf{x} = \sum_{i=1}^{n} \alpha_i \mathbf{u}_i \qquad (A.2)$$

for some $\alpha_1, \alpha_2, ..., \alpha_n \in \mathscr{F}$. Moreover, a set of elements $\mathbf{u}_1, \mathbf{u}_2, ..., \mathbf{u}_n \in V$ is said to be a **basis** of V if and only if (1) the set spans V and (2) the set is linearly independent.

The number of elements (i.e. vectors) that form a basis is called the **dimension** of V or, in other words, V is said to be n-dimensional if n linearly independent elements can be found in V, but any $n + 1$ elements of V are linearly dependent. If n linearly independent elements can be found in V for every n, then V is said to be infinite-dimensional; however, we will not consider such cases in this appendix (some examples of infinite-dimensional linear spaces are given in Chapter 2).

It must be noted that if some basis of a vector space V consists of a finite number (say n) of elements, then all bases have the same number of elements or, stated differently, although the dimension of a space is fixed, it is possible to construct many different bases.

If $\mathbf{u}_1, \mathbf{u}_2, ..., \mathbf{u}_n \in V$ (or, for short, $\{\mathbf{u}_i\}_{i=1}^{n} \in V$) is a basis for V then eq (A.2) holds for any vector $\mathbf{x} \in V$ and the coefficients α_i are called the **components** (or **coordinates**) of $\mathbf{x}$ relative to the basis $\{\mathbf{u}_i\}_{i=1}^{n}$; obviously, the components change with a change of basis.

Example A.3. Consider the space $\mathscr{R}^3$ and the 'standard' basis $\{\mathbf{e}_i\}_{i=1}^{3} = \{(1, 0, 0)^T, (0, 1, 0)^T, (0, 0, 1)^T\}$. Obviously, the vector $\mathbf{x} = 2\mathbf{e}_1 + 3\mathbf{e}_3$ has components $(2, 0, 3)$ relative to this basis. However, if we choose the basis $\{\mathbf{f}_i\}_{i=1}^{3} = \{(1, 1, 0)^T, (1, 0, 2)^T, (1, 1, 1)^T\}$ then it is left to the reader to show that (1) $\{\mathbf{f}_i\}_{i=1}^{3}$ is a set of three linearly independent vectors and that (2) the components of $\mathbf{x}$ with respect to this basis are $(1, 2, -1)$.

The way in which the components of a vector change under a change of basis can be obtained as follows. Consider two bases $\{\mathbf{u}_i\}_{i=1}^{n}$ and $\{\mathbf{v}_i\}_{i=1}^{n}$; each vector of the second basis can be expressed as a linear combination of the vectors of the first basis and we can write

$$\mathbf{v}_j = \sum_{k=1}^{n} c_{kj} \mathbf{u}_k \qquad (j = 1, 2, ..., n) \qquad (A.3a)$$

where the n^2 elements c_{kj} form the elements of the so-called transformation matrix. Now, since an arbitrary vector x can be written as

$$x = \sum_{k=1}^{n} a_k u_k$$

$$x = \sum_{j=1}^{n} b_j v_j$$

(A.3b)

where the coefficients a and b, respectively, are the components of x relative to the first and second basis, then – by taking eq (A.3a) into account – the second of eqs (A.3b) can be written as

$$x = \sum_{j=1}^{n} b_j v_j = \sum_{j=1}^{n} b_j \sum_{k=1}^{n} c_{kj} u_k = \sum_{k=1}^{n} \left(\sum_{j=1}^{n} c_{kj} b_j \right) u_k$$

from which it is evident (from the first of eqs (A.3b)) that

$$a_k = \sum_{j=1}^{n} c_{kj} b_j$$

(A.4a)

This relation, in turn, is equivalent to the matrix equation

$$a = Cb$$

(A.4b)

where

$$a = \begin{bmatrix} a_1 \\ a_2 \\ \cdots \\ a_n \end{bmatrix} \quad b = \begin{bmatrix} b_1 \\ b_2 \\ \cdots \\ b_n \end{bmatrix} \quad C = \begin{bmatrix} c_{11} & c_{12} & \cdots & c_{1n} \\ c_{21} & c_{22} & \cdots & \cdots \\ \cdots & \cdots & \cdots & \cdots \\ c_{n1} & c_{n2} & \cdots & c_{nn} \end{bmatrix}$$

Note that the components transform differently than the basis vectors (compare eq (A.3a) with (A.4a)).

We anticipated here some basic notions on matrices with which the reader should already have some familiarity; however, before turning our attention to matrices, two important concepts need to be introduced: the concept of inner product and the concept of linear operator on a vector space.

The structure of a vector space is greatly enriched by the introduction of a numerical function, the **inner product**, which allows us to generalize such notions as the length of a vector and the angle between two vectors.

Mathematically, the definition of inner product has already been given in Section 2.5: with reference to that section, given a complex linear space V, an inner product is a function defined on $V \times V$ (i.e. the set of ordered pairs of elements of V) with values in $\mathscr{C}$ (i.e. $\langle \cdot \mid \cdot \rangle$: $V \times V \to \mathscr{C}$) that satisfies axioms 1 to 4. If the underlying scalar field is $\mathscr{R}$, the relevant axioms to be satisfied are $1'$ to $4'$. With the notion of inner product at our disposal, two vectors $\mathbf{x}, \mathbf{y}$ are said to be orthogonal if

$$\langle \mathbf{x} \mid \mathbf{y} \rangle = 0 \tag{A.5}$$

where it is evident that this is just a generalization of the concept of perpendicular vectors encountered in basic physics. Particularly important in inner product spaces are the so-called orthonormal bases; an orthonormal basis being a set of linearly independent vectors which span the space, have unit length and are mutually orthogonal. Since a vector of unit length is a vector for which $\sqrt{\langle \mathbf{x} \mid \mathbf{x} \rangle} = 1$, an orthonormal basis $\{\mathbf{u}_i\}_{i=1}^n$ is a basis which satisfies

$$\langle \mathbf{u}_i \mid \mathbf{u}_j \rangle = \delta_{ij} \equiv \begin{cases} 1 & i = j \\ 0 & i \neq j \end{cases} \tag{A.6}$$

Given an arbitrary (nonorthonormal) basis in an inner product linear space, it is always possible to obtain an orthonormal basis for the same space. Although this task can be accomplished in many different ways, there exists a simple and far-reaching algorithm which is called the **Gram–Schmidt (orthonormalization) process**. Here, we will outline it briefly. Let $\{\mathbf{x}_i\}_{i=1}^n$ be an arbitrary basis and let us call $\{\mathbf{u}_i\}_{i=1}^n$ the set of mutually orthonormal vectors to be determined. Let us define $\mathbf{y}_1 = \mathbf{x}_1$ and choose

$$\mathbf{u}_1 = \frac{\mathbf{y}_1}{\sqrt{\langle \mathbf{y}_1 \mid \mathbf{y}_1 \rangle}} \tag{A.7a}$$

so that $\mathbf{u}_1$ is a vector of unit length. Next, define $\mathbf{y}_2 = \mathbf{x}_2 - \langle \mathbf{x}_2 \mid \mathbf{u}_1 \rangle \mathbf{u}_1$ (note that, with this definition, $\mathbf{y}_2$ is orthogonal to $\mathbf{u}_1$) and choose

$$\mathbf{u}_2 = \frac{\mathbf{y}_2}{\sqrt{\langle \mathbf{y}_2 \mid \mathbf{y}_2 \rangle}} \tag{A.7b}$$

which makes $\mathbf{u}_2$ a vector of unit length. Proceeding along this line of reasoning and assuming that $\mathbf{u}_1, \mathbf{u}_2, ..., \mathbf{u}_{k-1}$ have been determined, let

$$\mathbf{y}_k = \mathbf{x}_k - \langle \mathbf{x}_k \mid \mathbf{u}_{k-1} \rangle \mathbf{u}_{k-1} - \langle \mathbf{x}_k \mid \mathbf{u}_{k-2} \rangle \mathbf{u}_{k-2} - \cdots - \langle \mathbf{x}_k \mid \mathbf{u}_1 \rangle \mathbf{u}_1 \tag{A.7c}$$

so that $\mathbf{y}_k$ is orthogonal to $\mathbf{u}_1, \mathbf{u}_2, ..., \mathbf{u}_{k-1}$, and again choose

$$\mathbf{u}_k = \frac{\mathbf{y}_k}{\sqrt{\langle \mathbf{y}_k \mid \mathbf{y}_k \rangle}} \tag{A.7d}$$

The process is then continued until the desired n orthonormal vectors $\{\mathbf{u}_i\}_{i=1}^{n}$ have been obtained. Two things should be noted about this process:

1. At each step, the orthonormal vectors $\mathbf{u}_1, \mathbf{u}_2, ..., \mathbf{u}_k$ are a linear combination of the original independent vectors $\mathbf{x}_1, \mathbf{x}_2, ..., \mathbf{x}_k$ only.
2. It can be applied to any finite or countable – not necessarily linearly independent – set of vectors. If the set is not independent, the process will produce a vector $\mathbf{y}_k = 0$ for the least value of k for which $\{\mathbf{x}_1, \mathbf{x}_2, ..., \mathbf{x}_k\}$ is a linearly dependent set. In this case $\mathbf{x}_k$ is a linear combination of $\mathbf{x}_1, \mathbf{x}_2, ..., \mathbf{x}_{k-1}$. Substitution of $\mathbf{x}_{k+1}$ for $\mathbf{x}_k$ and continuation of the process can answer such questions as: what is a basis for, or the dimension of, or the span of $\{\mathbf{x}_1, \mathbf{x}_2, ..., \mathbf{x}_n\}$?

Let us now consider two vector spaces, U and V, over the same scalar field $\mathscr{F}$. Any mapping $A:U \rightarrow V$ such that

$$A(\mathbf{u}_1 + \mathbf{u}_2) = A\mathbf{u}_1 + A\mathbf{u}_2 \qquad \text{for all } \mathbf{u}_1, \mathbf{u}_2 \in U \tag{A.8a}$$

$$A(\alpha\mathbf{u}) = \alpha A\mathbf{u} \qquad \text{for all } \mathbf{u} \in U, \alpha \in \mathscr{F} \tag{A.8b}$$

is called a **linear operator** from U to V.

Among the set of linear operators in a vector space, many different classes can be distinguished according to some specific features. Here, we only mention the class of isomorphisms: an operator is said to be an **isomorphism** (from the Greek meaning 'same structure') if it is injective and surjective (see any book of linear algebra for these definitions). Broadly speaking, the importance of this particular class lies in the fact that two isomorphic spaces can be considered as equal for all practical purposes. Furthermore, in this regard, an important theorem states that two finite-dimensional vector spaces over the same scalar field are isomorphic if and only if they have the same dimension and it follows as a corollary to this theorem that any n-dimensional real vector space is isomorphic to $\mathscr{R}^n$ and any n-dimensional complex vector space is isomorphic to $\mathscr{C}^n$. More specifically, if V is n-dimensional over a scalar field $\mathscr{F}$ ($\mathscr{F} = \mathscr{R}$ or $\mathscr{F} = \mathscr{C}$) with specified basis $\mathscr{B} = \{\mathbf{u}_1, ..., \mathbf{u}_n\}$, then, since any element $\mathbf{x} \in V$ can be written uniquely as $\mathbf{x} = \sum_i \alpha_i \mathbf{u}_i$ ($\alpha_i \in \mathscr{F}$), we may associate $\mathbf{x}$ with the n-tuple $[\mathbf{x}]_{\mathscr{B}} = [\alpha_1, \alpha_2, ..., \alpha_n]^T$. In other words, for any fixed basis $\mathscr{B}$, the mapping $\varphi:\mathbf{x} \rightarrow [\mathbf{x}]_{\mathscr{B}}$ is an isomorphism between V and $\mathscr{F}^n$ and this is the reason why we can manipulate and operate on vectors by dealing with their components with respect to a given basis.

A.2 Matrices

Basically speaking, matrices are rectangular array of scalars from a field $\mathscr{F}$. When we speak of an $m \times n$ matrix the number m refers to the number of rows and the number n to the number of columns. If $m = n$ the matrix is said to be square. Matrix operations are probably quite familiar to the reader and we give here a brief account.

The addition of two matrices is written $\mathbf{A} + \mathbf{B}$, is defined entry-wise for arrays of the same dimension and inherits commutativity and associativity from the scalar field; moreover, the matrix whose entries are all zero is the identity under addition. In other words, given two $m \times n$ matrices $\mathbf{A} = [a_{ij}]$ and $\mathbf{B} = [b_{ij}]$ $(1 \leqslant i \leqslant m; \ 1 \leqslant j \leqslant n)$ we have

$$\mathbf{A} + \mathbf{B} = \mathbf{C} = [c_{ij}] \tag{A.9a}$$

where $\mathbf{C}$ is also a $m \times n$ matrix whose elements c_{ij} are given by

$$c_{ij} = a_{ij} + b_{ij} \qquad (1 \leqslant i \leqslant m; \ 1 \leqslant j \leqslant n) \tag{A.9b}$$

and $\mathbf{A} + \mathbf{0} = \mathbf{A}$.

Matrix multiplication of an $m \times n$ matrix $\mathbf{A}$ by a $p \times q$ matrix $\mathbf{B}$ is only possible when their sizes are compatible, that is when $p = n$ (the number of columns of the first matrix equals the number of rows of the second matrix). The result of multiplication is a $m \times q$ matrix $\mathbf{C}$, i.e.

$$\mathbf{AB} = \mathbf{C} = [c_{ij}] \tag{A.10a}$$

whose elements are given by

$$c_{ij} = \sum_{k=1}^{n} a_{ik}b_{kj} \qquad (1 \leqslant i \leqslant m; \ 1 \leqslant j \leqslant q) \tag{A.10b}$$

For example, the product of the 2×3 matrix

$$\mathbf{A} = \begin{bmatrix} 2 & -1 & 3 \\ 0 & 1 & 4 \end{bmatrix}$$

by the 3×2 matrix

$$\mathbf{B} = \begin{bmatrix} 1 & -1 \\ 3 & 2 \\ 5 & 7 \end{bmatrix}$$

is the 2×2 matrix **C** obtained as

$$\mathbf{AB} = \begin{bmatrix} 2 & -1 & 3 \\ 0 & 1 & 4 \end{bmatrix} \begin{bmatrix} 1 & -1 \\ 3 & 2 \\ 5 & 7 \end{bmatrix}$$

$$= \begin{bmatrix} c_{11} = 2 - 3 + 15 & c_{12} = -2 - 2 + 21 \\ c_{21} = 0 + 3 + 20 & c_{22} = 0 + 2 + 28 \end{bmatrix} = \begin{bmatrix} 14 & 17 \\ 23 & 30 \end{bmatrix} = \mathbf{C}$$

In general, matrix product is not commutative and $\mathbf{AB} \neq \mathbf{BA}$, provided that the two expressions have a meaning. However, within the set $M_{n \times n}$ of all square $n \times n$ matrices, the product can be commutative when restricted to certain subsets which are worthy of study in their own right.

Matrix multiplication is associative and distributive over matrix addition, i.e. when the following products are defined, we have

$$\mathbf{A(BC)} = \mathbf{(AB)C} \qquad\qquad\qquad\qquad (A.11)$$

and

$$\mathbf{A(B + C)} = \mathbf{AB} + \mathbf{AC}$$

$$\mathbf{(A + B)C} = \mathbf{AC} + \mathbf{BC} \qquad\qquad\qquad (A.12)$$

but it should be noted that, in general:

1. $\mathbf{AB} = 0$ does not imply $\mathbf{A} = 0$ or $\mathbf{B} = 0$.
2. $\mathbf{AB} = \mathbf{CB}$ does not imply $\mathbf{A} = \mathbf{C}$ (even if the reverse is true, i.e. $\mathbf{A} = \mathbf{C}$ does imply $\mathbf{AB} = \mathbf{CB}$ for all possible $\mathbf{B}$ for which the product is defined).

At this point, some basic definitions are in order. Given a general $m \times n$ matrix $\mathbf{A} = [a_{ij}]$ we call its **transpose** the $n \times m$ matrix $\mathbf{A}^T = [a_{ji}]$ which is obtained by interchanging rows and columns of the original matrix; for example, if

$$\mathbf{A} = \begin{bmatrix} 2 & -1 & 3 \\ 0 & 1 & 4 \end{bmatrix}$$

then

$$\mathbf{A}^T = \begin{bmatrix} 2 & 0 \\ -1 & 1 \\ 3 & 4 \end{bmatrix}$$

and a (square) matrix for which $\mathbf{A} = \mathbf{A}^T$ is called **symmetrical**.

The **Hermitian adjoint** of a matrix $\mathbf{A} = [a_{ij}]$ – denoted by the symbol $\mathbf{A}^H$ – is the matrix $\mathbf{A}^H = [a_{ji}^*]$ whose elements are the complex conjugates of $\mathbf{A}^T$; clearly, if the elements of $\mathbf{A}$ are all real then $\mathbf{A}^H = \mathbf{A}^T$. A (square) matrix with complex entries for which the property $\mathbf{A} = \mathbf{A}^H$ holds is called **Hermitian** (or self-adjoint) and it is evident that Hermitian matrices with real entries are symmetrical; by contrast, a symmetrical matrix with complex entries is not Hermitian.

Both the transpose and the Hermitian adjoint obey the 'reverse-order law' which reads

$$(\mathbf{AB})^T = \mathbf{B}^T\mathbf{A}^T$$

$$(\mathbf{AB})^H = \mathbf{B}^H\mathbf{A}^H$$

(A.13)

Note that, however, for complex conjugation there is no reversing, i.e.

$$(\mathbf{AB})^* = \mathbf{A}^*\mathbf{B}^*$$

Our attention will be mainly focused on square $n \times n$ matrices and in this light some other matrices which are given special names according to some property that they satisfy are as follows:

1. if $\mathbf{A} = -\mathbf{A}^T$ (i.e. $[a_{ij}] = -[a_{ji}]$) the matrix is called **skew symmetrical**.
2. if $\mathbf{A} = -\mathbf{A}^H$ (i.e. $[a_{ij}] = -[a_{ji}^*]$) the matrix is called **skew Hermitian**.
3. if $\mathbf{AA}^T = \mathbf{A}^T\mathbf{A} = \mathbf{I}$ the matrix is called **orthogonal**.
4. if $\mathbf{AA}^H = \mathbf{A}^H\mathbf{A} = \mathbf{I}$ the matrix is called **unitary** (a unitary matrix with real entries is an orthogonal matrix).

The symbol $\mathbf{I}$ indicates the unit matrix, that is the matrix whose only nonzero elements are all ones and lie on the main diagonal; for example, the 3×3 unit matrix is

$$\mathbf{I}_{3 \times 3} = \begin{bmatrix} 1 & 0 & 0 \\ 0 & 1 & 0 \\ 0 & 0 & 1 \end{bmatrix}$$

The matrix $\mathbf{I}$, in turn, is a special case of **diagonal** matrix, i.e. a matrix whose only nonzero elements are on the main diagonal. Also, **triangular** matrices are worthy of mention: a matrix $\mathbf{A} = [a_{ij}]$ is said to be upper triangular if $a_{ij} = 0$ whenever $j < i$ and strictly upper triangular if $a_{ij} = 0$ whenever $j \leqslant i$. Similarly, $\mathbf{A}$ is lower triangular (or strictly lower triangular) if its transpose is upper triangular (or strictly upper triangular).

A more general class of matrices is the class of normal matrices: a square matrix is called **normal** if it commutes with its Hermitian adjoint, that is if

$$\mathbf{AA}^H = \mathbf{A}^H\mathbf{A}$$

(A.14)

In this regard it is not difficult to show that:

1. All unitary matrices are normal (therefore all unitary matrices with real entries, i.e. orthogonal matrices, are normal).
2. All Hermitian matrices are normal (therefore all Hermitian matrices with real entries, i.e. symmetrical matrices, are normal).
3. All skew-Hermitian matrices are normal (therefore all skew-Hermitian matrices with real entries, i.e. skew-symmetrical matrices, are normal).
4. The matrix

$$\begin{bmatrix} 1 & -1 \\ 1 & 1 \end{bmatrix}$$

is normal but it does not fall in any of the above categories.

Before closing this section, we point out an aspect which is worthy of notice: the so-called **partitioning of matrices**. We mean by this term the fact that matrix manipulations are often simplified by subdividing a matrix into a (convenient) number of submatrices, where a submatrix is obtained by including only the elements of certain rows and columns of the original matrix. Consider for example the 3×5 matrix

$$\mathbf{A} = \begin{bmatrix} a_{11} & a_{12} & a_{13} & a_{14} & a_{15} \\ a_{21} & a_{22} & a_{23} & a_{24} & a_{25} \\ a_{31} & a_{32} & a_{33} & a_{34} & a_{35} \end{bmatrix}$$

if we take into account the only constraint that a line of partitioning must always run completely across the original matrix, one possible way of partitioning the matrix $\mathbf{A}$ is

$$\mathbf{A} = \begin{bmatrix} \mathbf{A}_{11} & \mathbf{A}_{12} \\ \mathbf{A}_{21} & \mathbf{A}_{22} \end{bmatrix}$$

where

$$\mathbf{A}_{11} = \begin{bmatrix} a_{11} & a_{12} \\ a_{21} & a_{22} \end{bmatrix} \qquad \mathbf{A}_{12} = \begin{bmatrix} a_{13} & a_{14} & a_{15} \\ a_{23} & a_{24} & a_{25} \end{bmatrix}$$

$$\mathbf{A}_{21} = [a_{31} \quad a_{32}] \qquad \mathbf{A}_{22} = [a_{33} \quad a_{34} \quad a_{35}]$$

Obviously, the same matrix can be partitioned in other different ways and the submatrices above, in turn, can be further partitioned. However, the point we want to make is that – besides the fact that partitioning has often the advantage of saving computer storage – using partitioned matrices we can add, subtract or multiply as if the submatrices were ordinary matrix elements, provided that the original matrices to be added, subtracted or multiplied have been partitioned in such a way that it is permissible to perform the prescribed operation. In other words, partitioning is a device to facilitate matrix manipulations and does not change any results. To this end, the reader is invited to calculate the simple product $\mathbf{AB}$ where, as an example

$$\mathbf{A} = \begin{bmatrix} 2 & 3 & -1 \\ 0 & 1 & 2 \\ 4 & -3 & -2 \end{bmatrix} \qquad \mathbf{B} = \begin{bmatrix} 1 & -5 \\ 4 & 0 \\ -4 & 2 \end{bmatrix}$$

and we partition $\mathbf{A}$ into the four submatrices

$$\mathbf{A}_{11} = \begin{bmatrix} 2 & 3 \\ 0 & 1 \end{bmatrix} \qquad \mathbf{A}_{12} = \begin{bmatrix} -1 \\ 2 \end{bmatrix}$$

$$\mathbf{A}_{21} = [4 \quad -3] \qquad \mathbf{A}_{22} = [-2]$$

and $\mathbf{B}$ into the two submatrices

$$\mathbf{B}_1 = \begin{bmatrix} 1 & -5 \\ 4 & 0 \end{bmatrix} \qquad \mathbf{B}_2 = [-4 \quad 2]$$

Then, it is not difficult to see that the above partitioning is consistent with matrix multiplication and to determine that

$$\mathbf{AB} = \begin{bmatrix} \mathbf{A}_{11} & \mathbf{A}_{12} \\ \mathbf{A}_{21} & \mathbf{A}_{22} \end{bmatrix} \begin{bmatrix} \mathbf{B}_1 \\ \mathbf{B}_2 \end{bmatrix} = \begin{bmatrix} \mathbf{A}_{11}\mathbf{B}_1 + \mathbf{A}_{12}\mathbf{B}_2 \\ \mathbf{A}_{21}\mathbf{B}_1 + \mathbf{A}_{22}\mathbf{B}_2 \end{bmatrix}$$

A.2.1 Trace, determinant, inverse and rank of a matrix

If we restrict our attention to square $n \times n$ matrices, two important scalar quantities which can be evaluated from the entries of a matrix are the **trace** and the **determinant**. The trace of a matrix is usually indicated by $\mathrm{tr}(\mathbf{A})$ and is the sum of its diagonal elements, i.e.

$$\mathrm{tr}(\mathbf{A}) \equiv \sum_{i=1}^{n} a_{ii} \qquad (A.15)$$

By contrast, the determinant of a (square) matrix $\mathbf{A}$ – usually denoted by det(**A**) or | **A** | – can be calculated by means of the so-called Laplace expansion

$$\det(\mathbf{A}) = \sum_{j=1}^{n} (-1)^{i+j} a_{ij} \det(\mathbf{A}_{ij}) = \sum_{i=1}^{n} (-1)^{i+j} a_{ij} \det(\mathbf{A}_{ij}) \qquad (A.16)$$

where the matrix $\mathbf{A}_{ij}$ is the $(n-1) \times (n-1)$ matrix obtained by deleting the ith row and the jth column of **A**. The first sum of eq (A.16) is the Laplace expansion along the ith row, while the second sum is the Laplace expansion along the jth column. For any choice of row or column, either expansion yields the determinant.

More specifically, the definition of determinant proceeds by induction by defining the determinant of a 1×1 matrix to be the value of its single entry, i.e. we have

$$\det[a_{11}] = a_{11}$$

$$\det \begin{bmatrix} a_{11} & a_{12} \\ a_{21} & a_{22} \end{bmatrix} = a_{11}a_{22} - a_{12}a_{21}$$

etc. Many properties and theorems are associated with the use of determinants and the interested reader can refer to a number of excellent books. For our purposes, one of the most important properties of determinants is that it is multiplicative, i.e. given two square matrices **A** and **B** we have

$$\det(\mathbf{AB}) = \det(\mathbf{A})\det(\mathbf{B}) \qquad (A.17)$$

which is a useful property in many numerical applications in which a given matrix is factorized – or 'decomposed' – into a product of (typically two or three) matrices. Also, it is straightforward to show that the determinant of a diagonal or triangular matrix is given by the product of its diagonal elements.

Now we turn our attention to the **inverse** $\mathbf{A}^{-1}$ of a (square) matrix **A**, and define $\mathbf{A}^{-1}$ as that matrix which satisfies $\mathbf{A}^{-1}\mathbf{A} = \mathbf{A}\mathbf{A}^{-1} = \mathbf{I}$. Not all square matrices possess an inverse and the ones that do are called **nonsingular** (and the inverse is unique!); by contrast, a matrix which does not possess an inverse is called **singular**. When the inverses of two matrices **A** and **B** exist, it is not difficult to prove that the 'reverse-order law' applies also in this case, i.e.

$$(\mathbf{AB})^{-1} = \mathbf{B}^{-1}\mathbf{A}^{-1} \qquad (A.18)$$

The existence of an inverse matrix is strictly connected both to the value of the determinant (more precisely, if it zero or different from zero) and to the **rank** of a matrix (indicated by rank(**A**)), this latter quantity being the largest number of columns of **A** that constitute a linearly independent set. This set of columns is not unique but the rank is, indeed, unique. Also worthy of notice is the fact that

$$\text{rank}(\mathbf{A}) = \text{rank}(\mathbf{A}^T) \tag{A.19}$$

so that the rank may be equivalently defined in terms of linearly independent rows. In other words, we can write 'row rank = column rank'.

As might be expected, the various ideas that have been considered up to this point are mutually interrelated and, for example, we can alternatively define the rank of a matrix $\mathbf{A}$ as the largest submatrix of $\mathbf{A}$ with a nonzero determinant. In the light of these interrelationships, we close this section by summarizing without proofs (which can be found on any book on matrices) a number of important results. If $\mathbf{A}$ is a square $n \times n$ matrix, the following statements are equivalent:

1. $\mathbf{A}^{-1}$ exists (i.e. $\mathbf{A}$ is nonsingular).
2. $\det(\mathbf{A}) \neq 0$.
3. $\text{rank}(\mathbf{A}) = n$ (i.e. the matrix has a **full rank**).
4. The rows of $\mathbf{A}$ are linearly independent.
5. The columns of $\mathbf{A}$ are linearly independent.

Also, referring to other aspects which will not be considered here or will be considered in later sections, for completeness it can be said that there are three other statements which are also equivalent to 1–5:

6. The only solution to the linear system $\mathbf{Ax} = \mathbf{0}$ is $\mathbf{x} = \mathbf{0}$ (where $\mathbf{x}$ indicates here an $n \times 1$ vector of unknowns).
7. The linear system $\mathbf{Ax} = \mathbf{b}$ has a unique solution for each $n \times 1$ vector $\mathbf{b}$
8. Zero is not an eigenvalue of $\mathbf{A}$.

A.3   Eigenvalues and eigenvectors: the standard eigenvalue problem

As before, let $M_{n \times n}(\mathscr{F})$ be the set of all $n \times n$ matrices on a given field of scalars $\mathscr{F}$, usually the real numbers $\mathscr{R}$ or the complex numbers $\mathscr{C}$. The following discussion is in general valid in the set of complex-entried matrices but it will seldom make a substantial difference if the material is interpreted in terms of real numbers. However, a major difference between $\mathscr{R}$ and $\mathscr{C}$ which should be kept in mind is that the complex field is algebraically closed and therefore it may often be useful to think of real matrices as complex matrices with restricted entries.

Given a square $n \times n$ matrix $\mathbf{A}$ and a nonzero vector $\mathbf{x} \in \mathscr{C}^n$, let us consider the equation

$$\mathbf{Ax} = \lambda\mathbf{x} \tag{A.20}$$

where λ is a scalar. If a scalar λ and a nonzero vector $\mathbf{x}$ satisfy eq (A.20), then λ is called an **eigenvalue** of $\mathbf{A}$ and $\mathbf{x}$ is called an **eigenvector** of $\mathbf{A}$ associated with λ.

The set of all $\lambda \in \mathscr{C}$ that are eigenvalues of $\mathbf{A}$ is called the **spectrum** of $\mathbf{A}$ and is often denoted by $\sigma(\mathbf{A})$.

Three observations can be made immediately: first, if $\mathbf{x}$ is an eigenvector of $\mathbf{A}$ associated with the eigenvalue λ, then any nonzero scalar multiple of $\mathbf{x}$ is also an eigenvector (meaning that eigenvectors are determined to within a multiplicative constant and some kind of 'normalization' is required); second, if $\mathbf{x}, \mathbf{y} \in \mathscr{C}^n$ are both eigenvectors associated to the eigenvalue λ, then any nonzero linear combination of $\mathbf{x}$ and $\mathbf{y}$ is also an eigenvector corresponding to λ and, third, $\mathbf{A} \in M_{n \times n}$ is singular if and only if zero is one of its eigenvalues, i.e. if and only if $0 \in \sigma(\mathbf{A})$ (see also statement 8 at the end of Section A.2.1).

It should be noted that eq (A.20) defines what is called a 'standard' eigenvalue problem, whereas a 'generalized' eigenvalue problem (the type of eigenvalue problem frequently encountered in engineering vibrations) involves two matrices and reads $\mathbf{Ax} = \lambda\mathbf{Bx}$. In what follows, in order to present a central set of fundamental results, we will only consider the eigenvalue problem in its standard form for three reasons: (1) a generalized eigenvalue problem can always be recast in standard form; (2) the fundamental concepts, ideas and results remain essentially the same; and (3) specific aspects of the generalized problem are given in the main chapters whenever needed in the course of the discussion.

If now we want to characterize the eigenvalues of a given matrix we can rewrite eq (A.20) as

$$(\mathbf{A} - \lambda\mathbf{I})\mathbf{x} = 0 \tag{A.21}$$

so that $\lambda \in \sigma(\mathbf{A})$ if and only if $\mathbf{A} - \lambda\mathbf{I}$ is a singular matrix, i.e.

$$\det(\mathbf{A} - \lambda\mathbf{I}) = 0 \tag{A.22}$$

From the Laplace expansion of the determinant, we note that $\det(\mathbf{A} - \lambda\mathbf{I})$ is a polynomial of degree n in λ – which is called the **characteristic polynomial** of $\mathbf{A}$ – so that from eq (A.22) it follows that the set of its roots (i.e. the roots of the **characteristic equation**) coincides with $\sigma(\mathbf{A})$. Then, from the fundamental theorem of algebra (a polynomial of degree n with complex coefficients has exactly n zeroes, counting multiplicities, among the complex numbers) it follows that every $n \times n$ matrix has, among the complex numbers, exactly n eigenvalues, counting multiplicities. Two observations are in order at this point:

1. The last statement depends heavily on the fact that the complex field is algebraically closed; for matrices in other fields (typically $\mathscr{R}$) little can be said about the number of eigenvalues in that field.

2. Here, the term multiplicity refers to the 'algebraic multiplicity' (a.m. for short), that is the number of times that a given eigenvalue appears as a zero of the characteristic polynomial; later, the concept of 'geometric multiplicity' will be defined.

Given the definitions above, it is now not difficult to prove the following results:

1. The eigenvalues of $\mathbf{A}^T$ are the same as those of $\mathbf{A}$, counting multiplicities (in fact, note that $\det(\mathbf{A} - \lambda\mathbf{I}) = \det(\mathbf{A} - \lambda\mathbf{I})^T = \det(\mathbf{A}^T - \lambda\mathbf{I})$).
2. The eigenvalues of $\mathbf{A}^H$ are the complex conjugates of the eigenvalues of $\mathbf{A}$, counting multiplicities (in fact, note that $\{\det(\mathbf{A} - \lambda\mathbf{I})\}^* = \det(\mathbf{A} - \lambda\mathbf{I})^H = \det(\mathbf{A}^H - \lambda^*\mathbf{I})$).

Note that nothing has been said about eigenvectors because, although $\mathbf{A}$ and $\mathbf{A}^T$ have the same eigenvalues, their eigenvectors corresponding to the same eigenvalue are, in general, different. In this regard, however, there exists an important property of eigenvectors which is worthy of notice; consider the eigenvalue problem (eq (A.20)) for the matrix $\mathbf{A}^T$ and let $\mathbf{y}_j$ be the eigenvector corresponding to the eigenvalue λ_j. We have $\mathbf{A}^T\mathbf{y}_j = \lambda_j\mathbf{y}_j$ or, equivalently

$$\mathbf{y}_j^T\mathbf{A} = \lambda_j\mathbf{y}_j^T \tag{A.23}$$

so that – owing to eq (A.23) – $\mathbf{y}_j$ is also called a left eigenvector of $\mathbf{A}$. On the other hand, let $\mathbf{x}_k$ be an ordinary (right) eigenvector of $\mathbf{A}$ corresponding to λ_k, where $k \neq j$. This means $\mathbf{A}\mathbf{x}_k = \lambda_k\mathbf{x}_k$. Now, premultiply this last equation by $\mathbf{y}_j^T$, postmultiply eq (A.23) by $\mathbf{x}_k$ and subtract one resulting equation from the other. The final result is

$$\mathbf{y}_j^T\mathbf{x}_k = 0 \tag{A.24}$$

meaning that $\mathbf{y}_j$ and $\mathbf{x}_k$ are mutually orthogonal. Equation (A.24) expresses the so-called **biorthogonality** condition stating that left and right eigenvectors belonging to distinct eigenvalues are orthogonal (note, however, that since $\mathbf{x}_k$ and/or $\mathbf{y}_j$ may be complex vectors, $\mathbf{y}_j^T\mathbf{x}_k$ is not an inner product as it is usually understood; see also Chapter 2). Furthermore, taking eq (A.24) into account, we can premultiply the equation $\mathbf{A}\mathbf{x}_k = \lambda_k\mathbf{x}_k$ by $\mathbf{y}_j^T$ and obtain the additional orthogonality condition

$$\mathbf{y}_j^T\mathbf{A}\mathbf{x}_k = 0 \tag{A.25}$$

which, in words, can be stated by saying that the eigenvectors $\mathbf{y}_j$ and $\mathbf{x}_k$ are also A-orthogonal.

Clearly, the biothogonality conditions (A.24) and (A.25) do not extend to eigenvectors corresponding to the same eigenvalue so that, for example, by enforcing the normalization $\mathbf{y}_j^T\mathbf{x}_j = 1$ we get

$$\mathbf{y}_j^T\mathbf{x}_k = \delta_{jk}$$

$$\mathbf{y}_j^T\mathbf{A}\mathbf{x}_k = \lambda_j\delta_{jk} \tag{A.26}$$

In the light of the above results, it now takes only a small effort to prove that

$$\mathbf{y}_j^H\mathbf{x}_k = 0 \tag{A.27}$$

where, since $\mathbf{x}_k$ and/or $\mathbf{y}_j$ are complex, $\mathbf{y}_j^H\mathbf{x}_k$ is an inner product in the usual sense.

Now, before turning our attention to symmetrical and Hermitian matrices which – besides deserving special attention in their own right – have played a major role in this book, we make some preliminary considerations.

First, as far as eigenvalues are concerned, it turns out that the theory becomes much simpler when the n eigenvalues are distinct. Let us denote them by the symbols $\lambda_1, \lambda_2, ..., \lambda_n$, with $\lambda_j \neq \lambda_k$ if $j \neq k$. The first result that can be shown (e.g. the classical reference by Wilkinson [1]) is that in this case each eigenvalue is associated with a unique (to within a constant arbitrary multiplier) eigenvector and, in addition, the eigenvectors are linearly independent. By the same token, the eigenvectors of $\mathbf{A}^T$ are also unique and linearly independent.

Second, since the n left eigenvectors can be arranged in an $n \times n$ matrix $\mathbf{Y}$ (in such a way that the components of $\mathbf{y}_j$ form the jth column of $\mathbf{Y}$) and the n right eigenvectors can similarly be arranged in an $n \times n$ matrix $\mathbf{X}$, the first of eqs (A.26) implies $\mathbf{Y}^T\mathbf{X} = \mathbf{I}$, i.e. $\mathbf{Y}^T = \mathbf{X}^{-1}$ and the second of eqs (A.26) implies

$$\mathbf{Y}^T\mathbf{A}\mathbf{X} = \mathbf{X}^{-1}\mathbf{A}\mathbf{X} = \text{diag}(\lambda_j) \tag{A.28}$$

This circumstance can be used to introduce two issues of fundamental importance in the theory: the concept of similar matrices and the concept of diagonalizable matrices. The definitions are as follows:

1. A $n \times n$ matrix $\mathbf{B}$ is said to be **similar** to a $n \times n$ matrix $\mathbf{A}$ when there exists a nonsingular $n \times n$ matrix $\mathbf{S}$ (the similarity matrix) such that

$$\mathbf{B} = \mathbf{S}^{-1}\mathbf{A}\mathbf{S} \tag{A.29}$$

so that the transformation $\mathbf{A} \rightarrow \mathbf{S}^{-1}\mathbf{A}\mathbf{S} = \mathbf{B}$ is called a similarity transformation and one often writes $\mathbf{B} \approx \mathbf{A}$ to mean that '$\mathbf{B}$ is similar to $\mathbf{A}$'.

In this regard, it is worthy of notice that similarity is an equivalence relation (i.e. a reflexive, symmetrical and transitive relation) on the set of all square $n \times n$ matrices $M_{n \times n}$. The direct consequence is that the similarity relation partitions $M_{n \times n}$ into disjoint equivalence classes, each equivalence class being the set of all $n \times n$ matrices similar to a given matrix which, in turn, is representative of the whole class and can be chosen in a particularly convenient (a so-called 'canonical') form. Then, noting that diagonal matrices are of a particularly simple form we say that:

2. If a matrix **A** is similar to a diagonal matrix, then **A** is said to be **diagonalizable**.

The considerations above can then be summarized in the following theorems:

Theorem A.2. If $A \in M_{n \times n}$ has n distinct eigenvalues, then it is diagonalizable.

Things may not be so simple if **A** has one or more multiple eigenvalues (i.e. eigenvalues with algebraic multiplicity > 1 or, in the terminology of physicists and engineers, if one or more eigenvalues are 'degenerate') even if it may still be possible that there is a similarity transformation which reduces **A** to diagonal form. In this regard, we state without proof the following theorem:

Theorem A.3. Let $A \in M_{n \times n}$. Then **A** is diagonalizable if and only if there is a set of n linearly independent vectors, each of which is an eigenvector of **A**.

If we examine this theorem more closely, we can introduce the concept of **geometric multiplicity** (g.m. for short) of a multiple eigenvalue λ_j as the dimension of the subspace generated by all vectors x_j satisfying the equation $Ax_j = \lambda_j x_j$ (or, in other words, the minimum number of linearly independent eigenvectors associated with λ_j) and note that, since always g.m. $\leqslant$ a.m., **A** is diagonalizable if and only if the algebraic multiplicity of each multiple eigenvalue is the same as its geometric multiplicity. In this case **A** is called **nondefective** whereas, when for some multiple eigenvalue of **A** we have g.m. $<$ a.m., the matrix is said to be **defective**. As an example, the matrix

$$\mathbf{A} = \begin{bmatrix} a & 1 \\ 0 & a \end{bmatrix}$$

is defective because the double eigenvalue $\lambda = a$ has a.m. $= 2$ and g.m. $= 1$. In this light, Theorem A.3 states that **A** is diagonalizable if and only if it is nondefective.

Now, the final result we need to close the present discussion is the following:

Theorem A.4. If $A, B \in M_{n \times n}$ and $B \approx A$ then they have the same characteristic polynomial.

The proof is easy and it is left to the reader but, more important, a corollary to Theorem A.3 is that

Corollary to Theorem A.4. If $A, B \in M_{n \times n}$ and $B \approx A$ then they have the same eigenvalues, counting multiplicity.

In other words, the eigenvalues of a matrix are invariant under a similarity transformation (however, note that having the same eigenvalues is a necessary, but not sufficient condition for similarity, i.e. two matrices can have the same eigenvalues without being similar).

A.3.1 Hermitian and symmetrical matrices

Symmetrical matrices with real entries arise in a large number of practical cases. As a matter of fact, they have also played a predominant role throughout this book whenever eigenvalue problems have been considered. From the point of view of the theory, however, it should be pointed out that symmetrical matrices with complex entries do not, in general, have many of the desirable properties of real symmetrical matrices. In order to extend the results to complex-entried matrices one must consider Hermitian matrices, i.e. matrices for which $A^H = A$, and not simply $A^T = A$. Conversely, one could present the discussion in terms of Hermitian matrices and note that real symmetrical matrices can be considered as Hermitian matrices with real entries.

The first important result is that the eigenvalues of a Hermitian matrix are all real. In fact, if $Ax = \lambda x$ then $x^H A x = \lambda x^H x$ where $x^H x$ (being an inner product) is always real and positive for any nonzero vector x; moreover, since $(x^H A x)^H = x^H A x$ implies that the scalar $x^H A x$ is real, it follows that $\lambda = x^H A x / x^H x$ is real also.

From this consideration it also follows that a real symmetrical matrix has real eigenvalues; however, note that real eigenvalues imply real eigenvectors for a real symmetrical matrix but, in general, it is not so for a complex Hermitian matrix. If now we take a step further it is not difficult to show that left and right eigenvectors of a Hermitian matrix are the same and we obtain the orthogonality condition

$$x_j^H x_k = 0 \qquad (j \neq k) \tag{A.30a}$$

which states that eigenvectors belonging to distinct eigenvalues are orthogonal.

This fact, together with the normalization $\mathbf{x}_j^H \mathbf{x}_j = 1$ (normalization, we repeat, is arbitrary, but this normalization to unity is one of the most common) leads to

$$\mathbf{x}_j^H \mathbf{x}_k = \delta_{jk} \tag{A.30b}$$

$$\mathbf{x}_j^H \mathbf{A}\mathbf{x}_k = \lambda_j \delta_{jk} \tag{A.30c}$$

At this point – in the light of the preceding section and of eq (A.30b) – we can conclude that if the Hermitian matrix $\mathbf{A}$ has n distinct eigenvalues it also has a set of n linearly independent and mutually orthogonal eigenvectors. In this case, we can arrange these eigenvectors as the columns of a $n \times n$ matrix $\mathbf{X}$ and rewrite eqs (A.30b) and (A.30c) in matrix form as

$$\mathbf{X}^H \mathbf{X} = \mathbf{I} \tag{A.31a}$$

$$\mathbf{X}^H \mathbf{A}\mathbf{X} = \mathrm{diag}(\lambda_j) \tag{A.31b}$$

which in turn imply:

1. $\mathbf{X}^H = \mathbf{X}^{-1}$, i.e. $\mathbf{X}$ is unitary (orthogonal if the same line of reasoning is followed by starting with a matrix $\mathbf{A}$ which is symmetrical with real entries).
2. The matrix $\mathbf{A}$ is unitarily ('orthogonally' for real symmetrical matrices) similar to the diagonal matrix of eigenvalues.

The reader should note that similarity via a unitary matrix is not only simpler ($\mathbf{X}^H$ is much easier to evaluate than $\mathbf{X}^{-1}$) than general similarity but also that unitary similarity is an equivalence relation that partitions $M_{n \times n}$ into finer equivalence classes because (see also next section) it corresponds to an **orthonormal** change of basis.

Example A.4.   The reader is invited to perform the calculations for this example. Let $\mathbf{A}$ be the real symmetrical matrix

$$\mathbf{A} = \begin{bmatrix} 1 & -4 & 2 \\ -4 & 3 & 5 \\ 2 & 5 & -1 \end{bmatrix}$$

Its characteristic polynomial $p(\lambda)$ is obtained from

$$\det(\mathbf{A} - \lambda \mathbf{I}) = \det \begin{bmatrix} 1 - \lambda & -4 & 2 \\ -4 & 3 - \lambda & 5 \\ 2 & 5 & -1 - \lambda \end{bmatrix}$$

so that $p(\lambda) = 0$ is the characteristic equation which, in our specific case, reads

$$-\lambda^3 + 3\lambda^2 + 46\lambda - 104 = 0$$

Then, the eigenvalues are the roots of this characteristic equation and we obtain, in increasing order

$$\lambda_1 = -6.5135$$

$$\lambda_2 = 2.1761$$

$$\lambda_3 = 7.3375$$

Next, the mutually orthogonal eigenvectors (normalized to unity, i.e. so that they satisfy the relations $x_j^T x_k = \delta_{jk}$; $j, k = 1, 2, 3$) can be determined; they are

$$x_1 = \begin{bmatrix} 0.4776 \\ 0.5576 \\ -0.6789 \end{bmatrix} \quad x_2 = \begin{bmatrix} 0.7847 \\ 0.0768 \\ 0.6151 \end{bmatrix} \quad x_3 = \begin{bmatrix} 0.3952 \\ -0.8265 \\ -0.4009 \end{bmatrix}$$

and the similarity matrix is therefore

$$X = \begin{bmatrix} 0.4776 & 0.7847 & 0.3952 \\ 0.5576 & 0.0768 & -0.8265 \\ -0.6789 & 0.6151 & -0.4009 \end{bmatrix}$$

At this point it is not difficult to verify that

1. X is orthogonal (i.e. $X^T X = XX^T = I$)
2. $X^T AX = \text{diag}(\lambda_j)$ where

$$\text{diag}(\lambda_j) = \begin{bmatrix} -6.5135 & 0 & 0 \\ 0 & 2.1761 & 0 \\ 0 & 0 & 7.3375 \end{bmatrix}$$

Now, the question could be asked as to what happens if A is Hermitian but has one or more multiple eigenvalue or, more specifically, what happens if A is defective. The answer to this question leads to one of the most important properties of Hermitian matrices: it does not matter if a Hermitian matrix has multiple eigenvalues because, even in this case, it is unitarily similar to the matrix $\text{diag}(\lambda_j)$. In other words, a Hermitian matrix is always nondefective.

If we want to be more mathematically rigorous, we can say that the statement above is one of the consequences of the following theorem.

Theorem A.5. If $A = [a_{ij}] \in M_{n \times n}$ has eigenvalues $\lambda_1, \lambda_2, ..., \lambda_n$ the following statements are equivalent:

1. A is normal.
2. A is unitarily diagonalizable.
3. $\sum_{i,j=1}^{n} |a_{ij}|^2 = \sum_{i=1}^{n} |\lambda_i|^2$
4. There is a orthonormal set of n eigenvectors of A.

The equivalence of 1 and 2 in Theorem A.5 is often called the **spectral theorem for normal matrices.** For our present purposes we recall that a Hermitian matrix is just a special case of normal matrix and we stress that – as expected – the statement of the theorem says nothing about A having distinct eigenvalues (and in fact, two or more eigenvalues could be equal).

Then, summarizing the results of the preceding discussion we can say that a complex Hermitian matrix (or a real symmetrical matrix) A:

1. has real eigenvalues;
2. is always nondefective (which means that – regardless of the existence of multiple eigenvalues – there always exists a set of n linearly independent eigenvectors, which, in addition are mutually orthogonal);
3. is unitarily (orthogonally) similar to the diagonal matrix of eigenvalues diag(λ_j). Moreover, the unitary (orthogonal) similarity matrix is the matrix X of eigenvectors in which the jth column is the jth eigenvector.

We close this section by briefly considering special classes of Hermitian matrices. A $n \times n$ Hermitian matrix A is said to be **positive definite** if

$$x^H Ax > 0 \tag{A.32a}$$

for all nonzero vectors $x \in \mathscr{C}^n$. If the strict equality in eq (A.32a) is weakened to

$$x^H Ax \geqslant 0 \tag{A.32b}$$

then A is said to be **positive semidefinite.** Moreover, by simply reversing the inequalities in eqs (A.32a) and (A.32b), we can define the concept of negative definite and negative semidefinite matrices.

Note that, if A is Hermitian, the definitions above tacitly imply that the term $x^H Ax$ – which is called the **Hermitian form** generated by A – is always a real number and so we can also speak of positive definite Hermitian form (eq (A.32a)) or positive semidefinite Hermitian form (eq (A.32b)).

The real counterparts of Hermitian forms are called **quadratic forms** and are expressions of the type $x^T Ax$, where A is a real symmetrical matrix. Quadratic forms arise naturally in many branches of physics and engineering, and – as we also saw throughout many chapters of this book – the subject of engineering vibrations is no exception. Clearly, the appropriate definition of a positive definite matrix reads in this case

$$\mathbf{x}^T \mathbf{A} \mathbf{x} > 0 \tag{A.33}$$

for all nonzero vectors $\mathbf{x} \in \mathscr{R}^n$. Similarly, the relation $\mathbf{x}^T \mathbf{A} \mathbf{x} \geqslant 0$ for all nonzero vectors $\mathbf{x} \in \mathscr{R}^n$ defines a positive semidefinite matrix.

For our purposes, the following result will suffice and we refer the interested reader to specialized literature for more details.

Theorem A.6.   A Hermitian matrix $\mathbf{A} \in M_{n \times n}$ is positive semidefinite if and only if all its eigenvalues are nonnegative. It is positive definite if and only if all its eigenvalues are positive (clearly, this same theorem applies for real symmetrical matrices).

Finally, it is left to the reader to show that the trace and the determinant of a positive definite matrix are also positive.

A.4   Matrices and linear operators

Some aspects of the strict relationship between linear operators on a vector space and matrices have been somehow anticipated in Section A.1. Given a basis $\mathscr{B}$ in an n-dimensional vector space V on a scalar field $\mathscr{F}$, the statement that the mapping $\varphi: \mathbf{x} \rightarrow [\mathbf{x}]_{\mathscr{B}}$ (i.e. the mapping that associates the vector to its components relative to the chosen basis $\mathscr{B}$) is an isomorphism constitutes a fundamental result which allows us to manipulate vectors by simply operating on their components. In fact, according to these developments, we saw in Section A.1 how the components of a vector change when we choose a different basis in the same vector space (in mathematical terminology, the sentence 'φ is an isomorphism but it is not a canonical isomorphism' translates this fact that φ is indeed injective and surjective, but the coordinates of a given vector change under a change of basis and therefore depend on the choice of the basis).

In a similar way, when we have to deal with linear operators on a vector space, it can be shown that – after a basis has been chosen in the space V – any given linear operator $T: V \rightarrow V$ is represented by a $n \times n$ matrix and it can also be shown that – given a basis in V – the mapping that associates a given linear operator with its representative matrix relative to the chosen basis is an isomorphism between the vector space of linear operators from V to V and the vector space of square matrices $M_{n \times n}$. Simple examples of such mapping are the null operator – i.e. the operator $Z: V \rightarrow V$ for which $Z\mathbf{x} = 0$ for all $\mathbf{x} \in V$ – which is represented by the null matrix and the identity operator – i.e. the operator $I: V \rightarrow V$ for which $I\mathbf{x} = \mathbf{x}$ for all $\mathbf{x} \in V$ – which is represented by the unit matrix.

In general, however, when a different basis is chosen in V, the same linear operator is represented by a different matrix. So, the question arises: since different matrices may represent the same linear operator, what is the relationship between any two of them? The answer to this question is that any two matrices which represent the same linear operator are similar. Let us examine these points in more detail.

First of all we must determine what we mean by a matrix representation of a given linear operator. To this end, let V be a n-dimensional vector space and let $T:V \rightarrow V$ be a linear transformation on V. If we choose a basis $\{u_i\}_{i=1}^{n}$ in the vector space, the action of T on any vector $x \in V$ is determined once one knows the vectors $Tu_1, Tu_2, ..., Tu_n$ because any x has a unique representation $x = \sum_{i=1}^{n} a_i u_i$ and linearity implies $Tx = T \sum_i a_i u_i = \sum_i a_i Tu_i$. Now, since every vector Tu_i, in turn, can be written as a linear combination

$$Tu_i = \sum_{k=1}^{n} t_{ki} u_k \qquad (i = 1, 2, ..., n) \qquad (A.34)$$

the n^2 coefficients t_{ki} can be arranged in a square matrix T, which is called the matrix representation of the operator T relative to the basis $\{u_i\}_{i=1}^{n}$. The entries of the matrix clearly depend on the chosen basis and this fact can be emphasized by indicating this matrix by $[T]_u$ so that, by choosing a different basis $\{v_i\}_{i=1}^{n}$ in V, we will obtain the matrix representation $[T]_v$ of T.

At this point, before examining the relationship between two different representations of T we need a preliminary result: we will show that – in a given n-dimensional vector space in which two basis $\{u_i\}_{i=1}^{n}$ and $\{v_i\}_{i=1}^{n}$ have been chosen – the 'change-of-basis' matrix is always nonsingular. In fact, since we can write

$$u_i = \sum_{j=1}^{n} c_{ji} v_j$$

$$v_j = \sum_{i=1}^{n} \hat{c}_{ij} u_i \qquad (A.35)$$

where $i = 1, 2, ..., n$ in the first equation (and the n^2 coefficients c_{ji} can be arranged in a square matrix which is the change-of-basis matrix from the basis $\{v_i\}_{i=1}^{n}$ to $\{u_i\}_{i=1}^{n}$) and $j = 1, 2, ..., n$ in the second equation (and the n^2 coefficients $\hat{c}_{ij}$ can be arranged in a square matrix which is the change-of-basis matrix from the basis $\{u_i\}_{i=1}^{n}$ to $\{v_i\}_{i=1}^{n}$). Then from eqs (A.35) we get

$$v_k = \sum_{i=1}^{n} \hat{c}_{ik} u_i = \sum_{i=1}^{n} \hat{c}_{ik} \sum_{j=1}^{n} c_{ji} v_j = \sum_{j=1}^{n} \left[\sum_{i=1}^{n} c_{ji} \hat{c}_{ik} \right] v_j$$

and since any vector can be expressed uniquely as a linear combination of the vectors $\{v_i\}_{i=1}^n$, the term within brackets must satisfy

$$\sum_{i=1}^n c_{ji}\hat{c}_{ik} = \delta_{jk} \tag{A.36a}$$

By the same token, it can also be shown immediately

$$\sum_{i=1}^n \hat{c}_{ki}c_{ij} = \delta_{jk} \tag{A.36b}$$

Equations (A.36a) and (A.36b) in matrix form read, respectively

$$C\hat{C} = I$$
$$\hat{C}C = I \tag{A.37}$$

meaning that $\hat{C} = C^{-1}$ (or, equivalently, $C = \hat{C}^{-1}$). Therefore, a change-of-basis matrix C is always nonsingular.

Also, with a slight change of notation, we can re-express the result of Section A.1 by noting that, since any vector $x \in V$ can be written as

$$x = \sum_{i=1}^n x_i u_i$$

$$x = \sum_{j=1}^n \hat{x}_j v_j \tag{A.38}$$

we can substitute the first of eqs (A.35) into the first of eqs (A.38) to obtain

$$\hat{x}_j = \sum_{i=1}^n c_{ji}x_i$$

which is equivalent to the matrix equation

$$[x]_v = {}_v[C]_u[x]_u \tag{A.39}$$

where the notation $[x]_v$ means that we are considering the components of x relative to the basis $\{v_i\}_{i=1}^n$. Similarly, $[x]_u$ indicates the components of x relative to the basis $\{u_i\}_{i=1}^n$ and ${}_v[C]_u$ indicates the change-of-basis matrix from $\{v_i\}_{i=1}^n$ to $\{u_i\}_{i=1}^n$.

The rather cumbersome (but self-explanatory) notation of eq (A.39) will now serve our purposes in order to obtain the relation between two matrix representations of the same linear operator. In fact, in terms of components the action of a linear operator T on a vector $\mathbf{x}$ can be written

$$[\mathbf{y}]_u = [\mathbf{T}]_u[\mathbf{x}]_u$$

$$[\mathbf{y}]_v = [\mathbf{T}]_v[\mathbf{x}]_v$$

(A.40)

where we defined $\mathbf{y} \equiv T\mathbf{x}$. Now, substituting eq (A.39) and its counterpart for the vector $\mathbf{y}$ into the second of eqs (A.40) yields

$$_v[C]_u\,[\mathbf{y}]_u = [\mathbf{T}]_v\,_v[C]_u\,[\mathbf{x}]_u$$

so that premultiplying both sides by the matrix $_u[C]_v = (_v[C]_u)^{-1}$ we get

$$[\mathbf{y}]_u = {}_u[C]_v\,[\mathbf{T}]_v\,_v[C]_u\,[\mathbf{x}]_u = (_v[C]_u)^{-1}[\mathbf{T}]_v\,_v[C]_u\,[\mathbf{x}]_u$$

which implies (compare with the first of eqs (A.40))

$$[\mathbf{T}]_u = (_v[C]_u)^{-1}[\mathbf{T}]_v\,_v[C]_u$$

(A.41)

that is, the matrices $[\mathbf{T}]_u$ and $[\mathbf{T}]_v$ are similar, the similarity matrix being the change-of-basis matrix $\mathbf{C}$.

Example A.5. As a simple example in $\mathcal{R}^2$, let us consider the two bases

$$\mathbf{u}_1 = \begin{bmatrix} 1 \\ 0 \end{bmatrix} \qquad \mathbf{u}_2 = \begin{bmatrix} 0 \\ 1 \end{bmatrix}$$

and

$$\mathbf{v}_1 = \begin{bmatrix} 1 \\ 1 \end{bmatrix} \qquad \mathbf{v}_2 = \begin{bmatrix} -2 \\ 0 \end{bmatrix}$$

Explicitly, the first of eqs (A.35) now reads

$$\mathbf{u}_1 = c_{11}\mathbf{v}_1 + c_{21}\mathbf{v}_2$$

$$\mathbf{u}_2 = c_{12}\mathbf{v}_1 + c_{22}\mathbf{v}_2$$

and we can immediately obtain $c_{11} = 0$; $c_{12} = 1$; $c_{21} = -1/2$; $c_{22} = 1/2$ so that we can form the change-of-basis matrix

$$_v[C]_u \equiv \mathbf{C} = \begin{bmatrix} c_{11} & c_{12} \\ c_{21} & c_{22} \end{bmatrix} = \begin{bmatrix} 0 & 1 \\ -1/2 & 1/2 \end{bmatrix}$$

Similarly, from the second of eqs (A.35) we obtain the change-of-basis matrix

$$_u[C]_v \equiv \hat{C} = \begin{bmatrix} \hat{c}_{11} & \hat{c}_{21} \\ \hat{c}_{12} & \hat{c}_{22} \end{bmatrix} = \begin{bmatrix} 1 & -2 \\ 1 & 0 \end{bmatrix}$$

so that, as expected (eqs (A.35)) $C\hat{C} = \hat{C}C = I$ or, according to the more cumbersome notation above,

$$_u[C]_v = (_v[C]_u)^{-1}$$

Now, consider the linear transformation $T:\mathfrak{R}^2 \to \mathfrak{R}^2$ which acts on a vector $x = [x_1 \ x_2]^T$ as follows:

$$T\begin{bmatrix} x_1 \\ x_2 \end{bmatrix} = \begin{bmatrix} x_1 + x_2 \\ x_2 \end{bmatrix}$$

(the proof of linearity is left to the reader). The representative matrix of T relative to the basis $\{v_i\}_{i=1}^2$ is obtained from the equations

$$Tv_1 = t_{11}v_1 + t_{21}v_2$$

$$Tv_2 = t_{12}v_1 + t_{22}v_2$$

from which it follows that

$$[T]_v \equiv \begin{bmatrix} t_{11} & t_{12} \\ t_{21} & t_{22} \end{bmatrix} = \begin{bmatrix} 1 & 0 \\ -1/2 & 1 \end{bmatrix}$$

and finally we get from eq (A.39)

$$[T]_u = \hat{C}[T]_v C = C^{-1}[T]_v C = \begin{bmatrix} 1 & 1 \\ 0 & 1 \end{bmatrix}$$

which is exactly, as can be directly verified from the equations

$$Tu_1 = \tilde{t}_{11}u_1 + \tilde{t}_{21}u_2$$

$$Tu_2 = \tilde{t}_{12}u_1 + \tilde{t}_{22}u_2$$

the representative matrix of T relative to the basis $\{u_i\}_{i=1}^2$.

If, in addition, the two bases that we consider in the complex (real) linear space V are orthonormal bases – this obviously implies that an inner product has been defined in V – the similarity matrix is unitary (orthogonal).

In fact, let for example V be a real n-dimensional linear space and let $\{\mathbf{u}_i\}_{i=1}^n$ and $\{\mathbf{v}_i\}_{i=1}^n$ be two orthonormal basis in V. Then, from the first of eqs (A.35) and from the orthogonality condition $\langle \mathbf{u}_j \mid \mathbf{u}_k \rangle = \delta_{jk}$ we get

$$\delta_{jk} = \langle \mathbf{u}_j \mid \mathbf{u}_k \rangle = \left\langle \sum_i c_{ij}\mathbf{v}_i \,\middle|\, \sum_m c_{mk}\mathbf{v}_m \right\rangle$$

$$= \sum_i \sum_m c_{ij}c_{mk}\langle \mathbf{v}_i \mid \mathbf{v}_m \rangle = \sum_i \sum_m c_{ij}c_{mk}\delta_{im} = \sum_i c_{ik}c_{ij}$$

so that the equality $\sum_{i=1}^n c_{ij}c_{ik} = \delta_{jk}$ reads in matrix form

$$\mathbf{C}^T\mathbf{C} = \mathbf{I} \tag{A.42}$$

which implies $\mathbf{C}^T = \mathbf{C}^{-1}$ and shows that, in a real linear space, we pass from one orthonormal basis to another orthonormal basis by means of an orthogonal change-of-basis matrix. In terms of linear operators, this means that two different matrix representations $\mathbf{A}$ and $\mathbf{B}$ of the same linear operator $T:V \to V$ are orthogonally similar and $\mathbf{B} = \mathbf{C}^T\mathbf{A}\mathbf{C}$, where $\mathbf{C}$ is the change-of-basis matrix.

Clearly, if V is a complex linear space, we get $\mathbf{C}^H\mathbf{C} = \mathbf{I}$ (i.e. $\mathbf{C}$ is unitary; recall that the inner product in a complex space is not homogeneous in one of the slots) and the matrices $\mathbf{A}$ and $\mathbf{B}$ are unitarily similar, that is $\mathbf{B} = \mathbf{C}^H\mathbf{A}\mathbf{C}$.

We will not go into further details here, but a final observation is in order: specific properties of linear operators are reflected by specific characteristics of the matrices which may represent such operators; these characteristics, in turn, are generally invariant under a similarity transformation. As an illustrative example of this situation, it can be shown that if a square matrix $\mathbf{A}$ is Hermitian, then $\mathbf{S}^H\mathbf{A}\mathbf{S}$ is Hermitian for all $\mathbf{S} \in M_{n \times n}$; this is because a Hermitian matrix represents a Hermitian operator and another matrix representing the same operator must necessarily retain this characteristic (the definition of Hermitian operator is beyond our scopes and the interested reader is referred to specific literature). Also recall the corollary to Theorem A.4 stating that eigenvalues are invariant under a similarity transformation: this circumstance reflects the fact that eigenvalues are intrinsic characteristics of a given linear operator and do not change when different matrices are used for its representation.

In the light of these considerations, we may recall the discussion on n-DOF systems (see Chapters 6 and 7, and also Chapter 9 for some important results on the characterization of eigenvalues) and note that the stiffness and mass of a given vibrating system can be envisioned as (symmetrical) linear operators on the system's n-dimensional configuration space. Then, the essence of the modal approach consists of finding the orthogonal basis – the basis of eigenvectors – in which such operators have a diagonal representation.

Solving the eigenvalue problem is the process by which we determine the basis of eigenvectors. The inconvenience of dealing with a generalized eigenvalue problem rather than with a standard eigenvalue problem translates into the fact that we have to diagonalize simultaneously two matrices instead of diagonalizing a single matrix. As stated before, however, this is only a minor inconvenience which does not significantly modify the essence of the mathematical treatment.

Reference

1. Wilkinson, J.H., *The Algebraic Eigenvalue Problem*, Clarendon Press, Oxford, 1988.

Further reading

Bickley, W.G. and Thomson, R.S.H.G., *Matrices – Their Meaning and Manipulation*, The English Universities Press, 1964.

Horn, R.A. and Johnson, C.R., *Matrix Analysis*, Cambridge University Press, 1985.

Pettofrezzo, A.J., *Matrices and Transformations*, Dover, New York, 1966.

Quarteroni, A., Sacco, A. and Saleri, F., *Matematica Numerica*, Springer-Verlag, Italy, 1998.

Shephard, G.C., *Spazi Vettoriali di Dimensioni Finite*, Cremonese, Rome, 1969 (In Italian). (Translated from the original English edition *Vector Spaces of Finite Dimension*, University Mathematical Texts, Oliver & Boyd.)

Voïevodine, V., *Algèbre Linéaire*, Mir, Moscow, 1976 (in French).

B Some considerations on the assessment of vibration intensity

B.1 Introduction

In a number of circumstances one of the main tasks of vibration analysis is to 'assess the vibration intensity'. This phrase, which is rather vague, can be interpreted as assigning to a specific vibration phenomenon a 'figure of merit' which can be used to predict the potential damaging effects, if any, of such a phenomenon. In these cases, one also speaks of 'assessment of vibration severity'.

Given the very large number of possible practical situations, it is obvious that the primary factors to be considered are, broadly speaking, the type, nature and duration of the excitation and the physical system which is affected by the vibration. Accordingly, there exist a number of specialized fields of investigation which study different aspects of the problem and consider, for example, the effect of shocks and vibrations on humans, buildings, various types of structures, electronic components etc. In this appendix, also in the light of the fact that it can be extremely difficult to categorize a complex phenomenon with a single number (as a matter of fact, there seems to exist no internationally accepted standard), we will obviously limit ourselves to some general considerations.

B.2 Definitions

In order to 'assess vibration intensity', the first definition we consider is the so-called **Zeller's power** (or strength) of vibration, which takes into account the acceleration amplitude a, in cm/s^2, and the frequency ν and is defined by the relation

$$Z \equiv \frac{a^2}{\nu} = 16\pi^4 \nu^3 x^2 \tag{B.1}$$

Zeller's power is in units of cm^2/s^3 and in the second expression on the r.h.s. of eq (B.1) we call x (in cm) the displacement amplitude.

From Zeller's power, another two quantities can be calculated: the first is the so-called **vibrar unit**, the strength S of a vibration in vibrar units being given by

$$S_{vibrar} = 10 \log_{10} \left(\frac{Z}{Z_0} \right) \tag{B.2}$$

where the reference value Z_0 is taken as $0.1 \ \text{cm}^2/\text{s}^3$. The second quantity is called the **pal** and the strength in pal (according to the original definition given by Zeller[1]) is calculated as

$$S_{pal} = 10 \log_{10} \left(\frac{Z}{Z_1} \right) = 10 \log_{10}(2Z) = S_{vibrar} - 7 \tag{B.3}$$

where the second and third expressions on the r.h.s. are obtained from the fact that $Z_1 = 0.5 \ \text{cm}^2/\text{s}^3$. Another definition of pal dates back to the German standard DIN 4150 of 1939 (current version 1986[2]) and defines the strength of vibration in terms of velocity ratios, i.e.

$$\tilde{S}_{pal} = 10 \log_{10} \left(\frac{v_{rms}}{v_0} \right) \tag{B.4}$$

where $v_0 = 0.1\sqrt{10}$ cm/s and v_{rms} is the root mean square value of the measured vibration velocity. Note that we use the symbol $\tilde{S}_{pal}$ to indicate the strength according to the DIN definition because this is different from Zeller's definition of eq (B.3).

The current German standard DIN 4150, Part 2 [2] deals with the effects of vibrations on people in residential buildings and considers the range of frequency from 1 to 80 Hz. In this standard, the measured value of principal harmonic vibration is used to calculate a factor of intensity perception KB by means of the formula

$$KB = d \, \frac{0.8v^2}{\sqrt{1 + 0.032v^2}} \tag{B.5}$$

where d is the displacement amplitude in millimetres and v is the principal vibration frequency in hertz. The calculated KB value (in mm/s) is then compared with an acceptable reference value which takes into account such factors as: use of the building, frequency of occurrence, duration of the vibration and time of day. For example, for small office buildings and office premises and a continuous or repeated source of vibration, the acceptable KB level is 0.4 mm/s in the daytime and 0.3 mm/s during the night, while the levels of 12.0 mm/s during the day and 0.3 mm/s during the night apply for a source of infrequent vibration.

If now, in the light of the above definitions, we turn our attention to the classifications that have been given, we can rate vibration phenomena in order of increasing effects on people and buildings. For example, a vibration with a Zeller power of 2 cm^2/s^3 is rated as 'very light', $Z = 50$ is 'measurable' and produces small plaster cracks, $Z = 250$ is 'fairly strong', $Z = 1000$ is 'strong' and defines the beginning of the danger zone, $Z = 5000$ is 'very strong' and produces serious cracking, $Z = 2 \times 10^4$ is 'destructive', $Z = 10^5$ is 'devastating' and so on. On the other hand, according to the DIN definition of pal, a vibration intensity up to 5 pal is 'just perceptible', 10 pal is 'clearly perceptible', 10–20 pal is 'annoying' and 40 pal is 'unpleasant'.

B.2.1 *Effects of vibrations on buildings*

As far the effects of vibrations on buildings are concerned, engineers are generally interested in the possibility of structural damage and the vibrar scale has been used by some researchers in order to give some general guidelines. So, a vibration up to 30 vibrar covers the 'light' and 'medium' ranges and no damaging effects should be expected; the 'strong' range is from 30 to 40 vibrar and there is the possibility of light damage (cracks in rendering, etc.); 40 to 50 vibrar is the 'heavy' vibration range with possible severe damage and more than 50 vibrar is the 'very heavy' range in which destructive effects should be expected.

Also, an alternative intensity unit called the **damage figure** (DF) has been proposed. This is expressed in mm^2/s^3 and is related to Zeller's power by

$$DF = 1.266Z \tag{B.6}$$

where Z, as usual, is expressed in cm^2/s^3. In this light, damage figures in the range 50–500 mm^2/s^3 are likely to produce small cracks in rendering (so-called 'minor' or 'cosmetic' damage), damage figures in the 500–2000 mm^2/s^3 range are likely to produce occasional light cracks in walls and damage figures in the 2000–7000 range produce serious cracks which extend to main walls. It is not difficult to determine that the damage figure and the vibrar strength are related by the equation

$$DF = (0.1266)10^{0.1(S_{vibrar})}$$

All these classifications – although useful in many practical circumstances – should clearly be used with judgement and the engineer should always refer both to his/her national standards and to his/her or other professional engineers' previous experience. For example, the German standard DIN 4150, Part 3 [2] deals with the effects of vibrations on structures and so does the Italian standard UNI 9916. (see also the list of further reading at the end of this appendix.)

Furthermore, it is worth noting that the most severe conditions to which a structure may be subjected are caused by earthquakes. In this regard, a **scale of macroseismic intensity** is a quantity which is used to evaluate the severity of shock on the basis of human perceptions and on the effects on humans and structures. Historically, many of such scales have been proposed through the centuries: the Gastaldi scale (1564), the Pignataro scale (1783) and the Rossi–Forel scale (1883). Nowadays, a modified version of the Mercalli–Cancani–Sieberg (MCS) scale is widely used in Europe: it is called the modified Mercalli (MM) scale and consists of 12 levels of intensity, from level 1 (hardly perceptible) to level 12 (total destruction). Other scales in use are the Medvedev–Sponheuer–Karnik (MSK), with 12 levels, and the eight-level scale of the Japanese Meteorological Agency. Since these scales are based on visible effects and on human perceptions, it is not always clear how they relate to one another and this fact has led to the definition of the magnitude M [3], which is the commonly adopted measure of the energy released during an earthquake. The basic definition is

$$M = \log_{10} A \tag{B.7}$$

where A is the maximum amplitude (in micrometres) recorded by a Wood–Anderson seismograph at a distance of 100 km from the epicentre. For practical purposes, however, different definitions are used because of local effects of the earth's crust on seismic waves. So, for example, for California earthquakes, one has

$$M = \log_{10} a + 3 \log_{10} \Delta - 2.92 \tag{B.8a}$$

while in Japan one has

$$M = \log_{10} a + 1.73 \log_{10} \Delta - 0.83 \tag{B.8b}$$

where a (in micrometres) is the ground amplitude of motion and Δ (in kilometres) is the distance from the epicentre. It should be noted that the Richter's magnitude does not apply for earthquakes which are measured at very large distances from the epicentre, where superficial waves are predominant and different relations are needed.

B.2.2   *Effects of vibrations on humans*

Although some vibration phenomena may not cause any structural damage, they can be annoying to the occupants of residential buildings, offices, etc. In this regard, it should be noted that the human body is extremely sensitive to vibrations and amplitudes as low as 0.1 μm may be detected by the fingertips.

Broadly speaking, some of the factors which influence 'human sensitivity' to vibrations are:

- position (standing, sitting, lying down);
- direction of incidence with respect to the spine;
- age and sex;
- personal activity (resting, working, walking, running, etc.);
- frequency of occurrence and time of day.

The 'intensity of perception' depends on:

- displacement, velocity and acceleration amplitudes;
- duration of exposure;
- vibration frequency.

Results from various researchers indicate that human perceptibility is proportional to acceleration in the 1–10 Hz range and proportional to velocity in the 10–100 Hz range.

In the low-frequency range (say 1–80 Hz) the human body can reasonably be modelled as an assemblage of masses, springs and dampers (whose characteristic values, however, are difficult to determine) and research has shown that there exists a resonant region in the 3–6 Hz range due to the thorax-abdomen subsystem and a further resonance in the 20–30 Hz range due to the head–neck–shoulder subsystem. Vibrations under 1–2 Hz, on the other hand, seem to affect the whole body and produce effects such as kinetosis (motion sickness), which are completely different in character from those produced at higher frequencies. For these vibrations, moreover, there seems to be a number of external factors (age, sex, activity, etc.) which have a significant influence on human reactions.

Above the threshold of about 80 Hz, the sensations and effects are extremely dependent on local conditions at the point of application (position, local damping due to clothing or footwear, etc.).

The International Standard ISO 2631 [4] applies to vibrations in both vertical and horizontal directions and deals with random and shock vibration as well as harmonic vibration. In this standard, three levels of human discomfort are considered: the 'reduced comfort boundary', the 'fatigue-decreased proficiency boundary' and the 'exposure limit'.

A completely different situation arises in the study of hand–arm vibrations induced by the use of working tools in heavy industry such as hammer drills, chainsaws, etc. High vibration levels and long exposure periods may lead, in the long run, to serious effects and also to permanent damage. In the hope of preventing such harmful effects, this important and interesting field of research at the boundary between engineering and medicine is currently being investigated in even greater detail.

References

1. Zeller, W., Proposal for a measure of the strength of vibration, *VDI, Zeitschrift*, 77, 323.
2. DIN 4150 (1986) Part 3, *Structural Vibration in Buildings: Effects on Structures*. (Part 1 (*Principles, Predetermination and Measurement of the Amplitude of Oscillations*) and Part 2 (*Influence on Persons in Buildings*) available in German.)
3. Richter, C.F., 'An Instrumental Earthquake Magnitude Scale', *Bull. Seismol. Soc. Am.*, **25**, 1–32, 1935.
4. ISO 2631 (1985) *Evaluation of Human Exposure to Whole-Body Vibration*; Part 1, *General Requirements*; Part 2, *Evaluation of Human Exposure to Continuous Vibration and Shock Induced Vibration in Buildings (1 to 80 Hz)*; Part 3, *Evaluation of Exposure to Whole-Body Z-Axis Vertical Vibration in the Frequency Range 0.1 to 0.63 Hz*.

Further reading

Allen, G.R., Proposed limits for exposure to whole-body vertical vibration 0.1 to 1.0 Hz, *AGARD – CP 145*, 1975.

ANSI S3.29 (1983) *Guide to the Evaluation of Human Exposure to Vibration in Buildings*, Standards Secretariat, Acoustical Society of America, New York.

Bachmann, H. *et al.*, *Vibration Problems in Structures*, Birkhäuser Verlag, 1995.

Boswell, L.F. and D'Mello, C., *The Dynamics of Structural Systems*, Blackwell Scientific Publications, 1993.

Dieckmann, D., A study of the influence of vibration on man, *Ergonomics*, **1**(4), 347, 1958.

Goldman, D.E. and Von Gierke, H.E., The effect of shock and vibration on man, No. 60–3, Lecture and Review Series, Naval Medical Research Institute, Bethesda, Maryland, USA, 1960.

Griffin, M.J. and Witham, E.P., Time dependence of whole-body vibration discomfort, *Journal of the Acoustical Society of America*, **68**(5), 1522, 1980.

Griffin, M.J., *Handbook of Human Vibration*, Academic Press, 1990.

Harris, C.S. and Shoenberger, R.W., Effects of frequency of vibration on human performance, *Journal of Engineering Psychology*, **5**(1), 1, 1966.

Harris, C.M., *Shock and Vibration Handbook*, 3rd edn, McGraw-Hill, New York, 1988.

Holmberg, R. *et al.*, *Vibrations Generated by Traffic and Building Construction Activities*, Swedish Council for Building Research, Stockholm, 1984.

ISO/DIS 5349.2 (1984) *Guidelines for the Measurement and the Assessment of Human Exposure to Hand-Transmitted Vibration*.

Studer, J. and Suesstrunk, A., Swiss Standard for Vibration Damage to Buildings, *Proceedings of the 10th International Conference on Soil Mechanics and Foundation Engineering*, Vol. 3, pp. 307–312, Stockholm, 1981.

Index